ART OR SOUND

ART OR SOUND

CURATED BY GERMANO CELANT

CA' CORNER DELLA REGINA

Fondazione Prada

In keeping with Fondazione Prada's aim to broaden its cultural scope, the exhibition "Art or Sound" addresses, for the first time, artistic creation that lies on the boundary between the visual arts and sound production, as well as the ways in which these two fields interact.

We began by investigating the Renaissance as it is the origin of a contemporary approach that, by extending its roots into the past, alters its cultural structures and categories. A closer review of what may at first appear secondary—like the decoration of a musical instrument or the suggested resonance of a painting—reconstructs a history, retold in its manifold forms, that demonstrates the ongoing interest in the convergence and exchange of ideas between disciplines.

An object of focus has been the ambiguity that arises out of the relationship between the object and sound, and whether the resulting artifact is considered to have the characteristics —depending on the moment, context and interpretation of the viewer—of a work of art or musical instrument.

In the exhibition at Ca' Corner della Regina, as in this book, art and music are not the only two elements interwoven. There are also notions of light and sound, writing and movement, sculpture and instrument, as well as the work of the artists and craftsmen who have produced hybrid, equivocal objects whose physical presence and sound production simultaneously activate our senses.

An *ensemble* of clocks and carillons, automata and musical machines, paintings and scores, sculptures and readymades, together with decorated, assembled, imaginary and silent musical instruments, transforms this exhibition into an orchestration to be seen, heard and experienced outside any previously established category.

Miuccia Prada and Patrizio Bertelli
Presidents of Fondazione Prada

ACKNOWLEDGMENTS

During the 20th century our understanding of art began to expand to include a multisensory dimension. It was no longer merely a privileged vision, but activated the other senses as well, from hearing to touch, smell and taste. It developed when the historical avant-gardes, from Futurism to Surrealism, attempted to make a creative transition that would embrace the complexity of a perception not monopolized by the eye, but open to all the body's sensory organs. Over the past few decades a wide range of theoretical studies and exhibition proposals have reinforced this opening, yet much remains to be done, especially with regards to sensorial totality.

"Art or Sound" at the Fondazione Prada in Ca' Corner della Regina in Venice is a continuation of these studies and investigations of the linguistic equivalence between artifact and sound, commencing its survey from the origins of this osmosis, an event that has led to the construction of aesthetic hybrids in which reciprocal modes of expression are fused and confused. It is a transition neither univocal nor unidirectional, and shuns all dichotomies in order to propose a complex and multiple perceptive structure within which difference and autonomy are intertwined, but devoid of any hierarchy, so that the sound is not subject to the image and vice versa but rather another identity is established, one that extends beyond polarities.

This sort of exploration, aimed at taking both a historical and a contemporary perspective in order to offer a stereoscopic view of the subject—at once temporal, spatial, visual and musical; and which embraces both the past and the present—could have not been achieved without the sensibility and drive of Miuccia Prada and Patrizio Bertelli, who have supported this project from the outset, enriching it with their own ideas, and have been involved both in the complicated research process conducted in the archives of museums and institutions, as well as in direct contacts with the artists. In order to reach a concrete and philosophical focus on the dialogue and encounter between art and sound, we have attempted to apply a trans-sensory strategy that can avoid discrimination and differences, bringing about real interaction between the different linguistic and cultural dimensions. This consideration, applied to objects and documents, artifacts and instruments, musical scores and performances, was only possible thanks to collaboration from museums, institutions, collections and individuals, all of whom demonstrated extreme generosity and receptiveness in making their knowledge, expertise and time available, as well as allowing us to use iconographic, digital and documentary materials from their libraries and archives. With this in mind my heartfelt thanks goes to: 303 Gallery (New York); Alanna Heiss Collection (New York); Archivio Maurice Henry (Milan); Corice Arman, The Arman Marital Trust (New York); Bauhaus-Archiv, Museum für Gestaltung (Berlin); Barbara Bertozzi Castelli Collection (New York); Beyer Clock Watch Museum (Zurich); Calder Foundation (New York); Castelli Gallery (New York); Centre Georges Pompidou, Musée National d'Art Moderne (Paris); Eric Ceputis and David W. Williams Collection (Chicago); Cité de la musique – Musée de la musique (Paris); Cittadellarte/ Fondazione Pistoletto (Biella); Collection of Historic Musical Instruments, Kunsthistorisches Museum (Wien); Paula Cooper Gallery (New York); Friedrich Christian Flick Collection, Hamburger Bahnhof (Berlin); FNAC (Paris); Fondazione Bonotto (Molvena); Fondazione Cassa di Risparmio (Bologna); Fondazione Marconi (Milan); Fundación Alumnos47 (Mexico City); Galerie Aanant & Zoo (Berlin); Galerie Chantal Crousel (Paris); Galerie EIGEN + ART (Leipzig/Berlin); Galerie Eva Presenhuber (Zurich); Galerie Xippas (Paris); Galleria Estense (Modena); Galleria Fumagalli (Milan); Galleria Michela Rizzo (Venice); Galleria Gian Enzo Sperone (Lugano); Hahn Collection (Cologne); Hauser & Wirth Gallery (New York); Het Noordbrabants Museum (Hertogenbosch); Kronos Quartet, Kronos Performing Arts Association (San Francisco); Lisson Gallery (London); Los Angeles County Museum (Los Angeles); Luhring Augustine Gallery (New York); Macedonian Museum of Contemporary Art (Thessaloniki); MART – Museo d'Arte Moderna e Contemporanea di Trento e Rovereto (Rovereto); Mitchell-Innes & Nash Gallery (New York); Moderna Museet (Stockholm); Moderner Kunst Stiftung Ludwig Wien (Vienna); Musée d'art contemporain de Lyon (Lyon); Musée des Instruments de Musique, Musées royaux d'Art et d'Histoire (Brussels); Musée historique de la ville de Strasbourg (Strasbourg); Museo di Castelvecchio (Verona); Museo Nazionale degli Strumenti Musicali (Rome); Museum für Kommunikation (Bern); Museum für Musikautomaten (Seewen); Museum Tinguely (Basel); Nouveau Musée National de Monaco (Monaco); Claes Oldenburg and Coosje van Bruggen Collection (New York); On Stellar Rays Gallery (New York); Marie-Madeleine G. Opalka Collection (Venice); Phil Lenihan, Orphic Gallery (Roxbury); Peruz Collection (Milan); Pierogi Gallery (New York); Pinacoteca di Brera (Milan); Prada Collection (Milan); Rheingold Collection (Düsseldorf); di Rosa Collection (Napa); Ruth Marten Collection (New York); Salon 94 Gallery (New York); Scriabin State Museum (Moscow); Seattle Art Museum (Seattle);

Speelklok Museum (Utrecht); Staatliche Museen zu Berlin, Nationalgalerie (Berlin); Sarah Hoe Sterling Collection (Hudson); Robin Sukhadia Collection (Los Angeles); T293 Gallery (Naples/ Rome); The Estate of Terry Adkins (New York); The Estate of Dennis Oppenheim (New York); The Estate of Wolf Vostell (Cáceres); The John & Maxine Belger Family Foundation (Kansas City); The Gilbert and Lila Silverman Fluxus Collection, The Museum of Modern Art (New York); The New York Public Library for the Performing Arts (New York); The Ochi Gallery (Ketchum); The Pace Gallery (New York); The Rachofsky Collection (Dallas); The Peggy Guggenheim Collection (Venice); The State Hermitage Museum (St. Petersburg); Rob Tufnell Gallery (London); Vehbi Koç Foundation (Istanbul); Pietro Verardo Collection (Venice); Claire Wesselmann Collection (New York); Max Wigram Gallery (London); Ethan Wiley Collection; Würth Collection (Künzelsau); Zimmerli Art Museum at Rutgers University (New Brunswick); as well as all those lenders who preferred to remain anonymous.

The task of bringing together sound artifacts from the different worlds of music and art could not have been accomplished without the enthusiastic cooperation of artists and their collaborators. I would like to thank them for their constant support and for the efforts they have made in tracking down and lending the fanciful yet functional sound objects they have designed, even though in many cases these are still being used for concerts and performances. Thus our gratitude goes to: Doug Aitken, Terry Allen, William Anastasi, Laurie Anderson, Athanasios Argianas, Tarek Atoui, Ay-o, Marco Bagnoli, Riccardo Beretta, Tore Honoré Bøe, Céleste Boursier-Mougenot, Ken Butler, Janet Cardiff and George Bures Miller, Maurizio Cattelan, Martin Creed, Paul De Marinis, Thomas Demand, Brian Dewan, Ruth Ewan, Neil Feather, Loris Gréaud, Subodh Gupta, Bart Hopkin, Rebecca Horn, Martin Kersels, Walter Kitundu, Milan Knížák, Jannis Kounellis, Bernhard Leitner, Maywa Denki, Walter Marchetti, Christian Marclay, Eliseo Mattiacci, Nicolas Bernier and Martin Messier, Haroon Mirza, Robert Morris, Bruce Nauman, Carsten Nicolai, Claes Oldenburg, Michelangelo Pistoletto, Edward J. Potokar, Pedro Reyes, Manuel Rocha Iturbide, Tom Sachs, Takako Saito, Anri Sala, Andreas Slominski, Akio Suzuki, Alberto Tadiello, Takis, Günther Uecker, Max Vandervorst, Yoshimasa Wada, Guido van der Werve and William T. Wiley.

The publication of a work on the relations between art and sound, viewed from a historical perspective with the aim of deepening people's understanding of their development, could not have hoped for better contributions than the ones produced by Jo Applin, Luciano Chessa, Christoph Cox, Geeta Dayal, Patrick Feaster, Christoph E. Hänggi, Bart Hopkin, Douglas Kahn, Alan Licht, Andrea Lissoni, Noel Lobley, Deirdre Loughridge, Simone Menegoi, Holly Rogers, Jonathan Sterne, David Toop, John Tresch, Eric de Visscher and Rob Young, who have extended and expanded our comprehension of the subject through their investigations of the theme and historical period spanned by "Art or Sound."

For display and graphic design, my gratitude goes to Michael Rock and the 2x4 studio in New York, who paid careful attention to the complexity of staging an exhibition in a building dating from the 18th century and relied on a thorough understanding of the relationship between art and music to create, with the help of Sungjoong Kim with Ryan Brooke Thomas and Jessica Dobkin, a well-organized and unprecedented architectural setting and presentation that enhances both visual and aural perception of the works. 2x4 also developed a surprising, engaging design for the publication, relying on constant and inventive collaboration from Liliana Palau Balada.

I also wish to thank Frédéric Sanchez, who paid careful attention to the relationship between space and sound that characterizes the show.

For coordination of Fondazione Prada's curatorial and operational teams in Milan and Venice, as well as for the professionalism with which she has overseen different aspects of staging the exhibition, I am grateful to Astrid Welter. The meticulous and detailed research conducted by Chiara Costa and Mario Mainetti has given the exhibition and the publication their form and structure. The sensitivity and accuracy shown in the design and realization of the display system was achieved thanks to Alessia Salerno, whose experience helped bring the operation to a successful conclusion. The decisive contributions made by all of these people has been supported by constant work and effort from the institution's staff members, including invaluable technical and professional contributions from Andrea Goffo for press relations; indispensable professional participation by Paola Magnanini; and intense and diversified assistance in dealing with organizational problems provided by Beatrice Boatto, Antonio Cascone, Francesca Cattoi, Camilla Finzi, Niccolò Gravina, Daniele Manfrè, Cornelia Mattiacci and Carlotta Predosin.

At Ca' Corner della Regina in Venice, the Fondazione Prada was able to rely on necessary technical and professional assistance from Maurizio Ciabatti, Giuseppe Zotti, Roberto Rigon and Romeo Scarpa. Mounting and displaying these precious works was made possible thanks to help from the Attitudine Forma team (Turin), as well as the conservation and restoration expertise of Luisa Mensi.

With regard to the publication, where knowhow and precision play such a vital role, we are grateful to the Buysschaert&Malerba studio, made up of Francesca Malerba, Laura Magda Barazza, Giacomo Serra, Zino Malerba, Aaron Maines and Timothy Stroud, for its constant input, dedication and concentration.

It would have been impossible to deal with the great number of requests for information, coordinating them with respect to the curator's schedule, without the patience and determination of Marcella Ferrari at Studio Celant, assisted in her research by Laura Conconi and Mattia Ruffolo, including valuable support provided by Lorrraine Dolot and Shirley Queiroz. Finally, I would like to thank Argento and Paris Celant who, with the greatest intensity, have been a constant presence at my side throughout this adventure.

Thanks to everyone

Germano Celant
Director of Fondazione Prada

ART OR SOUND

FONDAZIONE PRADA
Ca' Corner della Regina, Venice
7 June – 3 November 2014

EXHIBITION

CURATOR
Germano Celant

ASSOCIATE CURATOR
Mario Mainetti

RESEARCH
Chiara Costa
Mario Mainetti
with
Niccolò Gravina

PROJECT DIRECTOR
Astrid Welter

PROJECT MANAGER
Alessia Salerno

ORGANIZATION
Camilla Finzi
Daniele Manfrè
Carlotta Predosin
with
Cornelia Mattiacci

EXHIBITION DESIGN AND GRAPHIC DESIGN
2x4, New York
Michael Rock
Sungjoong Kim
Ryan Brooke Thomas
Jessica Dobkin
Ryan Weafer
Jaewon Seok
Minsun Eo

INSTALLATION
Alessia Salerno
with
Giuseppe Zotti

INSTALLATION CONSULTANTS
Attitudine Forma, Turin

VIDEO PRODUCTION
Show Biz, Milan

CONSERVATOR
Luisa Mensi
with
Matilde Dolcetti

PRESS OFFICE
Andrea Goffo
with
Beatrice Boatto

ADMINISTRATION
Paola Magnanini
with
Antonio Cascone

EDUCATIONAL ACTIVITIES & GUIDED TOURS
CoopCulture, Venice

SHIPPING
Apice Srl, Milan
Rubelli & C., Venice

INSURANCE
AXA Art Versicherung AG, Milan
Marsh, Milan

CONSTRUCTION MANAGEMENT
Maurizio Ciabatti
Roberto Rigon
Alessia Salerno
Romeo Scarpa

CONSTRUCTION
Augusto Capovilla Sas, Venice
Adriano Cincotto, Venice
Cantù Arreda Srl, Cantù
Contec Ingegneria, Verona
Esa Engineering Srl, Florence
Gemini luci, San Giuliano Milanese
Grifal Spa, Cologno al Serio
Gruppofallani, Venice
Gruppo Restauro Conservativo, Venice
Impresa costruzioni e restauri edili BCB, San Donà di Piave
Maser Group, San Donà di Piave
Openservice, Venice
Star Service Srl, Cernusco sul Naviglio
Tono Impianti Srl, Padua
Tramonte Elettroimpianti Engineering, Campagna Lupia
Zero4Uno Ingegneria Srl, Venice

THANKS TO
Bose Spa, Milan
Contec Ingegneria, Verona

SPECIAL THANKS TO
Frédéric Sanchez

BOOK

GENERAL EDITOR
Germano Celant

ASSOCIATE EDITOR
Chiara Costa

RESEARCH
Chiara Costa
Mario Mainetti
with
Niccolò Gravina

DESIGN
2x4, New York
Michael Rock
Sungjoong Kim
Liliana Palau Balada

PRODUCTION EDITOR
Buysschaert & Malerba, Milan
Francesca Malerba
Laura Magda Barazza
Giacomo Serra
Aaron Maines
Zino Malerba

TRANSLATIONS
Buysschaert & Malerba, Milan
Ilaria Ortolina
Antonella Santambrogio
Cristiana Spitali
Timothy Stroud
Language Consulting
Congressi srl
Anna Albano
Robert Burns
Christopher Evans
Marina Scotto di Carlo

PRODUCTION
COORDINATION
Marco Previtali
Mara Spinelli

PUBLISHER
Progetto Prada Arte, Milan

REPRO
De Pedrini, Milan

PRINTER
Nava Milano Spa

THANKS TO
Beatrice Boatto
Francesca Cattoi
Andrea Goffo
Daniele Manfrè
Cornelia Mattiacci
Carlotta Predosin
and
Studio Celant
Laura Conconi
Marcella Ferrari
Mattia Ruffolo

ISBN: 978-88-87029-56-7

TABLE OF CONTENTS

NOTE TO THE READER

This volume reconstructs and analyzes the relationship between investigation and experimentation in the fields of art, sound and music from the Renaissance through modern day.

In addition to a text by the curator, the volume contains essays by scholars, musicologists, theorists, artists, musicians, composers and art historians who conduct historical, critical, philosophical, technical and sociological surveys on the subject.

The essays are divided by sections with illustrations (complete with brief descriptive texts) that provide an exhaustive account of the exhibition "Art or Sound" by Fondazione Prada, curated by Germano Celant at Ca' Corner della Regina in Venice. Exhibited works (including musical instruments, mechanical objects, scores, paintings, sculptures, assemblages, installations and electronic devices) have been realized by artisans, craftsmen, artists, composers and musicians during the period stretching from 1520 to 2014.

All illustrated works (or the related authors/makers) mentioned in the essays are accompanied by a reference page number.

Presented in chronological order, each illustration is accompanied by a text describing the work and including author's/maker's name (when known), title, year (execution or production date; in case of reconstruction or replica the date is in parenthesis) and synthetic biographical or historic information on the author/maker (artisan or company, artist, composer or musician). Some of the texts refer to more than one image.

The works are described with regard to their relevant technical and artistic characteristics, and their relation to sound.

Information is based on historical documents, specialized publications, audio-visual documentation and cross disciplinary dialogue as part of the Art or Sound project, conducted with artists, scholars, archivists and other experts in museums, foundations and galleries.

The artist and musician citations included in this volume indicate bibliographic sources, except in those cases where oral testimony was gathered during the project's research phase.

The works mentioned in the text and followed by the abbreviation [exh.] are on exhibit but not illustrated in the book.

Further detailed information (media, dimensions, collections) can be found in the List of exhibited and illustrated works (p. 440).

GERMANO CELANT

ART OR SOUND

FROM THE MULTILINGUAL TO THE MULTISENSORY

The Museum has become a place of sensory withdrawal: a vision-centric territory that only permits an exploration centered on the eyes, in which the utilization of nonvisual senses is repressed. Questions are beginning to be raised about this focus on a single channel, which is based on multiple restrictions of the senses—those of smell, hearing taste and touch—and efforts made to find systems capable of stimulating the other senses. This shift in the sensory approach, though still rare and on a small scale, is leading in the direction of a multisensory display of art, in which the dictatorship of vision is giving way to a democracy that embraces smell, hearing, taste and touch. The intention is to augment the modes of exhibition by including elements and materials that do not ask for a response from the eyes alone, but also from other organs of the body; from the hands to the mouth, from the nose to the ears. Such an approach aspires to let visitors interact directly with the works, offering them an experience that involves an individual in his or her entirety. The possibility of hearing, touching, tasting and smelling brings visceral and sensual involvement. The feeling of involvement is amplified with divergent and plural characteristics that immerse the visitor in a more emotional and responsive universe, able to enrich the aesthetic perception and understanding of the artifact on display.

This practice of sensory presentation can also help to break down the normative schemes for the display and contemplation of art, entrusted today to restrictive systems that demonize participation and corporeal exchange between things and people. The relationship that is established between visitor and artifact in museums lacks the Oriental conception of a potential dialectic, in which the statue of the deity looks at its worshiper and the latter responds with offerings, gifts, candles, flowers and incense. A flow of energy from one pole to the other occurs in the space between the two, through which art primes the person and the individual activates the sculpture. In this way a relationship of reciprocity is established and the ceremonial aspect of finding oneself in front of a representation underpins a ritual of constant and endless experiential stimulus.

In stark contrast to Oriental culture, displays and exhibitions rely almost totally on the primacy of seeing and observing, of walking and exploring, as if the only source of understanding were observation of the image, accompanied at times by writing, since both—by ancient and philosophical tradition—are more dependable on the plane of reason and thought. The historical reasons for this privileged lexicon of knowledge transmission, entrusted to observing and reading, can certainly be traced back to the preference given to thinking over the senses, even though these have close links and important connections with intellectual processes and an awareness of reality.

The fact is that the Museum system and its techniques of display have extreme difficulty in including the senses and natural elements, whether they regard human beings or animals, heaven or earth. It excludes and confines fire and water, rain and snow, is terrified by the presence of any kind of fauna and flora, from the parrot to the horse, from the cactus to the tree, as if it wished to avoid any contact with organic and living elements. It perpetuates the Aristotelian view in which language, both visual and written, is seen as the distinction between human being and animal, out of which comes a rejection of any instinctive feeling, considered negative and inferior, vague and incidental, impure and muddled. It is a repudiation of the temptations of the other senses that is part of that inducement to purity professed by Western cultural idealism. An expression of the tendency to suppress physical and carnal pleasures as a source of bodily enjoyment antithetical to intellectual and spiritual hypersensitivity.

If we are going to avoid this process of reduction of the involvement of the senses, museum spaces can no longer be regarded as clinical, aseptic and sterile territories, places where we can devote ourselves to the conservation of the embalmed and frozen "bodies" of visual culture. Instead it is necessary to impel them to synchronize with life, with its fragrances and stenches, its noises and sounds, its severities and pleasures, its flavors and gustatory sensations. It is important for the Museum to be turned from a hospital or a house of the "dead" into a living place,[1] so that it can come closer to the real world, filled with sensory stimuli, and shake off the neurotic diktat to not touch, not listen, not taste and not smell.

However, these invisible and volatile ingredients continue to be absent from the systems used to communicate and display art. Indeed any interest in these dimensions of perception is viewed with suspicion and opposed, as if education, understood as cultural "treatment," does not allow for the dispensing of physical and sensory stimulation in its "pharmacies." And yet the smelly, the noisy, the hot, the sticky, the soft, the rotten, the organic and the decaying are *mementos* of living, as well as of dying. Perhaps this is why, beginning in the 1960s and continuing over recent decades, artistic production has not merely opened up to the industrial processes and chromatic language—Pop Art and Minimalism—, but has begun to include natural and bodily elements capable of supplying the mind with new kinds of information. The intention has been to prepare an interactive condition in which the inclusion of plants, animals, people and chemical and mineral processes, for their active and mobile character, is an invitation to consider the reciprocal situation of performance, as active observer of a vital and dynamic state.

With such historically recognized currents of research as Land Art, Arte Povera and Body Art, a holistic conception of the human being has developed in which intellectual and spiritual pleasures do not exclude the material and sensory ones that are determined by the wholly natural. This way of making art has affected viewer and visitor, placing them in a circular rather than frontal relationship with the work, so that they no longer focus their attention on one point, but on the entire environmental situation. They are stimulated to become involved, reverberating physically with respect to the object presented and shown. It is a change of perceptual direction based on the fact of producing sounds, noises, odors and sensations that are invisible but active entities. They come to occupy a space that, since it is not unidirectional, becomes enveloping and encompassing.

Such a 360-degree approach brings into question the centrality of vision, while perception is transformed into a whole range of sensory experiences located in a homogeneous space. This mode of comprehending, in its literal sense of taking together as one whole, is part of a series of emancipations from territorial constraints and limitations in art. It began in the 20th century when Italian and Russian Futurism and the historical avant-gardes began undermining the mythical metaphysical superiority of the major arts over the minor ones, of art and sculpture over photography, literature, music, theater and cinema. The hypothesis was to stop favoring one language over another, in order to arrive at a multilingual aestheticization of the real.

Here an immense and uncharted territory opened up, an ever-expanding space that came to include declarations and manifestos on sensual touch and sexual feelings. Art gave way to the arts, which showed themselves to be interested in an intense exchange between one another. Unexplored zones emerged, in which it was possible to get painting and dance, photography and theater, architecture and cinema, poetry and graphics, design and sculpture to work together... Boundaries tended to break down and the arts traveled from one visual-plastic world to another. Around 1968, the process fell in line with the upheavals created by the generation gap in a desperate search for an identity not marked by the past, in which everything was based on rigid and unequivocal, ideal and conservative coordinates. As we have seen, the rift caused by the introduction of a linguistic incandescence connected to natural organisms rendered the enclosures indistinct in order to bring out dormant energies. Along with vision, there was an activation of touch, hearing, smell and taste, plunging creation back into the melting pot of the senses.

The vertigo produced by these two historical moments converged in all-embracing projects with events that ranged—at the exhibition level—from "When Attitudes Become Form," curated by Harald Szeemann at the Kunsthalle Bern in 1969, to the series "Paris-New York," "Paris-Berlin" and "Paris-Moscow," curated by Pontus Hultén at the Centre Pompidou in Paris from 1977 to 1979. It was an accumulation of materials and expressions from the historical to the living, putting forth an idea of art in relation to the experiences and events of the time, to architecture and the urban and cultural spaces in which they were located. Within this vibrant exhibition activity, congealed into a set of objects and documents, artifacts and media, the aseptic gaze was intertwined with an eruption of materials and data. Forms and volumes precipitated into a container—the Museum—that succeeded in reestablishing contact with the life of today and yesterday. Material proved active. A contribution to this discovery has been made by all the instruments of the media, as well as by communication through the senses proposed by avant-gardes of the past and present, which in many cases have resorted to the performing and mutable aspects of things, both artificial and natural.

In the 1980s globalization disrupted the bipolar artistic axis of East and West, opening the doors to a "worldwide" consciousness capable of dealing with different expressive, communicative and aesthetic situations, as well as with different ethnicities and religions. After centuries of neutralization and dismissal of "other" linguistic territories, the art world opened up to non-dominant interfaces of identity: the female and the ethnic. The monolithic and univocal aspect of its system was short-circuited and local productions emerged or asserted themselves with the proliferation of communities that had post-patriarchal interests and postcolonial values, from "Women Artists: 1550-1950," curated by Ann Sutherland Harris and Linda Nochlin, Los Angeles, 1976, to "Magiciens de la Terre," curated by Jean-Hubert Martin, Paris, 1989. The problem of reconstructing a truly cosmopolitan vision, one open to all genres in an effort to redefine the term "universal," first reared its head.

This limited mapping of the various shifts in attitudes toward art does not aspire to define all the great changes in the modes of displaying and experiencing the artifact that have taken place from the 20th century to the present day. Rather it indicates a number of fundamental steps that have led to a move away from a hierarchy of techniques, genres, genders and ethnicities and toward a democratic equivalence of aesthetic languages and the institution of a multidirectional circularity alongside the unipolar order of the East-West axis. They are reformulations of the preestablished approach to art aimed at bringing it out of its dualism and into a more open and expansive outlook. Inserted into a truly cosmopolitan structure, made up of far-ranging and global references, the society of art as a multiplicity of languages and cultures has come to feed on new and unpredictable reticular differentiations that strike at the heart of a hierarchical and die-hard Western exhibition philosophy. Yet many pockets of resistance to these post-media and postcolonial openings remain, including that of the shift to a multisensory dimension, which cannot be reduced solely to the paradigm of vision but embraces the whole range of senses. A hierarchical subordination continues to hold sway. The eyes predominate, focusing attention on the outside rather than the inside of the experience. Thus the encounter with missing senses can represent another project of the opening and crossing of boundaries that will lead to further manipulation of the various possible relations between bodies and objects. A world that can be grasped through hearing and touch is more dynamic, complex and indeterminate.

With respect to vision, which establishes an explicit and managed territory, the other senses are implicit, and so more profound and permeating. This leads to the development of new sensations of things and one's own body, intensifying the awareness of self and reality. The reverberation created by hearing and touching makes it possible to experience oneself in a new way. Broadening the relationship with the world through the use of sensory instruments results in an intensification in which objects and artifacts are enriched with complementary aspects, creating an alternative landscape into which sound, smell, taste and touch have been inserted.

ART OR SOUND

What effect does the possible shift to a multisensory paradigm have on cognitive processes? First it means not yielding to the temptation of disguising the dominance of the visual by focusing on the experience of "castoff" senses. (A paradoxical condition that has recently been subjected to theoretical and practical criticisms in the name of experiential totality.) Today there is a tendency to speak of and practice a simultaneous presence and appurtenance in order to play on a mix of odor + sound + touch + taste + artifact in which the pluralism of experience is a constituent feature of environmental and cultural interpretation in the world of the arts. To overcome the unilateral vision of the process of perception it is necessary to produce an interaction which is unilaterality expanded and superseded, squeezing different sensory experiences into the same space. This is what has happened in the last few decades with regard to the attention paid to the relationship sound + art,[2] while much remains to be done for the combination and interaction of taste + touch + odor + art.[3] So the cohabitation of these polarities and pluralities, bringing together subjects that are still "in conflict," can create a place of mobile and unpredictable experience. The trans-sensory dynamic produces a flow that sucks all unilaterality into its vortex, giving rise to another difference, based on stimulation of all the senses. This sensory stereoscopy can be expressed today in a conjunctive as well as disjunctive sense, so that, from the perspective of the museum and the exhibition, the elements can coexist distinctly and indistinctly, in a play of mirrors between tactile, aural, visual and olfactory images. In welding together or connecting these closely intertwined elements, any supremacy tends to vanish in favor of the emergence of a unitary and total constellation of experience.

The exhibition "Art or Sound" aims to bring out the conflict as well as the association between these poles. Thus it underlines both respect for the identity of the individual language, art and sound, and the possibility that they interact to form a sum, art + sound, which converge in putting on display something within which their mutual influences and flows are put into effect. In addition, once their fusion and dialectic have been asserted, it tries to show the course of their historical development, offering a collection of aural artifacts from the 17th century to the present day and defining the display of a mnemonic, visual and sonorous experience. This passage from yesterday to today avoids the reduction of a pictorial and sculptural crossover between objects and music to the present in order to search for its roots in time.

Within the spaces of Ca' Corner della Regina in Venice the question of boundaries has been set aside in order to present a process that gives concrete form to both sides of a coin. The aim is to verify and document a tendency to level out the sensorial playing field, as had already happened in the 17th and 18th centuries through instruments like the clock or the automaton, and in a number of public collections, including those of the State Hermitage Museum and the British Museum. So it is a discourse on interde-
1 pendence and exchange, as well as on the autonomy of memory and the return of a mode of experience that was not kept separate in the Victorian age but left undivided, as visitors to collections and museums were allowed to touch, hear, taste, smell and look.[4]

The experience of sound animates the whole body, echoing in it and giving it a dynamic dimension. Something that comes from outside and is not produced from the inside, although arriving in it amplified and modified. The result of an agent or a producer of sounds and noises, as well as silences, that presents itself as an enveloping substance to be listened to, for a particular length of time and from a particular direction. Sound creates an interlocutory relationship that has both intentional and casual effects on the listener. This sound occupies a space in which it comes and goes in a feedback circuit. It radiates in an environment, spreading through it. It is amplified or attenuated in relation to the absence or presence of obstacles. The acoustic and sensory information provokes an amplified awareness of the tonal qualities of places and bodies, and therefore of the world.

The history of objects that develop their dimension, presence and color through sound, and thus animate spaces—framed in time but interweaving the senses and art to construct an autonomous core, formed out of the sum of sound with art and art with sound—, can be dated from the first attempts at mixing up the arts, conducted with the aim of uniting the human and the inhuman, the natural and the artificial, the organic and the industrial. This osmosis was first practiced[5] as early as the 13th century in the Arab world, and then developed in the Middle Ages and Renaissance, reaching its height in the 17th century when people set out to add an animated and acoustic dimension to artifacts, from clocks to androids, as those by Jaquet-Droz—the writer, the draughtsman, the musician—and Jacques de Vaucanson—flute player—which expanded on and surpassed the static and visual information of the time. It was a first step toward connecting and establishing a reciprocal relationship between sensory entities in which the invisible materiality of sound illuminated the way a person experienced the distance from and position of the object. At the same time it offered a different kind of memory from the visual image, which was temporary, fragmented and contingent.

The creation of an equivalence between ear and eye, visible and invisible, limited and unlimited establishes an echo of reciprocity that, in many cases, produces connections between one room and another, between a nearby object and a distant one, even when the visitor is not standing directly in front of the exhibit. A significant effect of interaction that is defined today as "interference," but is in fact a multimodal and multisensory manner of appreciating art, as well as its contingent surroundings. It is a challenge to the authoritarianism of the eyes, continued in the publication—catalogue or book—accompanying the exhibition, where the noises or sounds of things, not to mention their feel, taste and smell, are completely absent, given that it contains nothing but reproductions.

The choice to begin "Art or Sound" with hybrid objects in which the function is reflected in a fantastic *elsewhere* that is an interweaving of different senses, of sound and image, is clearly an indication of an aesthetic and iconographic articulation that has no intention of becoming fixed in a rigid structure, so that things are constructed by means of metamorphosis, amplification and reduction. The aim is to liberate unorthodox and marginal aspirations through which objects, from mechanical automata to musical instruments, are able to evoke the marvelous and the surprising, even if they reflect sounds in a traditional and classical manner. This allows the object to embark on a complex series of shifts. A violin or a guitar becomes a marble sculpture, as in the case of the Michele Antonio Grandi's work [50], while a buccin [70], a cornet [46] and a Russian bassoon [102] take on the appearance of a snake with the head of a dragon. Elsewhere a carriage turns into an organ [66] and a cage into a musical box [62], as if feeding on the chance collision that is a mark of the Surrealist experience. This choice of animal iconography may be linked to a desire for a pictorial and sculptural "realism" typical of the 19th century, when even the musical instrument was expected to say something about its presence in nature, as well as in the life of things. In this way it could connect with the type of music for which it was used, bourgeois or folk, liturgical or secular. It tended to take on a solemn or fantastical appearance in relation to the social context in which it operated, from the carriage organ to fairground organs [132]. It was the result of an iconography that represented the ancient world, like theaters with automata; or that drew on industrial discoveries and created equipment for the production of sound with gas, like Frédéric Kastner's pyrophone [114], as well as mechanisms that celebrated new industrial techniques, like the Stroh violin [180] invented by J. M. Augustus Stroh or Arthur Howson's phonofiddle [134] and the cylinder musical box developed by Paillard Vaucher & Co. [122].

So the choice of such objects becomes a way of highlighting a visual and acoustic, mechanical and material energy that cannot be unambiguously pigeonholed. The aim is not to prevent people from perceiving them with just the eyes, but to get people to think about them as emitters of sound that have undergone aesthetic and technical modifications in relation to the emergence of industrial culture, making them more and more like sculptures or artworks.

As if aspiring to create an effect of perceptual and plastic instability of visible and invisible, material and immaterial substances that can be subjected to an interchange of aesthetic properties and the effects of performance. A *perpetuum mobile* where the artifact or sound, like all other possible extensions of the senses including smell, touch and taste, leads to a multisensory "other-than-self." All instruments or sculptures that aspire to a synaesthesia in which the ear and the eye "listen" are historical examples of this—as the plastic representation of a fugue by Johann Sebastian Bach conceived by Henrik Neugeboren [178]—, because they host a sensation that mixes painting with music, the musical instrument with sculpture, thereby breaking down the boundaries traced by the univocality of seeing and perceiving. 2

An unlimited polysemy began to emerge with the historical avant-gardes of the 20th century, from Cubism to Futurism, in which the principle of metamorphosis generated and destroyed forms, from Giacomo Balla [162] to Pablo Picasso, and molded the object to give it a plastic connotation separate from its function. The same thing happened in the musical sphere, where the introduction of "nonmusical" instruments like the *intonarumori* [160], through which the artist Luigi Russolo brought audiences into contact with noise, suggested the breakdown or end of music. In the case of Duchamp the surprise was the "hidden noise," [164] foreshadowing John Cage's exploration of silence. The 20th century began under a contrast between noise and music,[6] partly in relation to the awareness of a clash between quiet countryside and deafening city, between the sounds of the rural world and the noise made by automobiles, streetcars and trains. It was also the consequence of developments in the history of music, from Claude Debussy to Igor Stravinsky, from Erik Satie to Edgard Varèse, from Henry Cowell to Karlheinz Stockhausen, whose musical works stemmed from a collage or montage of notes, harmonies and rhapsodies interwoven with sounds drawn from everyday life. Chaotic and aleatory ensembles that corresponded in painting to Wassily Kandinsky's dynamic use of color or Theo van Doesburg's structuring [166]. They were examples of chromaticism in sound, excited or calm, defining contrasting values and meanings from disorder to order.

The question of the interaction between art and sound is interwoven with the entire history of art from the 17th century to the present day as an aspiration to investigate a sensory space or territory that did not fit into the Western tradition and its schematic coordinates. It marks the whole course of modernity and the development of synesthetic relationships between different expressive languages of communication, with the aim of finding another, unconventional order. "Art or Sound" also ventures into the territories of history and memory, showing that an aspiration to the multisensory has existed for centuries as a methodological inquiry corresponding to a diversified mapping of sounds and musical effects, the results of which have simultaneously influenced both vision and hearing.

So rather than asserting the coincidence "art = sound" or the summation "art + sound," which might allude to a linguistic cannibalism in which everything is merged, their independence has been maintained in the expression "art or sound," fostering fluctuation and a free transition rather than antagonism: a space of backward and forward movement and exchange that serves to make art flow into sound and sound into art, but also distinguishes their different histories and paths.

The challenge taken up by the exhibition at the Fondazione Prada in Venice is the designation of a *theater of things* and events in which the connotation—Is it art or sound? Is it sound art or musical craft?—tends to lose its meaning. At times there is coincidence and attraction, at others refusal and rejection, and this serves to open up a gap between models of decipherment. The enigmatic approach adopted by artists like Man Ray [172, 174, 176] and Salvador Dalí [188], where what counts is the mutable multiplicity of an object, becomes many-sided and elusive, relying on unforeseen juxtapositions and unexpected appearances. Not so different from the nail violin [100] or Carlo Bugatti's tamburicorde [118] and the creations by Edward Kienholz [224], the Baschet brothers [190, 192], Christian Marclay [360] or Terry Adkins [364], which can be compared with the couplings and disruptions typical of Cubist, Dadaist and Surrealist experimentation.

The aspiration to embrace all senses that search for an acoustic sensibility of things, whether constructed or found, took shape with the linguistic discovery that led to the definition of an autonomous space of the arts in the post-World War II period. A necessity to make multisensory communication function irrespective of intentions related to the conscious or the unconscious, the social or the individual, the abstract or the real, so that the artifact is not reduced to the product of something "else" but is the almost tautological manifestation of the language of expression itself. Here the distinction between outside and inside breaks down, enabling the object to express various "projections" that make distinct trajectories of meaning and logic converge upon it.

Sculpture rediscovered an extraneous value like sound in the work of Alexander Calder [186], where the fragments of painted metal that make up his "mobiles" are numbered as if they were notes of a score. They produce—as in the case of Takis [284]—sounds when the parts, stirred by the wind, come into contact with or strike each other. In other cases the musical instrument retreats into its almost biographical interior, that of the world of Paganini, as in Maurice Henry's bandaged violin [184], or Ay-O [304], transforms the sound box of clocks and automatic musical boxes into a silent and oneiric container that alludes to voluptuous and yearned-for contents, however unfamiliar and distressing they may be. It is a muddling of experience in order to give in to the fluctuation of the senses, suddenly stripped of the certainty offered by optical discriminations with respect to physical and sensory ones. Since 1945 artists, scarred by the tragedy of war, have become well aware of the historical failure of a unique or unified understanding, denouncing its formalism and establishing the *reign of the formless* through Action Painting and Abstract Expressionism.[7] Everything crumbled and decomposed; lines of research expressed their fragility in the face of historical drama. Forms exploded with the sensation that it would be difficult to put them back together. It was a harrowing moment in which each language was dispossessed of its own nature, abandoning itself to gestures and actions that corroded and devoured every narrative or story, except that of an existential vacuity, which gave rise to nothing but emptiness or silence. All coordinates of reference were lost, suspended. It was inconceivable to plot a reliable identity for them.

Yet in this period of desiccated vitality, art, music, cinema and literature produced aggregates in which the extremes of life were crystalized, from noise to silence, from reality to abstraction, and the primary roots of painting, writing, documenting and narrating were brought back into circulation. This effervescent situation led to a total erosion of all linguistic limits. Subjecting materials to a strategy of fragmentation, immersing them in a sort of melting pot of the formless, the arts displayed a desire to "remake" themselves, to find a new foundation for all their polarities, and as a consequence reactivated all means of expression and exhibition, placing them on the same plane.

THE SOUND ARTIFACT

Since the 1960s, in addition to Andy Warhol's leveling of media, all the distinctions between floor, wall and ceiling have been lost. Demarcation lines between languages and senses have been disrupted, while the territory of the visible has come to include all sorts of bodies and objects, with a breakdown of differences between high and low, major and minor, full and empty, significant and insignificant: everything has acquired meaning. Alongside industrial noises, from the factory to means of transport, from the streetcar to the train, from the automobile to the motorcycle, that marked the beginning of the century, the perception of sounds was amplified by new technologies for acoustic transmission, from the radio to the gramophone [182].
It became possible to record concerts and pieces of music and broadcast them over the radio. In the wake of the World War II the soundscape expanded to include the din of war and the noises of natural events like eruptions and storms. It has incorporated film and theater scores, making the mingling of sounds total. We can thank John Cage [214, 216] for this placing of highbrow and lowbrow music, opera and Muzak, concerto and jukebox, silent music and radio on the same level. It is no wonder that Neo Dada and Pop artists, from Robert Rauschenberg [230] to Tom Wesselmann [238], included radio programs in their sculptures and paintings, describing them as "still lifes," while Jasper Johns inserted a windup clock into one of his paintings. What counted was the perception of the world, as it happens, through hearing, but also through the resounding chamber of one's own body, as in Nam June Paik's *Listening to Music Through the Mouth* [228].

In other cases the influence of introversion, which spread with knowledge of Zen culture, led to a search for one's own inner sound, which in the case of a sculpture or a painting lay in its active parts, as in Jean Tinguely's *Radio WNYR Nr. 15* [226].
A concept of individual illumination informed such combinations of art and sound in both an active and passive sense. In Milan Knížák [232] this found expression in an equivalence between life and death, between sound and silence, between working recorder and destroyed recorder.

Often in movement, all these indivisible wholes that incorporated different entities, found or constructed, and were not fixed onto a rigid structure—so that they brought into play traditional contrasts between dark and light, autonomy and heteronomy, mechanical and organic—had precursors in the 1920s in Russolo's *intonarumori*, in Alexander Scriabin's color piano [136] or Wladimir Baranoff-Rossiné's optophonic piano [170]. They cleared the way for Cage's modified or prepared pianos and accumulations of turntables, or Paik's sound ensembles and musical apparatuses, and found a sequel in the sound combines of Edward Kienholz and Wolf Vostell [252]. Mechanisms and instruments that presented strong affinities with historical clocks and automata opened the phase of artistic and sound "engines" used to modify and boost the effects of seeing and hearing. With them the art of seeing and playing music became a mixture, a contrivance that when set in motion issued entities in which subjective and objective could no longer be considered antinomic, thereby permitting continuity and integration between constructed and found, sculpture and readymade, noise and silence. More and more space was given to the language of art and sound as a protagonist of its iconic self-revelation, from Richard Artschwager [240], Arman [218, 344], Günther Uecker [236], and Walter Marchetti [314] to Claes Oldenburg and Coosje van Bruggen [378, 380], with works that seemed to renounce the acoustic component in favor of their chromatic, sculptural and environmental impact, but which formed and composed a visual and plastic silence.

European and American Neo Dadaists, especially those associated with Fluxus, from George Maciunas [270, 292] to Joe Jones [234], established a symbiosis between body, sound artifact and musical machine. They made it possible to turn the object into a prosthesis that connected the artificial and the human, bringing both of them out of a unique and fixed dimension. This passage closed the circle of reciprocal mutation of art and sound, establishing itself as an evolution that led to the birth of a being-machine or machine-being that, as object or as figure, is capable of autonomy in the production of sounds and images, in dancing and listening, as in Robert Morris [220] and Bruce Nauman [248], Joseph Beuys [310] and Stephan von Huene [244, 246], whose works are other examples of an inner and hidden sonority that has a connection with Cage's physical experience in the anechoic chamber, described in *Silence* (1961). Symbiosis and osmosis created a continual transition through which the acoustic instrument and the sculpture with or without sound, both linked to human action, sought an imagery with effects on seeing and hearing that seem to recall the dynamic forces defined by Michel Carrouges in his *Les machines célibataires* (1954).

On the plane of a machine that disturbs sensibilities and creates disquiet, owing to a series of shifts in meaning and insertions into its expressive values, we find the unstable and random dimension of an object that, brought into a succession of expectations on material and structure, is able to produce a fantastic and amazing "other." This visual and functional amplification has often been achieved, as was traditional in Cubism, Dadaism and Surrealism, through recovery of the remnant. This was taken up and reutilized for extraneous purposes, as with Ken Butler's musical instruments [298, 306], where everyday products from the axe to the suitcase were turned into guitars, violins or cellos. It is an attitude that recurs in William T. Wiley [274] and Pedro Reyes [432], for whom things, from metal cans to the firearms used by narcos, undergo a remodeling that produces an unexpected outcome, turning them into musical devices, while the instrumental results and effects remain traditional. And while 18th-century androids tended to imitate the living world, intent on bringing mechanisms to life and allowing them to evolve into human figures, their present day equivalent—the marionettes or robots of Dennis Oppenheim [282] and Maurizio Cattelan [368]—is also the opposite: the much-desired passage from artificial to organic, from art and music to bodily gesture.

It is a mirroring that highlights the vital component of art, which in Jannis Kounellis finds expression with the association between a visual element, such as a portion of a score (Igor Stravinsky's *Pulcinella*) and its repeated execution, to the point of exhaustion and withdrawal of the performers—a violinist and a ballerina [278]. The symbiotic association of object and subject by which they become two organisms that tend to swap roles is proffered again in painting, translated into vocal and acoustic material in the work of Roman Opalka [242] and Terry Allen [316], or in the musical instrument that Thomas Demand [406] turns into a photograph: it is almost a biotechnological evolution of sound in art. An osmosis between abstraction and performance, between musical tale and environment, between object and perception of sound that in the work of Laurie Anderson [296, 302], Takako Saito [294] and Céleste Boursier-Mougenot [428] requires audience participation, while in other cases the sound artifacts, alluding to Eastern and Western classical harmonies, or to Ethiopian and Brazilian music, to country or hip-hop, to the croaking of frogs and the ritual pealing of bells, can be activated by switches, timers or electrical connections—as in the work of artists like Bernhard Leitner [288], Rebecca Horn [354], William Anastasi [290], Eliseo Mattiacci [308], Marco Bagnoli [348], Manuel Rocha Iturbide [418], Janet Cardiff and George Bures Miller [430], Carsten Nicolai [350] and Anri Sala [408]—, and turning inside out and occupying space and time with emissions of sound that speak of nature, personal romantic adventures or bodily traversals of objects and places. Pictorial and sculptural prostheses that produce information and musical and acoustic narrations, constituting a new architecture of multisensory perception and comprehension whose meanings are related to the artist's autobiography.

Communication systems both within art and outside of it have continued to evolve since the 1990s, and enormous advances have been made in the use of data processing and electronics, the science of logic and artificial intelligence. Their effects have been amplified through the use of technological systems like the computer and the Internet, whose flows have had an influence on the structure, functionality, aesthetics and logic of objects and machines. With regards to hybrid artifacts, which interweave sound and art, the new acoustic creations increasingly resemble cyborgs. For Yoshimasa Wada [286] and Martin Kersels [416] they fuse with the human body, becoming its technological prostheses, albeit still primitive ones. The use of mechanisms seems to have also modified the mode of playing an instrument and producing sound, which changed its appearance. It is as if the configuration and iconography of the instrument were no longer important, and thus have been lost from view, and only emissions of sound count, for example in the abstract-technological and purely mechanical objects of Paul De Marinis [280], Tarek Atoui [426], Martin Creed [338], Brian Dewan [374], Ed Potokar [356], Bart Hopkin [346, 352], Haroon Mirza [400, 434] and Alberto Tadiello [422].

In other cases the evolution of mechanics has been rejected in the name of a memory of the past, when the instruments used to produce sound were still of a human dimension, and so we have seen a return to the 19th-century scale of Frédéric Kastner's pyrophone or Cocchi, Bacigalupo & Graffigna's street organ [124], as well as contemporary proposals like Tom Sachs's boombox [362] or Ruth Ewan's jukebox [370], which identify with the musical culture of rappers or political songs. Elsewhere constructions have grown more complex and been integrated into architecture. Rather than distancing themselves from the everyday environment they blend into it, proposing a double of it which can assume different linguistic connotations, from furnishings to functional use, as in the work of artists from Guido van der Werve [410] to Doug Aitken [420], Mirza, and Subodh Gupta [376]; or create a short-circuit that, as in the artwork by Athanasios Argianas [412], utilizes the architectural element in order to provide an unprecedented dimension to sound and images in relation to their internal and external urban locations.

The intensity of investigation of the sound artifact could remain limited to the polarity of art or sound, but we must not forget the active force of the performer, who represents the other *substance* in the passage from one to the other. The protocol of self-production, by manual winding or electric ignition, remains fundamental to understanding a sense of autonomous generation. But there is also a fusion between two polarities when these are activated by a person or human being. Here the fixity is translated into a physical, spiritual, organic and artificial journey that represents a vision of use and reflection on the object in question. The performer takes it, expropriates it and gives it another life. He or she brings the sound and the object out of their specificity and assigns another status to both. In this way its *virtuality* is enhanced, as in the case of Kounellis or Atoui, Anderson or Michelangelo Pistoletto [250], Akio Suzuki [272] and Maywa Denki [340]. The performer bolsters the potential of the musical and acoustic medium, be it picture or instrument, giving it real life, "embodying" it in a corporeity as in Loris Gréaud [372], Tore Honoré Bøe [436] and Neil Feather [358]. The involvement of the body renews its imagery by turning it into a real gesture, that of the concert. Here, however, music is not a staged event but the profundity of a regeneration of things found, composed and assembled that sometimes involves an alienation of gestures, in Kounellis again, or is expressed freely in the incandescence of the performance, as in Butler or Walter Kitundu [402, 404], Riccardo Beretta [424] or Reyes, Max Vandervorst [312] or Nicolas Bernier and Martin Messier [414]. In this case the mute and immobile dimension is subjected to a transmutation. Iconic silence gives way to objectified sound. Out of this come noises and musical flashes that reveal how the artifact is usually caught up in a trance that renders it immobile and paralyzed, when it can also be organic and dynamic. It comes back to life in waiting for the human gesture, with all its sensory complications, resounding with and smelling of its humors. Its pulsation generates circularity and endless gestation. Thus artistic research shows that it can radiate in all directions, forming a hinge between high and low, spirit and flesh, silence and noise, fragrance and stench, visible and invisible, between life and death. The aim is to opt for a labyrinth of the senses, for a geography of the formless and experientially indeterminate, in order to enter a multisensory cavity in which to brood a new birth.

1
Nina Levent, Alvaro Pascual-Leone (eds.), *The Multisensory Museum* (Plymouth: Rowan & Littlefield, 2014).

2
There is extensive literature on the relations between art and sound. Recent contributions from the perspective of the past and the present include both exhibitions and publications.

The main exhibitions worth highlighting include: "Sons & Lumières. Histoire du son dans l'art du XXe siècle," Centre Pompidou, Paris, September 22, 2004–January 3, 2005, exhibition catalogue edited by Sophie Duplaix, Marcella Lista (Paris: Centre Georges Pompidou, 2004); "Visual Music: Synaesthesia in Art and Music Since 1900," Museum of Contemporary Art, Los Angeles (CA), February 13–May 22, 2005 and Hirshhorn Museum and Sculpture Garden, Smithsonian Institution, Washington, D.C., June 23–September 11, 2005, exhibition catalogue edited by Olivia Mattis, Kerry Brougher (London/New York: Thames & Hudson, 2005); "SEE THIS SOUND. Promises in Sound and Vision," Lentos Kunstmuseum, Linz, August 28, 2009–January 10, 2010, exhibition catalogue edited by Manuela Ammer, Cosima Rainer, Stella Rollig *et al.* (Cologne: Walther König Publisher, 2010); "Sound Art. Sound as a Medium of Art," ZKM | Media Museum, Karlsruhe, March 17, 2012–January 6, 2013, exhibition catalogue edited by Peter Weibel (Karlsruhe: ZKM Publisher, forthcoming); "Soundings: A Contemporary Score," Museum of Modern Art, New York, August 10–November 3, 2013, exhibition catalogue edited by Barbara London (New York: The Museum of Modern Art, 2013).

Recent publications worth mentioning include: David Toop, *Haunted Weather: Music, Silence and Memory* (London: Serpent's Tail, 2004); Alan Licht, *Sound Art: Beyond Music, Between Categories* (New York: Rizzoli, 2007); Caleb Kelly, *Cracked Media: The Sound of Malfunction* (Cambridge, MA/ London: The MIT Press, 2009); Seth Kim-Cohen, *In the Blink of an Ear: Toward a Non-Cochlear Sonic Art* (New York/London: Continuum, 2009); Salomé Voegelin, *Listening to Noise and Silence: Toward a Philosophy of Sound Art* (New York/ London: Continuum, 2010); Patrick Feaster, *Picture of Sound: One Thousand Years of Educed Audio: 980-1980* (Atlanta, GA: Dust-to-Digital, 2012); Jonathan Sterne (ed.), *The Sound Studies Reader* (London/ New York: Routledge, 2012); Douglas Kahn, *Earth Sound Earth Signal: Energies and Earth Magnitude in the Arts* (Berkeley, CA: University of California Press, 2013).

3
Only a few museums, including the Victoria & Albert in London and the Art Gallery of Ontario, Toronto, have attempted to establish a relationship between modern art and odor, while the sensory interweaving of perfumes and aesthetic objects have been presented in exhibitions like "Five Holes: reminiSCENT," curated by Jim Drobnick and Paul Couillard, Toronto, 2003, and "Odor Limits," curated by Jim Drobnick and Jennifer Fisher, Philadelphia, 2005. For a bibliography see Jim Drobnick, "Reverie, Assault and Evaporating Presences: Olfactory Dimensions in Contemporary Art," *Parachute*, no. 89, Montreal, 1998, pp. 10–19; Anna Barbara, Anthony Perliss, *Architetture invisibili: L'esperienza dei luoghi attraverso gli odori* (Milan: Skira, 2006); Patrizia di Bello, Gabriel Koureas (eds.), *Art History and the Senses: 1830 to the Present* (London: Ashgate, 2010); Peter De Cupere, "The Use of Smell in Art, an 'Olfactology' Art Research," a lecture held at the College Art Association Conference, New York, February 16, 2013. The involvement of touch has often been confined to copies of artifacts or the manipulation of raw rather than worked materials. Here it is the preciousness of the object and its historical rarity that prevents its sensorial exploration. However, contemporary art ought to open itself up to this bodily approach, in order to offer experiences that are not merely visual. For a recent bibliography: Helen J. Chatterjee (ed.), *Touch in Museums: Policy and Practice in Object Handling* (Oxford, UK: Berg, 2008); Sandra H. Dudley (ed.), *Museum Materialities: Objects, Engagements, Interpretations* (New York: Routledge, 2010).

4
Constance Classen (ed.), *The Book of Touch* (Oxford, UK/ New York: Berg, 2005).

5
Gian Paolo Ceserani, Gli *automi. Storia e mito* (Bari: Laterza, 1983); Mario G. Losano, *Storie di automi* (Turin: Einaudi, 1990) and id., *Automi d'Oriente* (Milan: Medusa, 2003); Carlo Sini, *L'uomo, la macchina, l'automa* (Turin: Bollati Boringhieri, 2009); Bruno Corà, Pietro Bellasi (eds.), *Corpo, Automi, Robot: Tra Arte, Scienza e Tecnologia*, exhibition catalogue (Lugano/Milan: Museo d'Arte Città di Lugano/Mazzotta, 2009); Roberto Peverelli (ed.), *Gli automi sono tra noi* (Milan: Medusa, 2013).

6
Stefano Pivato, *Il secolo del rumore* (Bologna: Il Mulino, 2001); Kyle Gann, *No Such Thing as Silence* (New Haven, CT: Yale University Press, 2010).

7
Yve-Alain Bois, Rosalind Krauss, *Formless: A User's Guide* (New York: Urzone, 1997).

DEIRDRE
LOUGHRIDGE

MARVELOUS ILLUSIONS

VISUAL AND MUSICAL BEAUTY FROM THE RENAISSANCE THROUGH THE 18TH CENTURY

Beautiful sound should come from beautiful objects. Such was the basic assumption guiding musicians and instrument-makers in the 16th to 18th centuries. The fruits of this assumption can be seen not only in the marvelous instruments from the period preserved in museum collections—instruments like the harpsichord attributed to Vitus de Transuntinus (1560), held by the Berlin Musical Instrument Museum, which makes its twin aesthetic aims explicit in the motto inscribed above its keyboard: *Rendo lieti in un tempo gli occhi e 'l core* (I make the eyes and the heart happy at the same time). The precept is also evident in the standard forms of musical instruments that have been handed down to us from this time. Physicists are still working to understand how the contours and proportions of the *violin*—a shape that emerged in mid 16th-century Italy and reached perfection with the Amatis, Guarneris and Stradivarius—contribute to its sound quality. Using three-dimensional scanning lasers to map violins and their vibrational patterns, the Strad3D project has analyzed some of the world's most prized specimens, finding the placement and dimensions of the F holes and C bouts, for example, to be essential to determining the flexibility of the violin's top piece, and hence to the way it projects sound. As Samuel Zygmuntowicz explains, "a very lovely shape turns out to have a great deal of sophistication from a design point of view … and changing any one of those aspects will have an effect on the sound."[1]

If the unity of art object and sounding instrument was a basic condition of musical culture from the Renaissance to the turn of the 19th century, however, there were nonetheless also extraordinary cases —ones that called into question, or suggested new interrelations between visual and sonic, durable and ephemeral beauty. These centuries in fact can be seen as a period of transition during which a twinned concern with musical and visual beauty gradually dissolved, giving way to an assumption of separation and independence between these sensuous arts. This process can be traced through extraordinary cases that tested the assumption of unity: cases like the visually magnificent yet acoustically weak instruments of Michele Antonio Grandi [50] or Giovanni Battista Cassarini [52]; the perfect yet impossible instrument designs speculated by philosophers like Athanasius Kircher; Louis-Bertrand Castel's color keyboard and its ambivalent reception; and the musical automata of the 18th century, which located visual beauty not in the musical object alone but in its combination with a moving, human figure.

The beauty of an object stems not only from its shape, but also from the materials with which it is made. And so while the typical form of instruments like the violin, guitar and lute arose from a mutual concern with sound and appearance, these objects could be further enhanced through the use of precious materials. Gold, marble, gems—such materials increased the visual marvel of the instruments they comprised or adorned, making them especially attractive to princely collectors. But in such cases, visual beauty often trumped acoustic properties, resulting in unplayable or impossible musical instruments. Take, for example, the marble guitar thought to be made by Michele Antonio Grandi around 1687 for the House of Este. It is meticulously constructed in the shape of a Baroque guitar, with a peg box and tail-piece for attaching five double courses of strings, accurate frets, and rings carved into the marble for attaching a shoulder strap. But there is no sound hole in the body of the instrument; rather, a circular design appears on the marble, where the sound hole would be expected. Further, the nut is too low to lift the strings above the frets, and the first and last strings are placed too near the end of the fingerboard. As Gabriele Rossi-Rognoni, curator of instruments at the Royal College of Music, London, concludes, "it seems that the instrument is a marvelous illusion, created solely to please the eye with its extraordinarily accurate rendering of all the structural details of a guitar without, however, being —or ever having been—playable."[2]

While artisans, working for royal patrons, sometimes sacrificed acoustic resonance and playability for visual effect, philosophers envisioned musical designs based on mathematical ideals, conjuring speculative instruments that would not have behaved as they imagined. Many early modern thinkers believed the spiral to be specially related to sound. The shape features prominently in English philosopher Robert Fludd's *Temple of Music* (1618), where it represents the spiral movements of air thought to be caused by sound. The spiral was also commonly held to have sound-amplifying powers, based on the fact that the ear contains a spiral-shaped structure—the cochlea—which was believed responsible for making sounds louder. Based on this reasoning, the Jesuit polymath Athanasius Kircher described an instrument called the "Tubo Cochleato," which he maintained would greatly increase the sound of the voice when one spoke into it.

Perpetuating Kircher's fantastical design, Filippo Bonanni observed in his 1722 treatise on musical instruments, *Gabinetto armonico pieno d'istromenti sonori. Indicati e spiegati* (Musical Cabinet Full of Sounding Instruments, Shown and Explained), that one could hardly expect to make a working "Tubo Cochleato," since it would require inordinate skill to fashion the tube in a perfect spiral.

Another line of speculative design, however, was more readily put to the test: one based on an analogy between color and pitch. Isaac Newton popularized this analogy in his 1704 treatise *Opticks*. Observing the colors made visible when sunlight shone through a prism, he suggested that their relative width corresponded to the relative lengths of a string sounding the seven pitches of the diatonic scale. Based on this analogy, the French philosopher Louis-Bertrand Castel argued that it should be possible to create a new temporal art of colors—a music for the eyes. Though he was not an artisan by training, the interest aroused by his proposal drove him to build an ocular harpsichord in which keys activated colored silks or glasses. "The principal advantage of this new harpsichord," Castel wrote, "is thus to give to the colors … a certain vivacity and lightness which on an immobile and inanimate canvas they never have."[3] Visual beauty,
1 rather than being static and solid, could now take on the fleeting and dynamic quality of music.

Yet in practice, many found Castel's instrument unsatisfactory. The German professor Johann Gottlob Krüger agreed with the instrument in principle, arguing that, "our mind, in judging the pleasure or displeasure of the things that reach our eyes, follows the same laws and rules that are followed in distinguishing musical consonance from dissonance."[4] But he identified a problem at the conceptual level: Castel's instrument captured the sequential dimension of music but not its simultaneous aspect—a melody of colors without a harmony. Krüger thus proposed a color keyboard that would produce "color chords": its colored lights would appear in concentric circles, lower pitches being larger and higher pitches being superimposed within.

Others, meanwhile, found the very idea of music for the eyes untenable. As Jean-Jacques Rousseau argued, "to multiply the sounds heard at the same time or to develop colors one after another is to change their economy, to put the eye in the place of the ear, and the ear in the place of the eye."[5] For Rousseau and a growing number of thinkers, visual and musical beauty were incompatible: the former was based on simultaneities in space, the latter on sequences in time. As the arts were different in their essential nature, according to this view, it was folly to attempt to translate between them.

A similar tension surrounded the relation between the sound of music and the sight of performing musicians—a relation that 18th-century makers of musical automata explored in their art objects. As one maker of musical automata, Johann Carl Enslen, remarked of his pieces, "I have spared neither hard work, effort nor expense to fabricate a mechanical musical theater in which music, painting and sculpture combine, constituting the main objects, which appear in the form of life-like automata or artificial moving figures…. Viewers are entertained and surprised most agreeably."[6] As unions of mechanical, visual and musical art, musical automata reflected ideas about what a beautiful musical performance should look as well as sound like. Music teachers of the 18th century regularly counseled their students against grimacing, such disfigurement of the face—though a natural result of intense effort—being incompatible with ideals of grace. The French harpsichordist François Couperin recommended placing a mirror on the harpsichord, where the music would normally sit, in order to catch and correct oneself, and further taught that "one must have an air of ease at this harpsichord, without fixing his gaze too much on one object or looking too vague."[7] The makers of musical automata took such advice to heart, creating machines that mimicked the graceful appearance desired of performers. In doing so, they went beyond the work of artisans who made beautiful instruments, making the human figure part of the art object as well.

But if the ocular harpsichord had prompted Rousseau to articulate the difference between visual and musical arts, performers also taught him that visual and musical beauty did not go hand in hand. On a trip to Venice in 1743, he attended a church where he heard motets sung by girls who were concealed behind grilles. "I cannot conceive of anything so pleasurable or so moving as that music," he wrote. "What distressed me were the accursed grilles, which only let the sound through but concealed those angels of beauty—for the singing was worthy of angels—from my sight." He thus sought to meet the girls who produced such beautiful music. But upon being introduced to them, he discovered that rather than being possessed of angelic faces to match their voices, they were "hideous," one having only one eye and another being "disfigured by small pox." A couple of the girls seemed "passable" to him, but it turned out they were not the angelic soloists but sang only in the chorus.[8]

The account thus underscored the mismatch between musical and visual beauty—a mismatch that took listeners by surprise in the mid 18th century, and helped spur a parting of ways between visual and sonic arts.

Bartolomeo Cristofori, instrument-maker to the Grand Duke of Tuscany, invented the piano at the turn of the 18th century. Eleanor Selfridge-Field has suggested that Cristofori invented the piano not in response to musical needs or interests, but rather as a kind of scientific experiment—an application of Newton's new-found laws of force and motion.[9] The piano, in other words, was born out of an interest in the object of art—in material forms and properties—at least as much as it was born out of an interest in making music.

By the mid 18th century, however, much experimentation in instrument design was explicitly focused on tone quality and playability, with concerns for visual beauty increasingly taking a back seat. The glass harmonica, and its various incarnations, can be seen as part of this trend: an instrument celebrated for its celestial tone, it was also praised for letting listeners hear "the effect without the cause"—pure sound without any trace of material source.[10] Such an ideal emerged from the rifts between musical and visual beauty that accumulated over the 16th to 18th centuries, as artisans and philosophers pursued various ways in which they could, and could not unite. Ultimately, these rifts opened up the gap between art objects and sound objects that artists and audiences have continued to explore to this day.

1
"Strad3D: Video Tour," strad3d.org/demo/videoTour2.html (accessed March 7, 2014).

2
Franca Falletti, Renato Meucci, Gabriele Rossi-Rognoni (eds.), *Marvels of Sound and Beauty: Italian Baroque Musical Instruments*, exhibition catalogue (Florence: Giunti, 2007), p. 178.

3
Louis-Bertrand Castel, "Clavecin pour les yeux avec l'art de peindre les sons et toutes sortes de pièces de musique," *Mercure de France*, Paris, November 1725, p. 2573, trans. in Maarten Franssen, "The Ocular Harpsichord of Louis-Bertrand Castel," in *Yearbook for the History of Science, Medicine, Technology and Mathematics*, no. 3 (Utrecht: Institute for the History of Science, 1991), pp. 15–77: 21.

4
Johann Gottlob Krüger, "De novo musices, quo oculi delectantur, genere," in *Miscellanea Berolinensia ad incrementum scientiarum, ex scriptis Societatis regia Scientiarum exhibitis edita*, vol. 7 (Berlin: Johann Gottfried Michaelis, 1743), pp. 345–357: 352.

5
Jean-Jacques Rousseau, *Essay on the Origin of Languages and Writings Related to Music*, trans. and ed. by John T. Scott (Hanover, NH: University Press of New England, 1998), p. 325.

6
Wiener Zeitung, December 7, 1791, p. 3131.

7
François Couperin, *L'Art de toucher le Clavecin*, trans. and ed. by Margery Halford (Van Nuys, CA: Alfred Publishing, 1995), p. 30.

8
J.M. Cohen (ed.), *The Confessions of Jean-Jacques Rousseau* (Harmondsworth: Penguin Books, 1954), pp. 295–96.

9
Eleanor Selfridge-Field, "The Invention of the Fortepiano as Intellectual History," in *Early Music*, vol. 33, no. 1 (Oxford, UK: Oxford University Press, February 2005), pp. 81–94.

10
Anne Ford, *Instructions for Playing on the Musical Glasses*, London, 1761. On the 18th-century quest for ideal tone, see Emily I. Dolan, "E.T.A. Hoffmann and the Ethereal Technologies of 'Nature Music'," in *Eighteenth-Century Music*, vol. 5, no. 1 (Cambridge, UK: Cambridge University Press, 2008), pp. 7–26.

CHRISTOPH E.
HÄNGGI

SOUND, MUSIC, TIME AND MOVEMENT

ANTIQUE MECHANICAL MUSIC MACHINES

In ancient Greek and Roman scientific works, and in texts from the Middle East, scientists, technicians and mechanics, such as Ktesibios of Alexandria, Philon of Byzantium, Heron of Alexandria, the Muhammed Brothers, and Ahmed and Hassan Mûsà, frequently described organs, flute players, singing birds and chimes that were powered by water or wind. During the Middle Ages in Europe, knowledge of mechanical organs and other music machines waned, and no trace remains of the corresponding objects.

SOUND AND TIME

The oldest surviving mechanical music machines are carillons, which are controlled by cylinders fitted with teeth. The sound of bells enhanced the daily lives of the people and served to represent church dignitaries and medieval communities. In the 12th and 13th centuries, church clocks began to be fitted with bells, which developed into mechanical carillons in the 14th century. The oldest surviving carillon is the one that was fitted in the clock in the cathedral of Beauvais in 1310. Another early carillon was created for a similar clock in the cathedral of Strasbourg between 1352 and 1354. Arithmetic, geometry, astronomy and music—the ancient Quadrivium—underlay the construction of monumental astronomical clocks, which were sometimes not only furnished with carillons, but also other musical effects, such as cymbal crashes, rooster crows, organ piping and the like. Automated carillons, i.e. those powered by toothed cylinders, and those operated by hand quickly became known as carillons in what is now northern France and Benelux, where even today around 400 of them are still played daily in cathedrals and smaller churches, as well as town halls and other imposing buildings. However, the word *glockenspiel* remains the preferred term for this tradition in the German-speaking regions from Lübeck to Prague to Bern.

TIME AND MUSIC

The complicated mechanisms developed for clocks in the late Middle Ages and early Renaissance were also applied to the mechanisms used in operating musical instruments. As early as the start of the 14th century, clocks had found their way into the palaces of the gentry, the cloisters of the clergy and the dwellings of the wealthy bourgeoisie. When clockworks became smaller, so did carillons, resulting in domestic clocks in a variety of forms.

Developed in the 16th century, mechanical organs, also known as flute clocks, musically heralded the hours of the day in the chambers of kings and princes. These mechanical musical instruments with open or enclosed pipes were operated by a toothed cylinder, on which specially composed or arranged melodies had been prepared. An incorporated clockwork mechanism prompted the cylinder to rotate at regular intervals.

The wealthy city of Augsburg in southern Germany was the center of European clock-making during this period of the late Renaissance. Together with Nuremberg, it was also at the forefront of machine design and construction. In addition to carillons and mechanical organs, the masters of Augsburg and Nuremberg created table clocks and cabinet clocks fitted with mechanical string instruments, such as virginals, spinets and harpsichords, and trumpet mechanisms operated by toothed cylinders or wheels fitted into exquisite bureaus. Samuel Bidermann, Achilles and Veit Langenbucher, Matthäus Rungell and Hans Schlottheim were all master machine-builders, while the music for the cylinders of these instruments was composed by such highly regarded musicians as Hans Leo Hassler and Christian Erbach.

With the Baroque and Rococo, cabinet clocks and other pieces of furniture evolved and their musical content became more expressive. The number of pipes increased and the musical repertoire expanded by the newly developed opportunity to change the cylinders. Smaller instruments were incorporated into harp and flute clocks. In the second half of the 18th and throughout the 19th century, flute clocks remained very popular, and their artistic housings indicate that they were intended for the nobility. Nevertheless, these instruments, which were originally devised as gifts for emperors and kings, ultimately found their way into the circles of wealthy bourgeoisie as well as public art displays and dance halls. In part, flute playing was emancipated from the workings of clocks, as for example with the serinettes, which were made to imitate the chirping of birds.

In Great Britain, London was the center of flute clock production during the 18th century. It was here in 1718 that Henry Bridges constructed his clock called "The Microcosm," and Charles Clay, James Cox, James Newton, Eardley Norton and John Pyke and his son George Pyke all produced clocks with flute-playing mechanisms. In Switzerland, Pierre Jaquet-Droz [60, 62], his son Henri-Louis Jaquet-Droz, and Jean-Frédéric Leschot were all leading figures during the 18th century, as were Julien Leroy, his son Pierre Leroy and Jacques de Vaucanson in Paris. Vaucanson's famous flute player included its own music, and Jaquet-Droz also built flute-playing mechanisms into his machines as well as his clocks. But there were also important clock-making centers in the Netherlands and Sweden, with highly skilled clock-makers such as Cornelius Engeringh, Dietrich Nicolas Winkel and Per Strand. Flute clocks were also made in various German cities, such as Hannover, Göttingen and Neuwied am Rhein, where David Roentgen and his mechanic Peter Kinzing built the highest quality upright clocks with flute- and string-playing machines. In particular, a leading flute clock industry developed in Berlin, due in part to the initiative of Fredrick the Great, who established residences in the city for Swiss clock-making families from Neuchâtel and Geneva, and Huguenot clockmakers from France, thus spurring the city's native clockmakers. The most famous manufacturers of flute clocks in Berlin were Christian Ernst Kleemeyer and Christian Möllinger.

Famous composers like Georg Friedrich Händel, Franz Benda, Wilhelm Friedemann Bach and his younger brother, Carl Philipp Emanuel Bach, wrote original works for these instruments. Händel's compositions are also found on an instrument by Charles Clay, a London clockmaker who specialized in sophisticated pendulum clocks with bells and flutes during the first decades of the 18th century. But there is also a flute clock by Charles Pyke that features music by Händel, who had worked in London since 1709.

After Berlin, classical Vienna developed into a second center of flute clock production. [74] Around 1800, there were at least twenty manufacturers there, for instance, Joseph Primitivus Niemecz, who as a student and friend of Haydn was inspired by him to build a number of flute clocks. Niemecz was most likely also the manufacturer of the flute clock in Josef Graf Deym von Stržitéž's art display in Vienna, which played Mozart's *Fantasy for a Drum Organ*. Other names in Viennese flute clock manufacturing deserving of mention are Franz Aegidius Arzt, Eduard Heinrichssohn, Franz Erbs, Johann Nepomuk Mälzel (who also invented the metronome), Johann Georg Strasser, Leonhard Mälzel, Johann Adam Hoyer, Georg Anton Reinlein and his son Rudolf Reinlein, Johann Christian Heinrich, Joseph Franz Erbs and Joseph Arzt.[1]

In the second half of the 18th century, popular pieces of music were transcribed for flute clocks, including arias and famous instrumental works. Joseph Haydn, Antonio Salieri, Franz Anton Hoffmeister, Wolfgang Amadeus Mozart, Joseph Ignaz Pleyel, Luigi Cherubini and Ludwig van Beethoven all composed pieces directly for these mechanical instruments. As these composers all worked for a certain period in Vienna, where they could not have ignored the overwhelming demand for flute clocks, it is natural that they would compose for these instruments as well.

A number of Haydn's many short works for music machines are adaptations from his symphonies, quartets or piano works, however, many of them were composed specifically for the flute clock. Today they are among the best known compositions of this genre, and they are a brilliant reflection of the social and cultural currents of their time, such as stylized military music and newly fashionable coffeehouses. Or they simply represent the idleness of a bored wealthy elite that sat around amusing itself with music from machines as the accompanying clocks counted the hours. Haydn composed three series of works for flute clocks that he dedicated to different people. The first, a series of 16 pieces, was composed in 1772 for the baptism of the daughter of the choir master in Vienna, Florian Leopold Gassmann, which he dedicated to his friend's wife.

The second set, composed twenty years later in 1792, was written for the royal house of Lichtenstein. The third series was composed in 1793 for Prince Nikolaus Esterhazy, before his departure for London. Haydn used some of the pieces from the first series again in 1792/93, making a total of 32 pieces for flute clocks. It is unusual that the three instruments for which these pieces were composed still exist today, and they give us a valuable key to understanding the type of music played in that period.

Mozart's three surviving compositions for flute clocks were written for Count Deym von Stržitéž's cabinet of curiosities. In a scenario typical of the late 18th century, Count Deym returned from exile in 1770 to a glamorous Vienna under the name Joseph Müller. He ran the "Müllersche Galerie," an establishment similar to a wax figure cabinet, which offered amusements to the public. A section of this gallery was set up as a curiosity cabinet and from 1790 there was also a room that commemorated the late field marshal, Gideon Ernst Freiherr von Laudon. There were various flute clocks in this cabinet of curiosities, of which a few were created in collaboration with the clockmaker Niemecz. Mozart's *Fantasie KV 594* was composed in December 1790, having been commissioned by Count Deym as funereal music for the deceased field marshal to be incorporated in the largest of his flute clocks, and was first played on 23 March 1791.

The instrument included both the flute and reed registers and was therefore larger than a normal flute clock, whose range Mozart had found too restricting. The result was a programmatic work in remembrance of the high-ranking soldier. Mozart's other works for flute clock were also composed for Count Deym. On 3 March 1791 he completed the *Fantasie KV 608*, which was first played in public after the cabinet had been moved to another district of Vienna and after the mausoleum was remodeled into an Elysium for Field Marshal Laudon and Kaiser Joseph II, who died in February 1790. Lastly, the *Andante KV 616* was the final work Mozart composed in this series, intended for a much smaller flute clock.

Ludwig van Beethoven's works for mechanical musical instruments should without a doubt be considered 18th-century compositions. As a young composer, Beethoven was still very much under the influence of Haydn and Mozart, however, he was a friend of Count Deym, which allows us to suppose that his compositions in Vienna from 1799 were also commissioned by the count.

Flute clocks were replaced in the 19th century by clocks with modern music mechanisms. By the end of the 18th century, the Swiss had developed music boxes and the corresponding new method of tone production through the use of tone combs and steel plates. These new techniques were also applied to cabinet and wall clocks in the 19th century, replacing the cumbersome carillons and flute clocks, which were also more complex to manufacture. The most popular arias and best known instrumental pieces continued to be arranged for mechanical reproduction; however, they were now played using music-box mechanisms. Composers no longer wrote pieces specifically for music boxes, instead, arrangers adapted the most popular musical pieces of the period.

MUSIC AND MOVEMENT

The combination of music and movement was the goal of skilled machine-builders like Jacques de Vaucanson and Pierre Jaquet-Droz, both of whom were pioneers in the development of music machines, and this aim remains current today in many forms of contemporary art. In the art of clock making, another challenge was to incorporate moving figures—Jacquemarts—in church-tower clocks and Gothic wall clocks, to create the impression of bells ringing or anvils being hammered.

Shortly before the turn of the 16th century, Leonardo da Vinci experimented with machines that imitated human movement.[2] Similarly, more than a century later, René Descartes investigated the bodies of animals and humans, comparing them to machines and automatons. Technically accomplished machine-makers relied on moving figures in their constructions to impress the public and scientific circles alike.

For example, the automaton builder François-Joseph Camus designed a horse-drawn carriage with moving figures and presented it at the court of Ludwig XIV, and the French engineer Jacques de Vaucanson, and Swiss clockmakers Pierre Jaquet-Droz and Henri Maillardet, thrilled an enthusiastic public with their automatons during the century of the Enlightenment.

Vaucanson's "The Flute Player" marked the first successful synchronization of automated human-like movement with a mechanically reproduced piece of music. Born in Grenoble, Vaucanson devoted four years of his life to this, his first automaton, only completing it in 1737. He presented his Flute Player at an exhibition in Paris, where it enjoyed great success. Toward the end of 1738, however, because fewer people were visiting the work, Vaucanson decided to create two additional automatons—a pipe and tabor player and a mechanical duck.

Pierre Jaquet-Droz, from the Neuchâtel canton in Switzerland, had studied philosophy in Basel, but he switched to clockmaking and automaton building, developing the art to new heights. From his factory and branches in Geneva, Paris and London, Jaquet-Droz supplied a vast clientele of French and English nobility and bourgeoisie with his works. In 1758, with a carriage he had built expressly for the trip, he traveled to Madrid, where he was received by Ferdinand VI, who purchased six particularly valuable pendulum clocks.

Between 1768 and 1774 Jaquet-Droz, his son Henri-Louis Jaquet-Droz, and Jean-Frédéric Leschot together constructed his most famous automatons to date: the Writer, the Draughtsman and the Musician, which were presented in 1774 at La Chaux-de-Fonds in the Jura to enormous success. Each of the three automatons contains between 2,000 and 6,000 individual pieces and is a masterpiece of the genre. People began to travel to La Chaux-de-Fonds to see the works, as a result of which Jaquet-Droz decided to take them on tour. In Paris, the first stop, the precision of the mechanical movements of the automatons was met with great enthusiasm, after which the machines were taken to the royal courts in Brussels, London, Moscow and Madrid. Shortly before the outbreak of the French Revolution, the automatons were sold in Spain, thus they avoided the destruction in France.

As from 1812, they resumed their tour of Europe and went even further, traveling to China, India and Japan. In the 19th century they frequently changed ownership; however, rather than being publicized with their owner's name, it was that of Jaquet-Droz that remained engrained in the public's mind, which in turn brought international recognition to the Jaquet-Droz company's production of clocks, tobacco boxes and singing birds. At the beginning of the 20th century, all three automatons were returned to Switzerland.[3]

Jaquet-Droz built a fourth automaton that has since disappeared. It was known as Grotto, and represented an idyllic landscape with cliffs, cottages, mills, waterfalls, brooks, fountains, sheep, birds and human figures. A description of the automaton mentioned a shepherd, who may have mechanically played music like Vaucanson's flautist and may have included singing birds that functioned like those on Jaquet-Droz' snuff tins. There is also a corresponding illustration.[4] However, even if this piece was actually built, it no longer exists today.

Other noteworthy automaton builders of the late 18th century included Friedrich von Knaus, who built a Writer for the Habsburg court in Vienna, Abbé Mical, who produced talking heads and a Flautist, and Peter Kinzing and David Roentgen, who created a Cymbal Player for the French court in Versailles. The Swiss clockmaker, Henri Maillardet, a contemporary of Jaquet-Droz who worked closely with these men, constructed automatons with his brothers, Jacques Rodolphe and Jean-David, some of which still exist today. George III of England had a Writer built as a gift for the Chinese emperor. Another automaton could both write and draw, and various magicians are known. Lastly, there was also mention of Friedrich Kaufmann's automaton of a trumpeter constructed in Dresden between 1810 and 1812.[5]

In the 18th century, automatons built as uniquely individual pieces strongly impressed the public and were occasionally used for medical studies. Along with the construction of automatons and clocks, numerous automaton builders—including Jaquet-Droz and his partner Leschot—began to manufacture prosthetics, which sometimes even incorporated movements, such as knee bends.

The works made by the automaton builders were not only impressive at the time; they remain impressive even when viewed from today's perspective. Their goal was not just to reproduce human movement, such as the playing of the flute or piano, writing or drawing, but also to incorporate human physiognomy and breathing in order to create beings that so resemble people or animals as to stir a range of feelings from elation to fear in the viewer.

In a certain sense, the 19th century led to the industrialization of automatons. The constructions were simplified, so that the individual subjects such as a flautist, piper, drummer or draughtsman could be manufactured in larger numbers, though not yet mass-produced. With its many ateliers, Paris became the main center of production of this kind of automaton.

1
See Helmut Kowar, *Die Wiener Flötenuhr: "Sie spielt besser als das Orchester im Kärntnertor"* (Vienna: Technisches Museum Wien, 2001), and id., "P. Primitivus Niemecz (1750–1806): seine musikalischen Spieluhren und Maschinen," in *Das mechanische Musikinstrument*, no. 115 (Essen: Gesellschaft für Selbstspielende Musikinstrumente e.V., December 2012), pp. 7–13.

2
Claudio Giorgione, "Leonardo da Vinci e le macchine: forme del corpo e automatismi," in Bruno Corà, Pietro Bellasi (eds.), *Corpo, Automi, Robot* (Milan: Edizioni Mazzotta, 2009), pp. 65–69.

3
See also *Automates & Merveilles, Les Jaquet-Droz et Leschot*, exhibition catalogue, Musée d'Art et d'Histoire de Neuchâtel (Neuchâtel: Editions Alphil, 2012).

4
Charles Perregaux, François-Louis Perrot, *Les Jaquet-Droz et Leschot* (Neuchâtel: Attinger Frères, 1916).

5
See Rebecca Wolf, Friedrich Kaufmanns, *Trompeterautomat. Ein musikalisches Experiment um 1810*, in the series Beihefte zum Archiv für Musikwissenschaft, vol. 68 (Stuttgart: Franz Steiner Verlag, 2011).

Bartolomeo Veneto

WOMAN PLAYING A LUTE
1520

Bartolomeo Veneto (*c.* 1470s–*c.* 1540s) was an Italian painter who studied with Giovanni Bellini and worked in Veneto and Lombardy, where he was influenced by Leonardo da Vinci and Northern European art.

In the painting a woman is depicted in the guise of Saint Cecilia, the patron saint of music, while playing a lute. Beside her is a music book that shows part of a score from a frottola for three voices.

The lute existed in several sizes and had a variable number of strings. It was the most characteristic musical instrument in Renaissance Italian courts, and the most popular for accompanying secular songs.
At the end of the 16th century, lute-players changed the way they played in order to play Renaissance polyphony on a single instrument, abandoning the plectrum in favor of plucking strings with the fingertips, like the young woman posing for the artist.

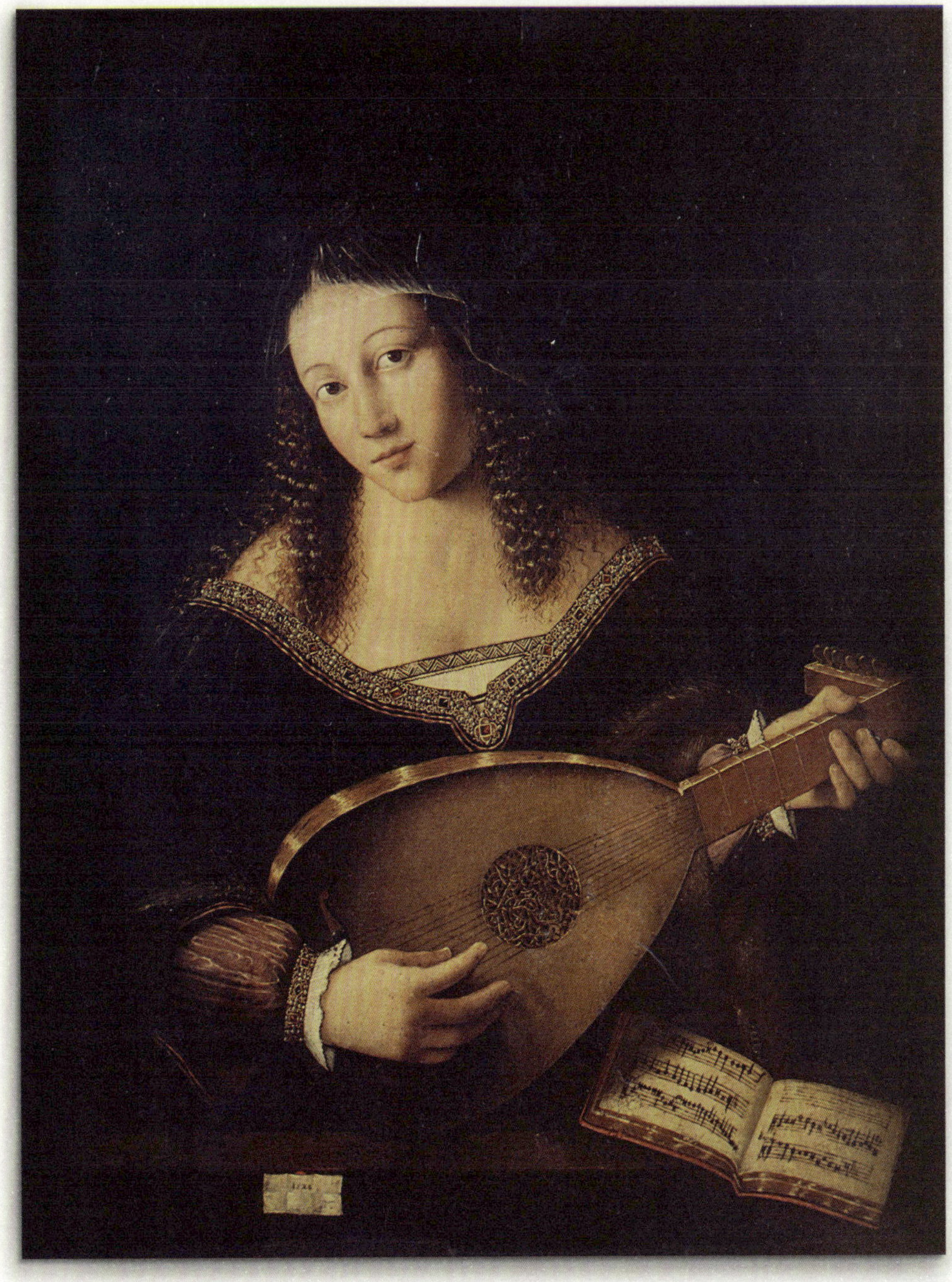

Nicola Giolfino

ALLEGORY OF MUSIC 1521–30

Nicola Giolfino (Verona, 1476–1555) was born into a family of wood-carvers and was a pupil of Liberale da Verona.

The fresco *Allegory of Music* was detached in 1873 from a room in the Theatine Convent of San Nicolò and is part of a cycle of 7 female figures representing the liberal arts now kept in the Museo di Castelvecchio. Here a young girl singing and playing the hammer dulcimer personifies music. A plaque bearing the first phrase of the soprano's part in the 1507 frottola *Ecco che per amarte* by Bartolomeo Tromboncino, a trombonist and composer from Verona, hangs from the branch of a laurel tree.

The frottola was a secular musical composition, generally for four voice parts, that became popular in Italy in the late 15th century. Towards the middle of the 16th century it developed into the more literary madrigal. Its principal characteristic was that it amalgamated an assortment of unrelated and sometimes quite senseless thoughts and facts in an irregular meter, characterized by both a casual sequence of verses of different tempo and an unpredetermined rhyming order.

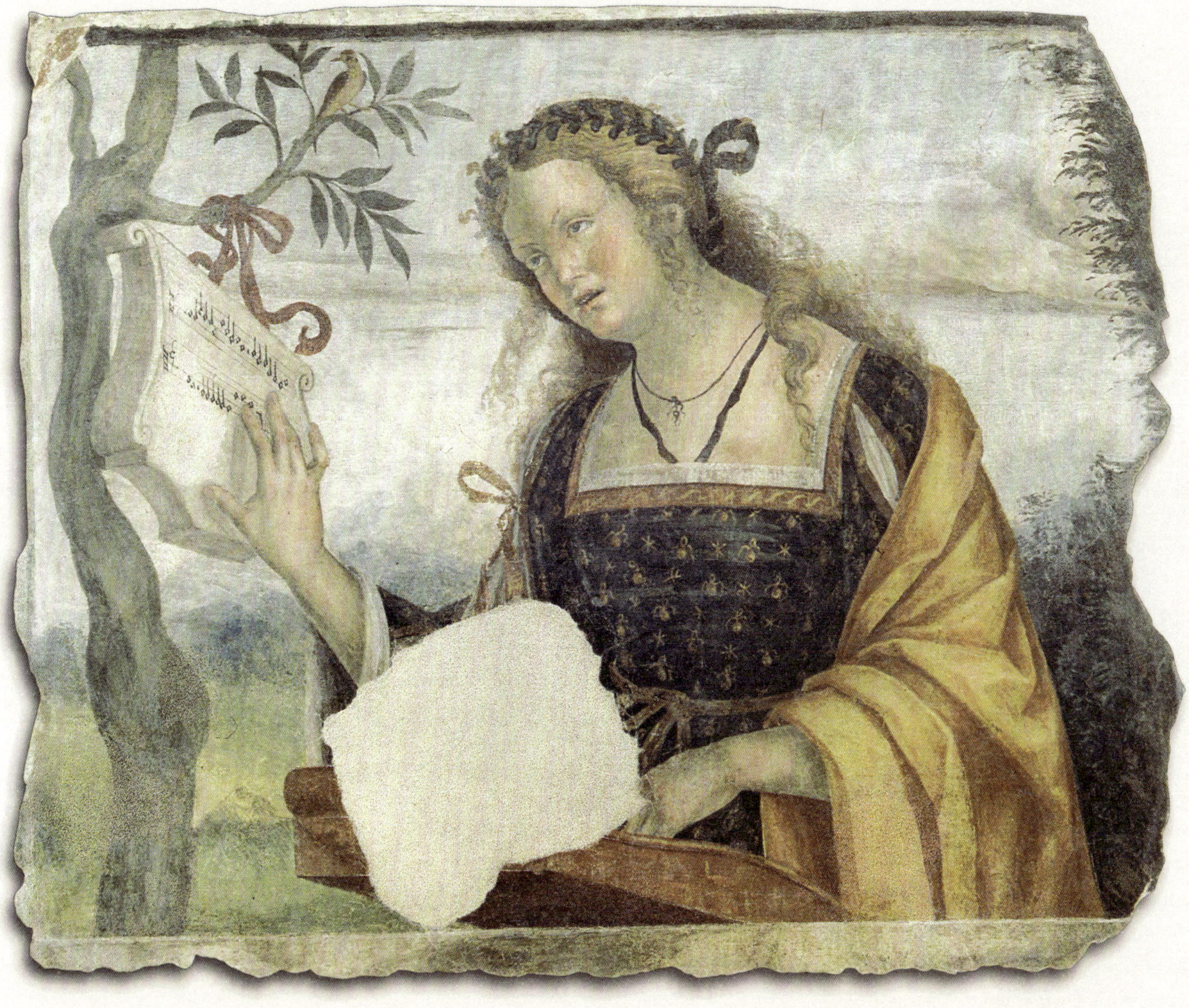

Unknown

LYRA-CISTER
16th–17th century

In creating this lyra-cister (or lyra-chitarrone), which is played by plucking 6 vibrating strings and 8 metal drone strings, the maker took his inspiration from ancient instruments for both the decorative volutes and tendrils and in the 2 arms of the blue and gold body, which resembles a lyre in which the yoke, or crossbar, has been replaced by 2 necks.

The unusual appearance and rarity of this instrument, of which only two known examples exist, are probably due to its use in theatrical and operatic productions set in the Classical age, typical of the late Italian Renaissance.

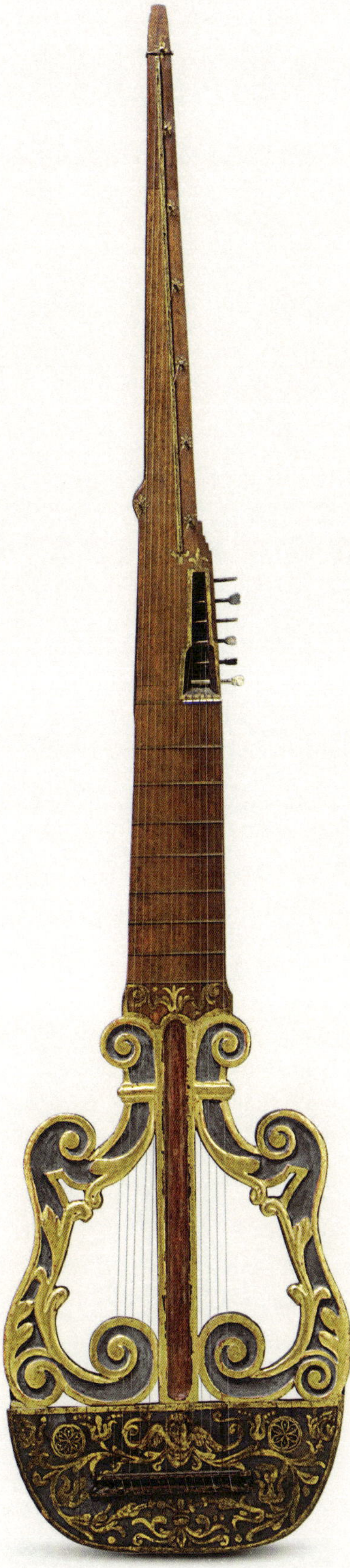

Unknown

TENOR CORNETT IN THE FORM OF A SERPENT WITH A DRAGON HEAD 17th century

The cornett was an instrument typical of Renaissance and Baroque music. It has a mouthpiece similar to that of a trumpet, and an eight-sided body of carved wood into which 7 holes are bored. The 2 hollowed out pieces of the body are usually entirely lined in leather.

The first examples of this instrument were made from animal horns, which may have inspired its sinuous shape. This zoomorphic cornett is in a double S that ends with the carved wooden dragon's head, with horns actually made from horn and teeth from ivory. A small cavity at the base of the palate suggests the instrument also had a metal tongue, today lost. The unusual form and quality of decoration make this a unique piece which, though with characteristics similar to the serpent family of instruments, produces a sound like that of a very mellow trumpet.

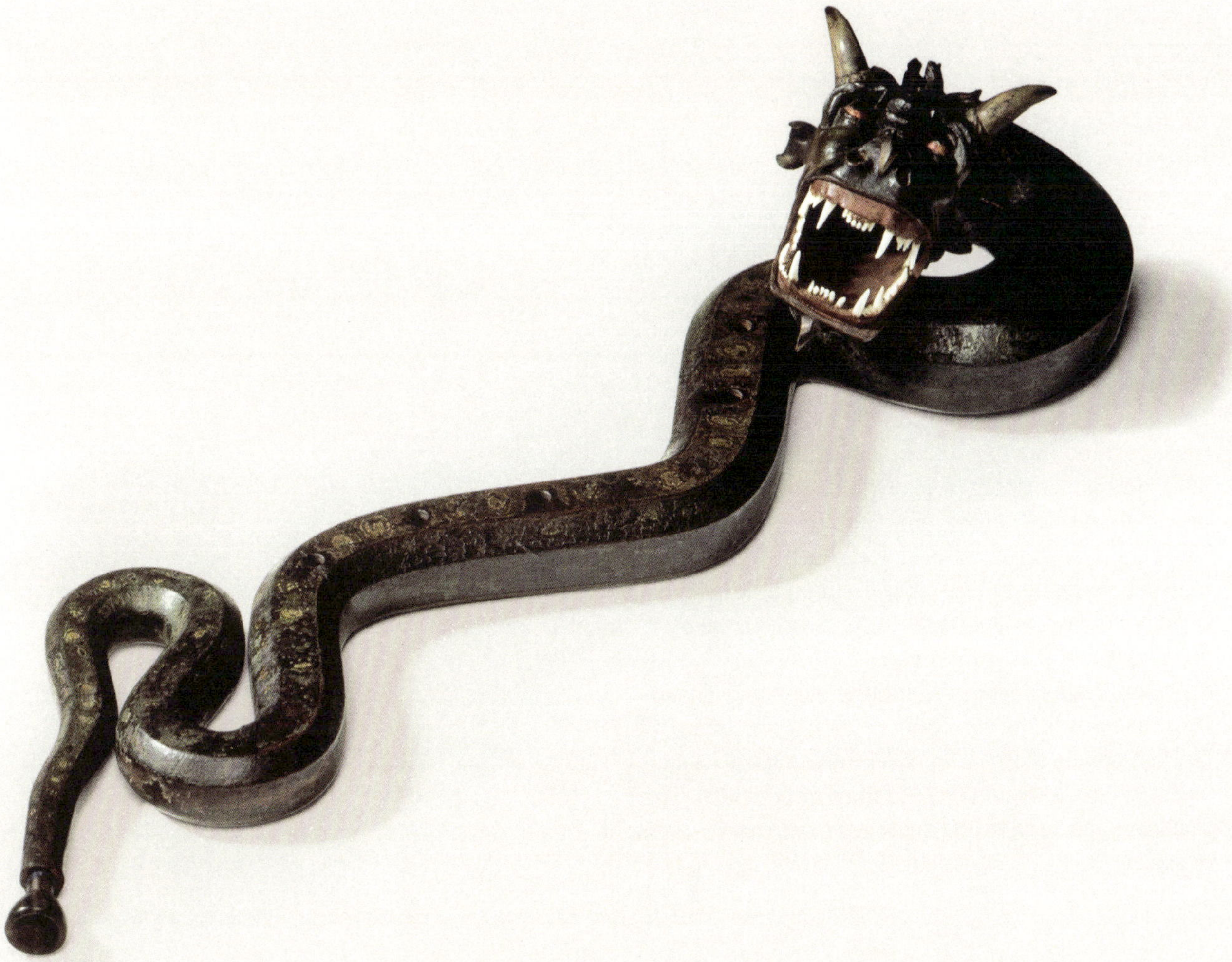

Jean Dubois au Puy

CHIMING CLOCK WITH IRON CASE
beginning of 17th century

48

The first mechanical carillons date to the Middle Ages. Incorporated in a mechanical clock, a carillon is activated at set times to play melodies "pre-recorded" on a cylinder containing pins that operate hammers. The carillons on town clocks were sometimes accompanied by "jaquemart" automata—human figures made from wood or metal—which signaled the time by striking bells with a hammer.

Built in France, this chiming clock in a closed iron case bears the signature of its maker, Jean Dubois au Puy, on one of its bells. The carillon has 9 bells: 8 play the melody and the 9th marks the hours. The melody is initiated every quarter hour by the movement of the clock.

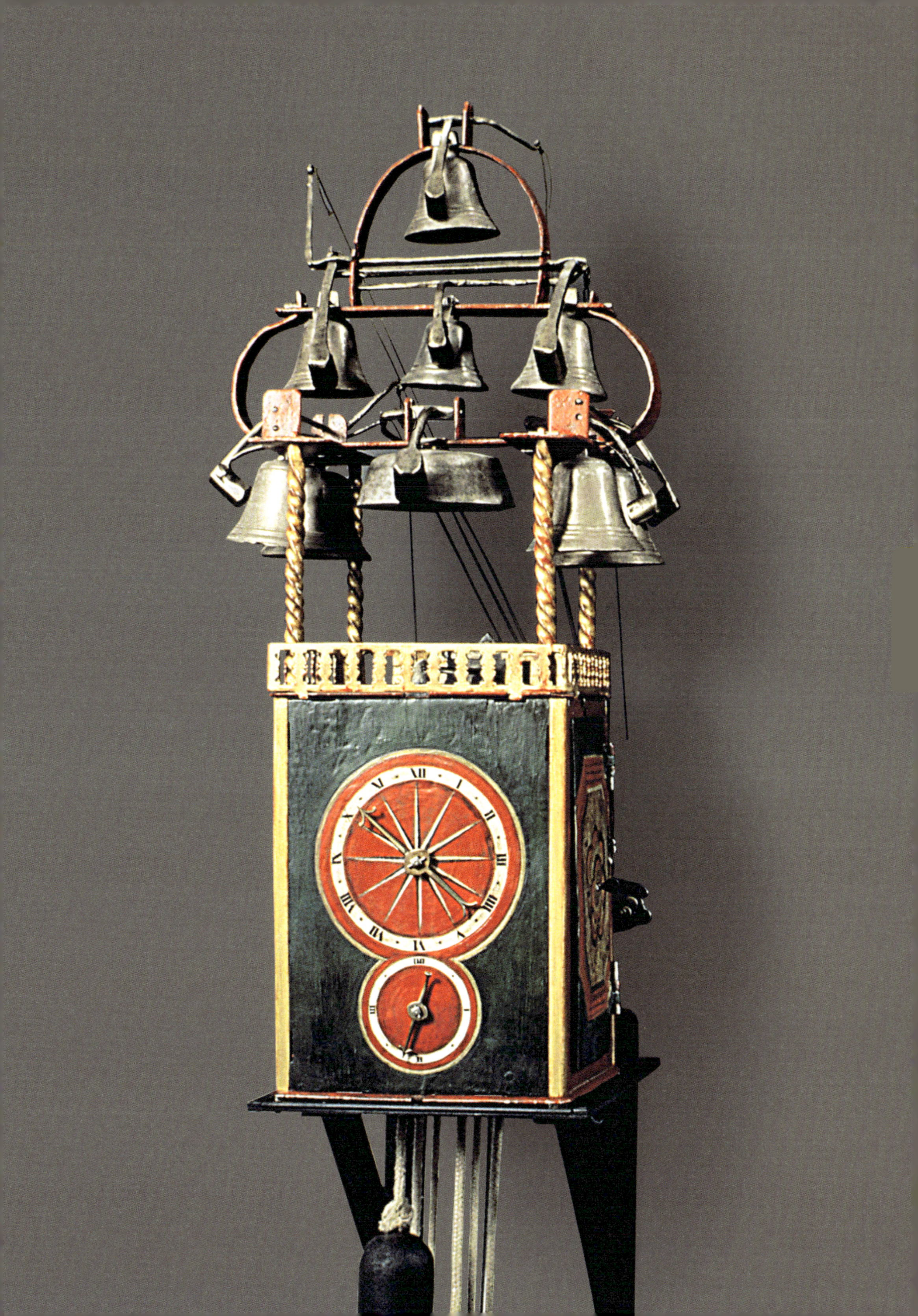

Michele Antonio Grandi

GUITAR
c. 1680

50

The guitar was made by the sculptor Michele Antonio Grandi (Carrara, 1635–1707) who, according to some sources, was close to Bernini. It was donated by Prince Cybo di Massa to Duke Francesco II d'Este in 1686 and taken from Carrara to Modena by Pietro Bertacchini, a musician from Carpi, who played it at the Estense court. The duke was very favorably impressed by the guitar and ordered Grandi to produce a harpsichord (also in marble), 4 flutes and a cornet. Originally fitted with strings, the white marble guitar has black marble-paste inlays of plant motifs and geometric decorations adorning the soundboard, fretboard and pegbox.

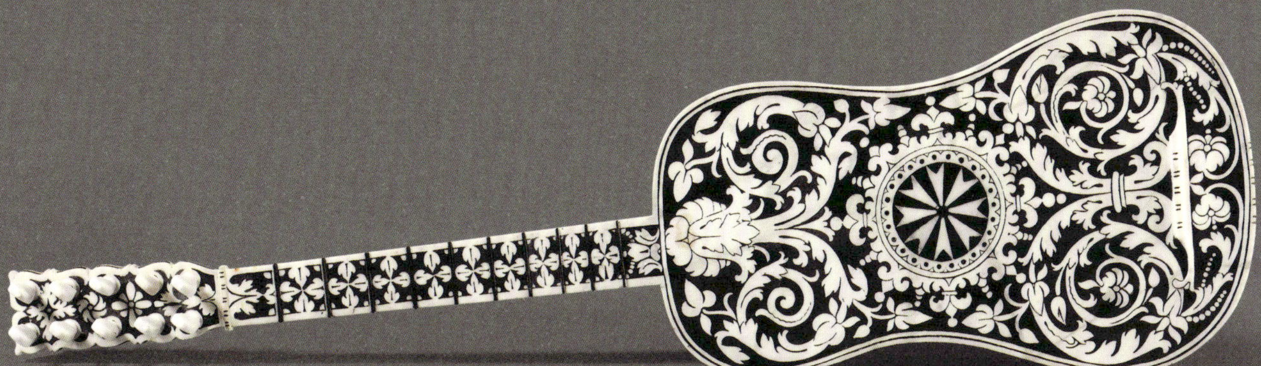

Giovan Battista Cassarini

CONTRALTO VIOLA
c. 1687

Made by Giovan Battista Cassarini (Carrara, 1643–1700), this viola was presumably commissioned by Duke Francesco II d'Este in 1687, as is suggested by the signature inside the soundbox: *D.O.M./Jo Baptista Casrini/Carrara 1687.* A music-lover, Francesco II retained a number of music masters at his court and collected works of extraordinary value. He placed emphasis on rare, curious and valuable pieces that would help his city earn recognition as a capital for music.

This white marble instrument is inlaid with marble-paste decorations in red and yellow Sienna. The entire case—on the top, sides and bottom—is decorated. The soundboard features a red fan-shaped design with plant volutes, tendrils and small flowers with open petals. The greater weight of the instrument's top end suggests a different thickness of marble was used in the upper sections.

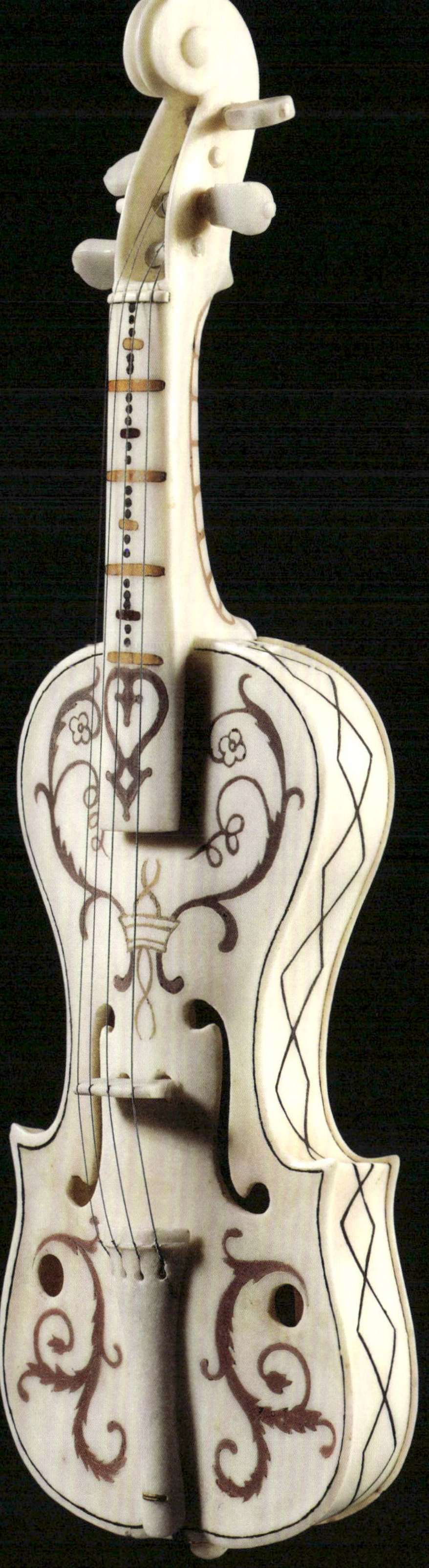

Onofrio Guarracino

RECTANGULAR SPINET
1692

The spinet was made and signed by Onofrio Guarracino (Naples 1628 – post 1698), the best known constructor of harpsichords and spinets in Naples during his lifetime. It features splendid Baroque decorative work. The soundboard is made of spruce, the moldings are in cypress, the case and bridge in walnut. It has 45 keys spanning 4 octaves; the 27 white keys have boxwood covers and the 18 black keys are veneered with ebony. The inside of the lid is decorated with a still life showing fruits, flowers and musical instruments. The still life is by the painter Nicola Casissa, who was active in Naples in the first half of the 18th century.

Like the harpsichord and virginal, the spinet is a plucked string keyboard instrument, though unlike the harpsichord, the spinet is quite small, making it easy to transport. For this reason it enjoyed a certain popularity until the 18th century, particularly for use at home.

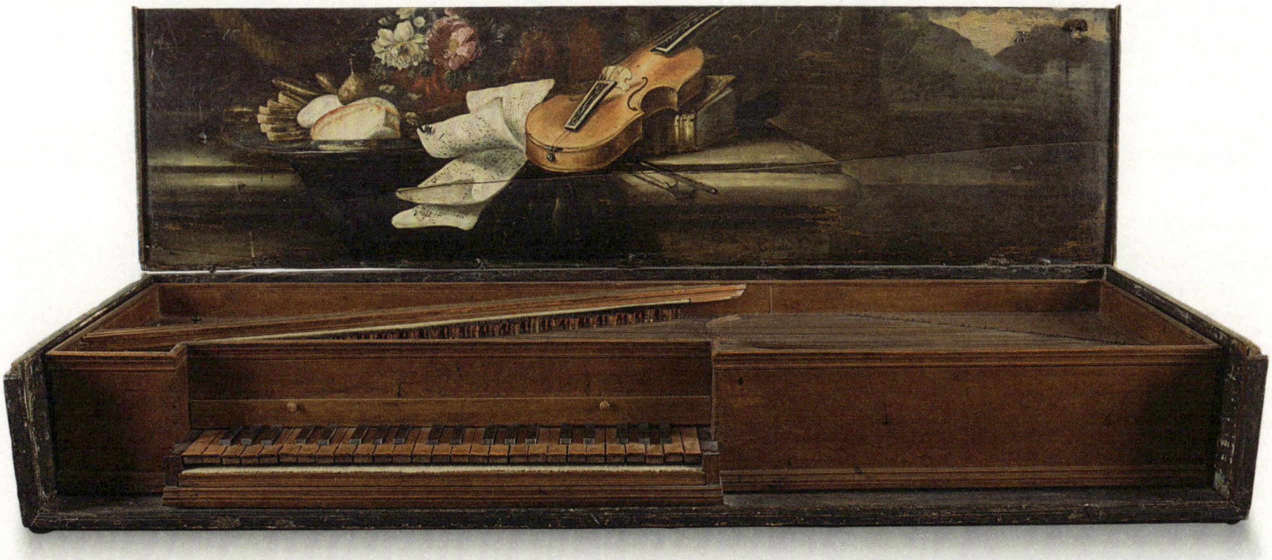

Atelier Pons (attr.)

LYRE-GUITAR
late 18th century

The lyre-guitar made its appearance in the Napoleonic era in conjunction with the fashion for Greek and Roman Antiquity that spread through French high society. Although similar to a lyre, this stringed instrument featured two characteristics of the modern guitar: a 6th string (the low E) and the use of single rather than double gut strings, which had been used until that time.

Designed for private use and closely associated with the notion of Neoclassical settings, the lyre-guitar was often fitted with an incorporated support that allowed it to stand independently as a decorative object. Supplanted by the piano as a salon instrument, it survived into the early years of the 20th century less as a musical instrument than as an object useful for suggesting Antiquity in paintings and photographs.

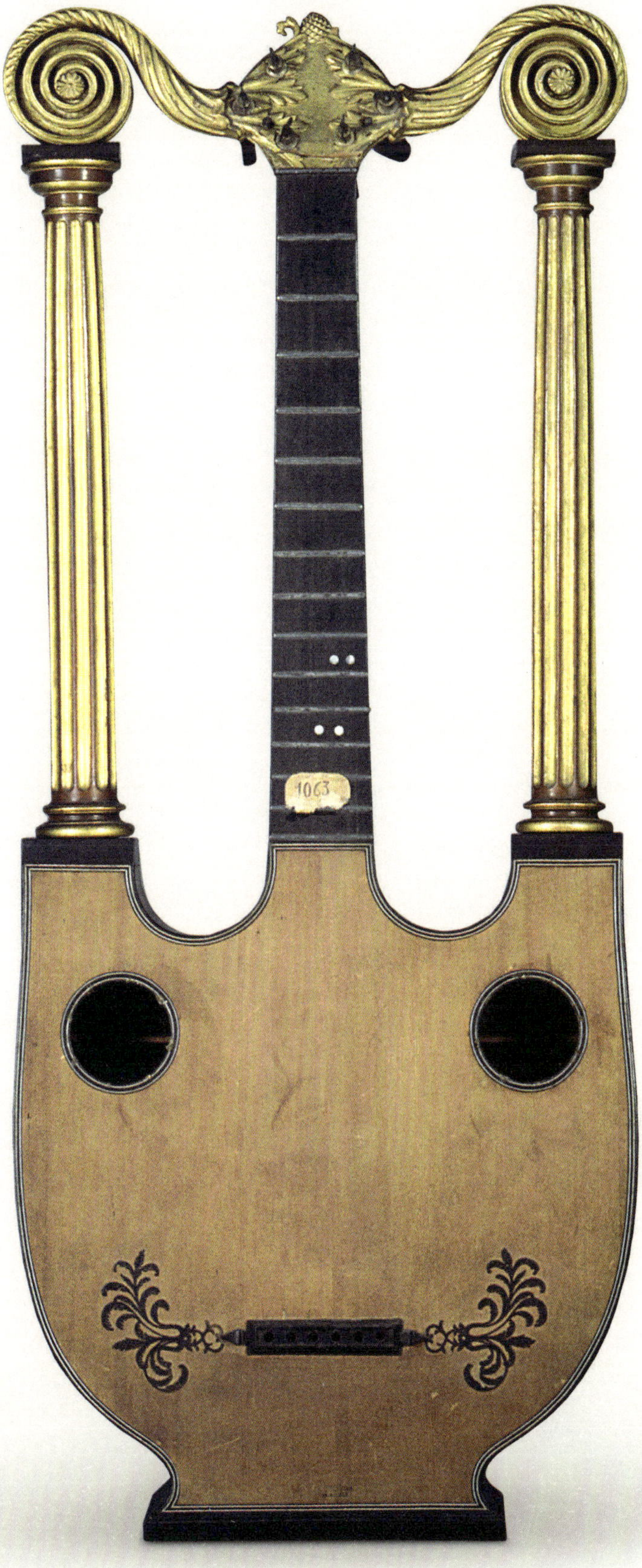
1063

Jean-Honoré Davrainville

BARREL ORGAN WITH FLUTES *c.* 1770

Davrainville Père et Fils was based in Paris and became renowned in the late 18th and early 19th centuries for the sophisticated mechanics of its watches and excellent sound of its automatic organs.

This organ was built by Davrainville père, who modelled it after the chapel organ at Versailles. It has a wooden cylinder and 7 tin pipes, all open, that sound at the double octave above. A lever opens the instrument's façade to reveal a thermometer and a Republican calendar. Another lever operates a bird organ that plays 5 different tunes.

[See also p. 106]

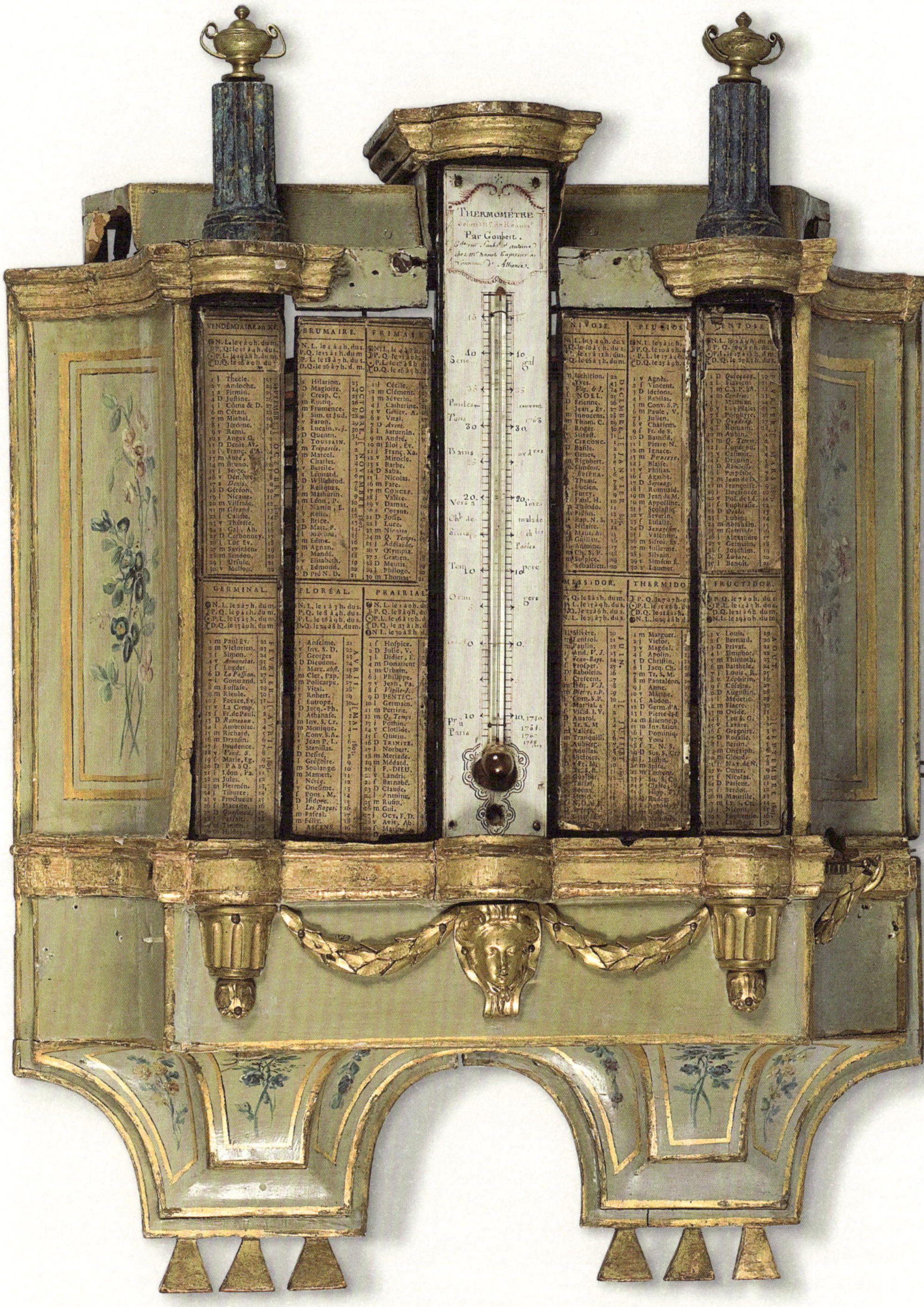
THERMOMETRE
Par Goubert.
BRUMAIRE
FRIMAIRE
GERMINAL
FLORÉAL
PRAIRIAL
FRUCTIDOR

Pierre Jaquet-Droz

CLOCK WITH CARILLON AND AUTOMATON 1780

Pierre Jaquet-Droz (La Chaux-de-Fonds, 1721 – Bienne, 1790) was born into a wealthy family of Swiss watchmakers and farmers. After studying philosophy in Basel, he was apprenticed to a clockmaker in the Neuchâtel mountain area, where this craft had begun to develop in the 17th century. In 1758 he traveled to Madrid to present his highly elaborate watches and automata to the royal court. Jaquet-Droz was received by Ferdinand VI and, fearing that he might be arrested as a sorcerer, invited the Grand Inquisitor to inspect the inner mechanisms of his devices, demonstrating how they were moved entirely by natural means. The king bought 6 clocks, and Jaquet-Droz invested all the profits in his enterprise, establishing an avant-garde company that specialized in the creation high-end technological parts.

The pendulum, in the shape of a box in Louis XIII or XIV style, contains a female musician automaton who conducts the music created by the mechanism, produced by 17 bells in 8 different melodies.

[See also p. 62]

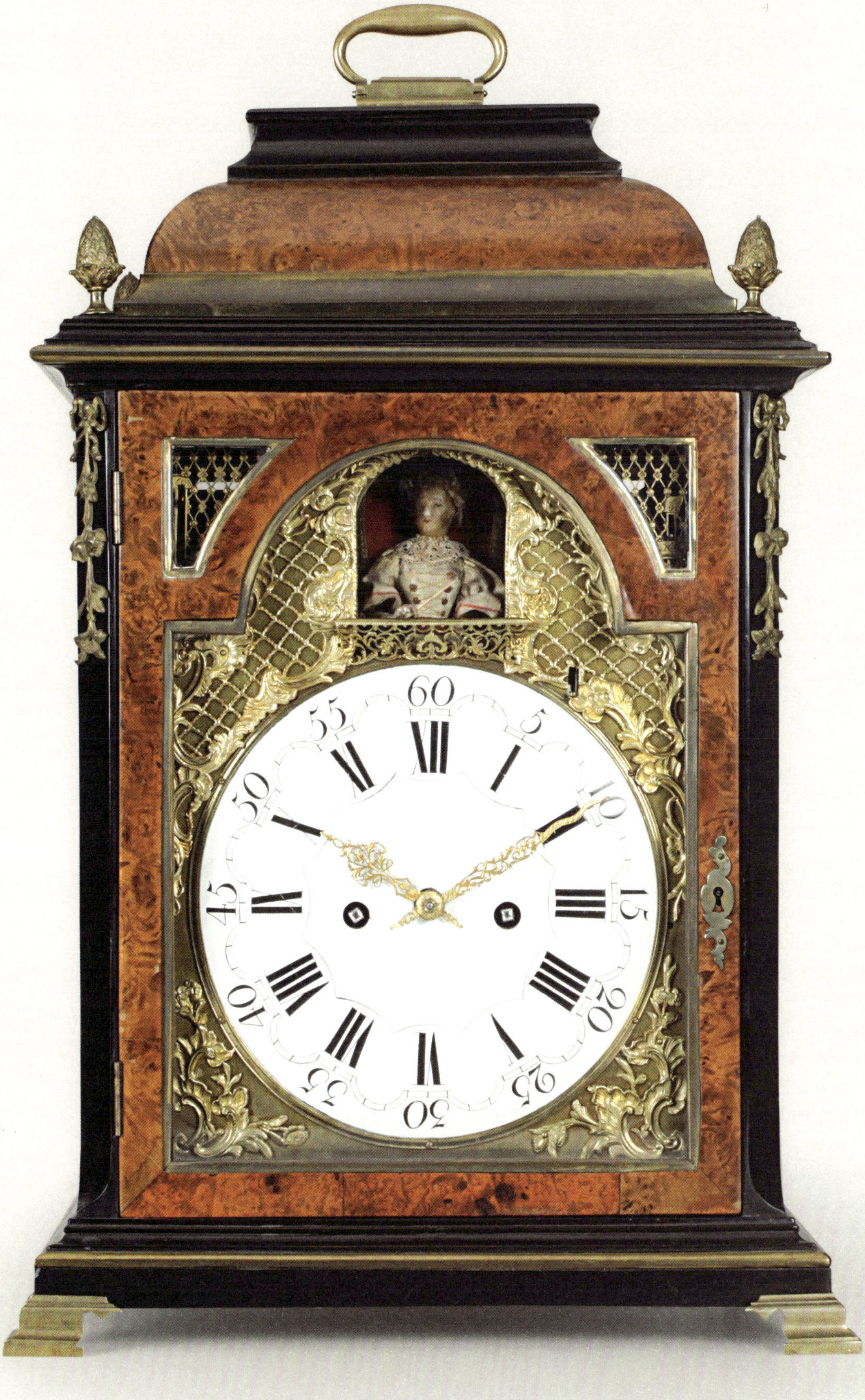

Pierre Jaquet-Droz and Henri Maillardet

SINGING BIRD CAGE WITH CLOCK
c. 1780

Pierre Jaquet-Droz

SINGING BIRD CAGE WITH CLOCK
c. 1785

Pierre Jaquet-Droz

SINGING BIRD CAGE WITH CLOCK
1790s

During the 18th century, the Swiss Pierre Jaquet-Droz (1721–90), his son Henri-Louis (1752–91) and their colleague Jean-Frédéric Leschot (1746–1824) acquired widespread renown in the art of clockmaking. Between 1768 and 1774, together they produced 3 automata—the writer, the draughtsman and the musician—that were applauded when premiered in La Chaux-de-Fonds, and won acclaim all over Europe for the unique way each encased the mechanism within the automaton's body. Henri Maillardet (1745–1830), another associate of Jaquet-Droz's, ran their London branch.

The bird cages, invented to reproduce birdsong as faithfully as possible, consist of a motorized movement that produces air using a bellows. A piston whistle allows the system to reproduce the melody while synchronizing the movement of the bird's beak, the flapping of its wings and its rotation. There are quarter-hour strikes with the bird automaton released every hour. Cages were usually made of chased fire-gilt bronze and brass, with china figures set at the base and a white enamel dial that could be seen on the bottom once the cage was hung.

[See also p. 60]

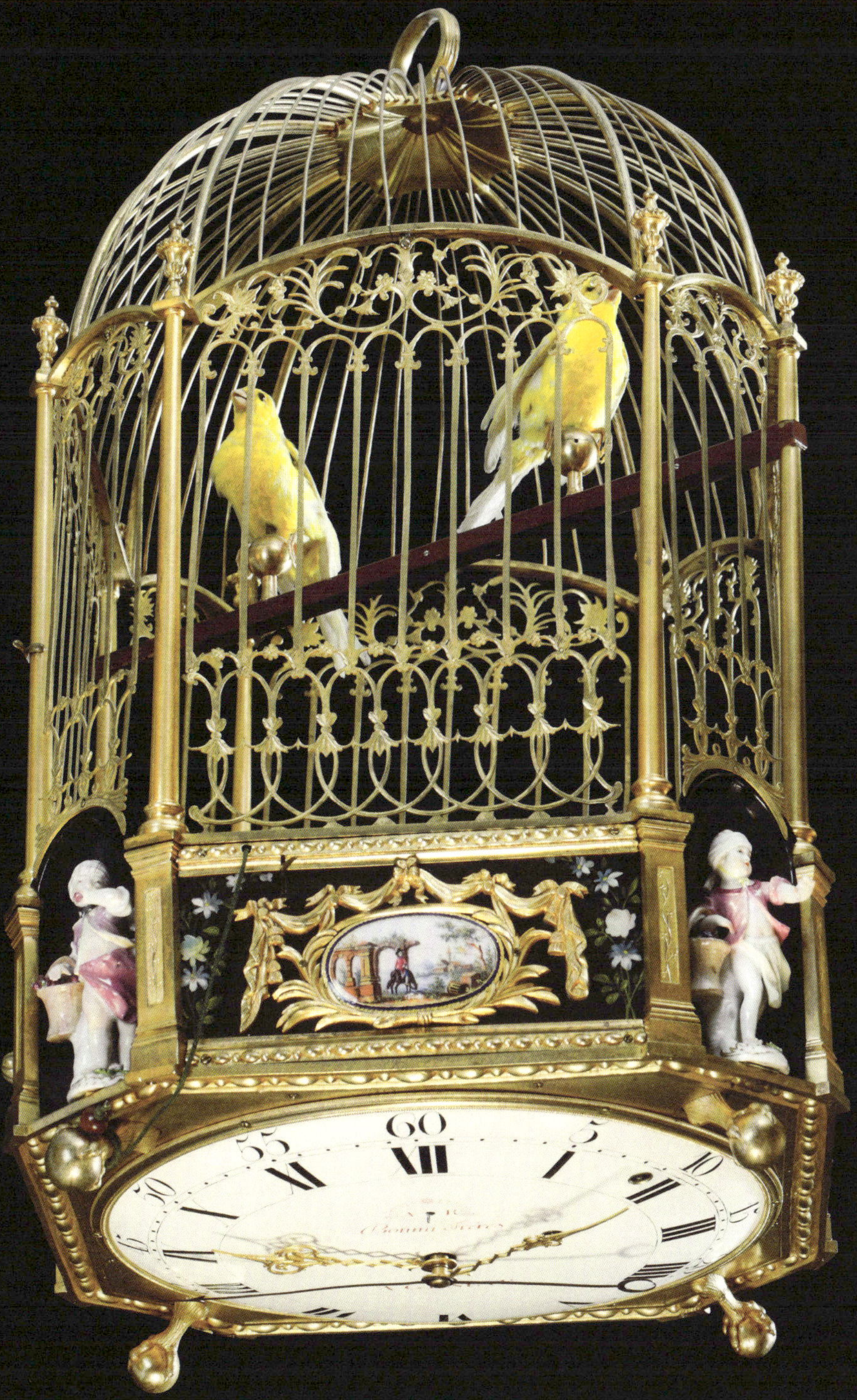

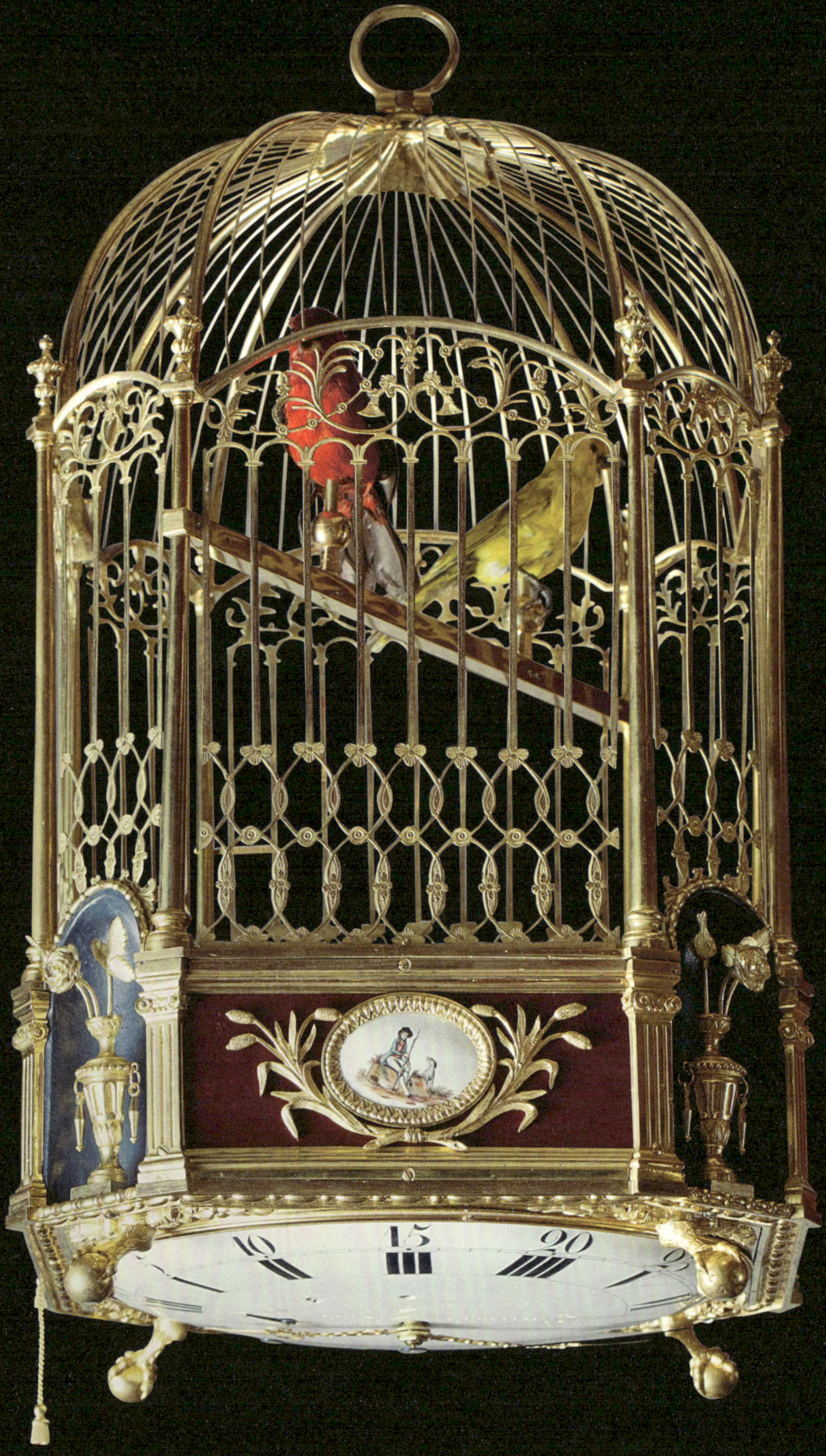
10
15
20

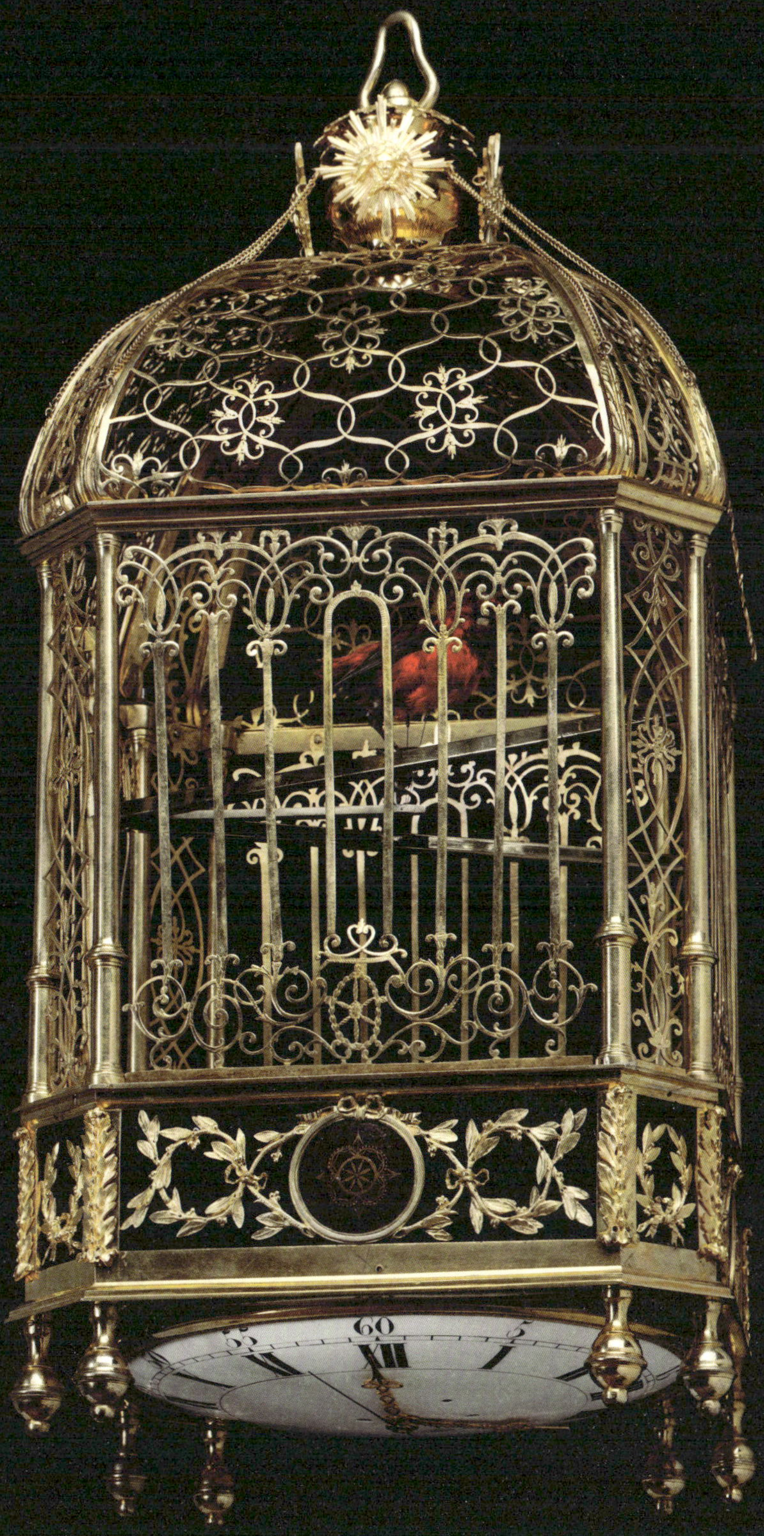

Egor Kuznetsov

CARRIAGE WITH BARREL ORGAN 1785–1801

The carriage was built by Russian inventor and clockmaker Egor Kuznetsov (1725–1805) for Empress Catherine II the Great and later offered to Marìa Fëdorovna, the widow of Catherine's son Paul I. It is equipped with an automatic organ that could be switched on during the ride to entertain passengers.

The sound comes from a wooden music cylinder and a series of stopped wooden pipes. The cylinder was programmed with 6 popular melodies that alternate automatically, and new cylinders could be made to keep up with the latest musical trends. The carriage also has a mechanism that is driven by the right back wheel and measures the distance covered. Every *verst*—a Russian unit of length equal to 1.0668 kilometers—is marked by a bell. Kuznetsov can be seen among the various telemeters on the back of the carriage, depicted with one of his watches.

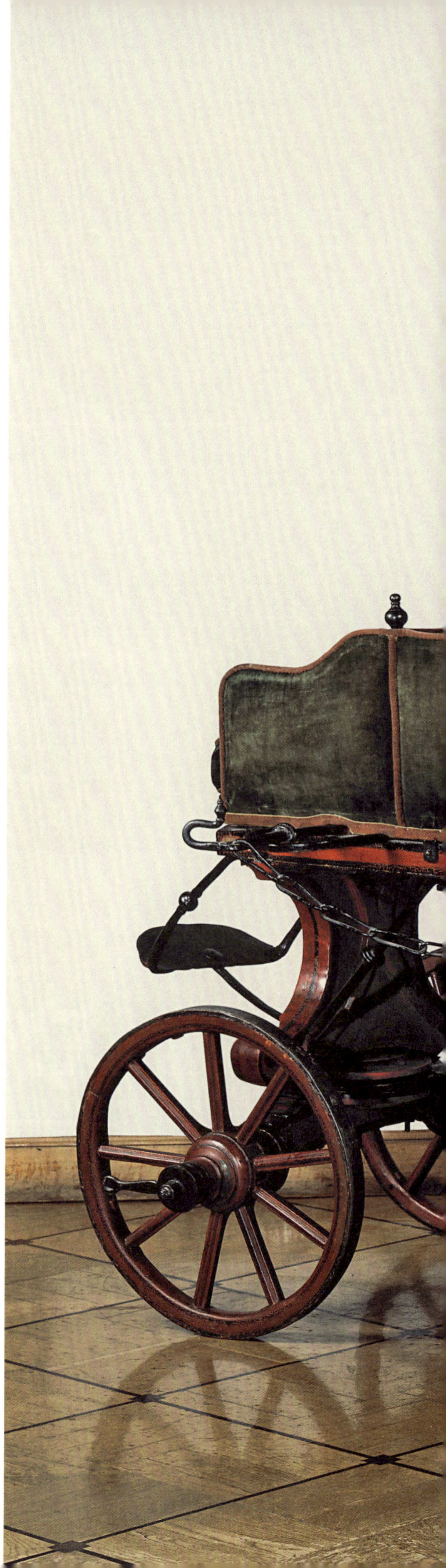

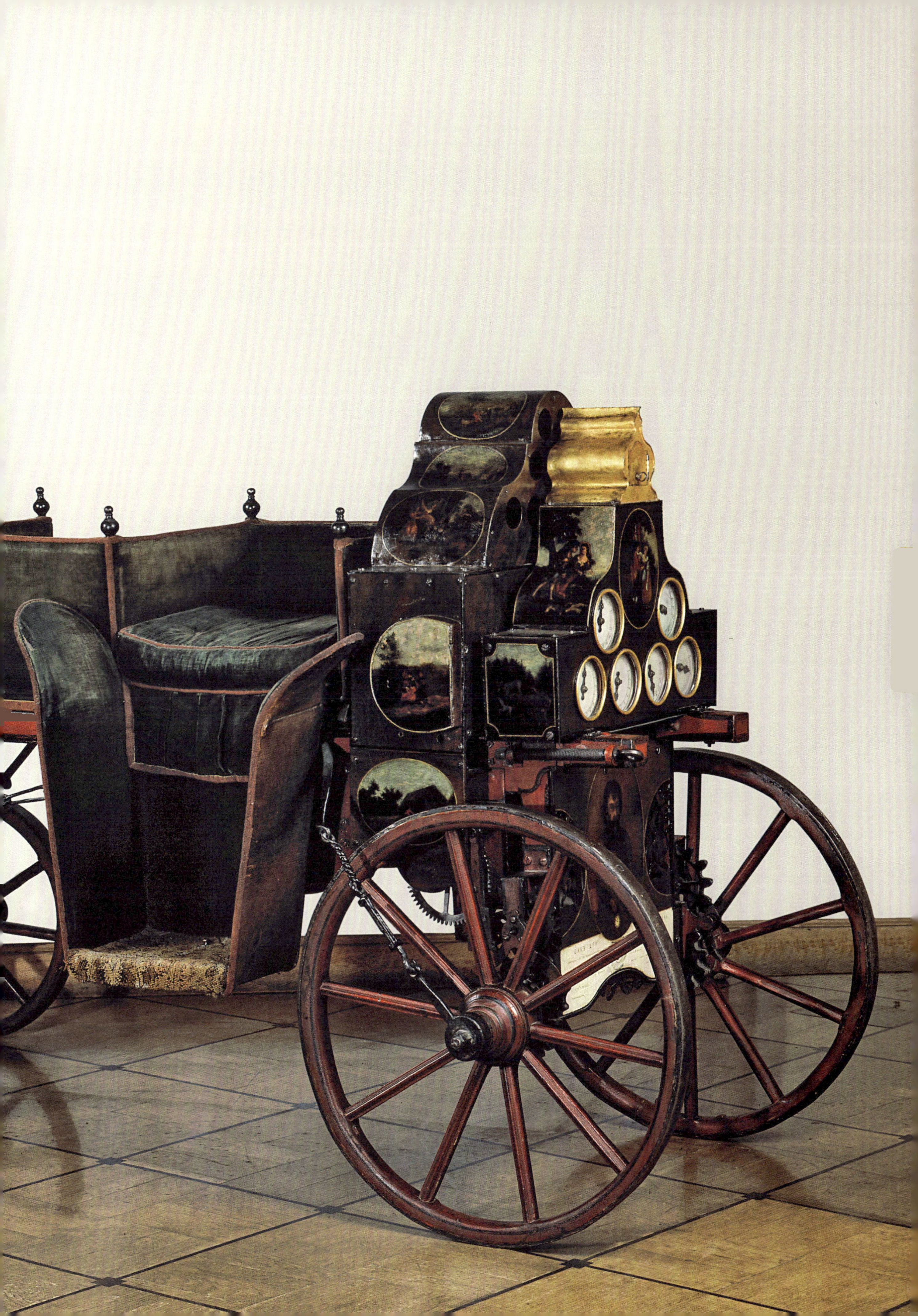

Unknown

CONCERTION
19th century

The concertion contains a cylinder—visible through an opening on the right side—with metal pins that pluck the teeth on a revolving steel comb. The very thin teeth are made of reed, wood or metal and fixed to the embouchure of several wind instruments that produce sound due to the vibration caused when the air column is agitated. The bellows and a small lever are operated in order to make this concertion produce sound, playing melodies including lieders, opera or dance tunes. The accordion is fitted with 3 cylinders.

The patent for this instrument can be found inside—"D.R. Patent (n. 25904)"—where D.R. stands for Deutsches Reich, indicating that the instrument was made in territories under German domination.

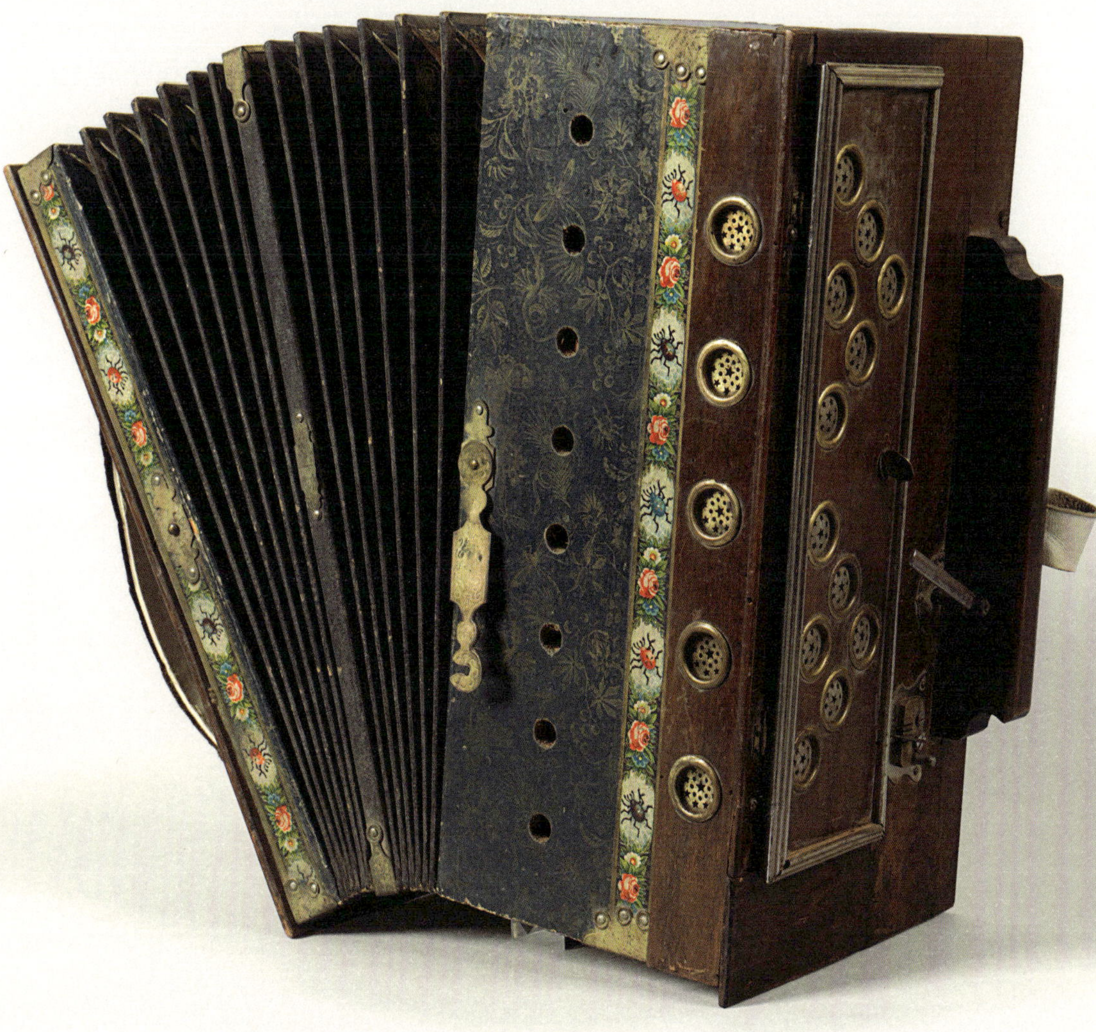

Jean Arnold Antoine Tuerlinckx

BUCCIN
c. 1800

François Antoine Sautermeister

BUCCIN
1809–30

B. Schott's Söhne

BUCCIN
c. 1820

The buccin is a type of trombone with a bell in the shape of a stylized serpent or dragon head that is often painted bright red, green or gold, and sometimes has a tongue made to vibrate by the passage of air.

The instrument's appearance is more significant than the quality of sound it produces. The latter is cutting but less vibrant than that of a conventional trombone, similar to that of a French horn. The instrument is linked to the spread of military bands in France and Belgium, whose trombones, serpents, bassoons and Russian bassoon with zoomorphic heads featured prominently in parades and popular festivals during the end of the 18th and first half of the 19th centuries.

[See also p. 102]

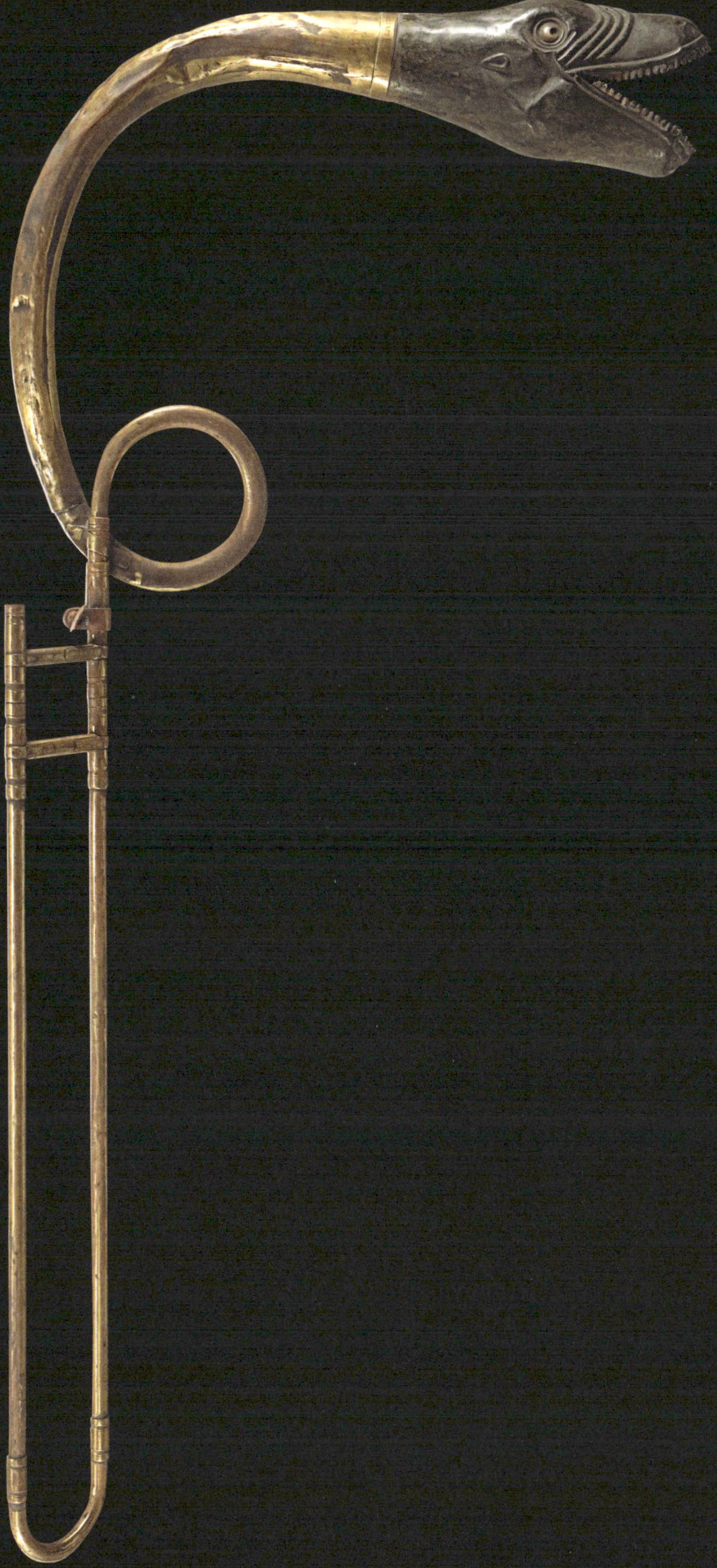

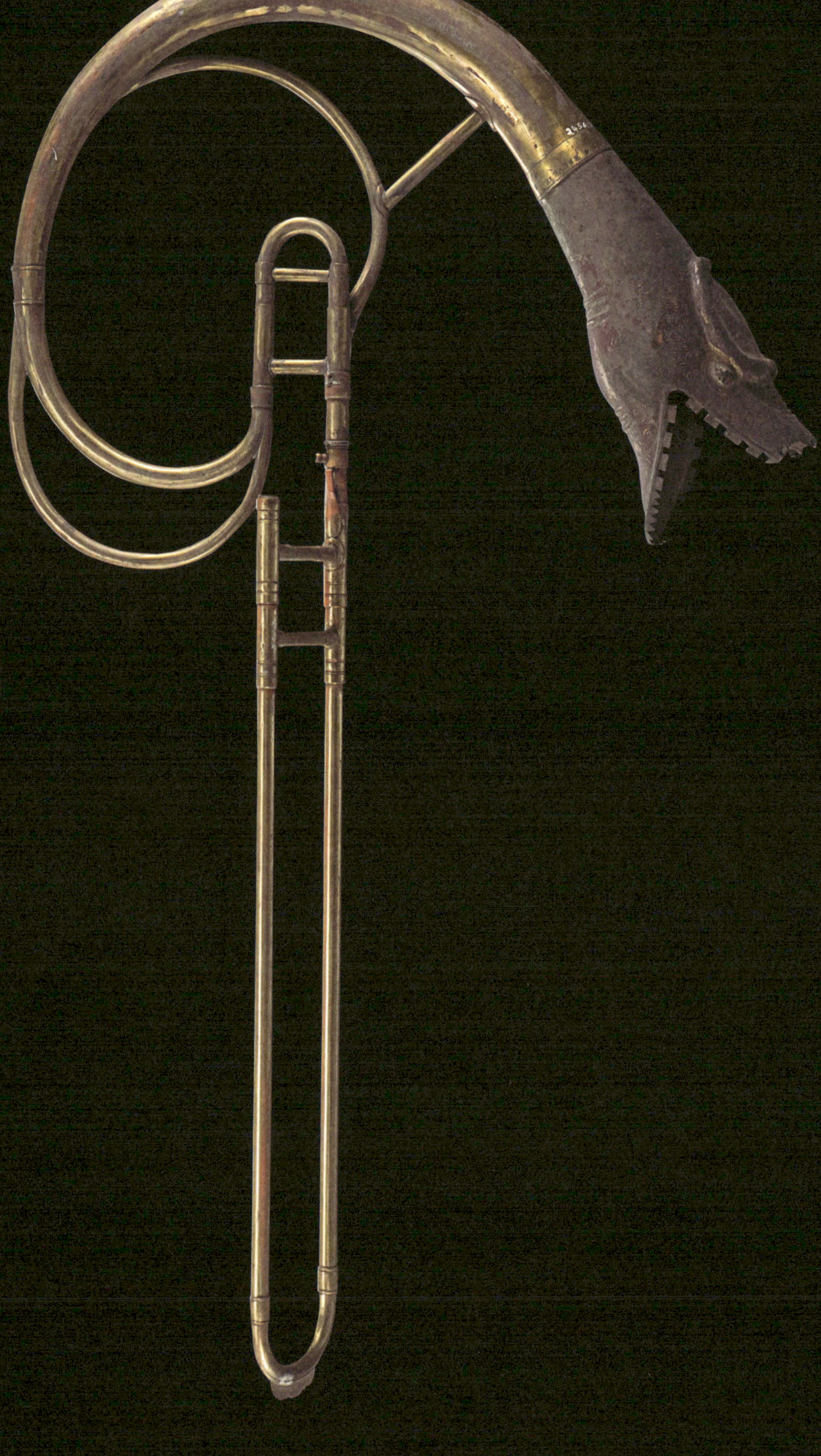

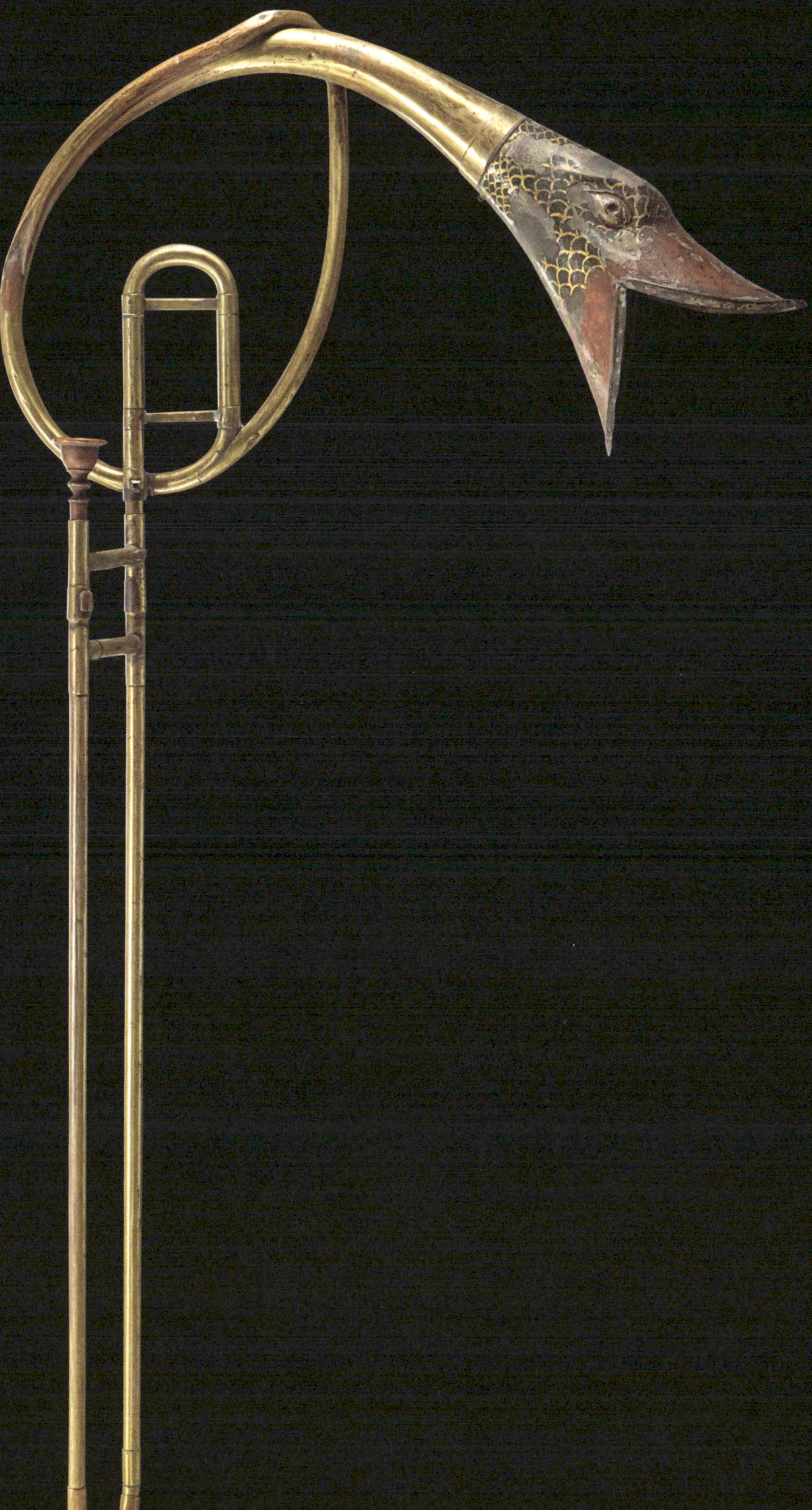

Unknown

CLOCK WITH FLUTES
c. 1800

Produced in Vienna, the clock is made of 26 pipes and an interchangeable, wooden cylinder with one melody. Flute clocks were enormously popular during the second half of the 18th century and throughout the 19th century. Originally designed as gifts for emperors and kings, these instruments found their way into the bourgeoisie and served as public entertainments in Vienna, which by the beginning of the 18th century had become a center of flute-clock production. Various works by Joseph Haydn, Wolfgang Amadeus Mozart and Ludwig van Beethoven were composed specifically for these mechanical instruments.

Flute clocks were replaced in the 19th century by clocks with modern musical mechanisms and composers no longer wrote pieces specifically for these boxes, although special arrangers adapted the most popular musical pieces of the period.

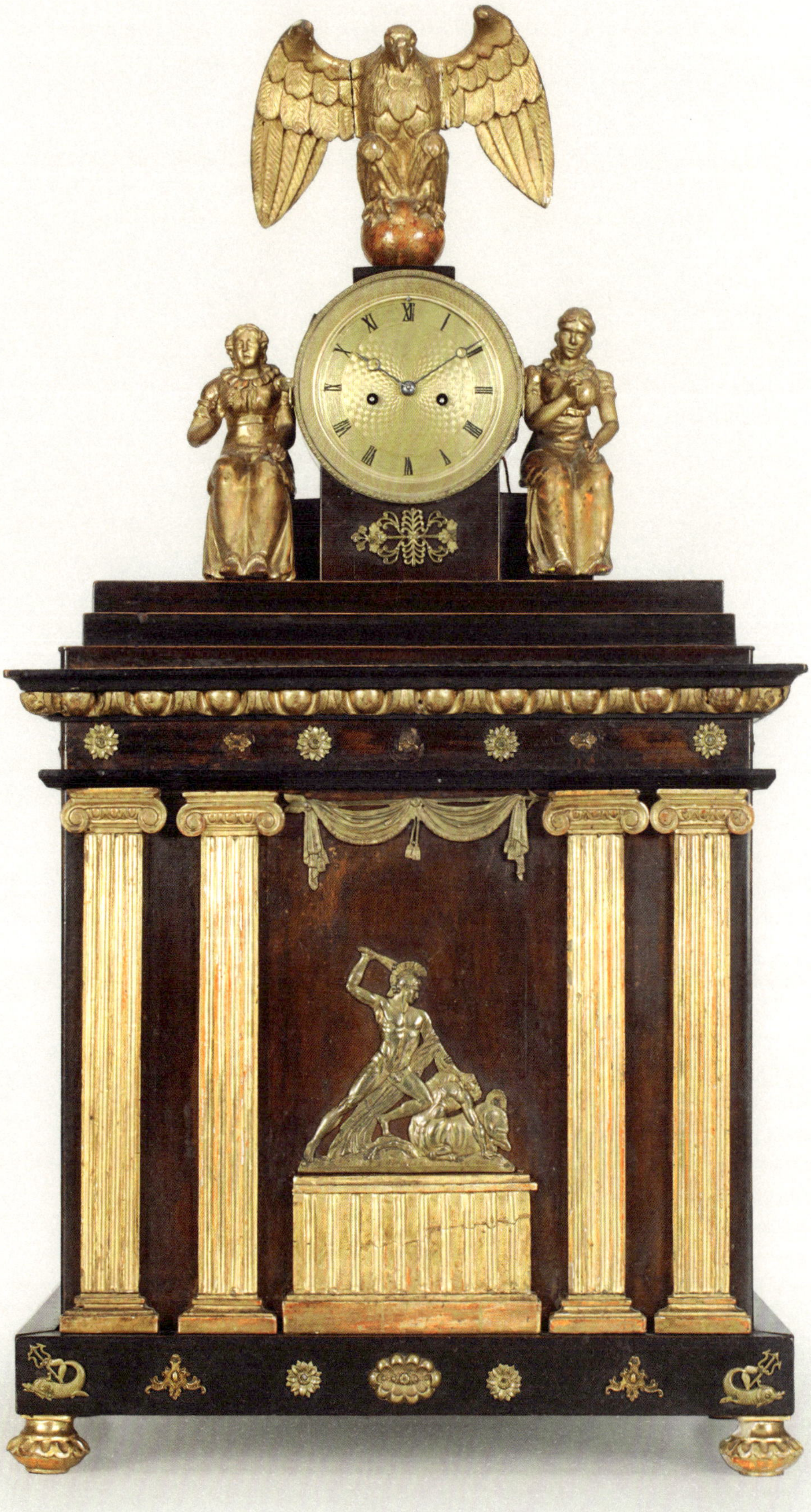

Unknown

SERPENT
c. 1800

Ludwig Embach

SERPENT WITH KEYS
1820–44

Joseph Dupré

MILITARY SERPENT
c. 1825

The serpent is a wind instrument invented during the 16th century as a member of the family of cornetts to accompany choirs in plain-chant. It began to spread across France during the 17th century, and became common in Belgium, Germany and England (where it was also used in theaters).

The double S shape is created by joining together 2 pieces of chestnut with strips of leather. It allows the player to support the instrument between the knees when seated and reach the six finger-holes without difficulty. The palms of the player's hands face downward.

The inclusion of the serpent in military bands in the 18th century required a development in the form of the instrument, which became more compact to permit portability, while the need to reach the finger-holes —now more distant—led to the adoption of keys, like in the instrument made by Ludwig Embach.

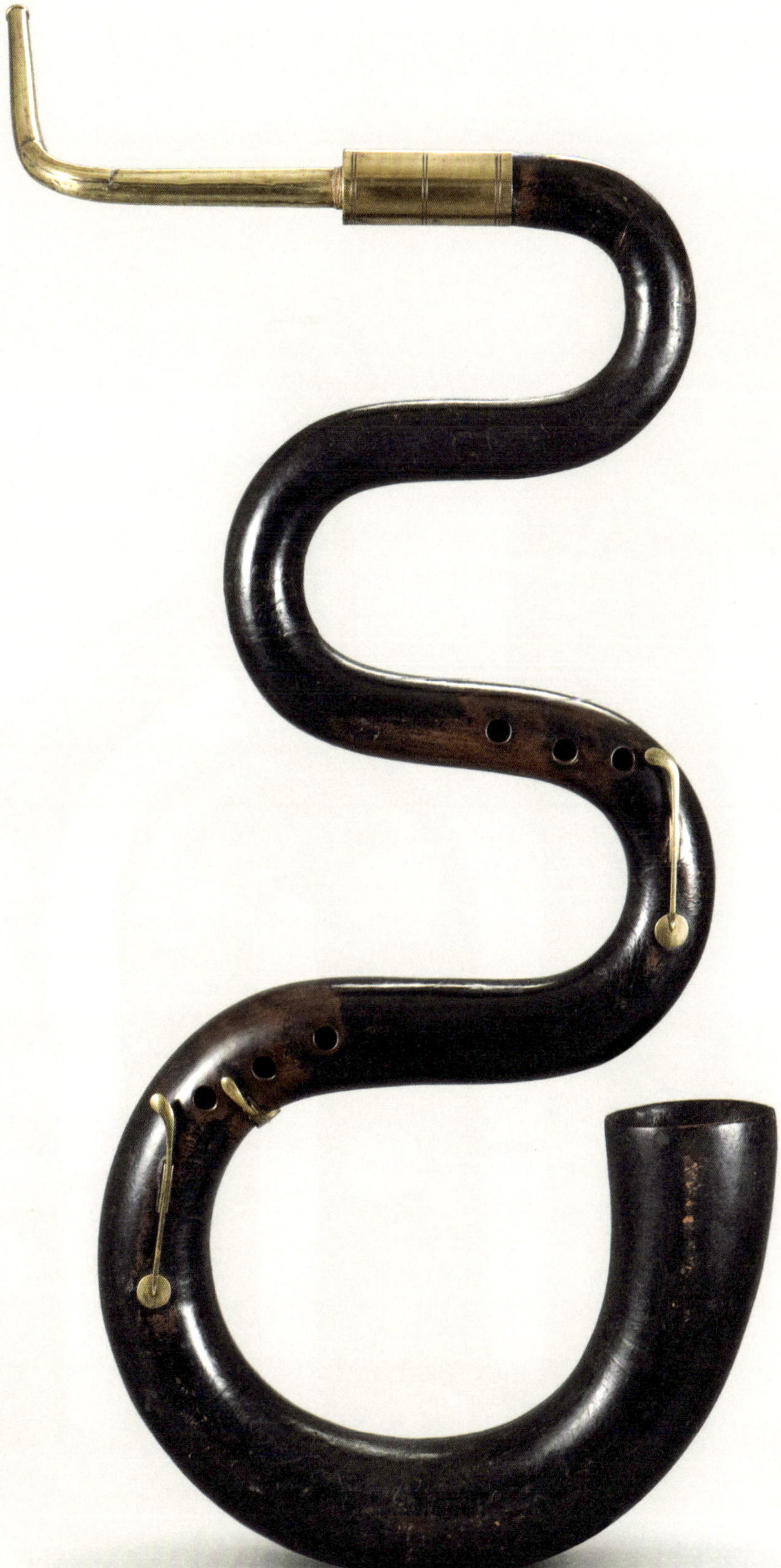

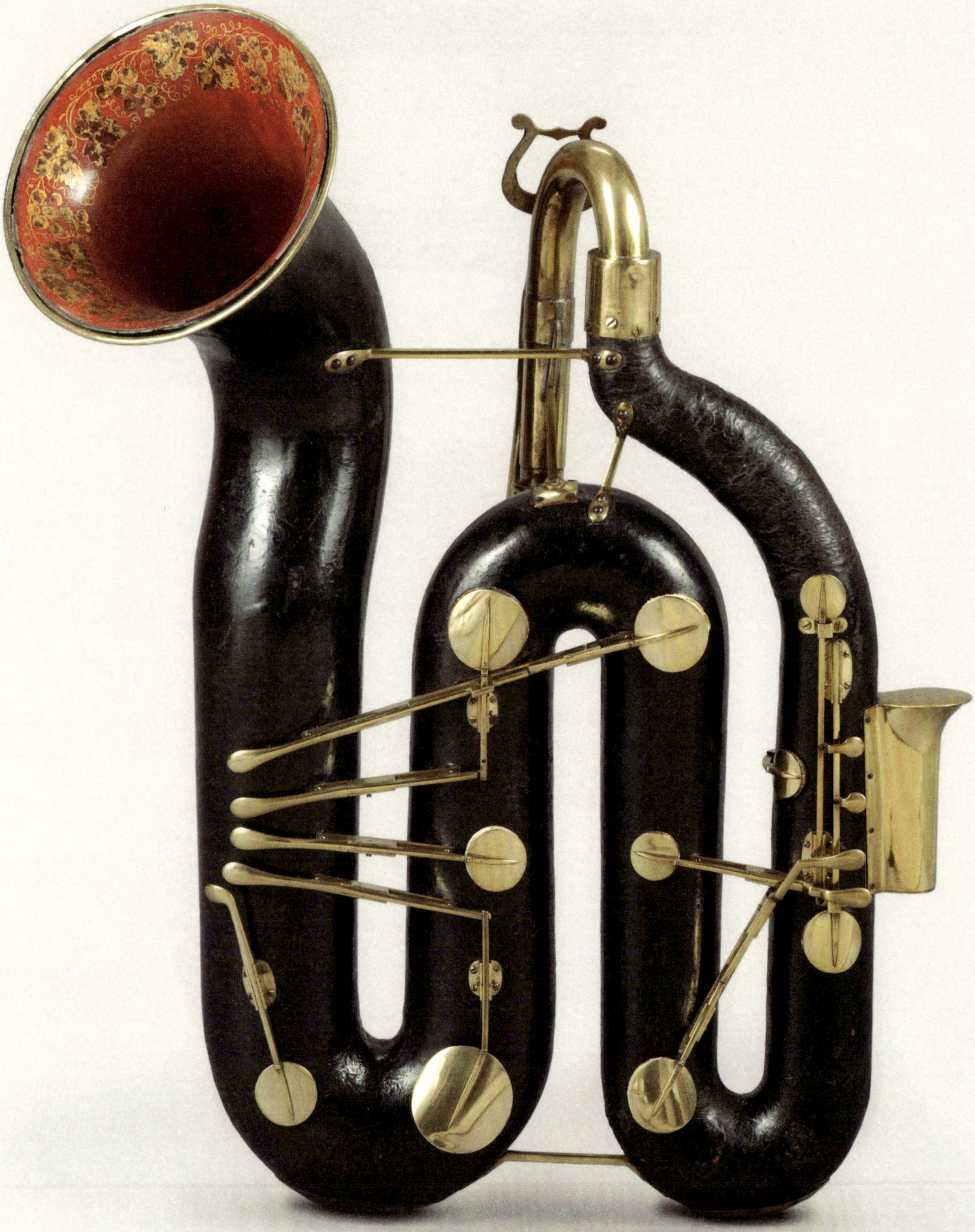

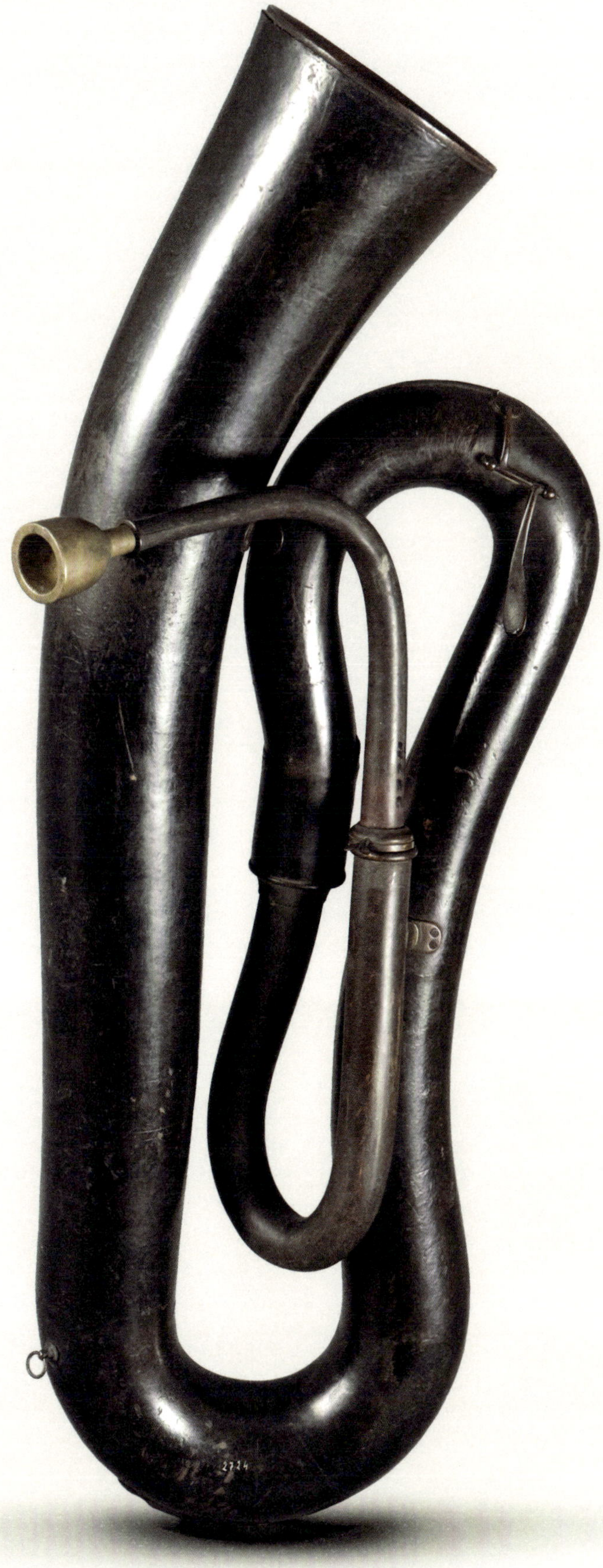

PATRICK FEASTER

ON SOUND-GRAPHS

A COORDINATED LOOK AT SONIC ARTIFACTS

Sound itself is evanescent, whether we regard it as vibratory motion or sensory perception, but enduring artifacts can still be said to embody sounds in a variety of ways.

First, there is the *instrument*: an object that a human performer uses to generate sounds at pleasure by exploiting its capacity for distinctive acoustic vibration, whether intrinsic or extrinsic (as with electrically-controlled speakers). Thus, a string might be said to embody the sound it makes when it's plucked or bowed, or a pair of coconuts might be said to embody the clip-clop of a horse's hooves behind the scenes in a theater. Second, there is the *score*: a writing that gives instructions for a human performer to actualize as sound, though perhaps not exclusively as sound. Such writings might be said to embody a poem for someone to read aloud, or a monolog for someone to declaim, or a musical composition for someone to perform. Third, there is the *program*: an object that, when operated upon by an appropriate device (an *automatophone*), will produce a sequence of sounds without further human intervention. This is the sense in which the pinned barrel of a barrel organ might be said to embody a musical composition, or a phonograph record might be said to embody a recorded speech or musical performance. And then there is the *sound-surrogate*: the visual or tactile form that represents sound through a sensory modality other than hearing, serving as an alternative or complement to the sound itself: the waveform displayed statically on a computer screen, for instance, or the KAPOW! in a comic book. Artifacts in this category embody sounds as intersemiotic translations of them; they "stand" for sounds even in the absence of actual sound.

Each of the four categories in my typology involves a distinctive relationship between artifacts and sounds, but there's nothing mutually exclusive about them, and they have continually interpenetrated and informed each other. A single enduring artifact might occupy more than one of these categories at once, or it might migrate freely between them, depending on how we use it. In this essay, I'd like to explore one particular approach to the embodiment of sound within the Western tradition that has transcended the distinctions between instruments, scores, programs and sound-surrogates: namely, the representation of sound as a *graph*, where meaning is tied to position along the x and y axes of a Cartesian coordinate system.

Certain graphic metaphors for sound are now so familiar that we tend to take them for granted as "natural," but they are emphatically culture-bound. Medieval European musical notation had already latched onto two of them. First, the spatial positioning of notes along an axis running from left to right served as a metaphor for the passage of time, following the convention of Latin alphabetic writing. Second, "up" and "down" served as metaphors for greater or lesser rapidity of vibration. We now use that same metaphor by default when speaking about the distinction between "higher" and "lower" notes, or about a pitch "going up" or "going down." But Classical Antiquity had instead used a metaphor of sharpness, according to which pitches had been understood as ranging from "sharp" to "flat" (*ὀξεῖα* to *βαρεῖα*; *acutus* to *gravis*). I'm not sure whether the new graphic metaphor of height in medieval musical notation gave rise to the linguistic metaphor or the other way around, but the two developments in tandem fostered a culturally distinctive approach to expressing pitch in spatial terms that is still with us today.

So medieval musical notation had already adopted a loosely graph-like logic centuries before Descartes, plotting time along the x axis and frequency along the y axis. Another graphic format that assigns x and y axes to those same parameters is the sound spectrogram that emerged from Bell Laboratories in the 1940s as a means of making audio signals visually intelligible, at first by running them through a series of narrow analog frequency filters and displaying their outputs—arranged from "top" to "bottom" of the spectrum—as a function of time. Nowadays, the "spectral" view in audio software performs essentially the same operation. Whereas medieval musical notation had been intended as a *score* for performance, sound spectrograms were originally conceived as *sound-surrogates*; indeed, their first projected use was to enable deaf people to use telephones as "a means for visual hearing."[1] But they have also been used as *programs* for automatophones.

One early example was the Pattern Playback unveiled by Haskins Laboratories in 1950, which exposed transparent or reflective spectrograms to modulated light for transduction into sound; today additive synthesis software commonly achieves the same effect digitally. The visual dimension of spectrograms has often retained its importance even under such circumstances: thus, the value of the Pattern Playback lay in allowing researchers to manipulate graphic representations of sound and then hear the results, and additive synthesis software has been used to hide spectrographic images within sound recordings, as with Aphex Twin's incorporation of a spiral shape into *Windowclicker* (1999) and a photograph of his face into the "equation" track on the B-side. The spectrographic format has also been used to construct synthesizer voices via graphic images (one of the advertised functionalities of Photosounder and MetaSynth), which draws it into the realm of the *instrument*. Time-frequency graphs thus span all four categories in my typology, and graphs created for different purposes are at least partially interchangeable: I've even found that we can use additive synthesis software to transduce digital scans of medieval musical notation directly into audio, and that the results have a recognizable resemblance to well-informed modern vocal performances of the same pieces.[2] The limited compatibility of medieval notation with 21st-century audio technologies strikingly illustrates a long-term continuity in our most basic cultural conventions for depicting sound.

There have been countervailing tendencies, to be sure. Western musical notation has usually tempered the "pure" time-frequency graph with other visually apprehensible conventions, such as sharps and flats and arbitrary correspondences between note-shape and duration (whole notes, half notes, quarter notes). But during the early modern period, certain technocultural developments prompted a return to the "pure" logic of the graph.

First, the manual preparation of *programs* for automatophones—such as barrel organs, bird-organs, musical clocks, and music boxes—introduced new inscriptive possibilities and requirements that differed from those of the human-readable score. Automatophones could exceed the capabilities of human performers in some respects, for example by enacting impossibly rapid trills—as embodied in pins or staples, perhaps, along a time axis that might be the circumference of a cylinder. However, their programs also needed to be "complete" in a way that scores did not because they couldn't rely on the discretion of a human performer immersed in a relevant performance tradition. Thus, Marie-Dominique-Joseph Engramelle—a leading 18th-century theorist and practitioner of automatophone programming—lavished considerable effort on encoding the "silences of articulation" which he believed human performers of keyboard music habitually inserted between notes, but which weren't shown in conventional notation.[3] The graphic approach lent itself well to such meticulous determinacy, much as similar graphlike scripts lent themselves to automation of the spectacles often associated with automatophones—think of the toothed wheels that control the corporeal movements of Jaquet-Droz's musical androids and mechanical singing birds [60, 62].

It was also in the 18th century that apparatus was first contrived in Europe for *recording* extemporaneous performances on keyboard instruments. The idea was that whenever a performer pressed a key, it would put a stylus in contact with a moving sheet of paper and leave behind a mark whose length corresponded to the time it was held down. John Creed, the first European known to have put forward a plan of this kind, articulated its advantages in terms that stressed its heightened precision: "Grave music from brisk, slow from fast, &c. will be better distinguished by this Machine, than in the ordinary Way by the Words *Adagio*, *Allegro*, *Grave*, *Presto*, &c. for, by these Words, we only know in general this must be slow or fast, but not to what Degree, that being left to the Imagination of the Performer; but here I know exactly how many Notes must be play'd in a Second of Time; *viz.* as many as are contain'd in 1 Inch of the Scroll.... Whereas, in the ordinary Way of writing Music, you have either no Character for Graces, or such as do not denote the Time and Manner of their Performance, here you have the minutest Particles of Sound that compose the most transient Graces mathematically delineated."[4]

Both of these projects—the programming of automatophones and the recording of keyboard improvisations—aspired to transcend the limits of ordinary notation, either capturing the features of live performance or substituting them. And both resorted to graph-like embodiments that had direct, causal relationships with the sounds they encoded—either they produced them, or they were made by the same actions that produced them. The graphic approach freed musical writing from some significant limitations of notational convention because a graph could in theory accommodate *any* gradation of pitch, *any* gradation of time, and *any* interplay between those two parameters, no matter how complex.

The graphic format used in the 18th century both for programming barrel organs and for recording keyboard performances was later inherited by the player piano, and the "piano roll" format has in turn carried over into digital tools such as MIDI sequencers. Composers thus continue to engage with this format on a regular basis, and some have found its distinctive graphic character to be significant in its own right. Jonathan Pitkin suggests that the "piano roll" format used for programming the Yamaha Disklavier—a digitally controlled piano—offers a "more faithful visible representation" of his compositions for that instrument than conventional notation. Part of the instrument's appeal lies in it "pressing down its own keys as if by magic," and Pitkin has sought to exploit this visual aspect of its performances: "I often found myself 'composing' the shapes and patterns that I wanted to see played out on the Disklavier's keyboard." Conventional notation would conceal these shapes and patterns, but the "piano roll" format clearly displays them.[5]

Beginning in the 17th century, an even more direct approach arose to embodying sound on the level of the individual vibration, ultimately giving rise to an audio waveform that graphs time against amplitude.

One objective of this approach was to create sound-*producing* devices. In the Savart wheel, sequences of equidistant teeth were made to strike cards or other objects; this device, recently revived by Bart Hopkin [346], was originated by Robert Hooke in the 1670s to demonstrate his belief that it was theoretically possible to "make all tunes by strokes of a hammer."[6] In other cases, sequences of equidistant holes were instead moved past a jet of air or water, as in the Cagniard-Latour siren; but either way, the speed of rotation determined the pitch of the sound. In the 1910s, some of Luigi Russolo's *intonarumori* [160] incorporated toothed wheels that worked on the same principle: "Certain noises obtained through a rotating movement can give us a complete ascending or descending scale through the speeding up or slowing down of the movement," he wrote.[7]

Meanwhile, the embodiment of sound on the level of individual vibrations was also used to create visible *sound-surrogates*. Sometimes this was done by plotting simple impulses or graphing complex waveforms at leisure, based on reasoned mathematical principles. At other times, it was done by automatically recording sound vibrations—say, by attaching a stylus to a tuning fork so that it could write its own oscillations on a moving surface. The graphing of sound vibrations as a function of time was initially perceived as a thoroughly *visual* art. This is nowhere more evident than in the aspirations of Édouard-Léon Scott de Martinville for his phonautograph, patented in 1857: an invention which substituted an eardrum-like membrane for the tuning fork in the above scenario so that it could be made to capture and write down sounds out of the surrounding air. Scott thought the wavy lines which his invention traced were attractive in their own right—he proposed to use it "to produce industrial designs for embroideries, filigrees, jewelry, shades"[8]—but what particularly excited him was the prospect of linking the aesthetics of sight and sound on a scientifically objective basis.

One of his favorite test recitations was a passage from Jean-François Ducis's *Othello* (1872), which he spoke repeatedly into his phonautograph to study correspondences between expressive vocal intonations and its "natural stenography." The code of waveforms turned out to be entirely different from that of the time-frequency graph (even though the one can be converted mathematically into the other through a Fourier transform): a rise in pitch yielded undulations that were more closely bunched together, for example, while an increase in volume yielded taller undulations. But once Scott had worked out these correlations, he drew up hybrid displays of the *Othello* text that superimposed features of his phonautograph's "natural stenography" on conventional alphabetic writing, apparently hoping that people in the future would learn to "read" great performances of music and oratory in this way.

In 1877, Thomas Edison's phonograph introduced playback into the field of automatic sound-vibration recording, reinventing it as a means of fixing sounds for repetition *as sounds*. Recorded waveforms now became programs for automatophones rather than sound-surrogates, prompting lawyers to argue over whether they were better understood as writings subject to copyright or as parts of machines subject
5 to patent law.[9] Edison's phonograph is often construed as an inherently documentary technology, a means of *copying* sounds more or less transparently, and some theorists have sought to differentiate analog phonograms from digital ones in terms of the perceived "authenticity" of their testimony: in the analog world, sounds are imagined to leave behind actual physical traces of themselves.[10] But there's nothing inherently documentary about analog sound-recording formats per se—they're simply another type of graph—and the phonograph was put to thoroughly creative uses from the start. Edison himself dabbled in sound synthesis during 1878 by substituting sequences of carefully drilled holes for the usual tinfoil record to create an "alarm for various purposes,"[11] aligning his invention with the older sound-producing tradition of toothed wheels and sirens.

In the years that followed, the hand-drawn phonogram became an object of perennial interest among avant-garde artists; in particular, László Moholy-Nagy envisioned gramophone grooves being "incised by human agency into the wax plate, without any external mechanical means, which then produce sound effects which would signify—without new instruments and without an orchestra—a fundamental innovation in sound production."[12] Something akin to that dream was realized beginning in 1929–30 with the manual creation of optical film sound tracks by Arkady Avraamov, Rudolf Pfenninger, Oskar Fischinger and others.[13] Sounds created in this way were derived in a direct, indexical way from images, based on how much light passed through each successive point on the film; but in many cases the visual artistry of the images took forms irrelevant to sound production per se, implying an interest in cultivating both the visual and aural dimensions of the new art rather than subordinating one to the other.

Compositions in that spirit took notational determinacy to an extreme, with the composer specifying the parameters of the musical work down to the level of individual wave shape. But visual and aural arts have also intersected at the opposite end of the spectrum with the creation of graphically innovative musical scores that are deliberately indeterminate, privileging new experiences over static, predictable ones in pursuit of the values articulated by John Cage [214, 216]: "My favorite music is the music I haven't yet heard. I don't hear the music I write; I write in order to hear the music I haven't yet heard."[14] A prime example of the kind of score Cage had in mind is Earle Brown's *December 1952*: a composition represented by a page of scattered lines and rectangles, to be "performed in any direction from any point in the defined space for any length of time and ... from any of the four rotational positions in any sequence."[15] As in the other cases we've examined, the graphic approach was intended here to free the writing of music from conventional limitations, but in this case the goal was to fix the composition *less* rigidly than usual: not to facilitate the automatic realization of music, but maximally to resist it.

And there's nothing contradictory about any of this. Graphs simultaneously permit hyperdeterminacy and hypodeterminacy of compositional form because they embody sound as raw data unconstrained by the arbitrary symbolism of tradition, their meanings tethered to axes that can be assigned and reassigned at will, narrowly or broadly, strictly or loosely. And it's precisely because of the heterogeneous potentialities of the sound-graph that I believe its uses offer us such an important window onto the history of sonic imagination.

1
Ralph K. Potter, George A. Kopp, Harriet C. Green, *Visible Speech* (New York: D. Van Nostrand, 1947), p. ix.

2
Cf. Patrick Feaster, *Pictures of Sound: One Thousand Years of Educed Audio: 980–1980* (Atlanta, GA: Dust-To-Digital, 2012), track 25k; and the performance of *Rex caeli domine* in the *Norton Recorded Anthology of Western Music*, vol. 1 (New York: W. W. Norton & Company Inc., 2009).

3
Dom François Bedos de Celles, *L'art du Facteur d'Orgues*, vol. 4 (Paris: L.F. Delatour, 1778), pp. 599–600.

4
"A Letter from Mr. John Freke," in *Philosophical Transactions (1683–1775)*, no. 483 (London: Royal Society, March–May 1747), pp. 445–450: 450.

5
Jonathan Pitkin, "Con Spirito," in Theresa Sauer (ed.), *Notations 21* (New York: Mark Batty, 2009), p. 177.

6
Henry W. Robinson, Walter Adams (eds.), *The Diary of Robert Hooke* (London: Taylor & Francis, 1935), p. 211.

7
Luigi Russolo, Robert Filliou (trans.), *Art of Noises, futurist manifesto, 1913* (ubuclassics, 2004), p. 9; www.artype.de/Sammlung/pdf/russolo_noise.pdf

8
Patrick Feaster (trans. and ed.), *The Phonautographic Manuscripts of Édouard-Léon Scott de Martinville* (Bloomington, IN: FirstSounds.org, 2010), p. 14.

9
See E. Fulton Brylawski, Abe Goldman (eds.), *Legislative History of the 1909 Copyright Act*, 6 vols. (South Hackensack, NJ: F.B. Rothman, 1976).

10
See in particular Eric W Rothenbuhler, John Durham Peters, "Defining Phonography: An Experiment in Theory," *The Musical Quarterly*, vol. 81, no. 2 (Oxford, UK: Oxford University Press, Summer 1997), pp. 242–64.

11
Reese Jenkins, Robert A. Rosenberg, Paul Israel *et al.* (eds.), *The Papers of Thomas A. Edison. The Making of an Inventor, February 1847–June 1873*, vol. 1 (Baltimore, MD: John Hopkins University Press, 1989), pp. 4–48.

12
László Moholy-Nagy, "Production-Reproduction," in Krisztina Passuth, *Moholy-Nagy* (New York: Thames and Hudson, 1985), pp. 289–90.

13
Andrey Smirnov, "Sound out of Paper," asmir.theremin.ru/gsound1.htm (accessed August 20, 2010); Thomas Y. Levin, "'Tones from out of Nowhere': Rudolf Pfenninger and the Archaeology of Synthetic Sound," *The Grey Room*, no. 12 (Cambridge, MA: The MIT Press, Summer 2003), pp. 32–79.

14
John Cage, "An Autobiographical Statement," newalbion.com/artists/cagej/autobiog.html (accessed March 2014).

15
Earle Brown, *Folio and 4 Systems* (New York: Associated Music Publishers, Inc., 1961).

JOHN TRESCH

THE ARCHAIC ELSEWHERE OF SKIES

THERMODYNAMICS AND ELECTRICITY, PANTHEISM AND VOID

"One might call it a din, a roar, a clamor, a hammering, a great shriek, a bellow of the gods, but noise is a simple word that serves just as well to describe what has no name."

Roberto Bolaño, *2666*[1]

CAPTURED SENSE

The "Art or Sound" exhibition walks us past objects we hear: devices that produce, transform or bury sound within them. These objects lead us to puzzle, across the centuries caught here, at the variable positionings of the seeable and the audible, and at the instruments that merge, exchange, juxtapose and decouple them.

The musical sculpture captures and converts. In it something is harnessed, held, transformed, redirected. Just as instruments capture air, movement, and the vibration of a sound, they can also capture the eye, hold the glance, rewarding it with a harmony, an intricacy, a reason to linger, a trap: two centuries of bird cages.

In wood and string instruments, the breath and motion (and skill and emotion) of the player are received, shaped and amplified: the action comes from without, from the human who plays. These contrast with automata, objects that move themselves. The difference between an *instrument* and a *machine* was a recurring 19th-century question, though many denied any essential divide: a skillfully played instrument appears to play itself, while the valves and pistons of tools of all sorts, including woodwinds and brasses from the 19th century onward, fold many simple machines into a single device.[2]

The visual forms of musical objects, as much as the sounds they emit, raise the question of the sources of action: who is in control, and through what means? Musical automata especially lead us to ask: who has wound the spring, stoked the furnace, patterned the circuits, flipped the switch? Such puzzles hover over the Aeolian harp, a stringed object—simple in principle, but frequently beautifully elaborated—played by the wind.

Coleridge's poem *The Aeolian Harp* used this instrument for a stealthy metaphysical provocation.[3] At sunset, the narrator and his wife sit wreathed by flowers, hearing the "soft floating witchery" from "that simplest Lute:" "a light in sound, a sound-like power in light." The poet is seized by the thought of a single substance uniting the universe, and dares to wonder "what if all of animated nature/ be but organic Harps diversely framed," caressed and animated by an "intellectual breeze" that is both God and soul. Then the speculative window snaps shut: his wife glances at him with "mild reproof," reminds him "to walk humbly with [his] God."[4]

The poet dutifully relinquishes his fantasies, yet it's hard not to suspect that the true prize of the poem, its beating heart and the source of fascination, is precisely the forbidden pantheist reverie, framed by the innocent proprieties of Christian and domestic dualism, at its center.

Coleridge's poem established a recurring link between innovative instruments, synaesthesia, and a metaphysical temptation. Behind the lure of synaesthesia, which merged senses, stood cosmological possibilities that merged matter, mind and elusive sources of activity. Synaesthetic instruments were a gateway drug to pantheism. Though such conjectures picked up on perennial debates, they gathered historical force in the 19th century, thanks to technological assemblages that unleashed unprecedented forces.[5]

CYMBAL, CYLINDER, CYGNE

Musicologists like to say that at the start of the 19th century Europeans heard two kinds of man-made sounds louder than any heard before: explosions in the Napoleonic Wars and Beethoven's symphonies. True or not, the anecdote points to a rupture and a connection, terrifying and sublime, between new technologies of destruction and creation. The disarming equation between strategically deployed artillery and the orchestra's technological assembly was frequently noted, with composers and conductors positioned as generals, ranked instruments as troops, symphonies as campaigns. Ludwig van Beethoven's *Eroica* (1802–04) symphony was inspired by Napoleon; his *Wellingtons Sieg* (1813) was composed not only to commemorate Waterloo, but as a proof-of concept for a new, gigantic orchestral automaton, with pounding drums, blaring fanfares, booming wind-up drums and crashing cymbals.[6]

The Panharmonicon for which Beethoven composed was designed by the same Johann Nepomuk Maelzel who invented the metronome, an artificial larynx, and toured with a chess-playing automaton. Though 18th-century automata were figures of bourgeois self-control, by the early 19th century many humans experienced such machines as rivals and an invasion: they provoked a creeping recognition of impersonal forces overtaking individuality.[7] Thundering in the background, too, they heard the clanging of mass-produced tools and weapons, soon joined by the groan and clatter of steam engines driving mills, presses and locomotives.

Overwhelming auditory experiences were matched by staggering visual experiences in the panoramas, which frequently used as subjects battles scenes and vertiginous views from towers and mountains, as well as the visual culture they spawned: whether enormous and confrontational historical paintings, or dazzling, shape-shifting backdrops for the theater and the opera. The railway journey produced a "panoramic vision" in which foreground details evanesced into a slowly shifting background, as well as disorienting shock through the train's violent commotions.[8] Battles, panoramas, steam engines and orchestras all involved sounds, sights, and concussions too powerful, vast or dynamic to grasp at once: they threatened to overtake, devour, encompass the body and mind of those who experienced them, to rubble the boundaries that kept humans separate from their surroundings.

Helmut Müller-Sievers has recently reframed 19th-century science, technology and arts through reference to a single geometric form, the cylinder, and the motions it made possible. The challenges posed by the steam engine were not limited to retaining heat and power—a capacity increased by James Watt's introduction of a second cylinder or condenser to the engine's primary piston—but translating rectilinear motion into circular motion and vice versa.[9] As motors, containers and tools (including the drums used to record physiological processes) cylinders ensured both the fixed order and the open progress demanded for consumption and creation. The cylinder was also an elemental form (however mutated) for new additions to the audio-visual arsenal: the brass tubes stretched and twisted by Adolphe Sax [110, 112] and Ludwig Embach [76], the voice-amplifiers for Giacomo Meyerbeer's demon chorus below the stage in Meyerbeer's *Robert le diable* (1831), opera glasses and panoramas themselves. It provided the logic for the senses' assembly and deployment over time: the unfurling of the moving panorama matched the sequence of the railway journey, music box and player-piano, the unrolling of the symphonic score.

An intriguing double movement appeared in studies of the physiology of the senses: the senses were separated in order to be later rejoined.[10] On the side of separation, Johannes Müller and Charles Bell argued that the nerves that carried auditory and visual signals were distinct, and that specific nerve energies applied to each.

The laboratory apparatus of Laplacean mathematical experimental physics was used by Félix Savart and Jean-Baptiste Biot to capture the distinct regularities of optical and acoustic phenomena, building mechanical replicas of both eyes and ears. Attempts to measure the speed of sound and light were undertaken with support from the scientific popularizer François Arago, a contributor to Alexander von Humboldt's cosmic and esthetic science of the earth and skies. To measure sound, a cannon was fired at the same time as a visual signal was raised: observers at a known distance measured the time it took to hear the explosion after seeing the sign. To measure the speed of light, Arago directed Léon Foucault and Hippolyte Fizeau to build a system of swiftly rotating mirrors that would cause a light beam to travel a vast distance in a contained space; the swiftly rotating mechanism was borrowed from a church organ's device for regulating the flow of air into pipes.

Yet profound analogies were seen to unite sound with light and the other "imponderable" fluids studied by physicists: some physicists assumed they were all susceptible to identical mathematical treatments, and others pressed for proofs of an underlying physical unity, seeing them as vibrations in a pervasive ether or as modifications of a fundamental force or power. Such theories were endorsed and enacted in romantic and fantastic arts, where the unity of senses suggested deeper metaphysical unity: Johann Wolfgang von Goethe's emotional and musical analysis of colors harmonized with his Spinozist nature; Charles Baudelaire programmatically announced his "Correspondences" (1857). The equivalence and interconversion between senses was an artistic goal and a critical common ground: painting by sound, the music of colors.

The mixture of senses suggested a mixture of essences; synaesthesia often surged forth as a first step to the assertion of a mystic unity between mind and nature—a vitalist monism or a pantheism. Both were responses to upheavals in the distribution of the sensible, a new technological sensorium in which analytic boundaries were relentlessly traced and relentlessly surpassed.[11] Synaesthesia resonated simultaneously with philosophical pantheism and the scientific and technological developments of the early 19th century: steam engines, electric networks, mass media.

This combination also produced new social visions. Followers of Charles Fourier propounded the analogies among the senses and the combinations made possible by the "keyboard of personalities;" new moons, new continents, new creatures and bodies would appear through the harmonic organization of labor. Saint-Simonianism, a religion invented by engineers for the spread of industry, combined conversions of steam, electricity, labor and the senses. The equivalence of heat and work in the steam engine was repeated in their view of a cosmic equivalence between mind and matter—two aspects of God. Thermodynamics, pantheism and synaesthesia went hand in hand.

In 1832, a Saint-Simonian denounced contemporary Paris as "a great satanic dance" of visual and auditory dissonance: the scientists of the Latin Quarter were face-to-face with the "howling animals" of the Jardin des Plantes; orphans screaming in the children's hospital disturbed astronomers at the Observatory. In "The New City of the Saint-Simonians," labor, knowledge, trade and pleasure would be first separated and then united in a single plan, distributed across the city, which would take the form of a man. Science filled his left side, where a university's giant pyramid and shining buildings "climb to the sky in crystals of light;" industry took the right, with "cylindrical chimneys opening jaws filled with flames," to "the rhythm of hammer blows and axes."

Between the man's legs lay buildings "consecrated to the ecstasies of the mind and the delirium of the senses," including "operas and theaters with their apparatus of instruments, costumes and décors." Separating functions between thought, action and feeling was a first step to their harmonious merger. At the head of this "new colossus," on the île de la Cité, was a new temple in the shape of giant woman. Staircases spiraled up her sides, like jeweled belts and garlands; music poured from an organ at her breast; behind her, the train of her dress was draped over an amphitheater. This hi-tech, synaesthetic fetish converted all the forces of nature into a melodious, striking device to unify humanity, its works, and the multifarious energies of a divinized nature.[12]

The Saint-Simonians' pantheist vision of a technologically improved and harmonized city was a key source for Georges Eugène Haussmann's transformation of Paris in the 1850s and 1860s. Napoleon III's remodeling, however, seemed to leave no place for quieter, more intimate harmonies. Like Victor Hugo, Baudelaire saw himself as an exile from this changed urban landscape, its scaffolds, clanging tools, clattering carriages and pavements. In *Le Cygne* (The Swan, 1861) he forged an allegorical bond between himself and a starving, desperate swan, escaped from a menagerie, that he saw walking near the Tuileries: "in the forest where his spirit is exiled"[13] he hears the notes of a horn play, recalling all others who are lost, away from home. The symbolic fusion of sound, sight and sense in this *cygne*/sign comes at the cost of the poet's expulsion from his surroundings. Sense and meaning are emphasized, while matter drops away, handed over to "modern fatuity."[14]

Baudelaire's phonetic capture of the eternal in the fleeting and of the heroism specific to the modern era resonated through the arts to come. Curious paradox: visions of a unity of sense, action and technique in the 1830s and 1840s had been shared among engineers, scientists, poets and artists— whom Saint-Simon baptized the "avant-garde." Yet in the second half of the century, many of the mystic aspirations of reformers were realized as merely material (and frequently exploitative) attainments. In response, artists redefined themselves and their domain, in many cases, as an escape—though a doomed one—to a realm of memory, sense, emotion and imagination alone. Baudelaire was the beacon for this shipwreck.

THE MATTER OF EMPTINESS

Wagner's *Gesamtkunstwerk* merged music, poetry and theater. His Parisian followers received his synaesthesic work in spiritualist terms: as a sensory, emotional and intellectual experience, but, significantly, as a spectacle that transcended its material supports. Art became an ideal medium that made it possible to leave matter behind. Wagner encouraged this dematerialization by hiding the orchestra: his musical and visual Leitmotifs now floated beyond instruments. The pantheist doctrines of material and spiritual unity from the first half of the century were frequently sublimated into a new form of dualism after 1850, one that resonated with table turning and spiritualism and with research into the "fourth dimension" of ether; a mercurial, weightless matter assumed to be at work in both séances and in gas-discharge tubes—precursors to today's fluorescent and neon lights.[15]

Nevertheless, material supports were needed to produce transcendent effects; the late 19th-century's spiritual art was inseparable from its material productions. Wagner was depicted as a practitioner of animal magnetism, in a musical arms race with his rival Hector Berlioz, who contemplated conducting via telegraph.[16] Networks of hydraulics and steam were busily accompanied by meshes of electric communication; telegraph lines raced along beside railroads, below and ahead of steamships; new synaesthetic equipment was launched. After David Brewster's hand-held Kaleidoscope (1815) came Frédéric Kastner's Pyrophone (1873) [114], its cylinders opening to dancing gas flames, followed by Alexander Wallace Rimington's Color-Organ (1893). Russian composer Alexander Scriabin [136] wove the multicolored electric lights of the Chromola device into his audiovisual symphonies.[17] Experimental psychology, with its multiplying apparatus of minutely accurate timekeepers, sirens and flashing lights, tried to catch up with synaesthesia: Théodule Ribot in *Psychological Emotion* (1896) and Frederic W. H. Myers in *Human Personality and Its Survival of Bodily Death* (1903) sought recurrent equivalences between specific sounds (vowels, notes, timbres), colors and forms through experiments on patients with the condition—including Scriabin himself, one of Myers's subjects.[18] Artists, composers and poets tuned in to these investigations. Paul Valéry later described artists of the 1880s developing "a doctrine of art derived from then fashionable theses of psycho-physics. The study of sensibility by the methods of physics, research into the (hypothetical) correspondence of sensations, the energetic analysis of rhythm," all impacted painting and poetry.[19]

Yet the submission of the senses to elaborate machines—whether to analyze their functioning or provoke new experiences—could enervate as much as elevate. Jean des Esseintes, anti-hero of Joris K. Huysmans' *À rebours* (1884),[20] immersed himself in the esthetic experiments of his avant-garde contemporaries and his mechanized age. From rooms transformed into a steamboat, his imaginary voyage to London without leaving the train station, and his bejeweled tortoise, the book details des Esseintes's desperate quest for new sensations. The pinnacle was his perfume-organ, a Wagnerian apparatus to compose symphonies by fragrance; after an evocative performance, he collapsed.

Stéphane Mallarmé, one of des Esseintes's idols, acknowledged the seductions of Wagner but stopped short "halfway up the sacred mountain," before reaching "the too lucid and too threatening absolute" that the cult of art was proclaiming.[21] For Mallarmé, a crowd transported by the total artwork, expositions or department stores revealed continuities with religious rites of the past. Yet partisans of the cult of art believed too earnestly and hastily that they could reach their barely-recognized goal of perfect communion through technical transubstantiation. He warned: "don't go and make the same mistake some preachers do, and lighten, through I don't know what dilution into the color of electricity and of the people, the archaic elsewhere of skies."[22] The wrong lighting, a music that was too audible, traded in counterfeit heaven.

Instead Mallarmé advanced a conception of poetry as primordial music, an art that gestured indirectly toward an unspeakable emptiness at the root of thought and things. Poetic language indicates what cannot be said, through its intrinsic inability to capture what it speaks. Words placed like the frozen moves of ballerinas could "evoke, with intentional vagueness, the mute object, using allusive words, never direct, reducing everything to an equivalent of silence."[23] In place of the too-literal fantasia of 19th-century spectacles in the color of electricity, he demanded more patient and obscure illuminations, the carefully positioned marks of a void.

Language spaced across the page became rarefied silence. To Mallarmé's devotée, Marcel Duchamp, gesture became a mode of inactivity; material traces became a species of emptiness. Avant-gardes followed off these cliffs, exploring automaticity, the specific limits of the medium, self-reflexive puzzles without solution [164].

Two centuries of networked signal and noise, of destruction choreographed by electric connection, divide the blasphemous speech of Coleridge's wind-strummed lute from the present. In Bruce Nauman's sculptures of word and light, perhaps we see and hear the fullness of pantheism that pulsed in romantic synaesthesia collapsing into a cosmic twin—silent dark matter undercutting "light in sound." In one of Nauman's enigmatic circuits (1981–82), neon hums in twisted tubes of lemon, orange and fuchsia; words flare, expire and repeat:
VIOLINS... VIOLENCE... SILENCE.

This essay is dedicated to
Robert Michael Brain

1
Roberto Bolaño, *2666*, trans. by Natasha Wimmer (New York: Farrar, Strauss, and Giroux, 2004), p. 795.

2
Helmut Müller-Sievers, *The Cylinder: Kinematics of the Nineteenth Century* (Berkeley, CA: University of California Press, 2012); John Tresch, Emily I. Dolan, "Toward a New Organology: Instruments of Music and Science," in *Osiris*, vol. 28, no. 1 (Chicago, IL: The University of Chicago Press, 2013), pp. 278–98.

3
Thomas L. Hankins, Robert J. Silverman, Willem Hackmann, *Instruments and the Imagination* (Princeton, NJ: Princeton University Press, 1995).

4
Samuel Taylor Coleridge, *The Eolian Harp*, vv. 13, 21, 29, 45–46, 48, 50, 53, in Ernest H. Coleridge (ed.), *The Complete Poetical Works of Samuel Taylor Coleridge*, vol. I (Oxford, UK: Clarendon Press, 1912), pp. 100–02.

5
Douglas Kahn, *Earth Sound, Earth Signal: Energies and Magnitudes in the Arts* (Berkeley, CA: University of California Press, 2013).

6
Emily I. Dolan, "The Origins of the Orchestra Machine," *Current Musicology*, n. 76, 2003, pp. 7–23; Nicholas Mathew, "History under Erasure: *Wellingtons Sieg*, the Congress of Vienna, and the Ruination of Beethoven's Heroic Style," *The Musical Quarterly*, vol. 89, no. 1 (Oxford, UK: Oxford University Press, Spring 2006), pp. 17–61.

7
Adelheid Voskuhl, *Androids in the Enlightenment: Mechanics, Artisans, and Cultures of the Self* (Chicago, IL: University of Chicago Press, 2013).

8
Wolfgang Schivelbusch, *Geschichte der Eisenbahnreise: Zur Industrialisierung von Raum und Zeit im 19. Jahrhundert* (München/Wien: Hanser, 1977), trans. by Anselm Hollo, *The Railway Journey: Trains and Travel in the 19th Century* (New York: Urizen Books, 1979), pp. 57, 137.

9
Helmut Müller-Sievers, *op. cit.*, pp. 23–40.

10
Jonathan Sterne, *The Audible Past: Cultural Origins of Sound Reproduction* (Durham, NC: Duke University Press, 2003); Jonathan Crary, *Techniques of the Observer: on Vision and Modernity in the Nineteenth Century* (Cambridge, MA: The MIT Press, 1992); Mara Mills, John Tresch, "Introduction: Audio/Visual," *Grey Room*, vol. 43 (Cambridge, MA: The MIT Press, Spring 2011), pp. 6–15.

11
Jacques Rancière, *The Politics of Aesthetics* (London: A&C Black, 2013); Todd H. Weir (ed.), *Monism: Science, Philosophy, Religion, and the History of a Worldview* (Basingstoke: Palgrave Macmillan, 2012).

12
Charles Duveyrier, "La Ville nouvelle, ou le Paris des Saint-Simoniens," in C. Ladvocat (ed.), *Paris, ou le livre des cent-et-un*, vol. 8 (Paris: Ladvocat, 1832), pp. 315–44.

13
[Editor's Note: Verse 49 from the poem *Le Cygne*. This poem is the 4th in a section of *Les fleurs du mal* entitled "Tableaux Parisiens" (Parisian Scenes), which was added to the 2nd edition of the book. First edition *Les fleurs du mal* (Paris: Poulet-Malassis et de Broise, 1857, 2nd ed. 1861).]

14
Charles Baudelaire, "Exposition universelle – 1855 – Beaux-arts," in Id., *Œuvres complètes*, vol. II (Paris: Michel Lévy frères, 1868), pp. 211–244: 219.

15
Falk Mueller, "Johann Wilhelm Hittorf and the Material Culture of Nineteenth-Century Gas Discharge Research," *British Journal for the History of Science* vol. 44, no. 2 (Cambridge, UK: Cambridge University Press, June 2011); Linda Dalrymple Henderson, "Etherial Bride and Mechanical Bachelors: Science and Allegory in Marcel Duchamp's 'Large Glass'," *Configurations* vol. 4, no. 1 (Baltimore, MD: Johns Hopkins University Press, Winter 1996), pp. 91–120.

16
Alison Winter, *Mesmerized: Powers of Mind in Victorian Britain* (Chicago, IL: University of Chicago Press, 2000).

17
Kerry Brougher, Judith Zilczer, Jeremy Strick, Ari Wiseman, Olivia Mattis, *Visual Music: Synaesthesia in Art and Music Since 1900* (London: Thames & Hudson, 2005).

18
Roy Porter, "Synaesthesia," in Adam Lowe, Simon Schaffer (eds.), *N01se*, exhibition catalogue (Cambridge, UK: Kettle's Yard, 1999).

19
Paul Valéry cited in Robert Michael Brain, "Genealogy of 'ZANG TUMB TUMB': Experimental phonetics, vers libre, and modernist sound art," *Grey Room*, vol. 43, (Cambridge, MA: The MIT Press, Spring 2011), pp. 88–117.

20
[Editor's Note: 1st edition published in Paris by G. Charpentier et Cie. Numerous editions of the book have been published in English under the titles: *Against Nature* or *Against the Grain.*]

21
Stéphane Mallarmé, "Richard Wagner: The Reverie of a French Poet," in Barbara Johnson (trans.), *Divagations* (Cambridge, MA.: Harvard University Press, 2007); Jacques Rancière, *Mallarmé: The Politics of the Siren* (New York: Continuum, 2011).

22
Stéphane Mallarmé, *op. cit.*, p. 246.

23
Ibid, p. 264.

CHRISTOPH COX

SEEING IS NOT HEARING

SYNAESTHESIA, ANAESTHESIA AND THE AUDIO-VISUAL

FROM *GESAMTKUNSTWERK* TO SYNAESTHESIA

Human beings are endowed with five senses, five different channels through which we gather information about our ever-changing environment. Communicating between inside and outside, four of these sensory channels originate in particular orifices—eyes, ears, nose and mouth—through which our bodies open out onto the world. The remaining sense, touch, is dispersed across the entire surface of the skin, an enveloping membrane that registers the minutest of stimuli.

The division of the arts largely accords with this sensory distribution and fragmentation, and also with the traditional hierarchy that separates the two "noble senses" (seeing and hearing) from their more vulgar associates (taste, touch and smell). Painting and sculpture address the eyes, music the ears. The former belong in a museum or gallery, the latter in the concert hall. Each has its own history, which only rarely overlaps with the other. And while there are surely mixed arts—theater and cinema, for example—we still unproblematically speak of "the visual arts" and distinguish them from music as apples to oranges.

However, despite this sensory fragmentation, physiological dispersal, and aesthetic hierarchy, we experience the world as a unity. We don't feel ourselves to be constantly coordinating and translating between heterogeneous streams of sensory data. Rather, the world seems to come to us undivided and complete. Acknowledging this mysterious emergence of unity from plurality, Aristotle posited a sort of sixth sense—a *sensus communis* or common sense—whose role was to coordinate the five sensory streams and ensure their agreement with one another.[1] Aristotle's hypothesis profoundly influenced his successors; and variants of it continue to inform current scientific research.[2]

If we experience the world as a sensory unity, shouldn't the arts affirm this union?

On this question, modernism was deeply divided. One of its most powerful and influential theorists, Clement Greenberg, argued that the distinctiveness of the modern consisted precisely in the segregation of the arts from one another and the autonomous development of each. In particular, Greenberg was concerned to purge the visual arts of every extra-visual sensation (tactility, for example) in order to render them "purely optical."[3] Yet, from the outset, prominent modern artists pressed an alternative position, advocating for sensory and esthetic unity. This call was boldly sounded by the composer and dramatist Richard Wagner in his 1849 essay *The Artwork of the Future*, which insisted that only a true synthesis of all the arts—a *Gesamtkunstwerk* or "total work of art"—could realize art's true purpose. Wagner wrote: "Each separate faculty of man is limited by bounds; but his united, agreed, and reciprocally helping faculties—and thus his faculties in *mutual love* of one another—combine to form the self-completing, unbounded, universal faculty of men. Thus too has every *artistic* faculty of man its natural bounds, since man has not *one only Sense* but separate *Senses*; while every faculty springs from its special sense, and therefore each single faculty must find its bounds in the confines of its correlated sense. But the boundaries of the separate senses are also their joint meeting-points, those points at which they melt in one another and each agrees with each: and exactly so do the faculties that are derived from them touch one another and agree. Their confines, therefore, are removed by this agreement; but only those that love each other can agree, and "to love" means: to acknowledge the other, and at like time to know one's self. Thus Knowledge through Love is Freedom; and the freedom of man's faculties is—*All-faculty*.

Only the Art which answers to this "all-faculty" of man is, therefore, *free*; and not the Art-*variety*, which only issues from a single human faculty. The Arts of Dance, of Tone, of Poetry, are each confined within their several bounds; in contact with these bounds each feels herself unfree, be it not that, across their common boundary, she reaches out her hand to her neighboring art in unrestrained acknowledgment of love. The very grasping of this hand lifts her above the barrier; her full embrace, her full absorption in her sister—i.e., her own complete ascension beyond the set-up barrier—casts down the fence itself. And when every barrier has thus fallen, then are there no more *arts* and no more boundaries, but only *Art*, the universal, undivided."[4]

In the decades that followed this proclamation, Wagner's *artistic* call for a union of the senses was bolstered by a burst of *scientific* interest in synaesthesia, the neurological phenomenon in which stimulation of one sensory modality induces sensation in another such that, for example, one sees sounds or hears odors. This phenomenon had fascinated philosophers and scientists for centuries, at least since 1690, when John Locke reported the case of a blind man who claimed to understand the color scarlet as the sound of a trumpet.[5] Yet, in the final decades of the 19th century, synaesthesia became a hot topic for neurological research.

It also sparked the interest of painters, composers and poets, who were drawn to the imaginative possibilities and spiritual resonances of sensory union. In his celebrated poem *Correspondences* (1857), Charles Baudelaire wrote of "odors ... sweet as flutes, and green as any grass."[6] Baudelaire's Symbolist compatriot Arthur Rimbaud drew equivalences between vowels, colors, textures and odors: the letter "A" invoked a "black hairy corset of shining flies/Which buzz around cruel stench;" "E," the "whiteness of vapors and tents," and so on.[7] As visual artists began to abandon figuration and approach abstraction, they often looked for inspiration to music, conceived as the highest and least representational of the arts. "All art constantly aspires toward the condition of music," declared the critic Walter Pater in 1877, a remark that would be echoed by many pioneers of pictorial abstraction.[8] Wassily Kandinsky's proto-abstract painting *Impression III (Concert)*, 1911, was directly inspired by the thrill of hearing Arnold Schoenberg's first atonal works at a concert in Munich. Others, such as Paul Klee, Marsden Hartley and František Kupka, took inspiration from Johann Sebastian Bach fugues, attempting a visual counterpoint akin to the fugue's rhythmic structure.

The late 19th-century craze for synaesthetic cross-wirings also caught the fancy of inventors, who set to work constructing machines that could directly connect sound with color. In the late 1860s, the Alsatian chemist and musician Frédéric Kastner developed an instrument he called the Pyrophone [114], in which a small keyboard produced both sound and colored light by igniting gas jets that lit 13 crystal pipes protruding from a console. Two and a half decades later, the British inventor and art professor Alexander Rimington built what he was the first to call a "color organ," an instrument that made no sound but used a standard organ keyboard to illuminate 14 arc lamps that glowed with different shades and intensities of color.

Kastner's and Rimington's inventions sparked widespread fascination with visual music, spurring artists and inventors to construct numerous variants on the color organ. In the first few decades of the 20th century, the artistic avant-garde was drawn to these inventions, not so much as musical instruments but as mechanisms capable of animating abstract forms through a sort of cinematic projection. The Russian Futurist painter Wladimir Baranoff Rossiné developed his optophonic piano (1920–23) [170], a complex keyboard-controlled contraption that produced sound while projecting light through an array of mirrors, filters, lenses and hand-painted discs. At the Weimar Bauhaus, Ludwig Hirschfeld-Mack constructed a machine that translated music into projections of mobile forms and colored light. The Danish-American artist Thomas Wilfred upgraded the color organ with his Clavilux (1922), which employed a bank of dials and sliding keys to rotate bulbs and mirrors that displayed ethereal, flame-like wisps of dancing color.

These live projections of visual music were paralleled by the exploration of musically-inspired abstraction on film. As in the work of Baranoff Rossiné and Hirshfeld-Mack, the "absolute cinema" of Hans Richter and Viking Eggeling aimed to animate the rhythmic and harmonic relationships inherent in non-representational painting, endowing them with the temporal dimension characteristic of music. The pulsing, growing, shrinking and layering rectangles in Richter's *Rhythmus 21* (1921) evoke the musical experience of volume, pitch, harmony and, above all, rhythm. In Eggeling's *Symphonie Diagonale* (1924), a series of comb-like forms alternate and succeed one another like chords, while curved and angled lines are drawn on the screen like silent melodies. Even more direct connections between cinematic sound and image were explored by Oskar Fischinger, who produced a set of visual "ornaments"—sawtooth patterns, stars, dots and waves—that could serve simultaneously as the visual content of a film and also as its soundtrack, given that a film's soundtrack is printed on the celluloid as a visual waveform to be read by an optical device that converts it into an audio signal.

This effort at a direct connection between sound and image has been a recurrent pursuit in modern and contemporary art, a project taken up in the early 1970s by Guy Sherwin's optical sound films, Steina and Woody Vasulka's video experiments, Alvin Lucier's *The Queen of the South* (1972), and again more recently in the work of Carsten Nicolai [350], Billy Roisz, Haroon Mirza [400, 434] and others.

SOUND/IMAGE

With the advent of sound film in the late 1920s, the dream of a synaesthetic art would seem to have been finally realized. Yet this development was not universally celebrated by filmmakers and other artists. In 1928, three of the most prominent representatives of Soviet cinema, Sergei Eisenstein, Vsevolod Pudovkin and Grigori Alexandrov, issued a statement warning that sync sound would surely be used merely to bolster cinematic illusion. "To use sound in this way," they noted, "will destroy the culture of montage." Instead, the trio advocated the non-synchronization of sound and image, a contrapuntal relationship between the two that would resist the subordination of sound to image, encouraging a tension that could thwart the naturalistic illusion.[9]

A few years later, Walter Ruttmann took up this call for a non-synchronous relationship between sound and image, and pushed it to the extreme. An associate of Richter, Eggeling and Fischinger, Ruttmann inverted their experiments in "visual music." His 1930 film *Weekend* consists solely of a soundtrack without images: an 11-minute collage of concrete noises—hammers, cash registers, sirens, voices, incidental music, etc.—that left the viewer to imagine corresponding visuals. The film might be seen as a complement to the silent visual music of *Rhythmus 21* or *Symphonie Diagonale*, yet the prevailing hierarchy of the visual over the sonic prevents this complementarity, insuring that the absence of the visual is experienced not as an invitation to synaesthesia but as a lack and disjuncture between sound and image.

This resistance to the seamless merging of sound and image is as prevalent in modern and contemporary art as is the desire for synaesthetic union. In part, it is based on the suspicion that any convergence of the senses is likely to retain the hierarchy that subordinates all other modalities to the visual. It is equally born of the desire not to eliminate the unique differences between the senses and the rich aesthetic tensions these differences generate.

This attitude has been prevalent throughout the history of 20th- and 21st-century art, from Duchamp's readymades through current practices in sound art. While Kandinsky, Klee, and so many other early modernists pursued the dream of *synaesthesia*, Duchamp instead championed *anaesthesia*. The selection of the readymades, he noted, "was based on a reaction of visual indifference with at the same time a total absence of good or bad taste…in fact a complete anaesthesia."[10] Indeed, throughout his career, Duchamp resisted the notion that art should be primarily "retinal," aiming instead to push it toward the conceptual and the verbal. In several works, this critique of retinal art took recourse to the sonic. Take, for example, *With Hidden Noise* (1916) [164], an "assisted readymade" consisting of a ball of twine sandwiched between two brass plates held together by bolts, with an unknown object inserted into the middle by Duchamp's patron Walter Arensberg. Though the work is in principle a sound-producing object or toy, when elevated to the status of an art object inertly displayed in a museum, it loses this auditory character, the trace of which remains only in the title, its sonic source now "hidden" or "secret."

A whole lineage of Duchamp-inspired art seeks to disjoin sound from image. Joseph Beuys' *The Silence* (1973), for example, is a sculptural object consisting simply of the 5 reels of Ingmar Bergman's 1962 film *The Silence*, the film's celluloid drenched in copper and zinc, and sealed in galvanized canisters. Eliminating both the sound and image of Bergman's film, Beuys' stack of canisters draws attention to film not as "talking pictures" but as mute material. This line of attack is furthered in Christian Marclay's *The Sound of Silence*, a photograph of Simon & Garfunkel's 1964 record *The Sounds of Silence*. Marclay's piece references Beuys and Duchamp but also René Magritte, whose famous 1929 painting *The Treachery of Images* boldly scribbled *Ceci n'est pas une pipe* (This is not a pipe) below a stylized image of a pipe. Magritte's painting highlighted the disjunction between word, image and object. Marclay's photograph extends this disjunction to sound, noting, like Beuys, that this mute image and object capture the silence to which Simon & Garfunkel's song and Bergman's film could only paradoxically allude.

Marclay's photograph is firmly within the lineage of classic conceptual art, which, often inspired by Duchamp, challenged the status of the work of art as both image and object. The search for non-retinal art led many conceptual artists to move toward the pure concept and toward language as its physical manifestation. Yet it led a number of conceptualists —among them Robert Barry, Christine Kozlov, Robert Morris and Bruce Nauman—toward sound as an ephemeral material at once invisible and powerfully physical. A pioneering work of both conceptual art and sound art, Robert Morris' *Box with the Sound of Its Own Making* (1961) [220] is precisely what its title describes: a wooden cube accompanied by a three-hour audio recording that documents the process of its construction. This dull visual object is nothing without its soundtrack, which supplies all the drama of the piece and offers an implicit critique of aesthetic reification: Art is not the result but the process, which is captured more fully through the temporality of audio than by the inert visual thing.

A number of Nauman's works from 1968 and 1969 explore the possibilities of sound to unsettle the visual and the object status of art. In *Six Sound Problems for Konrad Fischer* (1968), the visual element is simply a reel-to-reel tape recorder with a loop of audiotape stretched across the space and wound around a pencil attached to a chair. The 6 loops supplied with the piece register the sounds of simple activities performed by Nauman alone in Konrad Fischer's Düsseldorf gallery: walking, bouncing balls, playing violin, and combinations of these. Nauman had previously recorded several of these activities on video; here he presents them solely as sound—as "sound *problems*," because, once again, audio alone forbids us the documentary veracity of the visual, which more clearly registers the sources and causes of events. This capacity to sever sound from source is a characteristic feature of what the composer Pierre Schaeffer termed "acousmatic" listening, a form of auditory attention made possible by audio recording, which allows us to listen to sounds in the absence of their visual sources.

Schaeffer's *musique concrète*—audio works composed entirely from tape recordings of musical and documentary sound—explored the acousmatic possibilities of recorded sound. Nauman's *Concrete Tape Recorder Piece* (1968) [248] is a Duchampian pun on Schaeffer, homage to *With Hidden Noise*, and a response to *Box with the Sound of Its Own Making.* A 530-pound concrete block with a power cord protruding from it, the piece was described by the artist as a "tape recorder with a tape loop of a scream wrapped in a plastic bag and cast into the center of a block of concrete." Here, too, the noise is hidden, silent—or rather *silenced*, for Nauman's piece is more ominous than Duchamp's or Morris's. Its sound is entombed—repeated infinitely but inaudible, buried by the sculptural form that negates it. As such, the piece is perhaps an analog to Edvard Munch's silent *Scream* (1893) but more confounding, its sonic source denied all expression.

The search for sensory unity and the resistance to this union... These two tendencies (and the tensions between them) have driven modern and contemporary art for more than a century and a half. Since Wagner and Kandinsky, artists have pursued the dream of synaesthesia, a unification of the arts that could overcome their fragmentation and allow esthetic experience to affirm the unity of everyday experience—what Aristotle called the *sensus communis*, Wagner celebrated as the "universal faculty of men," and Brian Massumi has recently called the "amodal experiential confound," the "many-dimensional virtual whole of feeling" that characterizes our ordinary sensory lives.[11] Today, this drive toward sensory and artistic unity draws sustenance from the fact that art has become steadily more multimedia, intermedia, or post-medium, and that digital media facilitate data translation—an image into a sound, for example, or vice versa.

Yet this affirmation of sensory and esthetic fusion has been met by a powerful counter-discourse, articulated most emphatically by Clement Greenberg and Michael Fried, who championed a group of artists whom they saw as pushing in the opposite direction: toward the careful and rigorous exploration of specific media and sensory modalities.[12] Though this modernist camp would seem to have lost the battle against the hybrid and multimedia work that has dominated contemporary art, the prominent critic Rosalind Krauss recently renewed the Greenberg-Fried call for medium-specificity, maintaining that "the abandonment of the specific medium spells the death of serious art."[13] Much of the sound art that has emerged over the past two decades has implicitly aligned itself with these modernist critiques. Without rejecting multimedia and multimodal work, sound art prominently reminds us that seeing is not hearing and aims to thwart the imperial aspirations of the visual. Over the past century and a half, no work of art has fully satisfied either of these tendencies. But some of the most compelling works have responded to the provocation presented by these two poles, resisting both the assimilation and the segregation of the senses, operating in the fraught space between.

1
See Aristotle, *De Anima* III, 1–2.

2
See Richard Cytowic, *Synaesthesia: A Union of the Senses* (Cambridge, MA: The MIT Press, 2002), p. 75.

3
See Clement Greenberg, "Modernist Painting," in John O'Brian (ed.), *Clement Greenberg: The Collected Essays and Criticism, Volume 4: Modernism with a Vengeance, 1957–1969* (Chicago, IL: University of Chicago Press, 1993).

4
Richard Wagner, "The Art-work of Future," in W. Ashton Ellis (trans.), *Richard Wagner's Prose Works*, vol. I (London: Kegan Paul, Trench, Trübner and Co., 1892), pp. 97–98.

5
John Locke, Roger Woolhouse (ed.), *An Essay Concerning Human Understanding*, Book IV (London: Penguin, 1997), §11.

6
Charles Baudelaire, "Correspondences," in *Les Fleurs du Mal*, trans. by Richard Howard (Jaffrey, NH: David R. Godine, 1982), p. 15.

7
Arthur Rimbaud, "Vowels," in Wallace Fowlie (ed.), *Rimbaud: Complete Works, Selected Letters* (Chicago, IL: University of Chicago Press, 2005), p. 141.

8
Walter Pater, *The Renaissance* (New York: Modern Library, 1919), p. 111.

9
S.M. Eisenstein, V.I. Pudovkin, G.V. Alexandrov, "Statement [on Sound]," in Gerald Mast, Marshall Cohen, and Leo Braudy (eds.), *Film Theory and Criticism: Introductory Readings*, 4th edition (Oxford, UK: Oxford University Press, 1992), pp. 317–19.

10
Marcel Duchamp, "Apropos of 'Readymades'," in Michael Sanouillet and Elmer Peterson (eds.), *Salt Seller: The Writings of Marcel Duchamp* (Oxford, UK: Oxford University Press, 1973), p. 141.

11
See Brian Massumi, *Parables for the Virtual: Movement, Affect, Sensation* (Durham: Duke University Press, 2002), p. 169, and Id., *Semblance and Event: Activist Philosophy and the Occurrent Arts* (Cambridge, MA: The MIT Press, 2011), p. 88.

12
Greenberg's classic defense of medium specificity is "Modernist Painting," in John O'Brian (ed.), *op. cit.*, pp. 85–93. Fried's influential argument against the mixing of mediums is "Art and Objecthood," in Michael Fried, *Art and Objecthood: Essays and Reviews* (Chicago, IL: University of Chicago Press, 1998), pp. 148–72.

13
Rosalind Krauss, *Perpetual Inventory* (Cambridge, MA: The MIT Press, 2010), xiii.

Unknown

NAIL VIOLIN
early 19th century

Several friction idiophones were introduced during the 18th century, including Johann Wilde's nail violin in 1744, upon which tuned nails are bowed by a violin bow.

The flat wooden sounding-board is round on this instrument, rather than the usual half-moon shape, and the metal nails are mounted perpendicularly around its curved edge. As these nails diminish in height the notes rise in pitch, and the chromatic nails are distinguished by being slightly bent.

The nail violin has acoustic properties similar to a stopped organ pipe as opposed to one that is open at one end, having only odd-numbered overtones, and sounding one octave lower than a rod that is fixed or free at both ends.

Wilde said that he invented the nail violin after accidentally scraping the hair of his bow across the metal peg upon which he was about to hang it, producing a musical sound.

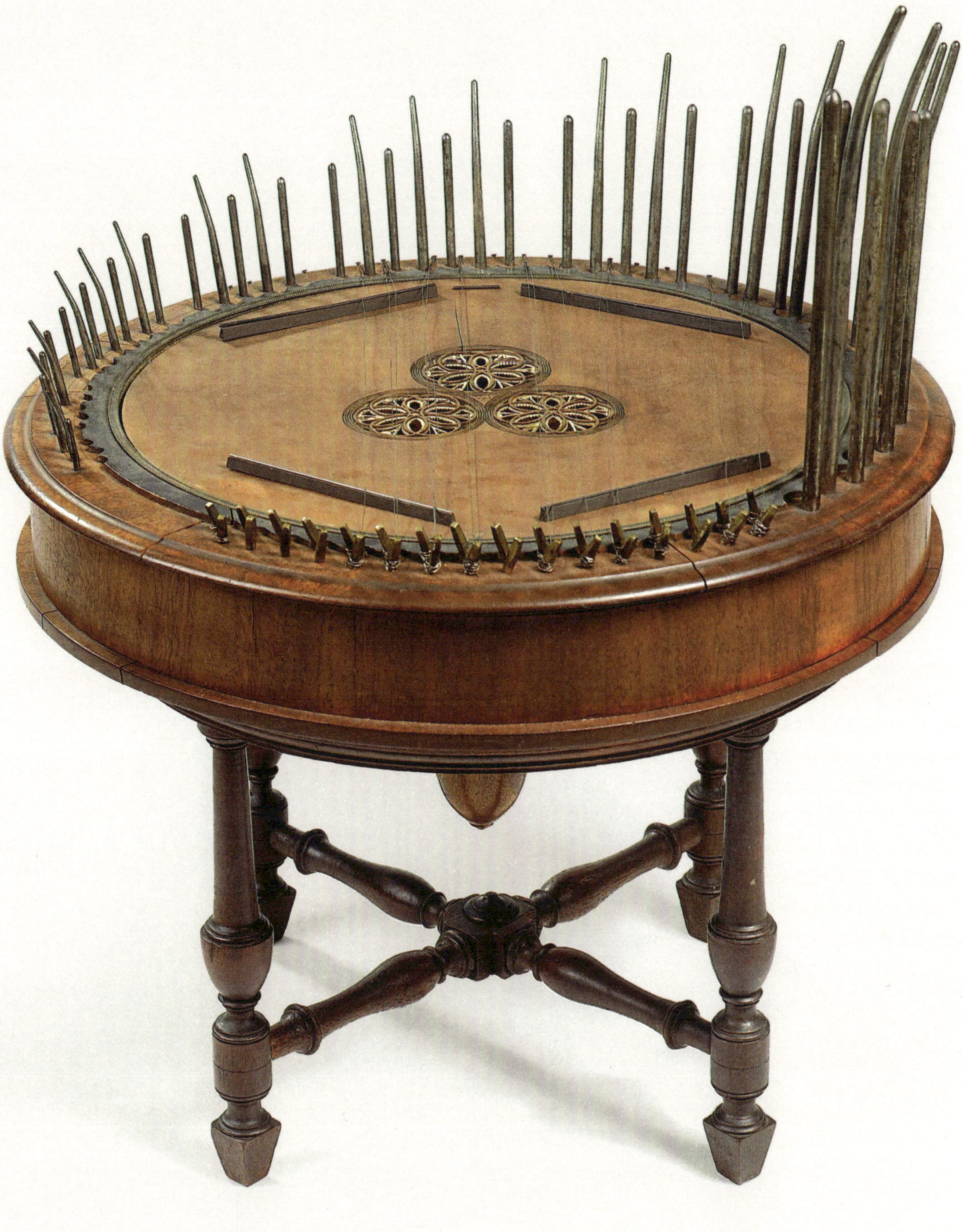

François Antoine Sautermeister

RUSSIAN BASSOON
1809–30

The Russian bassoon was a military instrument, and its name is thought to have been a linguistic evolution caused by the instrument's use in the Prussian military band. Designed by J.J. Regibo of Lille around 1784, it consists of 4 to 5 sections including butt and wing joints similar to the conventional bassoon, a conical bell column and dragon's head, and either a single coiled or swan-shaped bocal. The instrument, whose narrower bore dimensions permitted an expanded and clearer high register in comparison with other bass horns, such as its ancestor the serpent, was most popular in France and Belgium with a group of instrument builders in Lyon, like François Antoine Sautermeister (Wurtemberg, 1782–Lyon, 1830).

[See also p. 70]

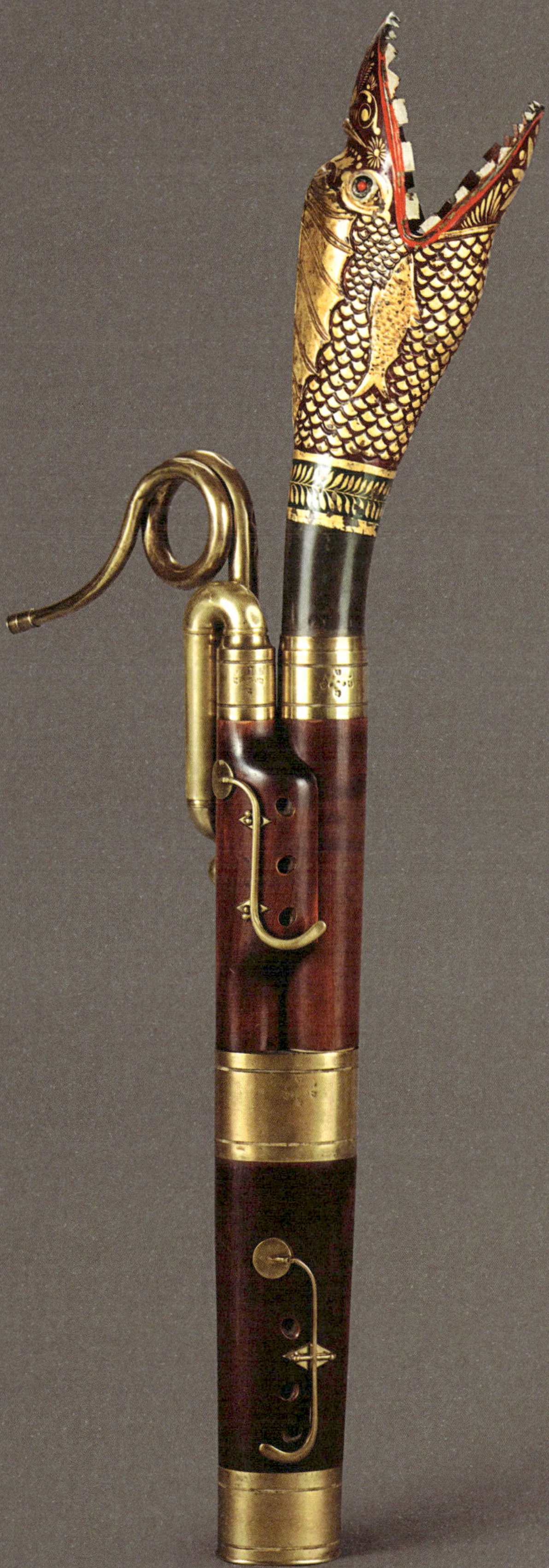

André Antoine Schmittschneider

HORN-SHAPED TRUMPET
c. 1823

The horn-shaped trumpet (also called a trompette-trombone), was invented and patented together with a new model of horn by the instrument-maker André Antoine Schmittschneider on December 13, 1821. The horn and trumpet were the result of a new manufacturing process that resulted in the two instruments receiving praise for the quality of their sound, particularly in the bass register. The hybrid conception of this copper instrument, which sounds like a trumpet and has a shape and mouthpiece similar to a horn, made it possible for horn players to play the trumpet as well.

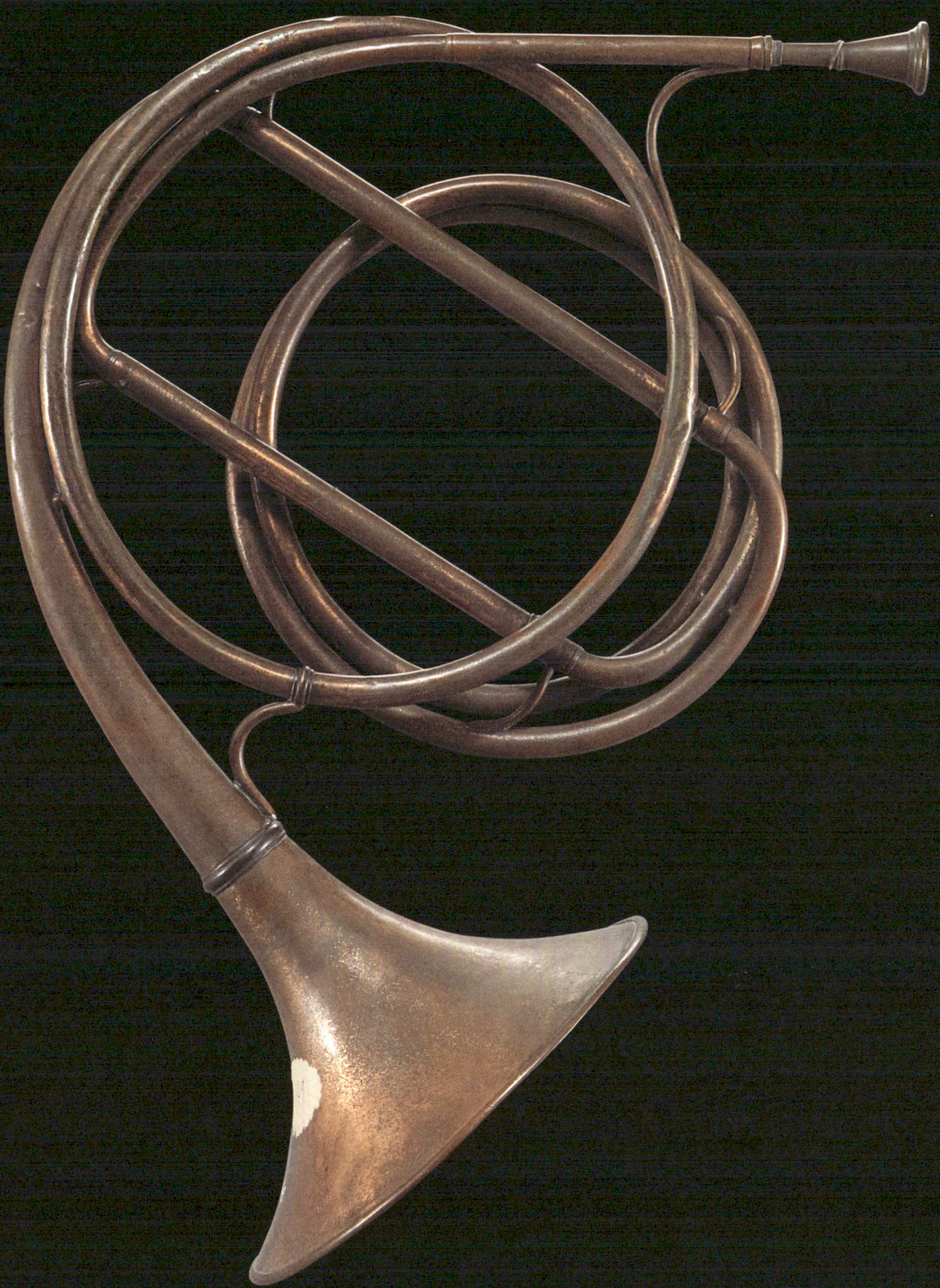

Davrainville fils

BARREL ORGAN WITH TRUMPETS 1828

The technology of the barrel organ—in which the continual rotation of a cylinder fitted with nails moves levers that operate a bellows which in turns blows air into the organ pipes—was in some cases applied to wind instruments as well.

In 1827 Davrainville designed the horns on omnibuses in Paris and, the following year, conceived this device to help teach cavalry trumpeters. Two interchangeable wooden cylinders can be turned to play the 32 musical signals used by the military through a set of 5 trumpets. Installed in the Célestin barracks until the July Revolution, it was later acquired by General Jean-Baptiste Ventura and ended up in Lahore, where this mechanical device served as the only instructor for 200 trumpeters of the Sikh army.

[See also p. 58]

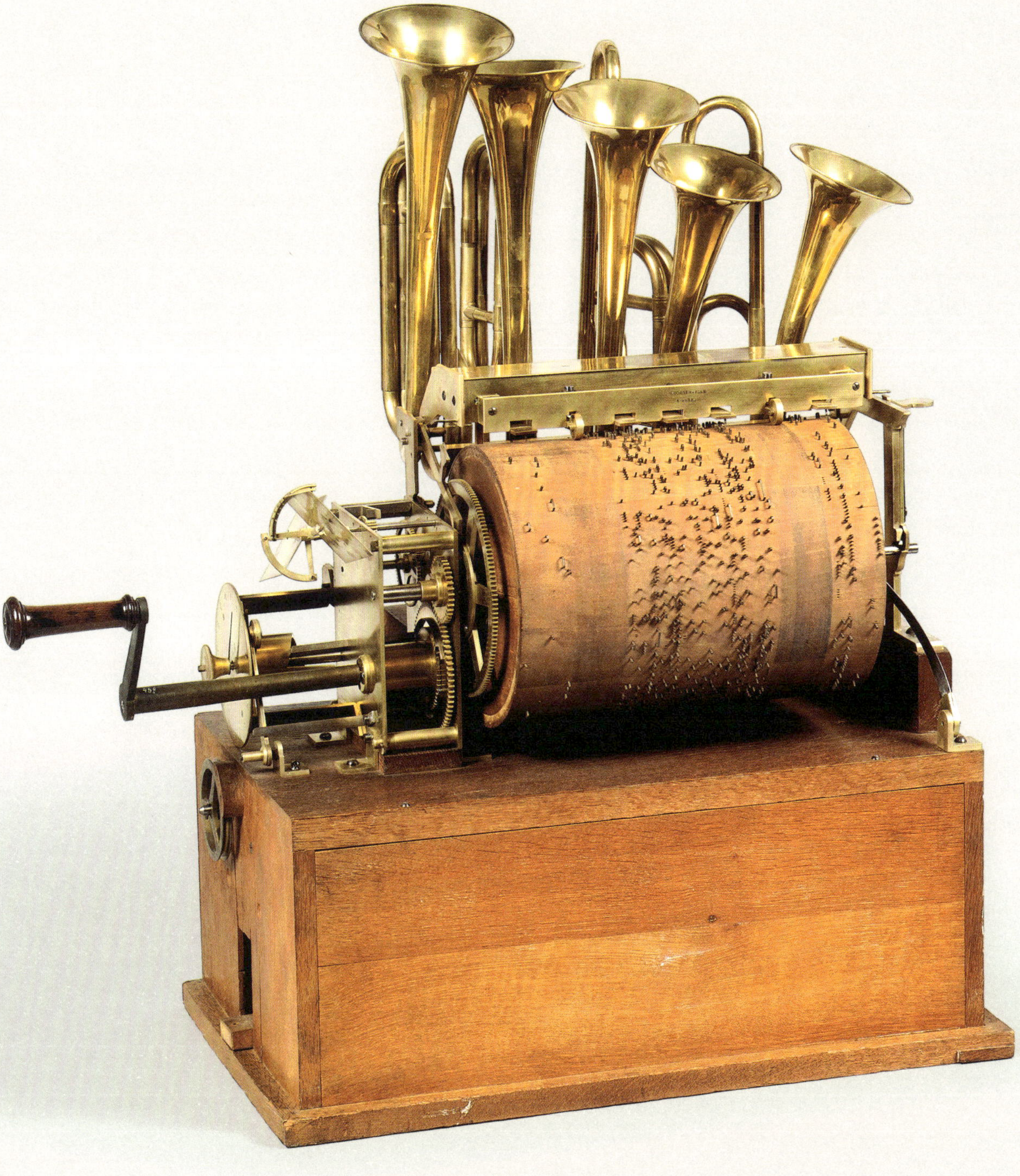

Unknown

DEMILUNE TRUMPET
first half of 19th century

108

The unusual half-moon shape of the demilune trumpet allows the player's hand to be placed in the bell to alter the pitch of the notes and make the instrument partially chromatic, thus giving it a greater musical range.

This technique was introduced by Anton Joseph Hampel, a horn player who had noted in 1775 that, by placing his hand in the bell of the instrument, he was able to lower the note by a semitone, or even a full tone. This was difficult in trumpets due to the length of the instrument, and was made pointless in the first half of the 19th century when keys and valves were added.

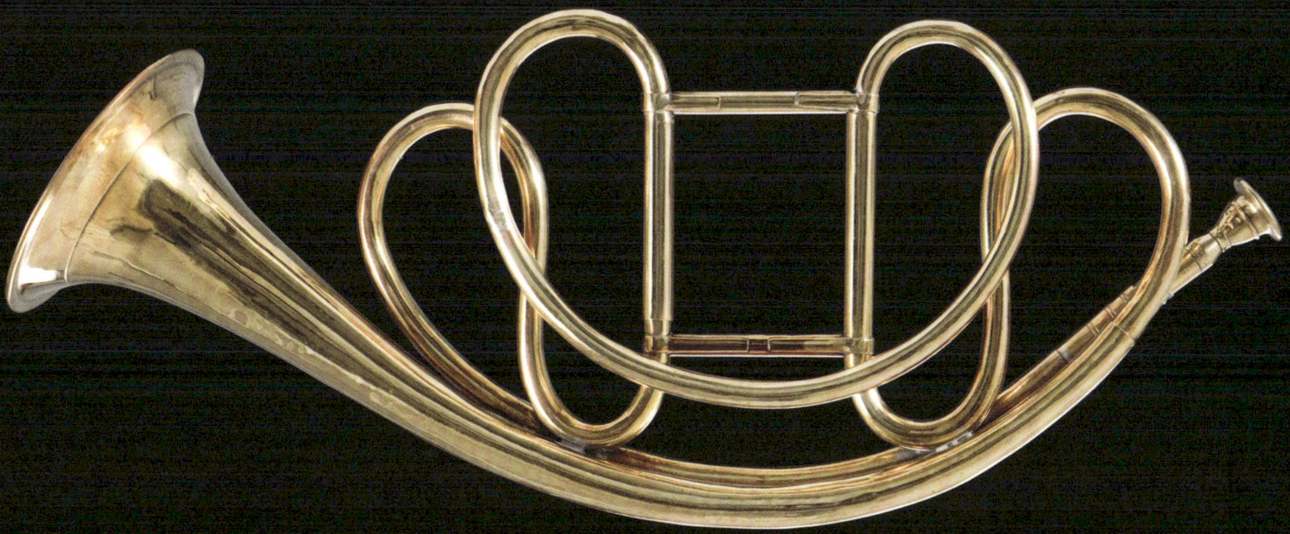

Adolphe Sax

NATURAL TRUMPET
1866–84

This flower-shaped natural trumpet was a stage instrument designed by the Belgian musician and instrument maker Adolphe Sax (Dinant, 1814 – Paris, 1894) specially for the first production of the opera *Sigurd* by Ernest Reyer at the Paris Opéra on June 12, 1885. Composed between 1862 and 1867 from the libretto by Camille Du Locle and Alfred Blau, the opera wasn't presented to the public until January 1884, when it debuted at the Théatre de la Monnaie in Brussels. It is based on the same principal characters and events in Richard Wagner's *Siegfried* and *Twilight of the Gods*, but the plot is closer to the Nordic sagas that inspired it than it is to *The Rhine Gold*.

[See also p. 112]

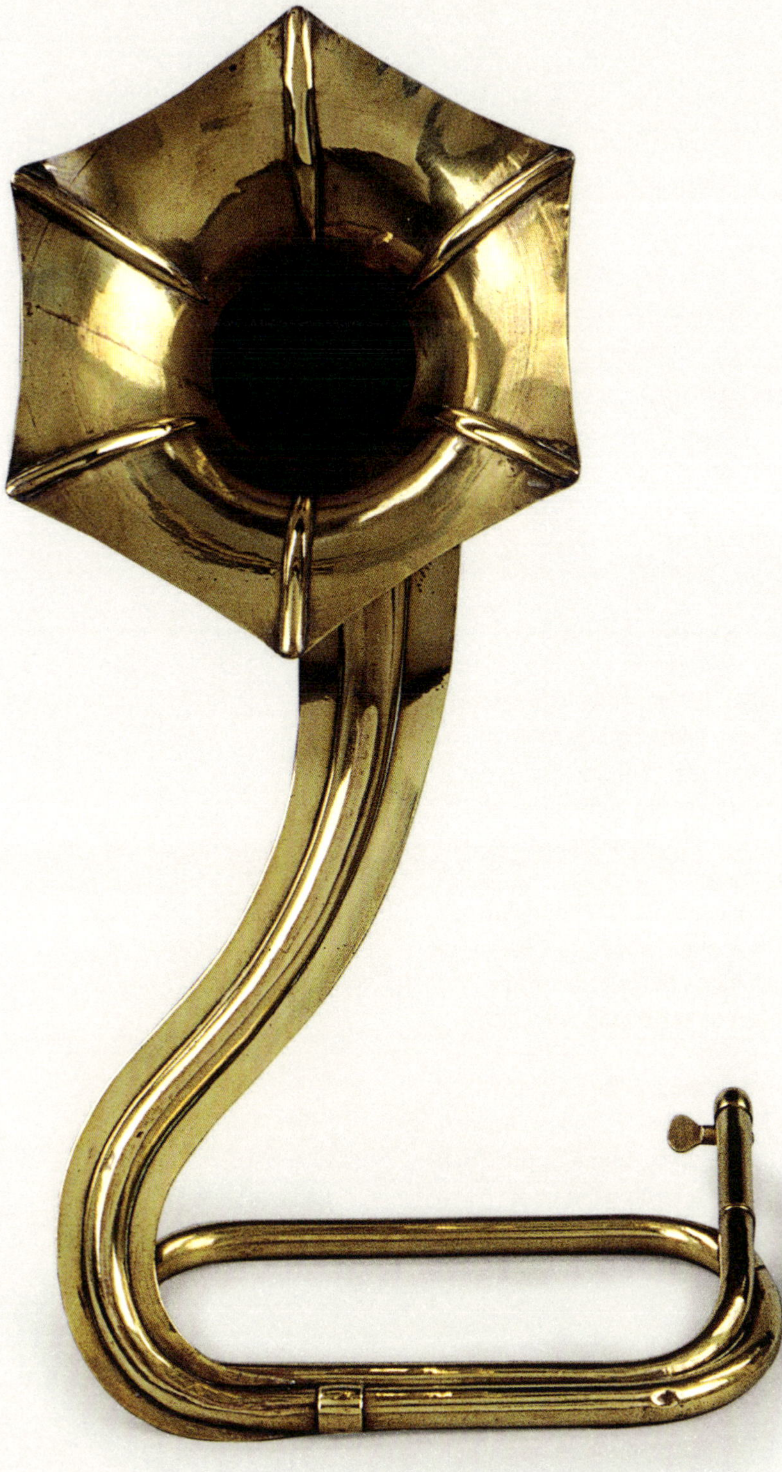

Adolphe Sax

SIX-VALVE TRUMPET
c. 1875

The principle of independent piston valves, invented by Adolphe Sax in 1852, and developed in 1859 and 1867, made it possible to build instruments of great precision, overcoming the defects in conventional instruments in which the simultaneous use of several valves reduced sound intensity.

With this new system, each of the piston valves produces its own series of sounds, allowing 7 harmonic series to be obtained with notes across the entire octave (if including the sounds produced without activating the valves).

The principle applies to trumpets, horns, cornets, tenor horns and trombones. Six-valve instruments were used in particular in the band of the Paris Conservatoire, conducted by Sax himself.

[See also p. 110]

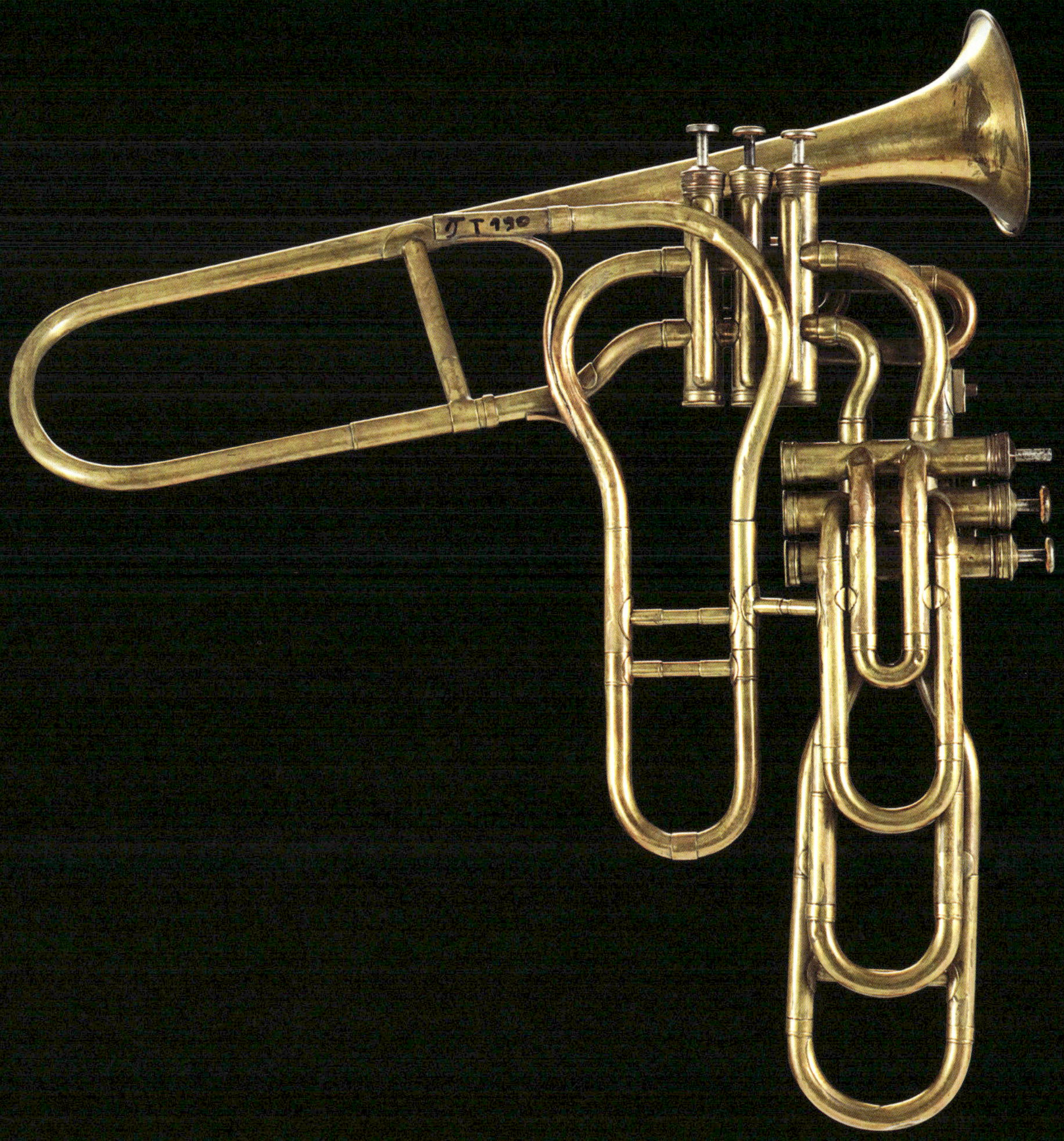

Frédéric Kastner

PYROPHONE
c. 1876

The pyrophone is a musical instrument invented by the Alsatian physicist, chemist and musician Georges Frédéric Kastner (Strasbourg, 1852–Bonn, 1882). It is visually similar to an organ with glass tubes, but within the pipes several jets of flame are arranged close together. When a key is depressed on the finger board, the jets in their corresponding glass tube are made to separate, creating the sound. Apparently, these sounds were similar to the human voice, piano or full orchestra.

The pyrophone is similar to the calliope, another gas-powered instrument, but its combustion is internal and produces light. For this reason the pyrophone is often compared to other coeval experimental devices such as the color organ created by the British painter Alexander Wallace Rimington.

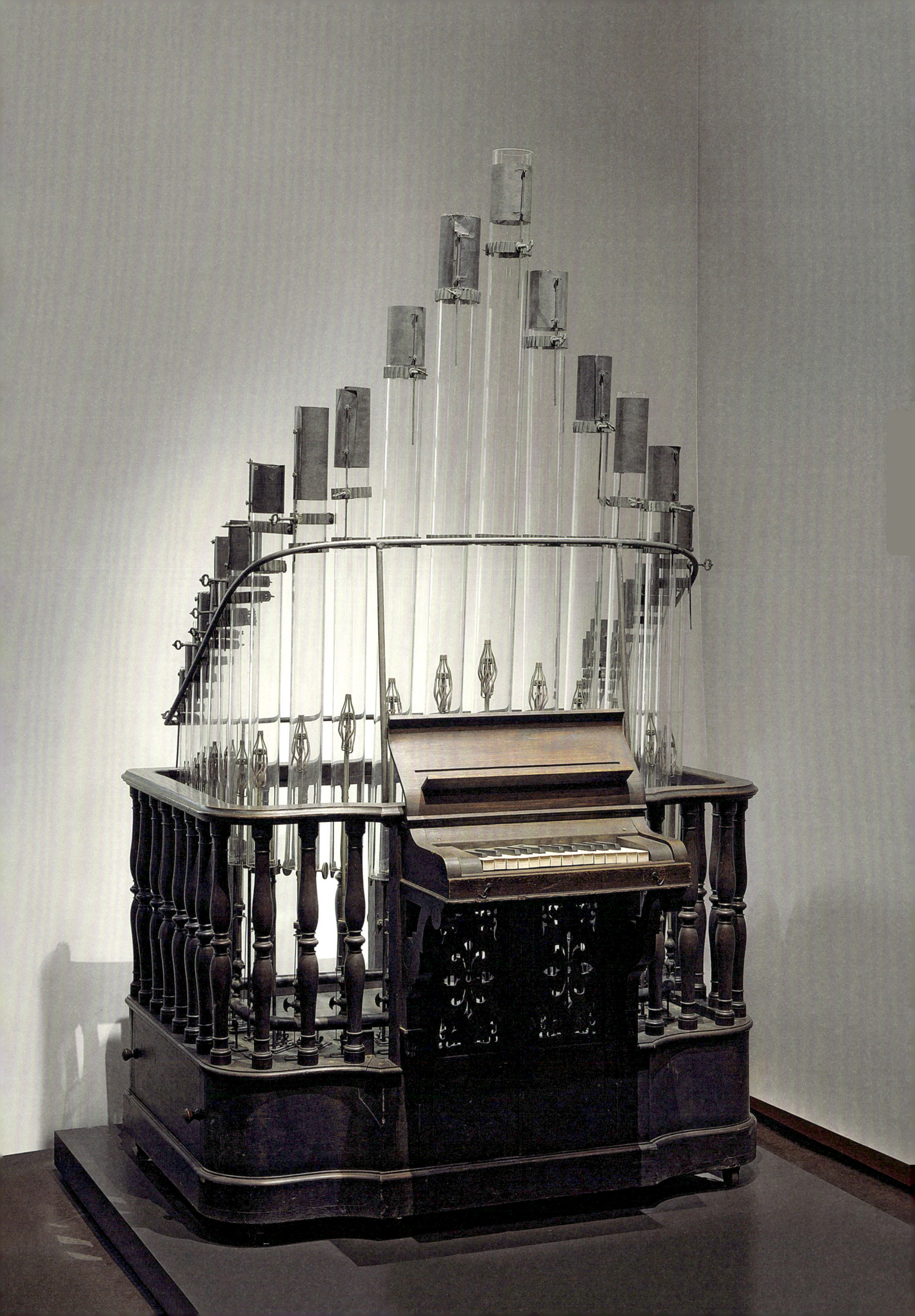

Charles-Abraham Bruguier

SINGING BIRD BOX WITH BUILT-IN CLOCK
c. 1880

Charles-Abraham Bruguier (London, 1818–Geneva, 1891) belonged to a family of clock builders from Geneva, the city where music boxes had been developed at the end of the 18th century, as well as a new method of tone production achieved with the help of tone combs and steel plates. His father, Charles-Abraham Bruguier the Elder, was also a renowned builder of singing-bird mechanical systems.

The case is in richly engraved silver-gilt with enamel medallions, and the cover above the bird is enameled on both sides. The bird revolves and turns its head, opens its beak and moves both wings and its tail. The automaton mechanism controls all movements by means of eccentric cam disks and levers. The bird's chirping is produced by a pair of bellows with a variable whistle. By pushing a small button at the side, the little bird pops out and starts moving and singing. It disappears with the closing of the lid.

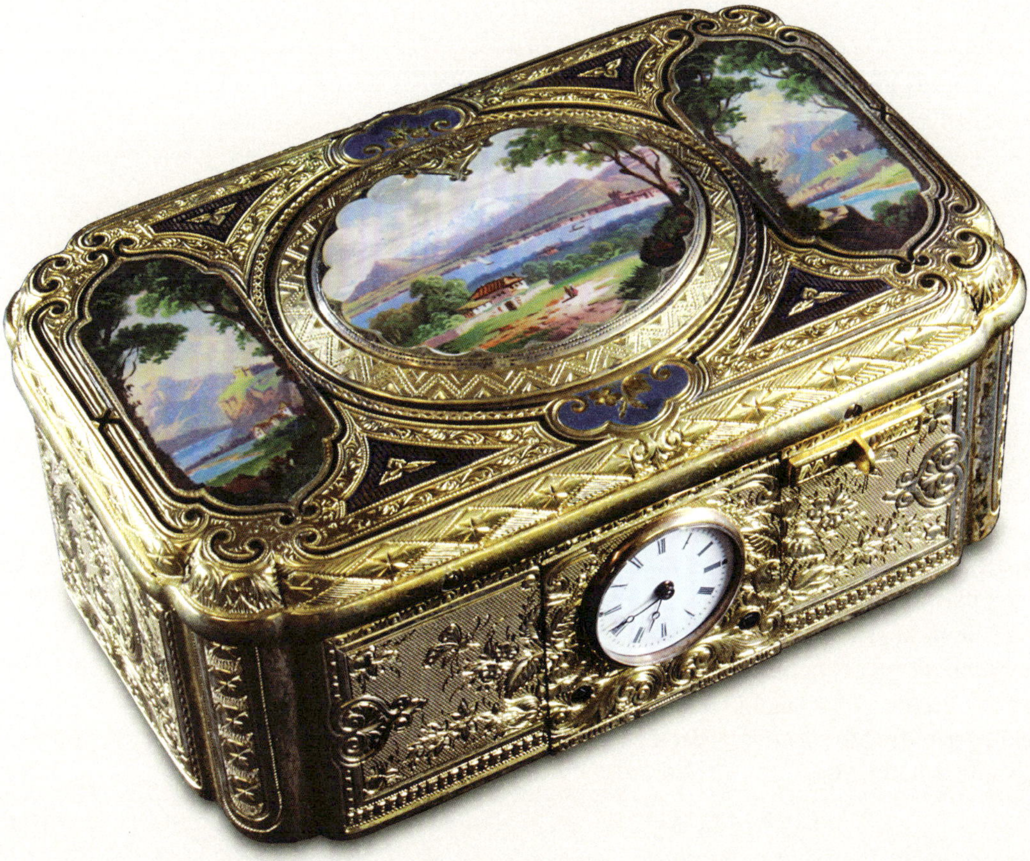

Carlo Bugatti

TAMBURICORDE
c. 1885

The tamburicorde was made by Carlo Bugatti (Milan, 1856 – Molsheim, 1940), an Italian decorator and cabinetmaker known for his eccentric furniture designs. He studied at the Brera Academy, where he joined Milanese artistic and musical circles that then centered around the music publisher Ricordi.

The instrument is eclectic, with 6 strings, a flat round soundboard, 2 stretched skin surfaces (the upper one decorated with painted flowers) and a long grip. When played, the strings vibrate over the stretched skin "soundboard" (rather than into a wooden case), attenuating the sound and giving it a more percussive quality. This tamburicorde, whose function was merely decorative, paired with a stool of the same oriental style.

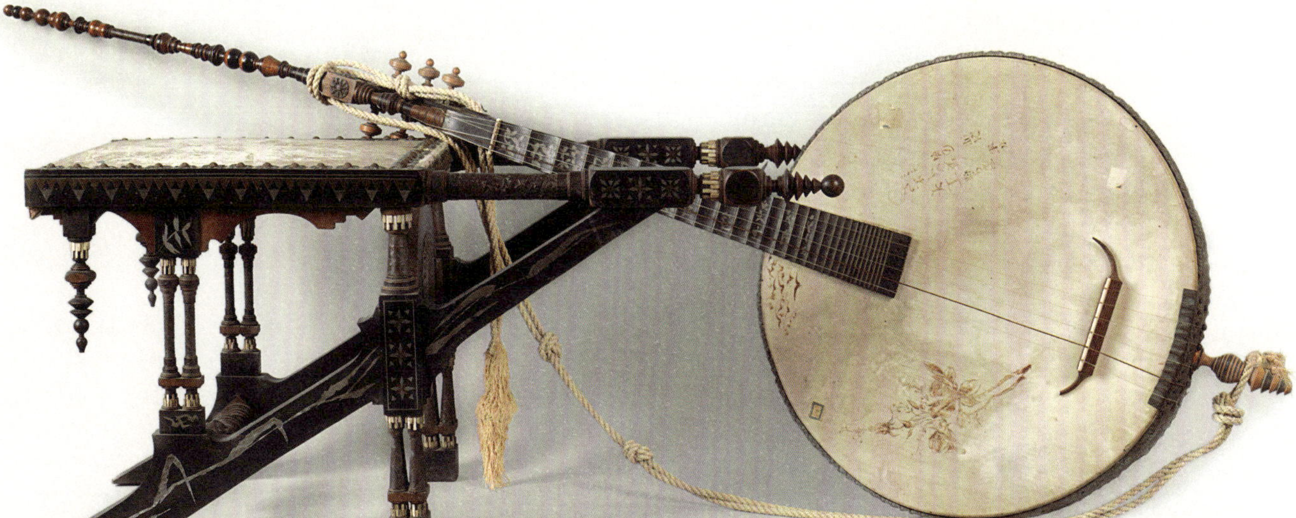

Roullet-Decamps

SNAKE CHARMER AUTOMATON 1890

Based in Paris, Roullet-Decamps was a toy producer famous for its automata. Founded in 1865, the company remained active until 1995.

One of the most famous automata produced by RD (this is the monogram that appeared on all the company's products) is Zulma, the snake charmer. The object was developed when music-hall shows featuring acrobats, wizards and dancers wearing ethnic costumes and accompanied by exotic animals had become fashionable in Paris. One of the most famous acts was a snake charmer known as Nala Damajanti who performed at the Folies Bergère.

Ten or so examples of this musical automaton were produced. The figure breathes, moves her head, closes her eyes, plays her trumpet, and moves a writhing snake close to, then away from her face by means of a spring mechanism hidden in the base and a series of connections inside the body. A development allowed Zulma to perform either nude or dressed in Oriental fashion, and a cylinder carillon could play 4 different melodies.

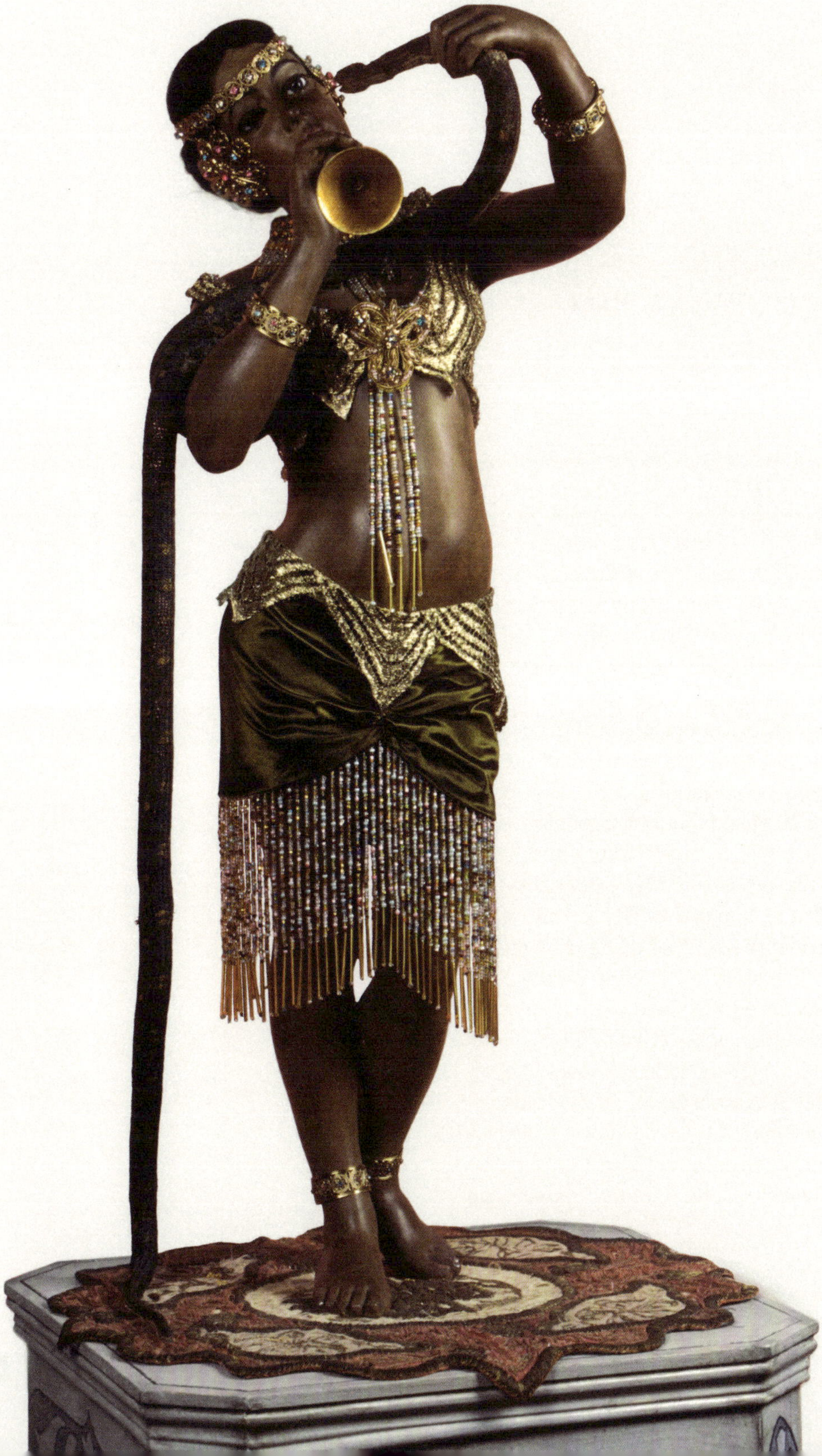

Paillard Vaucher & Co.

CYLINDER MUSICAL BOX
c. 1890

The box was built in Sainte-Croix, Switzerland, by Paillard Vaucher & Co., founded by Moïse Paillard (1753–1830) and active between 1814 and 1974. The company began producing musical boxes in 1825. The cylinder music box industry flourished in particular post-1850, when interchangeable cylinders were introduced. In the mid-19th century much of the top end of the business consisted of boxes that played opera overtures. The most popular composers were Bellini, Bizet, Donizetti, Rossini, Strauss and Verdi. In 1897 Ernst Paillard (1851–1922) began manufacturing the first gramophones outside the United States, which marked the start of the decline of musical boxes.

This box is veneered with bird's-eye maple and has an ebonized molding on the edges. The cylinder musical box has 6 interchangeable cylinders, 5 of which can be seen in the drawers. Every cylinder contains 6 melodies, for a total of 36. The musical code necessary to produce music is provided by the pinned cylinder (the pins pluck the teeth of the comb as the cylinder rotates).

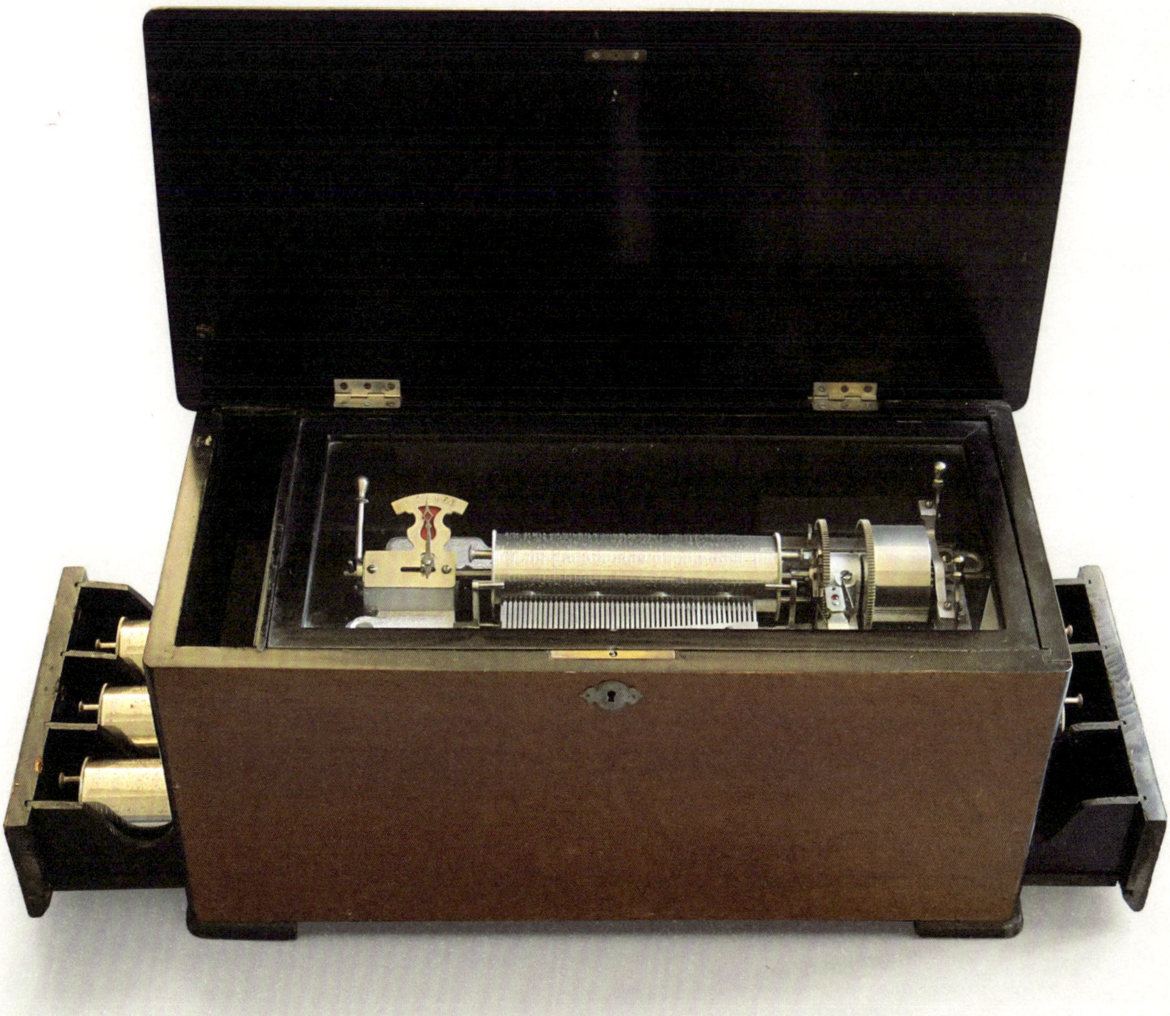

Cocchi, Bacigalupo & Graffigna

BARREL ORGAN
1893

Bacigalupo was a family of organ builders in Berlin who built barrel organs from 1879 to 1975 together with other affiliates. Giovanni Battista Bacigalupo (Berlin, 1889–1978) is considered the founder of the enterprise later named Cocchi, Bacigalupo & Graffigna.

A barrel organ (or roller organ) is a mechanical musical instrument consisting of bellows and one or more ranks of pipes housed in a case. The basic principle is the same as a traditional pipe organ, but the barrel organ is activated either by a person turning a crank, or by clockwork driven by weights or springs. The pieces of music are encoded on wooden cylinders, which are analogous to the keyboard of the traditional pipe organ.

This ebonized 50-key barrel organ with decorations of imitation gold has 8 registers and 10 tunes, and stands on a wooden pushcart. Barrel organs like this one were forerunners of the Dutch street organ. It was one of the last of its kind to be built before the organ book was invented in 1892, rendering instruments like this obsolete.

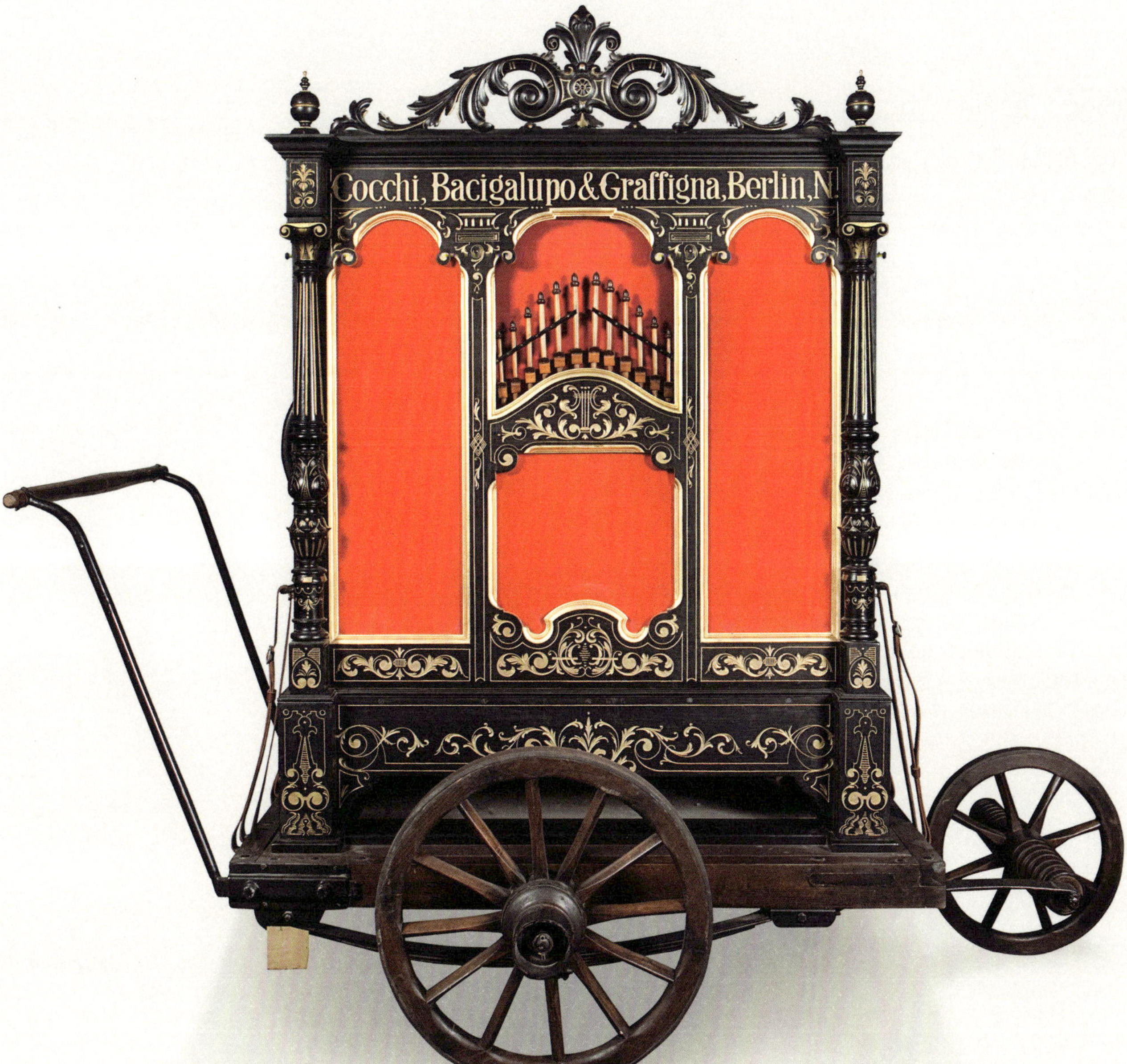
Cocchi, Bacigalupo & Graffigna, Berlin, N.

Victor-Charles Mahillon

EXPERIMENTAL TRUMPET before 1900

A manufacturer and collector of musical instruments, Victor-Charles Mahillon (Brussels, 1841 – Saint-Jean-Cap-Ferrat, 1924) was known for his studies on acoustics, which were published together in his *Éléments d'acoustique musicale et instrumentale* (1874). Wishing to demonstrate that curves—provided they are regular—in a column of air do not alter the timbre of a trumpet, he devised this experimental serpentine instrument, tuned to E flat, which produces a sound characteristic of a normal trumpet.

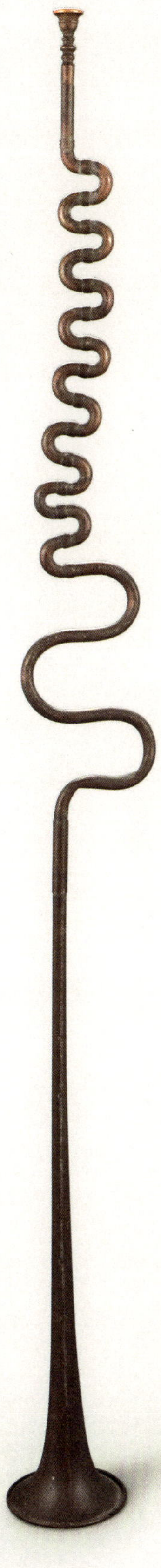

François Millereau

TROMPE DE LORRAINE
c. 1900

The *Trompe de Lorraine* is actually a very compact hunting horn in which the tube is rolled around the bell. The instrument was designed by Théodore Grégoire in Nancy in 1867, but was made by the Maison Raoux and its successors, including Herman Schoenaers and François Millereau (active 1861–1931). If unrolled, the instrument would be 4.60 meters long, which would make it impractical for the use for which it was created: hunting on foot.

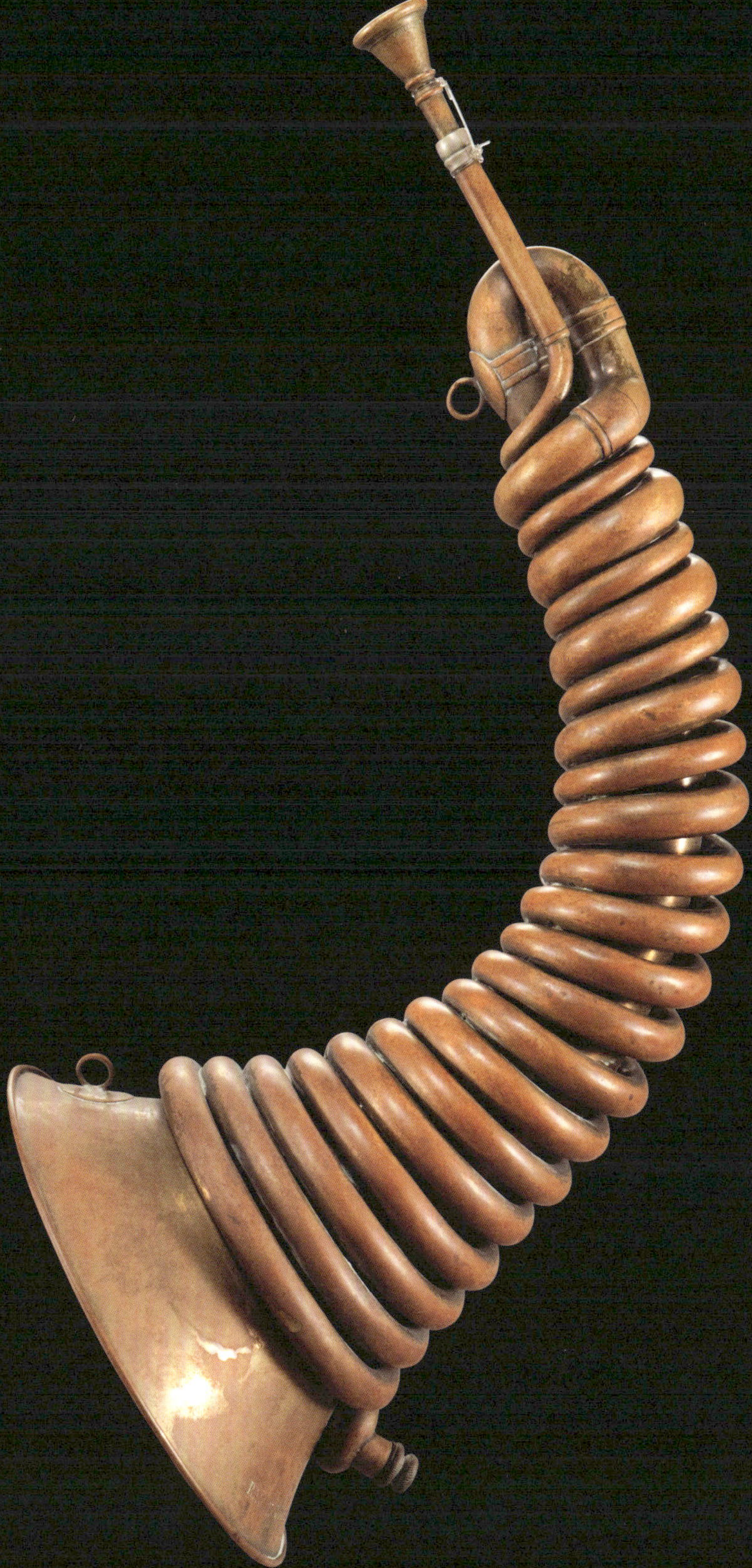

Kalliope Musikwerke

MUSICAL DISC BOX
c. 1900

1

During the 1890s and early 1900s, Kalliope Musikwerke produced a large number of disc-type music boxes. The firm started as a manufacturer of mechanical musical instruments (Kalliope Fabrik Mechanischer Musikwerke) in 1895, and was located in Leipzig. In 1898 it became a limited company and the name was changed to Kalliope Musikwerke.

This upright disc musical box has a double comb and a front-wind motor with cast-iron cover and walnut-veneered oak case. The music—overtures such as Verdi's *Aida*—is provided by a flat disc: the projections on the disc rotate star-shaped wheels that pluck the teeth of the comb.

The first disc musical boxes were produced in Germany with the purpose of overcoming the limitations of cylinder boxes, especially their high price. Disc musical boxes were made in a number of basic types: the auto change (the first jukeboxes), which were used in public spaces; the multiple disc, which played 2 or 3 discs simultaneously; and clocks, which included small or large hall clocks.

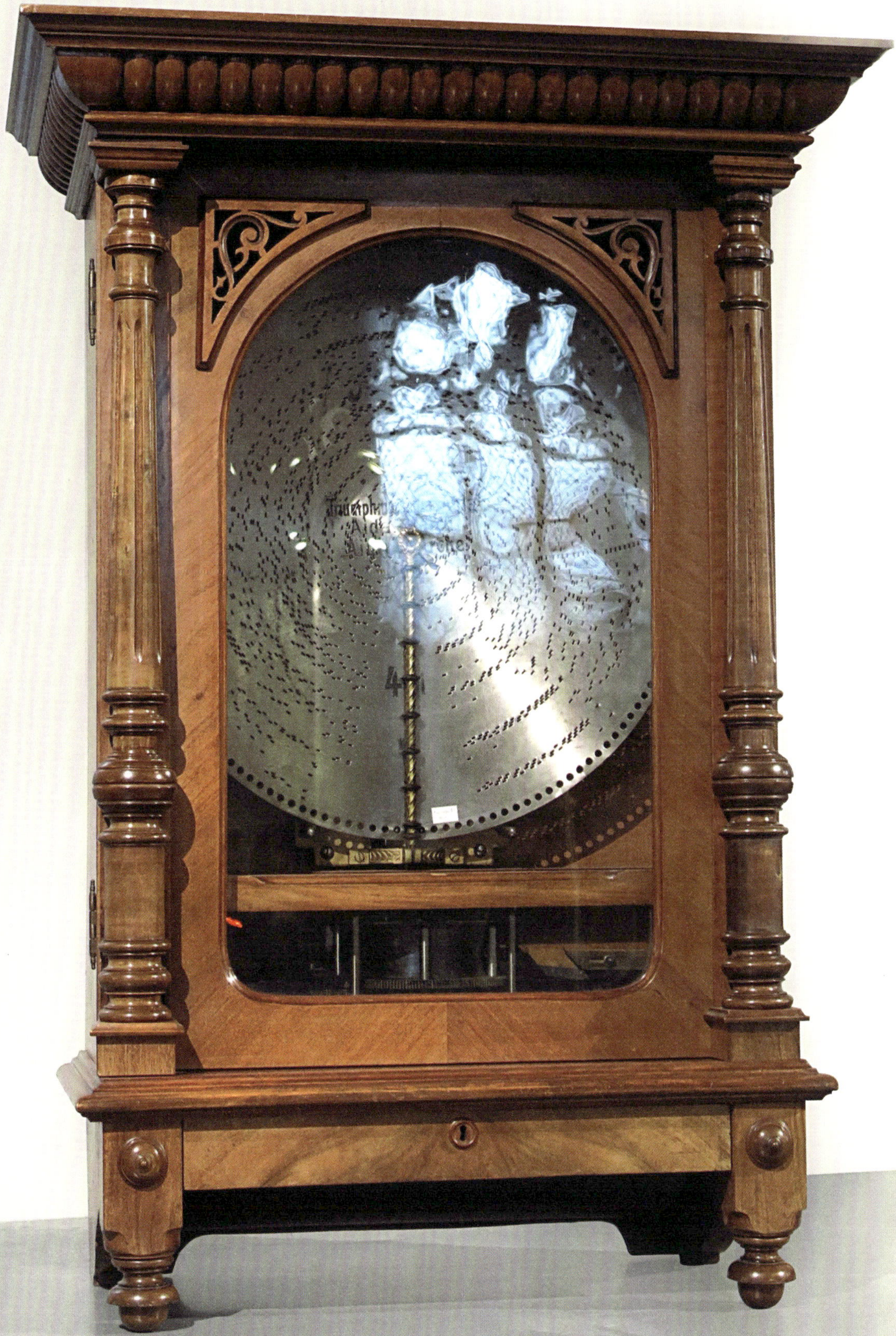

Gebrüder Wellershaus

FAIRGROUND ORGAN
early 20th century

Like many constructors of automata and mechanical devices, the German firm Gebrüder Wellershaus got its start making watches, in time adding these increasingly complex cylinder and roll organs.

This electrically powered fairground organ was made in the early 20th century. Its sounds are produced using compressed air and the music is read from perforated cardboard books using the system Anselmo Gavioli perfected in 1892.

The melody is not only played by organ pipes, whose powerful volume was designed for use in noisy environments like fairgrounds and country festivals, but also by a bass drum, snare drum and cymbal. The front is animated by a conductor and 2 women striking bells.

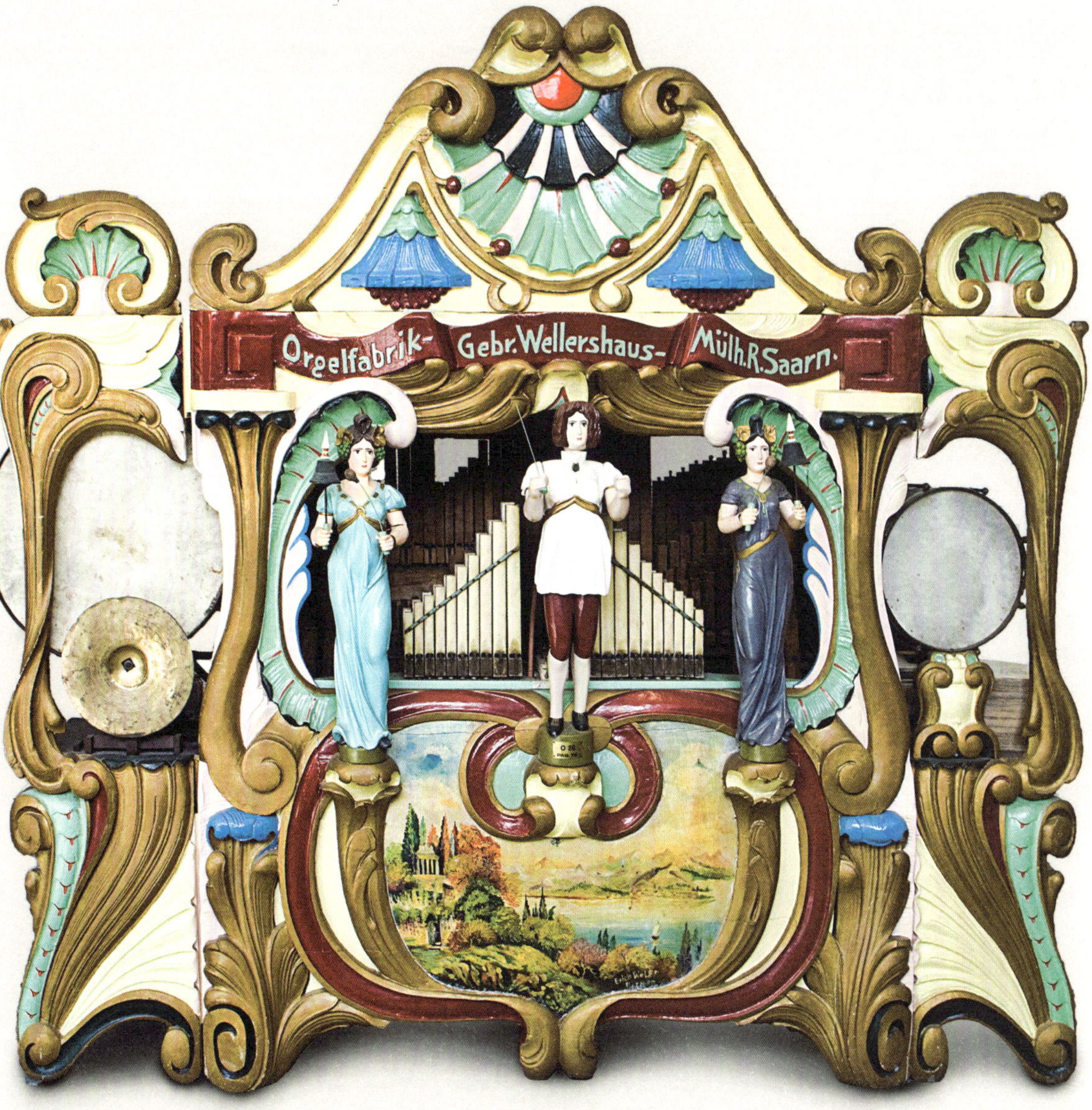
Orgelfabrik- Gebr. Wellershaus- Mülh. R. Saarn.

Arthur Howson

PHONOFIDDLE
1904–20

The phonofiddle is a musical instrument that emerged at the beginning of 20th century from the world of popular music-hall and variety entertainment in the United Kingdom. Invented in London by musical instrument maker Arthur T. Howson (1866–1928) in 1904, it applies to the earlier Japanese fiddle—in fact a monochord of western European origin—J.M. Augustus Stroh's idea of using a phonograph type reproducer as a voice-box to amplify string instruments.

Howson also replaced Stroh's metal cone diaphragm, producing sound with mica Isinglass sheets, reducing the cost of production and enabling the phonofiddle to be successful in a popular context. Thanks to its louder and more directional sound, known minstrels, such as George H. Chirgwin, adopted the instrument in their show, playing it seated as for a cello.

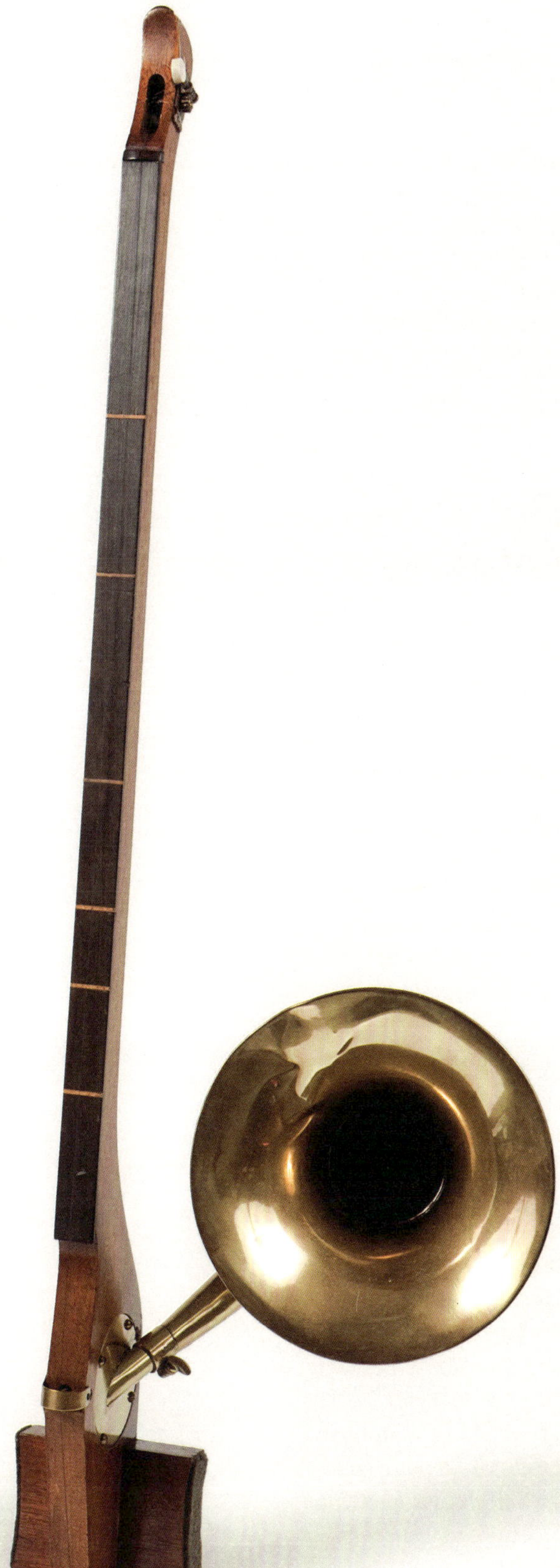

Alexander Scriabin and Alexander Moser

COLOR PIANO FOR THE PART OF LUCE IN SCRIABIN'S PROMETHEUS 1911 (2014)

The symphony *Prometheus: The Poem of Fire* (1910) by Russian pianist and composer Alexander Scriabin (Moscow, 1872–1915) was the first composition in history to include notation for lights, interlacing color with sound to create multi-sensorial harmony.

Scriabin envisaged the use of a color keyboard to transpose the "Luce" part of the score into colored light projections for his symphony, associating the 12 tones of the chromatic scale with 12 colors that changed in conjunction with the basic tones of the harmonies.

A specially-designed multicolor light projector with colored light bulbs was made—from this model by physicist Alexander Moser—for the first performance of *Prometheus*, which was given in Moscow on March 15, 1911, but the machine that was supposed to perform the lighting part did not operate.

The first public presentation of Scriabin's symphony accompanied by colored lighting occurred a year later at Carnegie Hall in New York. The color portion of the score was played with a color-projection instrument, later named the Chromola, made by Preston S. Millar, a specialist in electrical lighting.

DOUGLAS KAHN

REVELATION OF HIDDEN NOISES

IMPROBABLE MUSICAL INSTRUMENTS AND SOUND DEVICES IN EARLY 20TH-CENTURY LITERATURE

The French writer Raymond Roussel was the most formidable musical instrument builder that the 20th century produced. That his instruments and musical moments were built in words alone should not distract from the character of their sounds, ingenuity of construction and organological significance. An odd orchestrion fueled by the expansion and contraction of a newly discovered thermally sensitive metal, bexium; the pitched wheels of twelve chariots performing popular melodies; a large worm that drops beads of heavy water on the strings of a cithara performing a melody from memory much like a biological player piano; two brothers juggling twelve balls between them by batting them with drums; a one-legged man who plays a flute carved from his own tibia; a man who can sing four parts simultaneously from different areas of his gigantic mouth; a limbless one-man band; a child who could imitate "the sounds of a train getting up speed, the cries of domestic animals, a saw grating on a free-stone, the sharp pop of a champagne cork, the gurgling of liquid as it is poured out of a bottle, the fanfare of hunting horns, a violin solo and the plaintive notes of a 'cello'," with incredible veracity; a rodent hair that produces two notes when bowed.

These descriptions appeared in his novel *Impressions of Africa* (1910)[1]. To call it a novel may be misleading, since its narrative structure is more like a processional in a Federico Fellini film, the ecclesiastical fashion show in *Roma* (1972), or panning past open doors onto isolated scenes in *Satyricon* (1969), except what is on display is much more varied, obsessively detailed and bizarre, as though a watchmaker were having a fevered dream in which a convoluted Jules Verne fused with William Burroughs. There has been a shipwreck on the Atlantic coast near the equator, an imaginary event buttressed by French colonialism in Africa. The passengers just so happen to be an unusually talented group who, while awaiting release by Emperor Talu, stage a gala that overlaps with the spectacle of Talu's own series of rites, performances, decrees, public tortures and executions.

I have yet to mention the incredible scenes from his other major novel *Locus Solus* (1914)[2]. Taken together they establish Roussel as a prolific master builder of what I have called *conceptual instruments*, although he did imagine several that were merely improbable rather than impossible, as the one I shall describe below.[3] This instrument, for instance, could be built as long as one accepted its inherent cruelty. 1
Roussel undoubtedly belongs at the center of any discussion of musical instruments and the arts in his own right, in part because of the influence he had over such key figures as Marcel Duchamp, Guillaume Apollinaire, André Breton and many others in literature and the arts.[4]

More concretely, Duchamp and Apollinaire attended a theatrical performance of *Impressions of Africa* together with Francis Picabia, and Duchamp would later state that Roussel was "responsible" for nothing less than Duchamp's major work, *The Large Glass*. *Impressions of Africa* is written all over Guillaume Apollinaire's short story *The Moon King* (1916), including the scene where the narrator navigating mountain caverns happens upon the eponymous player of a special keyboard. Each key is attached to a microphone somewhere around the world. "The flawless microphones of the king's device were set so as to bring in to this underground the most distant sounds of terrestrial life. Each key activated a microphone set for such-and-such a distance. Now we were hearing a Japanese countryside. The wind soughed in the trees—a village was probably there, because I heard servants' laughter, a carpenter's plane, and the spray of an icy waterfall."[5] Another key socialist form of labor and hot springs in New Zealand are heard, and so on at different sites around the world. Apollinaire described an improbable if not impossible musical instrument for the time that would not be difficult to construct and perform one century later.

I will focus here on one instrument encountered in the early pages of *Impressions of Africa*. It is sometimes called the "musical braids," pointing to how the pomp and circumstance of formal decoration on a military uniform might be accompanied by its own music. However, naming the instrument in this way misses how the sound is generated: from the lung still inside a living person. Moreover, we hear reports of the sound of "musical breathing" long before seeing the braids or understanding any other aspect of the instrument. Instead, everything but the sound was hidden within a small, dark, cabin-like structure. The cabin had four black walls and no windows or doors except for a small, covered egress. Roussel most often introduces a strange scene and leaves it unexplained, disregarding its internal behaviors and mechanisms, if not meanings, until it appears a second time. And that is what I will do with my own description of this instrument. Its details will remain hidden for now and revealed later.

Roussel's instruments were the products of an unlikely craft that itself was fabricated from ostentatious wealth and an obsessive personality. What other fantastically rich obsessive has been so uniquely skilled? He was driven by an expectation of universally recognized genius facilitated by an aristocratic level of entitlement. But instead of reclining into leisure, he foisted an arbitrarily arduous writing discipline upon himself. His odd genius was eventually recognized among artists and the intelligentsia, but during his life he expected to be acclaimed as a popular genius like those he so admired: Victor Hugo and Jules Verne.

Roussel was pressed into musical discipline early on by his mother, but didn't earn admission to the Conservatoire until he was 16. He eventually fared well, living up to the high standards of the institution, as one might with a teacher nicknamed "the hundred-handed pianist."

Mme Roussel spared no expense for his lessons, nor any other activity at the mansion on Rue de Chaillot where an elevator lined with exotic wood led to the floor where Raymond lived; each child, as his biographer François Caradec tells us, "had his or her own floor and separate suite of rooms, and their own servants, horses and carriage, lodged in the building of the inner courtyard."[6] The family had a box at the opera, their mansion was filled with music, musicians frequented his mother's salon, and he was regularly taken along to concerts, popular music events, and musical theater, although no amount of privileged experience could explain the fertility of his musical imagination.

His life in literature began when the verse he set to song overtook the music, and we can characterize his achievement in part as an application of the complex rules governing certain types of musical composition applied to writing. In keeping with his obsessive character and self-described "rule-o-mania" he developed a technique that at once constrained his writing and opened up the floodgates of the imagination. He would take a sentence and change but one letter that would change the meaning of one word. That change would then rattle through the sentence with puns and other forms of transformation, creating a wildly divergent second sentence. He placed one of these sentences at the start of the story, placing the second at the end of a story he was required to create. This very slight variation produces a substantial non sequitur that demands the resolve of an intricate, fugue-like exposition.

Think of this involution of language as a burl on a tree. A small implanted imperfection creates an alternative reality to which it is attached. Most mutations in nature die, but burls live as a chaotic expression of their hosts, the grain veering from linear to a turbulence that we happen to perceive as beautiful. For Roussel, wordplay begat worldplay. The transformative effect of one letter rattled changes through a sentence, generating surface appearances, actions and behaviors that then step aside for a detailing of mechanism. It is a disobedience of code that generates an informational machine. *Impressions of Africa* was the first major fruit of this device.

Roussel's literary device presaged similar mechanistic techniques throughout the century, e.g., the twelve-tone system, Dada and Surrealist techniques and games, John Cage's various ways for deriving chance and indeterminacy, William Burroughs' cut-ups and many others, including the productive constraint encountered in the early digital arts.[7] Such arbitrary devices were designed with many intentions in mind, primarily to distance one from the normal intentions and habits of expression. Michel Leiris put it best:

"The fact of making oneself the voluntary slave of a hard and complicated rule, on which it is necessary to concentrate one's attention and ignore one's surroundings, causes a freeing of the unconscious which is more real than in the pure and simple abandonment of [Surrealist] automatic writing. Perhaps because of this, concentration of the attention on some futile point lifts inhibitions more effectively than passivity or the attempt to empty one's mind."[8]

The drive for technique could also be compared to Ferdinand de Saussure's search for anagrams in Latin verse, in reverse, but not quite. Roussel only revealed his secret in his posthumously published *How I Wrote Certain of My Books* (1935).[9] Only from the grave, in other words from within his crypt, did he provide the key to decrypt his writings, to reveal what was hidden in the innermost mechanism. He died without achieving the popularity he felt he deserved, but by revealing his secret relegated his writings to examination in the obsessive spirit in which he wrote them, to all the Saussures in the future.

Although he thought he wrote for a wide audience, Roussel has joined other esteemed writers and artists whose works were made to be studied: James Joyce's *Finnegans Wake* or Marcel Duchamp's *The Large Glass* come immediately to mind. Intended neither as a popular outing nor hermetic quest, such works display their requirements at the outset. But unlike puzzles that may have a solution, studying only unlocks greater complexity. No destination will be reached; no inner sanctum will be breached, as well it should be. Solving mysteries is a violent act because it destroys their mystery, their generative power. It is better that they elude destruction by changing shapes right under our noses, and by creating new vistas of knowledge right under our gnosis.

Both Joyce and Duchamp constructed a great deal of their hermetic works by moving pun functioning toward codes and cryptograms. Puns have a bad reputation for being painfully obvious, oblivious or just plain painful. Yet even at their worst they are saying at least two things at once; they evidence an economy of action and space. They never speak with just one tongue, but also with a tongue in-cheek, as though they were speaking out of both sides of the mouth. They are not unlike the singer Ludovic with the enormous mouth in *Impressions of Africa* who trained himself "to separate his lips and his tongue into sections which functioned independently of one another, and to pronounce, at one and the same time, without difficulty, several conflicting parts, with different tunes and different words...."[10] The more one mouth or word can say or sing at once, the better.

"Wheatstone's magic lyer," was a musical pun in James Joyce's *Finnegans Wake*. Joyce was perhaps the greatest word wizard to have ever lived; just ask any of the scholars who have devoted their lives to his *Wake*, where it seems every word has been buried in a fruitful conflict with itself, much like a *harpsdischord*.[11] Joyce is calling Sir Charles Wheatstone, the Victorian scientist and inventor, a liar for the deceit involved in his unusual musical instrument, the Enchanted Lyre. The lyre was suspended from the ceiling and appeared to produce music by itself, and not simply lyre music, when in fact the sound was transmitted from instruments in the room above, their vibrations conducted through the means of suspension. It was a matter of deceit, much like Wolfgang von Kempelen's *Mechanical Turk*, which pretended to be a chess-playing automaton when in fact a real chess player hid in its cabinet. It made sense that Wheatstone would later develop cryptography and ciphering machines. Wheatstone had another word for his Enchanted Lyre, calling it the Acoucryptophone, literally *hearing a hidden sound* (*acou* as in hearing, *crypto* as in hidden, *phone* as in voice or sound).

Marcel Duchamp's well known work *With Hidden Noise* (1916) [164] is a ball of twine sandwiched between two metal plates secured by four long narrow bolts that also act as posts that support the whole piece. Inside is an unidentified object that makes a noise when rattled about. Duchamp's friend Walter Arensberg placed an object in the piece and was not to tell anyone what the object was, including Duchamp himself. There are cryptic inscriptions on the top and bottom plate, three lines with three words in each line, with periods punctuating the absence of letters within the words. Substituting letters will generate English and French words easily enough, but the possible meanings of those words pose their own problems. Duchamp remarked that the inscriptions were like letters missing on a faulty neon sign that render the sign meaningless. There could be no final determination as to the identification of the source of the sound or the meaning of the inscriptions; they were all destined to remain hidden. The art historian Seymour Howard has punned upon the punster, calling this tendency in Duchamp a *hidden naos*, an inner sanctum where an occluded authority rests.[12] Interpretation, in this regard, becomes labyrinthine with pleasure in the pattern, like life in the living, while the noise/naos hides like the crypt in cryptography, the dead enumerating a rule-o-mania from the afterlife.

Among Duchamp's genres of readymades, *With Hidden Noise* is often associated with *Fresh Widow* (1920), a small-scale window with eight panes of glass covered with black leather, with the inscription "FRESH WIDOW COPYRIGHT ROSE SELAVY 1920." The title itself is a pun on *French window*, accompanied by the *Eros, c'est la vie* of Duchamp's familiar pseudonym. There are many twists and turns in possible interpretation, but both pertain to hiding and masking, to decipherment of inscriptions with little or no illumination. Combined, *With Hidden Noise* and *Fresh Widow* suggest an object and musical instrument introduced and described in the early pages of Raymond Roussel's *Impressions of Africa*. Like *With Hidden Noise*, there existed a hidden noise, but one that would be revealed.

We return to the fictional nation of a real French Equatorial Africa in Roussel's *Impressions of Africa*. There is an esplanade that frames a square; nearby stands a row of statues, and near them a small cabin. It is an unusual cabin: it has no door, only four walls made of thick black canvas, much like the blackened panes of glass of *Fresh Widow*. Its roof looks thatched, but is made instead of the yellowing pages of a Sir Walter Scott novel. It would be totally dark inside if not for a judas window—like the term implies, a place of surveillance, hiding, betrayal and clandestine escape. Without a door or regular windows in the cabin, this would be the only means of egress. But it too is covered with the pages of the novel, masking the ability to eavesdrop from the inside and, instead of being entirely dark, the pages over the window must (the narrator speculates) cast a diffuse yellow light inside the cabin. What is not masked is a sound emanating from the cabin, "a chord, recalling the tone of brass, but very much fainter, sounded at regular intervals from the cabin, giving the exact sensation of a musical breathing."[13]

The absence of windows and masking of the judas window is reminiscent of Roussel's ecstasy at the time his writing first took hold of him. He "was surrounded in rays of light, I would close the curtains for fear the shining rays that were emanating from my pen would escape through the smallest chink; I wanted to throw back the screen and suddenly light up the world."[14] The black canvas of the cabin were like curtains drawn with no light escaping, or better yet coming in. No pinhole of light shines into this room; only a sound emanates from this closed, shuttered *camera obscura*. The existence of a judas window on a blackened space might recall the small window on his mother's coffin lid. This was not such an eccentric idea, for she was the one who, on a long trip to the Indian Ocean, brought along a coffin should anything untoward happen and her embalmed body require a suitable berth back home.

He accompanied her on that trip, and he was the one who had the window on her coffin installed so that he would be able to see her face up to the last moment. However, *Impressions of Africa* was written before both her trip and her demise: crypt followed cryptography.

The source of this sound remains mysterious for many more pages, between the hidden sound and its revelation, much like the original sentence and its variation that completes the story. In the interim, the cruel Emperor Talu approaches the square with his large entourage while the inhabitants of the town gather, anticipating a show. They are treated to a series of bizarre events, tortures and executions. Another musical moment occurs when ten of Talu's wives—he has more—perform a dance after eating an unknown substance that will become known only through its effects. It animated their song and dance, from "a few slow steps, mingled with lithe, sinuous movements" increasing eventually to a frenzied and contorted delirium, while their song came in the form of a barking chorus of loud belches, a sound made more intense through their competition for the greatest volume and flamboyance. The performance culminated in a deafening climax, after which it ended, "accompanied by a sustained final chord which gradually fell into silence."[15] Again, we have the musical breathing.

But nothing stays hidden for long in a Roussel novel. The judas window opens and up steps a young woman, Louise Montalescot, in an officer's dress uniform. On the right shoulder, denoting status in the French military, is a gold braided and knotted cord. Hollow aglets with vibrating plates inside each hang from the knot. These small sounding bodies emitted the musical breathing that could be heard emanating from inside the cabin.

The sound was driven "by means of a surgical connection, linking the base of the lung and the arrangement of coiled braids," such that "at each contraction of the lung, a part of the air expelled passed through the multiple pipes and, setting the plates in motion, produced a harmonious resonance." When Montalescot saw the body of one of the people who had been executed, "she wept hysterically, her bosom shaken with terrible sobs which set in motion the cords of her shoulder-tags and increased their sound."[16]

Technically, the instrument sits somewhere between bagpipes and the voice. The lung is a bag or bladder like those used in the bagpipes, and similarly distributes the generation of tones across several sounding devices. However, as a reservoir it would empty too quickly, if not have a direct expressive effect, as with the voice. The instrument has a certain similarity with the chorus of belching wives, since their song bypasses the lung power in order to tap the storage of air using a bladder more amenable to bagpipes: the stomach. It may be that the tube that connects Montalescot's lung with the public display of an officiated music is umbilical. This cord connecting with the chord of musical breathing is available in the autobiographical Roussel's dependency on and devotion to his mother and her musical influences, just as her later drug addiction echoed the mysterious ingested substances that animated Talu's wives in a bizarre song and dance.

The pleasures of these interpretations lie in the patterns, which may or may not exist in language alone. Luckily this is not an important distinction in literature, nor should it be in the design and construction of musical instruments. In the above, there is no discernible relation between James Joyce, Charles Wheatstone's Enchanted Lyre and Raymond Roussel, apart from a proclivity of puns and concealed musical sounds. Marcel Duchamp, on the other hand, is indelibly stamped with the specter of Roussel, but whether he is a ghost that haunts the specific works or a noise that remains hidden, remains hidden.

In *Impressions of Africa* Roussel has generously let us know what was making the hidden sound of musical breathing. This was part of the narrative structure of a processional of many little stories of disclosure that existed in place of a grander story that one might recognize as a novel. We may not end up knowing *why*, but through the self-revelatory process of the novel, a dual action of presentation and eventual explication, we at least gain a better notion of *how*, the pure surface of *how* versus the noises of *why*. What remains hidden is the shift in the one letter that created this instrument. And if that were revealed, what would be known?

1
Originally published as *Impressions d'Afrique* (Paris: Lemerre, 1910). English edition *Impressions of Africa*, trans. by Lindy Foord, Rayner Heppenstall (London: John Calder, 1983).

2
Originally published as *Locus Solus* (Paris: Lemerre, 1914). English edition *Locus Solus*, trans. by Rupert Copeland Cunningham (London: Calder and Boyar, 1970).

3
Douglas Kahn, "Conceptual Instruments," *Experimental Musical Instruments*, vol. 6, no. 3, October 1990, pp. 17–20.

4
Along with Alfred Jarry and his Pataphysics, Roussel was also the patron saint of Oulipo (*Ouvroir de littérature potentielle*), a group of writers who relished rules of constraint through which to unleash their writing, and of the novelist Alain Robbe-Grillet, who argued for a what-you-see-is-what-you-get surface, rather than ulterior motives seething beneath the surface.

5
Guillaume Apollinaire, "The Moon King," in *The Poet Assassinated and other Stories*, trans. by Rod Padgett (San Francisco: North Point Press, 1984), pp. 70–83.

6
François Caradec, *Raymond Roussel*, trans. by Ian Monk (London: Atlas Press, 2001), p. 26.

7
Oliver Grau (ed.), "Between a Bach and a Bard Place: Productive Constraint in Early Computer Arts," *MediaArtsHistory* (Cambridge, MA: The MIT Press, 2007).

8
Michel Leiris, *Roussel l'Ingénu* (1987), cited in Caradec, *op. cit.*, p. 64.

9
Originally published as *Comment j'ai écrit certains de mes livres* (Paris: Lemerre, 1935). English edition *How I Wrote Certain of My Books*, trans. by Trevor Winkfield (New York: Sun, 1975).

10
Raymond Roussel, *Impressions of Africa*, cit., p. 64.

11
James Joyce, *Finnegans Wake* (London: Penguin Classics, 1999), p. 13.

12
Seymour Howard, "Hidden Naos: Duchamp Labyrinths," *Artibus et Historiae*, vol. 15, no. 29, 1994, pp. 153–180.

13
Raymond Roussel, *Impressions of Africa*, cit., p. 12.

14
Quoted in Pierre Janet, "The Psychological Characteristics of Ecstasy," in Alastair Brotchie *et al.* (eds.), *Raymond Roussel: Life, Death and Works* (London: Atlas Press, 1987), pp. 40–41.

15
Raymond Roussel, *Impressions of Africa*, cit., p. 21.

16
Ibid., p. 25.

LUCIANO CHESSA

A META-PHYSICAL ORCHESTRA

RESEARCHING AND RECONSTRUCTING THE INTONARUMORI

In 1913 the Italian painter Luigi Russolo launched a Futurist manifesto entitled *L'arte dei rumori* (The Art of Noises) with the intention of radically updating the concert music of his time by forcing it to come to terms with a contemporary, noisy, mostly urban sound world.

Abandoning painting and engaging full-time in the activities of instrument building, composing and conducting, Russolo concretely and effectively offered his own revolutionary version of what music in the 20th century should be like.

Russolo's sound revolution, first ridiculed and loathed, came to be increasingly regarded as seminal, while Russolo himself, a subject of musicological attention since at least the 1970s, currently enjoys a considerable cult following.

However, when in 2001 I started working on a doctoral dissertation on Russolo's aesthetics,[1] I was surprised to discover that the field of music history still lacked a substantial hermeneutic effort both on the art of noises, and on the noise-intoner instruments (the *intonarumori*) [160] Russolo built to showcase it.

The pioneeristic archival works published in the mid 1970s by Giovanni Lista and Gianfranco Maffina,[2] for the most part based on documents that Russolo's widow Maria Zanovello had preserved and catalogued, had not been followed by any systematic interpretation of Russolo's actual work on sound. And though in secondary literature Russolo's name was often dropped, most of the writing on his musical activities had remained anecdotal, while the meaning of such activities had not been investigated in depth.

This state was even more surprising if one considers that plenty of research was being done on Russolo as a painter—especially surprising for those of us who regard his contribution in the field of music to be more relevant than his contribution in the field of painting.

That said, the importance of Russolo's training as a painter in the context of his philosophical interests and, of course, in the context of his experiments on sound should not be underestimated. The key to understanding his work on music lies in fact in understanding what motivated or, rather, animated, his painting.

My research began by studying Russolo's creative process *vis-à-vis* formulations of early Futurist aesthetics, to which he first contributed *as a painter*. By reviewing the many secondary sources on Futurist art, my attention was caught by research that revealed influences of occult theories within the Futurist movement.

Chief among them was a seminal article by Germano Celant entitled *Futurismo esoterico*.[3] This article, together with portions of Maurizio Calvesi's volume *Il futurismo. La Fusione della vita nell'arte* of 1967,[4] served as a basis for all my subsequent findings.

As a result of these readings, I discovered that under an apparently modernist surface the Futurists were interested in occult arts. Following this line, I realized how Russolo's work was influenced by Renaissance art and technology as they blended in harmony in the proto-futurist figure of Leonardo da Vinci.

Subsequently, it became clear to me how the whole of Russolo's work had originated from a genuine interest in the occult, an interest Russolo demonstrated continuously throughout his life, from his earliest canvases (the 1908 *Autoritratto con teschi*, or the 1910–11 *Autoritratto con doppio eterico*) to his late "classico-moderno" paintings of the 1940s.

My firsthand reexamination of Russolo's writings and work revealed the connection he had found between painting and music through synaesthesia. Russolo and the Futurists understood synaesthesia in an occult way: as the field of possible arcane interconnections between different sensory fields.

His notion of synaesthesia was derived from a specific late 19th-century branding of the occult arts: Theosophy. Russolo and his Futurist comrades had likely absorbed Theosophy through their fascination for late 19th-century French thought, and in particular for French Symbolism, an art movement that had inspired Futurism since the start.[5] All early Futurism was soaked in Symbolism. As a result, all early Futurism was soaked in the occult via Theosophy, and the keyword was synaesthesia.

If we consider Futurism's earliest members, beginning with the founder of the movement, Marinetti, pretty much all these artists and intellectuals had one way or another flirted with Symbolism before turning to Futurism. Marinetti's own poetry in French is surely inspired by Symbolist poetry, while early works by Boccioni and Russolo present themes commonly found in Symbolist painting.

How did these occult interests influence Russolo's art of noises? I believe that the art of noises can be read as a process of conjuring the spirits articulated in two phases: in the first phase noise becomes spiritualized; in a second phase, spirits materialize. I argued that Russolo first painted the process in 1911–12 in the large canvas titled *La musica*, and subsequently put this process into practice a year later by formulating the art of noises and by building the *intonarumori*.[6]

A mere glance at *La musica* should be enough to convince us that we are in front of a highly symbolic work. *La musica* is a work clearly soaking in theosophical doctrine, and in particular in that eminently synaesthetical theory of *sound forms* first introduced by Charles Leadbeater in 1904[7] as an expansion of his and Annie Besant's better-known theory of *thought forms*.

At a deeper hermeneutical level, the painting shows a performer/medium in a trance-like state, and thus able to evoke spirits and to produce materializations of *thought* and *sound forms*. I identified the restless masks in the painting, which show a full range of facial expressions, as figurative materializations of *thought forms*, as well as a key concept in early Futurist painting, that of *stati d'animo* (states of mind). The sinuous blue ribbon that occupies a significant portion of the painting I identified as Russolo's own representation of the appropriate *sound form* for a sound source that Russolo would have defined as *enharmonic*, i.e. a sound source that moves continuously across the pitch-space in gliding glissando, uncorrupted by any sense of intervallic subdivision.

The art of noises, which Russolo formulated a year later, and the *intonarumori* orchestra he constructed to materialize it, could also be read within this same occult frame of reference. Russolo's occult persuasion led me to understand the *intonarumori* not merely as cleverly engineered musical instruments but rather as philosophical instruments, and thus as part of an illustrious tradition whose lineage dates back to Pythagoras and his experiments on the monochord.

The *intonarumori*, also essentially monochords, first generate noise as raw matter. Thanks to the lever interface, this raw matter is then spiritualized as it is first tuned, and then made to exist, beyond any tone divisions, in the fluid "enharmonic" space. By producing glissandos, the *intonarumori* proves that pitch is an infinitely divisible quantity, and therefore that pitch-space is continuous. Russolo called this gliding feature of the *intonarumori* "dynamic continuity."[8]

Though the word "*intonarumori*" can be used both in the singular and in the plural, Russolo didn't think of the *intonarumori* as an individual instrument; an individual sound module. He was committed to the creation of an orchestra, which, as a whole, could synthesize the totality of available noises of the natural and mechanical world.

Having introduced a taxonomy of such noises in the Art of Noises Manifesto, in the summer 1913 he proceeded to build the instruments that could produce them all. For Russolo, the *intonarumori* orchestra recreates the world through its sounds: first as a simultaneous chaos and then—thanks to the energy and ability of a conductor/medium—as a substantial unity. When the conductor/medium reached the point of supreme inspiration, spirits of the dead could be conjured, awoken, and even made to materialize.

The principle of materialization/awakening, omnipresent in theosophical doctrine, is found in the work of Marinetti and other Futurists, including Ardengo Soffici, Boccioni and of course Russolo. And far-fetched as my aforementioned interpretation of the art of noises' intention may seem, it is corroborated from a coeval source by Futurist Paolo Buzzi, who often explained Russolo's art of noises precisely as an occult operation of conjuring spirits: first in a brief 1919 text titled *Russolo ferito* (Russolo wounded), then in a poem titled *Russolo*, likely written in 1919 but published in the 1922 collection entitled *Poema dei Quarant'anni*, and finally in his 1924 novel, *La cavalcata delle vertigini*.[9]

Expanding an interpretation of the art of noises as a process of conjuring expressed in the 1919 texts, *La cavalcata delle vertigini* features Russolo, under the name Marzio, as the main character, and includes a full chapter on Russolo/Marzio conducting/conjuring.

Though not directly and literally, a deeper understanding of the art of noises as an ambitious metaphysical operation was the fundamental principle that guided my reconstruction of the *intonarumori* orchestra. This reconstruction was possible thanks to a commission I received from RoseLee Goldberg of Performa, New York City's Biennial of Performance Arts.[10]

For Performa 2009 I ended up producing the first complete modern reconstruction ever attempted of Russolo's earliest (and lost) *intonarumori* orchestra.

The model for my reconstruction was in fact the orchestra of 16 *intonarumori* that was first unveiled on August 11, 1913, at a press concert that took place in Marinetti's Casa Rossa, the Milanese headquarters of the Futurist Movement; this orchestra disappeared during the German retreat in World War II.

When I began work, a few reconstruction projects had already been produced. The main one, by Pietro Verardo and Mario Abate, was originally commissioned by the 1977 Venice Biennale. Verardo was the first to try to reconstruct some of Russolo's *intonarumori* to the extent of what the available patent would allow. Obviously, I felt that I needed to produce something other than what was already available: I was in search of an identity, as well as a unifying concept for my reconstruction project.

The intention to rebuild the entire first orchestra did not come to me right away. As I started thinking about it, it became clear that what I really wanted to do was build a set of instruments that would allow me to perform Russolo's fragment from *Risveglio di una città*—a 1913 fragment that first appeared in the Futurist periodical Lacerba in 1914, and had not been heard live in almost a hundred years—as close as possible to the original score.[11]

When I began the reconstruction, no *intonarumori* ensembles could perform that piece live. And since I knew from the start that I wanted to rebuild the *intonarumori* to use them in live performances, the score for *Risveglio di una città* became my guiding model.

When Performa contacted me in 2008, I had been working on Russolo for about seven years. My work up to that point was mostly theoretical. I had investigated Russolo's aesthetics for my doctoral dissertation, which focused entirely on Russolo's art of noises.[12] The information I gathered to support my argument included Russolo's patents and patent drawings for the *intonarumori*. Since Russolo did not patent each individual instrument, only the basic *intonarumori* mechanism, these drawings were only helpful to a certain extent, though they could still provide plenty of information. More importantly, thanks to my earlier research, I felt I had a grasp on what Russolo was hoping to achieve when he decided to build these instruments.[13]

In December 2008 I contacted musical instrument builder Keith Cary, a friend of a friend living in Winters, California. I asked him if he wanted to help me with this project, initially by building a prototype. This turned out to be an *Ululatore medio*, the most straightforward of Russolo's *intonarumori* to build, since it has the most direct relationship to the patent drawing.

Once the prototype was built, I had to deal with a very difficult question: how to reconstruct the other 15 instruments, many of which I knew to be considerably different from the one basic mechanism Russolo had presented in the patent. I began gathering all the evidence I could find: in addition to the patents, there were Russolo's articles and books, his charts detailing the ranges of the instruments, texts by Futurists and their associates, reviews, etc.

Photographs of the instruments provided valuable help. The most useful may be a pair taken in July 1913 at Russolo's studio in Milan. These are the famous photographs of Russolo and his assistant Ugo Piatti surrounded by 16 instruments.

These two images were extremely important for my reconstruction. They show the very orchestra of 16 *intonarumori* I planned to reconstruct, the one unveiled at the August 1913 press concert. Since the
1 instruments shown in the photos were still under construction and stacked all over the room, a number of useful details not seen in later photographs can be observed: the position of cranks in relation to levers; the position of cones; construction tools on a table, etc.

The photographs of live performances were also useful: some from the London concerts Russolo conducted in June 1914; some from the 1921 Paris concerts, etc. All these photographs show the exteriors of *intonarumori*. One photograph, likely taken in 1914, showed the inside of the *Ronzatore*. This photograph proved decisive.

The photograph reveals the insides of an instrument fitted with an electric motor. We know that Russolo's first *intonarumori* orchestra contained 16 instruments, 10 of which were hand-cranked and thus not powered by electricity; 6 were electric. In fact, this is precisely the initial distinction that can be made among the *intonarumori*: some were hand-powered and others ran on electricity.

Having a sense of how both typologies looked was very important for the purpose of reconstructing them: from the patent I could find out more or less how the cranked instrument looked; from the *Ronzatore* photo, I had a sense of how one of the electric instruments looked inside. I could then apply this knowledge to other electric instruments.

Sadly, we don't have photographs of the insides of all the other instruments. This would have made reconstruction of the whole set much easier, and in all likelihood a replica of it could have been completed much earlier than mine.

The evidence I gathered during the reconstruction process included information related to previous *intonarumori* reconstructions, all of which had been partial reconstructions. The first one—the one for the 1977 Venice Biennale—was the most useful of all.

Pietro Verardo worked with firsthand documentation from the very beginning, and spent decades perfecting his instruments, carefully selecting the materials, etc. I was able to visit his studio with composer and radio personality Robert Worby and a crew from the BBC in the summer of 2009. There were nine instruments in the studio, some of which I was able to play.

The other reconstructions I knew of (one made in Japan, one in the Netherlands, etc.), were far more limited in scope, and included only a few hand-cranked instruments that more closely resemble the instrument featured in the patent. Regardless, getting as much information on these earlier reconstruction projects was unquestionably helpful for me.

While it is true that my understanding of Russolo's aesthetics fueled my reconstruction, it is also true that—having been in charge of a reconstruction project and thus benefiting from the opportunity to concentrate intensely on minute construction details—it wound up influencing my understanding of Russolo's aesthetics as well.[14]

Luigi Russolo, whose eclectic personality constantly oscillated between the theoretical and the practical, would have been very pleased by this process.

I would like to thank Todd Lerew for his help in reviewing this essay

1

1
Luciano Chessa, *Luigi Russolo and the Occult*, PhD dissertation (Davis, CA: University of California, 2004). In 2012 my dissertation became the book titled *Luigi Russolo, Futurist. Noise, Visual Arts, and the Occult* (Berkeley, CA: University of California Press 2012).

2
Giovanni Lista, "Russolo, peinture et bruitisme," in Luigi Russolo, *L'art des bruits* (Lausanne: l'Age d'Homme, 1975); Gianfranco Maffina, *Luigi Russolo e l'Arte dei rumori. Con tutti gli scritti musicali* (Turin: Martano, 1978).

3
Germano Celant, "Futurismo esoterico," *Il verri* 15, nos. 33–34, Milan, October 1970.

4
Maurizio Calvesi, *Il Futurismo. La fusione della vita nell'arte* (Milan: Fratelli Fabbri Editori, 1967).

5
It can be argued that Futurism's founding act, Filippo Tommaso Marinetti's 1909 manifesto in *Le Figaro*, was modeled not so much on Karl Marx's 1848 *Communist manifesto*, but rather Jean Moréas' *Symbolist manifesto*, also published in *Le Figaro*, in 1886.

6
For a detailed explanation of this process, see Luciano Chessa, *Luigi Russolo, Futurist. Noise, Visual Arts, and the Occult*, cit., chapters 7, 8, 11.

7
Charles Webster Leadbeater, "L'aspect occulte de la musique," *Le Lotus Bleu: la revue théosophique*, no. 10, vol. 15, December 1904.

8
Luigi Russolo, *L'arte dei rumori* (Milan: Edizioni Futuriste di Poesia, 1913).

9
See Luciano Chessa, *Luigi Russolo, Futurist. Noise, Visual Arts, and the Occult*, cit., pp. 209–18.

10
For the 2009 edition of the Biennial, RoseLee Goldberg chose the Futurist manifestos as the curatorial theme to unify the over one hundred events that made up the Biennial's impressive program. Thanks to a suggestion from a friend of mine, Tom Welsh, Performa's General Manager Esa Nickle contacted me in 2008 and asked if I could reconstruct Russolo's *intonarumori.*

11
Luigi Russolo, "Grafia enarmonica per gl'intonarumori futuristi," *Lacerba*, March 1, 1914.

12
See note no. 1.

13
2008 was the perfect time for me to switch to a more hands-on research approach, directing my interest from aesthetics to instrument construction. I'd filed my dissertation at UC Davis in 2004, and then worked in other directions for a while (writing on Italian composer Giacinto Scelsi, for instance). 2008—roughly four years after putting that dissertation aside—was a good time to return to Russolo.

14
After this experience, I was able to return to the draft of *Luigi Russolo, Futurist* with newfound enthusiasm and understanding. The experience helped me complete the book in 2012.

ROB YOUNG

SO RUDELY FORCED

THE POSTWAR TURN FROM MUSIC TO SOUND

You could say the moment music began to metamorphose into "sound" occurred on a ski slope. In early 1948 Pierre Schaeffer, a Radio France producer who had worked with the station during World War II, was feeling disillusioned with the state of music and was restlessly pondering the stagnation of traditional musical forms. Clinging to a ski lift in the French Alps, he found his ears tuning in to the creaks and metallic groans of the machinery, rendered even stranger by the unfamiliar acoustic backdrop of powdery snow. He was, perhaps, the first human being to acknowledge that such concrete noises could be enjoyed aesthetically for their own sake and heard as, if not a composition, then at least a form of musical experience.

This trip opened Schaeffer's ears anew and within a few weeks he had secured a room at his workplace in which to set up a makeshift sound chamber where he began experimenting with turntables, locked-groove vinyl sides, tape reels and echo/reverb effects. Over the next few years the results of this work emerged as new musical works, broadcast on French radio or performed "acousmatically" (anonymously through sound systems), including the *Études de bruits* (1948), *Les paroles dégelées* (1952), and *Symphonie pour un homme seul* (1952, in collaboration with Pierre Henry).

Schaeffer's new practice forced very different working practices from the previous model of the music composer. Rather than generating harmonic and melodic combinations of instruments while sitting at a desk in front of stave paper, his quest for new sounds made him engage physically with technical equipment such as turntables and tape machines, as well new devices he built himself. Embarking on a quest for the *objet sonore*—the sound object—he roamed the Radio France HQ assembling a collection of bits and pieces, percussion instruments and metal objects, an old piano, and then went further afield, scavenging abandoned organ pipes from a bombed church, and arriving at the Gare de Batignolles with a truck containing portable recording equipment and microphones with which he taped the noises of locomotives standing by the platforms in preparation for one of the early masterpieces of electronic music, the *Étude aux chemins de fer* (1948).

To fully grasp the paradigm shift that electronic music undertook during this first decade following the World War II, it's necessary to place ourselves on the same journey taken by composers during these years, and abandon the notion of music as a construction of harmony, melody and rhythmic elements. In his introduction to one of the first histories of electronic music, the writer Paul Griffiths states: "most of our music makes use of only very narrowly defined selections from the enormous range of sounds we can hear ... the palette of instruments available to the composer, though comparatively variegated in the modern symphony orchestra, again confines his imagination in quite particular ways."[1]

The development of electronic music in all its forms was in effect an unspoken assault on the limitations of a musical medium that had remained composed of roughly the same elements for centuries. The main characteristics of electronic music in all its forms include the combination of frequencies, microtones, noise, interference, concrete and found sounds, and the manipulation and distortion of instrumental timbre. For his part, Pierre Schaeffer was even dissatisfied with the course that contemporary music had taken thus far, dominated as it was by the atonalism and serialism pioneered at the turn of the century by the Second Viennese School of Arnold Schönberg, Alban Berg and Anton von Webern. "After the war," he wrote, "in the 1945–48 period, we had driven back the German invasion but we hadn't driven back the invasion of Austrian music, twelve-tone music. We had liberated ourselves politically, but music was still under an occupying foreign power: the music of the Viennese school."[2] Schaeffer and the whole *musique concrète* movement he set in motion acted as a resistance movement committed to undermining the Teutonic axis that had conquered modern music in the first half of the 20th century.

Schaeffer began recombining snatches of noise on vinyl and tape loops, spinning them backwards, varying their speed and dreaming up three-dimensional playback systems. Until then, music had no plastic qualities—composers conceived music in the mind, then wrote marks on paper that corresponded to the stable frequencies of musical notes, intended for performance. Concrete music, on the other hand, could not pre-exist, but was a product of its constituent recorded elements. "As long as meaning predominates, and is the main focus, we have literature and not music," writes Schaeffer. "But how can we forget meaning and isolate the in-itself-ness of the sound phenomenon?"[3]

On May 15, 1948, he announced: "I have coined the term *musique concrète* for this commitment to compose with materials taken from 'given' experimental sound in order to emphasize our dependence, no longer on preconceived sound abstractions, but on sound fragments that exist in reality and that are considered as discrete and complete sound objects..."[4] The *objet sonore* (sound object)—a fragment on tape or shellac—allowed him to analyze and transform sounds at a magnification never possible before, and Schaeffer talked of vivisecting atoms of sound to fuse unheard audio elements.

This notion of the violence of electronic music was alluded to in a 1964 piece for voice and synthesizer by the American composer Milton Babbitt. *Philomel* is based on the Greek myth of Philomela, an Athenian princess kidnapped and "so rudely forced"[5] by her brother-in-law, Tereus, who cuts out her tongue so that she is unable to broadcast the news that he has raped her. Together with her sister Procne (Tereus's wife), Philomela eventually takes revenge by forcing Tereus to eat the flesh of his own son, while Philomela herself is transformed into a nightingale, the bird most revered for the beauty of its song. Babbitt's *Philomel*, with a female soloist mirroring herself on tape and the voice distorted by the synthesizer, enacts in mythological form electronic music's own "rude force" upon what has gone before. Echoing contemporary fears that music based on tape and electronic machines would kill off the familiar concert-going musical experience, the voice—the most original and essential musical instrument—is first "devoured" by the synthesizer, then metamorphoses and takes flight in a transmuted form. It is a pertinent work that, by merging serialist procedures with a live soloist, tape mirroring and synthesis, confronts and transcends many of the problematics in the contemporary music of its time.

One of Schaeffer's most striking techniques in the late 1940s and early 1950s also involved a kind of severing of the tongue of a sound. By using a vinyl groove or a magnetic tape to isolate a fragment of sound—perhaps the striking of a bell, or the chime of a piano note—he could go microscopically deep into the sound object's atomic structure and slice off its opening milliseconds, also known as the "starting transient." This turned a percussive noise into a gently haunting texture, which seemed to float out of the silence instead of appearing with an explosion. Like Philomela after her tongue was mutilated, sounds transcended reactionary forces and were eventually allowed to transform and fly free of their physical origins.

In this new world, the "composer" had to be more than a desk-bound individual inscribing notes on manuscript paper; it forced a very different, often physical engagement with the materiality of sound. The American composer Harry Partch, who ended up designing and building his own instrumentarium to realize his wholly other world of sound, once described himself as "a philosophic music man seduced into carpentry."[6] The composer could no longer entrust the production of music to a third party (soloist, ensemble, orchestra, conductor) but often, like La Monte Young, David Tudor, Morton Subotnick, Pierre Henry, Alvin Lucier, Pauline Oliveros and many others, found him- or herself forced into the position of self-performer. Schaeffer himself was involved in the creation of the *pupitre d'espace* (space desk), a gesture-controlled system of electromagnetic "halos" for moving sounds around a performance space, the *phonogène* (sound generator), which varied the pitch and speed of tape playback, and the morphophone, an early effects box whose multiple tape heads produced primitive delay and echo. One critic described the results as akin to "discovering a sound continent as virgin as Robinson Crusoe's island."

Although he moved away from performance to a more theoretical approach during the 1950s and 1960s, the composers and musicians of Schaeffer's "school" who clustered around the GRM (Groupe de Recherches Musicales, established in 1958) took up his methodologies and turned them to their own individual purposes. These included Luc Ferrari (1929–2005), who declared an hour-long recording of sunrise over a French coastline to be music (in *Presque rien no 1 – le lever du jour au bord de la mer*, 1967–70), foreshadowing the ambient tendency, and the plethora of field recordists active today. Another, Eliane Radigue, was celebrated in a major London retrospective in summer 2011. Radigue, now in her eighties, adopted synthesizers in the late 1960s, pressing their malleability and extreme frequencies into service as signifiers of the "eternal now" she sought through Buddhism. Still composing today, she now investigates similar transcendent qualities in acoustic instruments such as the cello and harp. The soundworld may be very different from Schaeffer's earliest "noise studies" but the interpenetration of matter and form, and the quest for a realm of sound beyond conventional "music," survive.

Electronic music's expansive qualities seemed to demand new space for reception. Poorly adapted to the formal concert hall (even today, audiences and critics can be skeptical about concerts involving playback, sound diffusion or use of a laptop), electronic music's natural habitat was the radio broadcast, the long-playing record, or the spatial installation. Another landmark in that significant year was the Philips Pavilion built for the Brussels Expo 58. Commissioned by the Dutch electronics firm (Philips) that was at the forefront of developing home hi-fidelity systems and recording devices, the architect for the structure was nominally Le Corbusier, but the building was actually designed from a Corbusier sketch by one of his studio assistants, Iannis Xenakis. This thirty-six year old Greek-born architect was also a composer in the making who was interested in applying mathematical and architectural principles to the process of composing music.

Le Corbusier's conception of the pavilion, which resembled an obscure Platonic solid undergoing high levels of torsion, was that it should act as a kind of digestive system, with the audience entering through a mouth, being bombarded with sound, color, light and image while in the stomach, and being expelled from the exit hole after a seven-minute multi-sensory experience. Xenakis's *Concret PH* was at the time one of the most abstract pieces of music composed to date, a proto-noise continuum constructed from a recording of burning charcoal cut into one-second fragments and stitched together to form a granular texture that was played as the audience entered and exited the space. Once inside, they were bombarded with a work entitled *Poème électronique* by Edgard Varèse, a collage of percussive noise, cut-up and manipulated tapes and sonic transformations. Varèse had been one of the premier avant-garde composers since well before World War II, a prophetic figure whose ideas were always too advanced for the available technology to realize. As early as 1916 he was already adopting a zealously Futurist attitude to new sonic potential, when he told a newspaper interviewer: "I refuse to submit myself to sounds that have already been heard. What I am looking for are new technical mediums which can lend themselves to every expression of thought and can keep up with that thought."[7] By the late 1950s, when he had reached his seventies, the machinery was beginning to catch up with his radicalism. Diffused via a net of speakers embedded in the walls, *Poème électronique* utilized the full breadth of the architectural space, and was operated via a system of telephone dials controlling the waves of sound in three dimensions.

At the same time, in Cologne others were focused on different kinds of musical machines that could keep up with the speed of human thought. The Westdeutscher Rundfunk (WDR) Studios, steered by Herbert Eimert, had taken what they called "the first step in musical control of nature,"[8] employing very different methods from Pierre Schaeffer and his cohorts. Instead of transforming recorded actualities *à la musique concrète*, the Germans generated electronic tones with oscillators, ring modulators and primitive synthesizers.

A 1951 concert for the radio was described as "a sound-world of music;"[9] the phrase has more recently become a journalistic cliché, but its use at the time suggests a desire to create a hermetic space for a new evolutionary strain of music. Like Varèse, Eimert was transfixed by the possibilities of an entire reinvention of the language of music through technology. He called for a "subterranean revolution in hearing"[10] and claimed that classical music, when frozen in time in the form of a gramophone record, "may be the surrogate and electronic music the true music."[11]

Eimert's pupil-assistants included Karlheinz Stockhausen, who would go on to be the figurehead of Germany's postwar contemporary music field. If large-scale compositions, such as *Gruppen* (1955–57) for three orchestras, showed an early interest in stretching the limitations of formal classical music, he soon pushed himself into non-traditional settings, such as the ring-modulated choirboy on *Gesang der Jünglinge* (1955–56), the amplified gong of *Mikrophonie I* (1964), and massive conceptual electroacoustic pieces such as *Telemusik* (1966) and *Hymnen* (1966–67). Epic in scope, Stockhausen's works increased in dimension over the 1960s and into the 1970s. In 1970 a descendant of the Philips Pavilion —the German Pavilion at the Osaka Expo 70—, broadcast a five-hour selection of his work in a geodesic dome fitted with 7 rings of loudspeakers. *Hymnen* opened up an imaginary acoustic space in which all the nations of the world were represented by distorted versions of their national anthems, the first composition to attempt a pan-global scale; while *Telemusik* played with telematic ideas of closeness and distance while utilizing ethnic music samples as its raw materials. Music broke its boundaries to reach across a planetary audience, in a cosmic gesture that Stockhausen would later augment with statements of beliefs that his music had its origins in the distant star system of Sirius.

Elsewhere, music was expanding like a white dwarf star. In 1948 John Cage had already proposed an unprecedented space in which all sounds were permitted, in the notorious and highly un-silent piece *4'33"*. His compositions for tapes, such as *Williams Mix* (1953) and *Fontana Mix* (1958), were not so much about the actual sounds produced than they were about giving performers a democratic choice as to how and what elements they would choose to present the score. Cage's use of chance operations, indeterminacy and interchangeable sheets of transparent score paper wrenched total control of a musical work away from the composer and placed many, or all, of the decision-making either with the faculties of the performer(s), or even left to chance [214, 216].

Cage's 1969 response to the space age was *HPSCHD*, a multimedia happening that combined computer music (a collaboration with Lejaren Hiller) with visual material from the NASA space program. Like the Xenakis/Varèse pavilion, *HPSCHD*—premiered one month before the Apollo Moon landings—drenched the viewer in sound and image, with the insistent timbre of the baroque harpsichord backed with computer-generated tapes as views of planet Earth taken from space provided an immersive and weightless environment. The installation continued for five hours and was visited during that period by around 6,000 people.

Such long durations were becoming increasingly common in the arts of the late 1960s as forms stretched elastically, and music was no exception. The films of Andy Warhol and Michael Snow documented the unfolding of real-time events in all their unmediated banality, creating conditions in which the passing of time became undetectable. Similar effects were audible in the early music of former Fluxus member La Monte Young. His Theater Of Eternal Music collective picked up the long durations of Indian and Asian classical music as a structuring device, as well as using drones and frequencies from bowed instruments, chants, tambura drones and resonant objects. The group included John Cale in its line-up, who went on to influence the world of avant-garde rock in The Velvet Underground (a group set up under the auspices of Warhol's Factory), and violinist Tony Conrad, who went on to pursue his own course exploring alternative tuning systems and just intonation. For his part, Young and his wife Marian Zazeela established another of the important physical spaces in contemporary sound art, the Dream House in Manhattan, New York.

Here they installed a permanent, ever-changing series of rooms containing resonant drones and morphing colored light effects, designed to have a meditative and hypnotic effect. Young and Zazeela continue to live according to their own circadian rhythm, out of sync with the standard twenty-four-hour clock, underpinned by a music connected to universal forces outside the human realm.

Other examples of sound artists exploring space include Alvin Lucier, whose *I Am Sitting in a Room* (1970) harnessed the feedback accumulating around a voice rerecorded multiple times on tape, and played back into a room, thus producing a sounding of a unique space. Max Neuhaus's *Public Supply I* (1966) threw open the telephone network as an input node, mixing contributions from the ether in real time. Maryanne Amacher's *Sound-joined Rooms* (1980) connected unrelated spaces by transmitting their miked-up interiors to each other across long distances; while Christina Kubisch's *Electrical Walks*, begun in the late 1970s, allows members of the public to roam a city wearing wireless headphones that amplify electromagnetic pulses, revealing a hidden electroacoustic dimension of human habitations. A process that began with Pierre Schaeffer roaming the city has evolved into the listener being invited into a similar flâneur's role.

All these examples serve to demonstrate that although music's harmonic tongue may have been cut out after the war, it continued to soar on broad wings.

1
Paul Griffiths, *A Guide to Electronic Music* (London: Thames & Hudson, 1979), p. 7.

2
Pierre Schaeffer, *In Search of a Concrete Music*, trans. by Christine North and John Dack (Oakland, CA: University of California Press, 2012).

3
Ibid.

4
Ibid.

5
Taken from T.S. Eliot's poem *The Waste Land* (1922): "... The change of Philomel, by the barbarous king / So rudely forced; yet there the nightingale / Filled all the desert with inviolable voice / And still she cried ...," in Id., *The Waste Land and Other Poems* (New York/ London: Penguin, 2003).

6
Harry Partch, *Genesis of a Music* (Cambridge, MA: Da Capo Press, 1947).

7
Malcolm MacDonald, *Varèse: Astronomer in Sound* (London: Kahn & Averill, 2002).

8
Paul Griffiths, *op. cit.*

9
Ibid.

10
Ibid.

11
Ibid.

Luigi Russolo

INTONARUMORI
1913–14 (2002–09)

Luigi Russolo (Portogruaro, 1885–Cerro Laveno, 1947) was active in the Futurist movement, first as a painter, then as a musician and composer. In March 1913 he wrote the manifesto *The Art of Noises*. At a Futurist soirée at the Teatro Storchi in Modena in June the same year, Russolo presented *intonarumori*, a series of musical machines he'd invented that could modulate noises and reproduce a great variety of sounds typical of the city and modern machines. His *intonarumori* were grouped by timbre, just as in an orchestra, and included gurglers, cracklers, screamers, roarers, screechers, whisperers and beaters.

Reconstruction of these instruments, which were destroyed in a bombing raid during World War II, was initiated in 1977 by Pietro Verardo, an expert in music and the history of Futurist music, acting on an invitation from the Archivio Storico delle Arti Contemporanee of the Venice Biennale, the head office of which was then located in Ca' Corner della Regina. Verardo based his work on the drawings patented by Russolo in January 1914.

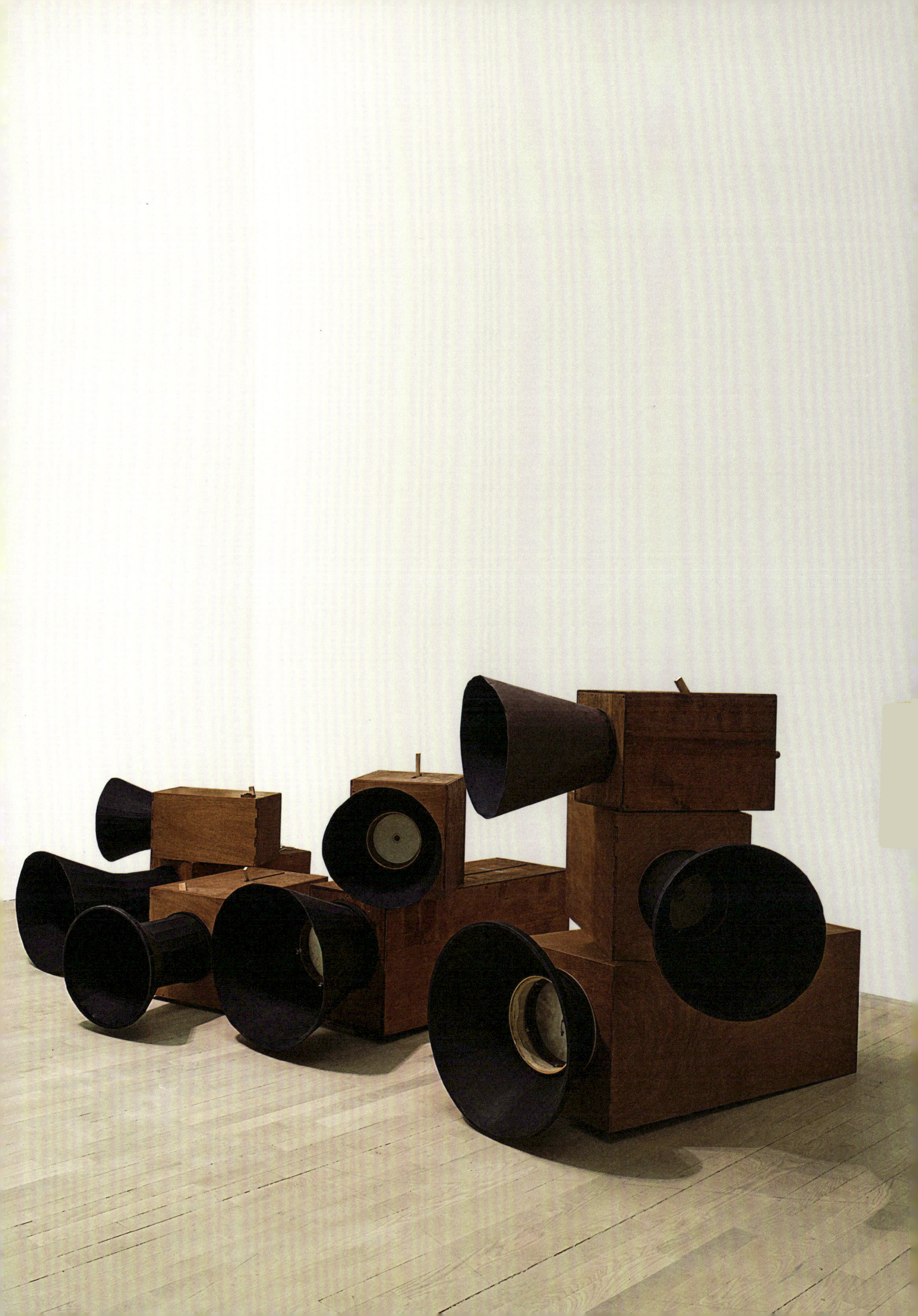

Giacomo Balla

CIAC CIAC
1914

In 1915 Italian painter, sculptor, set designer and photographer Giacomo Balla (Turin, 1871–Rome, 1958) drafted the manifesto *The Futurist Reconstruction of the Universe* with Fortunato Depero. The notion of a form of art apparent in all aspects of life, however banal, was a recurrent characteristic of Balla's output, which ranged across many forms of expression, including music. For the production of *Piedigrotta* (1914), a work by Francesco Cangiullo, Balla built various instruments for producing noises typical of the Neapolitan carnival, among which a *scetavajasse*, also known as a *ciac ciac* for the sound it made.

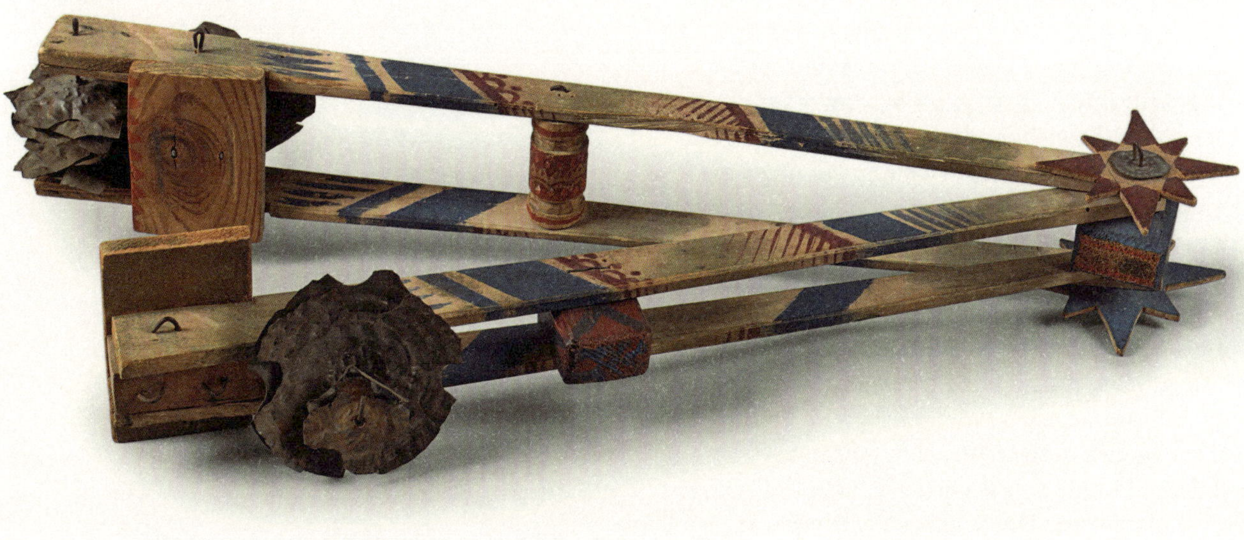

Marcel Duchamp

WITH HIDDEN NOISE
1916 (1963)

With Hidden Noise (*À bruit secret* in French) came as a result of a collaboration between French artist Marcel Duchamp (Blainville-Crevon, 1887 – Neuilly-sur-Seine, 1968) and his friend Walter Arensberg. On Easter Day 1916, Duchamp set a ball of nautical twine between 2 brass plates. He then asked Arensberg to put an unknown object inside—and never to tell him what it was—before he clamped the readymade shut with 4 long screws. The 2 brass plates are engraved on the top and bottom with a mysterious inscription of English and French words on both sides; in order to decipher it the readymade has to be turned over repeatedly. The title alludes to the rattling sound the hidden object makes when the readymade is shaken.

This replica of the 1916 readymade was constructed in 1963 by Swedish art critic Ulf Linde in Stockholm for the Duchamp retrospective at Moderna Museet. In Milan the following year, Duchamp produced an edition of 8 copies (with the hidden object placed by his wife Alexina Duchamp).

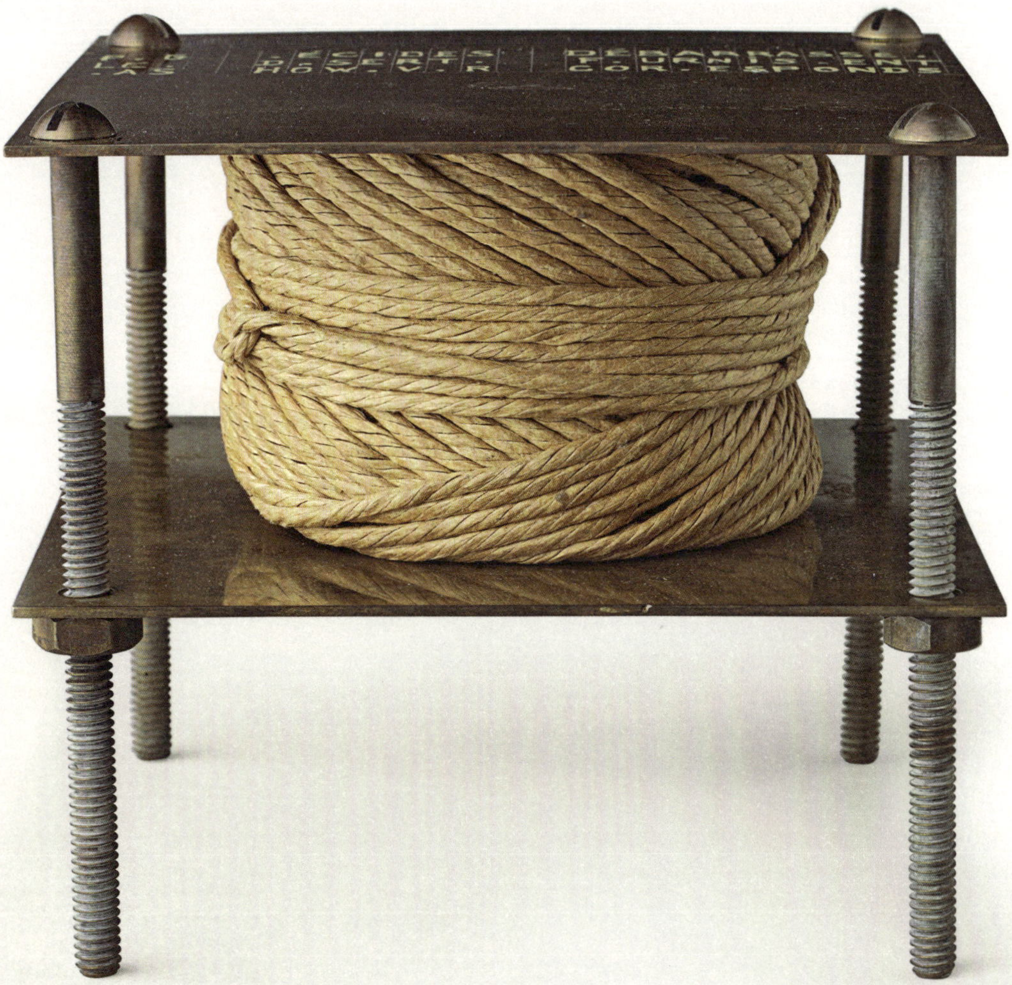

Theo van Doesburg

COMPOSITION IN GRAY (RAG TIME) 1919

In October 1917, Dutch architect Theo van Doesburg (Utrecht, 1883 – Davos, 1931) founded *De Stijl* magazine with Piet Mondrian and other artists and architects. The name was soon used for the whole De Stijl group, which sought to create an autonomous, universal aesthetic consciousness using geometric forms, vertical and horizontal lines, and a limited palette that included only the primary colors combined with the three non-colors, as theorized by Mondrian, in order to depersonalize artwork.

Composition in Gray is an abstract representation of a couple dancing ragtime, a musical genre that began as dance music in the red-light districts of African American communities in St. Louis and New Orleans years before being published as popular piano sheet music. Its main characteristic is its syncopated or "ragged" rhythm. This music was very popular in the second decade of the 20th century and composers of the time made several references to it, among them Erik Satie, who included the piece *Ragtime du paquebot* in his *Parade – Ballet réaliste* (1917).

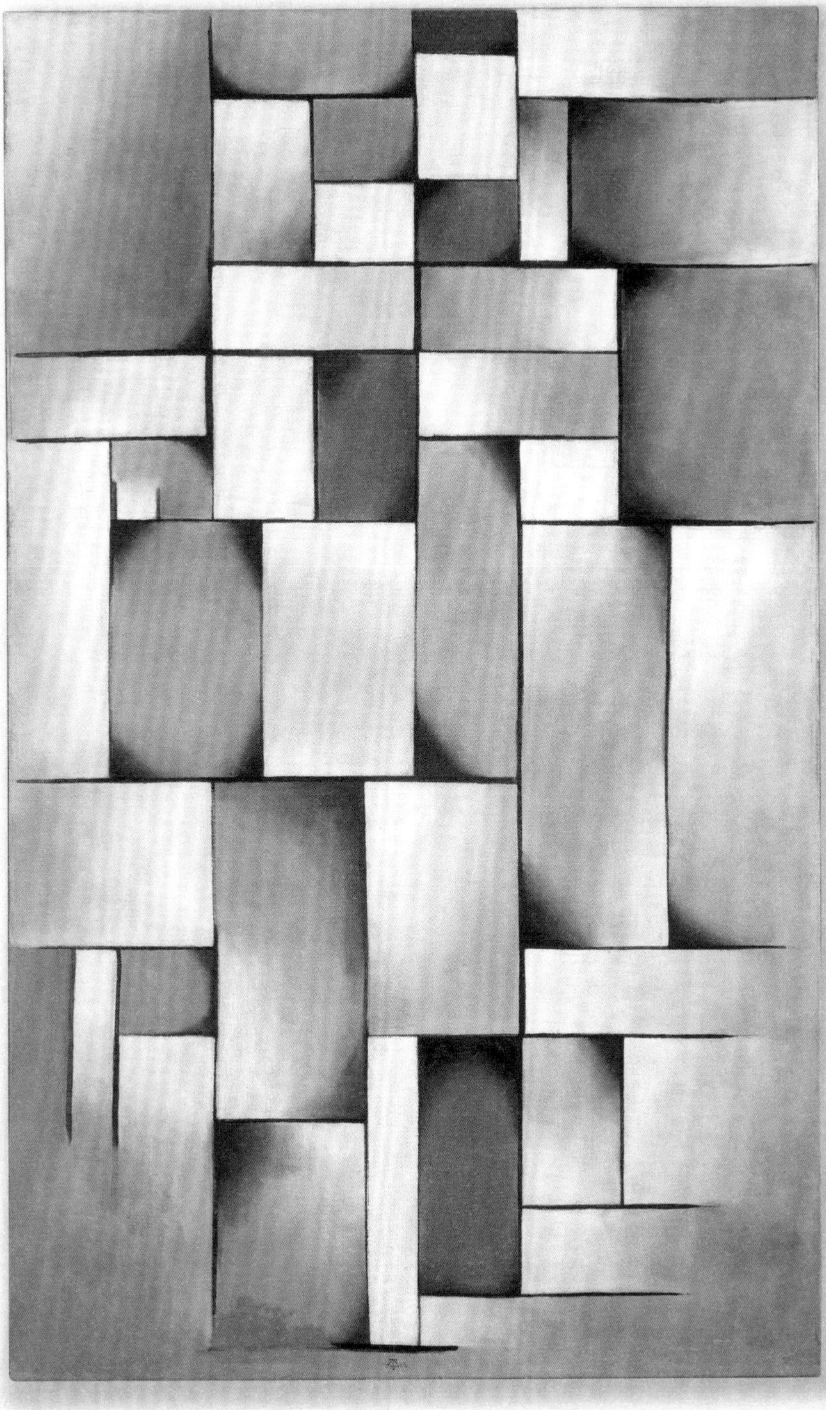

Amelotti

ORCHESTRION ACCORDEO JAZZ
c. 1920

Originated in the middle of the 18th century in Germany but developed in particular from the mid-19th century as a result of the work of the German Michael Welte (1807–80) and his Freiburg-based company, the orchestrion was an automated instrumental ensemble featuring a real organ, to which percussion instruments were added, and occasionally others, such as upright pianos and violins.

In the case of the *Orchestrion Accordeo Jazz*, built by Amelotti, a company in Nice, an accordion made by Paolo Soprani & Figli of Castelfidardo played music recorded on perforated cardboard together with a cymbal, bass drum, snare drum and castanets, which were moved by a compressed air system.

Unlike fairground organs, which they resemble, orchestrions were designed to be used in indoor spaces, like houses and ballrooms, and generally produced a better quality of sound.

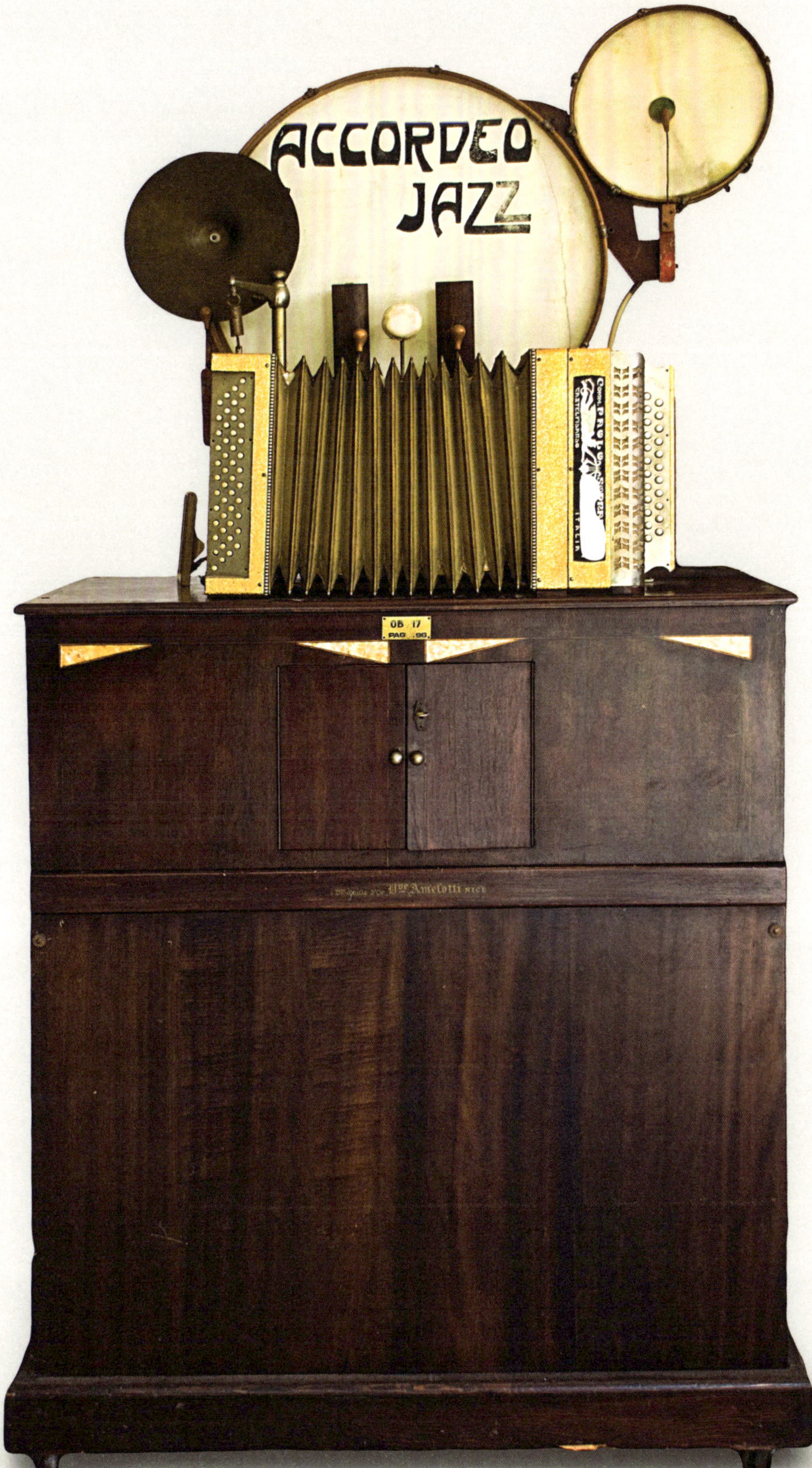
ACCORDEO
JAZZ

Wladimir Baranoff-Rossiné

OPTOPHONIC PIANO
1920–23 (1971)

The Russian Jewish musician, painter and sculptor Wladimir Baranoff-Rossiné (Kherson, 1888 – Auschwitz, 1944) settled in Paris in 1910, joining Abstractionist and Cubist circles. In 1912 he began to work on a synaesthetic instrument included in a series of "light concerts" in 1916 in Sweden, which he perfected in the years that followed.

On April 16, 1924 he presented the optophonic piano at the Meyerold Theater in Moscow, where he performed an "opto-visual colored concert" in which music was transformed into visual art.

The apparently normal piano contained not strings but a mechanical device composed of painted disks, prisms, lenses, mirrors and a light source. Moving images were projected onto a screen in harmony with the rhythm of the music on the disks.

The original instrument was destroyed, but was rebuilt in 1971 by Jean Schifrine with Eugène Baranoff-Rossiné, the artist's son. It interprets a recording of the prelude to Act IV of Edvard Grieg's *Peer Gynt*.

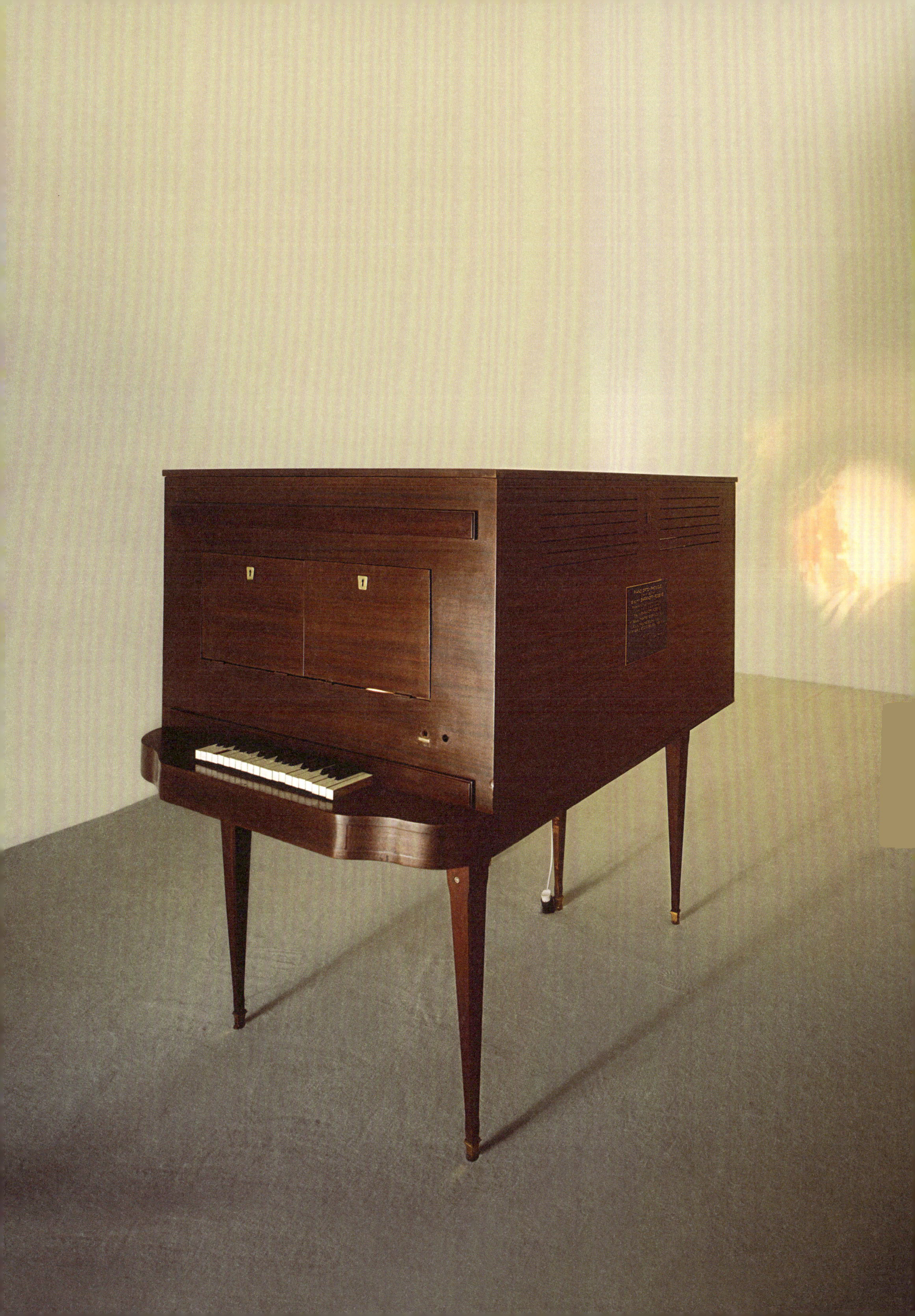

Man Ray

INDESTRUCTIBLE OBJECT
1923 (1965)

American Surrealist artist Man Ray (Philadelphia, 1890–Paris, 1976) created in 1923 *Object to be Destroyed*, later renamed *Indestructible Object*. It is a photograph of an eye attached with a paper-clip to the swinging arm of a metronome. Ray said that he used the object so that he could work to the rhythm of the ticking, and added the eye because a painter needs an audience. "Like the pianist sets it going when he starts playing, its ticking noise regulated the frequency and number of my brushstrokes. The faster it went, the faster I painted; and if the metronome stopped then I knew I had painted too long, I was repeating myself, my painting was no good and I would destroy it."* Ray smashed the object one day when it stopped ticking.

During the following decades, he did many replicas, the first of which had a cut-out of the eye of Lee Miller, a photographer and Ray's former lover, with whom he had a violent breakup.

[See also pp. 174, 176]

* From Arturo Schwarz, *Man Ray: The Rigour of Imagination* (New York: Rizzoli, 1977), pp. 205–06.

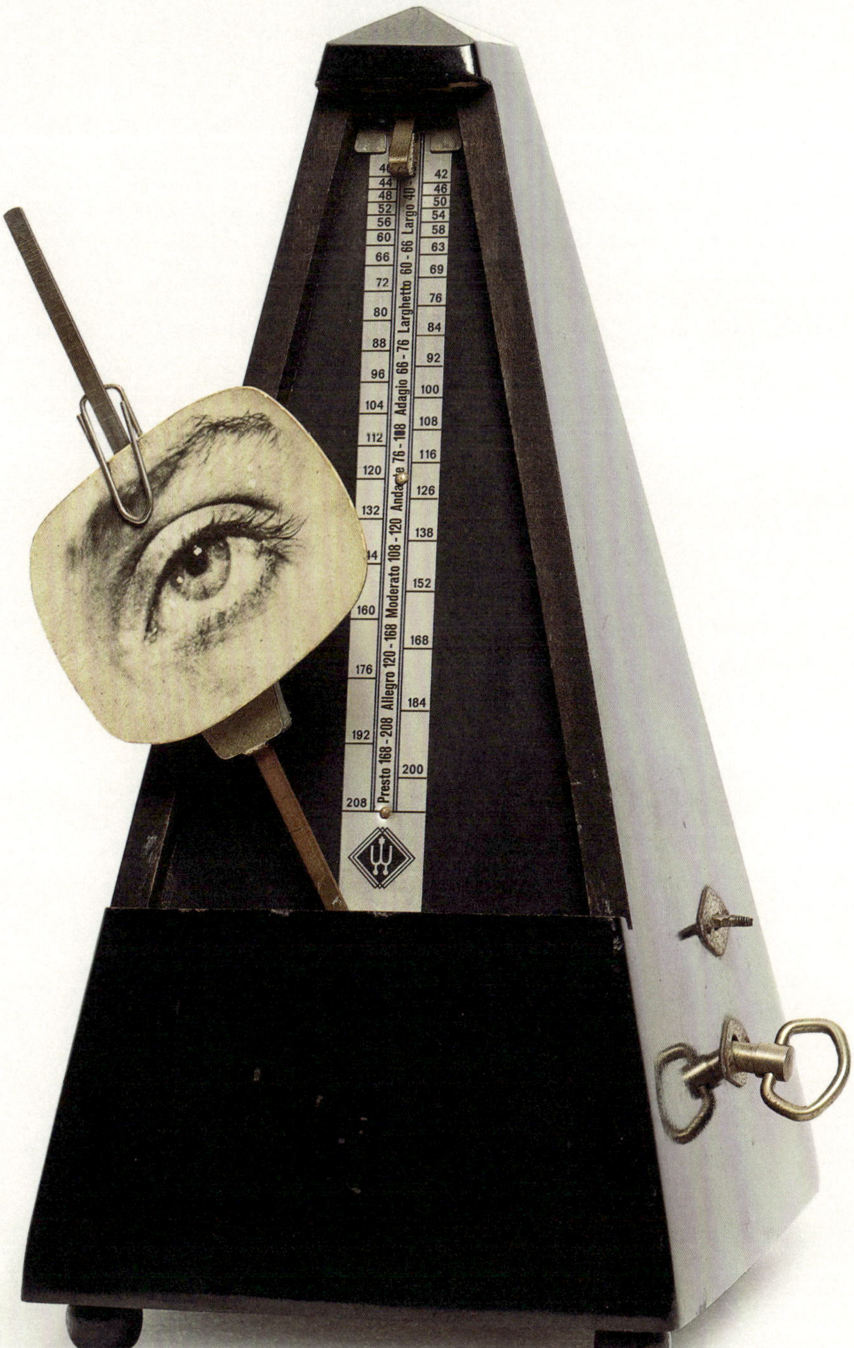
Presto 168 - 208 Allegro 120 - 168 Moderato 108 - 120
Adagio 66 - 76 Larghetto 60 - 66

Man Ray

LE VIOLON D'INGRES
1924 (1980)

1

Man Ray was an admirer of the paintings of Jean-Auguste-Dominique Ingres and made a series of photographs inspired by his figures of nude women at Turkish baths.

Le violon d'Ingres (Ingres's violin) is based on a photograph of Ray's frequent model Kiki wearing a turban. He painted the f-holes of a stringed instrument as symmetrical orifices on the photograph of the model's back, and then re-photographed the print.

The title comes from the French idiom "violon d'Ingres," which refers to a person's second skill beyond the one for which they are more commonly known, since Ingres played second violin at the Orchestre du Capitole de Toulouse as a teenager.

By simply adding 2 signs, the female body becomes a sensual musical instrument to be played, as if it were Ray's personal "violon d'Ingres."

[See also pp. 172, 176]

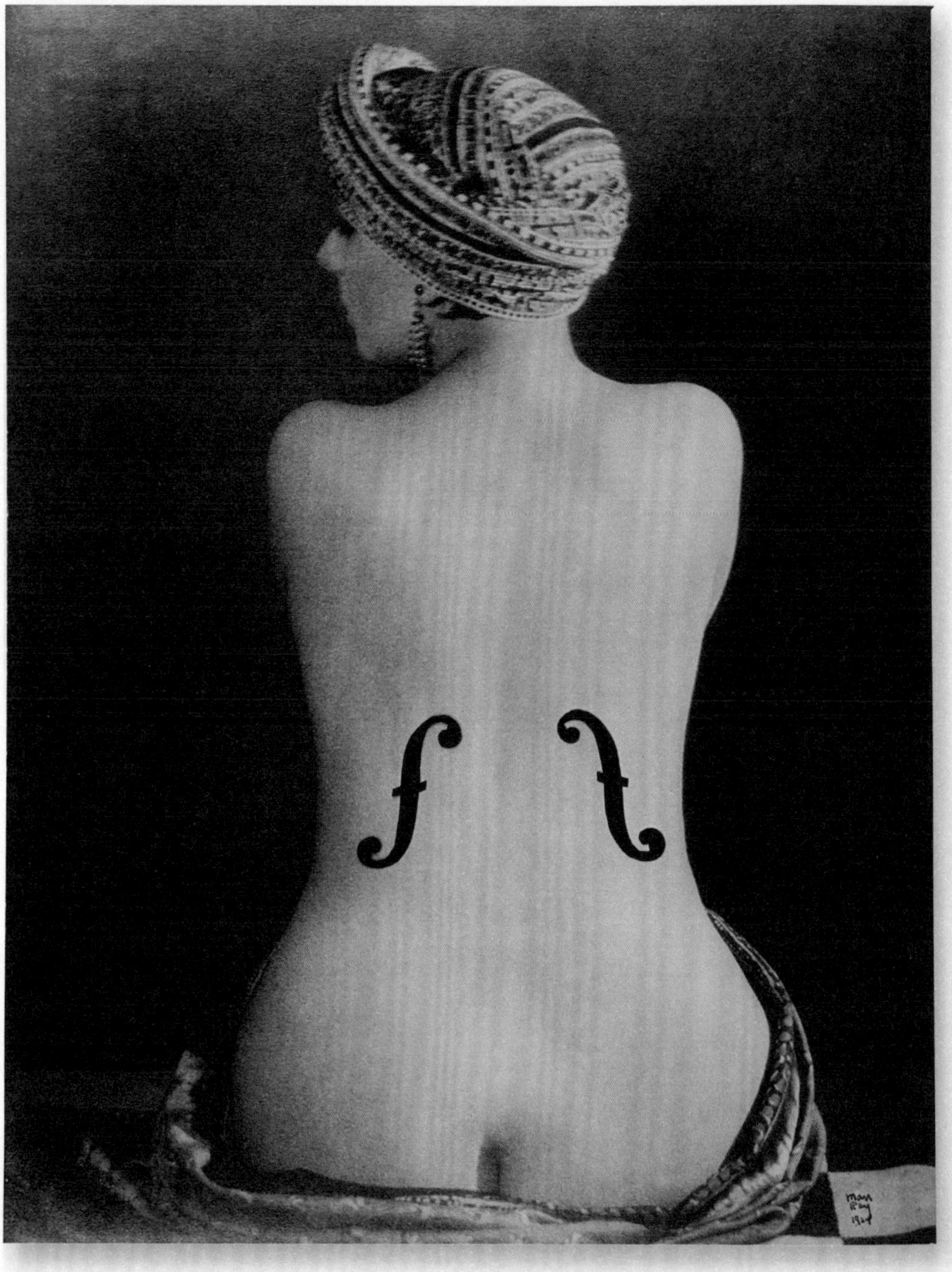
man
Ray
1924

Man Ray

EMAK BAKIA
1926 (1970)

Emak Bakia consists of the fingerboard of a cello that has had its strings replaced by light gray horsehair, which is the material typically used to make a cello bow. The original work was built in 1926 but was lost (like many other of Ray's early readymades) so it was remade in 1970. *Emak Bakia* was the name of the villa in Biarritz where the artist spent the summer of 1925, as well as the title of an avant-garde film he made that same year. The name means "leave me alone" in Basque and, as its title suggests, the object refers to the impossibility of producing music. It would also appear as a sort of self-portrait in Ray's painting *The Artist's Home* (c. 1931), where it is placed next to a female head, a rendition of a photograph of Lee Miller.

[See also pp. 172, 174]

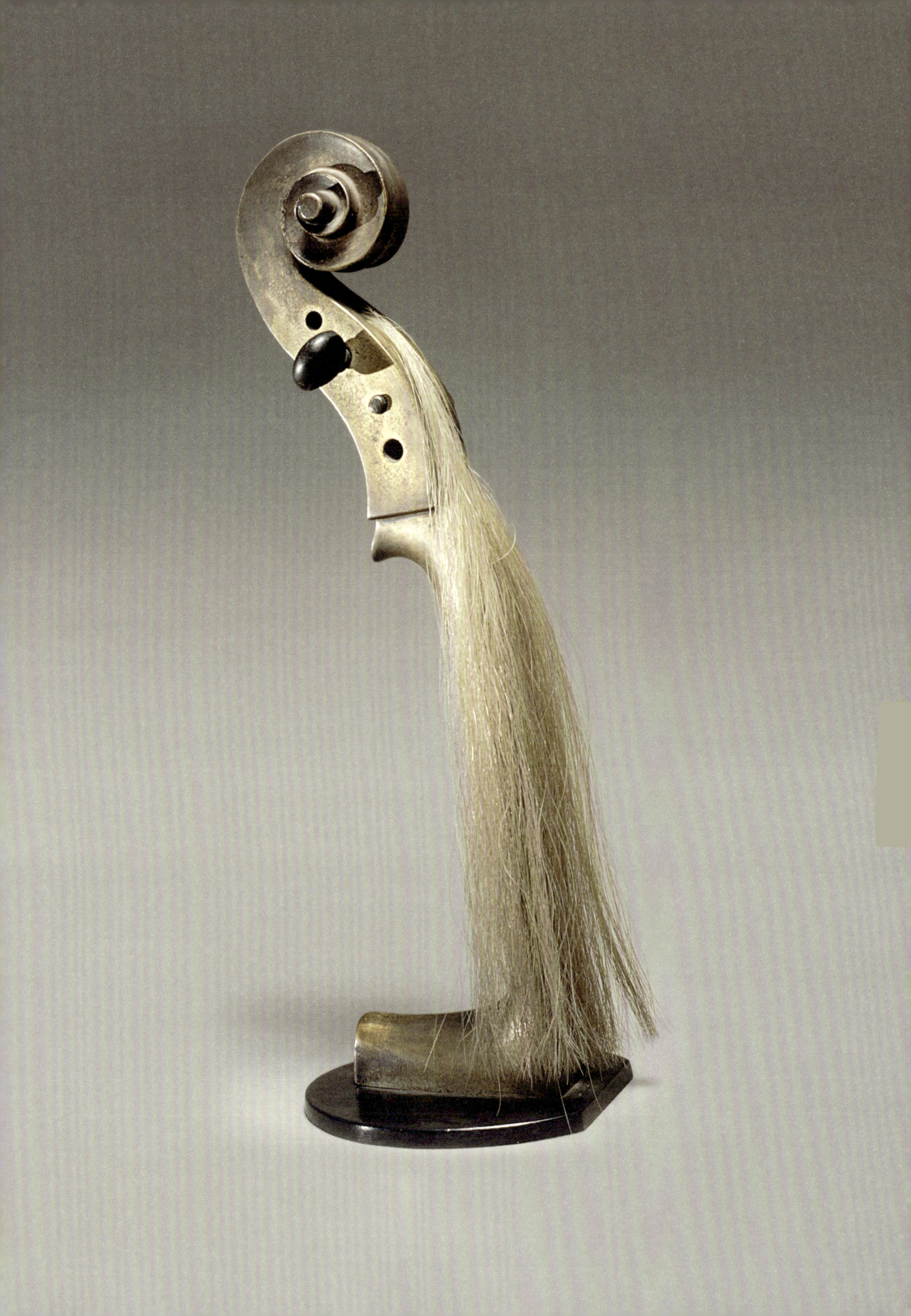

Henrik Neugeboren

SCULPTURAL REPRESENTATION OF BARS 52-55 OF E-FLAT MINOR FUGUE BY J.S. BACH. PROPOSAL FOR A BACH MONUMENT 1928 (1966)

In 1928 the Romanian artist and composer Henrik Neugeboren (known as Henri Nouveau, Brasov, 1901 – Paris, 1959) was intrigued by the concept of synaesthesia discussed at the Bauhaus by Wassily Kandinsky and Paul Klee, whereupon he decided to create a plastic representation of a fugue by Johann Sebastian Bach. His aim was to clarify the spatial and temporal aspects of the work by producing a sculptural version of it that allowed the perceiver not only to hear the temporal and spatial development of the music, but also to see it.

Neugeboren's intended that the work should not be a personal interpretation influenced by sentiment, but a scientifically exact transformation of the music into another system. According to the artist the model "has such an architectural and monumental effect that the question must be asked whether a representation of this nature might not be a worthier monument to Bach." *

* From an article published in *Bauhaus* magazine, no. 1, January 1929.

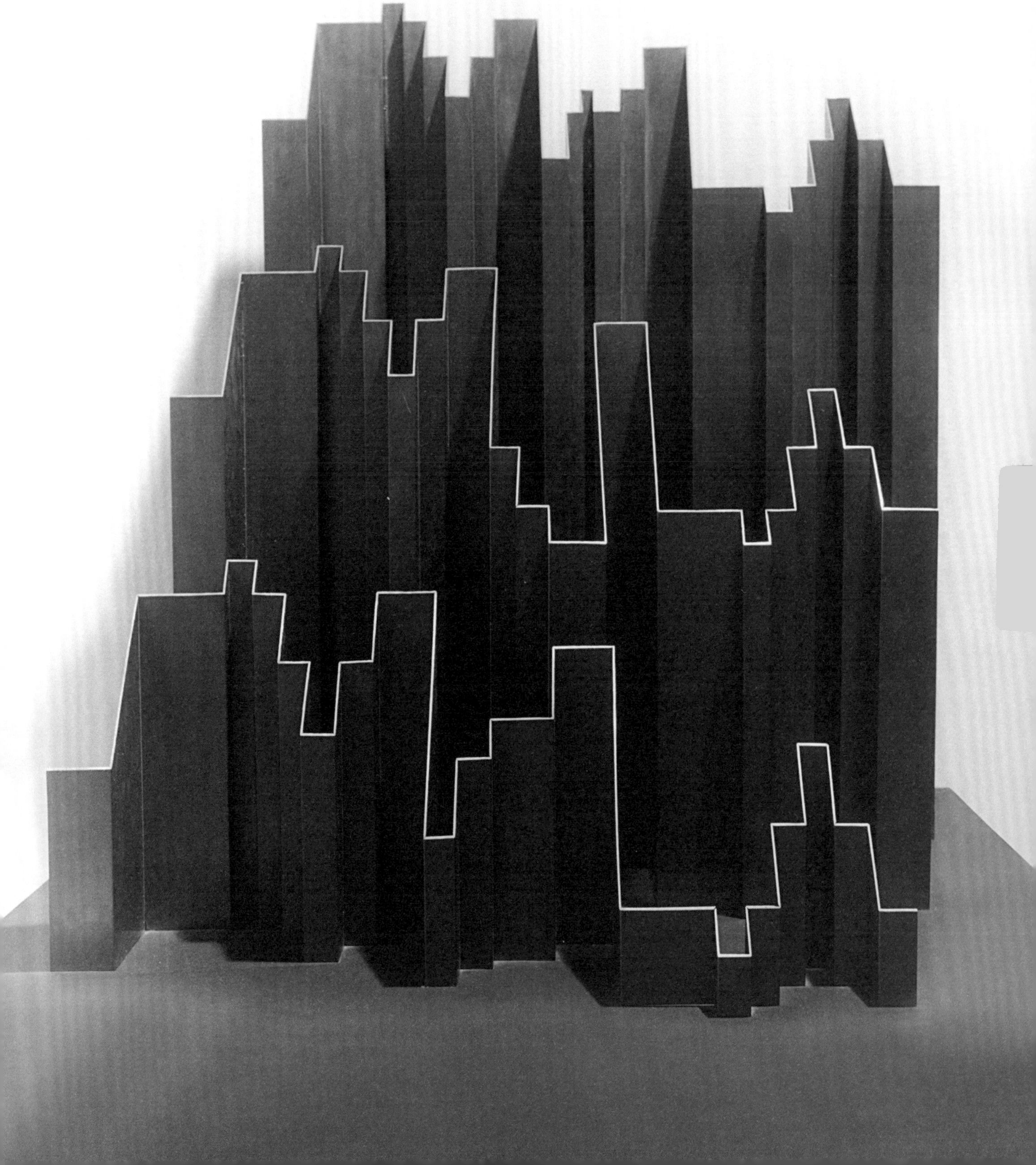

Georg Evans & Co.

STROH VIOLIN
1930

Horned string instruments were developed at the turn of the 20th century because the recording equipment on wax cylinders of the time required a loud and directional sound from the performer.

In 1900 engineer J. M. Augustus Stroh (Frankfurt am Main, 1828 – London, 1914) invented a violin with a patented diaphragm arrangement, then considered the best instrument available for recording. Stroh's idea was to replace the usual wooden body with a metal resonator in order to produce a louder, more penetrating sound. The metal horn at the end of the fingerboard directed this sound either into the recording horn or toward the ear of the singer, while the smaller horn enabled the musician playing to hear the sounds more distinctly. Later violins, violas, cellos, basses, ukuleles, mandolins and guitars using this patented diaphragm were also produced.

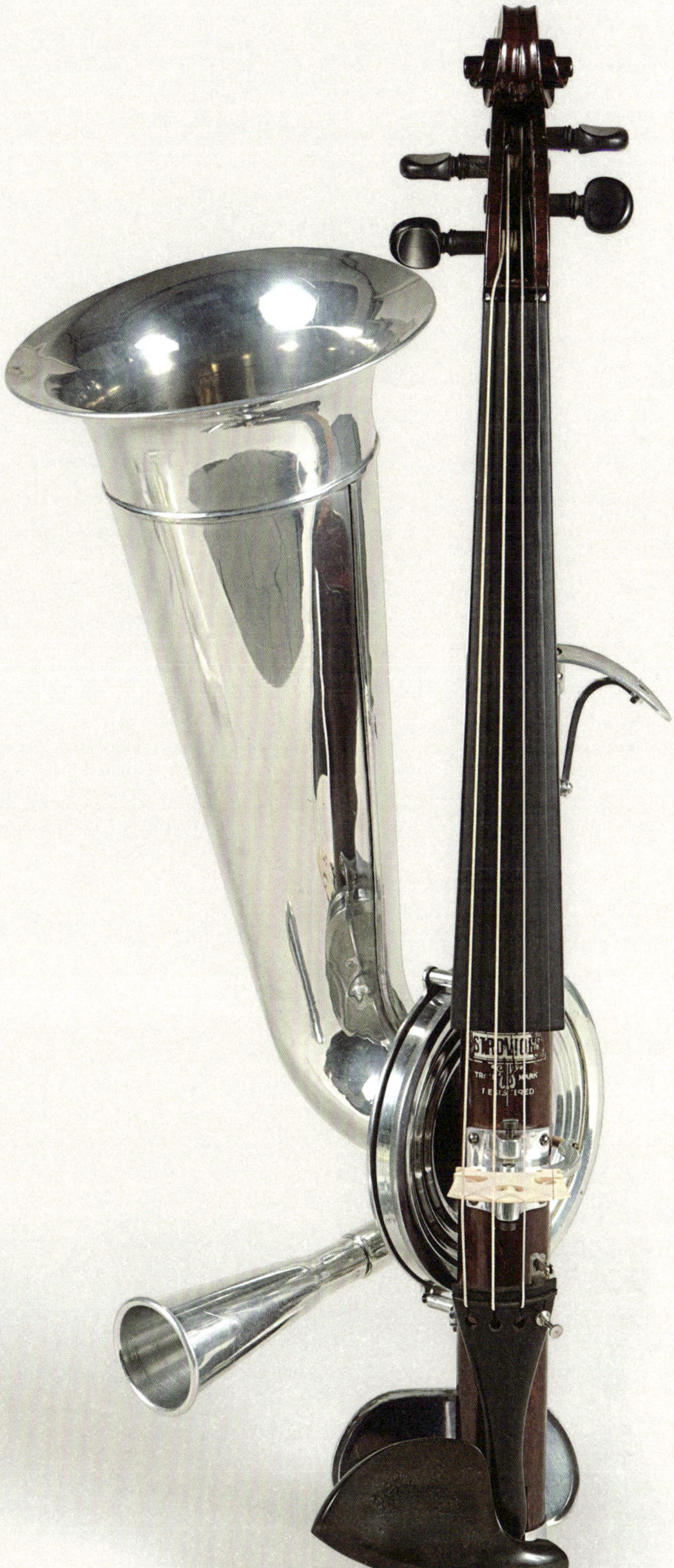

His Master's Voice

DOUBLE HORN DISC GRAMOPHONE
c. 1930

Until the invention of the phonograph by Thomas Alva Edison in 1877, and the invention of the gramophone by Emile Berliner 10 years later, the reproduction of sound was essentially the replication of a melody by mechanical means.

The phonograph and then the gramophone were able to reproduce the human voice and associate singing with music produced by musical instruments recorded on the surface of disks.

The potentially infinite increase in the number of melodies that could be reproduced simply by changing disks, which cost substantially less than rolls or cylinders, rendered all earlier music machines obsolete, particularly those for private use. An affordable music box had been invented, but it was also a new object for the home, and its trumpet horn, sometimes fitted in a pair, were decorated accordingly.

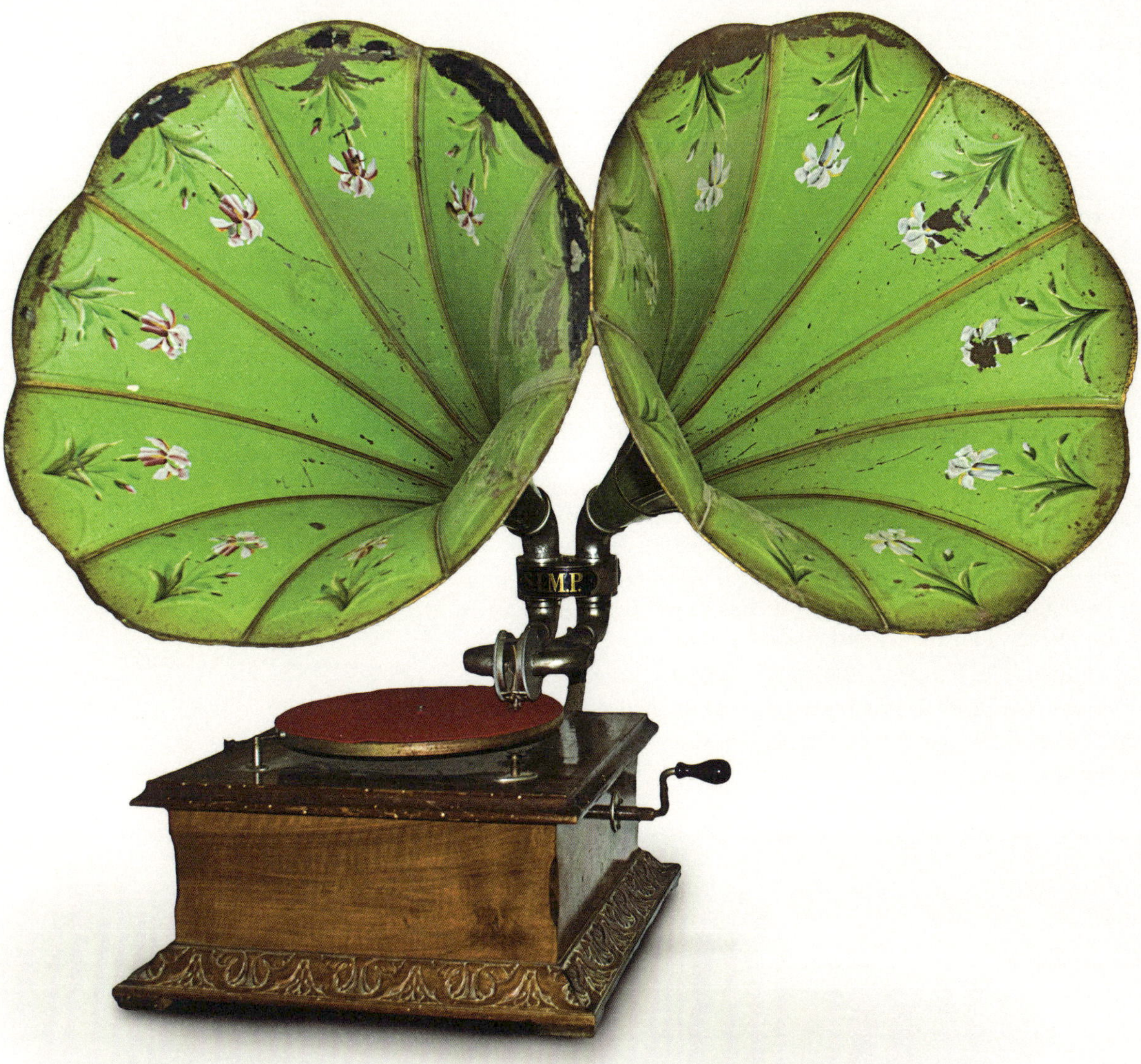

Maurice Henry

HOMMAGE À PAGANINI
1936 (1968)

A French poet, illustrator, painter, sculptor and a great music lover, Maurice Henry (Cambrai, 1907 – Milan, 1984) co-founded the group Grand Jeu in 1926, then joined André Breton's Surrealist group in 1933, also contributing to the magazine *Surréalisme au service de la révolution.*

In May 1936 he exhibited in the "Surrealist Exhibition of Objects" at the Charles Ratton gallery in Paris, which traveled to London and New York the same year. Henry showed a small violin that was tightly wrapped in a "Velpeau" bandage (named after the doctor who invented it) and placed on a grassy ground. The object was recreated by Henry in 1968 as an artist's edition.

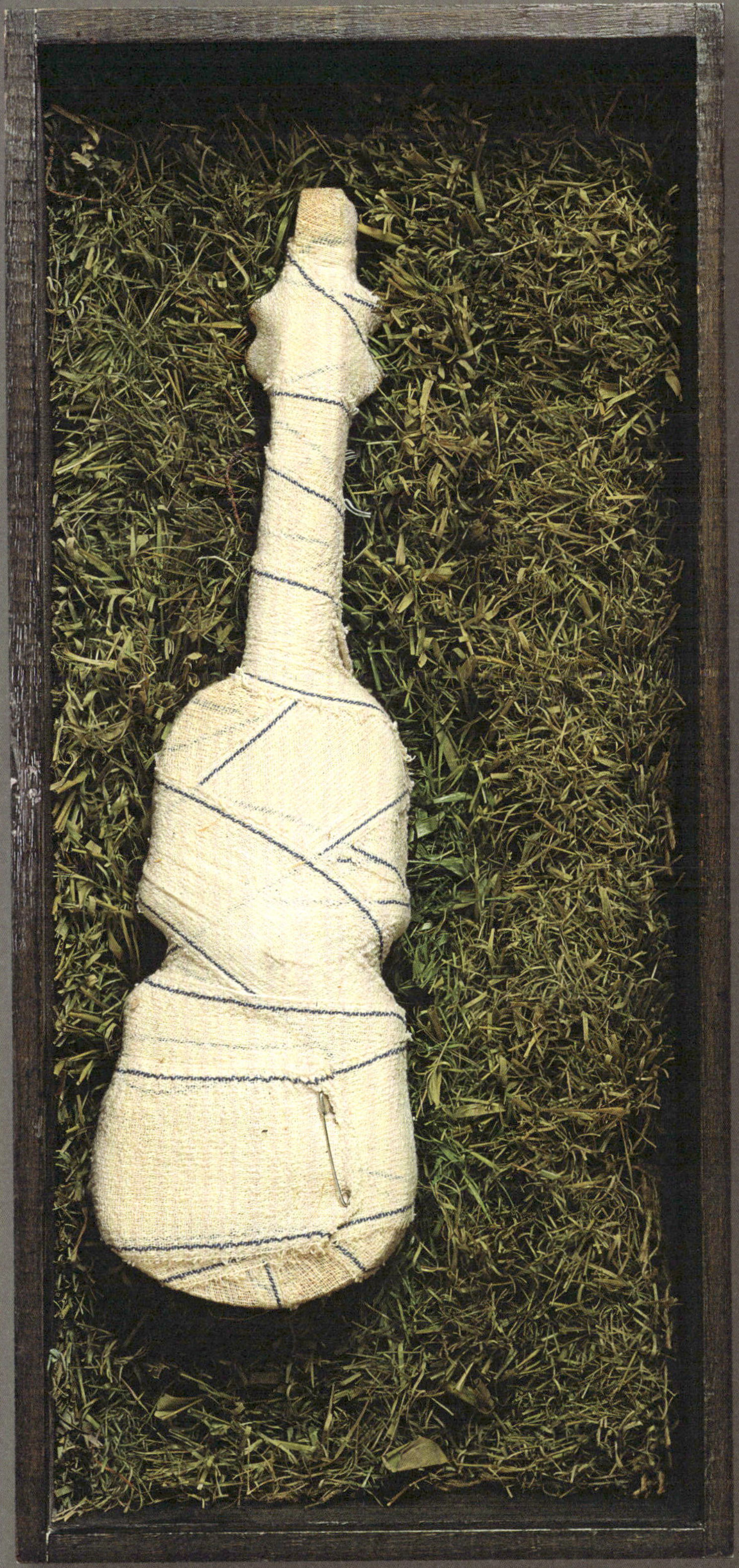

Alexander Calder

UNTITLED
c. 1940

Alexander Calder (Lawton, 1898 – New York, 1976) engaged the composition of motion in his "mobiles," kinetic sculptures made with balanced components that respond to motor power or air currents. Beginning with his first hanging "mobile" in 1932, which invites the viewer to initiate avant-garde music, the American artist went on to create an exotic breed of mobiles that incorporate sonorous elements. "To me the most important thing in composition is *disparity*," he wrote in 1943.* Indeed, the unpredictable component of sound within a kinetic sculpture yields unbounded variation.

Untitled is one of many gifts from Calder to French painter Jean Hélion over the course of their friendship, which began in Paris in the early 1930s. In this standing mobile, 3 black angular elements are arranged within a shared proximity on the same plane, colliding with one another and providing the viewer with a distinct aural and visual experience.

* Alexander Calder, "A Propos of Measuring a Mobile," manuscript (Archives of American Art, Smithsonian Institution, 1943).

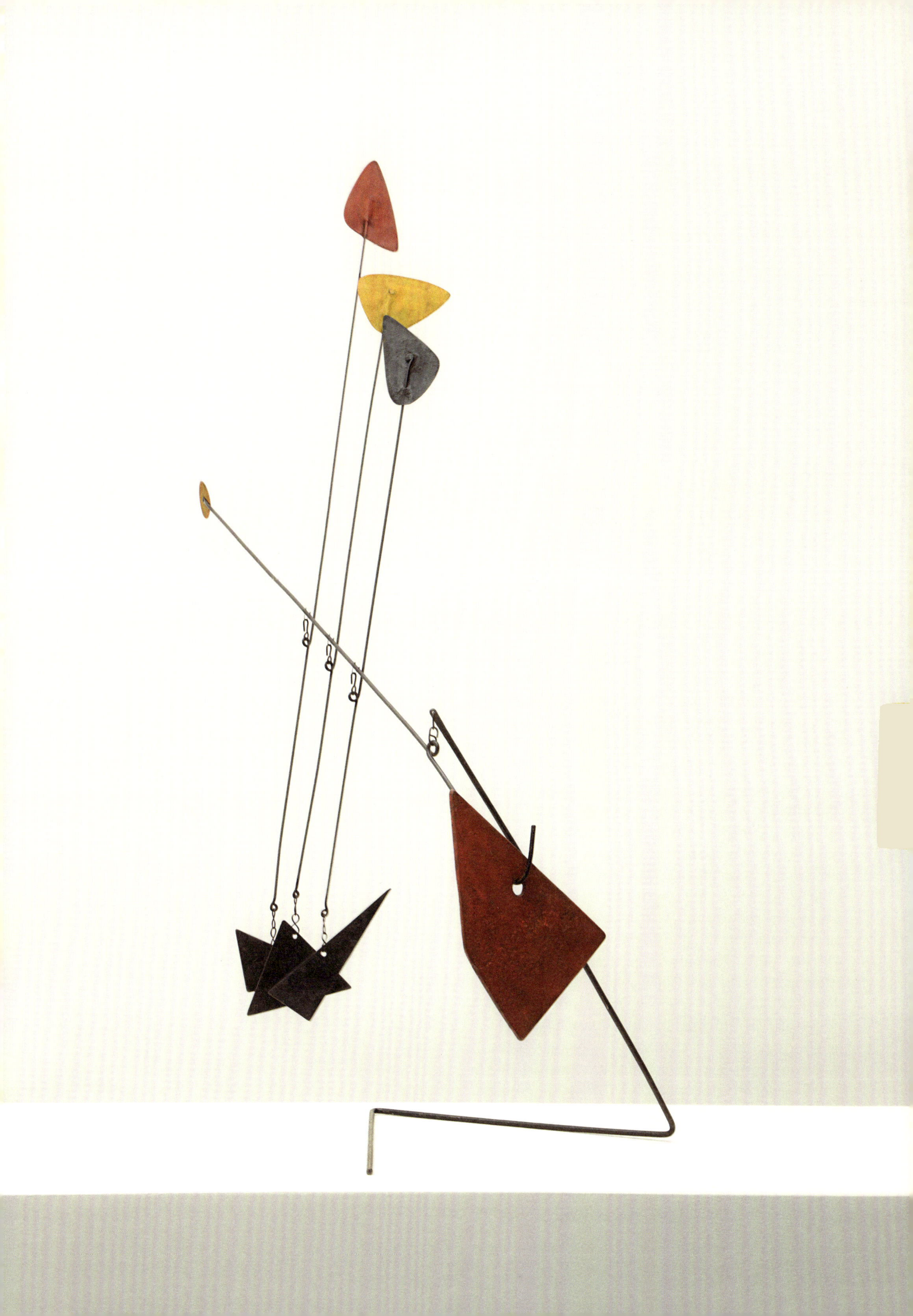

Salvador Dalí

MÉTRONOME
1944

Salvador Dalí (Figueres, 1904–89) was drawn to the metaphysical painting of Giorgio De Chirico and Carlo Carrà, and in 1928 became a member of André Breton's Surrealist group. He met Marcel Duchamp and Man Ray, whom he frequented regularly. His link with Man Ray is clear in this artwork, bringing to mind Ray's *Indestructible Object* (1923) [p. 172]. Dalí was always attracted to the theme of time, as in *The Persistence of Memory* (1931), in which mechanical time, which can be measured by watches and clocks, and is thus regularized, is overthrown by the perception of time in human memory, which renders watches "soft." The only difference from Man Ray's metronome is the eye, which is not photographed but drawn in charcoal. Eyes recur frequently in Dalí's imaginative world—painted and otherwise—as doorways to the soul. Consider the opening dreamlike scene in *Un chien andalou*, the 1929 film directed by Luis Buñuel together with Dalí, in which a woman's eye is cut open with a razor.

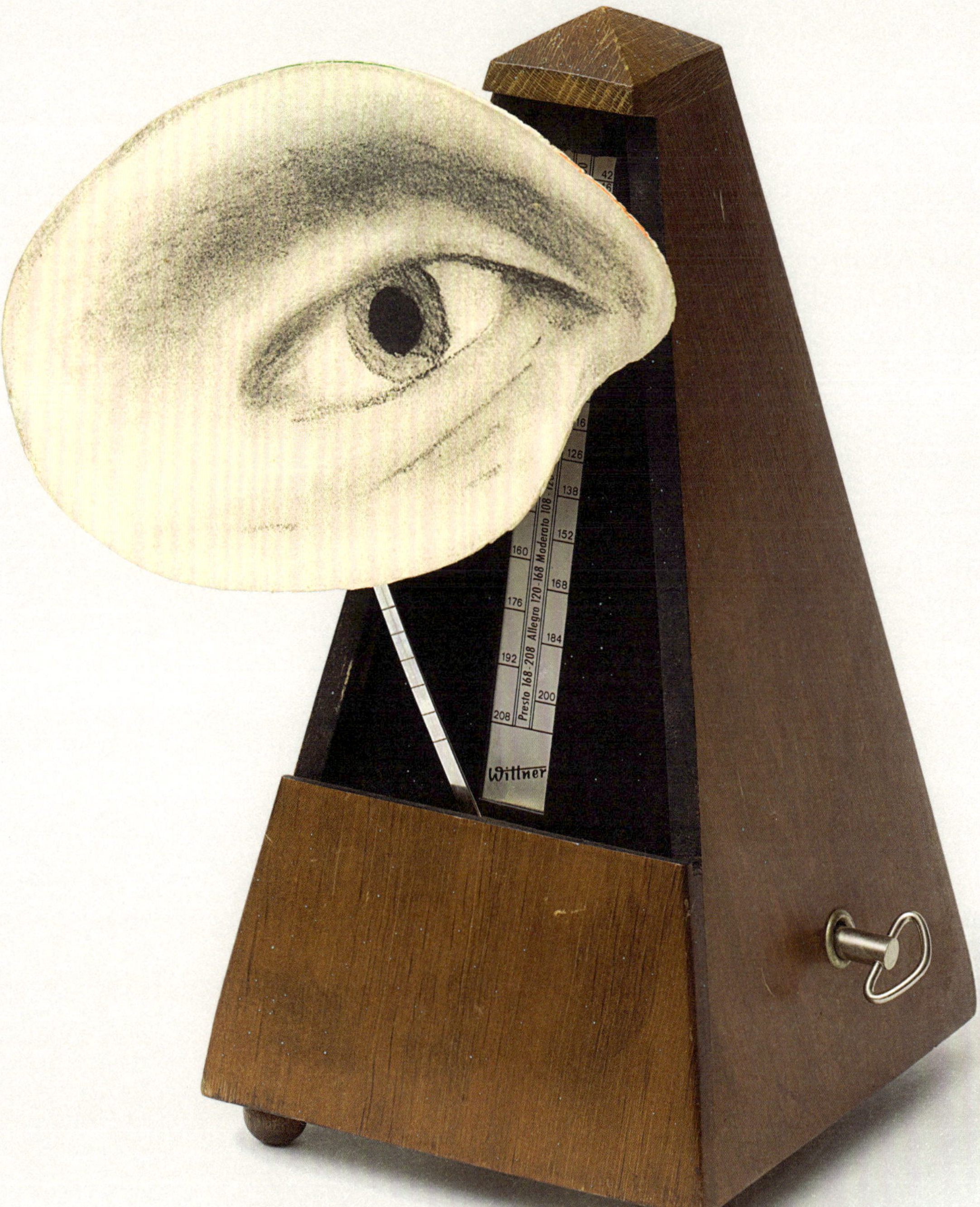
Presto 168-208 Allegro 120-168 Moderato 108
Wittner

Bernard Baschet and François Baschet

FOLDING AND INFLATABLE GUITAR 1952

1

In 1952 Bernard Baschet (Paris, 1917), who studied *musique concrète* under Pierre Schaeffer, began researching acoustics with his brother, the sculptor François (Paris, 1920–2014).

The first instrument devised by the Baschet brothers, this folding and inflatable guitar seems to have been the playful outcome of a practical problem: how to travel with a guitar. "When traveling, it is an excellent idea to get yourself invited everywhere. The best way is to learn 10 songs and carry a guitar. They say the world is vast. It's true but it's not so large if you carry a guitar, the landscape shrinks." *

The solution was simple: replace the body of a classical guitar with an inflatable plastic cushion so it could be blown up when needed. The folding inflatable guitar—renamed Gonflandoline—can be compared to a banjo in which the membrane is expanded with air inside a plastic cushion, rather than being mounted on a rigid frame.

[See also p. 192]

* François Baschet, *Les mémoires sonores* (Paris: Éditions l'Harmattan, 2007).

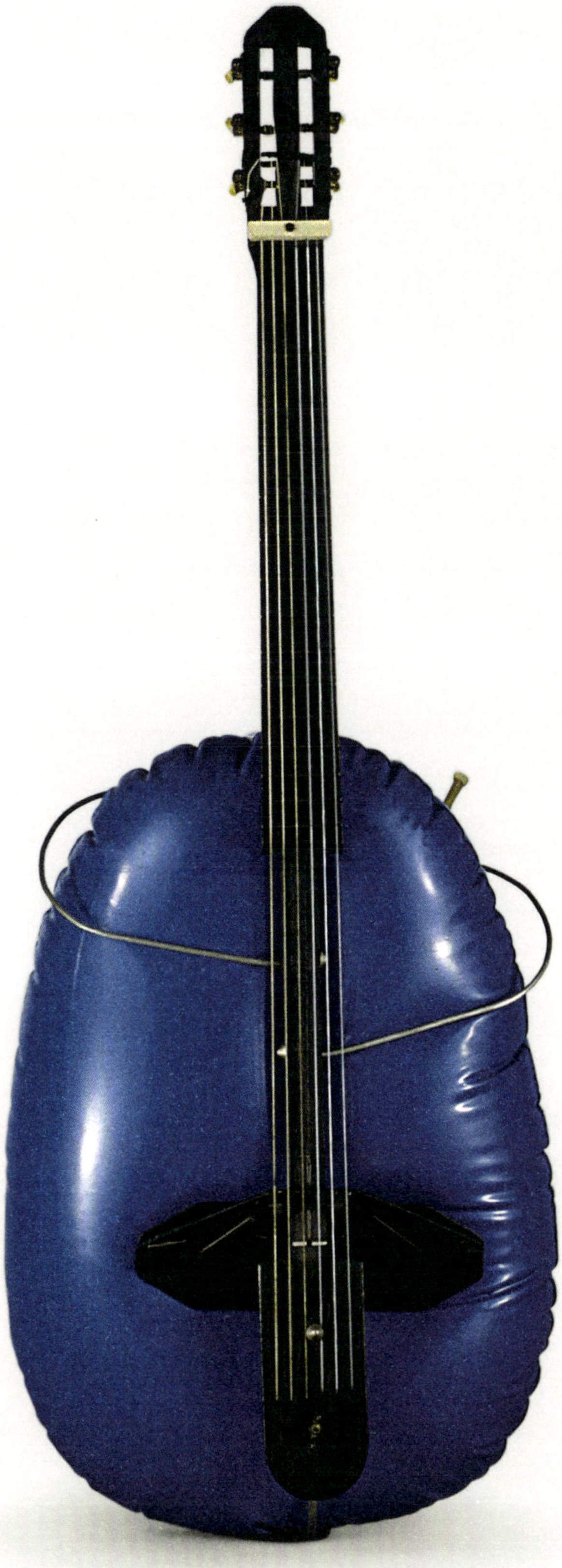

Bernard Baschet and François Baschet

CRISTAL
1952 (1980)

Bernard and François Baschet produced sound sculptures, meant for museums and art galleries, and sound structures like the Cristal, which are essentially instruments meant for musicians.

Conceived during the 1950s as an acoustic synthesizer, the Cristal was developed in different shapes and compositions by the Baschet brothers throughout their investigation into the propagation of metal vibrations. The musician plays the instrument by rubbing moistened fingers over glass cylinders set in a metal plate, producing amplified sounds similar to electroacoustic noises but without using electricity or electronics.

[See also p. 190]

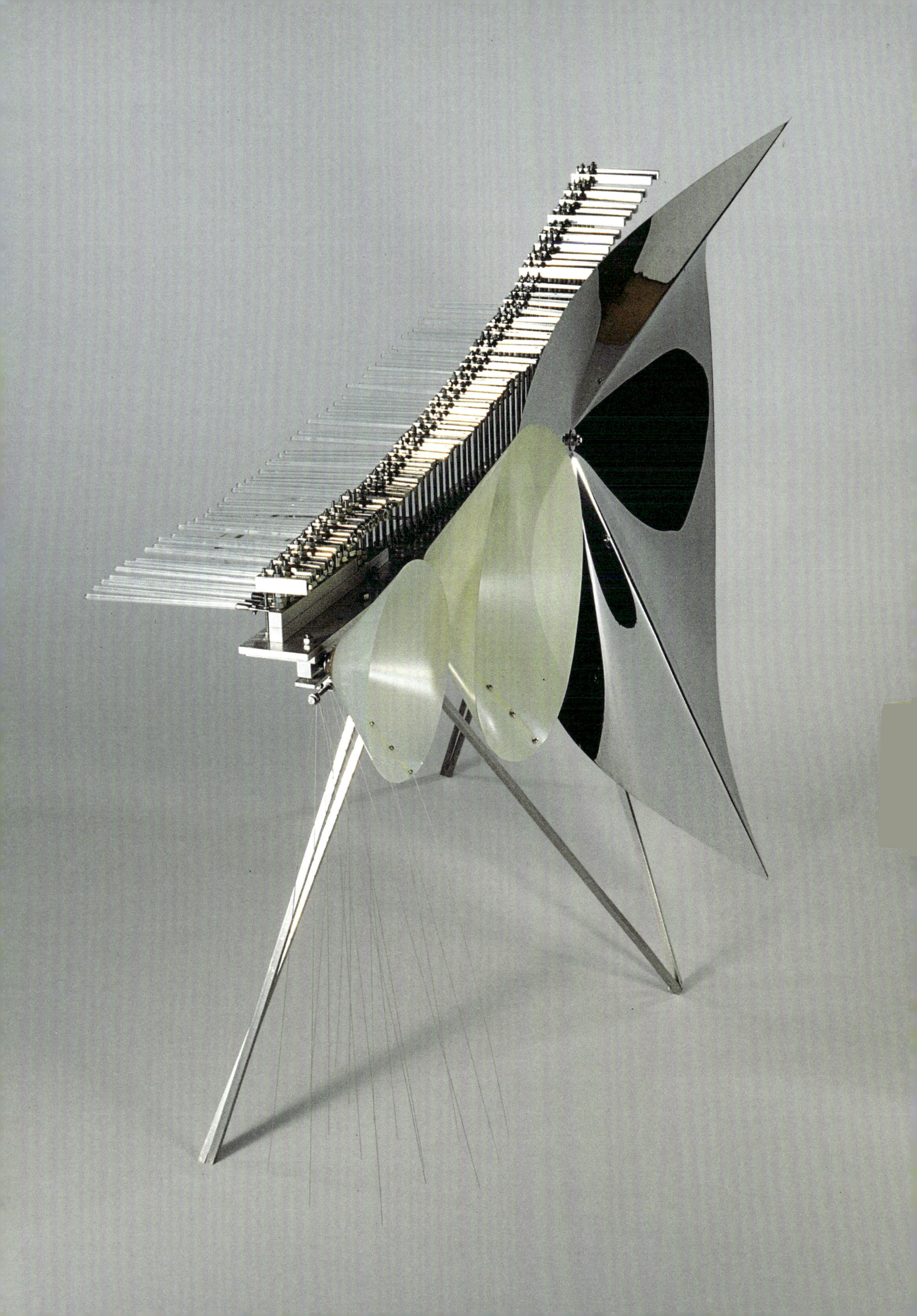

ERIC DE VISSCHER

SILENCE AS A MUSICAL SCULPTURE

JOHN CAGE AND THE INSTRUMENTS IN 4'33"

At the boundary between art and sound, where music and silence, space and time, instruments and concepts continually intermingle, the career of the composer John Cage is an obligatory point of passage. The silent piece *4'33"* occupies a central place in his oeuvre, as affirmed by the artist himself. The work was conceived and developed over a period of four years (1948–52), precisely at the core of the 20th century; it thus radiates outward over the entire period, its roots drawing on the avant-garde movements at the beginning of the century and its influence extending over the fifty or so years that have followed it. But what meaning does this silent work have today and how does it relate to instruments and technology?

Born in 1912, Cage began to develop creatively in the early 1930s, which may be considered his formative years. At the time, he made an important trip to Europe, passing through Paris and Germany, where he visited Bauhaus. After returning to Los Angeles, he quickly found a place for himself in the avant-garde, which was mainly constituted by German immigrants bent on exploring music as much as the visual arts. It was in this artistic and intellectual milieu that he met a host of personalities, such as Galka Scheyer, Pauline Schindler, Oskar Fischinger and, naturally, Arnold Schoenberg, a tutelary figure who was both feared and worshipped. Heeding the advice of Henry Cowell, composer, editor of the *New Music Quarterly* and teacher at the New School of Social Research (where Cage would teach twenty years later), Cage went to New York to spend a year studying with the pianist Adolphe Weiss before approaching Schoenberg, the grand master. While in the city, he met other pillars of modern music, such as Edgard Varèse and Virgil Thomson. Cowell, along with Schoenberg, was a structuring element in the gestation of a new way of thinking: from the former, Cage would gain an immense openness to the most experimental music of the time (the percussion works of Varèse, Chavez, Amadeo Roldán and many others, the first forays into electronic sounds...); from the latter, he would learn the need for structure and rigor in a strongly chaotic universe.[1]

From the first "serial" works to the percussion pieces, which would be followed by the so-called "invention" of the prepared piano, Cage's thinking on music would take a good dozen years to jell, from the *Sonata for Clarinet* (1933) to the various *Constructions* (*in Metal*) or *Imaginary Landscapes*, of which the performance at the New York Museum of Modern Art in February 1943 would be a defining moment. In parallel, Cage was very active in writing about his music, the most important texts being "The Future of Music: Credo" (1937) and "For More New Sounds" (1942). These constituted at once the synthesis of the creative ferment that was shaking up the United States in the wake of the Europe avant-garde and a work plan that allowed the composer to create the Varésian definition of music: organized sound. Music thus had to be opened up to all sounds, including noise, as well as to silence, considered to be an integral part of the continuum of sound. To organize 1
this, a structure is necessary, but this structure can only be based on time and not on tones, because only duration is common to both sound and silence.

Like other composers, Cage focused on percussion to achieve "more new sounds," but he did so in a way that was far removed from the well defined purview of the "percussion instrument." While Varèse had already included fire-truck sirens and industrial anvils among the instrumentation of *Ionisation* (1929–31), Cage took things further by adding automobile parts and other found objects. But percussion was only a stage in this process of opening up: "I believe that the use of noise to make music will continue and increase until we reach a music produced through the aid of electrical instruments, which will make available for musical purposes any and all sounds that can be heard. Photoelectric, film, and mechanical mediums for the synthetic production of music will be explored."[2]

In "For More New Sounds," Cage elucidated his quest for new technical means and tools for composition. Alluding to the fact that new machines were developing as much within the field of radio as in the then nascent film industry, Cage demanded that composers seize the opportunity: "Many musicians, the writer included, have dreamed of compact technological boxes, inside which all audible sounds, including noise, would be ready to come forth at the command of the composer. Such boxes are still located somewhere in the future.

At present the choice is either to wait and lament the fact that they aren't available now for experimental and musical purposes, or to continue to work with what 'axes and buckets' can be found or made."[3] The sense of frustration betrayed by this quote followed the failure of his various attempts to set up a center for experimental music. And just as his work with the Columbia Workshop of Chicago, which produced the radio piece titled *The City Wears a Slouch Hat* in 1942, turned out to be a disappointment for Cage, the means at his disposal also did not measure up to his ambitions. As for the "axes and buckets," these were the quite simple instruments he used in his first electronic pieces, such as *Imaginary Landscape No. 1* (1939). He used three variable-speed phonographs from the radio studio of the Cornish College of the Arts in Seattle, where he was teaching at the time, to play test-tone recordings. The variations in pitch were achieved by varying the speed of the turntable, while the rhythm came from lifting and lowering the stylus. In the following two works in the series (*Imaginary Landscape No. 2* and *No. 3*, both dated 1942), Cage further elaborated his "electro-acoustic" technique: in *No. 2*, he used an amplified coil of wire attached to the phonograph tone arm, while in the third, he added an oscillator, a recording of a generator, a buzzer and an amplified *marímbula*. In these three works, the new sounds are always intermixed with percussion instruments, just like another piece from the same year, *Credo in US*, which incorporated live radio broadcasts.

The need for a rhythmic articulation for these sounds and "non-sounds" became evident starting in the 1940s and would obsess Cage into 1952. It was precisely the evolution of his thinking on silence during that period that changed both his relationship with technology and his conception of form, most notably his acceptance of the element of chance in his compositions. This development is well known. It was first seen in a text dated 1948 but discovered much later, "A Composer's Confessions," in which Cage talks about writing a piece made of "uninterrupted silence," with the idea of selling it to the Muzak company:[4] "It will be 3 to 4½ minutes long—those being the standard lengths of 'canned' music—and its title will be *Silent Prayer*."[5]

It is essentially a reactionary work, fighting against the invasion of the environment by sound, borrowing the very tools of commercial music, namely, the short-duration hit. Yet Cage harbored a fixed conception of silence as the absence of sound. This idea prevailed even when he talked about the structure being able to accommodate both sound—all sounds—and "non-sounds," silences. However, his position evolved in the early 1950s under the simultaneous influence of Zen Buddhism ("every something is an echo of nothing"), Robert Rauschenberg's *White Paintings*, and the soft sounds of the music of Morton Feldman. The spatial, and thus plastic, dimension of silence became clear when Cage wrote that "Silence surrounds many of the sounds so that they exist in space unimpeded by one another and yet interpenetrating one another...."[6] The famous experience in the anechoic chamber at Harvard put another stone in the edifice, allowing Cage to state that absolute silence did not exist because our body is always producing sound. And so silence becomes the set of all non-intentional sounds and is expressed as an attitude of listening and openness, doubtlessly exaggeratedly positive, to all sounds coming from the surrounding space. This led to the premiere of the silent piece *4'33"* at the Maverick Concert Hall in Woodstock in 1952, in the middle of a piano recital by David Tudor.

This development reveals, on the one hand, a shift from the polemical view in *Silent Prayer* (the absence of sound as an alternative to noise pollution) toward a conception that is apparently more philosophical and individualistic (listening and attention to non-intentionality). On the other, it betrays a constant and growing recourse to technology to get the concept across. In themselves, the anechoic chamber and the acoustic piano, both mute and contributing to the performance of *4'33"*, give material form to something that might appear to be purely conceptual. Electrical technology, which was so characteristic of the *Imaginary Landscapes*, was used in 1951 to approach silence: in the fourth of this series of landscapes, written for 12 radios controlled by 24 operators, with no other instrumentation, the score provides a structure of durations and tunings for each of the radios. It is told that at its premiere on May 10, 1951, due to the late hour of the performance, many local radio stations were no longer broadcasting and the piece was a long string of silence and static. The radios were detectors of unheard or inaudible sounds, a function that Cage gave to all the technology that he would use from them on.

As has been aptly shown by Douglas Kahn,[7] one of the logical consequences of *4'33"* is that technology no longer serves exclusively to produce new sounds. It is there to give sound to silence—or more exactly, to the unheard, that which we cannot hear with our ears but which is always there. The microphone and loudspeaker were the indispensable tools in this plan, spreading amplified sounds (or silences) live through space. Starting in 1952, the successive versions of the silent piece always involved the dimension of amplification, strongly influenced by the liberatory ideas of Marshall McLuhan on decentralization, de-hierarchization, and anarchism brought about by technology. This process of the swelling and bursting of audible matter was also heard in other pieces from the same period, particularly the series of *Variations* [214], *Cartridge Music* or those using fixed supports such as recording tape (*Williams Mix*, *Rozart Mix*) or disks (*33 1/3* for a minimum of 12 turntables and a large number of records, operated by the audience).

But the series of silent pieces offers particularly interesting insights into the Cagean relationship with technology and its political and social implications. Because ten years later, in 1962, Cage returned to that moment of illumination by proposing *0'00" (4'33" No. 2)*, a radically different piece, a "solo to be performed in any way by anyone." The score consisted simply of a main phrase and some qualifications: "In a situation with maximum amplification, perform a disciplined action." We thus go from a shared order, that of the score that tells the musicians what they must play—or not play—, to much more explicit and directed performance. And Cage kept up the momentum with *0'00" No. 2*, which appeared in his *Song Books* (1970): "On a playing area (e.g., table, chessboard) equipped with contact microphones (four channels preferably, speakers around the audience, highest volume without feedback)—play a game with another person (e.g., chess, dominoes) or others (e.g., Scrabble, bridge)." Again there was the maximum amplification, but this time in the form of a social action (implying a plurality of participants), one that is musical because amplified, and necessarily gamelike.

Because if music is not a game, we still say that we play a musical instrument and that certain rules of the game can certainly be used in composing music—as Marcel Duchamp aptly demonstrated. For that matter, one of the most famous performances of *0'00" No. 2* took place in Toronto in 1968—prior to the publication of the score—with the title *Reunion*: over the course of the evening, Marcel and Teeny Duchamp, together with Cage, played chess in public on a chessboard made by Lowell Cross that transformed the movement of the chess pieces into sounds transmitted through loudspeakers and filling the space of the auditorium.

A little more than twenty years later, in 1989, Cage performed a last silent piece, *One*^{3} = *4'33" (0'00")* + 𝄞. The references in this title are as numerous as they are arcane: *One*^{3} is one of the "Number Pieces," late works where the base indicates the number of musicians required to perform the piece and the exponent refers to the order of the works with the same number. The title cites the two main silent works, followed by a G-clef, making reference to a remark made to Cage by the Russian composer Sofia Gubaidulina, who was not fond of measured time in a musical work. The title thus refers to three different types of time: chronometric time, zero time that negates in some way the concept of duration and flow, and lastly a musical tempo taking form in the movement of sounds.... The only performance of the piece took place in Tokyo. Cage sat in the hall while microphones on the stage captured and amplified the sounds of the hall to a level just short of feedback. After an amount of time not specified in the score (it seems that more than 12 minutes passed), Cage got up, went onto the stage and stopped the amplification.[8]

What should we think of this evolution of the silent work over nearly 40 years? A number of elements are shared by the different versions: the acceptance of the unwanted sounds, a form of theatricality that transgresses the framework of a "musical" performance to spill over into other social situations (such as games), and lastly the recourse to a technical element (the musical instrument or the electroacoustic device) conceived as a sort of prosthesis to the body of the performer.

This technology, taken in the broad sense, is necessary, according to Cage, to distinguish the performance of the silent work within a well-defined situation (a concert, for example) from the simple possibility of listening to the silence at any given moment once one is in a situation of auditory attention and openness to all that may come. For Cage, the silent work exists at all times: "… in nature, and at every moment. 'Distinct' means that there is amplification. It's a work on a work—like all my indeterminate works! … That's how I act, with the aid of technology."[9] Technology reveals sounds that we could not perceive without it, but it also contributes to the staging of the work, creating an event that is different from our daily experience. In some way, technology generates the work, it does not supplant it. This was the conclusion Cage reached at the end of his life, when he realized that there was a real potential danger from the improper use of the technology: "it's now a technological silence … now silence includes technology. In a way that is not necessarily … good," he said in an interview with Joan Retallack.[10]

Going from *4'33"* to *One*^{3} thus meant for Cage that the silence of 1990s is not the same as that of 1952, because our environment has necessarily been modified, especially under the influence of the increasing presence of technological tools. If we can no longer, as in "Silent Prayer," impute the omnipresence of sound in our environment to the Muzak company, it is technology itself, once seen as bringing freedom, that seems to pose the problem by being invasive or oppressive.

A handwritten note on the score of *Sculptures Musicales*, a work Cage composed in 1989 based on a note by Marcel Duchamp for his piece *The Green Box* ("Musical Sculpture: Sounds lasting and leaving from different points and forming a sounding sculpture which lasts," 1934), seems to confirm this diagnosis. Composed for a dance piece with Merce Cunningham, the score, presented exclusively in text form, dictates the production of a series of envelopes composed of a minimum of three sounds from different spatial sources, separated by silences that are open to external sounds, that is, those of the dancers…. The term "silence" is written in quotation marks, indicating the ambiguity with which Cage viewed the concept (is it a "non-sound"? Or a sound filled with noise that he cannot control?), whereas one of the printed scores includes an annotation in the composer's own hand: "the dark side of silence."

This hidden side of silence betrays a certain disenchantment on Cage's part, referring to a more pessimistic vision of silence, that of the poet Edmond Jabès, who associates it with "*listening to a sinking ship*." Perhaps he too came to the idea that the anechoic chamber, which opened so many possibilities in revealing the permanence of sound, may also signify the loss of senses, the constraint of silence, and thus the opposite of freedom.

Cage already sensed this form of resignation in 1980 when he created a mesostic based on the word SILENCE, admitting the ineluctability of the concept, the need to convert our spirit to a form of anarchy and "chaos."[11]

While the term "anarchy" is found throughout Cage's political thinking since 1960, "chaos" appears less frequently. Cage refers to it in speaking of the mathematical theory by the same name, which seeks to describe the complexity of certain situations, mixing determinism and incertitude, which likens "chaos" to "chance," a manner of composition used by Cage since 1951. But we can also think of the chaos in Greek mythology, that state of violent confusion out of which the world was born. In these two cases, we are faced with a perception of complexity, not free of danger but heralding a future. And it is the same complexity of our relation to the world that characterizes Cage's view of silence. The acknowledgement of a hidden side of silence does not negate the possibility of a positive anarchy, it enriches it. The potential danger of omnipresent technology does not necessarily keep it from playing an extremely liberating role in how we experience the world. And it is this ambivalence, this acceptance of complexity, that makes Cage's thinking still relevant today. Anyone can listen to silence today, and they can also perform *4'33"* via an iPhone app and communicate that experience to others by means of this new technology.[12] Anyone can cut themselves off from the world around them by clapping their headphones over their ears, but they can also create their own "imaginary landscape" mixing recorded music and external sounds.

Thinking of silence today means constructing our relationship with the world based on hearing, and distinguishing it from other ways of apprehending reality. Silence may also offer us the possibility to think of the future, using our ears as receivers that can be open or closed, as wrote the Japanese novelist Yôko Ogawa: "If you want to know the future, you have to plug your ears. Then you will hear your story far into the future."[13]

1
Regarding his apprenticeship years, see Thomas S. Hines, "'Then Not Yet Cage': The Los Angeles Years, 1912–1938," in Marjorie Perloff, Charles Junkerman (eds.), *John Cage. Composed in America* (Chicago, IL: University of Chicago Press, 1994).

2
John Cage, "The Future of Music: Credo," in Id., *Silence* (Middleton, CO: Wesleyan University Press, 1961), pp. 3–4.

3
John Cage, "For More New Sounds," in Richard Kostelanetz (ed.), *John Cage* (New York: Praeger, 1970), p. 65.

4
The term "Muzak," inspired by the success of the name Kodak, was invented by George Owen Squier, an engineer who sold and broadcast music over electric power lines during the 1920s. Squier's fee system was quickly overtaken by radio, and the inventor turned his attention to broadcasting music in large public areas, department stores and businesses. Based on the concept of "Stimulus Progression," the programs he offered through Muzak aimed at conditioning the listener (workers or consumers), and broadcast different programs – sometimes stimulating, other times relaxing – depending on the time of day. Although it has been strongly criticized for its manipulative aspects, the concept of Muzak has nevertheless attracted the interest of musicians like John Cage and Brian Eno. Erik Satie's "Musique d'Ameublement," (Furniture music) can be considered an involuntary precursor of this concept.

5
"A Composer's Confessions" was published in 1991–92 simultaneously in the magazines *Musicworks* (no. 52, Toronto) and *MusikTexte* (nos. 40–41, Cologne) at my initiative and that of Gisela Gronemeyer, editor of *MusikTexte*, and later in Richard Kostelanetz (ed.), *John Cage Writer* (New York: Limelight Editions, 1993), p. 65.

6
John Cage, "Julliard Lecture," in Id., *A Year From Monday* (Middleton, CO: Wesleyan University Press, 1967), p. 100.

7
Douglas Kahn, *Noise Water Meat. A History of Sound in the Arts* (Cambridge, MA: The MIT Press, 1999), p. 192.

8
Regarding the different versions of the silent piece, including the versions for television and cinema (*WGBH-TV, One*[11],...), as well as its contemporary transpositions, see Dieter Daniels, Inke Arns (eds.), *Sounds Like Silence. John Cage. 4'33". Silence Today*, exhibition catalogue (Leipzig: Spector Books, 2012).

9
John Cage (in conversation with Daniel Charles), *For the Birds* (London: Marion Boyars, 1981), p. 210.

10
John Cage (in conversation with Joan Retallack), *Musicage. Cage Muses on Words, Art and Music* (Middleton, CO: Wesleyan University Press, 1996), p. 77.

11
John Cage, *X—Writings '79–'82* (Middleton, CO: Wesleyan University Press, 1983), p. 117.

12
The *4'33"* app is available for iPhone and iPad; see www.johncage.org

13
Yôko Ogawa, *Chinmoku Hakubutsukan* (Tokyo: Chikuma Shobo, 2000), trans. from the French edition *Le Musée du Silence* (Arles: Actes Sud, 2003).

GEETA DAYAL

EXPERIMENTAL MUSIC AND PERFORMANCE

DAVID TUDOR, JOHN CAGE AND MERCE CUNNINGHAM

Today, finding and listening to music seems easier than ever before. Right now, you're just a few clicks away from rapidly downloading or streaming a musician's entire discography. The difficult part seems to be sorting through it all—contending with the massive digital deluge. The easy availability of rare artifacts—say, obscure recordings of live electronic music, free jazz outtakes and B-sides—also makes it harder to contend with genres of music that may have never been suited to recording in the first place. One problem is that a lot of older experimental and avant-garde music was first presented with a strong visual element, something we tend to forget now. Sometimes dance was involved, or theater or performance art. Perhaps there was a film flickering behind the performers, or the music itself was a score that was tightly married to film. Perhaps there were extravagant costumes, or sculptures, or paintings. One thing is for certain—the music wasn't originally intended to be presented as a long list of MP3s. YouTube gets a little closer, when the footage exists, but there's still a lot missing from the experience.

In the fascinating book *Records Ruin the Landscape* (2014), the musician and professor David Grubbs discusses how many emerging forms of avant-garde and experimental music in the 1960s weren't well-suited to being preserved and listened to in recorded form at all. Grubbs understandably focuses a large portion of his book on John Cage, using Cage's fraught relationship with recordings as a lens to discuss free improvisation, minimalism, varied Fluxus happenings and performances, live electronic music, sound installation art and many other genres that were budding at the time. "Cage's complex relationship to recorded sound became a touchstone in this decade, and it's an attitude that one can see refracted through each of those genres," Grubbs writes. "Fluxus artists created text scores that demanded to be experienced in performance. Minimalists made works that could last for hours or, in La Monte Young's case, weeks or months or years. Live electronic music and free improvisation defined themselves against the portability and repeatability of recorded music. Visual artists who worked with sound did so in performances and installations that emphasized the sculptural qualities of sound reproduction."[1]

I experienced the strong sensation that I was missing out on a broader picture some years ago, while trying to contend with Cage's voluminous output by listening to CDs. While I was struck by much of what I heard on the recordings and read in books, it wasn't until I saw an extensive series of live performances of Cage's compositions that I started to develop a deep personal appreciation for his work. In a small room, with the right musicians, Cage's work became three-dimensional to me; in some cases, it became so vivid as to be perhaps seven-dimensional or nine-dimensional. I was deeply moved by the performances; the feelings I felt for Cage's work were profoundly visceral and emotional, not simply intellectual and theoretical. These were experiences that punched me in the gut. I still have strong memories of running out of a room holding my ears during a performance of Cage's *Cartridge Music* (1960); it was such an intense and deafening experience that my earplugs did nothing to stem the rising tide of overpowering feedback. After reorienting my ears, I ran back into the room, eager to continue listening. In addition to being a tremendous sonic experience, it was a fantastic, understated work of theater, with the performers straight-facedly wielding Slinky toys and other implements, along with the amplified cartridges.

This rich and fascinating visual element was also what was missing for me with Harry Partch; much of Partch's great work predated the 1960s. When you listen to Partch recordings, you can't help but feel that you're missing out on something big. An understanding of Partch's music goes deeper when you're able to see his whimsical one-of-a-kind musical instruments, such the Chromelodeon, built in 1945. Partch also composed much of his greatest work for the stage, another piece of the puzzle that gets lost in translation while listening to audio recordings. *Delusion of the Fury*, one of Partch's greatest works, received its onstage debut in 1969 in Los Angeles. It incorporated mimes and dancers, West African myths and folklore, and Japanese Noh drama. According to the 1971 liner notes for the CBS recording, the costuming included "a huge pair of pantaloons," "a poncho-like garment," "imaginative make-up," and "a fantastic headpiece" for each musician.[2]

The liner notes also make reference to the instruments as sculptures, as works of art that were meant to be experienced as part of the *Gestalt* of the whole production. Even the names of the instruments used in *Delusion of the Fury* were suffused with potent mystery: the Chromelodeon, with its colorful painted keys, signifying a 43-tone scale system; the Diamond Marimba, the Cloud Chamber Bowls. "The Partch instruments are not only performed upon; they, too, perform," Eugene Paul writes in the CBS liner notes. "Besides being co-conveyors of his music, along with singers and dancers, they are also striking sculptural works employed in the stage design, not only in 'Delusion,' but also in all of Partch's performed compositions. They are the vehicles of his creativity, around which Partch has designed his own notation system."

One of the first pieces of avant-garde electronic music I ever listened to, *Poème électronique* by the great composer Edgard Varèse, was from a CD. Played on a decent stereo, it sounded beautiful—all 8 minutes of it—but it wasn't the tremendous multimedia experience I had read about in books. *Poème électronique* was exhibited at the World's Fair in Brussels in 1958, in the Philips Pavilion. The dramatic sonic experience—spread over many, many speakers—was meshed with innovative architecture, film, lights and projections. Films by Le Corbusier were projected while the music played, in the unique architecture of the pavilion that Le Corbusier was contracted to design. (Le Corbusier was traveling at the time—most likely in the Indian city of Chandigarh—so the Greek composer Iannis Xenakis, who was then working for Le Corbusier, oversaw the project.) A composition titled *Concret PH* by Xenakis also played intermittently in the pavilion. Xenakis was a polymath—a mathematician, engineer, architect, composer and artist who embodied the cross-pollination between visual art and electronic music. His numerous architectural drawings, graphic scores and schematics were works of art unto themselves.

Cartridge Music (1960) was one of the rare Cage works that did actually translate reasonably well to a recording, as Grubbs notes in his book. Seeing *Cartridge Music* performed live took some of the abundant mystery of the recording away—I could link the objects I saw to the sounds that I was hearing—but it also offered a rich layer of visual information.

Cage's partner Merce Cunningham took *Cartridge Music* a step further, using it to soundtrack a dance piece he choreographed in 1973 entitled *Changing Steps*. Charles Atlas designed the vibrant costumes for the dancers. The Walker Art Center in Minneapolis, which has the costumes in its collection, lists that Atlas's costumes included "cotton jumpsuits with front zippers (some containing front pockets) in a variety of colors including cream, green, blue, and brown."[3] The Walker Art Center also holds many of the colorful costumes that Robert Rauschenberg created for Merce Cunningham's dance company, beginning in the 1960s.

Rauschenberg's alliance with Cage and Cunningham was longstanding. This alliance also extended to others in Cage's circle, including David Tudor. Tudor's collaborations with Cunningham and Rauschenberg in the 1960s showed how profoundly interdisciplinary the budding genre of live electronic music could be. Many of Tudor's works—most famously *Rainforest* (1968)—were striking examples of how sculpture and dance could be seamlessly incorporated into trailblazing electronic music.

Tudor was best known earlier on in his career as a virtuoso pianist, aligned strongly and most famously with Cage. Cage often said he wrote his works with Tudor in mind.

Tudor studied piano with Stefan and Irma Wolpe from 1944 to 1950 before breaking out on his own. His association with Cage, beginning in the 1950s, led to a long and legendary collaboration and lifelong friendship. Many Cage compositions were not just written with Tudor in mind; they were enabled by Tudor's incredible skill and his unique intuition for interpreting complex scores.

In the 1960s, Tudor reinvented himself, shifting from acting as being the key interpreter of other people's compositions to being a composer of his own live electronic music. Tudor created soundscapes and sound installations, working for many years with the Merce Cunningham Dance Company, and with painters including Robert Rauschenberg and Jasper Johns.

A curious performance piece by Rauschenberg, unveiled in Paris in 1961, is interesting to think about with regards to Tudor and his varied interests. The piece was titled *Homage to David Tudor*. Rauschenberg enlisted the talents of Tudor, Jasper Johns, Jean Tinguely, and Niki de Saint Phalle. Rauschenberg painted in front of the audience (the work he made was entitled *First Time Painting*), but with the back of the canvas facing the audience so that they couldn't watch what he was doing.

The art historian Jeffrey Saletnik writes that "Tudor performed Cage's composition *Variations II* (1961); Johns crafted a target out of flowers; de Saint Phalle included one of her tir, or 'picture-shoot,' paintings; and Tinguely contributed a motorized sculpture."[4] Rauschenberg stopped painting when an alarm clock went off; the painting was summarily taken off the stage.

In 1964, Rauschenberg and Tudor performed at an event at the Moderna Museet in Stockholm, as part of a series called *5 New York Nights*. The piece of music Tudor created for the night was spontaneous and site-specific; it had the curious—and literal—title *Fluorescent Sound*, because it was based on a system of fluorescent lights. (An exhaustive recent box set of Tudor's electronic music on New World Records doesn't include *Fluorescent Sound*, which is considered to be Tudor's first composition. It appears that the piece was never recorded.) Rauschenberg's contribution was called *Elgin Tie*; according to photographs and eyewitness accounts, he was slowly lowered from the ceiling into a barrel of water, and held his breath as long as he could, and was then lifted out. A cow was also involved.

It was tough to compete with Rauschenberg's plans—the scene was certainly dramatic. "I walked around the museum and thought 'what am I going to do?'" Tudor said in an interview in 1988 with the journalist and critic Teddy Hultberg. "I noticed that there were—it must have been—a thousand fluorescent light bulbs. One day I was in the room when someone was turning on the fluorescent lights and they didn't know which to turn on and all of a sudden there was the most beautiful music. "I thought, ok, I put some contact microphones up there from the bulbs to see if the sound can be made really audible.... So then I had to get up on a large ladder and place the contact microphones which actually took nearly three days. Fortunately I had help to do it because there were, I believe, 75 switches with three light bulbs on each switch. It was a big job. It worked very beautifully. The only problem was that Bob Rauschenberg was to come in from the side entrance of the museum riding a Brahma cow, which he did, and they asked me to stop the music. (laughs) That was my first composition."[5]

Judging from the scant documentation that exists, the night sounds like a marvel—a merger of performance art, lights, entertainment and uniquely strange and interesting music. And a cow.

A bit more documentation exists of *Bandoneon! (A Combine)*, a work Tudor made in 1966, which was shown as part of the legendary *9 Evenings* series at the Armory in New York. The word "combine," of course, was a reference to Rauschenberg and his series of combines. Tudor's use of the term is interesting, and indicative of his connection to Rauschenberg—at the very least, it points to a mutual admiration and understanding.

Rainforest, which premiered in 1968, was Tudor's most famous work. *Rainforest* had several different iterations over the years with different numbers, but in general, the piece used a system of transducers attached to objects to create sounds. Despite the abundance of electronic circuitry, *Rainforest* never sounded overtly "electronic," regardless of the objects that were used. *Rainforest* was oddly natural—it sounded almost like birds chirping in a lush, verdant environment.

In Tudor's vision, each speaker or resonating object could be thought of as an individual instrument, with its own personality. "I felt for many years that loudspeakers are an obscenity, or at any rate, it's a false notion—they don't exist," Tudor said in 1972. "Because an attempt was made to create something to reproduce sound. To me, a loudspeaker is a vibrating membrane, and is capable also of creating sound, by its vibration. And so I was dreaming one day, and I thought how nice it would be to have a whole forest, or orchestra of loudspeakers, each one with a quite different sound. Which means that they should have the sound that they have, rather than reproducing some other sound which is in somebody's mind ... a loudspeaker is an instrument."[6]

Tudor's electronic music stood on its own, but on stage it looked like magic. The set design included the work of Andy Warhol, who provided his luminous, minimalist *Silver Clouds* (1966) as a backdrop. Cunningham's choreography was beautifully spare and concentrated, and the ripped, nude-colored costumes worn by the dancers were striking—it turned out that Jasper Johns had designed them.

Johns explained the situation many years later. "I had asked Andy [Warhol] to design costumes to go with the pillows, but his only suggestion was for the dancers to be nude, an idea that had no appeal for Merce," explained Johns to Alastair Macaulay in a 2013 interview in the *New York Times*. "Merce showed me an old pair of his tights that were ripped and torn. I imitated these."[7]

Rainforest—by way of Cunningham and his dancers, Johns, Warhol and Tudor's pioneering score—had become something much larger than any of its individual parts. Archival footage of the 1968 performance of *Rainforest* is transfixing to watch; it is a living, breathing work of art.

Tudor's further iterations of *Rainforest* in the 1970s were more like sculptural, interactive sound installations, without the dance component. It was a powerful experience, but in a different way.

"In 1973 I made *Rainforest IV* where the objects that the sounds are sent through are very large so that they have their own presence in space," Tudor explained to Hultberg in 1988. "I mean, they actually sound locally in the space where they are hanging as well as being supplemented by a loudspeaker system ... it makes, I thought, quite a harmonious and beautiful atmosphere, because wherever you move in the room, you have reminiscences of something you have heard at some other point in the space .... It's important that each person makes their own sculpture, decides how to program it, and performs it themselves."[8]

I thought about this concept of making my own sculpture as I put on a recording of *Rainforest IV*, and left the vinyl record spinning while I did other work. Played on a standard stereo, I don't get to fully experience Tudor's dream of an "orchestra of loudspeakers," nor do I get to see a poetic Merce Cunningham dance production, *circa* 1968. There were no Andy Warhol silver clouds, nor artfully ripped Jasper Johns costumes, nor sculptural sound installations. But I could see all of those things in my imagination, if I tried, in a dream of sound.

1
David Grubbs, *Records Ruin the Landscape: John Cage, The Sixties and Sound Recording* (Durham, NC: Duke University Press, 2014).

2
Liner notes by Paul Eugene, accompanying Harry Partch's *Delusion of the Fury* (New York: CBS recording, 1971).

3
Costumes for *Changing Steps*, Walker Art Center collections, Minneapolis, walkerart.org/collections/artworks/costumes-for-changing-steps

4
Jeffrey Saletnik, "Trophy IV (for John Cage)," Rauschenberg Research Project, July 2013. San Francisco Museum of Modern Art, www.sfmoma.org/explore/collection/artwork/25848/essay/trophy_iv_for_john_cage

5
Teddy Hultberg, "I smile when the sound is singing through the space," interview with David Tudor, Dusseldorf, May 17–18, 1988; davidtudor.org/Articles/hultberg.html

6
David Tudor, "From Piano to Electronics," *Music and Musicians*, no. 20, August 1972.

7
Alastair Macaulay, "Cunningham and Johns: Rare Glimpses Into a Collaboration," *The New York Times*, January 7, 2013.

8
Teddy Hultberg, interview with David Tudor, cit.

JO APPLIN

OPTICAL NOISE

THE SOUND OF SCULPTURE IN THE 1960s

During the 1960s, Modernism's privileging of the visual over other sensory responses to the work of art, an assumption that has held sway since the 19th century, came under numerous assaults. While influential modernist critics such as Clement Greenberg and Michael Fried may have insisted upon the purely optical aspect of Modernist painting—its unique address to "eyesight alone"—as its defining critical feature, at the hands of a number of painters, sculptors, musicians, dancers, filmmakers and writers, other kinds of environmental, external, lived experiences and encounters began to seep back into the production and reception of art. A number of discordant notes and excessive, noisy and at times violent bodily interventions marked the art world of the 1960s. For example, Dick Higgins's *Danger Music #17* (1962) instructed performers to "Scream! Scream! Scream!"[1] while Bruce Nauman made a cassette recording of a woman screaming which he subsequently silenced by encasing it inside a solid block of concrete for *Concrete Tape Recorder Piece* (1968) [248]. The only evidence that Nauman's work contained sound was the redundant electrical flex protruding from the side. Nauman's work foregrounded the role of media, the obdurately material supports for recorded sound, but also highlighted their limits and failures. Other artists associated with the international Fluxus, Zero and Nouveau Réaliste movements enacted various attacks on postwar commodity fetishism by destroying everyday and familiar objects, and often musical instruments too, assaulting directly the elite values embodied in the orchestral instrument. For some artists, music became a contemporary motif through which to think about the historical context within which they were working, while others sought not to destroy but to remake the very idea of sound art to radically far-reaching ends: John Cage's infamous "silent" orchestral arrangement *4'33"* (1952) in which the only sounds were those ambient ones produced inside the concert hall such as coughing, page turning, shuffling and breathing, was a signal work that underscored many artists' engagement with sound in the 1960s.

There is something definitive about the act of destroying, attacking, silencing or breaking a musical instrument. The sharp disjuncture between the sound of an instrument being played, and the sound it makes when attacked and subsequently silenced, was explored by artists keen to challenge the stranglehold of the visual over other expressive means of art-making. Iconoclastic gestures such as these also represented a very public gesture of refusal—a refusal to listen, to learn from, incorporate, repeat or perform the sounds of another musician, composer or instrument. In his 1964 work *Piano* [236], Günther Uecker hammered hundreds of sharp nails into the surface of an upright piano, an act echoed formally by George Maciunas's *Piano Piece* [270] in 1970, although in the latter case Maciunas reversed the procedure, placing each nail upright on the individual keys as though to substitute fingers, invoking, yet at the same time serving as a barrier to, touch. In a similar key, Arman's series of *Colères* (from 1961) [218] or "rage" works, in which he destroyed musical instruments before mounting and framing them as sculptural assemblages, performed an aggressive silencing which also bore close parallels to Nauman's *Concrete Tape Recorder Piece*. Meanwhile, Philip Corner's *Piano Activities* piece, performed at the Wiesbaden Fluxus festival in 1962, involved a group of Fluxus artists collectively smashing a piano to smithereens. Each of these works is reliant upon the embodied participation or intervention of another body, whether they are required to play, scream, attack or remain mute. Sound is not so much silenced as remade, drawing on a dissonant set of environmental noises that exist beyond the written score to draw in the world outside.

Such practices, which invoke a sonic register at the level of both production and reception, prompt consideration of the consequences of substituting a language of the auditory for the more conventional visual responses to the work of art. What might it mean to "listen to" rather than to "look at" an artwork, including those that do not make an audible noise?

I pose this question in order to think about how the conceptual and formal connecting tissue linking one work of art to another across time and history might be mapped, not by tracking visual similarities and differences, but by paying attention to an auditory register in which artworks might be described as "sounding" like others. How might certain sculptures be understood both visually and aurally, and what happens when these two sensory fields become confused or intertwined?

To approach this question, we are required to address not only the role of sound in sculpture at the level of figuration or intent (although these are important topics) but more the collaborative, fugitive means by which sound is produced, projected, displaced or otherwise invoked by the sculptural object. In *The Life of the Mind* (1978) Hannah Arendt noted how often "thinking has been thought of in terms of seeing," an assumption which trumped earlier tropes of the storyteller "whose stories are being listened to" (Arendt recalls the early modern motif of the "blind bard"). "One may wonder," Arendt continued, "why hearing did not develop into the guiding metaphor for thinking."[2] A number of three-dimensional works from the 1960s attempted to render hearing, and sometimes an inability to hear, central to intellectual activity and material practice.

Broadcast (1959) and *Oracle* (1962–65) [230] are two assemblage works by the American artist Robert Rauschenberg that incorporate radio sets. They are works of art that demand viewers listen as well as look.[3] *Broadcast* is a wall-mounted Combine painting, the surface of which comprises a chaotic mismatch of newspaper and other printed materials. Concealed beneath the surface are three radios, only two knobs of which are visible: one to control the volume; the other to change the station. The radios are rigged so that however the viewer turns and twists the dials, they will never be tuned to the same channel, producing instead a cacophony of overlapping sounds and ghostly static interference, which appears to emanate from the surface of the work.

Oracle is the more ambitiously scaled of the two works, a fully three-dimensional installation that recalls Luigi Russolo—the Futurist artist and author of *L'arte dei rumori* (The Art of Noises, 1913)—and his *intonarumori* or "noise-machines." [160] Its significance lies not in its formal similarity to Russolo's machines, but rather in its attempt to fuse the noises and cacophony of everyday life with the work of art. *Oracle* contains five concealed, operating radios among its array of found objects, which include a series of ventilation ducts, a wooden window frame, a crushed metal car door placed on a typewriter table, and a range of assembled items such as wire baskets and electronic components.

The hidden radios are controlled by a series of variable-speed motors that, as with *Broadcast*, produced clashing noises whose flux and openness was in tune with the cacophonous visual effect of the assemblage. *Broadcast* and *Oracle* not only invoke the material and affective properties of Rauschenberg's contemporary media landscape, but trouble any sense that its interpretation might be a matter for visual contemplation alone. Thinking here—both for Rauschenberg and for the viewers of these works—requires that one listen.

OPTICAL NOISE

The sounds produced by Rauschenberg's Combines are deliberately distracted, and the works' physical edges rendered indeterminate. Sound leaks. These objects are, in every sense, environmental. The sensory feedback they provide as you wander around the work is multiple, akin to walking down a busy city street as they draw in and on the world outside (for Russolo the human ear had to rapidly adapt to the onslaught of new sounds thrown up by the metropolis). It was this aspect of Rauschenberg's work the critic Leo Steinberg picked up on in his famous essay *Other Criteria* (1968–71). Steinberg claimed that Rauschenberg's work produces "optical noise," offering in the process a powerful corrective to Modernist criticism's insistence on artworks' purely optical address to the viewer. For Steinberg a shift occurred in the "psychic address" of the work, reorienting it so it was no longer "the analogue of a visual experience" but a horizontal "plane" offered as a lateral spread or matrix of information.[4]

For Steinberg this shift was particularly visible in Rauschenberg's use of a flatbed printing press to make his transfer works, in which imagery, text, paper and newsprint were printed and overlaid on top of one another in a horizontal position to produce a palimpsest of information or visual "noise." Steinberg argued that the overlaid images were analogous to the way sound is transmitted, except that instead of clear communication the mismatched snippets of information "interfered" with one another, leaking to create a kind of visual static. They resembled the "waste and detritus of communication—like radio transmission with interference, noise and meaning on the same wavelength, visually on the same flatbed plane."[5] While Steinberg's comments were limited to Rauschenberg's two-dimensional works, his argument also extends sideways, to describe the new kinds of three-dimensional objects that similarly incorporated aspects of "optical noise."

HIDDEN NOISE

Another kind of "optical noise" was produced by two sculptures made at different moments in Modernism's 20th-century trajectory, in 1916 and 1961 respectively, that are best described not in terms of how viewers "listen" to the objects, but rather in terms of how the objects "listen" to each other. The first is Marcel Duchamp's *With Hidden Noise* (1916) [164] and the second, Robert Morris's homage to Duchamp, *Box with the Sound of Its Own Making* (1961) [220]. Both works offer formal and aural echoes of one another (as well as Nauman's later concrete box, too). There is a temporal play to both works that, like a ticking metronome, clicks back and forth between the finished "object" and the sound of its earlier making (a rattle in the case of Duchamp; the sounds of the workshop in that of Morris).

This model of exchange suggests the establishment of a relationship based not on vertically oriented patriarchal inheritance, in which one object leads inexorably to the next (say, from Higgins to Nauman, Russolo to Rauschenberg, or Duchamp to Morris). Rather, these relationships might be understood to be alternative models of kinship, articulated via an auditory register in which earlier artists and artworks "listen" out for the echoes and resonances of others.

Just as Michel Foucault argued that "genealogy" should not simply be considered in terms of a linear "unbroken continuity" from one thing to the next, but as "an unstable assemblage of faults, fissures, and heterogeneous layers than threaten the fragile inheritor from within or from underneath," so too the relationships between artworks are not always determined via purely formal, visual means; an aural reminder, just like a half-remembered smell, can produce other kinds of recall not tied solely to a visual regime.[6]

Building on Steinberg's claims, we might instead imagine a model of affiliation and connection aligned not with the forward march of chronology but with the lateral and diachronic field of practices out of which any one artist works at any one time.[7] For example, in 1967 Nauman made an ear from a knot of rope and some molded wax. The object, titled *Knot an Ear*, was part of a series of sculptural works in homage to the work of Henry Moore and H.C. Westermann—two artists alongside whom Nauman was attempting to position himself in a fraught and contradictory relationship. Through the trope of the ear Nauman cast himself as "listening" to the work of his peers, while with its punning title the work invoked the contingencies of spoken language. However, *Knot an Ear* also called up the frustrations, deafnesses and non-linear character of such a model of inheritance. Such "fragile inheritances" could, in another context, be described as "phantom effects"—after psychoanalysts Nicolas Abraham's and Maria Torok's description of the delayed effects of earlier encounters, memories, sights and sounds—upon later generations. These effects are enfolded, or "encrypted" within the other; like the "hidden noise" in Duchamp's work, or the memory of music when looking at Uecker or Maciunas's broken pianos, these works speak back to each other, and sometimes across one another, to frequently unexpected ends.

HISTORY HEARD

In his 1933 essay *The Lamp*, Walter Benjamin wrote about the historical task in auditory terms, describing how his approach involved pressing his ear against the shell of the 19th century.[8] Invoking the ways in which historians glean their information from a variety of sources that are not necessarily the most obvious and overt, Benjamin understood the historian's task as seeking out barely discernible traces and almost-forgotten only-just-audible episodes, figures and motifs.

This "aural" model in which works of art are considered as not merely invoking looking but also "listening," might be aligned with ways of thinking about history in auditory terms. In another context T.S. Eliot famously described this as the "auditory imagination" demanded by poetry which, Eliot wrote, "works through meanings, certainly, or not without meanings in the ordinary sense, and fuses the old and obliterated and the trite, the current, and the new and surprising."[9]

To "listen" to a work of art suggests a mode of attending to aspects of the object that are not articulated visually, but which resonate instead through less clear-cut, perhaps more ephemeral, or at least differently mediated signals. Sometimes these might ring out loud and clear, but at other points the effects might be subtle and opaque and take the form of a vibration, a crackle, a hum or even a haunting, silent scream. The sound might be discordant, troubling even—a disruptive noise signaling a moment of breakage or rupture. There was a moment during the 1960s when sound became, for a range of artists working with three-dimensional objects, both a material and a concept or metaphor. Listening to works of art was both a literal requirement and a historical problem. Evaluating a time when the visual and the aural came together in unexpected and confusing ways, we might well return to Arendt's call for hearing as a "guiding metaphor" in order to think about how these artists used sound and its reception to challenge the longstanding privileging of the "visual" within modernism in general and Modernist discourse in particular.

1
See Michael Fried, "Three American Painters: Kenneth Noland, Jules Olitski and Frank Stella," in Id., *Art and Objecthood: Essays and Review* (Chicago, IL: University of Chicago Press, 1998), pp. 213–65: 224; Clement Greenberg, "Modernist Painting," in John O'Brian (ed.), *Clement Greenberg: the Collected Essays and Criticism*, vol. 4 (Chicago, IL: Chicago University Press), pp. 85–93. For two very different accounts of the visual in modernity see Jonathan Crary, *Techniques of the Observer: On Vision and Modernity in the Nineteenth Century* (Cambridge, MA: The MIT Press, 1992) and Martin Jay Downcast, *Downcast Eyes: The Denigration of Vision* (Berkeley, CA: University of California Press, 1994). [Editor's Note: The complete title was *Danger Music # 17: Scream! Scream! Scream! Scream! Scream! Scream!*]

2
Hannah Arendt, "Metaphor and the Ineffable" in *The Life of the Mind: Thinking* (New York: Harcourt Brace, 1981), pp. 110–11. See also Angela Frattarola "Developing an Ear for the Modernist Novel: Virginia Woolf, Dorothy Richardson and James Joyce," *Journal of Modern Literature*, vol. 33, no. 1, (Bloomington, IN: Indiana University Press, Fall 2009), pp. 132–53, p. 132.

3
Brandon Joseph, *Random Order: Robert Rauschenberg and the Neo-Avant-Garde* (Cambridge, MA: The MIT Press, 2007) p. 187.

4
Leo Steinberg, "Other Criteria" [1968/1971], in Id., *Other Criteria: Confrontations with Twentieth Century Art* (New York: Oxford University Press, 1972), pp. 55–93: 84.

5
Ibid., p. 88.

6
Michel Foucault, "Nietzsche, History, Genealogy" [1971], in Donald F. Bouchard (ed.), Donald F. Bouchard, Sherry Simon (trans.), *Language, Counter-Memory, Practice* (Oxford, UK: Blackwell Publishing, 1977), p. 146. The intimate entanglement of the senses provides the opening motif for Marcel Proust's novel *In Search of Lost Time*, "Swann's Way," vol. 1, trans. by Lydia Davis (London: Penguin, 2004).

7
See chapter 5 of my *Eccentric Objects: Rethinking Sculpture in 1960s America* (New Haven, CT/London: Yale University Press, 2012) for a discussion of alternative models of non-linear relationships between artists and artworks in which an aural register of "listening" and "mis-hearing" is explored.

8
Walter Benjamin, "The Lamp" [1933], trans. by Rodney Livingstone, in Michael W. Jennings, Howard Eiland, Gary Smith (eds.), *Walter Benjamin: Selected Writings, vol. 2, part 2, 1931–1934* (Cambridge, MA/London: The Belknap Press/Harvard University Press, 1999), p. 692.

9
T.S. Eliot, "Matthew Arnold," in *The Use of Poetry and the Use of Criticism* (London: Faber and Faber, 1933), pp. 118-19. Jonathan Sterne also cites this well-known passage in Jonathan Sterne, "Sonic Imaginations," in Id. (ed.), *Sound Studies Reader* (New York: Routledge, 2012), pp. 1–19: 5–6.

John Cage

VARIATIONS I
1958

John Cage (Los Angeles, 1912–New York, 1992), best known for his 1952 composition *4'33"*, which consists of 4 minutes and 33 seconds of silence, was an American avant-garde composer, pioneer of electro-acoustic music and indeterminacy in music, and an inspirer of important movements such as Fluxus.

The score of *Variations I* consists of 6 transparent squares: one with 27 points of 4 different sizes, 5 with 5 lines each. The squares are to be combined in any way, with points representing sounds, and lines used as axes of various characteristics of these sounds: lowest frequency, simplest overtone structure, etc. Such characteristics are obtained by tracing perpendiculars from points and measuring these perpendiculars.

As can be read on the title page, the work is dedicated to David Tudor, an American pianist and composer of experimental music, on his birthday "(tardily)." The artist himself indicates that parts had to be prepared from the score, with or without the extra materials, for any number of performers, and any kind and number of instruments.

[See also p. 216]

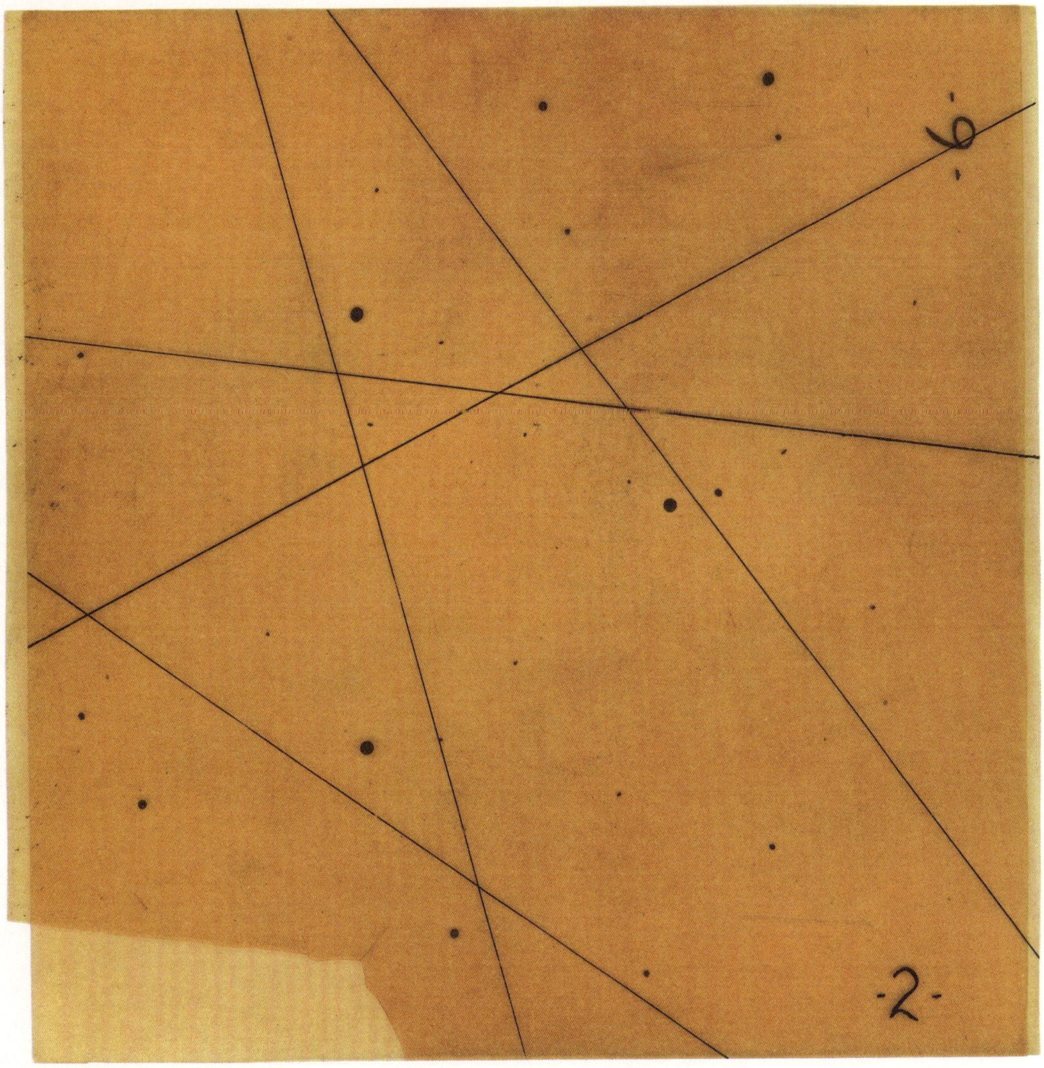
-2-

John Cage

WATER WALK
1959

In 1958 John Cage spent a few months in Milan, having been invited by composer Luciano Berio to work in the Italian state television recording studios. He also participated as an expert on mushrooms in a popular TV quiz called “Lascia o Raddoppia,” where he performed one of his “chance music” compositions live, whose elements are left to chance or to the determination of the performer.

Conceived during his stay in Milan, *Water Walk* is a score with a list of properties and instruments to be used, a floor plan for placing instruments and objects, a timeline with the actions and a list of notes on the 27 actions to be made in order of occurrence, such as “Explode bottle up and over piano” or “Place ice in glass (preparing a drink).” The materials required are mostly related to water—bathtub, toy fish, ice cubes—but Cage also calls for a grand piano and 5 radios. The timings are not accurate: “Start watch and then time actions as closely as possible to their appearance in the score.”

[See also p. 214]

1

WATER WALK

FOR SOLO TELEVISION PERFORMER

JOHN CAGE
MILANO 1959

Arman

NBC RAGE
1961

The son of an antique dealer and a cello player, the French artist Arman (Nice, 1928–New York, 2005) joined the nascent art movement Nouveau Réalisme in 1960. His fascination with found objects began in 1959 with his *Poubelles* (Trash Cans) series, and was followed by *Accumulations*, collections of identical items squeezed into glass vitrines. In 1961, Arman made his debut in the United States and, during this period, explored creation via destruction. The *Coupes* and the *Colères* featured sliced, burned or smashed objects arranged on canvas, often using musical instruments, mainly violins and saxophones.

NBC Rage is of the *Colères* series and was made as a result of a performance in Paris during filming by the television network NBC. The artist stated, "My approach is conceptual, rational, not at all instinctive. Chance, despite its appearance, is secondary. Only in two cases does it play a role: when I break an object, like a cello, the pieces stay right where the cello fell." *

[See also p. 344]

* Interview with Michel Giroud, in *Kanal*, monographic issue, July 1993.

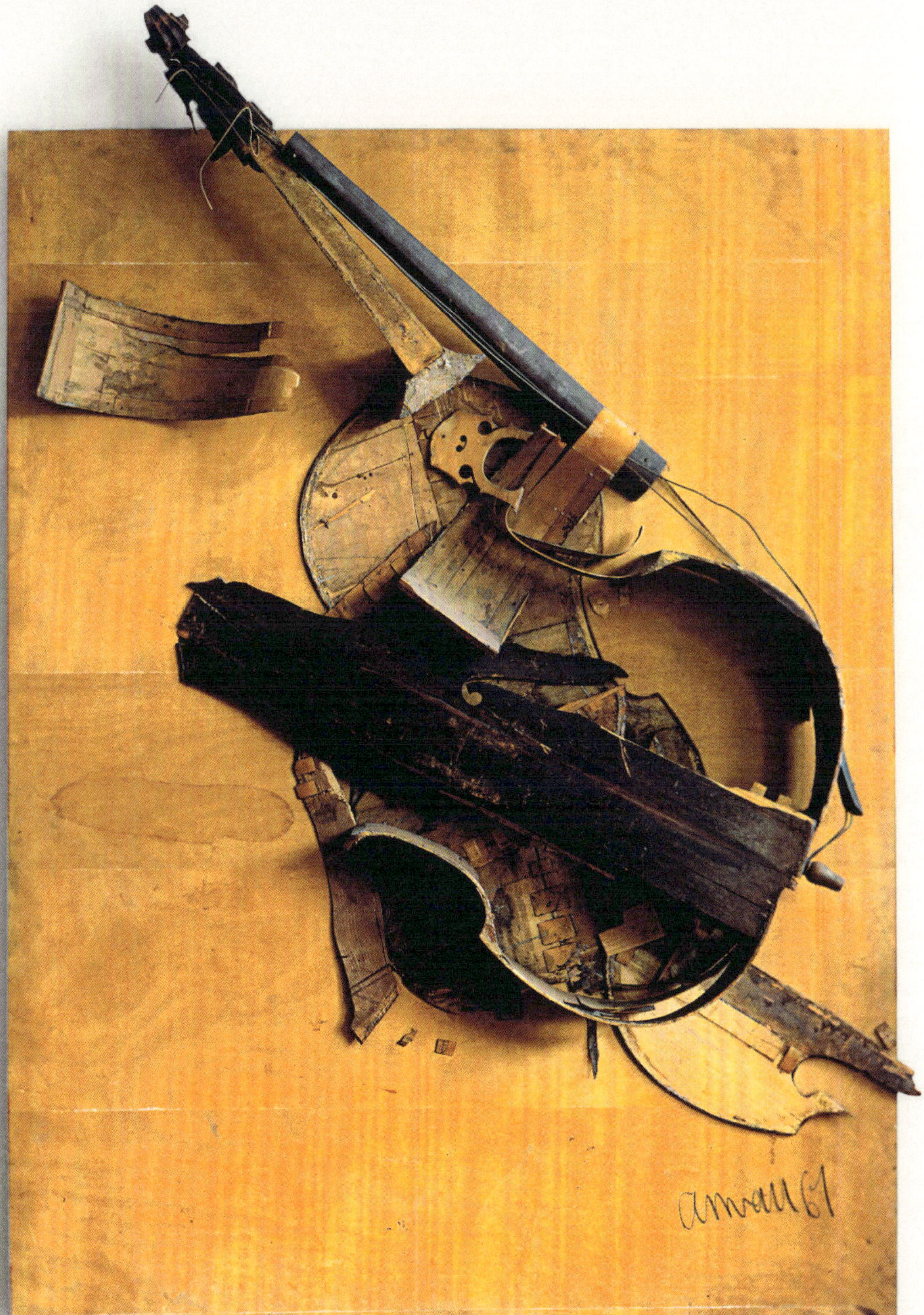
arman 61

Robert Morris

BOX WITH THE SOUND OF ITS OWN MAKING 1961

American artist Robert Morris (Kansas City, 1931) built *Box with the Sound of Its Own Making* in New York in January 1961. The wooden cube is made of 6 pieces of nutwood joined by Morris, and a small loudspeaker inside that reproduces the carpentry sounds (of a hammer and saw) made while constructing it. The tape recording lasts approximately 3 hours. This performative element is a reference to Marcel Duchamp's *With Hidden Noise* (1916) [p. 164], but instead of hiding a secret, the box reveals its soundtrack: the process of making an artwork. John Cage, visiting Morris in his studio, sat in front of the box and listened to the whole duration of the tape, as if it were a theater performance. As Morris said, "I quit painting for a particular reason—certain problems I couldn't solve. There was a kind of ontological character to painting I couldn't accept. Because on the one hand you were involved in some activity, on the other hand you ended up with an object. …One of the first objects I made was *Box with the Sound of Its Own Making*, which does resolve the problem." *

* From Thomas Krens, "The Triumph of Entropy," in Rosalind Krauss, Thomas Krens, David Anti *et al.*, *Robert Morris. The Mind/Body Problem*, exhibition catalogue (New York: The Guggenheim Museum, 1994), p. XIX.

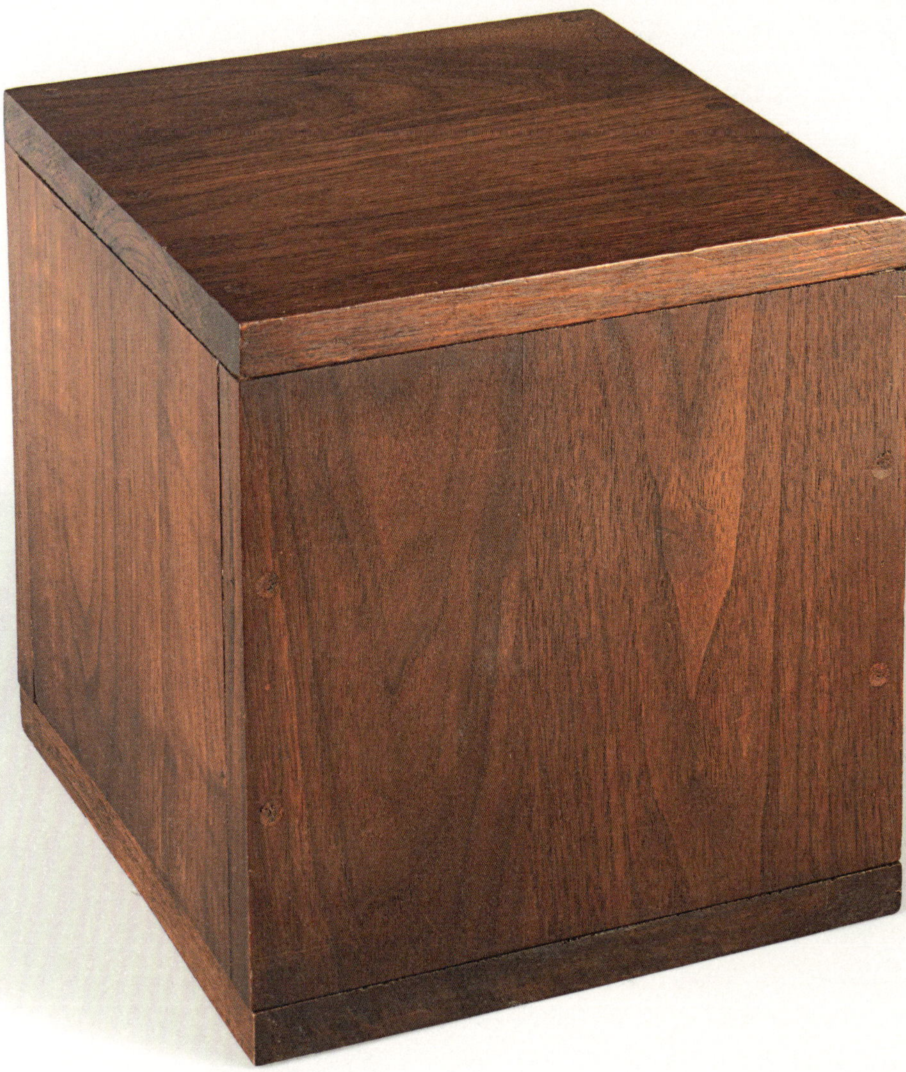

Nam June Paik

URMUSIK
1961

During his studies in music history in Germany, the Korean artist Nam June Paik (Seoul, 1932–Miami, 2006), who had long trained to be a classical pianist, met the composers Karlheinz Stockhausen and John Cage, and subsequently specialized in the study and composition of contemporary electronic music.

As the title of this work—*Urmusik* (Primeval Music)—suggests, when the wooden case is played freely and in an improvised manner, it operates as a simple sounding board: the taut wires in the upper section can be plucked, and a metal can on the side containing a marble can be rotated. The creation of musical instruments employing commonly used objects was a recurrent theme in Paik's work, such as in *TV Cello* (1971): in this, televisions placed on top of one another to suggest a cello were played with a bow by the American performer and cello player Charlotte Moorman, with whom Paik enjoyed a long lasting artistic relationship.

[See also p. 228]

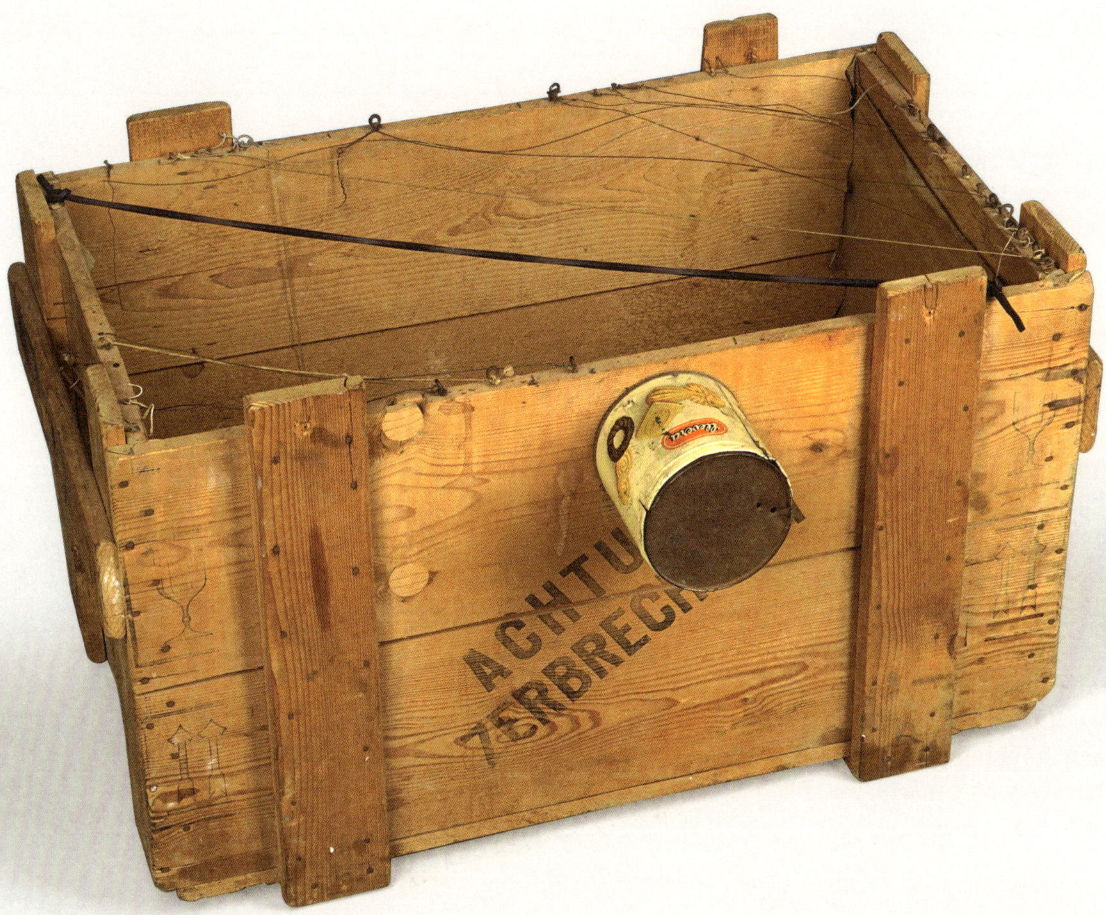

Edward Kienholz

THE CARNIVORE
1962

American artist Edward Kienholz (Washington, 1927 – Hope, 1994) was born in Washington State and spent his youth on his family's farm. Kienholz moved to Los Angeles in 1952 and in 1957 he and Walter Hopps started the Ferus Gallery, where Kienholz held his first solo exhibition in 1960. Throughout the 1960s and 1970s he produced large-scale tableaux from assembled and constructed objects that criticized the politics of the period.

His skills in carpentry, metalwork and auto repair are a constant element of his sculptures, such as the early *The Carnivore*, an abstract construction made of a cello out of which a disquieting, furry creature emerges. Kienholz's use of musical instruments in assemblages returns in his sculptural work, *The Tadpole Piano Pool with Woman Affixed Also* (1971), consisting of a concert grand piano with a pregnant woman lying on the keyboard. The inside of the piano houses a pool in which a biological interplay is supposed to take place.

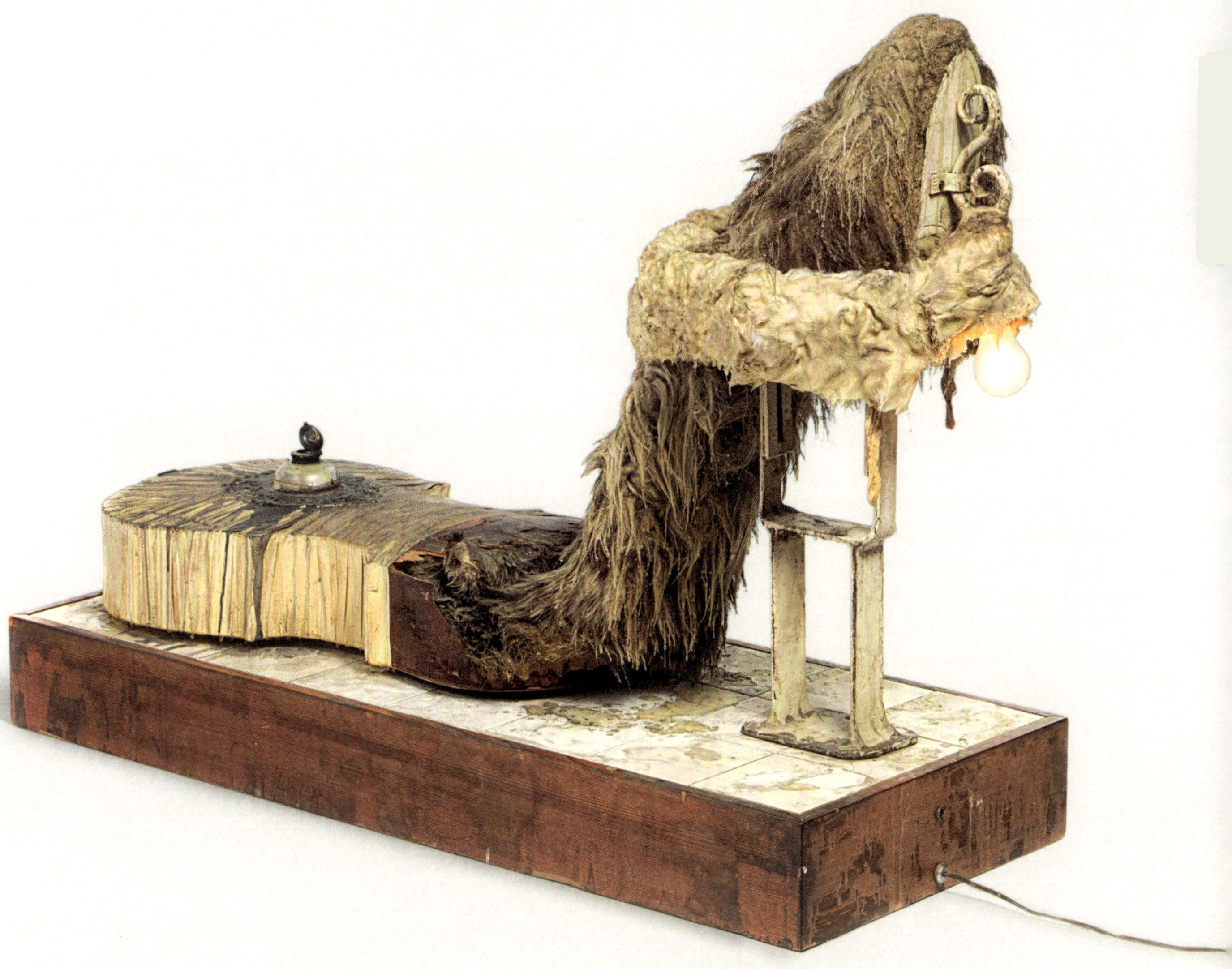

Jean Tinguely

RADIO WNYR NR. 15
1962

In 1962 Swiss artist Jean Tinguely (Fribourg, 1927 – Bern, 1991), a leading figure in the Nouveau Réalisme group, started working on a series of works involving assemblages where the functioning raw materials of disassembled radios were mounted behind Plexiglas to make them visible. A small motor mounted on a radio continually turns the frequency dial, thus constantly switching channels. The incomprehensible radio sounds are therefore determined by chance. These works were made in New York—where, during the same period, Robert Rauschenberg [p. 230] and Jasper Johns incorporated radios in their works—and presented at the Alexandre Iolas Gallery in November 1962. The title is an allusion to the New York radio station WNYC, one of America's earliest non-commercial stations.

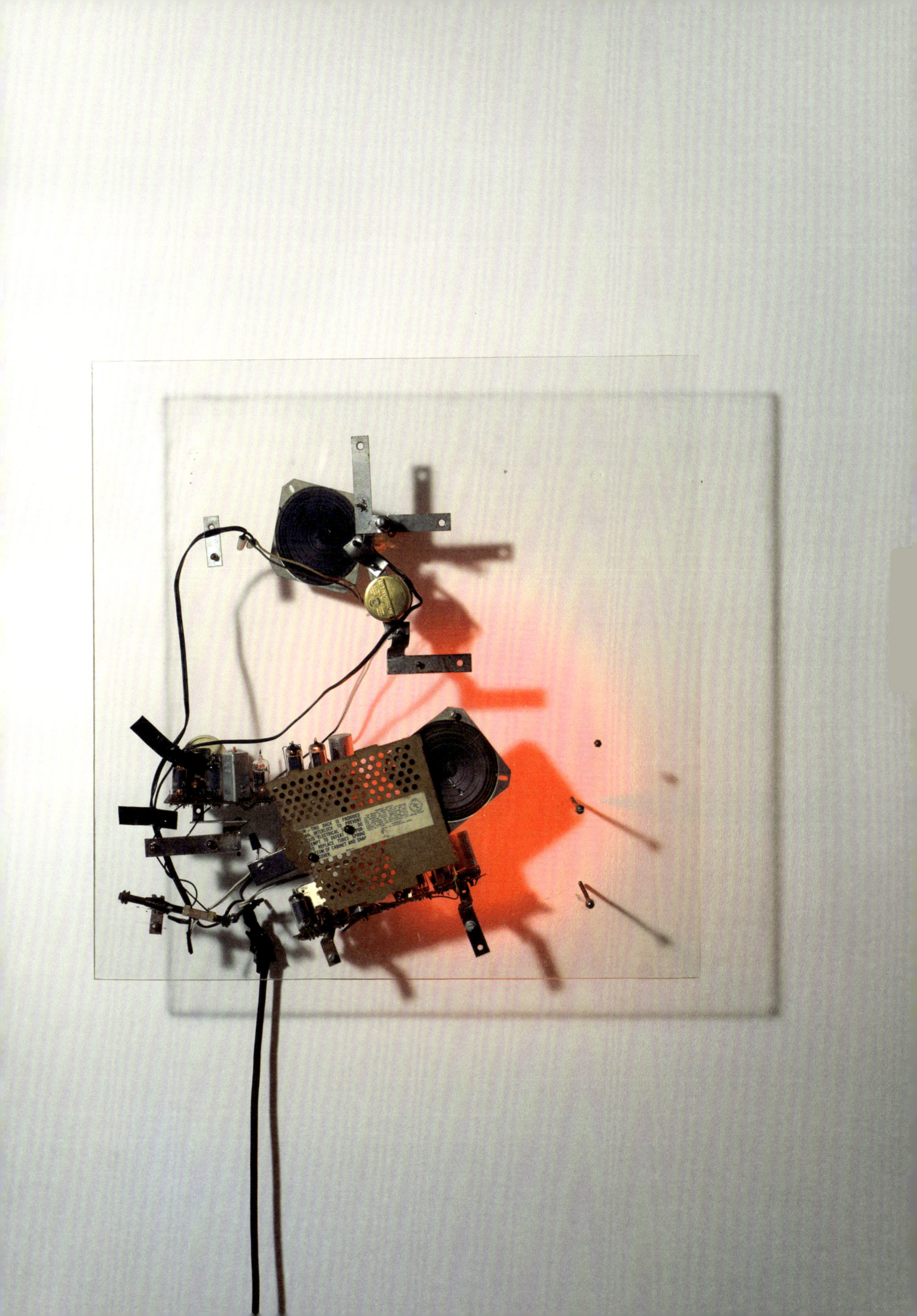
-THIS BACK IS PROVIDED
AN INTERLOCK TO PREVENT
ELECTRICAL
EMPT TO DEFEAT
TO REPLACE TUBES SPRING
TTOM OF CABINET AND SNAP
COVER.

Nam June Paik

LISTENING TO MUSIC THROUGH THE MOUTH 1962–63

Nam June Paik's works often call for interaction between the artist and the objects. In his March 1963 performance, on the occasion of the "Exposition of Music–Electronic Television" at the Galerie Parnass in Wuppertal, the artist "listened" to a disk using his mouth rather than his ears, the "sound" of which was diffused through a phallic object. The sound was thus experienced as a material object through the vibrations that this produced in his mouth, with his body consequently acting as a sounding board. Oral auscultation substituted its aural equivalent using headphones.

The anti-traditional method of using musical devices and instruments in this work makes reference to the "Prepared Piano"—a piano that had its sound altered by placing objects between or on the strings—used by John Cage from 1938, and is evident in other of Paik's works, such as *Violin with String* (1961), in which Paik makes a violin produce sounds by dragging it along the ground by a string.

[See also p. 222]

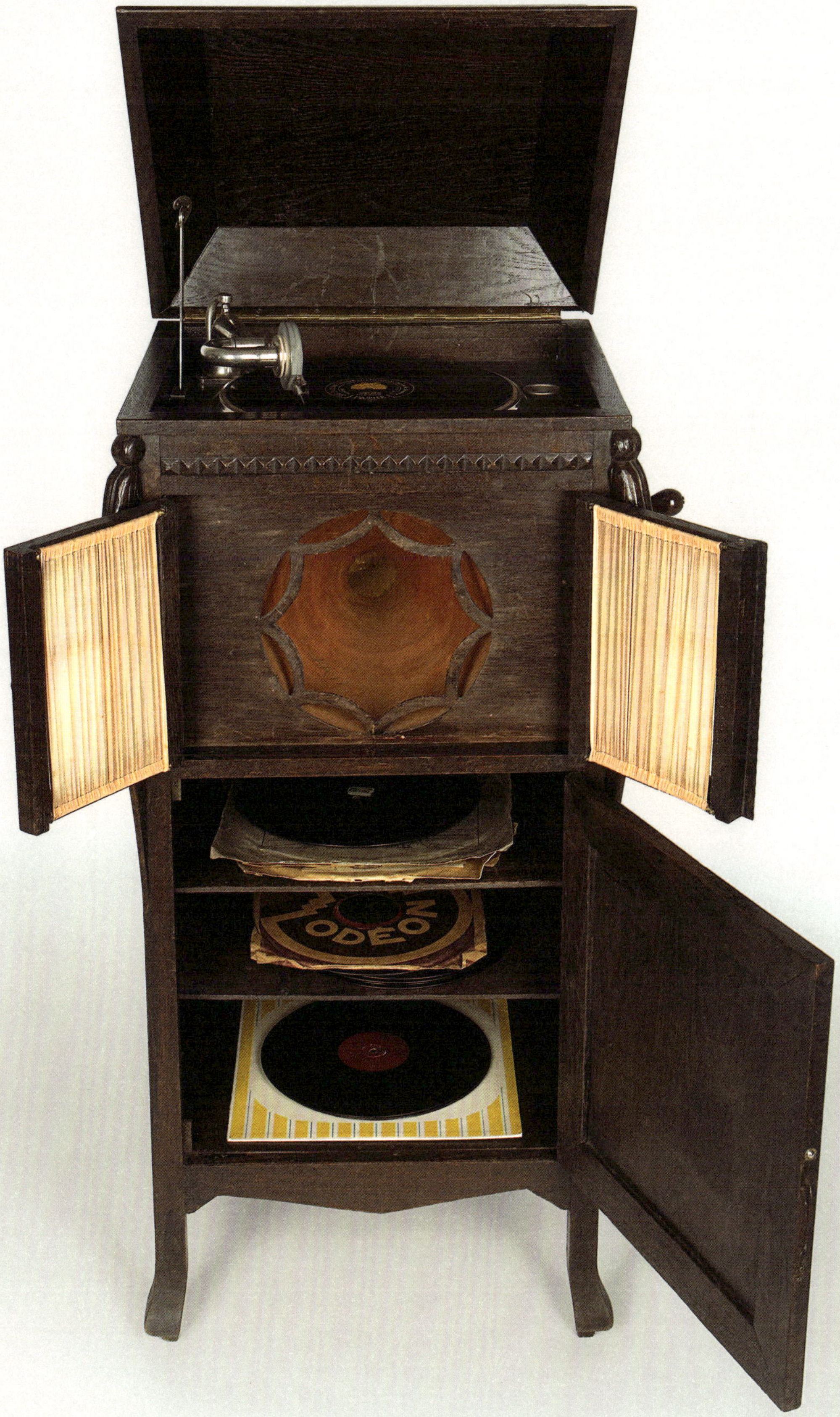
ODEON

Robert Rauschenberg

ORACLE
1962–65

Research into the field of sound by the American artist Robert Rauschenberg (Port Arthur, 1925–Captiva Island, 2008) began in 1953 with *Music Box.* A tribute to Marcel Duchamp's *With Hidden Noise* [p. 164], this wooden box had several old nails driven into the inside and 3 stones added, and was representative of Rauschenberg's interest in the interaction between a work of art and the observer. The box had to be shaken to produce sounds in the same way that the 3 radios in his next sound work, *Broadcast* (1956), needed to be controlled by the observer. This work was one of a series of combine paintings in which a random sound was added to mixed techniques on canvas.

In *Oracle*, which was presented to the public at the Castelli Gallery in New York in 1962, the sounds of 5 radios, of which the volume and rate of change of the radio signals received were varied, were added to the sound of running water from the shower into the bathtub, bringing life to an environment built with found objects. *Oracle* was created with the help of engineer Billy Klüver, with whom Rauschenberg was to found Experiments in Art and Technology (E.A.T.) in 1967, an organization designed to promote the relationship between art and technology.

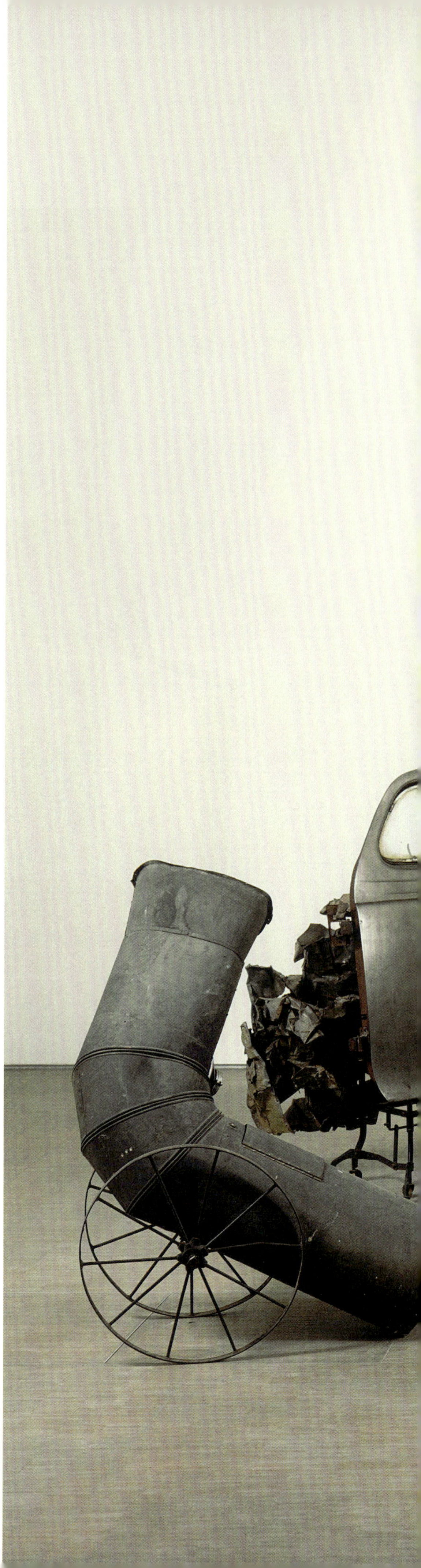

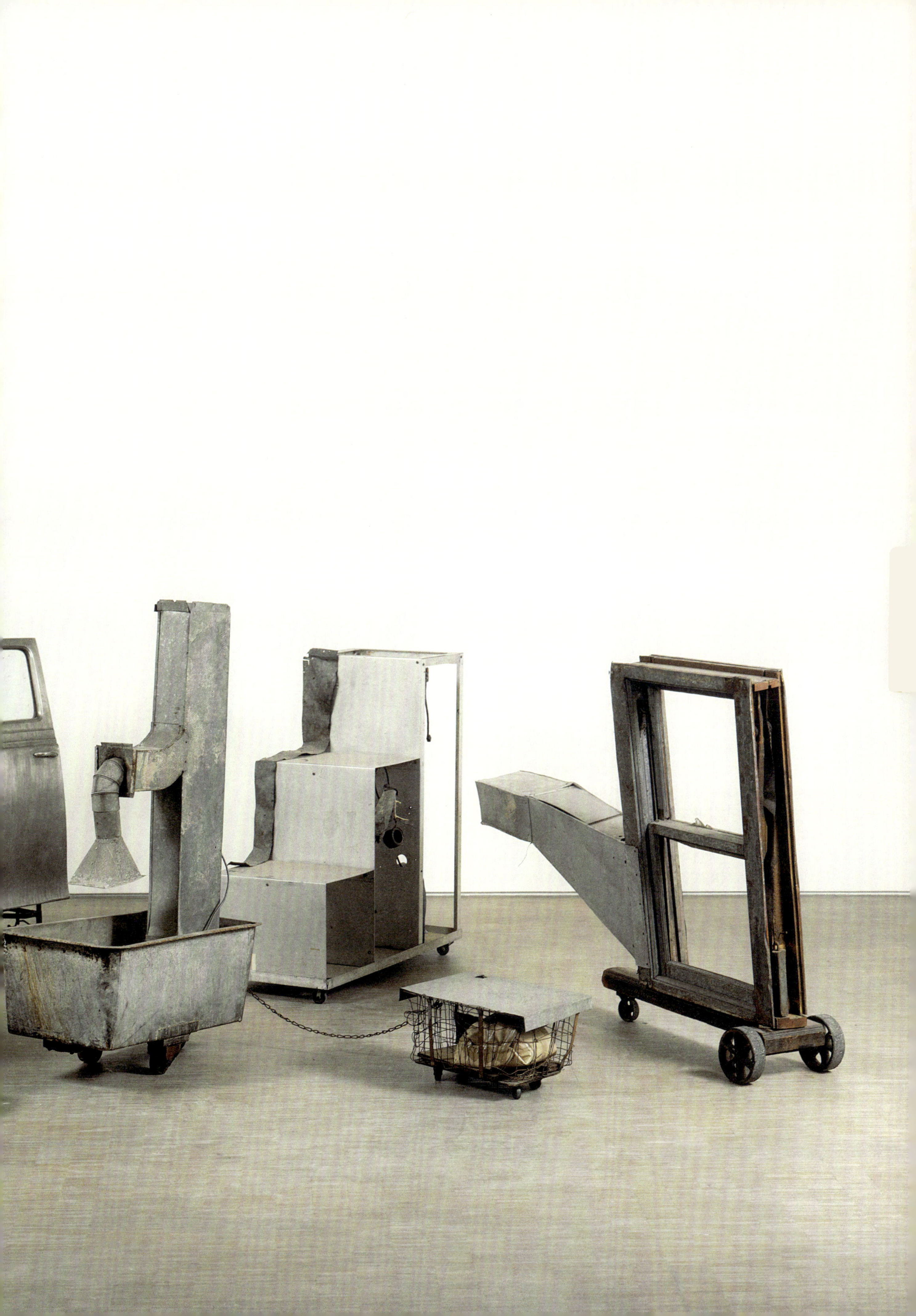

Milan Knížák

DESTROYED MUSIC
1963 (1975)

Milan Knížák (Pilsen, 1940) is a Czech visual and performance artist. At the beginning of the 1960s, he created what he called "Ceremonies" and "Demonstrations"—happenings, ceremonies, installations and various environments that were performed in the streets of Prague. He described as follows the moment he took up the practice of destroying art objects that was then becoming established in the Fluxus milieu: "In 1965 I began destroying records: scratching them, puncturing them, breaking them. Playing them—which ruined the needles and sometimes the whole record player—created a whole new type of music, one that was surprising, jarring, aggressive and funny." *

Milan Knížák's *Destroyed Music* [exh.] resulted from gluing, painting over and mechanically modifying records. Playing over this altered vinyl record produces an entirely new music that is unexpected and tempestuous but also humorous.

* From Caleb Kelly, *Cracked Media: The Sound of Malfunction* (Cambridge, MA: The MIT Press, 2009) p. 144.

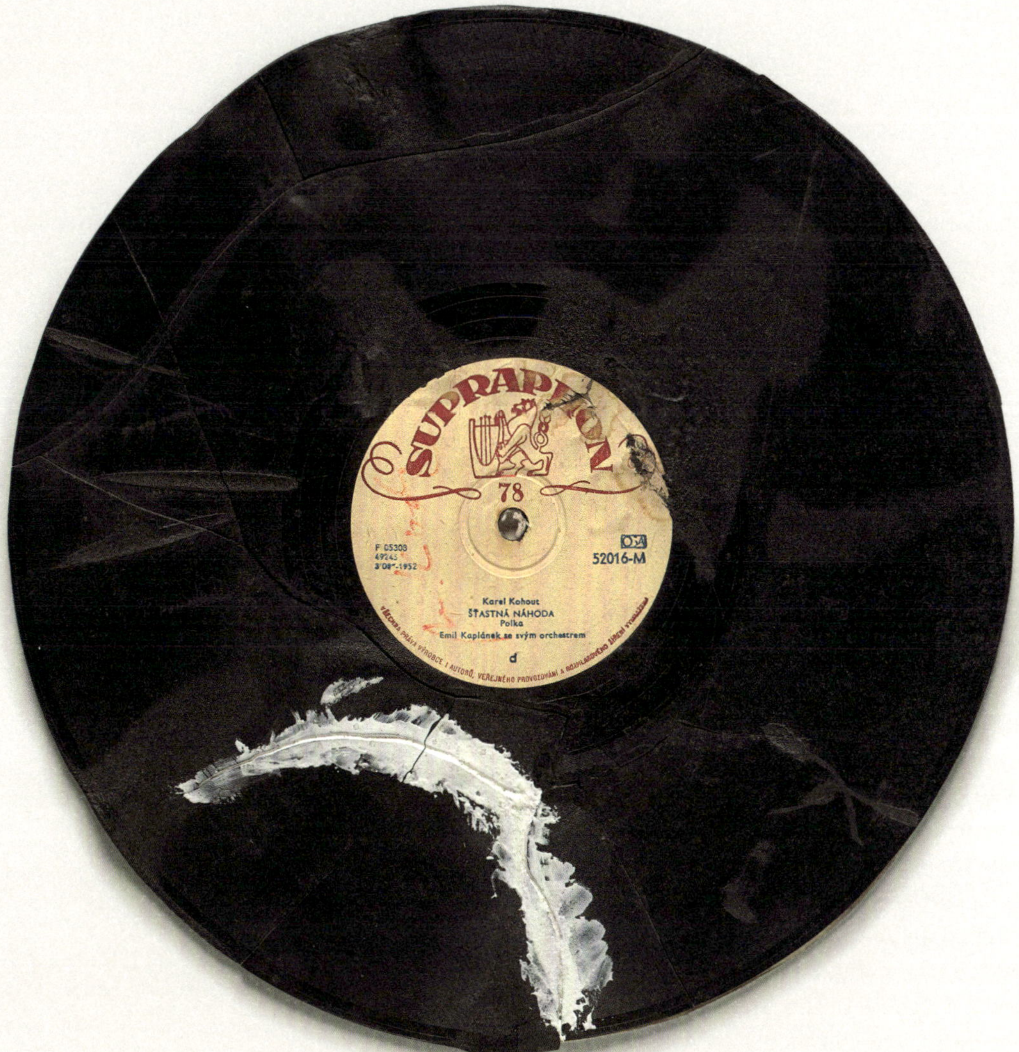
78
F 05308
49245
52016-M
Karel Kohout
ŠŤASTNÁ NÁHODA
Polka
Emil Kaplánek se svým orchestrem
d

Joe Jones

BIRD CAGE
1964

2

Joe Jones (New York, 1934 – Wiesbaden, 1993) was an avant-garde American artist and musician on the fringes of Fluxus and known above all for the creation of music machines.

He began his experimentation in the musical field in the early 1960s, making mechanical instruments and sound objects powered by an electric motor that would make them vibrate like musical instruments. He constructed a series of such objects and created, with George Maciunas, *Mechanical Flux Orchestra* (*c.* 1964) composed of a violin, 3 bells and 2 aerophones.

Bird Cage is a tribute to the American composer John Cage, who had a great influence on Jones. The metal cage contains a painted violin, various objects and a battery-powered motor. When the sculpture is in operation, the motor stimulates a rotary motion that plays the violin strings in a random manner.

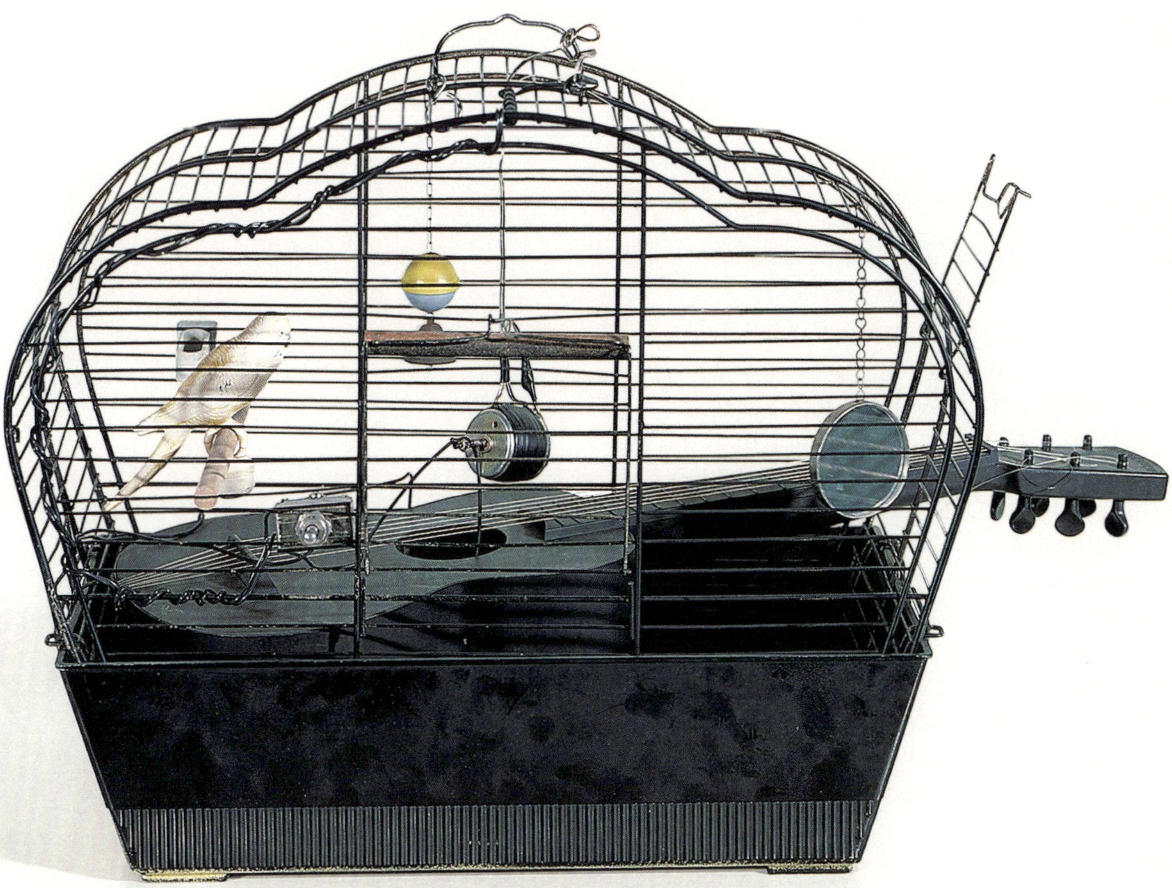

Günther Uecker

PIANO
1964

Günther Uecker was born in Wendorf (1930) in what was to become East Germany. Early on he moved to Berlin, West Germany, and in 1961 joined the ZERO group. The movement entered into dialogue with the contemporary interest in multimedia experimentation. In 1963 Uecker began to hammer nails into items of furniture like chairs, television sets and tables. In this work, the nails were repeatedly hammered into the surface of a piano.

The concept of rhythm, linked with the principles of Zen philosophy, was influential on Uecker's work. Uecker has noted that he regarded nails "as structural elements," and was "trying to create vibrations by using these materials in an ordered relationship to one another, one that disturbs the geometric order and may even irritate." *

* Günther Uecker, "Die Schöenheit der Bewegung" (1961), in Stephan von Wiese (ed.), *Schriften* (St. Gallen: Erker, 1979), p. 105.

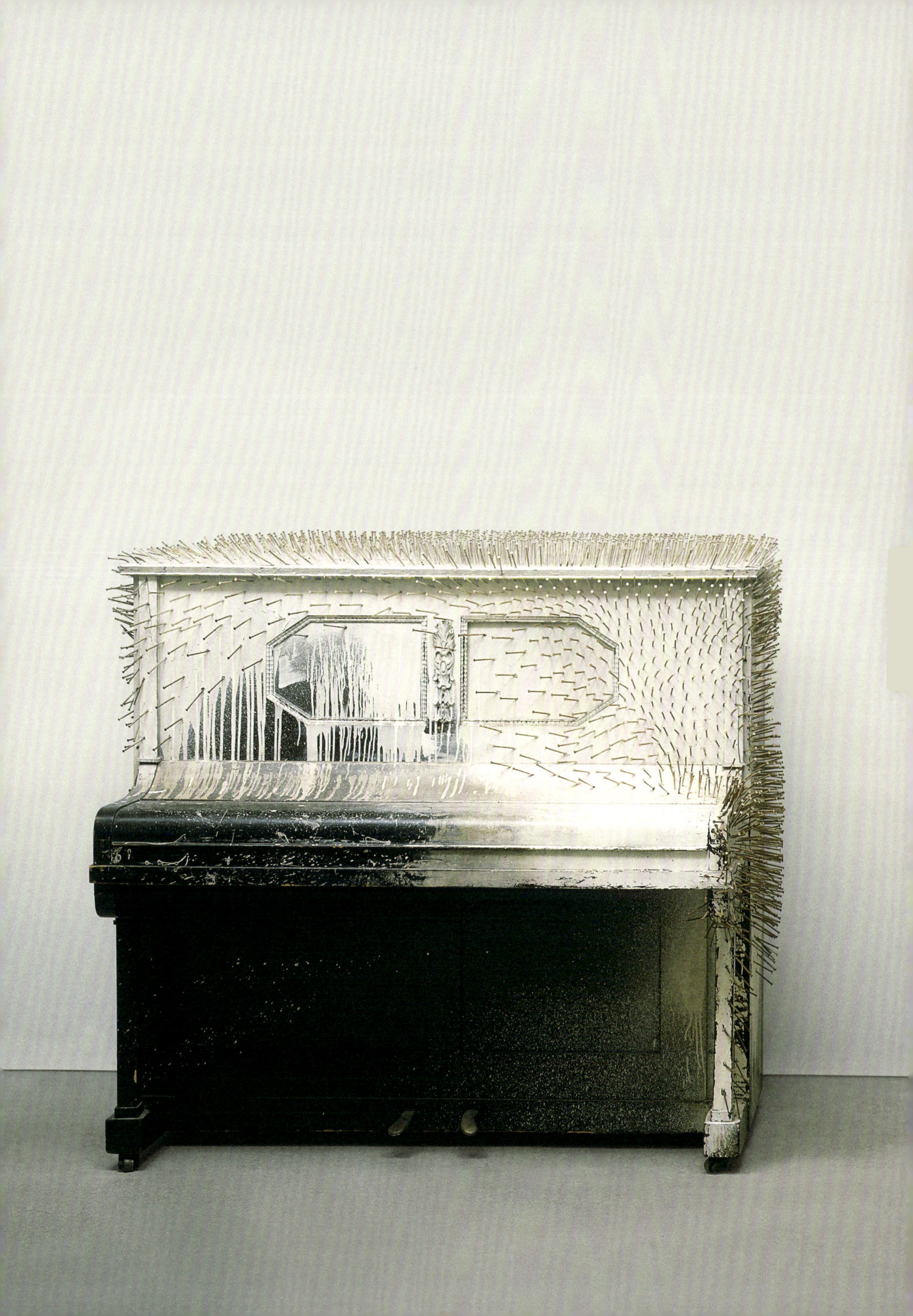

Tom Wesselmann

LITTLE STILL LIFE #18
1964

In the early 1960s, American Pop artist Tom Wesselmann (Cincinnati, 1931 – New York, 2004) started using sound, light, and later also televisions in his still life paintings, thus making them 3-dimensional. *Little Still Life #18* is a small-scale work, featuring flat and 3-D objects on a protruding shelf. It is an assemblage work including 2 Guinness beer bottles, plastic oranges and plastic flowers, and a carved wooden image of a radio, as well as a working transistor radio. A photograph of a vase and tablecloth are collaged onto cut-out shapes. A small reproduction of a master drawing resides in a metallic oval frame that hangs above the radio.

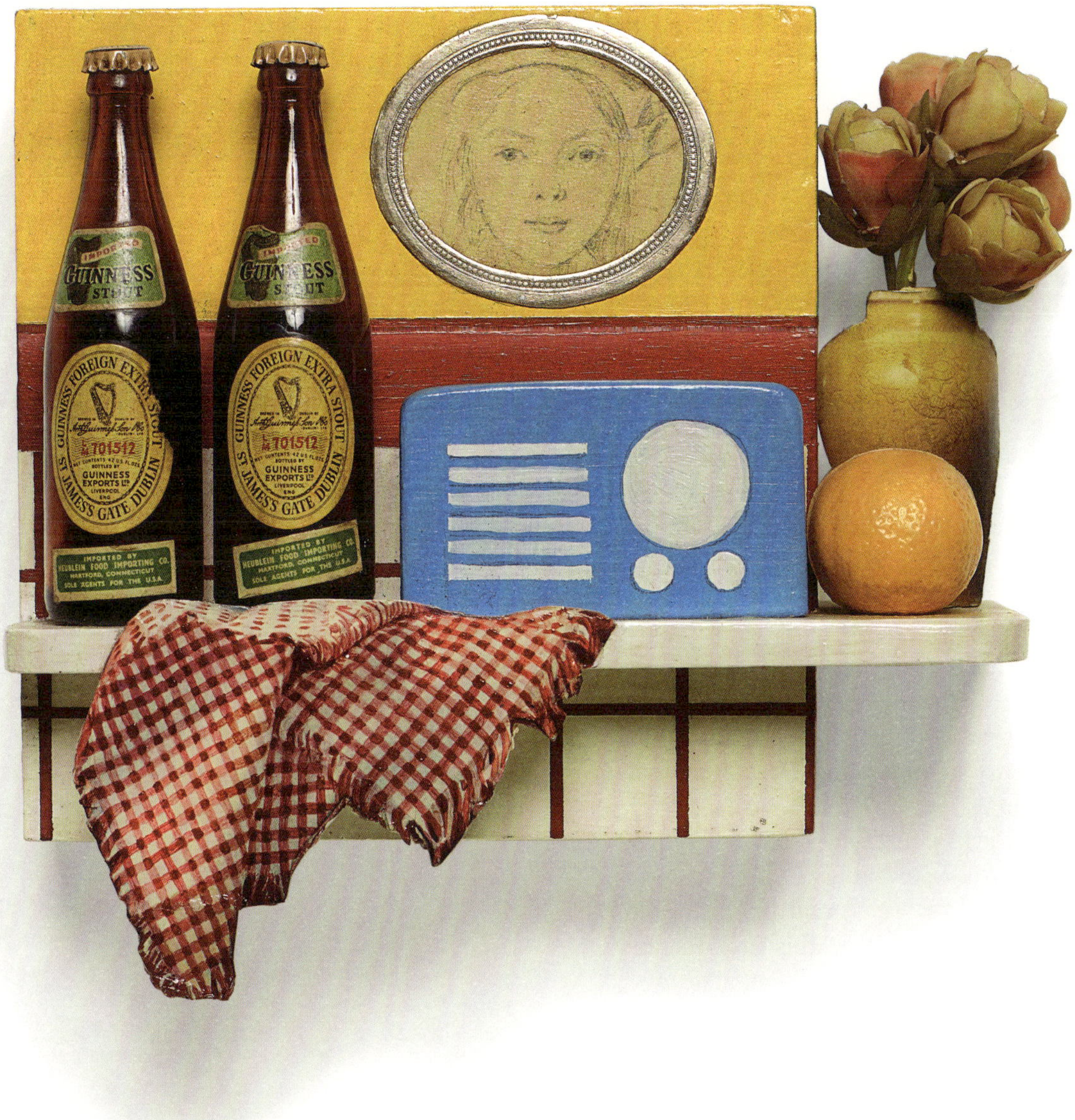
GUINNESS STOUT
GUINNESS FOREIGN EXTRA STOUT
ST. JAMES'S GATE DUBLIN
GUINNESS STOUT
GUINNESS FOREIGN EXTRA STOUT
ST. JAMES'S GATE DUBLIN

Richard Artschwager

PIANO
1965

Richard Artschwager (Washington, D.C., 1923 – Albany, NY, 2013) was an American painter and sculptor. His work is associated with Pop Art, Minimalism and Conceptual Art.

From the end of the 1950s Artschwager created sculptural reinterpretations of ordinary objects, furniture in particular, that question the processes of perception and comprehension of the space that the objects occupy.

Piano was the first in a series that ended in 2012. It explores the boundaries between painting and sculpture and perceived reality and art, deliberately creating confusion in the viewer. It is close to a real piano in size but the proportions and number of keys—accurate reproductions that allow the notes to be recognized—are unreal. The piano seems to be made of wood but is in fact Formica laminate, an imitation of natural material.

The piano is clearly not a real musical instrument but at the same time it continually calls to mind aspects of reality, such as toy pianos, everyday materials and familiar proportions, and prompts a desire to play it even though this is not possible.

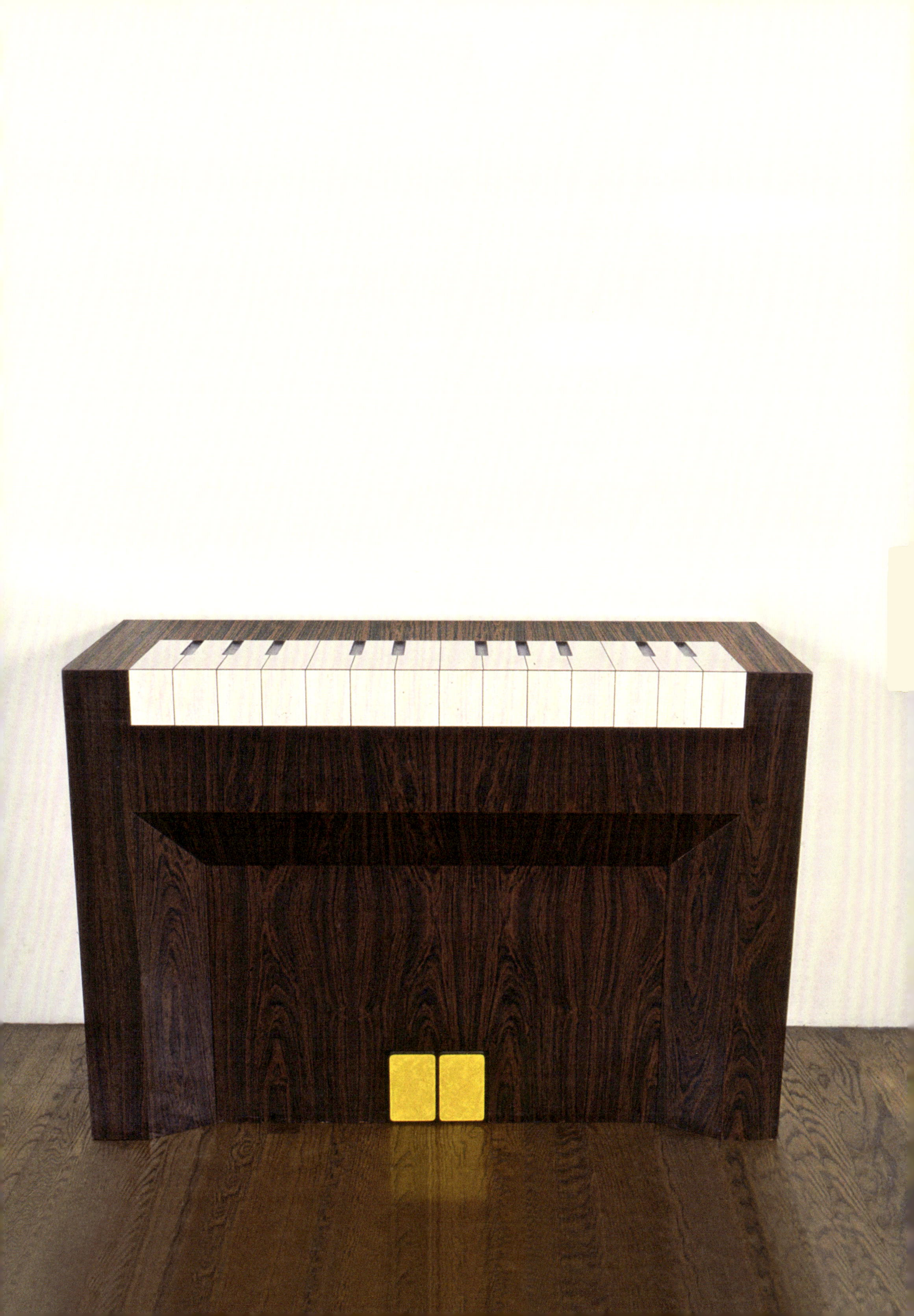

Roman Opalka

OPALKA 1965 / 1 – ∞
DÉTAIL 2450737 – 2473067
1965

Born in France to Polish parents, in 1965 Roman Opalka (Abbeville, 1931 – Chieti, 2011) began to transfer the process of counting from one to infinity to a painting medium.

Starting in the top left-hand corner of the canvas and finishing at the bottom right, Opalka began painting tiny white numbers in horizontal rows on a black background. Each new canvas—which he called a *Détail* (Detail) and dated 1965—took up counting where the last left off. He altered this system in 1968, changing the background to gray and introducing a recording, speaking each number in Polish into a microphone as he painted it. Then in 1972 he decided to make the background of each painting about 1% whiter each time. *Détail 2450737 – 2473067* has a gray background and includes the recording.

Using the non-finite to define the finite, Opalka kept a ritual bookkeeping of time's passage, creating an endless process measured against its goal of infinity. Opalka's mission was to find a language that could reflect the inherent "problem that we are, and are about not to be."

Stephan von Huene

KALEIDOPHONIC DOG
1967

2

In the mid-1960s American artist Stephan von Huene (Los Angeles, 1932 – Hamburg, 2000) turned from abstract expressionist paintings to sculpture and began creating surrealist objects. Initially making them from traditional materials like wood and leather, he eventually began including mechanical elements that brought sound and movement to his pieces. The first of a series of musical sculptures that included *Rosebud Annunciator*, *Tap Dancer* and *Washboard Band*, this work, which makes reference to ancient automatic organs, in particular the work *Tipu's Tiger* exhibited at the Victoria and Albert Museum, is controlled by a perforated ribbon and a compressed air mechanism that allows the figure of the dog to move, thus producing a continuous and variable cycle of movements and sounds. The name of the work came about by error. The artist stated, "I always thought that the kaleidoscope was just colliding patterns … . Other people make music of phasing, I make music of collisions." * *Kaleidophonic Dog* is an expression of the need to leave room for chance happenings, as opposed to the notion of planning.

[See also pp. 246, 342]

* Marvin Altner (ed.), *Stephan Von Huene: Split Tongue | Essays and Interviews* (Munich: Hirmer, 2012), p. 96.

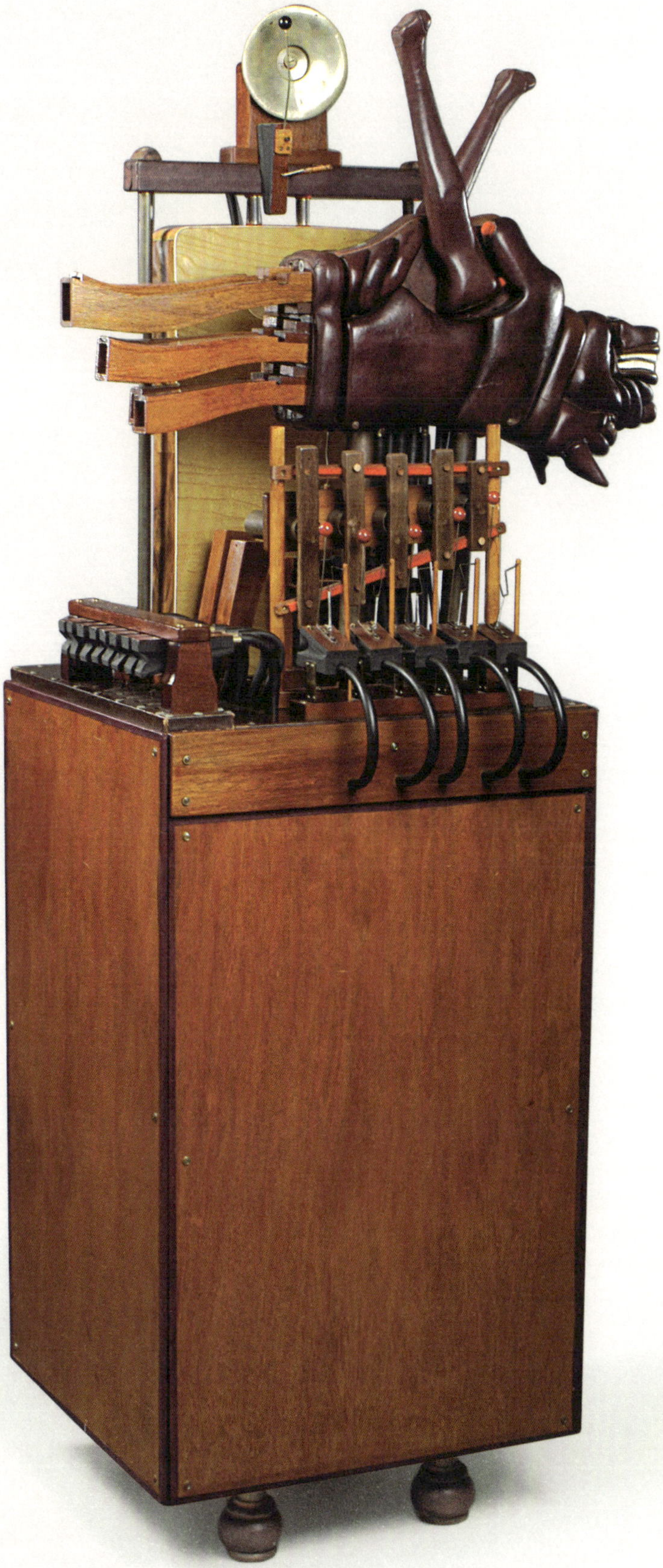

Stephan von Huene

WASHBOARD BAND
1967

2

The washboard was a percussion instrument used by African-American street musicians who played on the streets of New Orleans during the first half of the 20th century.

In *Washboard Band* Stephan von Huene makes use of a mechanical and pneumatic control system to produce sounds using a washboard, gong, plates and a cowbell. When the sculpture is operating, it automatically produces a musical performance. The table is inscribed, "NO MAN (the person being crossed out)." The work arose from the artist's wish to create a machine able to eliminate the need for a live performer by creating an automated process. As von Huene claimed, "*Washboard Band* is where I killed the man in the one-man-band."*

[See also pp. 244, 342]

* Marvin Altner (ed.), *Stephan Von Huene: Split Tongue | Essays and Interviews* (Munich: Hirmer, 2012), p. 60.

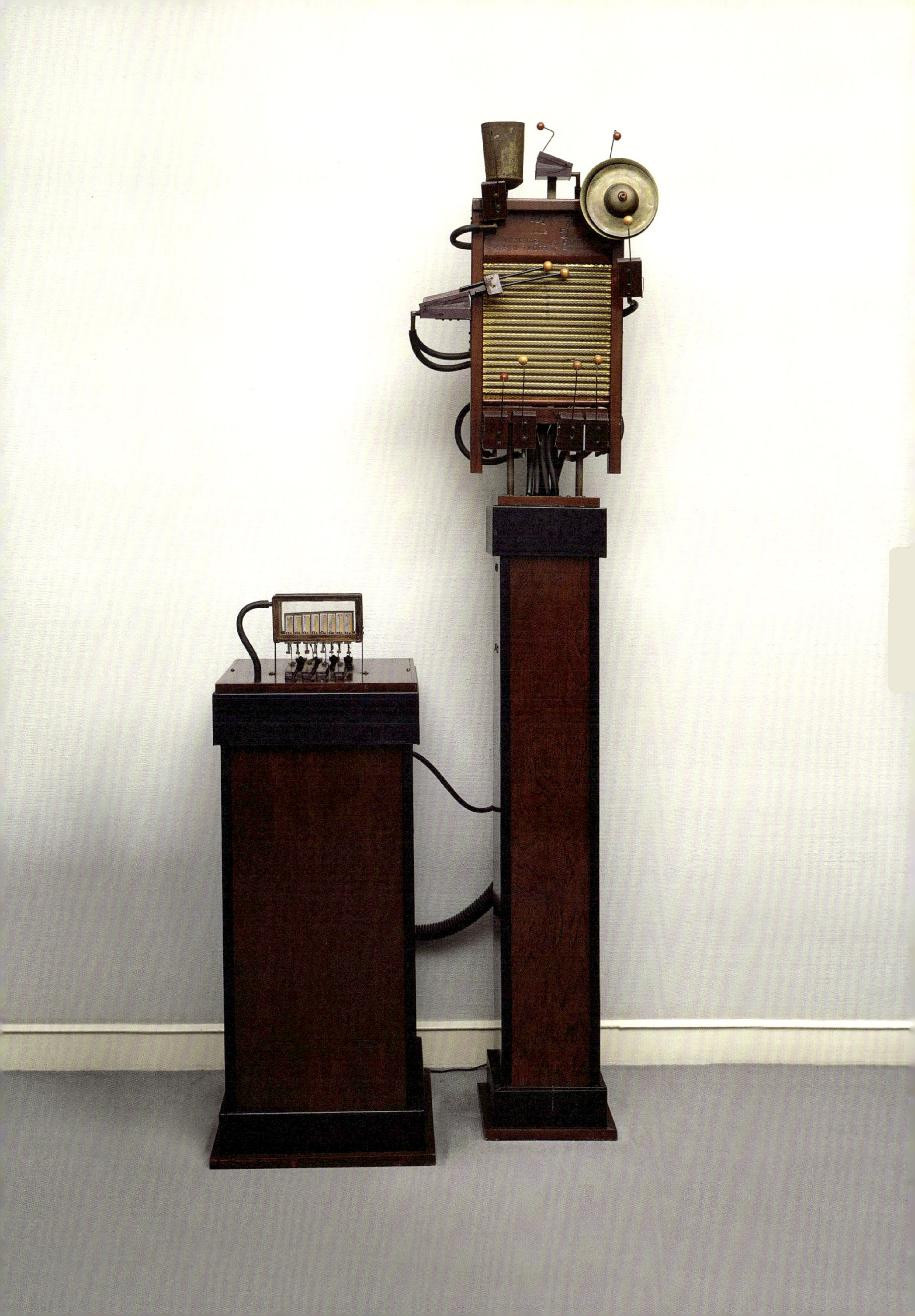

Bruce Nauman

CONCRETE TAPE RECORDER PIECE 1968

Since the 1960s, the works by American artist Bruce Nauman (Fort Wayne, 1941) have involved a broad range of media. After studying mathematics and then the arts, he stopped painting and began to make sculptures based on the backs of objects or molded from parts of his own body, as well as neon pieces with words and films of his performances.

Concrete Tape Recorder Piece consists of a tape recorder with a tape loop of a screaming woman, wrapped in a plastic bag and cast into the center of a block of concrete weighing about 240 kg. The electric cord and plug coming out of one side of the block constitute the only evidence that the work is supposed to contain a sound, although it would be inaudible at any volume of playback. By putting the recording of a scream inside a concrete block, Nauman reduces the scream to silence.

As in Robert Morris's work *Box with the Sound of Its Own Making* (1961), the reference to Marcel Duchamp's *With Hidden Noise* of 1916 [p. 164] is evident in the suggestion that a sound has been located inside a box.

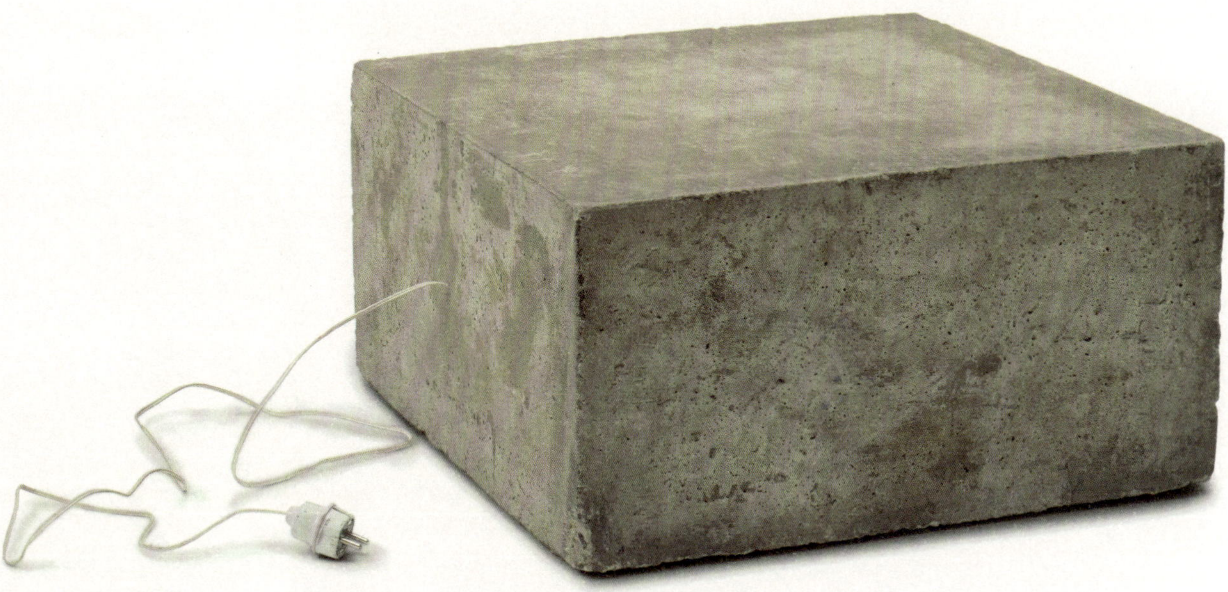

Michelangelo Pistoletto

LE TROMBE DEL GIUDIZIO
1968

Together with *Oggetti in meno* (The Minus Objects, 1965–66), Italian artist Michelangelo Pistoletto's (Biella, 1933) mirror paintings explored the theme of the continual and dynamic reflection of reality in art, one that would return in his early collective actions in the milieu of Arte Povera. Working with Lo Zoo, a group of people from different artistic disciplines (music, literature, theater, the visual arts), Pistoletto staged theatrical productions: "It is not a question so much of involving the public, or making them participate, but of impinging on their freedom and imagination, of triggering liberating mechanisms in them." *

Le trombe del Giudizio (The Trumpets of Judgment) was the name first given to a performance staged in 1968 by Pistoletto and Maria Pioppi in the courtyard of the artist's studio, when the pair were photographed playing 3 old trumpets once used as loudspeakers to amplify radio broadcasts or speeches made in public squares. Beginning in 1991, the trumpets were used on various occasions in public performances of Pistoletto's *Concerto per megafoni* (Concerto for Megaphones).

* Interview with Guido Boursier, "Far scattare nella gente meccanismi di liberazione," *Sipario*, no. 276, Milan, April 1969, p. 17.

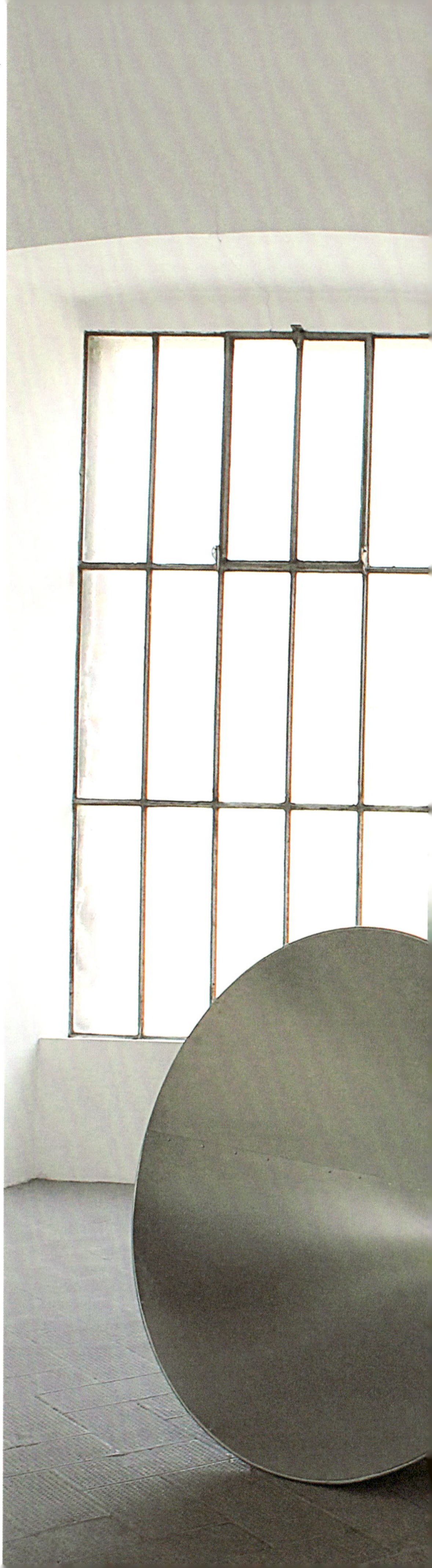

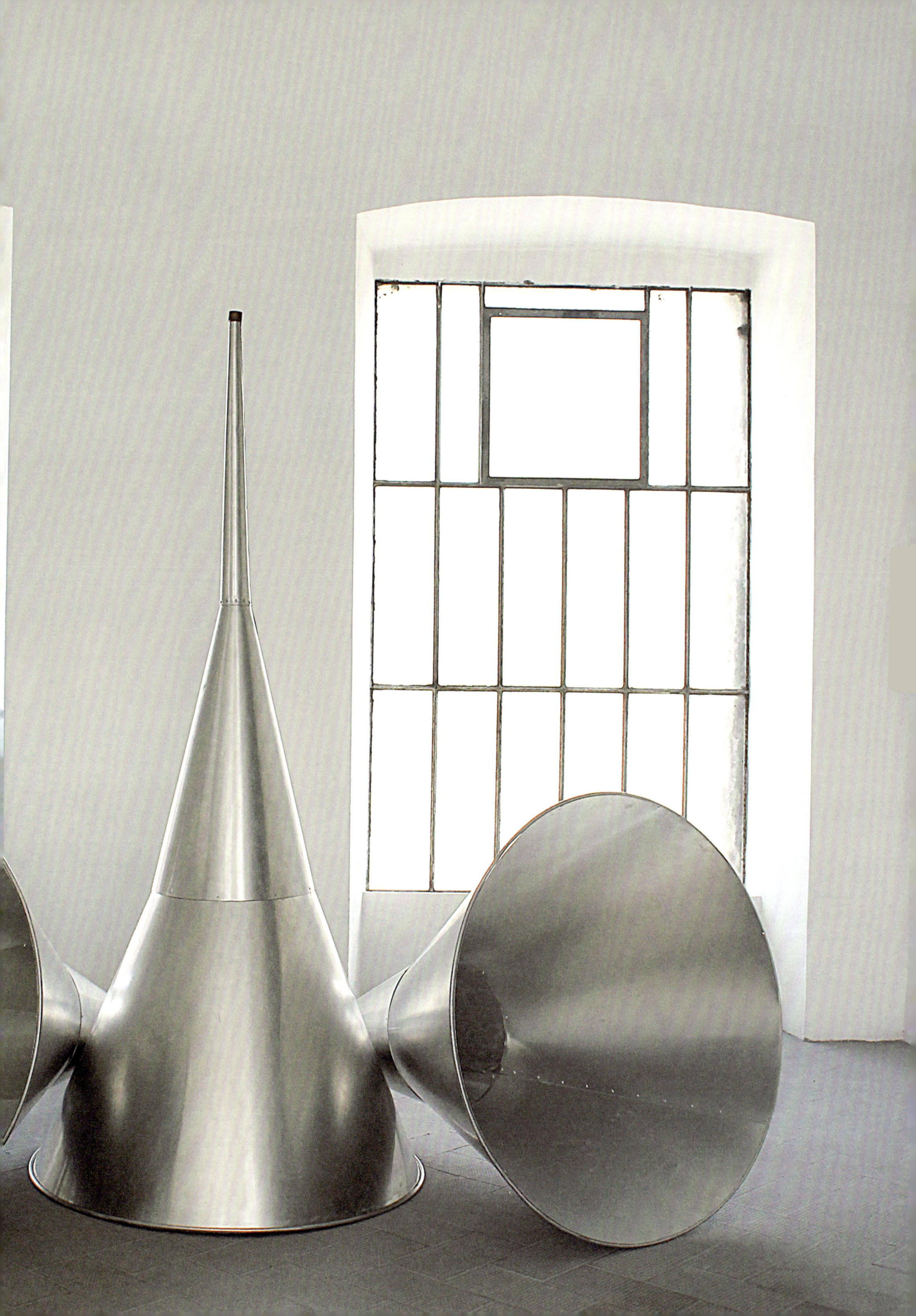

Wolf Vostell

RADAR ALARM
1969

Wolf Vostell (Leverkusen, 1932 – Berlin, 1998) was a German musician and artist and one of the foremost exponents of the Fluxus movement. Among the first to use the technique of décollage, Vostell laid claim to the artistic nature of the most ordinary objects and elementary gestures and advanced the existence of the creative act in the flow of daily life, with a particular focus on sound. His production includes *Fluxus-Symphony for 40 Vacuum Cleaners* (1976) and *Le Cri* (The Cry, 1990), a concert for 90 performers, in which 200 disparate objects acted as instruments.

Radar Alarm consists of a bicycle with a television and a series of 40 alarm sirens. The TV screen shows a normal television program, while the alarm sirens are connected to a timer that switches them on for 10 seconds every hour.

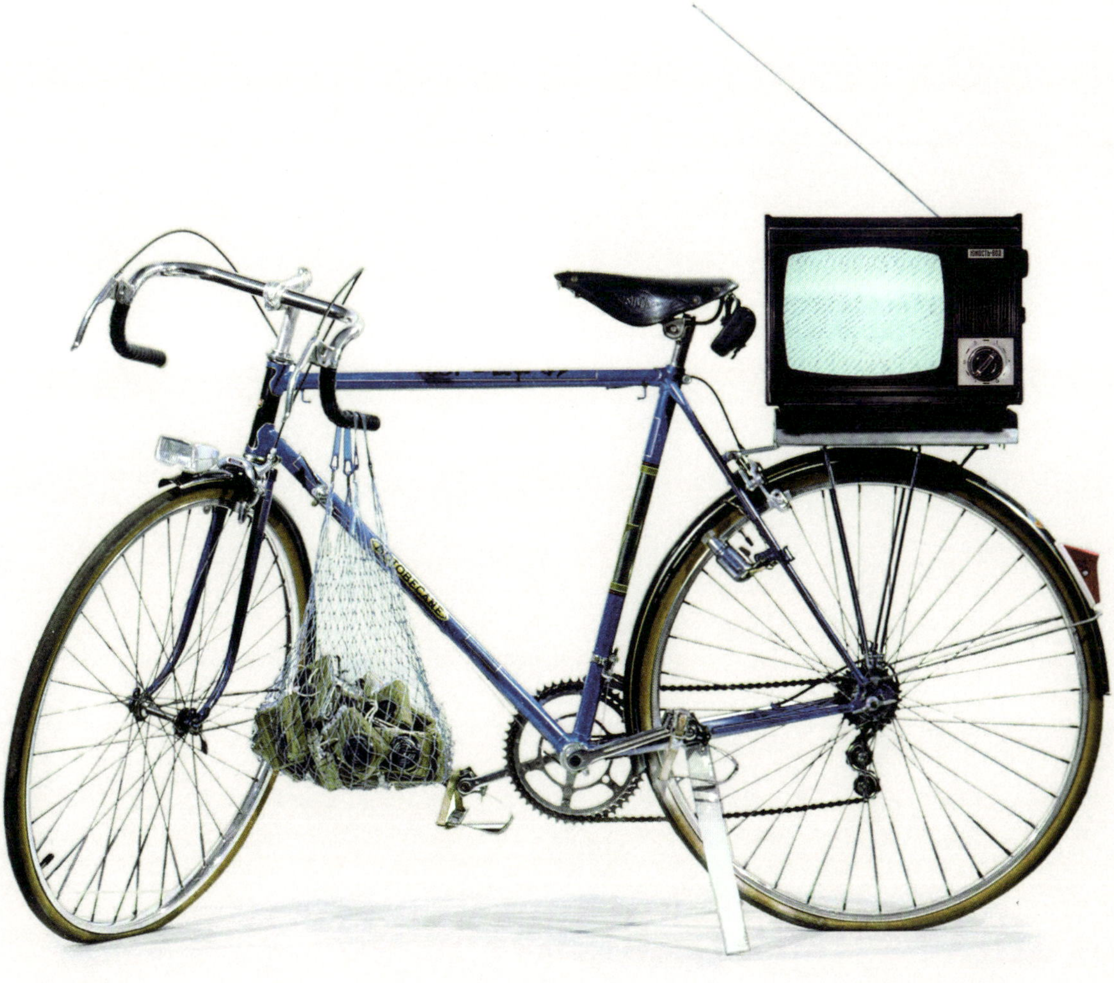

ALAN LICHT

THE NOISE OF SURFACE

MAKING ART AUDIBLE IN THE 20TH CENTURY

As ethereal as sound may seem, its existence is predicated on the interaction of matter. Sound is the auditory result of physical contact, whether water dripping from a faucet into a sink or an electronic tone being pushed through the speaker of an amplification system. A musical instrument is also based on touch, fashioned with the specific intent to generate sound by causing one or more of its parts to vibrate. Customarily, the instrument is intended to supply pitches that can then be manipulated by the player.

Sound has a role as an indicator of presence. Even an object that does not intrinsically have a sound can be *made* to sound by being touched. It's a key part of identity, not only for animals that emit sound, whether it's a whale singing or a fly buzzing, but for motorized objects (e.g. a car, a lawnmower) and elements of nature (rain, wind and fire). An instrument is an extension of a human being's innate sound-producing ability. In the case of most wind or string instruments, the instrument becomes something of a substitute for the voice, while percussion instruments relate more to the heartbeat and body movement. Any instrumentalist uses music as a second language, a form of communication and expression that is nonverbal (except for singing) but still subject to equivalent notions of grammar.

The plastic arts are fundamentally an extension of sight, not speech or hearing. They are mute: there is no aural component. And the viewer is forbidden to touch the work—the engagement is strictly visual, not tactile. Silence is the general condition in which the work is observed and contemplated. It was only in the 20th century that artists began to conceive visual artworks that would also have sounds. Marcel Duchamp made *With Hidden Noise* (also known as *À bruit secret*) in 1916 [164], a ball of yarn with a mysterious object inside meant to be shaken to produce a sound. In the 1950s, Kinetic artists like Jean Tinguely [226], Takis [284], Len Lye and Nicholas Schöffer forged motorized sculptures in which separate parts would strike each other. Two artists of the conceptual era of the 1960s juxtaposed objects with their sounds: Robert Morris created *Box with the Sound of Its Own Making* in 1961 [220], enclosing a three-hour recording of himself constructing a box inside the box itself, while William Anastasi [290] made a series of "sound objects," reimagining the readymade by including the sound of its operation, such as a radiator accompanied by a tape recording of it clanking.

Bruce Nauman's sculpture *Musical Chairs* (1983) put two steel bars and a chair, suspended from the ceiling, on a collision course with each other in order to "actually make noise—make music." More recently, in 2002 Stephen Vitiello attached microsensors to Donald Judd sculptures in Marfa to see what vibrations, ordinarily inaudible, would be revealed. In each case, the artwork is given a voice, a sonic individuality that goes beyond the visual surface.

Harry Bertoia and the Baschet Brothers were pioneers in making sculpture that was conceptualized as both exhibitive and playable. Bertoia heard two pieces of wire collide while making a metal sculpture and began making "tonal sculptures" of thin vertical rods fastened to a base that, when activated, knock together to produce shivery, bell-like chimes. The sound varies according to size and the type of metal used, and when used collectively the sculptures can be almost orchestral in range. Bertoia recorded eleven albums of himself playing the sculptures and gave concerts. The sculptures can be found in numerous museum collections as well as public outdoor installations. The Baschet Brothers also made sounding sculptures [190, 192] using metal or glass rods that are combined with other objects (glass, plastic, piano wire, balloons) that interact when the sculpture is vibrated with wet fingers. The Baschets collaborated with Jacques and Yvonne Lasry, who composed and performed with the sculptures, and other composers including Toru Takemitsu and Jean Michel Jarre have used their work in scores. The sculptures themselves have exhibited internationally, starting with New York's Museum of Modern Art in 1965.

Parallel to these developments in the art world and the advancement of an aesthetic of abstraction, at the dawn of Modernism there was an interest in recalibrating a definition of music and what could be considered a musical instrument. The classical music avant-garde moved to shatter taboos, at first with the increasing prevalence of atonality, dissonance and more complex rhythms. But the truly radical ideas for a new music came from the Futurists, who exhorted taking dissonance to the extreme by incorporating machine-age noises into music, as in Luigi Russolo's manifesto *Art of Noises* (1913).

The sound poetry of Futurist F.T. Marinetti and Dadaists Hugo Ball and Raoul Hausmann broke down language vocally, retaining a sense of structure but substituting gibberish and guttural noises for speech, hinting that noise could likewise replace pitch in instrumental music (which, as I have already noted, is inwardly linguistic). Kurt Schwitters, in fact, composed a sound poem in musical terms (scherzo, cadenza, etc.) for non-pitched vocal sounds, *Die Sonate in Urlauten* (1922–32). Piet Mondrian called for "a new order of sounds and nonsounds (determined noise)" in his essays on Neo-Plasticism from 1921 and 1922, and László Moholy-Nagy looked at phonographs as potential music makers, rather than merely sound reproducers in his own 1923 essay *New Form in Music: Potentialities of the Phonograph.*

Some composers took notice: Edgard Varèse likewise proposed music as an "organization of sounds." George Antheil scored his *Ballet Méchanique* (1927) for an eccentric array of player pianos, airplane propellers, xylophones, electric fans, a siren, bass drums and a tam tam. John Cage used turntables in *Imaginary Landscape No. 1* (1939). Classical instruments were also reassessed: Henry Cowell began scraping the strings of a piano in his compositions, while his former student Cage "prepared" a piano with various nuts and bolts between the strings, expanding its tonal palette and underscoring its structural rigging as a percussion instrument. In the 1940s, *musique concrète* tape composers like Pierre Schaeffer and Karlheinz Stockhausen made tape and the recording studio viable instruments in their own right, twisting recorded instrumental and natural sounds into alien soundscapes via electronic processing. Cage's *Cartridge Music* (1960) brought the amplification and mixing of non-instrumental sounds (of furniture and other quotidian odds and ends, fitted with phonograph cartridges or contact mics) to a live concert situation.

Cage's now-famous visit to an anoechoic chamber in 1951 certified, in his mind, that there was no such thing as silence. If *everything* has a sound, it then followed that *anything* could be enlisted in the service of music composition.

New instruments were fabricated in the spirit of forward-looking music. David Tudor, a premier interpreter of modern piano works, began making his own electronic devices in the 1960s, composing and performing with them. Composer Harry Partch built special instruments in order to play his own microtonal music rather than the equal-tempered music favored in Western classical music since the 15th century. (Western classical music has 12 notes to the octave, while Partch's instruments can have up to 43 notes to the octave). In addition to "adapted" guitars and viola, Partch invented his own versions of an organ (the Chromelodeon), marimbas (the Boo, a bamboo marimba; the Diamond Marimba and the Mazda Marimba, made of 24 tuned light bulbs), and singular creations (Spoils of War, a pile of junk retrofitted for use as percussion, including several shell casings; the Kiharta, which resembles a harp but is played like a slide guitar; and Cloud Chamber Bowls, glass gongs that hang from a wooden frame). In the early 1980s Glenn Branca devised his own unusual guitars (harmonic guitar, mallet guitar), which are played horizontally with mallets and constituted of planks of wood with additional bridges and strings, born of a similar desire to accommodate a special expanded tuning system based on the harmonic series.

Some artists also investigated homemade instruments, chiefly for their own performance purposes. Russolo's advocacy of noise music led him to develop the *intonarumori* ("noise machines"), wooden sound boxes with a single speaker that were activated with a crank and equipped with a lever to control pitch. He had 27 different types, each designed to produce specific sounds, e.g. the *Ululatore* (which howled), the *Crepitatore* (which crackled) and the *Rombatore* (which roared). Much later, in the neo-Dada Fluxus movement of the 1960's, came Joe Jones' music machines [234] (instruments played automatically via motors, including guitar, violin, drums, bells, gongs, etc., occasionally grouped together as the Mechanical Fluxorchestra) and Nam June Paik's *TV Cello* (1971), played extensively by Charlotte Moorman.

In the late 1960s, sculptor Robert Rutman introduced his *Steel Cello* and *Bow Chime* (both fashioned from metal and played with a bow). A decade later Yoshi Wada put together long horns out of plumbing pipe, while Akio Suzuki concocted his *Analapos* (1970) [272] designed to study echoes, comprising two metal cylinders and a spiral coil. More recent entries include Ken Butler's hybrid instruments [298] (stringed instruments whose bodies consist of found objects, e.g. a violin made from a rotary phone, a skateboard turned into a guitar). The music the artists produce with these instruments is often somewhat hard to differentiate from a demonstration of their capabilities. Paul De Marinis' *The Pygmy Gamelan* (1973) [280], a gadget that would play five notes in response to whatever electromagnetic field it was situated in, was conceived both as a stand-alone exhibitive sculpture and something that the *viewer* could play (at one point De Marinis even envisioned marketing them as a kind of alternative radio). Laurie Anderson's *Handphone Table* (1978) [296], an exploration of the combination of a table and human bones as sound conductors, was also geared to a visitor's experience rather than an artist's performance: the table is wired to taped sounds, which are heard/felt by placing elbows on the table and cupping hands over the ears.

Occasionally artists have harnessed a regular musical instrument for something other than its original function. In his installation *Playing the Building* (2008) David Byrne took an old organ, attached long cables to the keyboard instead of pipes, and connected them to various points in an old building—pipes, plumbing, beams and other built-in metal works. Visitors were invited to press the keys, which would elicit jarring clangs, whistling flute-like tones and low rumbles (arranged according to the low, middle and upper registers of the keyboards in a nod to the instrument's usual capabilities). The piece turned the room into a sound sculpture, but also transformed the organ into a non-electronic sampling keyboard. It was an installation, but without continuous sound.

Manuel Rocha Iturbide's *I Play the Drums with Frequency* (2007) disperses a drum set in different parts of a room, planting a speaker on each piece. A computer feeds sinusoidal waves to the speakers, which then make the drumheads and cymbals vibrate, resulting in a timbrally complex sound environment that is latent within a drum set but not evident in the usual manner of playing the kit with sticks. In his video *Guitar Drag* (2000), Christian Marclay plugs a new Fender Stratocaster electric guitar into an amplifier on the back of a pickup truck, tying it to the vehicle with a rope at its headstock. The truck drags the guitar behind it as it traverses fields and highways in Texas. As the strings are all broken within minutes, the guitar's pickups become contact mics on a chunk of wood and metal in friction with earth and pavement; it is the guitar body that now vibrates, not the strings, with the changing terrain providing much of the differentiation in sonic texture. Initially inspired by the gruesome murder of James Byrd Jr. in a similar manner in Texas, Marclay also cited a Nam June Paik performance in which he dragged a violin behind him with a rope, as well as rock guitarists Pete Townshend and Jimi Hendrix's decimation of guitars onstage. That Townshend, Hendrix and Marclay's guitars were still roaring with noise even after being rendered "unplayable" speaks to the unintended consequences of amplification and how deeply the Modernist re-think of music's parameters affected both instruments and musical idioms in the last half century.

In his art school days in the early 1960s, Pete Townshend heard Gustav Metzger lecture on auto-destruction in art, while another British art student of the time, the improvising musician and painter Keith Rowe, was arguably the first to lay a regular guitar down flat on a table to be played. Following Jackson Pollock's lead in terms of what could be done by dripping paint on a canvas laid on the floor, and with an awareness of Cage's prepared piano, Rowe dropped various objects on the supine guitar, attaching contact mics to the body and never play the instrument in the traditional fashion (he also never tuned it).

Rowe was among the first practitioners of what was dubbed "free improvisation," produced in the 1960s and 1970s by a small but ultimately influential group of European musicians who attempted to do real-time atonal, aleatoric music without any predetermined score, under the influence of both jazz improvisation and the open scores of modern classical music. Because the emphasis was on texture rather than harmony, the idea of unorthodox extended technique (like Rowe's) was embraced by improvising musicians, who had to find ways of continually surprising themselves, their co-players and audiences. Although less thoroughly radical than Rowe's rejection of the usual method of playing his instrument, percussionists like Han Bennink, Tony Oxley and Jamie Muir still heavily modified their drum kits with various metallic and wooden objects and even electronics (in Oxley's case) for a greatly widened sonic scope (Bennink and Muir were also visual artists). Other improvisers built their own acoustic and/or electronic instruments, such as Paul Burwell, Max Eastley, Hans Reichel and Hugh Davies.

However, predating the rise of free improvisation among experiment-minded musicians were the musical investigations carried out by artist Jean Dubuffet in his wildly cacophonous 1960 record-
▸9 ings with Asger Jorn. The two tackled piano, violin, cello, trumpet, recorder, Saharan flute, guitar and tambourine, instruments with which they'd had no prior experience. The Artists Jazz Band & the Nihilist Spasm Band, both emerging in Canada in the 1960s, were also comprised of musically untutored visual artists who embraced free improvisation as a form of music making (several members of the Spasm Band made their own instruments, specifically electric guitars and electric kazoos).

By eliminating many of the accepted systems of playing music and preferring to approach instruments "improperly," free improvisation was open to both untrained *and* trained players. Marclay, also a non-musician, joined the ranks of the international improvisers in the 1980s. He used records and turntables as his instruments, and more importantly made "recycled records" glued together from shards of other broken discs that not only served him in performance but were exhibited and sold as artworks.

If Marclay, Bertoia and the Baschets blurred the line between sculpture and instrument, then Bill Fontana's use of the term "sound sculpture" for his sound installations provides yet another perspective. Fontana extensively utilizes field recording as a way of perceiving a given location's sound from every possible vantage point, much as one would do while walking around a sculpture. Fontana has written, "In the ongoing sculptural definition of my work I have used different strategies to overcome the ephemeral qualities of sound, that seem to be in marked contrast to the sense of physical certainty and permanence that normally belong to sculpture and architecture."[1] He notes his 1976 work *Kirribilli Wharf*, a recording he made by the sea in Australia: "The percussive wave action at Kirribilli Wharf had continuousness and permanence about it. This 8-channel tape was not a recording of a unique moment, as with the total eclipse, but was an excerpt from a sound process that is perpetual."[2]

Fontana's approach is to listen first, and then do the sculpting. In sound art, of which Fontana's work is a prime example, listening itself is the seed; a completed work operates on a listener-to-listener relationship. Composed music, by contrast, has a player-to-listener relationship. At Kirribilli Wharf the sound is already there, freestanding; it existed before, during, and after Fontana made his field recording. It is not "played" by Fontana, and is hearable in its natural state by anyone. Much sound art springs from a desire to give sound a continuity, as found in an almost systematic environmental soundscape like those explored by Fontana.

The relative regularity of the activity at the wharf locale allows for immersive, in-depth listening, rather than the type accorded to following the narrative arc of a composition. Music itself is episodic, lasting for a given interval and then ending. An instrument is played, then falls silent when put down. When a craftsman designs a musical instrument, a tool is made: it is intended to be durable and eye-catching, but not ornamental. When artists make objects that are supposed to be audible, they have produced an embodiment of the paradox between the finality of a work of art and the transience of sound.

2

1
Bill Fontana, *Resoundings*, resoundings.org

2
Ibid.

SIMONE MENEGOI

SOUND ART?

VISUAL ARTS AND SOUND AT THE TURN OF THE 1970s–1980s

The late 1970s and early 1980s marked a significant historical shift in the relationship between visual arts and sound in the 20th century. In this relatively brief span of time, a series of thematic exhibitions in Europe and the Unites States presented, as a unitary phenomenon, the work of the artists who had begun over the previous two decades to use sound in conjunction with other media typical of the visual arts, some even going so far as to use it exclusively, ascribing an importance to the spatial dimension of sound at least comparable to its temporal dimension. This marked the official debut in cultural institutions of the phenomenon that would take the name "sound art" in those years, one that continues today to be the subject of exhibitions and critical debate, and that poses, now more than ever, problems of definition, categorization and presentation.

If we look at the chronology of thematic exhibitions dedicated to the relationship between art and sound since the 1960s, the phenomenon we have mentioned is strongly represented.[1] The beginning of the 1960s was marked by a burgeoning of events and exhibitions associated with the Fluxus movement—that great laboratory of interbreeding among disciplines. However, after this initial burst, the later 1960s and 1970s brought a relative paucity of events (among the few that did take place, of particular note is "Record as Artwork, 1959–73" curated by Germano Celant, Royal College of Art Gallery, London, 1973). Interest in the relationship between art and sound as an exhibition medium resurfaced in the second half of the 1970s, culminating during the late 1970s and early 1980s in a proliferation of small and large exhibitions, some immediately recognized as milestones.

Europe witnessed "Für Augen und Ohren. Von der Spieluhr zum akustischen Environment – Objekte, Installationen, Performances" (Akademie der Künste, Berlin, 1980, organized by René Block), a large collection of objects, instruments and machines for producing sound: from the musical automata of the 17th and 18th centuries to the vinyl-engraved cacophony of Fluxus artist Milan Knížák [232].

On a smaller and somewhat more obscure scale, but predating the Berlin exhibition, another German exhibition, "Sehen um zu hören: Objekte und Konzerte zur visuellen Musik der 60er Jahre" (Städtische Kunsthalle, Düsseldorf, 1975, curated by Inge Baecker), pivoted around the Fluxus movement. Meanwhile, in a crescendo of dimensions and ambitions, the United States produced "Audio Works" (Artists Space, New York, 1978, organized by Helene Winer), which included the works on magnetic tape by conceptual artists such as Barbara Kruger, John Baldessarri, Robert Barry, Guy de Cointet, Bernard Venet and Lawrence Weiner that had been featured in a slightly earlier exhibition ("Narrative Themes/Audio Works," Los Angeles Institute for Contemporary Art, 1977, organized by Robert Smith); "Sound. An Exhibition of Sound Sculpture, Instrument Building and Acoustically Tuned Spaces" (Los Angeles Institute of Contemporary Art and subsequently, in an expanded version, at PS1, New York, 1979, curated by Robert Smith and Bob Wilhite); and "Soundings" (Neuberger Museum, SUNY at Purchase, New York, 1981, organized by Suzanne Delehanty), presented as "the first American survey of visual artists' use of sound, music and acoustical phenomena from 1900 to the present."[2]

The broadest and most significant shows, such as "Für Augen und Ohren" and "Soundings," which undertook the titanic enterprise of summarizing an entire century (and more: the Berlin exhibition also included objects dating back four centuries) of art-sound relationships, shared common ground with more narrowly focused events: the works of the 1960s and 1970s that occupied a no-man's land between sound and vision. There was now quite a bit of material of sufficient variety and elaboration as to constitute a critical mass—to the point that its authors, harking as much from visual and performing arts as from experimental music, had begun to develop specific artistic conceptions and terminology to talk about their work. If we look at the phenomenon of today's visual arts, it is rather clear that the emergence of this new area of exploration in artistic theory and practice prompted, in a more or less openly declared manner, the exhibitions we have listed (and many others would follow in the ensuing years).

Even in large historical art exhibitions, one of the main motivating factors for organizers was to lay out and establish within 20th-century culture a background for this recent (and stimulating) artistic phenomenon. Suzanne Delehanty's introduction to "Soundings" in the homonymous exhibition cata-
3 logue is exemplary in this sense. She framed (within a historical and cultural perspective spanning two centuries, the 19th and 20th), the work of artists emerging in the 1960s and 1970s who used sound, alone or mixed with other media, within the visual arts—from Nam June Paik [222, 228] and Harry Bertoia to Max Neuhaus and Bruce Nauman [248]—and that of composers of experimental music whose works, in terms of instrumentation, format and reception, had years ago abandoned the concert halls—from Takehisa Kosugi and John Cage to Alvin Lucier and David Tudor—and organized them into two main areas of pursuit.

The first regarded technological innovation—in this case innovations allowing sound to be recorded and reproduced, or allowing it to be associated with images, such as cinema and video—and its impact on the arts, both in terms of providing new means for expression and in creating and defining new artistic conceptions. The second regarded attempts to create fusion among the arts: the recurrent utopia in 19th- and 20th-century culture of a *Gesamtkustwerk* that could recapture the original unity that the arts had had, for example, in ancient Greek drama; the search for synaesthetic correspondence between sound and color as seen in Symbolism and early abstract painting; and the work with graphics, meaning and sound of the poetry of Stéphane Mallarmé, Guillaume Apollinaire and the Futurists. Overall, it was a solid but traditional critical approach, concerned less with describing a new trend in progress than with giving it a cultural genealogy (the latter certainly a worthy undertaking!). The organization of the text faithfully reproduced that of the exhibition, effectively recapitulating, through works, objects and documents, the works of the first half of the 20th century before continuing to discuss the creations of the artists cited above.[3]

Conversely, another major American exhibition in those years exploring the relationship between art and sound, "Sound" (1979), focused precisely on recent artistic production, mustering over forty artists, most of whom had come onto the scene in the 1970s. The catalogue writings also concentrated on the latest trends. The most important essay, by Peter Frank,[4] essentially delineated three areas of exploration, all of them represented in the show: the first (represented by Connie Beckley, Charlemagne Palestine, Peter Gordon, Laurie Anderson [296, 302] and the duo Christina Kubisch and Fabrizio Plessi) is strictly related to the dimension of time, and included the physical presence of the artist themselves, making sounds with their voice or an instrument, as a more or less constant component.

The second is located somewhere between sculpture and the building of musical instruments, in which Frank identifies two subsets: "sculptural instruments," created mainly by musicians who wanted to move beyond the limits of the instrumental legacy of the 19th century and featuring a sculptural element (Susan Rawcliffe, Bob Bates and Emmet Chapman, along with veterans such as Harry Partch and the Baschet brothers [190, 192]), and "sounding sculptures," created mainly by visual artists intent on exploring an additional, dynamic and temporary dimension of sculpture (among others, Bob Wilhite, Jim Pomeroy, Tom Recchion, Stephan von Huene [244, 246]).

The third area of exploration, one that had a more radically "intermedia" character, difficult to see even as a sum of different disciplines in that it appears to constitute a realm of exploration completely in its own right, is the "acoustically tuned spaces" (the exhibition subtitle) or "sonic installations," as Frank calls them. They are environmental works where the accent is less on the timespan of the sound and more on the space it occupies, in a phenomenological acceptation of their relationship. (The names in this case are Michael Brewster, Max Neuhaus, Liz Phillips, Bill Fontana, Marvin Torffield, Karen Wolff and William Klingsbury).

At the conclusion of the text, as a sort of recap and to underscore the complexity and variety of the artistic area whose confines he is trying to map out, Frank adds a small list of the most singular works, apparently completely independent of one another, that he has cited in his text: "A musical score that looks like Mondrian. One-word instructions for concert pieces. A room full of clicking. Rock and country bands in art performances. Musicians playing sculpture. An artist playing flute through a gas mask. A fragmentary opera gleaned from the words of a sodden Victorian novel."[5]

As heterogeneous as the catalogue is, it could have embraced even more variety. For example, it could have included the art tapes of the conceptual artists who were the vital force in the exhibition "Narrative Themes / Audio Works," or the singular audio performances aired in those years on Close Radio, a show hosted by John Duncan and Neil Goldstein on KPFK Los Angeles.

Words are important. Naming is an act of creation—or at least of appropriation. And so, the efforts of critics and curators in those years were not just to circumscribe the area shared by music (or sound, pure and simple) and visual arts, but also to find the words to do so. We have already read some of them: "sculptural instruments" and "sounding sculptures," "sonic installations" and "acoustically tuned spaces," to which we must add "musical art"—the catchall term Peter Frank used in referring to the art-sound phenomenon in 1979.[6] The two German exhibitions mentioned above, "Sehen um zu hören" and "Für Augen und Ohren," gave us definitions such as *visuelle Musik*—which well suited the Fluxus pursuits that innervated the 1975 Düsseldorf exhibition—and *akustischen Environment*.

In a review of "Soundings," another critic named Frank, this time Robert Frank, used an expression that would enjoy greater fortune in the upcoming decades becoming in English-speaking circles, "the" definition of a certain type of artistic realm: "sound art." The expression is not found in Suzanne Delehanty's text or in the title of the exhibition. Robert Frank used it because it had been in the air for a number of years. Paternity is usually attributed to William Hellermann, a New York artist (not included in "Sounds," but cited in the exhibition catalogue) who created the SoundArt Foundation in the late 1970s, and who, in 1983, organized another important exhibition for the period at the Sculpture Center of New York titled "Sound/Art."

This exhibition included, among others, eminently visual/auditory artists who had already been featured in "Sounds," alongside other fairly well established names who were not primarily associated with sound but who nevertheless used it, such as Vito Acconci, Hannah Wilke and Keith Sonnier.

It is certainly no coincidence that Hellermann was an artist. One of the organizers of "Sounds," Bob Wilhite, was also an artist. It was the artists' efforts to find a new identity for their emerging quest that generated the phenomenon that the exhibitions of the 1970s and 1980s sought to encapsulate, addressing it from different angles. Another of the artists mentioned above, Max Neuhaus, was the one who first used the term "sound installation" to describe a cycle of works, begun in 1967, that sought to liberate sound from the spatial logic of the concert hall, and from the temporal logic of the piece of music with a set beginning and end.

Some thirty-five years after those artistic forays and the exhibitions that documented them, what is the final reckoning for the transition years from the 1970s to the 1980s? And what sort of theoretical framework can be applied?

At the time, the search for an understanding and definition of certain works oscillated between an appeal to a theoretical framework that had originated over ten years earlier with Dick Higgins in the Fluxus movement and was by then familiar, i.e. intermedia, and the attempt to create new and specific terminology and categories (not always perfectly well focused); between a long-term vision drawing on the synaesthetic quests of the 19th and 20th centuries, and on the experiments of the historical avant-garde movements and an occasionally over-meticulous and at times gap-riddled mapping of the events and actors that populated the scene.

Today, with the benefit of hindsight, it may be worthwhile to explore other routes. For example, we may try rethinking the artistic experiences in sound and vision and the effort of definition (and self-definition) that they entailed using the terms of one of the most interesting philosophers of esthetics to come onto the international scene in the past decade, Jacques Rancière. In particular, we may examine the terms with which Rancière reinterpreted all of artistic modernity from the 18th century to the 1950s in *Aisthesis. Scènes du régime esthétique de l'art* (2011).

In the introduction to the fourteen chapters, each one dedicated to a work and its reception by contemporaries, Rancière pointed out that what he has produced might even be considered a "counter-history" of the modern period in the arts by virtue of the degree of its opposition to the prevailing criteria by which this period has been identified and described: "Influential histories and philosophies of artistic modernity identify it with the conquest of autonomy by each art, which is expressed in exemplary works that break with the course of history, separating themselves both from the art of the past and the 'aesthetic' forms of prosaic life. Fifteen years of work have brought me to the exact opposite conclusions: the movement belonging to the aesthetic regime, which supported the dream of artistic novelty and fusion between art and life subsumed under the idea of modernity, tends to erase the specificities of the arts and to blur the boundaries that separate them from each other and from ordinary experience. These works only create ruptures by condensing features of regimes of perception and thought that precede them, and are formed elsewhere."[7]

The idea of reading the emergence of "sound art" (necessarily in quotes) as a case study—excluded from Rancière's treatment because of the chronological limits set by the philosopher—of the essential logic of arts in the modern era, i.e. their tendency to embrace on the one hand the means and ends of other arts and on the other aspects of the real world hitherto excluded from the sphere of the arts, belonging instead to that of ordinary experience, is certainly alluring. As regards the first aspect, there is no use to be gained from the point: the trend of which we have been speaking, whatever we may wish to call it, emerges by constitution as a hybrid, beholden as much to the arts of vision and space as to those of hearing and time. As regards the second aspect cited by Rancière, i.e. the tendency to annex aspects of the real hitherto excluded from the artistic sphere, the art/sound pursuits are no less exemplary. Cage, leaving the score of *4'33"* (1952) pristine and untouched, had already focused the audience's attention on the web of sounds that accompany us everywhere, even in the presumed acoustic neutrality of the concert hall, the equivalent of the "white cube" in the visual arts.

In a series of acoustic events entitled *LISTEN*, beginning in 1966 and continuing, in various forms, until 1978, Max Neuhaus had brought things full circle: his "concerts" involved leading the audience out of the designated listening place—whether theater or museum—and having them walk along the streets of their city, stopping in places that the artist thought were particularly interesting in terms of acoustics. These were places with completely ordinary sounds, and at times, such as the case with traffic sounds, even sounds that were unpleasant to the average listener. And this is only one example among many.

One of the stakes in the sound-vision experiences of the 1970s and 1980s was certainly the conquest of an integral acoustic dimension capable of appropriating and using any sound, any space and any interaction between the two.

There is another criterion elaborated by Rancière for interpreting modernity that offers some stimulus in the sound-art quest, marking a path that some have already begun. In *Aisthesis*, the philosopher does not speak only of works (painting, sculpture, cinema, theater, photography) that are unanimously considered to be "major." Just as often, he dedicates his attention to "minor," at times even obscure, works and events—pantomimes by clowns, exhibitions of decorative arts—following a compelling intuition: that those works and events have contributed to creating the "perceptual regime," the substrate of ideas and intuitions on arts and the real from which the "major" works arose. Applied to "sound art," Rancière's intuition counsels examining not only what happens in museums, galleries and festivals, but also what goes on around them. This includes what takes place in cinema, where the work of Walter Murch in Francis Ford Coppola's film *The Conversation* (1974) gave birth to the figure of the "sound designer" (and, in Murch's case, a sound designer who applied the techniques of the contemporary musical avant-garde to the sound fabric of the film). And it regards the emergence of an ecological conscience applied to sound, of which R. Murray Schafer's book *The Tuning of the World* (1977)[8] represents the first manifesto. These are, as I have said, quests that have already begun. Indeed, both references are cited in a very brief and useful profile of "sound art" written by one of the most discerning scholars in the field (and sound artist himself), Alan Licht.[9]

A final consideration. Already at the beginning of the 1980s, the realm of art and sound appeared to the best informed critics to be so complex and heterogeneous as to generate doubts about the possibility (and the wisdom) of giving it a unitary image in either an exhibition or a text.[10]

Over thirty years later, the doubt has become, in my opinion, a reasonable certainty. Retrospectively, the differences are even more evident: differences between those who used sound occasionally and others who made it the pivot of their work; differences between artists who looked to sound as a way of extending their visual work and who were quite at home in museums and galleries, and avant-garde musicians whose quest, due to the prevailing conservatory-oriented bent of the institutions of music, ended up finding refuge in the institutions of art for lack of any other alternative; differences between those who focused on the spatial, plastic properties of sound in relation to architecture and those who, like conceptual artists, conceived of sound purely as a vehicle for verbal information.

In the 1980s, all things considered, these differences might have seemed sufficiently negligible with respect to a unifying element—the use of sound in the visual arts—that was still fairly recent and little assessed at the institutional level, to make a unitary approach perfectly acceptable. Over time, however, the differences have become points of divergence, both theoretically and in terms of the practices of artists and curators. At the same time, as in other media (video first and foremost), the use of sound became so ubiquitous in the work of contemporary artists that it lost much of its power as a thematic glue.

In 2004, echoing the resurgence of interest in experimentation in sound and vision between the late 1990s and early 2000s, the magazine *Artforum* called a round table meeting via web with artists, curators, critics and philosophers who had shown a particular interest in the matter at hand or were directly active in it: Cristoph Cox, Carl Michael Von Hausswolff, Anthony Huberman, Branden W. Joseph, Steve Roden, Marina Rosenfeld, David Toop and Stephen Vitiello.[11] Reading their exchanges, the first thing that strikes you is the Babelian divergence of opinion on even the most basic questions of how to circumscribe the realm of discussion. Neuhaus, whose polemically titled essay "Sound Art?"[12] (the question mark challenges the very legitimacy of the term) is cited by Cox, argues that most of that which is categorized under the label in question is simply contemporary music, and concludes that it is disorienting to define it as anything else or place it within art institutions. Von Hasswolff agrees with this premise, but differs from Neuhaus in being happy that museums and galleries have enough slack in their disciplinary confines to welcome and accommodate new music. Christoph Cox proposes seeking the discriminating factor between music and sound art in the way that the dimension of time is addressed: linear and having a beginning and an end for music; and tending to be open-ended in sound art. However, he receives a lukewarm response from the other participants in the debate, particularly from the artists, typically reluctant to relinquish opportunities to express themselves for theoretical reasons.

And so on it goes. The discussion offers a palpable representation of the fact that grouping artists on the mere basis of the medium "sound," with no further qualifications (and qualifications that must be as sophisticated as the medium is ubiquitous) has long been, more than difficult, senseless. For that matter, as Douglas Kahn quipped several years ago, "Many artists who have been using sound for a long time would rather be called *artists* than *sound artists*."[13]

1
The material we consulted, which was incomplete but nevertheless quite useful, was compiled and made available online (joy.nujus.net/SAET) by two musician-artists and scholars of the intersections between visual arts and sound, the Frenchman Jerôme Joy and the American Thom Holmes. When we consulted the material (March 28, 2014) the chronology covered the period from 1964 to 2004 and sought, especially for the first fifteen years, to apply an inclusive rather than an exclusive criterion, including any "collective public presentation of sound and musical works (which overlap the concert framework such as installations, performances, etc., i.e. sound and music events which use another public configuration than the presentation on a stage, and which haven't a beginning nor an end) under a topic, a theme or a specific question and problem expressed by the title of the event."

2
As reported in Robert Frank's review of the exhibition "Soundings at SUNY," *Art Journal*, vol. 42, no. 1, New York, Spring 1982, pp. 58–62.

3
Suzanne Delehanty, "Soundings," in Suzanne Delehanty, Dore Ashton, Germano Celant, Lucy Fischer (eds.), *Soundings*, exhibition catalogue (Purchase, NY: Neuberger Museum, 1981), pp. 21–38.

4
Peter Frank, "The Sound and the Theory. A Short History of the Art-Music Interface," in Robert Smith, Robert Wilhite (eds.), *Sound. An Exhibition of Sound Sculpture, Instrument Building and Acoustically Tuned Spaces*, exhibition catalogue (Los Angeles/New York: Los Angeles Institute of Contemporary Art/Project Studio 1, 1979), pp. 9–16.

5
Ibid., p. 16.

6
"One result of this open acceptance in the art world of new musical and fused musical-artistic forms is the emergence of musical art of all kinds," in *Ibid.*, p. 12.

7
Jacques Rancière, *Aisthesis. Scenes from the Aesthetic Regime of Art*, trans. by Zakir Paul (London: Verso Books, 2013), p. XII. The philosopher's target is not hard to find, being explicitly named on the following page in extraordinarily harsh terms: Clement Greenberg, the foremost exponent of the "monument of retrospective modernism [which,] failing to found any important art, ... would however succeed in imposing the golden legend of the avant-gardes and rewrite the history of a century of artistic upheavals to its advantage." In *Ibid.*, p. XIII.

8
Raymond Murray Schafer, *The Tuning of the World* (Toronto/New York: McLelland and Stewart/Knopf, 1977).

9
Alan Licht, "Sound Art: Origins, Development and Ambiguities," in *Organised Sound. An International Journal of Music Technology* (Cambridge, UK: Cambridge University Press, April 2009), vol. 14, no. 1, p. 8.

10
In the above-cited review "Soundings at SUNY," Robert Frank wrote: "*Soundings* is only the most recent, and most formally museological, in a series of sound art shows of all shapes, sizes, forms and focuses. This series is likely to continue, although future sound art exhibitions here are likely to concentrate on specific aspects of the polymedium (sound-producing object, say, or the visual representation of music)," p. 58.

11
"Audio Files. Sound Art Now: An Online Symposium," 2004, artforum.com/symposium/id=6682 (last access March 27, 2014).

12
Max Neuhaus, "Sound Art?," liner notes for the audio CD *Volume: Bed of Sound* (New York: P. S. 1 Contemporary Art Center, July 2000). Now available at www.max-neuhaus.info/bibliography (last access March 28, 2014).

13
(Italics added). Douglas Kahn, "The Arts of Sound Art and Music," *The Iowa Review Web*, vol. 8, no. 1, February/March 2006, iowareview.uiowa.edu/TIRW/TIRW_Archive/feb06/kahn.html (last access March 27, 2014).

George Maciunas

PIANO PIECE
1970

Fluxus appeared in the early 1960s as a loose, international network of artists, composers, musicians and designers such as George Brecht, Dick Higgins, La Monte Young, Yoko Ono, Nam June Paik, and its founder George Maciunas (Kaunas, 1931 – Boston, 1978), a Lithuanian-born American artist.

The destruction of musical instruments as a creative process was part of the Fluxus repertoire, as for example Philip Corner's *Quiet Work of Destruction*, the slow dismantling of a piano during the 1962 first Fluxus Festival at Wiesbaden.

Piano Piece is also known as *Carpenter's Piano Piece*, and is the result of a performance during which the keys of a piano are nailed down one by one. The content of the work is the sound produced while nailing, bashing and smashing, together with the resonance made by the piano during this action. Nam June Paik, to whom the piece is dedicated, is also known for instrument destruction, such as in *Klavier Integral* (1958–63), which involved the ornamentation and then destruction of a piano.

[See also p. 292]

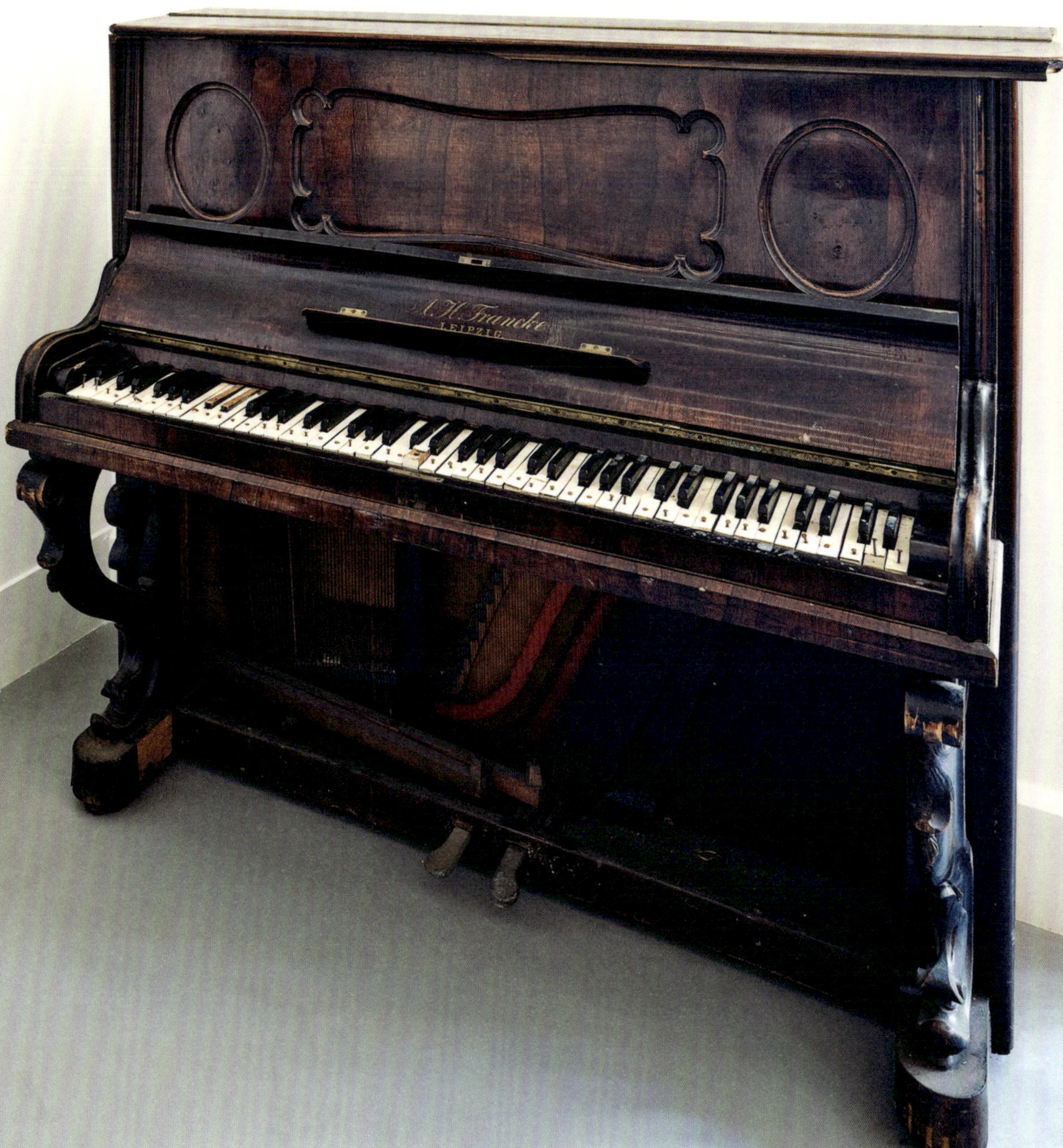
A. H. Francke
LEIPZIG

Akio Suzuki

ANALAPOS
1970

Visual and sound artist Akio Suzuki (Pyongyang, 1941) grew up in Aichi, Japan. Since the 1960s, Suzuki has been investigating the acoustic quality of selected locations and creating corresponding topographies. His intensive involvement with the phenomenon of pulse and echo led him to develop his own instruments in the 1970s. One of these is the spiral echo instrument *Analapos*, which consists of a coil spring and 2 iron cylinders that function as resonating chambers and is played with the voice or by hand. The name *Analapos* combines the terms "analog" and "postmodern" and stands for analogue listening in the digital age.

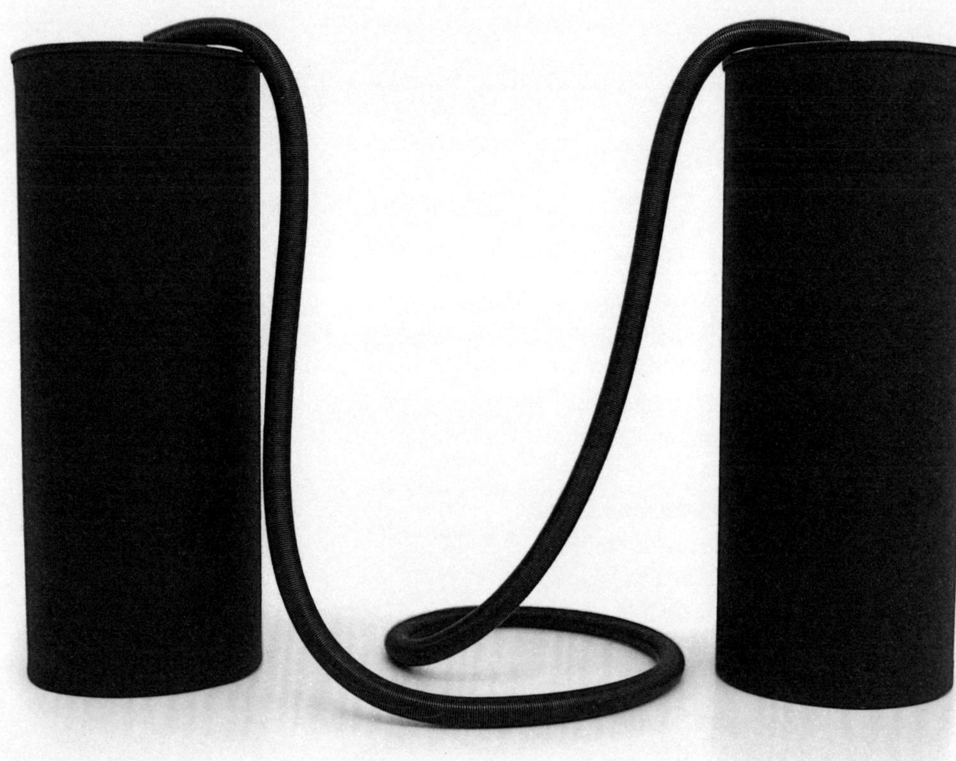

William T. Wiley

CANJO
1971

BANJO FOR J.B.
1985

DEBILSLIED
1986

During his career, American artist William T. Wiley (Bedford, 1937) has created drawings, paintings, sculptures, installations and films. His works share the aim of challenging viewers' certainties by layering meanings and intents, using self-references, references to current political affairs, simple decoration and autobiographical confession, in which every artistic language he uses—verbal, visual or musical—is equivalent to the others, and their combination increases their potential.

Wiley plays a range of musical instruments—from Jew's harp to didgeridoo—and the guitar and banjo (or their fantasized equivalents) seem to be the artist's favorite symbols, his metaphors for a metaphor.

Canjo is a playable banjo that extends the artist's pun to the material—cans, wire and a stick; *Banjo for J.B.* is a tribute to Joseph Beuys, an artist admired by Wiley and with whom he shares several practices, such as the use of symbols referring to universal meanings; *Debilslied* (1986)—a 5-string guitar—is a complex allegory of a possible apocalypse caused by man.

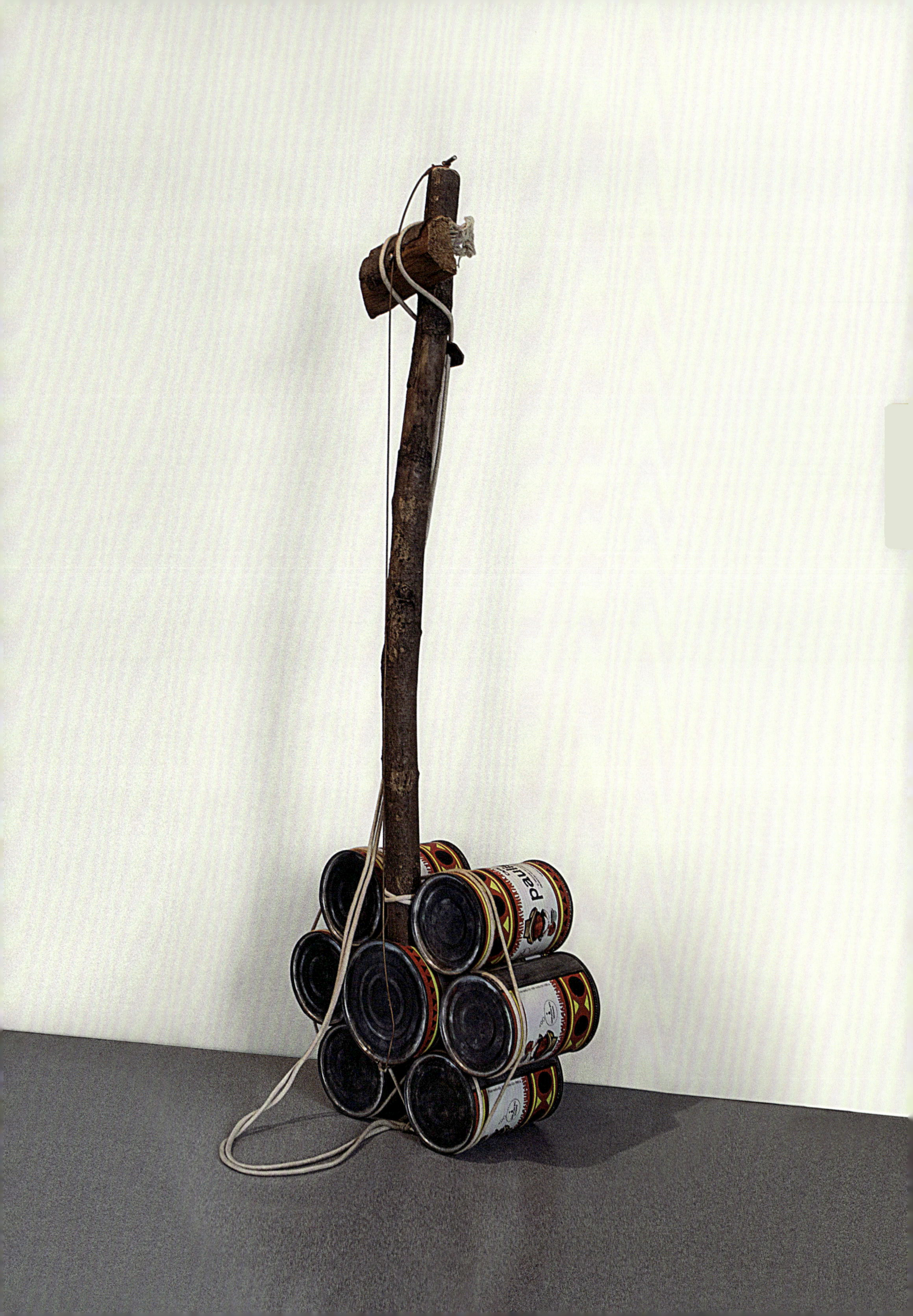

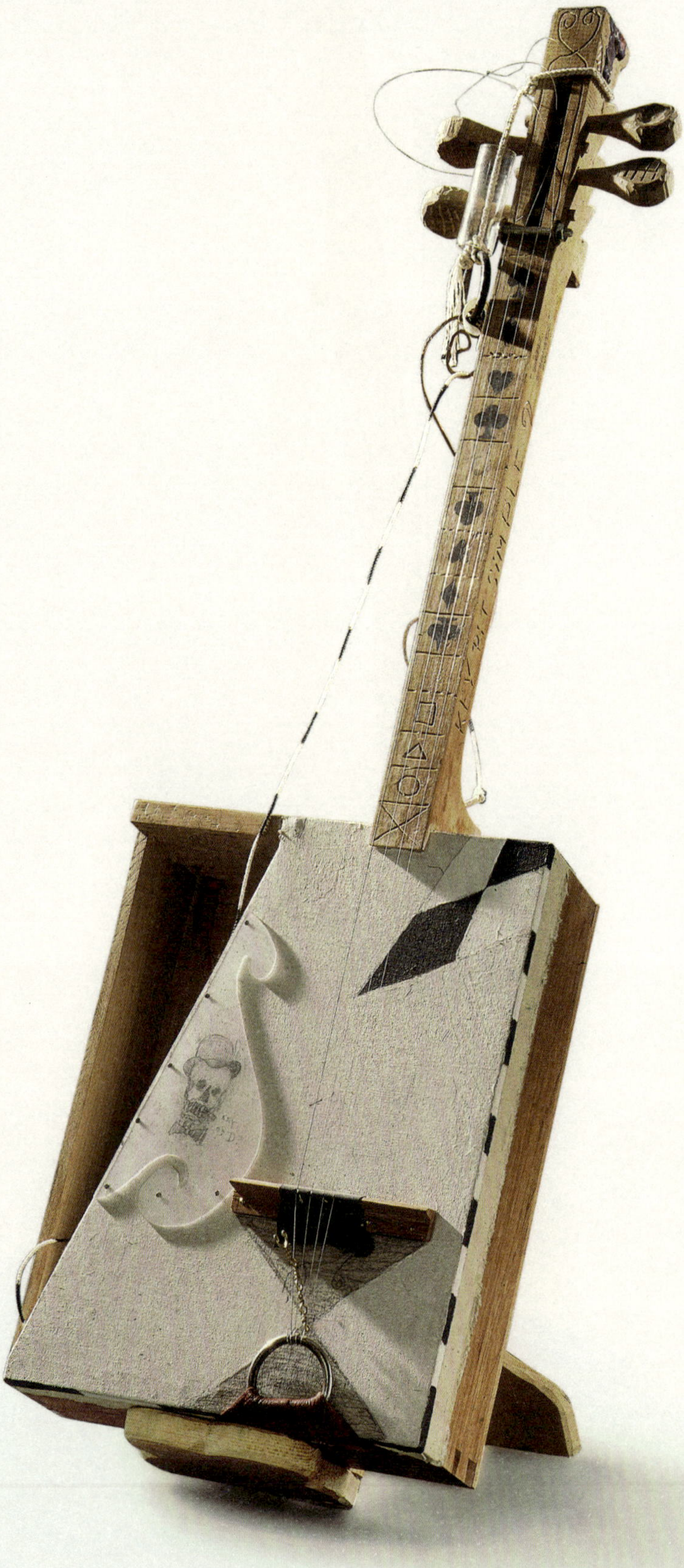

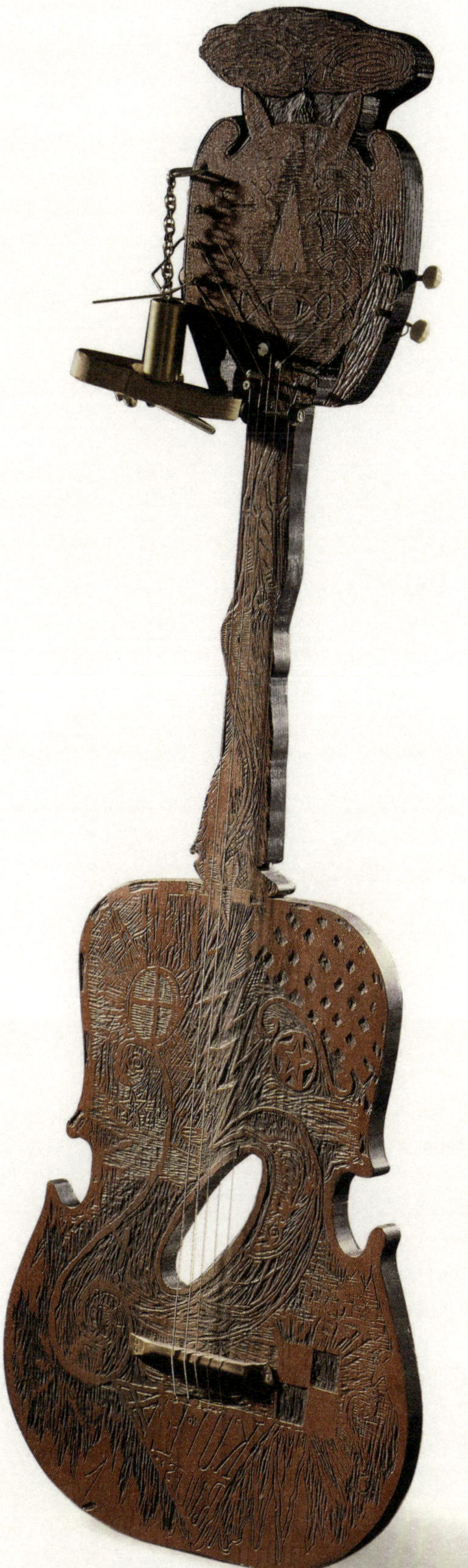

Jannis Kounellis

SENZA TITOLO (DA INVENTARE SUL POSTO) 1972

Greek artist Jannis Kounellis (Piraeus, 1936) is one of the leading exponents of Arte Povera. From his earliest works in the 1960s he has explored the contrast between natural (plants, animals and fire) and industrial elements (metals, processed fabrics), thus taking reality into the artistic domain. Some of his works have included performances of music by Mozart, Verdi, Bach and Stravinsky in which the sounds, closely associated with actions, become uncontrollable energy.

For documenta 5 in Kassel, curated by Harald Szeemann in 1972, he presented *Senza titolo. (Da inventare sul posto)* (Untitled. To be Invented on the Spot), in which 3 elements—a large painting, a violinist and a dancer—are each indispensable to the others. The painting reproduces extracts from the score of the ballet *La Pulcinella* by Igor Stravinsky, which are partially covered by pink paint. The violinist plays the score, while the ballerina improvises a dance in front of the painting. After a short pause, the operation is repeated identically and for an indefinite time.

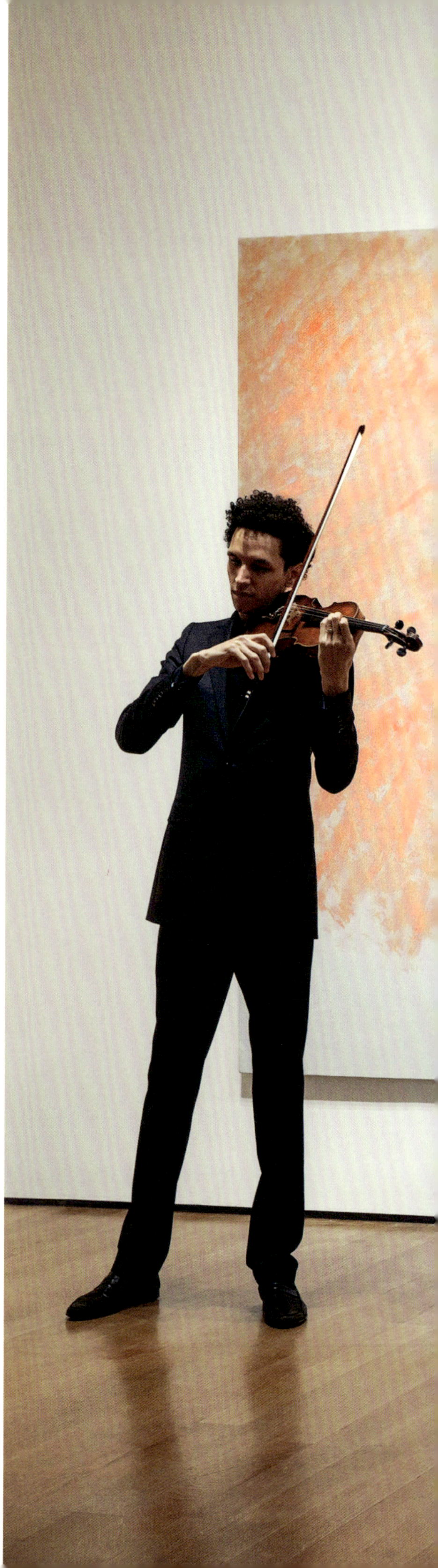

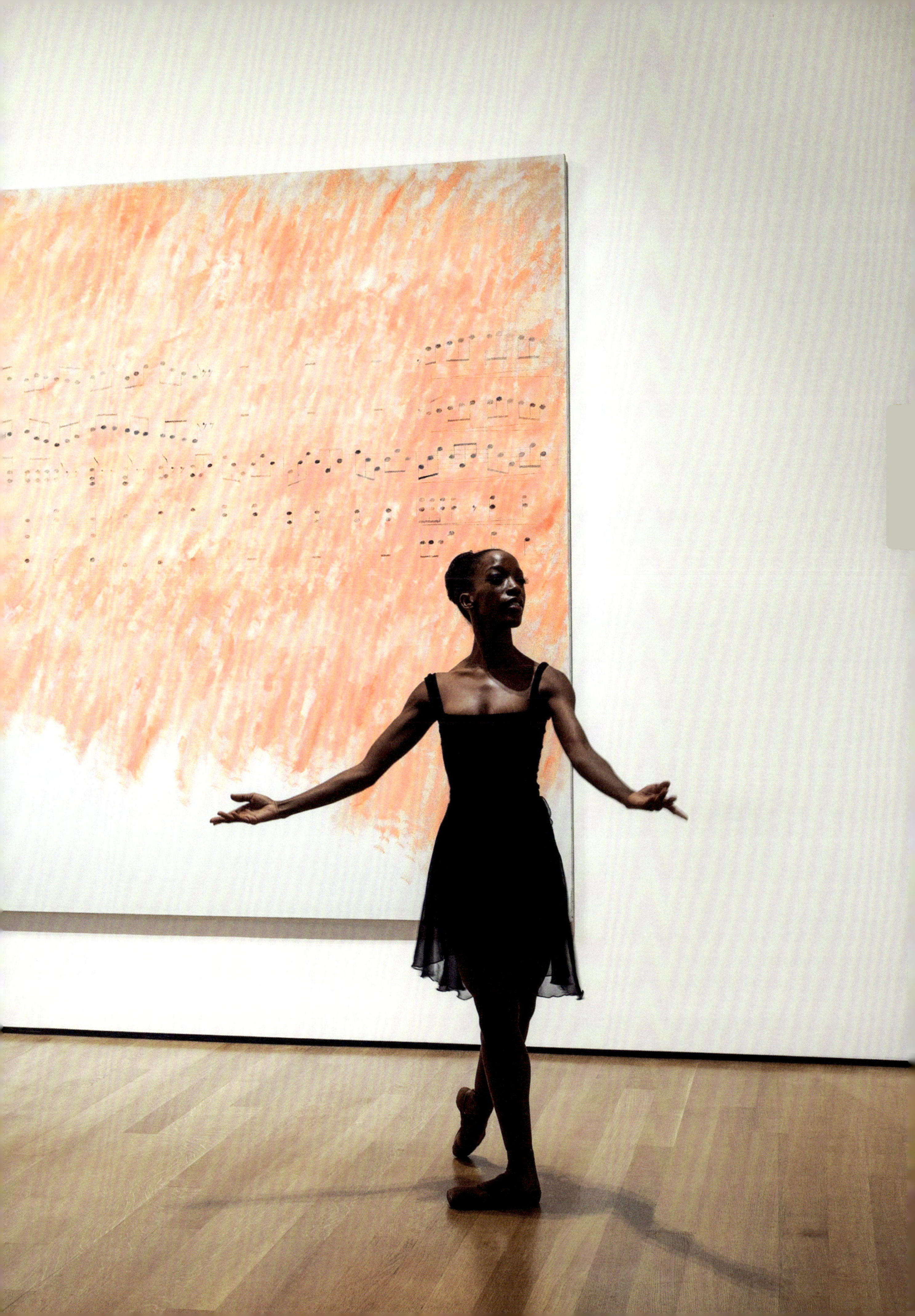

Paul De Marinis

THE PYGMY GAMELAN
1973

Paul De Marinis (Cleveland, 1948) is an American electronic music composer and a sound, performance and computer-based artist. During college, De Marinis studied music with composer Robert Ashley, who experimented in the field of electronics and technologies applied to opera and theatrical works. De Marinis's works consider the ideas and dreams that technical artifacts and mechanisms embody, addressing points of deep contact between technology and culture, as well as between science and superstition.

The Pygmy Gamelan is a nondescript device that amplifies ambient radio waves: it is made up of 5 to 10 little electronic circuits that respond to electrical fluctuations in the galaxy by improvising around 5-note phrases in the style of a Gamelan orchestra (a traditional Indonesian musical ensemble)—the incorporated sensors are in fact capable of reflecting a person's electronic field.

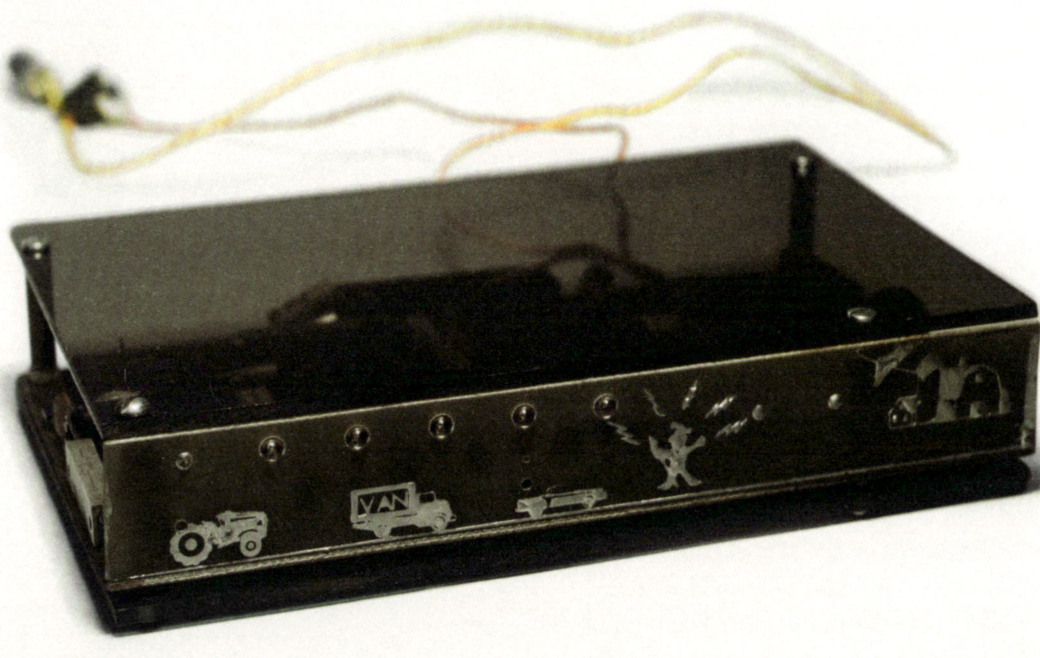
VAN

Dennis Oppenheim

THEME FOR A MAJOR HIT
1974

The early works of American artist Dennis Oppenheim (Electric City 1938 – New York, 2011) were associated both with performance art and the early land art movement. Inspired by architectural forms and grandiosity, his sculptures incorporated moving industrial parts, casts of animals, uprooted buildings, water effects and fireworks, often documented through photography. Oppenheim described *Theme for a Major Hit* as the first of a series of "surrogate performance pieces" which he referred to as "post-performance pieces." Oppenheim developed a form of narrative where the body of the artist is replaced by a figure that performs for him because "it's a device or agent that can perform in such a way that the individual cannot." In this case, the artist is the absent performer. The puppet with the artist's face is mechanically activated and executes an endless dance to a rock soundtrack whose lyrics refer to going back to motivation and seeking the boundaries of art, as well as the motivations that make artists make art: "It ain't what you make, it's what makes you do it." *

* Quotes from Kim Levin, "Dennis Oppenheim, Post-Performance Works," *Arts Magazine*, September 1978, pp. 123–25.

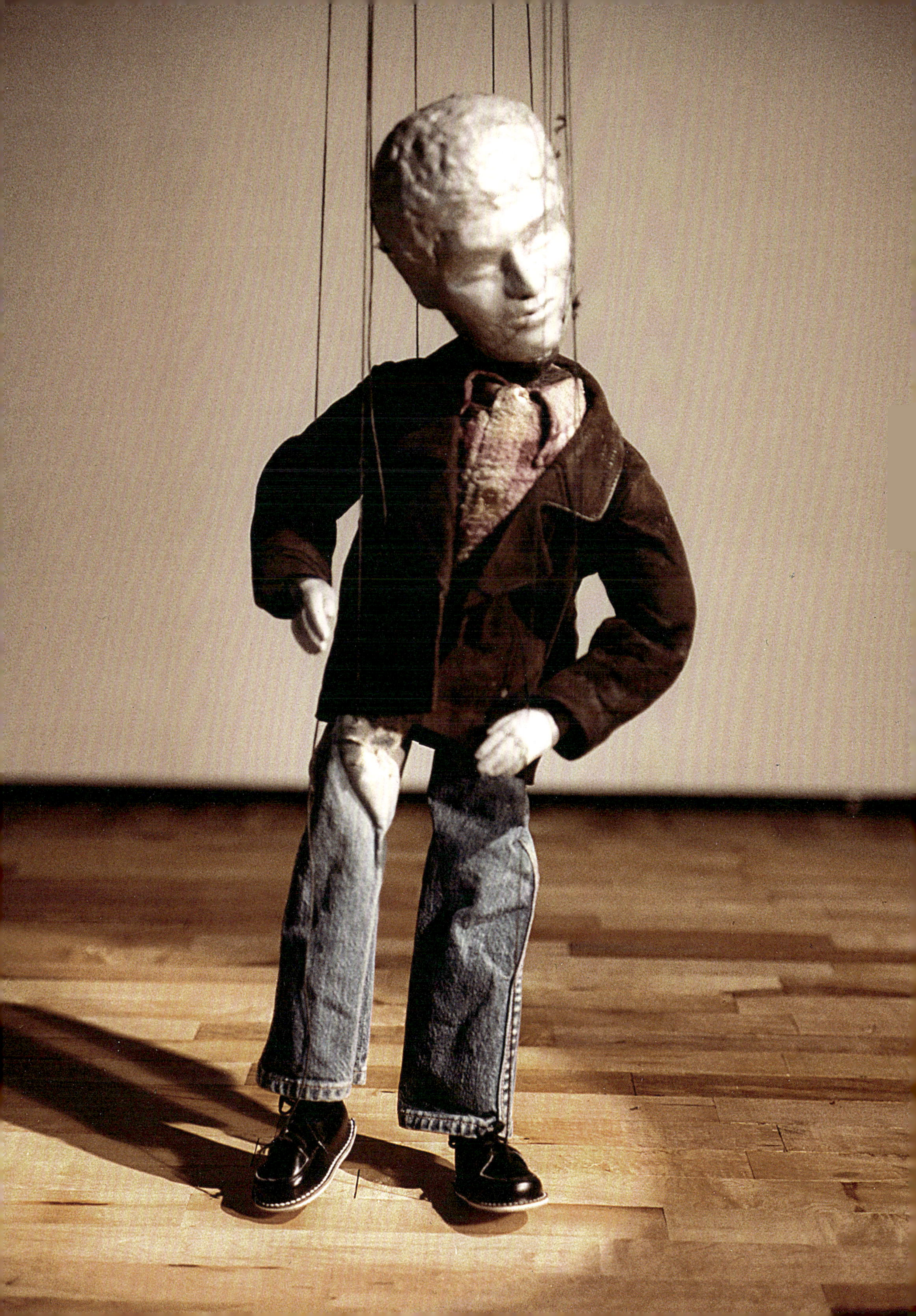

Takis

SIGNALS
1974

The research of Greek artist Takis (Athens, 1925) focuses on the exploration of energy as something intangible, and introduces magnetism, sound, light and motors into the static world of sculpture.

His first *Signals* were rods consisting of piano wires that created musical vibrations as they struck against each other in the wind. These sculptures— made of found objects and later modified to launch firework rockets or contain colorful light bulbs—produce a musical expression in which natural forces call forth sounds in an unprogrammed manner. The artist says they were inspired by the technological landscape of a train station in Calais: "Eyes of monsters lit up and dimmed, tracks, tunnels, an iron jungle. Multicolored signs, railings, passages." *

* From *Collection art contemporain – La collection du Centre Pompidou, Musée national d'art moderne*, directed by Sophie Duplaix (Paris: Centre Pompidou, 2007).

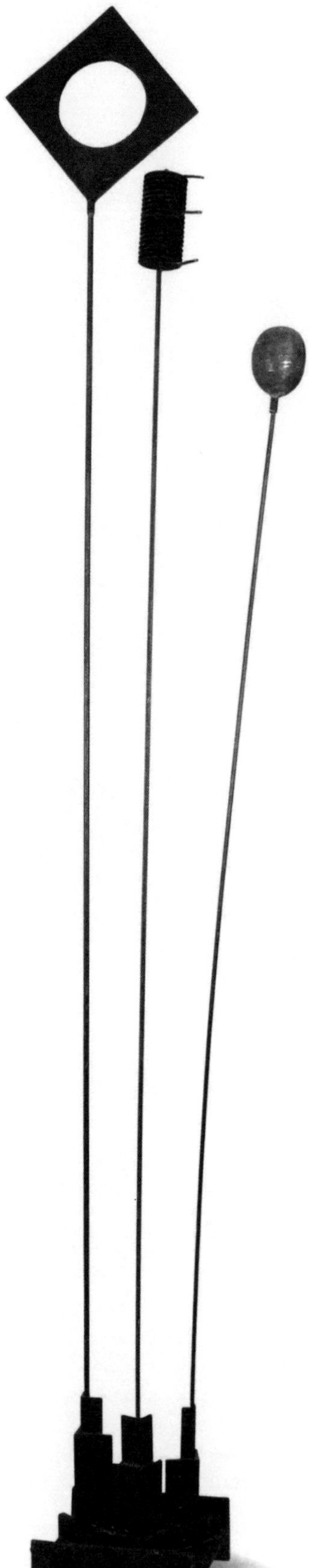

Yoshimasa Wada

EARTH HORN
1974

Yoshimasa Wada (Kyoto, 1943) is a Japanese sound artist and musician close to Fluxus. After taking lessons from the Indian composer and instrumentalist Pandit Pran Nath, the artist began to include drones in his own works. Drones are harmonic or monophonic sounds typical of classical Indian music in which a note or chord is played continuously for much of a composition. His musical works make use of instruments invented by the artist himself, such as pipe-horns, composed of long hydraulic pipes, or large wind instruments that include several pipes taken from bagpipes. Wada's works are usually performed at very high volume, allowing for the overtones within the sound to be heard clearly.

Yoshimasa Wada worked as a plumber in New York in the early 1970s, when he made *Earth Horn*, a wind instrument built from iron pipes of different size. The set of pipes, held together by hydraulic joints, stands vertically from the ground level.

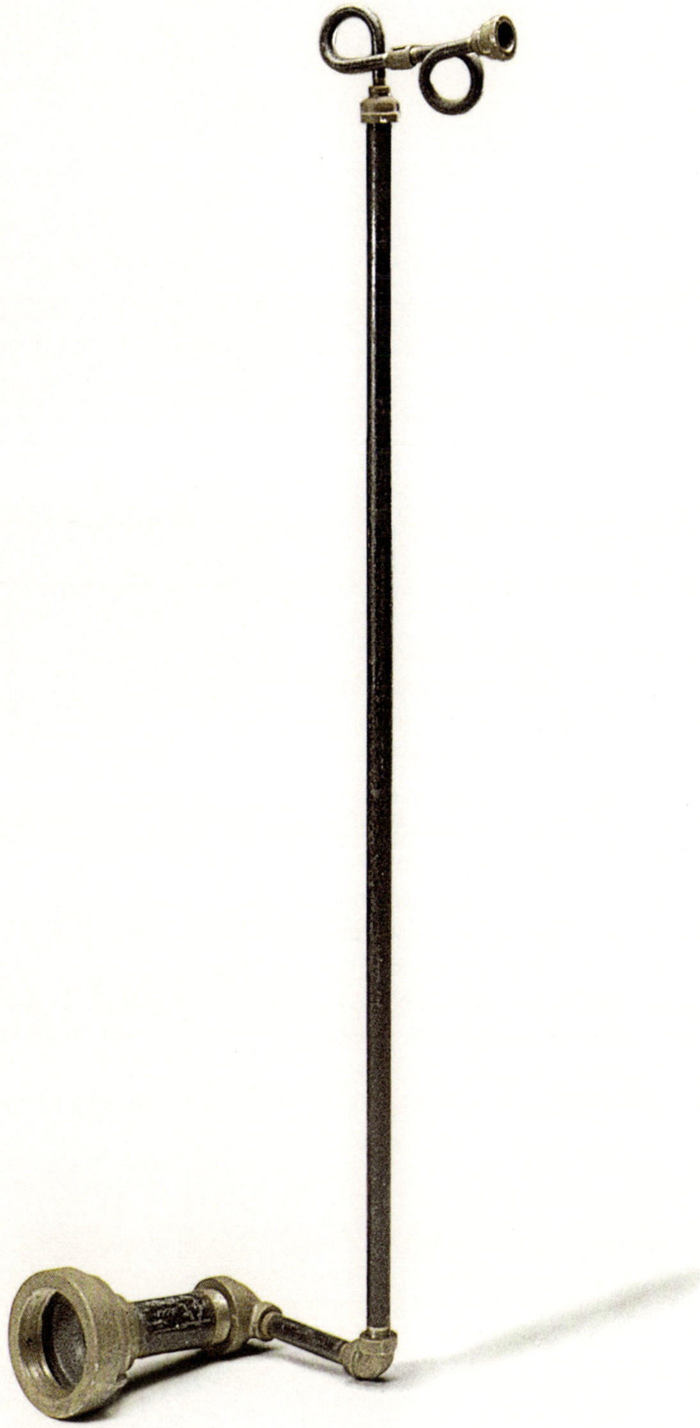

Bernhard Leitner

TON-LIEGE
1975 (2006)

After earning a degree in architecture, Austrian artist Bernhard Leitner (Feldkirch, 1938) moved to New York, where he lived for about twenty years, also working as an urban designer. Since the late 1960s he has investigated the relationship between sound, space and body: "My sound architecture and functional objects ... are really instruments. These instruments, in their specific forms, make it possible to program and compose sound spaces. ... The visual aesthetic of the instrument is part of the aural experience." *

In *Ton-liege* (Sound-chair) the listener's body lies in the axis of sonic motion; 2 loudspeakers under the lower back and calves of the reclining body define the sound space. Processed cello and horn sounds sweep the body, moving back and forth so that the sound shifts into the body.

* "Conversation between Wolfgang Drechsler and Bernhard Leitner," in Bernhard Leitner, *Sound : Space* (Ostfildern: Cantz Verlag, 1998).

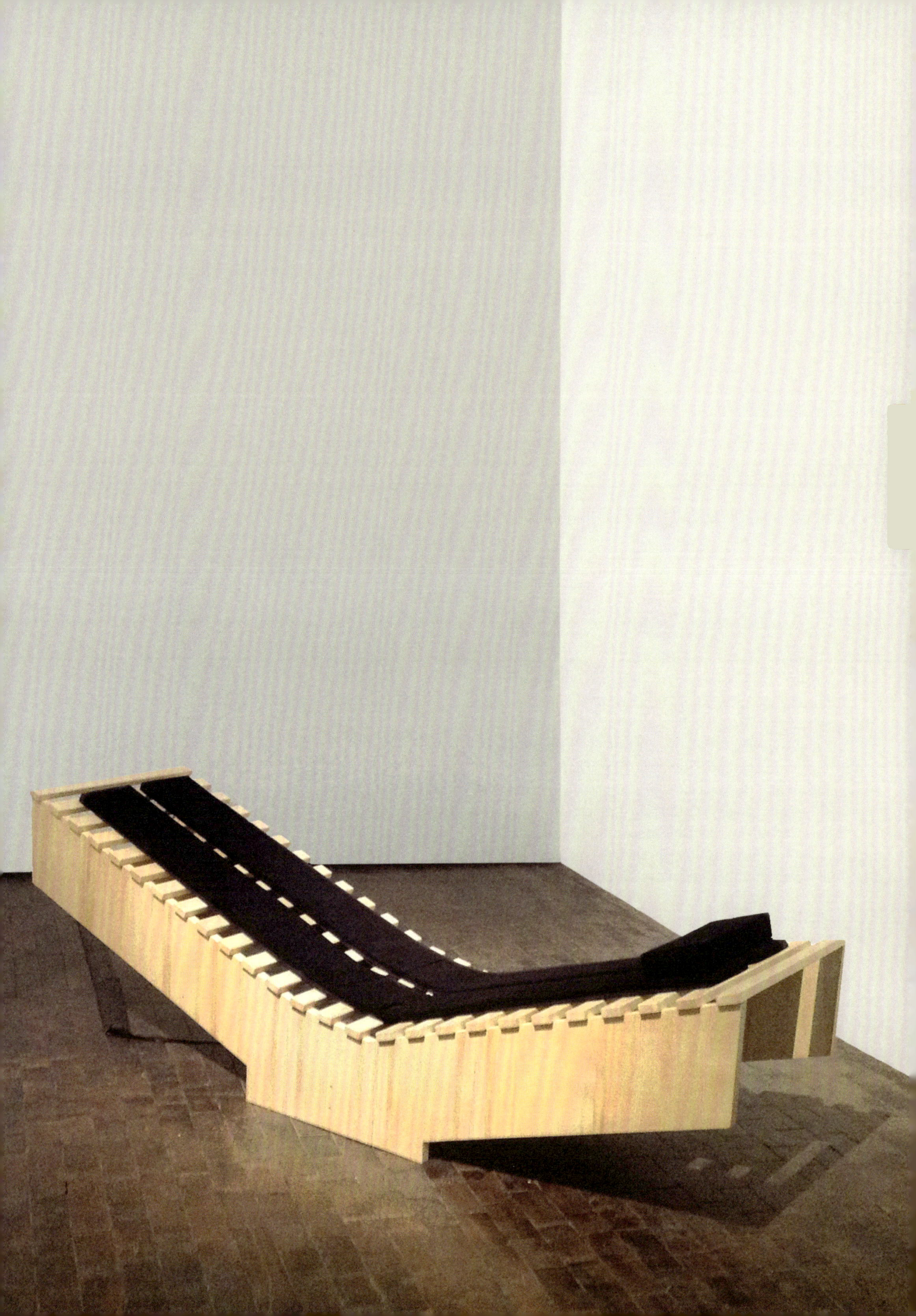

William Anastasi

THE WORLD'S GREATEST MUSIC
1977

American artist William Anastasi (Philadelphia, 1933) questions the relation between art and perception in his early works, which often refer to the Cagean concept of the impossibility of silence. He describes *Microphone* (1963), which is a tape recorder placed inside a closet and left to record its own sound, as a "recording of the recorder recording the recorder." In 1966 the Dwan Gallery presented his *Sound Objects*, tools and appliances displayed alongside recordings of their sounds. *The World's Greatest Music* consists of 3 children's turntables playing records from the series "World's Greatest Music," which was launched in 1938 as a newspaper circulation program by the Publishers Service Company and sold millions of copies. The records of music by Brahms, Wagner and Mozart become stuck at the end of the disk and spin eternally, producing a crackling sound.

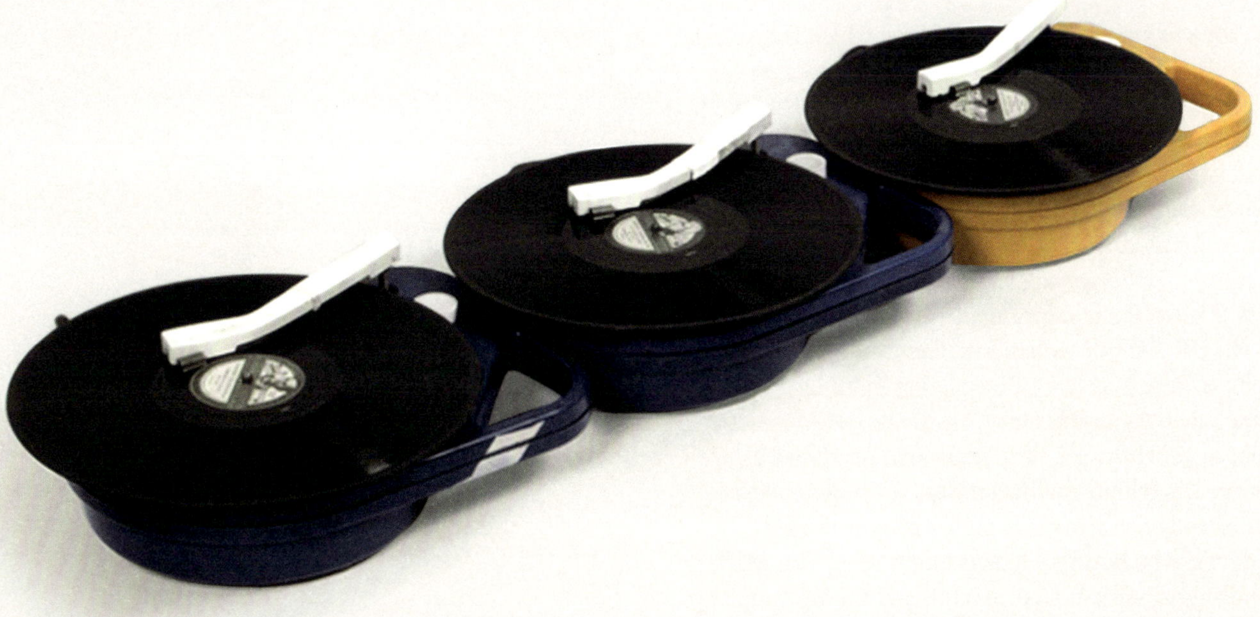

George Maciunas

JASMINKA BAN. SPELL YOUR NAME WITH THESE OBJECTS 1977

The idea of art and life as a game in which the artist reconfigures the rules is central to Fluxus and refers to the works and theories of Marcel Duchamp and John Cage. Many of the objects these artists created involve direct, playful participation by the viewer. George Maciunas assembled boxes and issued many different works in boxed format. For example, *Gift Box for John Cage: Spell Your Name with These Objects*, was made around 1972 using an old US Army medical kit. The items in this box include objects that can be used to spell Cage's name, such as a pine Cone, an Acorn, a Glass bottle stopper and an Egg. Maciunas also produced a gift box for his friend Jasminka Ban, a German artist connected with Fluxus, and his lawyer Jerald Ordover, who handled leases and helped him with building purchases that established the first artist-owned coops in New York's Soho.

[See also p. 270]

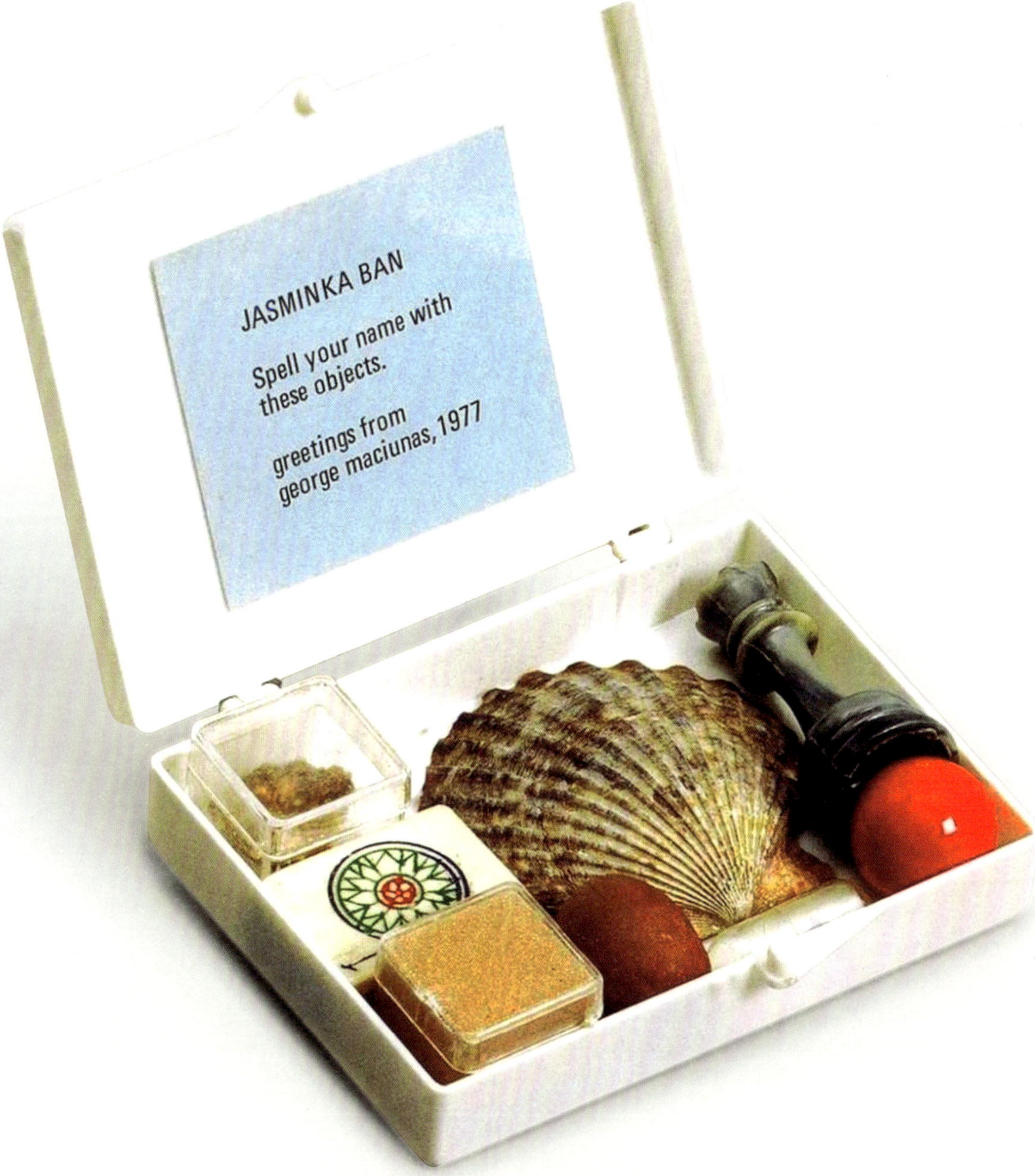
JASMINKA BAN
Spell your name with
these objects.
greetings from
george maciunas, 1977

Takako Saito

SOUND CHESS
c. 1977

Takako Saito (Sabae-Shi, 1929) is a Japanese artist associated with Fluxus. She joined the movement in 1963 when she moved to New York from Tokyo to join her friend Ay-O, who introduced her to George Maciunas.

Sound Chess is an artist's multiple originally conceived in 1964–65 to be sold in Maciunas's Flux Shop in Soho. It is part of a larger series of objects known as *Flux Chess*, where Saito challenged the visual aspect of the game by replacing the set with pieces that relied on the use of alternative senses (smell, touch, sight…) to play.

In the case of *Sound Chess*, traditional pieces are substituted by identical sealed boxes containing different objects. Originally Maciunas asked Saito to create a chess set as an homage to Marcel Duchamp, a passionate chess player (like Maciunas himself); *Sound Chess* pays a further homage to Duchamp, and to his readymade *With Hidden Noise* (1916) [p. 164] in particular.

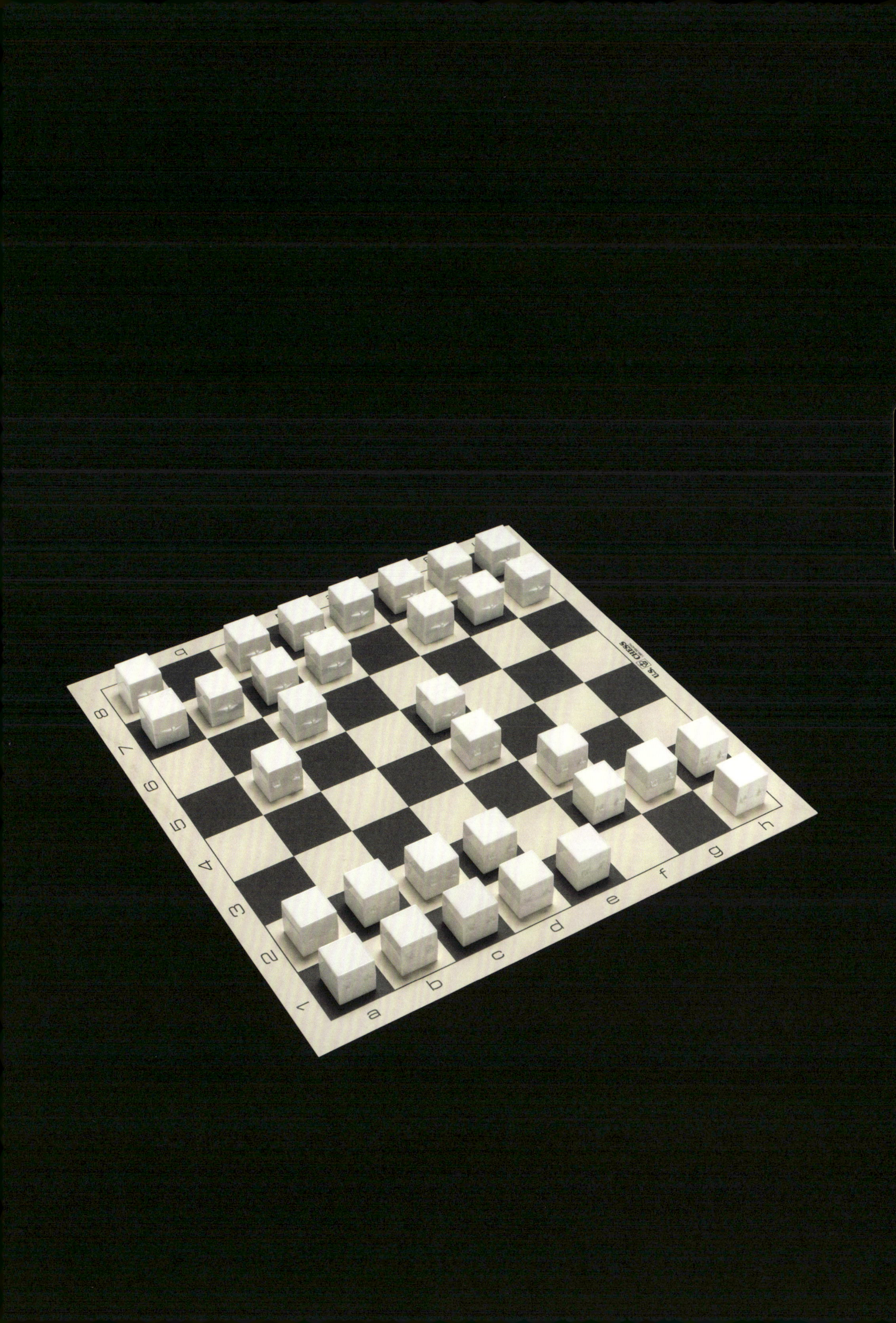

Laurie Anderson

HANDPHONE TABLE
1978

American performance artist and musician, Laurie Anderson (Glen Ellyn, 1947) conceived *Handphone Table* when she rested her head in her hands and discovered that her forearms were conducting sound. The wooden table conceals a playback system of wires connected to nodes under 2 concave disks where the viewer's elbows are to be placed. The 2 participants are invited to sit with their hands over their ears, and to activate 2 different lyrics at the opposite end of the table: *Now You in Me Without a Body Move* and *And I Remember You in my Bones*. A photograph on the walls shows a previous listener with a handwritten sentence: "The Way You Moved Through Me." Talking about this work, Anderson notes: "I loved the symmetry: you heard the final piece using the same gesture involved in its invention, closing the circuit, your head in your hands, listening, listening through your bones."*

[See also p. 302]

* Laurie Anderson, *Stories from the Nerve Bible: A Retrospective, 1972–1992* (New York: HarperPerennial, 1994), p. 48.

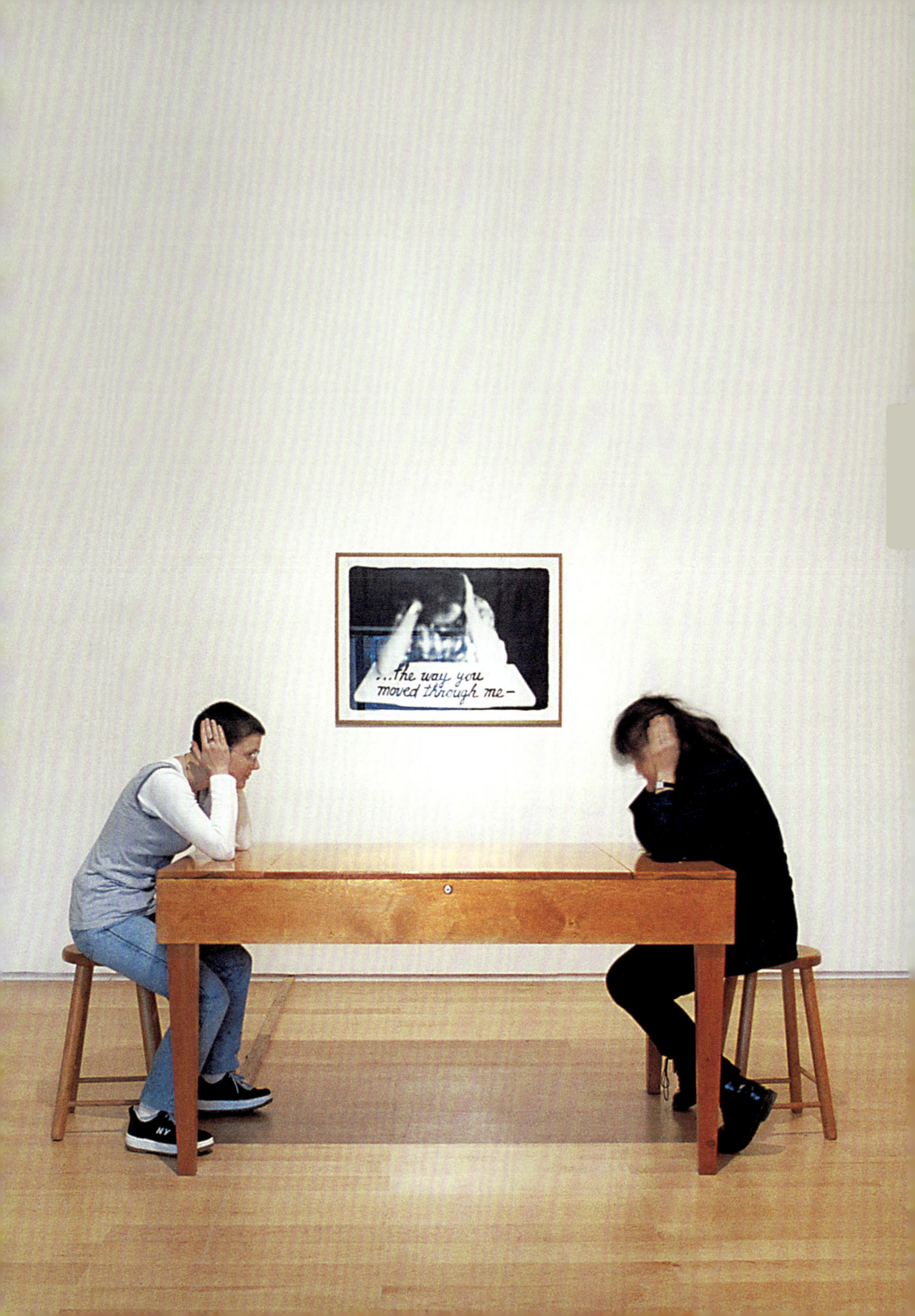
...the way you
moved through me-

Ken Butler

AXE VIOLIN
1978

BRIEFCASE GUITAR
1980

BENT ARM VIOLIN
2002

After studying viola as a child, American artist Ken Butler (Bethesda, 1948) maintained an interest in music while studying visual arts during the 1970s.

His hybrid instruments are constructed from readily available consumer objects designed to perform a different function and, when amplified, are shaped with cutting-edge sound processing equipment that enables artful musical sounds and expression.

When on a concert stage or viewed in an exhibition, these assemblages of hammers, hockey sticks, tennis rackets, golf clubs and brooms turn into violins, guitars and cello-like instruments with multiple playing surfaces and a diverse range of percussive and assorted sounds, in addition to those produced by the strings.

[See also p. 306]

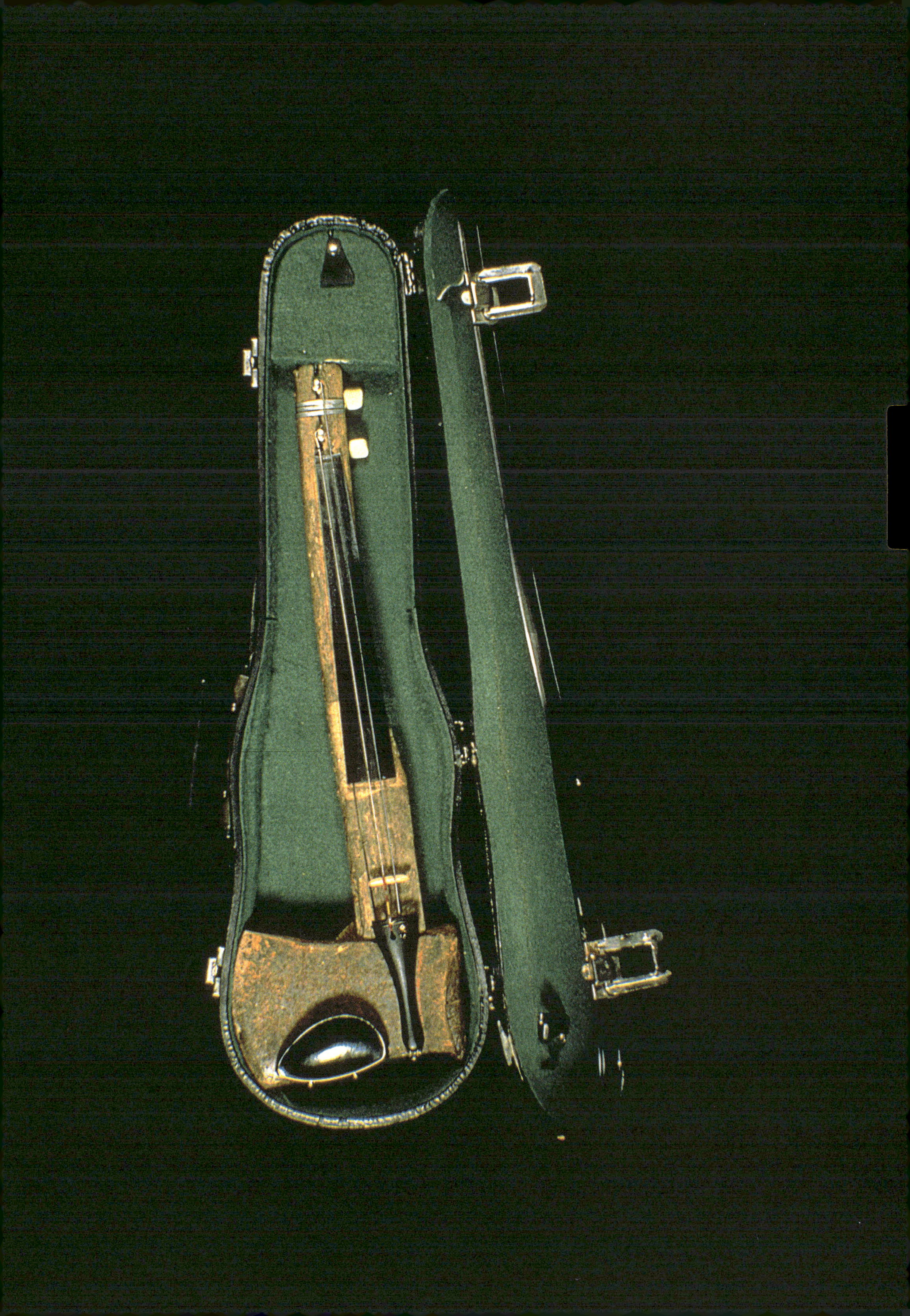

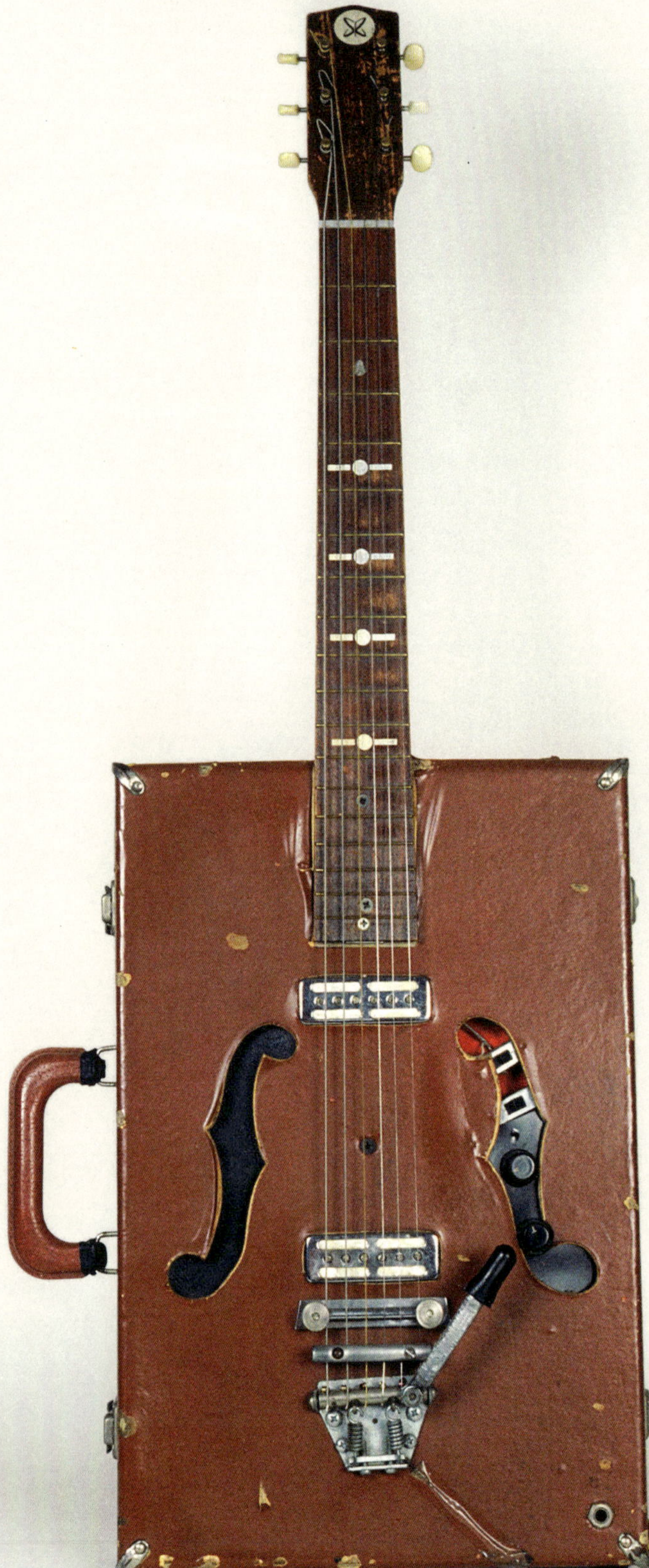

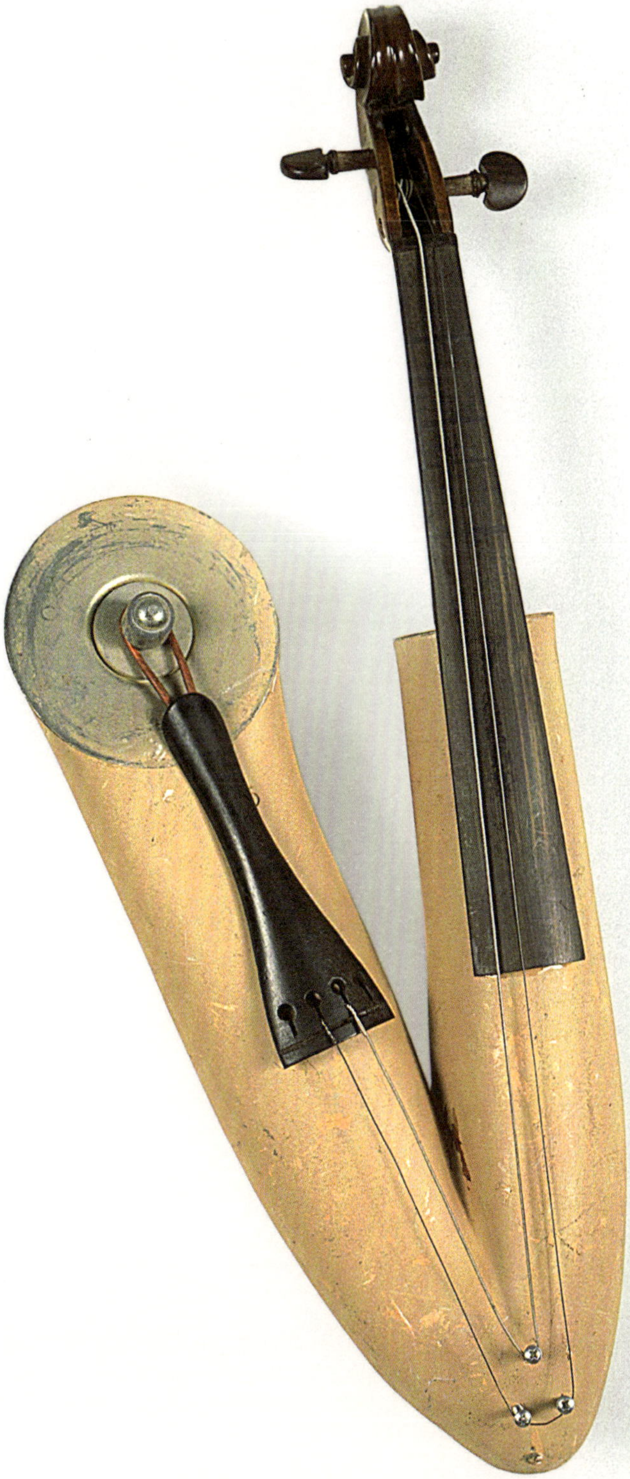

Laurie Anderson

NUMBERS RUNNERS
1979

Laurie Anderson's *Numbers Runners* is an altered public telephone booth. People enter the booth, lift the phone from its hook and hear 2 voices: Anderson's message on tape and their own words delayed by ¾ of a second. By interacting with the public's input, the telephone booth enables the listener and Anderson to participate in a dialogue. Lapsed-time sequences had already been included in the musical performance *New York Social Life* (1977), for which Anderson alternates 2 distinct voices between telephone and microphone, performing the words very rapidly as one long run-on sentence.

With this work the artist explores the performing potential of the telephone as a plastic artificial conductor of real relationships between people: "… the question is: Is the phone alive? What's alive and what doesn't have life?" *

[See also p. 296]

* From the performance *United States* (1979) as cited in Laurie Anderson, "Hotline," in Id., *Stories from the Nerve Bible: A Retrospective, 1972–1992* (New York: HarperPerennial, 1994), pp. 62-64: 62.

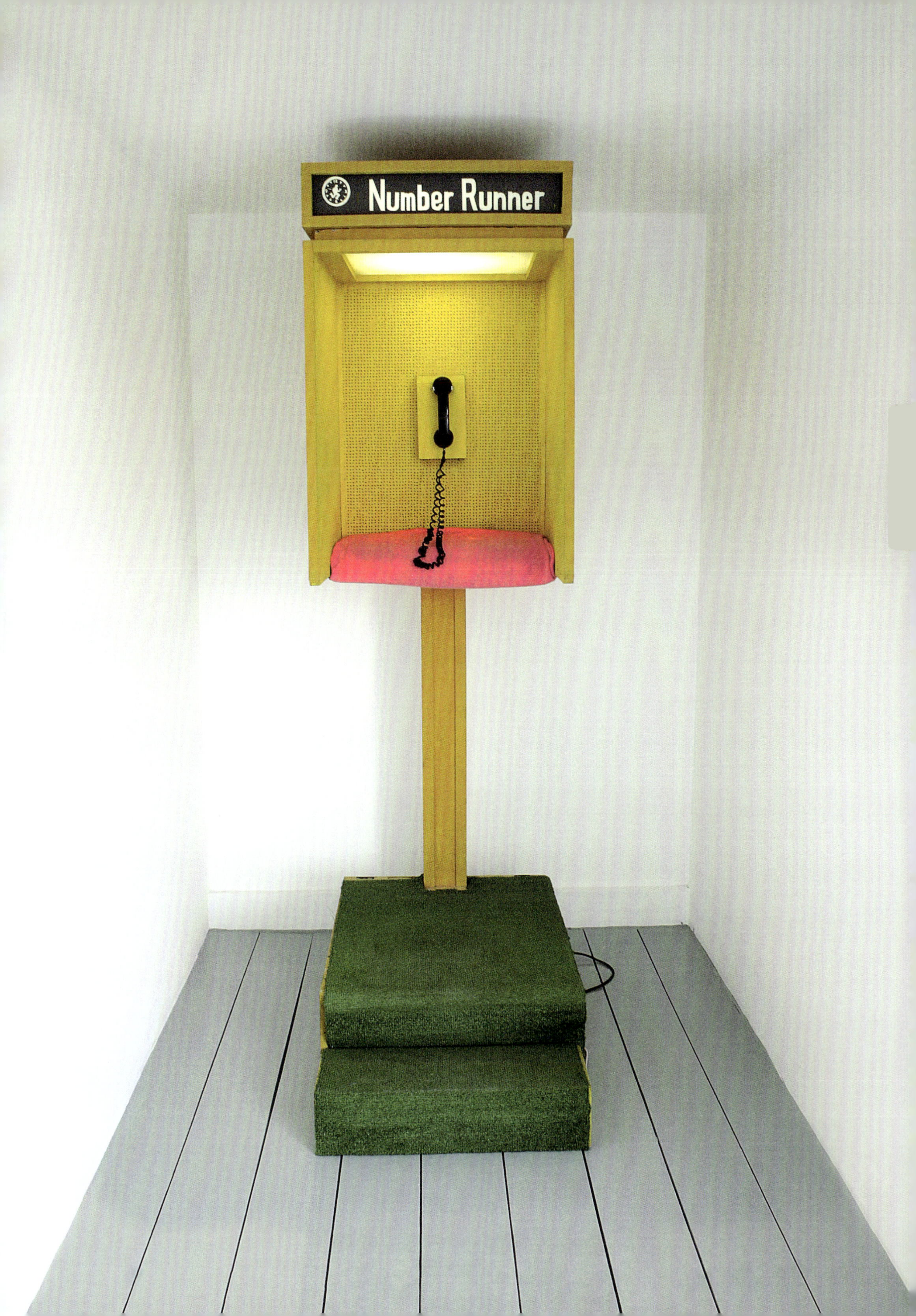
Number Runner

Ay-O

AY-O'S AND JOE'S SOUNDBOX
1979

Known as "Rainbow Man" for the use of color in his works, Ay-O (the artist name of Takao Iijima, born in Ibaraki, 1931) is a Japanese artist and performer who was a member of Fluxus. In 1955 he joined the Demokrato Artists Association (Demokurato Bijutsuki Kyōkai), founded by Ei-Q, which promoted the freedom and independence of art.

Ay-O was introduced to George Maciunas in 1961 by Yoko Ono. At the time Ay-O was beginning to produce his *Finger Boxes*, a series of wooden boxes devised for the public to insert their fingers into.

Ay-O's and Joe's Soundbox is a reconstruction of a work from the early 1960s. A large wooden cube has various openings into which hands and fingers can be inserted. Inside are a number of mechanical and electric sound sources that can be activated. The work can be touched and explored by several people at the same time, thus creating a collective performance.

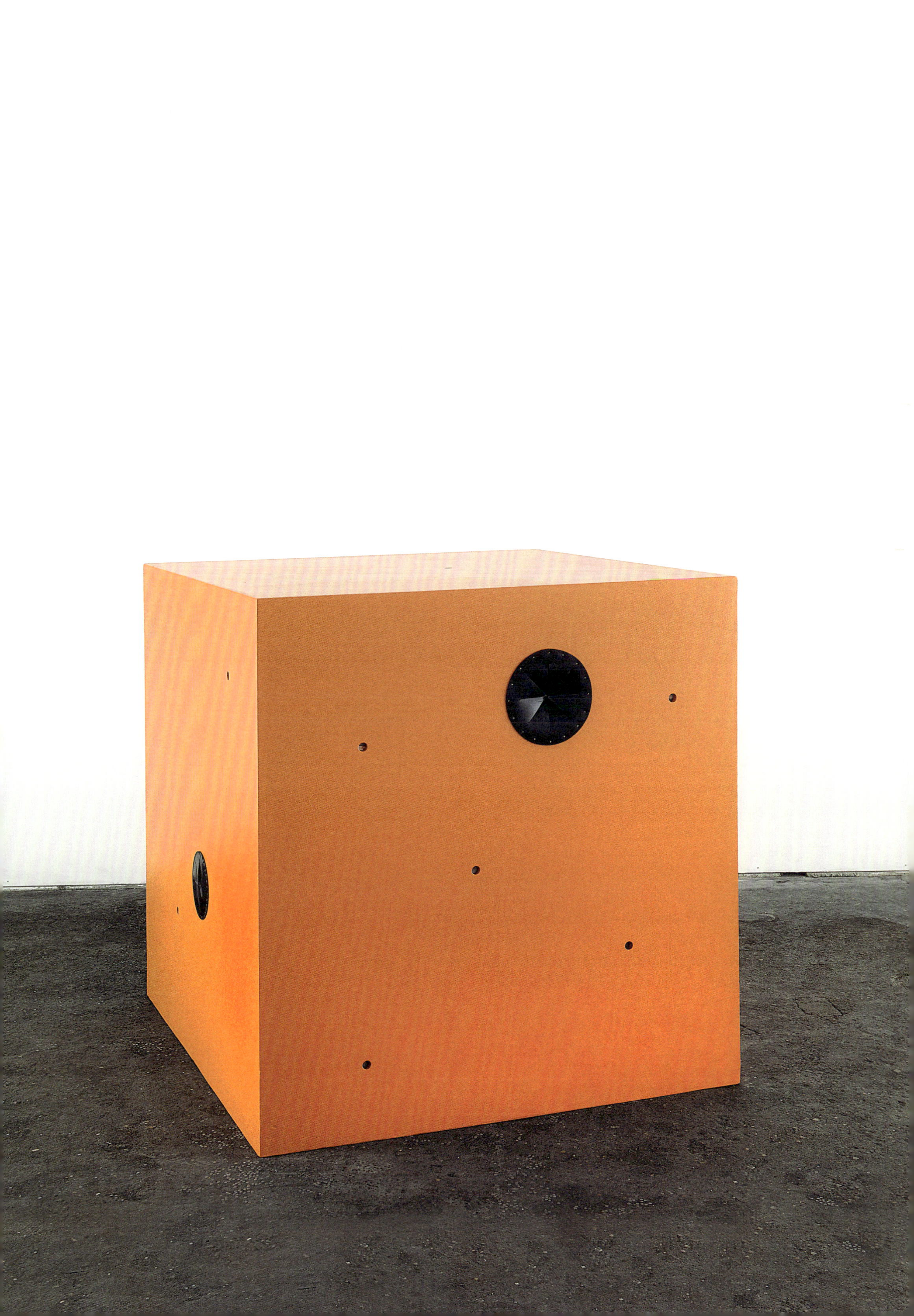

Ken Butler

K-BOARD
1983

Ken Butler's keyboards take the idea of the hybrid instrument beyond that of a sound producer. When they are connected to radios, fitted with lights or projectors, or have objects attached, they become a contemporary version of experiments carried out by 20th-century avant-garde movements in their attempts to unify the arts.

K-Board is an audio-visual keyboard with 110 AC outlets into which lights and other electrical devices can be plugged and activated by touching the keys.

A single amplified string directly beneath the keys is strummed by a motorized bicycle wheel rimmed with guitar picks, producing an intricate droning rhythm and microtonal scales as the keys are depressed.

[See also p. 298]

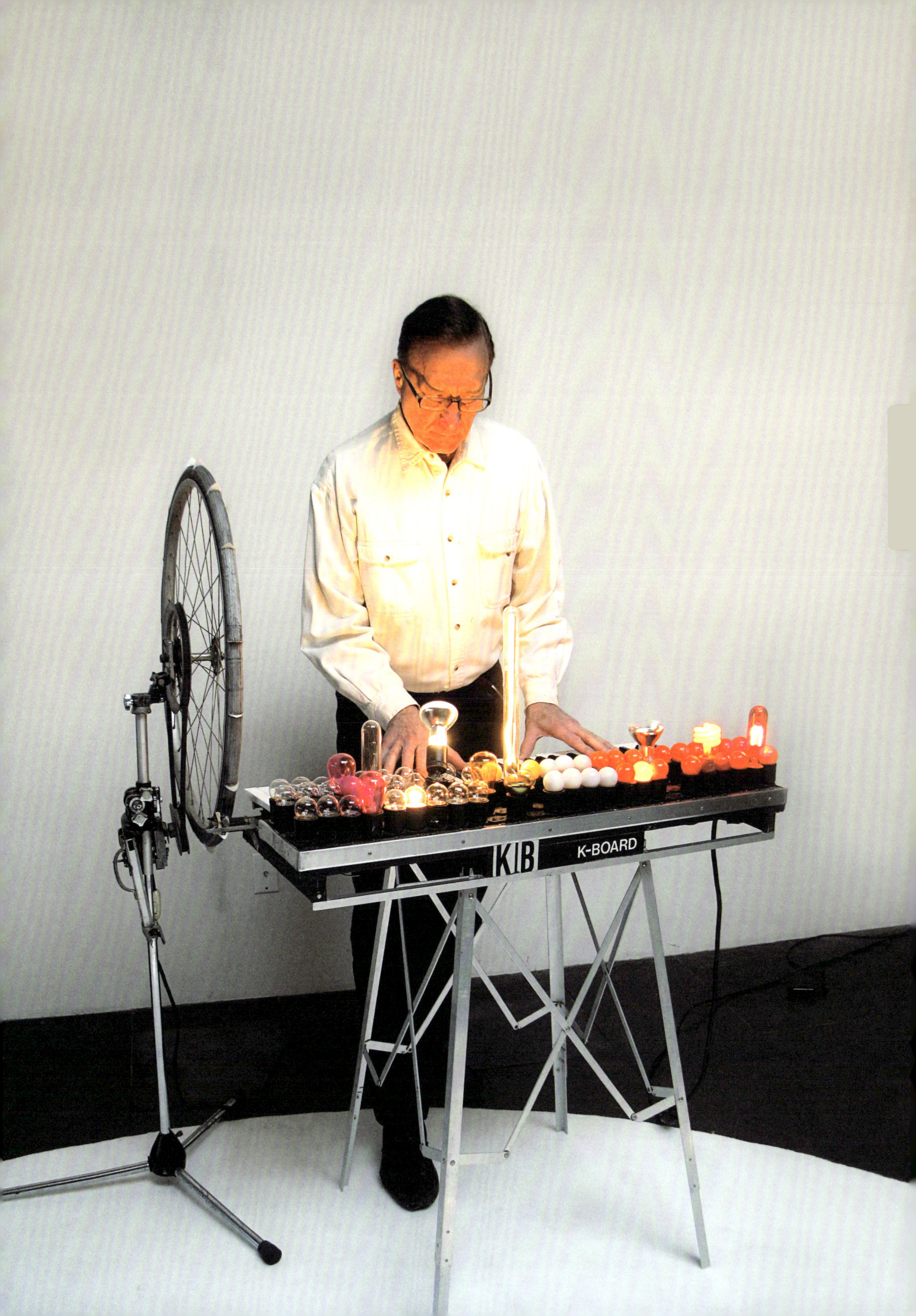
K|B
K-BOARD

Eliseo Mattiacci

ECHI DI SUONI E CANI CHE ABBAIANO 1983

Italian artist Eliseo Mattiacci (Cagli, 1940) moved to Rome in 1964, where he frequented the artistic environment of Arte Povera and experimental performance.

The artist used sounds from the human and natural domains (the artist breathing, sounds of animals) from the 1970s up until the mid-1980s. *Echi di suoni e cani che abbaiano* (Echoes of Sounds and Dogs Barking) comprises 11 metal elements culminating in cymbals and a recording of the striking of a gong, followed by dogs barking. It was created during the artist's stay in New York and inspired by "the rhythm of the sounds in the background that, mixed with the barking of dogs, is a constant noise in American cities." * Mattiacci's fondness for metals led him to use different types of bronze in the cymbals. The gong is a recurring element in his artwork, and featured in his 1995 work *Ascolto di vuoto* (Listening to the Void), in which an enormous iron disk was suspended between rocks by the Cava di Sant'Anna and struck by a brass ball, allowing the sound to reverberate into the quarry.

* From Germano Celant (ed.), *Eliseo Mattiacci* (Milan: Skira, 2013), p. 208.

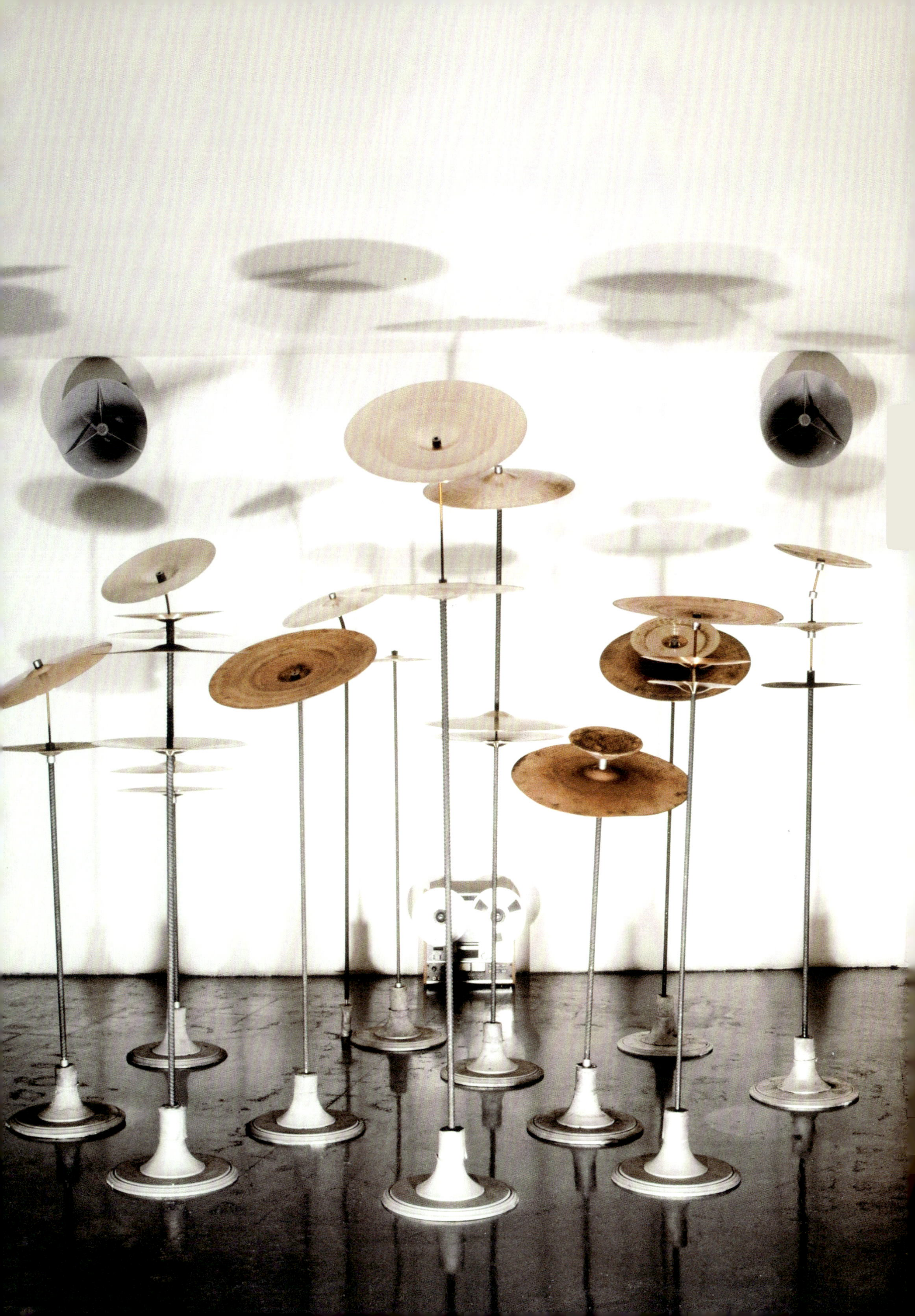

Joseph Beuys

KLAVIER OXYGEN
1985

Joseph Beuys (Krefeld, 1921 – Düsseldorf, 1986) took piano and cello lessons in his childhood, and his works, whether sculptural or performed, often involved sound and musical instruments, especially pianos, as in *Infiltration homogen für Konzertflügel* (Homogeneous Infiltration for Grand Piano, 1966), a piano covered with felt; and *Revolutionsklavier* (Revolution Piano, 1969), a glass display case containing a piano and roses. The last time the German artist featured a piano was in Hamburg on November 29, 1985, during the opening of the Art of Peace Biennale, curated by René Block. Beuys was supposed to perform with Nam June Paik and Danish composer Henning Christiansen. There were 3 pianos on stage, but as Beuys was absent (he was to die 2 months later), a phone was placed on the top of his piano and an oxygen tank beneath. The oxygen was released upon Beuys' request, and a sentence he spoke by phone was written on a blackboard: "If you want to free the work resulting from ability, it's logical to first free the ability."*

* "Bei einem wesensgemäßen Beschreiben des Geschehens zur Befreiung der von der Fähigkeit getragenen Arbeit ist es doch logisch, daß das Tragende zuerst befreit werden muß," trans. by Greg Lutz, surround.noquam.com/nature-and-culture

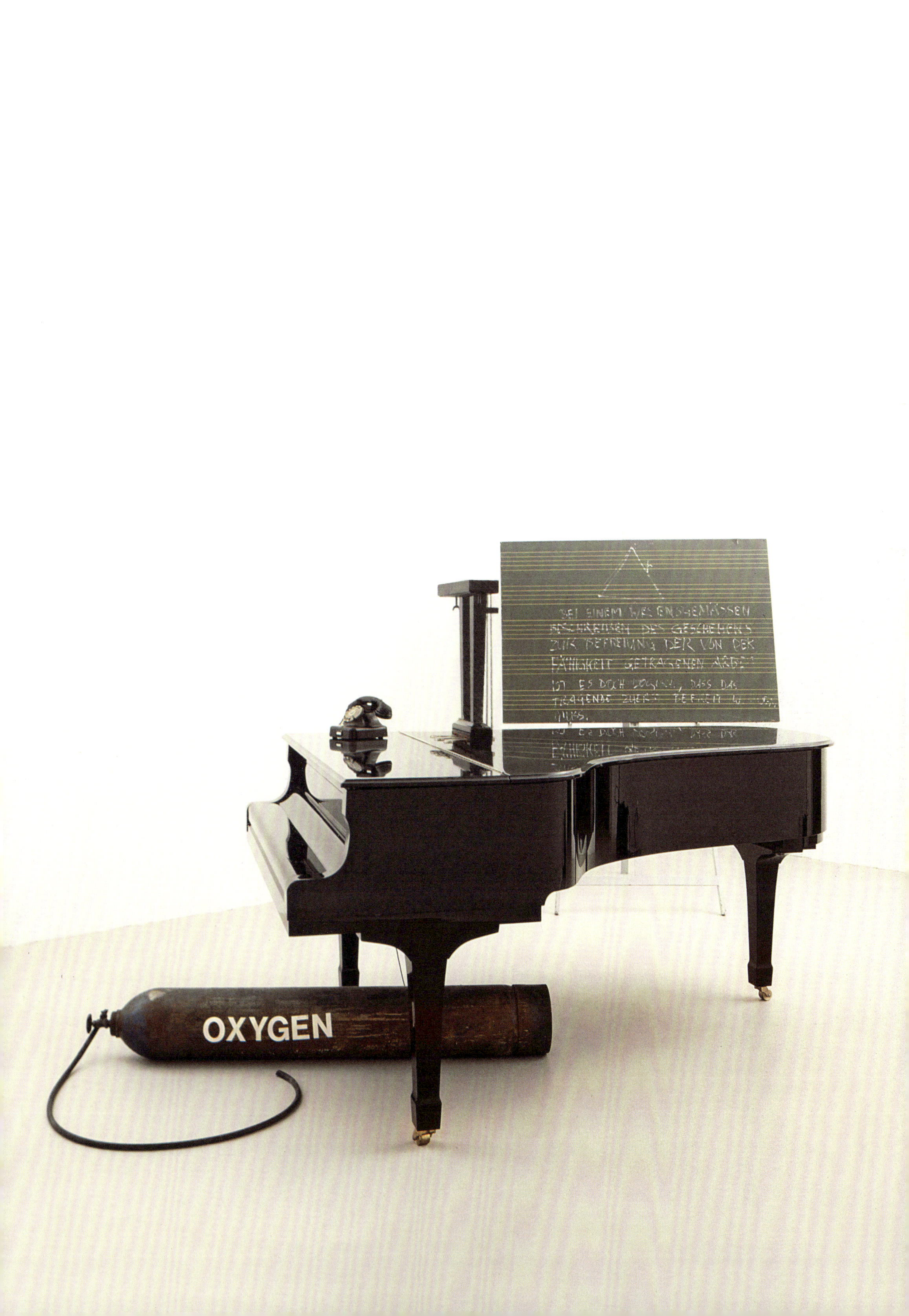
BEI EINEM WESENSGEMÄSSEN
BESCHREIBEN DES GESCHEHENS
DER VON DER
GETRAGENEN
ES DOCH
DASS
MUSS.
OXYGEN

Max Vandervorst

SAXOSOIR
1988

Max Vandervorst (Schaerbeek, 1961) is a Belgian musician and composer who has presented performances whose protagonists are musical instruments created using ordinary objects not usually considered suitable for making sounds.

The *Saxosoir* is formed by the combination of a 5-liter watering can and a tenor sax whose ebony mouthpiece is fitted into the semiconical spout of a watering can perforated with 7 holes. The resulting sound is unique, somewhere between a clarinet and a saxophone.

The instrument is part of a show called "Symphonie d'objets abandonnés" (Symphony of abandoned objects) conceived by Vandervorst in 1988, and is significant of what the artist calls "Lutherie Sauvage" (literally Wild Instrument-Making), an activity that attempts to "broaden the strict boundaries of the musical domain to embrace other fields of expression, such as poetry, theater, sculpture…." *

* Max Vandervorst, *Nouvelle lutheries sauvages* (Paris: Alternatives, 2006), p. 5.

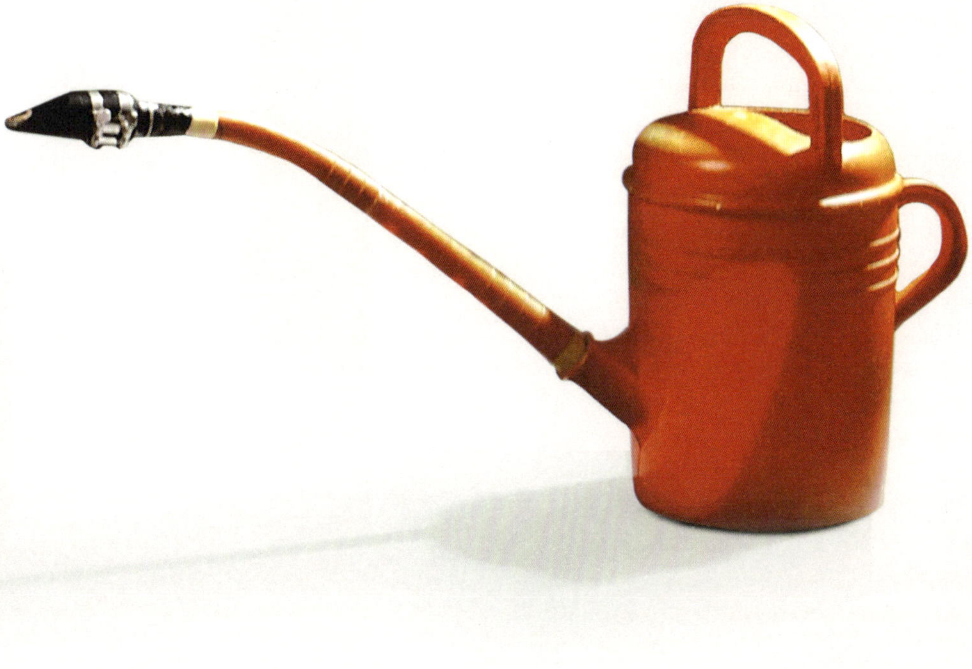

Walter Marchetti

MUSICA DA CAMERA N° 211
1991

Italian artist Walter Marchetti (Canosa di Puglia, 1931) studied composition as an autodidact. In 1958 Marchetti met John Cage, evolving his practice in strong relation to the Fluxus movement. He was a founding member of the art group ZAJ, which was formed in 1964 in Madrid and focused on music and the performing arts. Beginning in 1990 and parallel to the composition of experimental music, Marchetti has focused on the *Musica da camera* series. *Musica da camera n. 211* is the result of a ZAJ music action and installation during a festival in Milan. He created the entire volume of a piano by neatly stacking toilet paper rolls. The instrument is present through the medium of ordinary objects and materials, and its sound potential is investigated in albums recorded by the artist, such as *De Musicorum Infelicitate: Ten Variations of Musical Action* (2001–03) where several pianos play at once, or a piano is recorded several times on different channels.

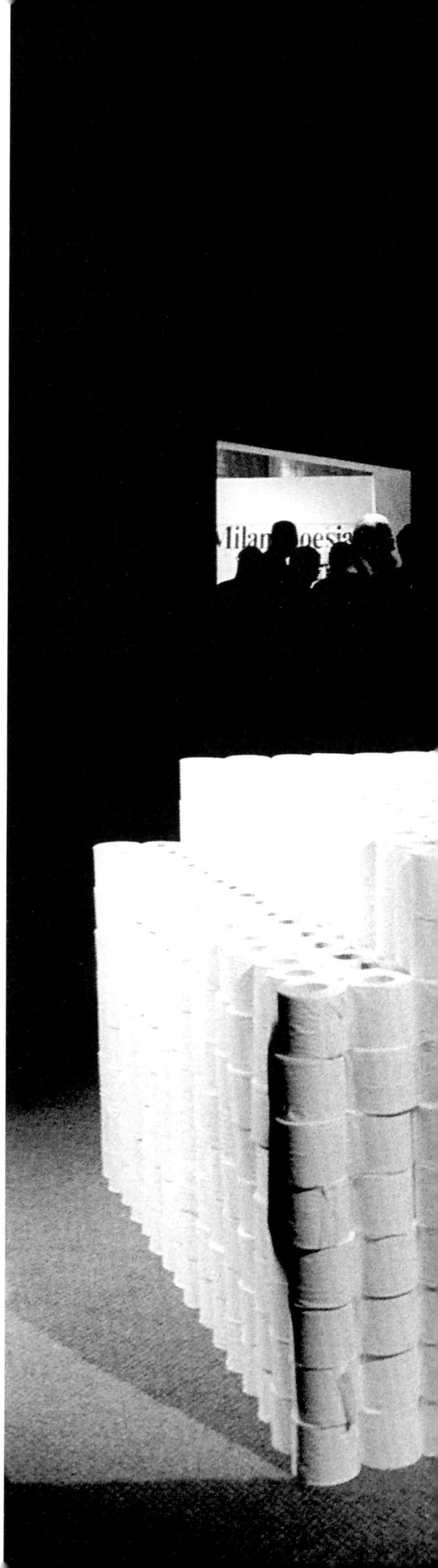

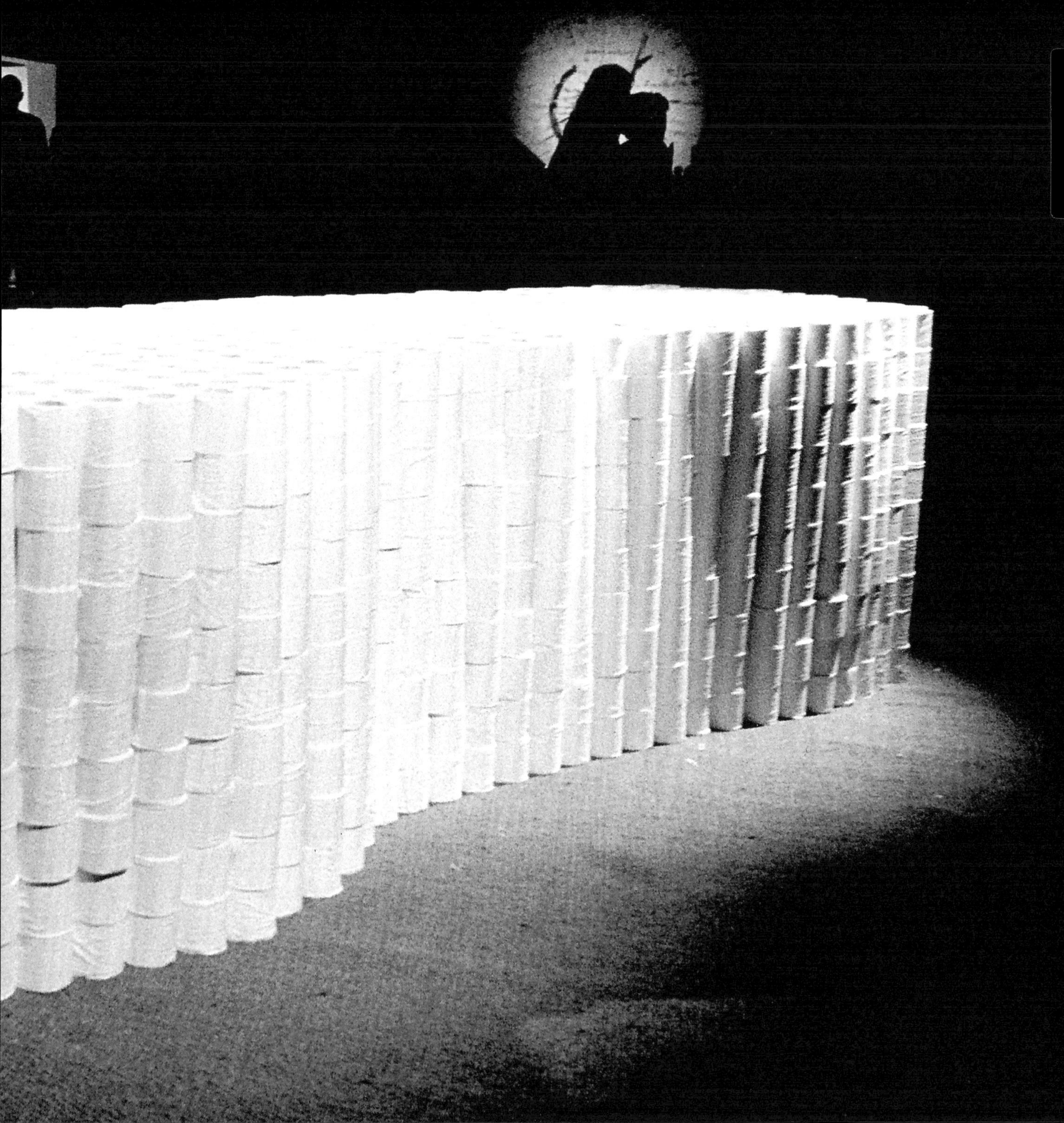

Terry Allen

MONTEZUMA 6-CASTLE
1992

American artist Terry Allen (Wichita, 1943) has been working as a musician, writer and visual artist since 1966. Raised in Texas, for Allen the folklore, culture and landscapes of the land along the border with New Mexico constitute an element of continuing inspiration in his artwork. In *Montezuma 6-Castle*, a landscape drawing in pastels is hung on a wall and connected via audio speakers and cables to a used shoe and audio components arranged on the floor. The shoe and the drawing broadcast the recording *Montezuma Castle* in a loop.

Music composed by Allen is often inserted into his installations, just as albums set his stories to music, in an interaction between multiple languages that evolve into performances, shows and theater pieces.

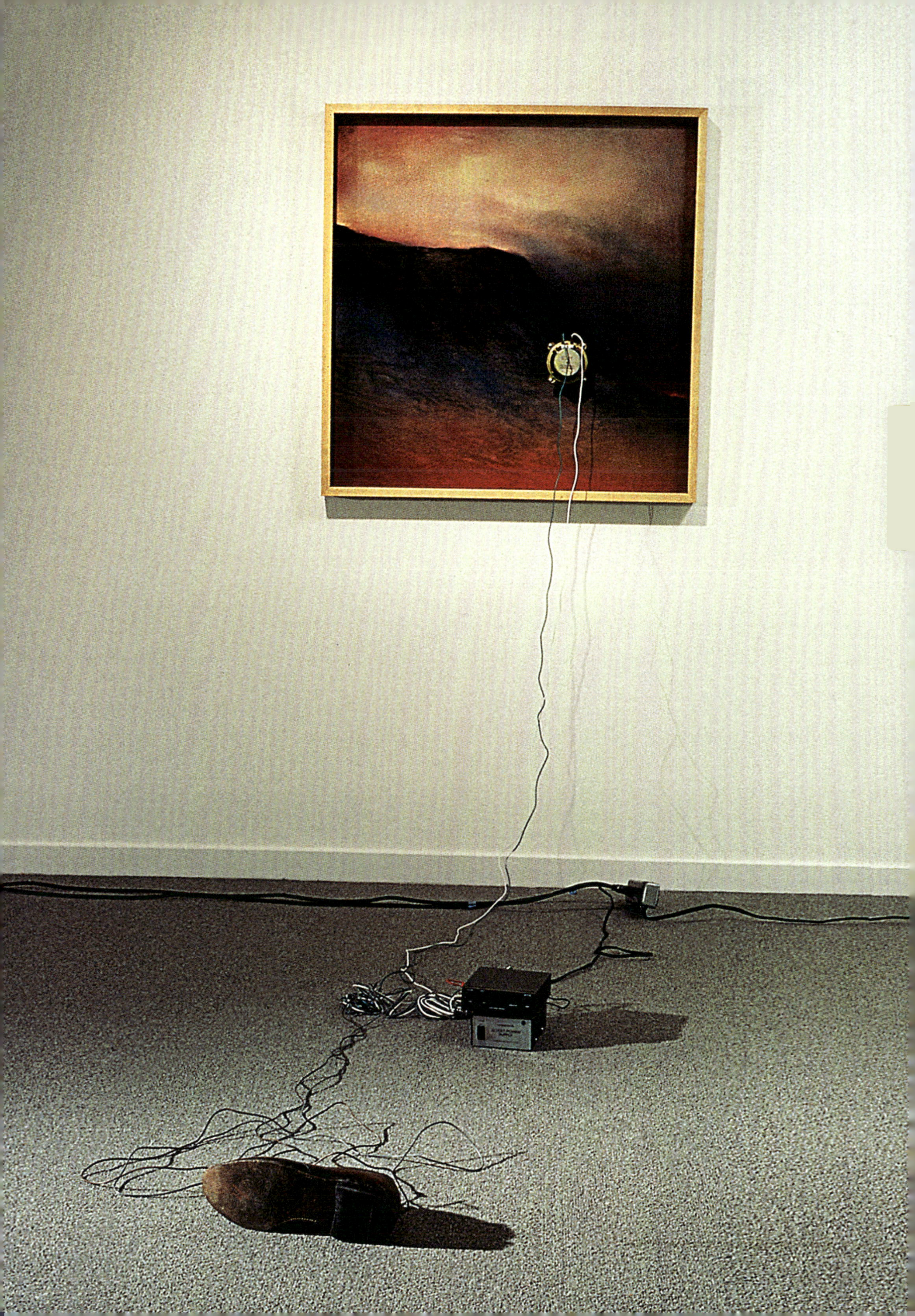

DAVID TOOP

INSTRUMENTS OF NON-EXISTENCE

(THROUGH WHICH HEAVEN AND EARTH SEEK RECONCILIATION)

In the *Sound (Acoustics)* section of Joseph Needham's *Science & Civilisation in China* (written by Kenneth Robinson), a strange instrument is described. A single-roomed building is built, in fact three rooms built concentrically, like a Russian doll. The walls are plastered to cover any cracks; doors prevent the ingress of draughts from outside. Pitch-pipes are mounted on wooden stands, further sealed from external air within a tent of orange-colored silk, each pipe arranged around the compass points in their proper positions. The upper ends of the pitch pipes are packed with reed ashes. Once prepared, they are watched according to the calendar. As the Han expert on astronomy, acoustic and music Tshai Yung wrote: "When the emanation (*chii*) for a (given) month arrives, the ashes (of the appropriate pitch-pipe) fly out and the tube is cleared."[1]

Noting that in the 21st century, cosmological machinery such as this might crassly be described as "sound art," can the arrangement of rooms and their esoteric contents be described as a musical instrument? Of course not, but then again its existence as an instrument of observation, of "waiting for the *chii*," was predicated upon the use of bamboo tubes to determine tuning in relation to the proportionate, cyclical, physical world and to *chii*, the mysterious, rarefied emanation which, rising from earth to the heavenly ancestors and from ancestral heaven back down to earth, could give life to all phenomena. So bamboo tubes were divinatory instruments, "humming-tubes" as they are called in Needham, used by shamanic music-masters to predict outcomes and influence events. "Every man has within his body his own *chii*," writes Robinson. "The diviner uses his to set up a disturbance in the outside world when he blows through his humming-tube. One *chii* will then 'by a kind of mysterious resonance' react on another *chii*, just as one musical instrument will touch off another which is in tune with it."[2]

So *chii* (*chi*, *ch'i* or *qi*, as it is better known today) and sound can be understood as closely related emanations within ancient Chinese belief, just as such divinatory actions can be contextualized within a wider category of spirit or ancestral voices "speaking" through voice disguisers, flutes and other esoteric instruments, a form of extra-human communication once widely practiced in regions of Papua New Guinea, Central and West Africa. Even in 1930, when Henry Balfour was collecting material in Nigeria for his paper *Ritual and Secular Uses of Vibrating Membranes as Voice-Disguisers*, such "mysteries," as Balfour described them, "had already been shorn of much of their former significance and terror,"[3] and yet the principle remains consistent throughout the global history of sound-producing devices.

The demonstrably physical yet alluringly intangible properties of sound can be shaped, modified, projected, transformed and given life through constructions that belong to the prosaic earthliness of matter on earth. No matter how beautiful to all the senses, they remain abject. Though fetishized, monetized and anthropomorphized, musical instruments are the waste products left behind after sound. By representing sound through matter as a static design for the eyes, they represent the betrayal of music, particularly from the radical Taoist point of view, that "the greatest music has the most tenuous notes." In James Joyce's *Ulysses* (1922), Leopold Bloom ruminates on sound and the outer limits of the musical instrument: "a blade of grass, shell of her hands, then blow."[4] Such an instrument can be found illustrated in Jaap Kunst's volume, *Hindu-Javanese Musical Instruments*,[5] described as a rice-stalk aerophone, the instrument that has no archive, becomes without sound, air, lips, the cupped hands, simply what it was before its sounding—described by Kunst as "a remarkably powerful shawm-like sound"—an abject wisp of discarded organic matter.

In Bloom's museum of streaming consciousness there is comb and paper, hunting horn, shepherd's pipe, policeman's whistle, the bestiary of the instrumentarium—"Brasses braying asses through uptrunks ... Semigrand open crocodile music hath jaws"—the holes of the female body to be played like a flute, the chamber music of Molly Bloom pissing in her chamber pot: "It is a kind of music I often thought when she. Acoustics that is. Tinkling. Empty vessels make most noise. Because the acoustics, the resonance changes according as the weight of the water is equal to the law of falling water ... Drops. Rain. Diddle iddle addle addle oodle oodle. Hissss."

Even Joyce's attempts here to transcribe sound into language are simplistic, as ignominious as the action itself, yet in their audition of an echo they echo Rembrandt's etching of 1631, *Man and Woman Pissing*, both examples gazing without judgment (though in both cases a degree of sexual interest is suggested) on a hidden history of abject, rough music, one of the origins of music perhaps (though never mentioned in those speculative texts that imagine early humans listening to birds and imitating their cries).

ORGAN(OLOGY) WITHOUT A BODY

"Composing means: building an instrument,"[6] Helmut Lachenmann wrote. In response: Building an instrument means composing (if composing is taken in its broadest sense, a gathering of materials).

An instrument may question the fixity of boundaries between all that it is to be human and all that it is to be not human. Many musical instruments are based on animal forms. They are channels, reservoirs, surfaces, chambers out of which sound is projected into the world through varieties of violence—friction, expulsions of breath, shaking, striking—as if empty animals whose cries can be transmuted into human desire.

In dreams, hallucinations, the generative machine of Raymond Roussel's fiction, alchemical symbology and in paintings we find ensembles which broach the boundary between human, animal and heaven, proposing a reversion to the garden of Eden: a hybrid creature depicted in the 14th century alchemical work, *Aurora Consurgens*, its instrument a lobster bowed by what may be a snake, a harmony created from the elemental chaos of Materia Prima: the chorale of ass and angels singing to the heavens in Piero della Francesca's *The Nativity*; Caravaggio's *Rest on the Flight into Egypt*, in which an angel plays violin, Joseph helpfully holding up the score of Nöel Baulduin's setting of the Song of Songs while rubbing one foot on the other to stay awake, Madonna and child asleep, a donkey's nose close to the violin within this intimate scene of breathing both sacred and profane, the sound of grasses and leaves, friction of small stones underfoot, a melody: "How fair and pleasant you are O loved one, delectable maiden."

This silent music is closest to the ideal of an instrument that has become dematerialized to fully allow music its heavenly status. Music without sound was a Taoist ideal. One note would only spoil the last one or the next; the most exalted of instruments, the *ch'in* (translated by Robert Hans van Gulik[7] as 'lute') could reduce earthly instruments to silence, its heavenly sound only audible to supernatural beings: "Now when the lute is played, the sounds of the instruments made of metal and stone die out, and the breath blown into those made of gourd or bamboo ceases. Wang Pao breaks off his chant, Ti Ya loses his gift of taste." In this resounding silence a more subtle silence rouses immortals, a phoenix pair and heavenly fairies.

In search of regeneration (or heaven), music apostatizes its instruments, relinquishing agency to the forces of nature or automata, hypothesizing (to reverse the Antonin Artaud formulation of a free body without organs) a (free) organology without body.

Mississippi-born R&B singer Bo Diddley, creator of guitars whose shapes and textures—rectangular, rocket-shaped, covered in pink fur—celebrated the non-guitar nature of the solid body electric guitar, identified the catalyst as an instrument without body: "There was this dude called Sandman that used to carry a board, an' a bag of sand an' a swish broom, you know, one of them 'sweep-ups'. He'd put the board down on the ground, an' dump this sand on it an' spread it around, and then he would sing an' sand—it was kinda like a shuffle, but it was called 'sandin.' It carried a rhythm with it, like a tap-dancer taps, an' not too many people around here could do it. When he finished his act, he'd sweep up the sand an' put it back in his little sack."[8]

This modulation between the (literally) earthly and immaterial is reminiscent both of the sand mandalas created by Tibetan llamas over days, then ceremonially destroyed and discarded in water, and the Japanese *dōtaku*, bronze bells displayed in the British Museum. These bells were made by the Yayoi people between 500 BC and AD 250, and often buried at the edge of fertile agricultural land. Decorated with images of animals valued because they preyed upon the insects that destroy crops, their ritual absence from sight and activation suggests a profound relationship between the invisible, hence mysterious energies of sound, the observable workings of nature and the unfathomable proclivities of the gods. A disappearance is enacted; the presence or physical sounding of the object is unnecessary for the magic to take place.

Such sounding devices and actions-at-a-distance cling precariously to the outer edges of our contemporary idea of the instrument. They are closer to an extra-human zone beyond culture, to the involuntary and uncontrollable, to meteorology, bioacoustics, cavernous reverberations and inexplicable submarine and subterranean noises, to supernatural conceptions of auditory phenomena. They arise as if from the deep, often treated as living beings, symbolic of afterlife and speaking their own language.

The huge garamut slit gongs recorded by Ragnar Johnson in Papua New Guinea in 1976 were decorated with long-nosed ancestor figures carved as a projection at each end of the garamut slit. Made from single logs, these wooden gongs were used both to send messages over long distances and to accompany the sacred flutes whose voices were the cries of spirits. "In the past," wrote Johnson, "when a new garamut had been carved and brought into a village, the villagers would dress up and hold a feast, decorating the garamut with feathers and leaves, and give it offerings of sago and coconut milk."[9]

Wooden slit gongs played a pivotal role in Bamum society in Cameroon,[10] their function to send out signals from within the royal court in times of crisis. By the time German colonists reached the Bamum kingdom for the first time in 1910, they found eight of these spectacular instruments, each hollowed out from a central slit, decaying on the palace ground like felled giants from a mythical past. Two French colonial administrators stationed in the Bamilike region noted that these gongs were carved when a king was enthroned; when he died they were abandoned in the market square and left to rot as if undergoing the same process of decomposition as the king himself. One Bamilike gong photographed in 1911 was carved in the shape of a buffalo, bigger than life, its body decorated with relief carvings of a chameleon, a lizard, a toad and the iron double gong used by secret societies. Consecrated by sacrifice, a gong like this might be used by all-male secret societies to summon members for a funeral. Torn from earth, the sound of these tree gongs gave voice to the dead and the otherworld of the deep. Thus it is with these unknown devices that speak in chthonic tongues from James Joyce's "rambling undergroands" of the lower depths.

As music of the European diaspora ripped itself away from the laws of elegant harmony to become a work of listening, sacrifices became central to the musical ceremonies of the 20th century. To plunge into a deeper well of time, far darker and deeper than the formalism of counted meter, was in one sense to engage with death.

In 1966, artist Raphael Montañez Ortiz took part in the Destruction In Art Symposium in London. He destroyed a piano, and while he was not the first artist to murder this object so charged with symbolism, nor surely the last, he locates his action in a moment when "many people were searching for a non-European aesthetic framework and many Native American and African American people were searching for an identity outside of the existing colonial ones." For Ortiz it was a matter of bridging the gap between his Catholic background and Yaqui ancestry "that took in a shamanic sense of reconciliation through sacrificial processes."[11]

The themes of concealment, sacrifice and the organ without a body are insistent throughout the 20th–21st centuries. In 1964, improvising musician, composer, inventor and scholar Hugh Davies performed in the ensemble that recorded Karlheinz Stockhausen's *Mikrophonie I*, a piece in which the surface activation of a large tam-tam with a wide variety of implements (many of them domestic) is amplified via microphones.[12] These, in turn, are transformed by other operators using filters and sound distribution. The technique of activating and amplifying the tam-tam had been developed by Stockhausen together with his technician, Jaap Spek. In this case the "instrument" has no location in space. Its body is nowhere to be found in an ecology of relations between the violence of actions (scraping, crunching and grunting are some of the terms Stockhausen used to describe the sounds of activation), the abject objects of activation, the microphones, electronic processing and sound distribution, and the flat surface of a tam-tam and its secret potential. In his preface to the score, Stockhausen described the process as a kind of auscultation; like a doctor probing with a stethoscope, he was able to reveal to the world those sounds hidden within metal.

Hugh Davies returned to London with the knowledge he acquired from Stockhausen (working as his assistant) and Jaap Spek, and from that moment developing a practice based on the amplification of small or hidden sounds using contact microphones. The sounds themselves came from humble artifacts to be found in a kitchen, bathroom cabinet, garage or garden workshop—springs, kitchen egg slicer, rubber glove, 3–D postcard, vegetable peeler and so on. These were collected together in assemblages known as Shozygs, multi-functional, tactile and idiosyncratic workstations requiring an intimacy of touch, a sensitivity of listening, their purpose both sculptural and microaudial, their secret possibilities becoming apparent through improvisation (Davies performed in Music Improvisation Company between 1968 and 1971 with Evan Parker, Derek Bailey, Jamie Muir and Christine Jeffrey, then with many other improvisers until his death in 2005).

In the 1970s Davies was part of a loosely affiliated group of inventors, all asking the questions: what is a musical instrument, what are its boundaries, what might its music become? Anna (now Annea) Lockwood, a colleague of Davies, published 15 copies of her *Sound-Hat* through Henri Chopin's Revue-Disque *OU* in 1970.[13] The instructions written by Chopin for this simple hat of hanging bamboo and shells were to turn the head "slowly from left to right, from north to sound, from west to east" as if divining the future through sound and its correspondence to points of the compass. "Think that Anna, who created the hat," he wrote, "wishes to do away with the cult of the composer, by asking you—all of you—to compose your own music."

If this was, in some way, a new vision for instruments (instruments of the European diaspora at least), it was no longer predicated upon music or composition. Even human agency was sacrificed. Artists such as Max Eastley built instruments operated by meteorological forces like wind and water, or by motors that functioned according to their own complexity. Their activation is not generated by human violence but by the unpredictability of nature or the fugitive scratching of a wire. The instrument was becoming disembodied, skeletal, a place of emanation not dissimilar to the mysterious Chinese room of pitch pipes packed with ashes.

To be within the "instruments" of Haroon Mirza [400, 434], Tetsuya Umeda,[14] Samson Young and Rie Nakajima[15] (to choose four contemporary artists who have inherited the proposition of organ(ology) without bodies) is to feel a dynamic part of sound fields whose autonomous volition works through unstable boundaries. As if holding a mirror to 21st-century life, their materials are abject and (de)composed, an elemental chaos of disparity; out of modest, fragile existence, questions of heaven and earth hang in the air, other sounds.

1
Joseph Needham (with Kenneth Robinson), *Science & Civilisation in China*, Volume IV: 1 (Cambridge, UK: Cambridge University Press, 1962), pp. 187–88.

2
Ibid., p. 138.

3
Henry Balfour, "Ritual and Secular Uses of Vibrating Membranes as Voice Disguisers," *Journal of the Royal Anthropological Institute of Great Britain and Ireland*, vol. 78, nos. 1–2, London, 1948, p. 47.

4
James Joyce, *Ulysses* (London: Picador, 1997), references taken from the Sirens section.

5
Jaap Kunst, *Hindu-Javanese Musical Instruments* (The Hague: Nijhoff, 1968), p. 29 (and illustration, after Curt Sachs, fig. 103).

6
Helmut Lachenmann, *Über das Komponieren* (1986), in Joseph Häusler (ed.), *Musik als existentielle Erfahrung* (Wiesbaden: Breitkopf & Härtel, 1996), p. 73.

7
Robert Hans van Gulik, *Hsi K'ang and his Poetical Essay on the Lute* (Tokyo: Sophia University, 1968), p. 118.

8
George R. White, *Bo Diddley: Living Legend* (Chessington, UK: Castle Communications, 1995), p. 27.

9
Ragnar Johnson, *Sacred Flute Music from New Guinea: Madang*, Quartz 001 LP (London: Quartz Publications, 1977).

10
Christraud M. Geary, *Slit Gongs in the Cameroon Grassfields: Sights and Sounds of Beauty and Power*, in Marie-Thérèse Brincard, *Sounding Forms: African Musical Instruments* (New York: American Federation of Arts, 1989).

11
Raphael Montañez Ortiz, quoted in "Art Attack at Tate Britain," *The Guardian*, September 28, 2013, gu.com/p/3j4tb

12
Hugh Davies, youtube.com/watch?v=wPT9A0IsGgs

13
Annea Lockwood, "Sound-Hat," *OU, Revue de Poésie Evolutive*, nos. 36–37, 1970.

14
Tetsuya Umeda, youtube.com/watch?v=TYptaKc-LMo

15
Rie Nakajima, rienakajima.com/_work/_installations.html/carpet.html

BART HOPKIN

TYPES AND ONE-OF-A-KINDS

HOW MUSICAL INSTRUMENTS FUNCTION SOCIALLY AND AESTHETICALLY

Here's an odd linguistic convention: people speak of something called "the violin," as in the sentence "she plays the violin." The oddness is in the use of the definite article *the*. Which violin are we referring to? Wouldn't it be more accurate to say "she plays *a* violin"?[1]

The usage reflects the fact that people tend to think of musical instruments as types. There's the violin, the bassoon, the waterphone, the trumpet marine, the snaggleodeon (OK, I made the last one up) and many more. Each of these represents a taxonomic category defined by certain characteristics—the violin, for example, is the category of musical instruments with four strings, normally played with a bow, having something close to a certain shape and size… and more: central to violin-ness is the existence of a surrounding culture of violin. This includes a large repertoire of composed works in the classical European tradition, plus association with other popular, ethnic or folk traditions. It also includes an available pool of highly-trained players. More subtly, it includes a bundle of ideas people have about what a violin should sound like and what kind of music it's good for. An individual violin, like the one that our friend plays, is a single representative of the category. While this particular violin inevitably has its own distinctive characteristics, in overall form it fits the standard description closely enough to qualify.

The reason that people think this way—conceiving of musical instruments as types, and reflecting this in language—is simply that, for the most part, it's an accurate reflection of the existing situation. The great majority of musical instruments in the world are recreations of pre-existing types, having only minor individual variations. It doesn't have to be that way, of course. If you decide to make a musical instrument, you're free to make anything you want, including an uncategorizable one-of-a-kind, unrepresentative of any pre-existing type. But let us acknowledge that there are good reasons for the "typing" of musical instruments. All those things mentioned above—the pool of skilled players, the existing repertoire, and (more important than it might seem) a shared sense of the instrument's sound possibilities—have proven to be immensely valuable in the real world of music-making. This accumulation of cultural capital around established instrument types has created fertile ground for much of the most wonderful music the world has enjoyed.

One of the factors at play in the typing of musical instruments may lie in the nature of human progress as reflected in their historical development. It can be seen that in musical instruments a pattern of development exists that is remarkably parallel to biological evolution. With musical instruments as in biology, different types develop gradually over time, building upon previous progress, with a process similar to natural selection ensuring that the most suitable types survive and are increasingly perfected. It is quite possible, as mentioned above, that some creative individual will create an instrument that is quite new and unconnected to previous lineage, and indeed that occasionally happens…but not often. More the norm is an evolutionary development in which people build upon and refine ideas inherited from predecessors. As a result, we find ourselves in a world with a seemingly finite number of nicely developed "species" of musical instruments.

It's interesting to consider the role of the individual in all this. Across most of human history and in most cultures, material development in musical instruments as in other fields has tended to be not only gradual and evolutionary, but also collective and anonymous. The idea that the individual should innovate and claim credit has been less celebrated historically than it is currently in the Western world. A caveat is needed here, however: to some extent this view of a traditionalist, collectivist past may reflect a facile and possibly inaccurate perspective on the part of us moderns. Since we don't know the individuals responsible for particular innovations, it's easy to credit those ideas to some vague essence called "tradition" or "culture," perhaps seen through a haze of sentimental primitivism. This might be a bit irksome to the person who, sometime in the unrecorded past, actually did come up with a valuable innovation and would like a bit of credit for it, but unfortunately that person is no longer around to tell us about it. That said, the tension between the individual and the collective is real in human life, and it does appear, in instrument-making as in other fields of endeavor, that celebration of the individual inventor or artist has been greater in the last few centuries in the West than in other epochs and places.

So if someone were to ask me today whom I admire most as an instrument maker, instead of giving the name of someone from 17th-century Cremona, or perhaps one or another of the wonderfully creative artists working today, I might look into that pool of people who never got credit and say "The person who developed the key-mounting system used in mbira! That person was a genius!" Or, if asked tomorrow, I might say "whoever thought of placing the sound post under one foot of the violin bridge! Very clever idea!" But of course I have no idea what the names of either of those people might have been. And in fact these technologies may have been the result not of an individual's work but indeed of collective and evolutionary processes.

The exhibit catalogued here is full of the work of creative individuals who have somehow sidestepped the idea that musical instruments and other sound devices should be made in accordance with pre-existing types. And as the exhibit shows, once you step away from the need to standardize, interesting doors begin to open. When you begin to look at the core processes of sound-making, avenues of exploration arise not only in sound, but in the ways that sound-work gives rise to visual form, in the ways that the underlying physics gives rise to both, and in the ways that sound-art can reflect broader conceptual perspectives.

A first example: Musicians and artists have long been intrigued with the concept of synaesthesia, the idea that the perceptions of one of the senses may be perceived through another. Most prominent has been in the cross-channeling of music and the visual. Can the pitches, timbres and time patterns of music translate to patterns of color and form?

Among early ventures in synaesthesia was Alexander Scriabin's color piano [136], a color organ using electric lights conceived as part of the score for the symphonic poem *Prometheus*, constructed by Alexander Moser for the first performance in 1910. Mack & Schwerdtfeger's *Farblichtmusiken* device was another early effort in translating musical sound into patterns of light. In more recent work, Henrik Neugeboren's *Sculptural Representation of Bars 52-55 of E-Flat Minor Fugue by J.S. Bach for a Bach Monument* [178] is a less literal and more poetic attempt at something similar.

As a contrast, let us note a similar effort made in the realm of the olfactory: The Smell Organ, proposed in some detail but apparently never built by the French chemist Dr. Septimus Piesse in the 1920s, was a keyboard instrument designed to release scents into a room in a manner analogous to the way music fills the air. The plan included an impressive-looking organ console and scores in music notation. One surviving drawing depicts a haughtily handsome keyboard virtuoso performing before a rapt audience in a grand concert hall.

Central to many such cross-channeling efforts is the question of mapping: Given any musical event or pattern, how do you decide what sort of visual (or olfactory) pattern it should translate to (or map onto)? Some people working in these areas have postulated specific and presumably universal mappings, seeming to suggest for instance that the note C is somehow inherently blue or G inherently red. Others have thought in terms of more personal or subjective associations between musical tones and visible color or form. Going the other way around, some artists have found ways to map visual phenomena into musical patterns, for instance turning the pattern of an urban skyline or birds on a telephone wire into musical compositions. Other sorts of mappings have been explored as well; for instance, mapping the light spectra of chemical compounds into corresponding sound timbres, or transposing radio signals from space (occurring as electromagnetic radiation at frequencies outside the human hearing range) into audio.

The question of mapping is also at the core of a particularly important area of contemporary exploration: what are the possibilities for translating human body gestures into patterns of sound? To see the significance of this, recall for a moment the early days of commercialized electronic musical instruments (think of electronic organs and synthesizers of the 1960s and following decades; the name *Casio* may come to mind). The natural thing to do, it seemed at the time, was to attach every new advance in electronic sound production to a piano-style keyboard. In subsequent years it has increasingly dawned on people that there could be other, potentially interesting ways for a human being to interact with an electronic sound-generating device.

As digital sound technology has become increasingly rich in potential and flexible in operation, it has become possible to frame the question of the player interface in this way: How can we map human movements into musical outputs? The keyboard, with its system of associating white and black keys with particular musical notes, is the interface for one mapping. Many others are possible. Dance controllers? Multi-player interactive controllers? Sensors to read and digitize other sorts of movement in three-dimensional space? The engineer's imagination is just about the only limit in defining the ways the person can interact with the device to manage audible output. Don Buchla has been a pioneer in this area, designing devices which not only recognize a wide and creative variety of playing gestures, but provide a high degree of flexibility and control in how one chooses to make the mappings from gestures to sounds, sound manipulations, and interactions.

Let us take a step beyond the constructs of artists and engineers and toward the underlying source of it all: nature and her ways. A number of sound artists in recent years have been fascinated with the physics and the physicality of sound, seeking not only to understand these things, but to make them visible; to highlight their beauty or their fascinating strangeness. A term has been coined for the effort to make sound vibration visible to the eye through physical and acoustical means: *cymatics*, suggested by the Swiss physician and scientist Hans Jenny (1904–72). Borrowing some ideas from the earlier German physicist and musician Ernst Chladni (1756–1827), Jenny created fascinating visual patterns using plates and membranes sprinkled with sand or other granular substances, and subjecting the plates to vibrations at different frequencies. Inspired by the cymatics idea, the German photographer Alexander Lautenwasser has produced stunning visual work with containers of water agitated into beautiful forms by sound vibrations.

Pyrophones [114], in which an audible tone is set up by means of a flame in a tube, represent an early thrust in the direction of sound made visible and physics made audible. Early pyrophones, and some contemporary ones as well, are set up to function almost like organs for playing conventional music. Other recent explorations, like those of the French artist Michel Moglia, have taken shape as wild and potent demonstrations of the pyrotechnical forces at play in both sight and sound.

Alvin Lucier has been a leading visionary in bringing forth the physicality of sound. In this exhibit, his *Music for Pure Waves* uses lightweight pendulums bouncing off of drumheads excited by electronically-generated sound waves, creating observable patterns of movement in response to the acoustic and mechanical factors in play. Carsten Nicolai's *Modell zur Visualisierung* (2001) displays sound patterns as contortions in a visible electron beam.

Among many other contemporaries whose work explores the underlying physics, we can mention the long-string works of Ellen Fullman. She works with strings dozens of meters long vibrating not in the transverse mode that is usual for musical strings, but longitudinally. This, it turns out, has intriguing physical implications which translate not only into an exotic and rather beautiful music, but also into a natural sort of choreography: the primary playing gesture consists of walking along the strings, drawing rosined fingers along them in passing, while the choice of walking paths among the strings functions almost as a score for the music—composition made visible. The physics of musical scale theory likewise become observable on a grand scale, manifest in the lengths of the strings or the placement of tuning weights along the strings, and in distances walked.

As ideas about the physical structure of musical instruments open up, so do ideas about the personal and social structures surrounding music-making. As an alternative to seeing instruments primarily as tools for concertizing, artists can explore alternative notions of how sound may function personally or socially.

I recall a passing comment made to me by Sharon Rowell, the maker of triple-chambered globular flutes exquisite in both appearance and sound. It was not a groundbreaking observation, but it stayed with me: her instruments, she said, were made primarily as personal instruments; private and suitable for solo meditation. She mentioned playing them alone among the dunes at a beach near her home. While her instruments have often been played to richly moving effect in concert settings, it's the image of Sharon alone among the dunes that abides.

On the other hand, sound instruments can also serve as mediators of interaction between pairs or groups of people. We touched on this earlier, in connection with computer systems that monitor input from several players and generate new musical material from the interplay. Much more physical sorts of interplay can be seen, for example, in sound set-ups in which several people interact with multiple conjoined long strings, jointly varying the tension and sound output with their pulling and leaning motions. Some multi-player instruments take the form of games, like the French group Les Phones' ping pong table in which the table surface is made of closely spaced metal bars producing different tones as the ball bounces randomly among them. For a more intimate interaction, think of a single wind instrument played jointly by two players as the sound varies with the mingled give-and-take of their breath pressure within the instrument. In Oliver DiCicco's *Due Capi* (1994) this takes place, with the optional addition of a third player to manipulate valves.

At the opposite end of the social spectrum, one can think of instruments and sound devices played by no one… or at least, by no human hands. The idea has never gone away, and has seen a resurgence with recent advances in computerized robotics.

From among the many working with computer-controlled musical robotics, we can mention the impressive automatons of Compressor Head, a robotic rock band, or the EMMI group (Expressive Machines Musical Instruments, consisting of three composer-builders), which produce more abstract musical styles. As a counterpoint to the interest in computer robotics, an interest in old-fashioned mechanics also remains, as exemplified in the low-tech constructions of the French maker Pierre Bastien or the Australian Ernie Althoff.

Music itself is a powerful message-maker; it naturally shapes the instrument maker's work. But as one learns to see sound-construction in a less type-restricted manner, another perspective opens up in the opportunity to think about pure sound independent of musical considerations. An important idea in modern sound art suggests that we can hear all sound as music; we need only learn to listen with open ears. My own contrarian sense is that the way people respond to music and the way people respond to other sorts of sounds are very different, and it's ultimately an unproductive exercise to conflate the two. This leaves plenty of room for an aesthetic appreciation of sound which does not concern itself with musical parameters. And indeed, many contemporary artists have found the idea of pursuing sound as sound itself very liberating. For the purest of abstract beauty in sight and sound, from among many that that could be cited, here's one representative: the Italian-American designer Harry Bertoia (1915–1978), with his Sonambient sculptures. These pieces—he made quite a few of them—are small groves of upright metal rods mounted on bases of wood or another material, sometimes weighted at the top. When set moving back and forth by gentle brushing, they strike one another to produce a shimmery, softly swaying aspect accompanied by sustaining, shifting, swishing sounds, as if breeze-touched rushes could be made of metal.

Placed in the context of a museum exhibit with an emphasis on the visual, musical instruments and other sound constructions have the potential to function a little differently from other sorts of creative art. Consider the question of abstract versus representational/narrative/programmatic.

Musical instruments seemingly aren't abstract, in that one does not consider abstract form alone in looking at them. There's a sense in which they engage the viewer by having a visual story to tell—particularly one-of-a-kinds that are not recognized by type. But it's a particular kind of story. It has to do with function, purpose, what they might do and how they might do it, why they are formed the way they are and how this particular manifestation arises from the intended objective. It's liberating in a way because the message, while engaging, is not fraught with social, political or aesthetic import; it's just about how things do what they do.

At the same time, musical instruments are unlike other machines that might make enjoyable viewing, because the instruments presented here as art are themselves devoted to creating art (of an audible sort). Design decisions are based both on that ultimate output and on the look and feel of the object itself. The process has an intensely human aspect, because central to the intent is an intimate physical interaction between the instrument and its player-to-be.

This much is true: in musical instruments form follows function. The forms in this case often turn out to be beautiful or, if not conventionally beautiful, then intriguing in unusual and engaging ways.

1
This convention—use of the definite article in our sample phrase "She plays the violin"—seems to hold in most of the Latinate languages, including Italian. Interestingly, it doesn't seem to hold in most Germanic languages for the phrase in question (e.g., "Sie geige spielt").

NOEL LOBLEY

ANIMATING ETHNOGRAPHIC OBJECTS

EVERYDAY AND EXTRAORDINARY MUSICAL INSTRUMENTS

"The artfulness of the ethnographic object is," according to Barbara Kirshenblatt-Gimblett, "an art of excision, of detachment, an art of the excerpt."[1] When musical instruments are collected, extracted from the fields and cultures from and for which they make sound and music, what practices and concepts also become detached and buried? Which are made visible and become revealed? How does an experience of an object and its local resonances help us locate the art of sound production?

In "Art or Sound" I can see and imagine hybrid instruments, pygmy gamelans and phonoharps; guns, TV screens and oxygen canisters inserted into instruments; idiophonic sound boxes, locked grooves and melting guitars. When thinking about an exhibition that explores the relationships between musical instruments, sound and contemporary art objects, we might look beyond the collections, galleries and displays of Western musical instrumentation during the last three centuries in order to develop wider concepts of what musical instruments and sound objects may or may not be. The very concept of "musical instrument," when considered cross-culturally, is one that asks us to locate the notion of musical artfulness and artistry in various, and at times hidden, places within and beyond material objects.

COLLECTING AND DISPLAYING ETHNOGRAPHIC INSTRUMENTS

Objects in collections of ethnographic instruments in museums typically range from the everyday to the extraordinary. The Pitt Rivers Museum at the University of Oxford, for example, includes objects as diverse as children's toy kazoos and Chinese classical qin (zithers) whose body and strings encourage philosophical and cosmological contemplation. Ethnographic collectors were encouraged to look beyond their own aesthetic notions of music and art, and to include as musical instruments "any object whether natural or artificial, and however simple, which is employed for the purpose of producing sound (whether 'musical' in an aesthetic sense or not)."

Henry Balfour, polymath and first curator of the Pitt Rivers Museum in Oxford, offered this advice in 1929 to anthropologists engaged in the collection of musical instruments, advice which seems prescient when we consider that distinctions between sound and music are continually dissolved and re-articulated in contemporary thinking about musical, sonic and art objects. Balfour was as interested in decorative art, tools and weapons as he was in music and everything else ethnographic. Like many curators, he recognized the value of starting with objects, for the appreciation and study of their material and visual properties and their wider cultural meanings and resonances.

What do we learn from the display of sound objects behind glass or in galleries, especially when musical instruments are often personified as being vital essences of the creative act itself? Objects of ethnography, Kirshenblatt-Gimblett reminds us, "become ethnographic by virtue of being defined, segmented, detached, and carried away by ethnographers" with the result that "like the ruin, the ethnographic fragment is informed by a poetics of detachment."[2] The process of asking gallery visitors to look closely at musical objects when their value often lies far beyond their visual appearance is part of this detachment. It is also the very process that can enable such objects to move in and through the realms of curio, specimen and art object, "though not necessarily in that order."[3] The collection, display and investigation of objects representing the material store that is the ethnographic collection of musical instruments and related sound recordings, an at times undifferentiated mass of objects and documents, explore these intersections between musical instruments, sound and art objects.

African musical instruments and sound objects continue to be widely featured in ethnographic museums, music exhibitions and shows related more widely to African traditional and contemporary art. Their presentation has often been highly varied and multi-faceted.

For example, New York's Museum for African Art's traveling exhibition "Dogon Now: Masks in Motion" explored the performance of music and masquerade through theatrical displays of both ancient and new Dogon sculpture, painting and drawing, textiles and clothing, masks and masquerade costumes;[4] Victor Gama, the Angolan composer and instrument builder, explores interactive instruments fusing ancient African and Eastern musical principles with Minimalism and further music for today;[5] and the Beninese artist Calixte Dakpogan inserted broken African 7 inch vinyl records among salvaged car parts to create his anthropomorphic standing figure *Perroquet (Parrot)* (2005).[6] These examples all explicitly explore the important relationship between visual and sonic communication within sounding objects. "Sounding Forms: African Musical Instruments" was an exhibition at four museums in the US and Paris showing African instruments as art objects in order to address the "scholarly silence on the subject of African musical instruments as sculpture."[7] It was probably the first concerted attempt to address this conspicuous gap within a Western exhibition format.

LOCALIZING THE CONCEPTS OF MUSICAL INSTRUMENTS

Recognizing that cultures around the world have developed their own formal and informal ways of conceptualizing musical instruments, ethnomusicologist and organologist Margaret Kartomi claims that "to classify groups of objects such as musical instruments is a fundamental principle of human thinking."[8] She refers to case studies drawn from many different communities throughout the globe in order to explore the properties of instruments and their relationship to people and their art. "Unlike the Hellenic and Hellenistic views that musical instruments are inanimate objects," writes Kartomi, "West African peoples such as the Dan in the Ivory Coast and neighboring Liberia (where they are known as the Gio) and the Kpelle of Liberia adhere to a human-centered concept of instruments."[9] "To the Dan or Gio," she continues, "human beings are not the exclusive owners of the instruments that they play, nor consequently of the music itself [because] most instruments are believed to have originally been owned by nonhuman beings such as spirits, masks, sorcerers, or animals."[10]

The Ghanaian ethnomusicologist J.H. Kwabena-Nketia has analyzed the aesthetic dimensions of many African musical instruments in order to explore the relationship between objects, power, speech and methods of visual communication. Nketia points out that some instruments are revered as religious symbols of political office. Among some Akan communities in Ghana, he notes, the hierarchical structure of chiefship is regulated by drums, the chief swears oaths to drums and "in the past, a chief could be rendered powerless or reduced in status by forcibly seizing his drums."[11] There can be, Nketia claims, a "tendency to regard some musical instruments as 'entities' with inherent vitality (analogous to that of a silent inactive person), which is made manifest when they are sounded."[12] He gives by way of example from Nigeria the Yoruban *iya ilu* drum, which is built from a tree that grows along the side of a road where people regularly pass by, because it is believed that the tree will have consequently learned the ability to 'speak' by absorbing the sounds of its environment in action. Conversely, when drums are hung silent outside some Wagogo homes in Tanzania to symbolize that an initiation ceremony is taking place within a family, it is clear that instruments also retain the art of communicating silently.

Musical instruments are also regularly personified as animate objects and bodies. Brincard notes that "some Africans view [instruments] as amplifications or extensions of the vibrations of the performer's soul and body."[13] For the Bakongo communities of Lower Zaire, Angola, and the Congo Republic in Central Africa, Robert Farris Thompson claims that "music is a living entity, as vitally present as a moving human body," and that the "Bakongo also believe that musical instruments are human beings."[14]

Visual symbolism and representation can be extremely powerful. In her analysis of some of the varieties of harps of Central Africa—many of the traditions of which are thought to be at least 5000 years old—Sue Carole DeVale argues that the traditionally high status of the harp is usually marked by very elaborate decoration and symbolism. The Bwiti *ngombi* in Gabon, for example, has a resonator that symbolizes the stomach of the female while the neck symbolizes the male, and the body is usually painted red and white to represent both blood and semen, while four of its strings are considered male and four female. Elaborate ritual procedures surround the instrument, which, itself considered the power of night, must not be exposed to the light of the sun. Accordingly, at the end of night ceremonies the *ngombi* might be accompanied by specific exit songs before being stored in darkness securely away from daylight.[15]

ZIMBABWEAN *MBIRA* CONCEPTS

Paul Berliner's landmark study of the music and traditions of Shona *mbira* (lamellaphone) music from Zimbabwe paints a very rich picture of the fluid concepts surrounding musical instruments.[16] Surveying some of the origin myths and uses of a variety of *mbira* instruments, he observes that although some instruments might be little more than toys, other instruments achieve very high status, often being restricted to specific players and ancestral ceremonies. The association of the *mbira* with the world of the ancestors is often called the rule or law of *mbira* and "some musicians consider that the 'law of *mbira*' obliges them to recognize the ancestors formally as owners of their instruments and to care for these instruments with due respect: the theft of an *mbira* was traditionally considered in Shona culture to be a very serious crime."[17]

Berliner shows how it is often the relationship between the instrument and the player that is paramount, indeed that "some performers imply … that the instrument has a strong will of its own" and that it in effect calls you.[18] Some *mbira* players refer to their instruments with personifications at both the level of the entire instrument and also for individual keys. Today, many players explain the relationship between the different keys in term of pitches, but older musicians such as Mubayiwa Bandambira, a respected elder at the time of Berliner's original research in the late 1970s, have a much more detailed system for describing family relationships both between registers of keys and between individual keys. The three registers of keys in an *mbira dzavadzimu* ("*mbira* of the ancestors") —bass, middle and high—are referred to, respectively, as "old men's voices," "young men's voices" and "women's voices."[19] Bandambira also eventually details an intricately interwoven system naming and personifying each of the 22 keys on his *mbira dzavadzimu*, classifications which "can be regarded in large part as praise names, names that praise each key for its musical quality and for its effect upon listeners."[20] Bandambira detailed the sharp voice of the *Benzi* ("mad person") keys as being the ones that frighten you or make you feel wild and excited; the *Duri* ("mortar") keys as providing the rhythmical sounds like stamping millet; the *Nhiura* ("big drum") keys as the ones that lift up your lungs and make you feel like shaking your body; and the *Shumba* ("lion") keys as the animal in which the powerful spirits live.

Hundreds of *Mbiras* hang on the walls of the International Library of African Music in Grahamstown, South Africa, and dozens are displayed under glass in the cases of the court of the Pitt Rivers Museum in Oxford, but how does the exhibiting of such muted objects relate to the personified and animated relationships between musicians, materials and spirits? Such a question is being explored through painting by Zimbabwean artist and musician Moses Bikishoni.

The *mbira* is commonly played inside a calabash or fiberglass gourd (*deze*) studded with buzzing bottle-caps, for both amplification and the addition of complex rhythms, pitches and timbres. Over time, and in the hands of an expert player, repeating melodic cycles can spiral into fantastically woven textures that reveal complex hidden sound worlds of water flowing and gushing, and of children singing and laughing amid huge creaking and crackling structures. Bikishoni's paintings, such as *Gwariva – Inside the Gourd* (2014), mix cubism and impressionism with idiosyncratic visions of bone-like spirals to represent these intricate relationships between *mbira* instruments, sound, and communication with an ancestral spirit world.[21]

HUNTING AND GATHERING MUSICAL INSTRUMENTS

Whenever we get too comfortable with a classification, pattern, or relationship, we can always look for cultural comparisons and critiques from anthropology. Anthropology advances cultural relativism, asking us to question the predominance or naturalness of any system or idea, critiquing whatever might be considered normative.

BaAka hunter-gatherer communities, semi-nomadic pygmies inhabiting the Central African Republic and parts of Northern Congo, have created an astonishingly rich musical system based on an ingenious relationship between instrumentation, bodies and the environment. The musical instrumentation for rainforest ceremonies such as *boyobis* extends to include the cathedral-like acoustics of the rainforest canopy, the use of the most cursory of utensils such as jerry cans, discarded food tins and upturned bowls, and the illusive sound and art of dancing masked spirits to create beautifully complex webs of interwoven polyphonic and polyrhythmic relationships.

BaAka people are intimately aware of the sonic properties of the materials and objects in their environments. In the rainforests, young male youths sometimes select, bend and then tie a young sapling to form an arc that is fastened into a resonating pit dug underground. The resulting improvised earth bow (*ngbindi*) is then plucked and stopped with thumbs and forefingers to create a powerful rasping and roaring bass motif that, when listened to closely, can sound like heavily processed electronic dubstep. The *ngbindi* is simply discarded when the BaAka move to another camp or settlement. Male youths might then create the most stunning cross-rhythms from drumming their bodies and the buttress roots and trunks of selected trees using their palms, fists and sticks. The trees are usually selected as instruments for the way the slapping, cracking and thudding of hands against wood and human bodies echo, resonate and repeat throughout the environment.

At other times, a musical instrument can be made out of a river. BaAka women and children bathing in rivers might spend time cupping their hands into various controlled shapes to coax both rhythm and melody from the plunging percussive rippling of the water's surface, splashing cross-rhythms into joyous entertainment.

The art of a musical instrument can, tragically, also be lost. The *mbyo* is a four-hole notched flute that was commonly heard among BaAka communities at Yandoumbé up until relatively recently, near the town of Bayanga in the southwest of the Central African Republic. Expert players like Mindumi, Mobila and Contreboeuf could play variations lasting until dawn, listening to the pulsing sounds of forest insects, playing along with the warm pattering sheets of night rainfall, exploring and interlocking their notes with the suspended overtones resonating in the rainforest canopy. As soothing soundtracks for sleeping camps, this was quite literally music played to enter people's dreams. As the last expert player died and no children picked up the instrument, the last known *mbyo* was being preserved by Louis Sarno, a recorder and researcher who lives with the BaAka at Yandoumbé. In March 2013, during a coup d'état in the Central African Republic, Séléka rebels ransacked homes at Yandoumbé. The last *mbyo*, made from a local plant that can be crushed if not held carefully, was stamped into the ground by army boots.[22]

CONCLUSION: THE ART IN SEEING SOUND OBJECTS

The ethnographic lens of collected and displayed musical instruments often privileges the visual dimension and aesthetics of sound objects, asking us to look closely instead of listening to the art of sound relationships. Examining the varying concepts of musical instruments in several different African cultures reveals that sound objects may be considered materials owned by gods or spirits; networks of relationships between humans, objects and environments; ephemeral natural materials that are discarded after use along hunting pathways; or visual symbols of power and status that are mastered by some individuals, strictly forbidden to certain social groups and even destroyed by others.

1
Barbara Kirshenblatt-Gimblett, "Objects of Ethnography," in Ivan Karp, Steven Levine (eds.), *Exhibiting Cultures: the Poetics and Politics of Museum Display* (Washington, DC/London: Smithsonian Institution Press, 1991), pp. 386–443: 388.

2
Ibid., p. 387.

3
Ibid., p. 392.

4
africanart.org

5
victorgama.org

6
caacart.com

7
Marie-Thérèse Brincard (ed.), *Sounding Forms: African Musical Instruments* (New York: The American Federation of the Arts, 1989), p. 15.

8
Margaret Kartomi, *On Concepts and Classifications of Musical Instruments* (Chicago, IL/London: The University of Chicago Press, 1990), p. 3.

9
Ibid., p. 241.

10
Ibid., p. 243.

11
J. H. Kwabena Nketia, "The Aesthetic Dimensions of African Musical Instruments," in Marie-Thérèse Brincard (ed.), *op. cit.*, pp. 21–29: 22.

12
Ibid., p. 23.

13
Marie-Thérèse Brincard (ed.), *op. cit.*, p. 17.

14
Robert Farris Thompson, "Body and Voice: Kongo Figurative Musical Instruments," in Marie-Thérèse Brincard (ed.), *op. cit.*, pp. 39–45: 39.

15
Sue Carole DeVale, "African Harps: Construction, Decoration, and Sound," in Marie-Thérèse Brincard (ed.), *op. cit.*, pp. 53–61.

16
Paul Berliner, *The Soul of Mbira: Music and Traditions of the Shona People of Zimbabwe*, reprint (Chicago, IL/London: University of Chicago Press, 1978, reprint 1993).

17
Ibid., p. 236.

18
Ibid., p. 128.

19
Ibid., p. 56.

20
Ibid., p. 58.

21
mosesbikishoni.com

22
This section is based on research into an extensive archive of BaAka music recorded by Louis Sarno and curated by the Pitt Rivers Museum, Oxford. For further information see: web.prm.ox.ac.uk/reel2real/index.php/collections-sarno.html

Martin Creed

WORK NO. 97
(A METRONOME WORKING
AT MODERATE SPEED)
1994

Martin Creed (Wakefield, 1968) is a British artist and musician working across a multitude of disciplines, including installation, sculpture, music, dance and film. Creed has also been an experimental musician since 1994, the year he established his first band, Owada, which performed until 1999, and in 2011 he launched his own label, Telephone Records. Sound has also featured in his gallery-based work, with pieces using doorbells and metronomes.

In *Work No. 97*, a metronome produces a percussive piece of music while also marking out time—the ticking of the metronome actually echoes throughout the space, its unique identity becoming part of the natural environment in spite of its monotone clicks. In 2011, Creed further developed his fascination with metronomes in *Work No. 189*, an installation comprising 39 metronomes set to different tempos or speeds, assembled as a chorus that juxtaposes rhythmic precision with dissonant cacophony.

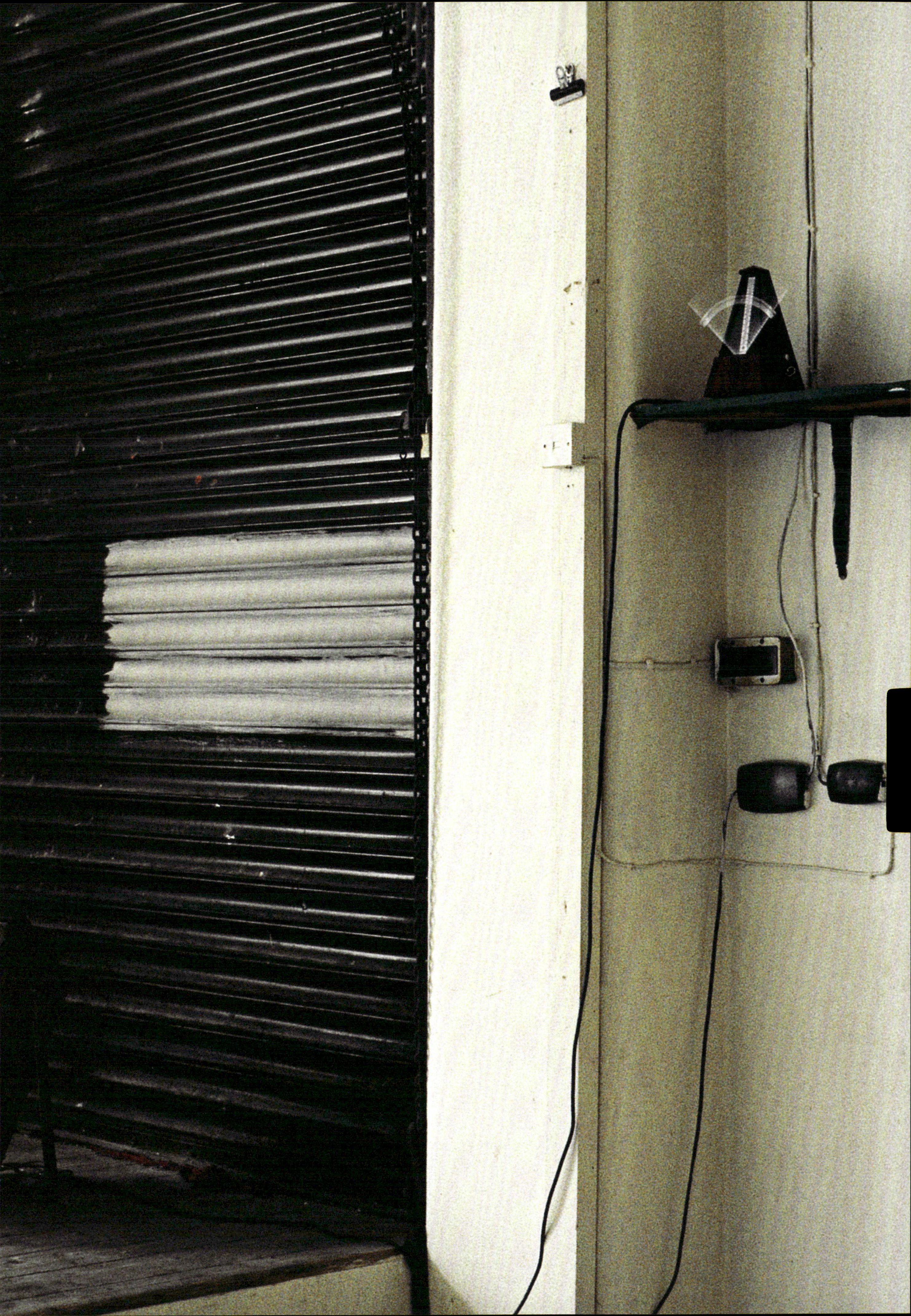

Maywa Denki

GUITAR-LA
1994

Maywa Denki is an art unit that mimics a company. It was formed in 1993 by two Japanese brothers, Masamichi and Novumichi Tosa (1965 and 1967, Hyogo Prefecture) and named after the small manufacturing business their father ran in the 1970s. Instead of pieces of art, Maywa Denki sells "products," and exhibitions or live performances are referred to as "product demonstrations." Maywa Denki is renowned for their "product lines" *Naki*, nonsensical machines featuring fish motifs, and very popular *Tsukuba*, which are tiny working instruments driven by motors or electromagnets.

Guitar-la consists of 5 guitars arranged to form a *sensu*, a traditional Japanese fan, which can be controlled through the pedal organ with a built-in controller. The instrument works by using organ keys that operate 100V switches to strike the guitar strings.

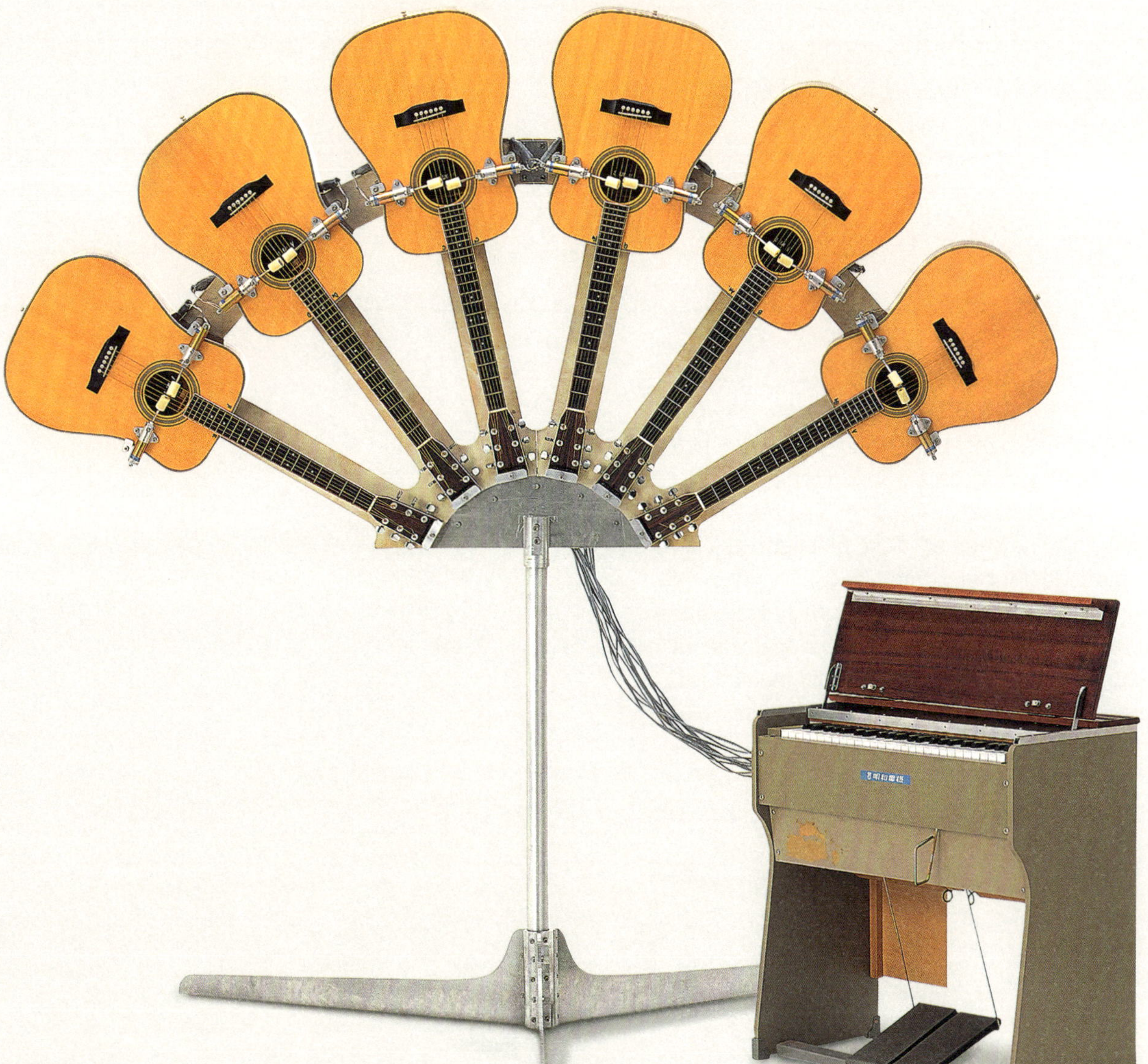

Stephan von Huene

DER MANN VON JÜTERBOG
1995–96

Stephan von Huene conceived his first musical sculpture in 1967. Entitled *Tap Dancer*, the piece comprises 2 mechanical legs that produce a loud noise by tap-dancing on a wooden chest.

In *Der Männ von Jüterbog* (The Man from Jüterbog), the movement of the mechanical legs is accompanied by a recording of a text written by German-American writer Reinhard Lettau, played through a loudspeaker. The automaton presents the text through the language of the mechanical sculpture. Each sound is numbered and associated with a movement. When the story (narrated by the writer himself) suddenly halts, the leg of the sculpture remains fixed in the air before restarting its movement. The drum adds a dimension to the work by accompanying individual words with either a single beat or a drumroll.

[See also pp. 244, 246]

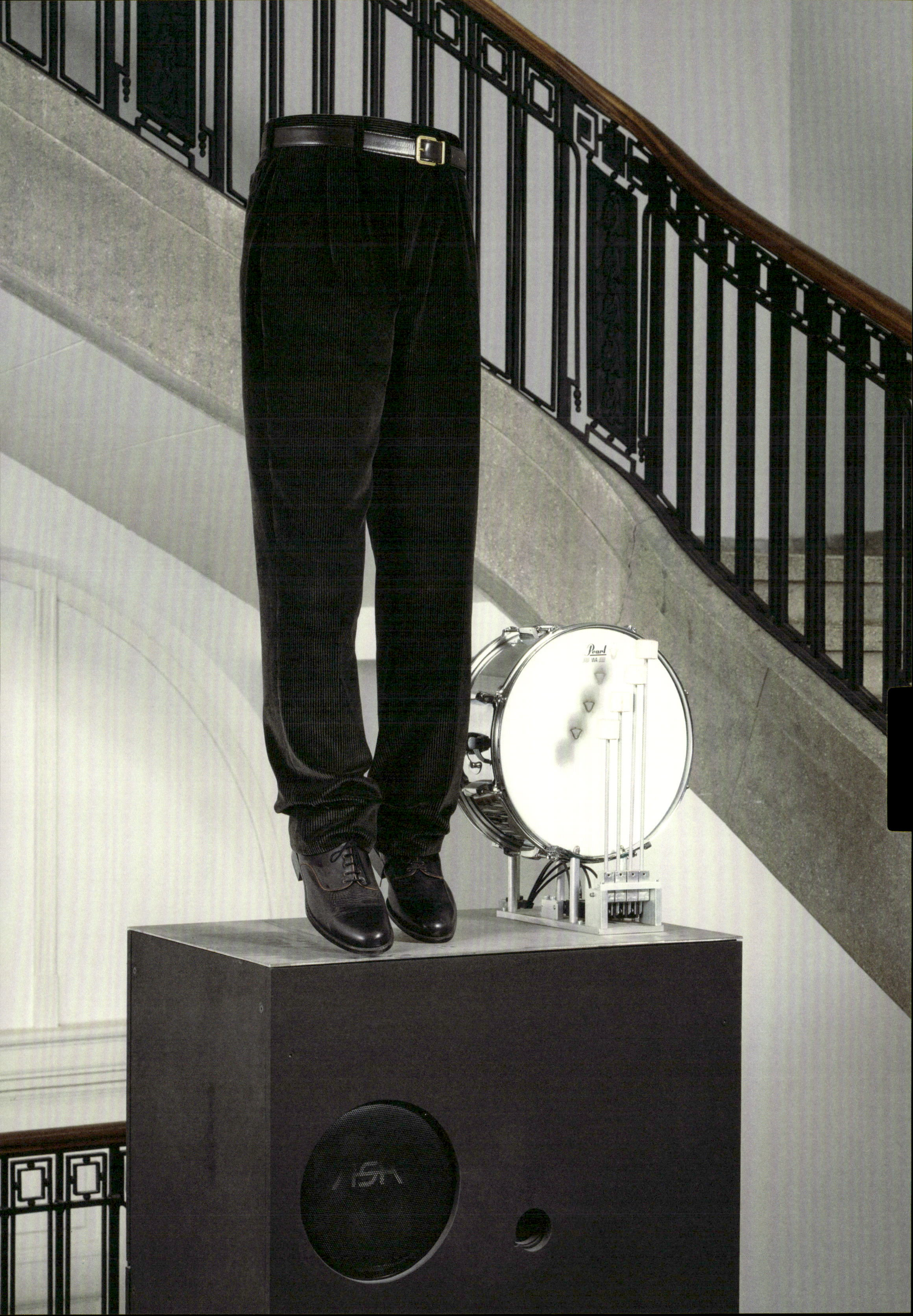
Pearl

Arman

THE SPIRIT OF YAMAHA
1997

French artist Arman is associated with the practice of artistic creation through the destruction of objects, often musical instruments. *The Spirit of Yamaha* was the first work in the *Sandwich Combos* series, in which the object is not altered through smashing followed by recomposition on a wall panel—as occurred with the double bass in *NBC Rage* (1961)—but by slicing and inserting within it another object of completely different and contrasting nature. In this case a baby grand piano has been sliced into 3 parts and 2 Yamaha motorcycles sandwiched between.

As an exponent of Nouveau Réalisme, Arman identified a composite object, like the piano, with the multiple aesthetic potential that its plastic destruction causes, producing unconventional images and forms in an everyday object.

In his next series, *Fragmentation*, Arman sliced a piano, separated the parts, and then recomposed the fragments with the title *Concerto for Four Pianos* (1998).

[See also p. 218]

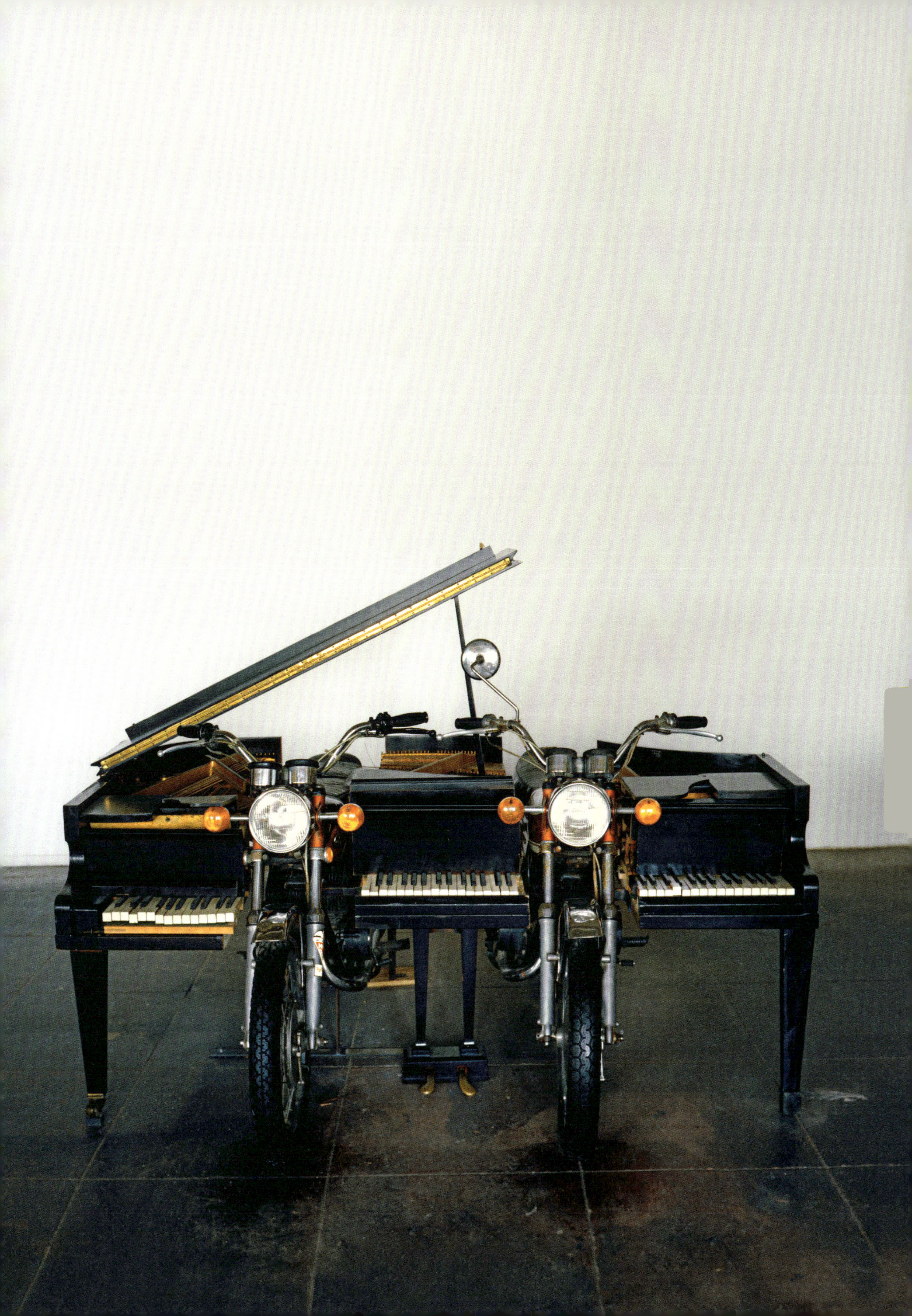

Bart Hopkin

SAVART'S WHEEL
1997

Over the years American artist and musician Bart Hopkin (Evanston, 1952) has been active in many facets of music-making, including performance and recording, composition and arrangement, music education, theater music, ethnomusicological research and instrument making.

Savart's Wheel is a tuned rasp consisting of 30 wooden disks of graduated size on a motor-driven spindle. The edge of each disk is surfaced with ridges spaced about 3 millimeters apart. As the spindle turns, the player holds a plectrum against the ridged edge of one of the rotating disks. The number of ridges that pass against the plectrum per second is the frequency of the sound. With disks of different diameters, the number of ridges-per-second changes, so by holding the plectrum against different disks you get different frequencies, producing different notes. The plectrum is not like a typical guitar pick; the plastic piece that does the picking is attached to a polystyrene cup that acts as a soundboard, projecting the sound into the air.

[See also p. 352]

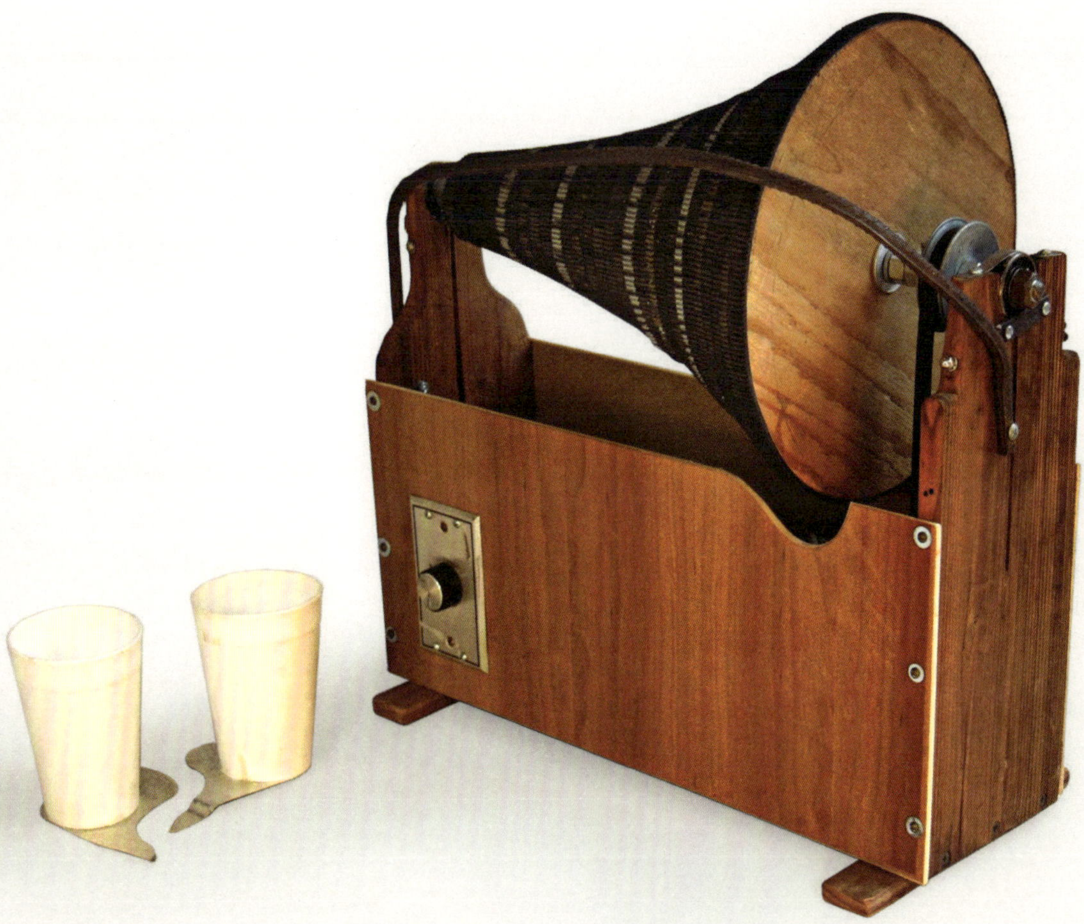

Marco Bagnoli

SONOVASORO
1997 (2000)

After studying chemistry, around the end of the 1970s Italian artist Marco Bagnoli (Empoli, 1949) became interested in the relationship between art, science and spirituality. He presented his *Sonovasoro* for the first time in 1997, collaborating with Giuseppe Scali for the sounds. An essential part of the work was the inclusion of sounds produced by didgeridoos.
A new version was presented in 2000, in which frog croaks were included for purposes of harmonics and their strong connection with primordial nature (an evocative and highly symbolic aspect of the work). The decision to include the punctiform sound of frogs croaking, and thus its rhythmical scansion as well, was deliberate, and refers to the significance these animals hold in Classical mythology, in particular to a citation by Aristophanes in which he describes them as being “especially dear to the gods.”

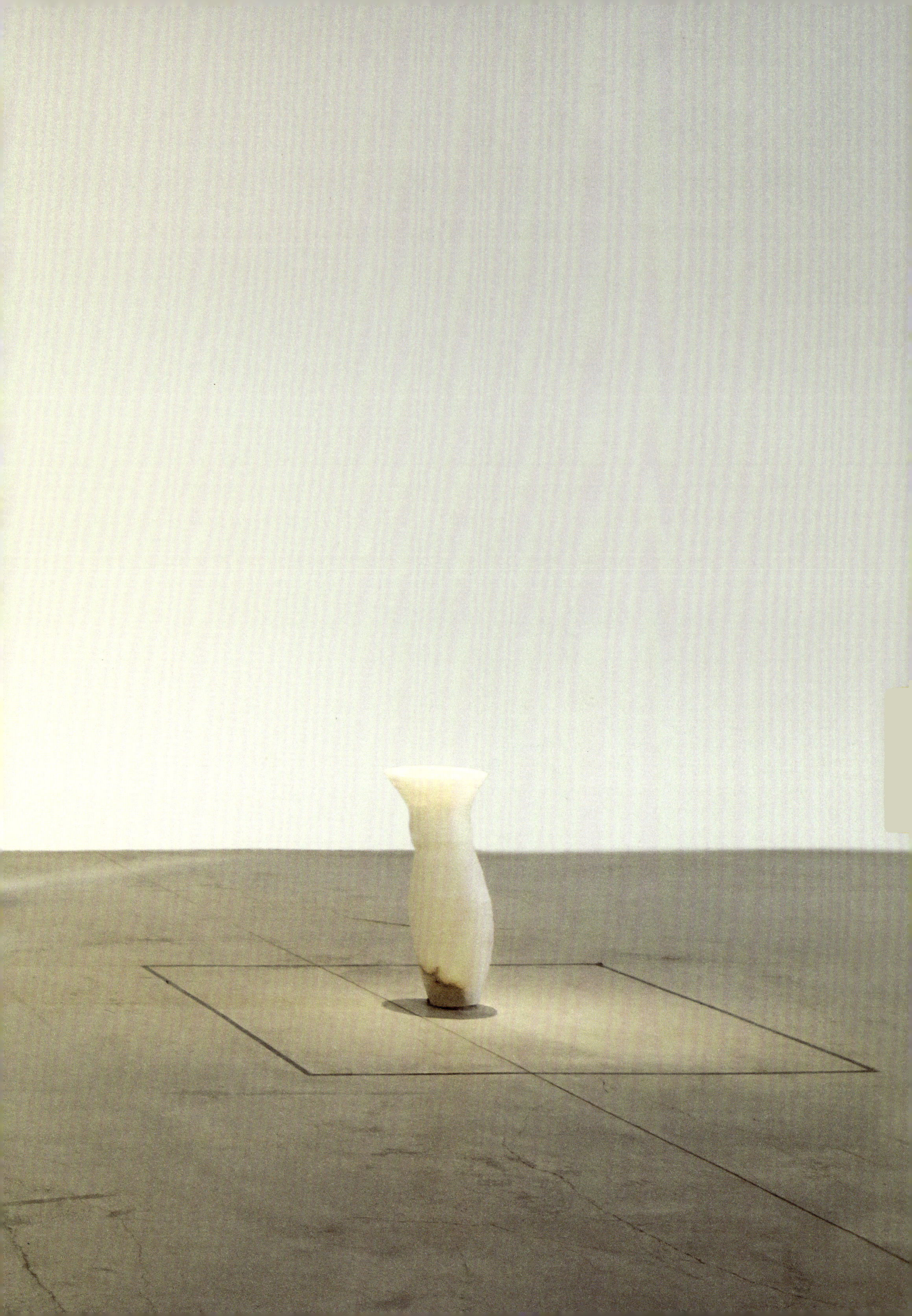

Carsten Nicolai

BAUSATZ NOTO ∞
1998

Also known as Alva Noto, German artist Carsten Nicolai (Chemnitz, 1965) is a sound artist whose work lies at the intersection of sound, art and design. He studied architecture and landscape design before founding the electronic music label Raster-Noton and pursuing art.

Nicolai describes his installations as visualizations of frequencies that allow the works to create their own voices. In the case of *Bausatz noto ∞*, 4 turntables integrated into a table give viewers an opportunity to play several sound-loops, layering and superimposing them to create endless new combinations while attentively following the emergence of the sound surface through headphones.

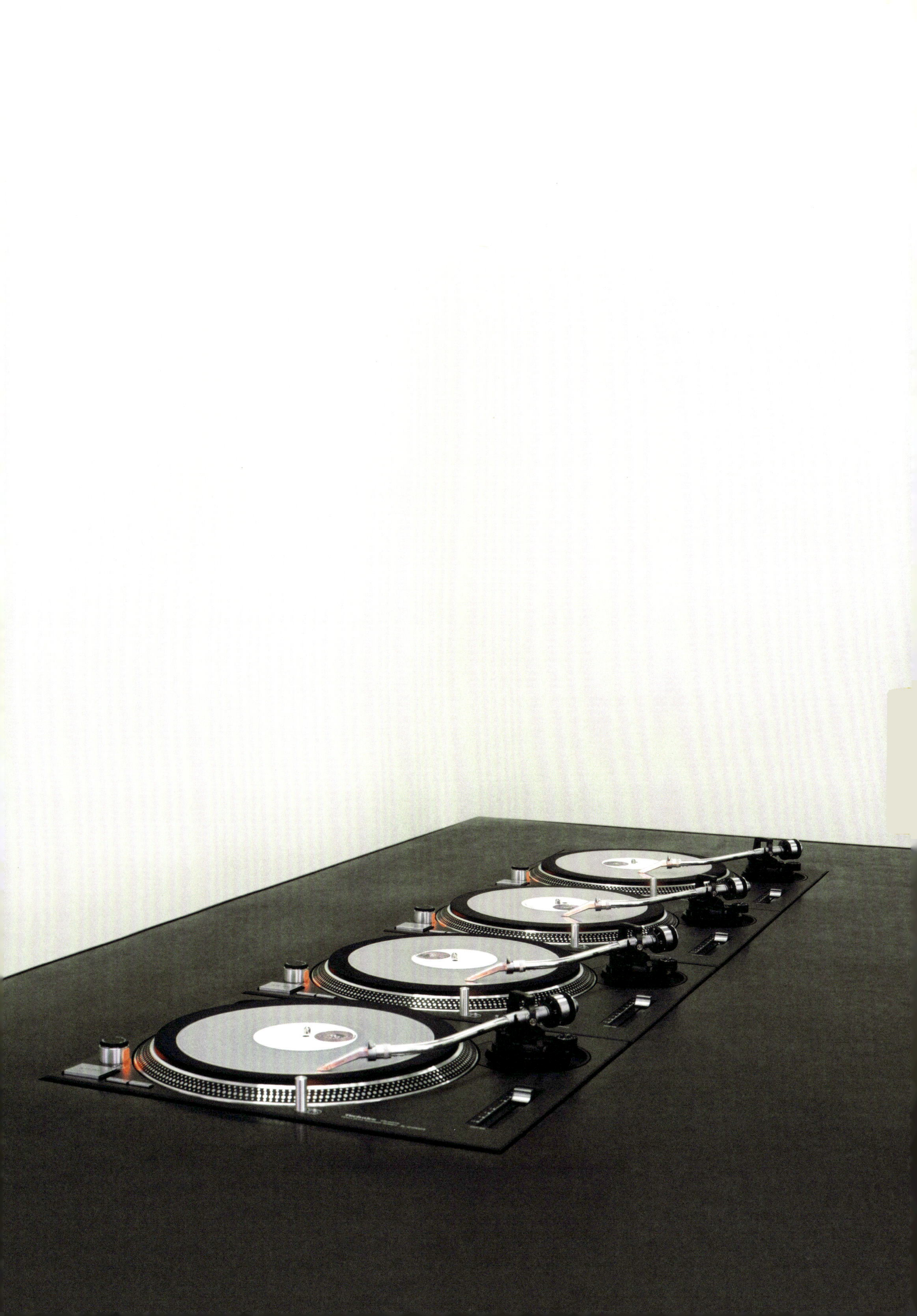

Bart Hopkin

OVERHANGING
1999

Thanks to its external amplification, Bart Hopkin's *Overhanging* reveals the inner vibratory life of a narrow spring-steel rod whose sounds would otherwise be nearly inaudible.

It consists of a simple system for holding the rod firmly enough to make vibration possible while allowing continual variation of the active length. The rod can be activated by plucking, bowing or percussion. Positioned beneath the rod is a magnetic pickup, allowing the player to monitor the vibrations.

Overhanging produces fascinating, odd and comical sounds, and can demonstrate many unexpected and interesting acoustic phenomena.

[See also p. 346]

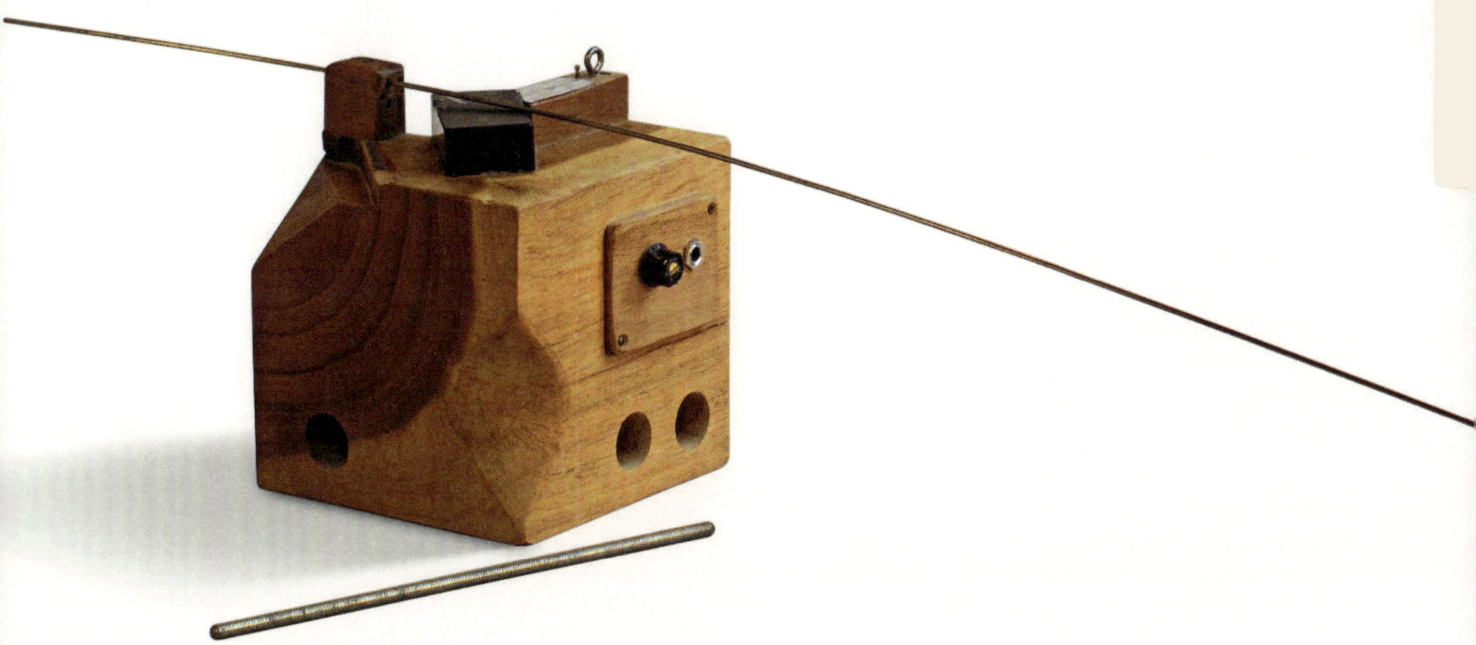

Rebecca Horn

CONCERT FOR THE PATH OF THE BEES 1999

Rebecca Horn (Michelstadt, 1944) is a German artist. Her artistic practice includes body art, performance, installation, sculpture and film. In her early works she attempted to turn her body into a machine by attaching various objects and instruments to it; then the human body slowly disappeared, substituted by mechanical sculptures, as in *Concert for Anarchy* (1990) in which a grand piano is suspended upside down from the ceiling, and a timed mechanism pushes the piano keys out of the keyboard and opens the piano's lid to reveal its chords.

A Concert for the Flight Path of the Bees and *Sighs of the Winds* [exh.] are mechanically activated instruments—a cello and a violin, respectively—that can be considered further steps in investigating the theme of the absent performer replaced by the public through its participation: "...I simply allow the consequences of whatever the machine is doing to meet with the particular reality or mood each person brings to it."*

* From an interview with Germano Celant in "The Bastille Interviews 1 Paris 1993" in Various Authors, *Rebecca Horn*, exhibition catalogue (New York: Guggenheim Museum, 1993), pp. 14–23: 18.

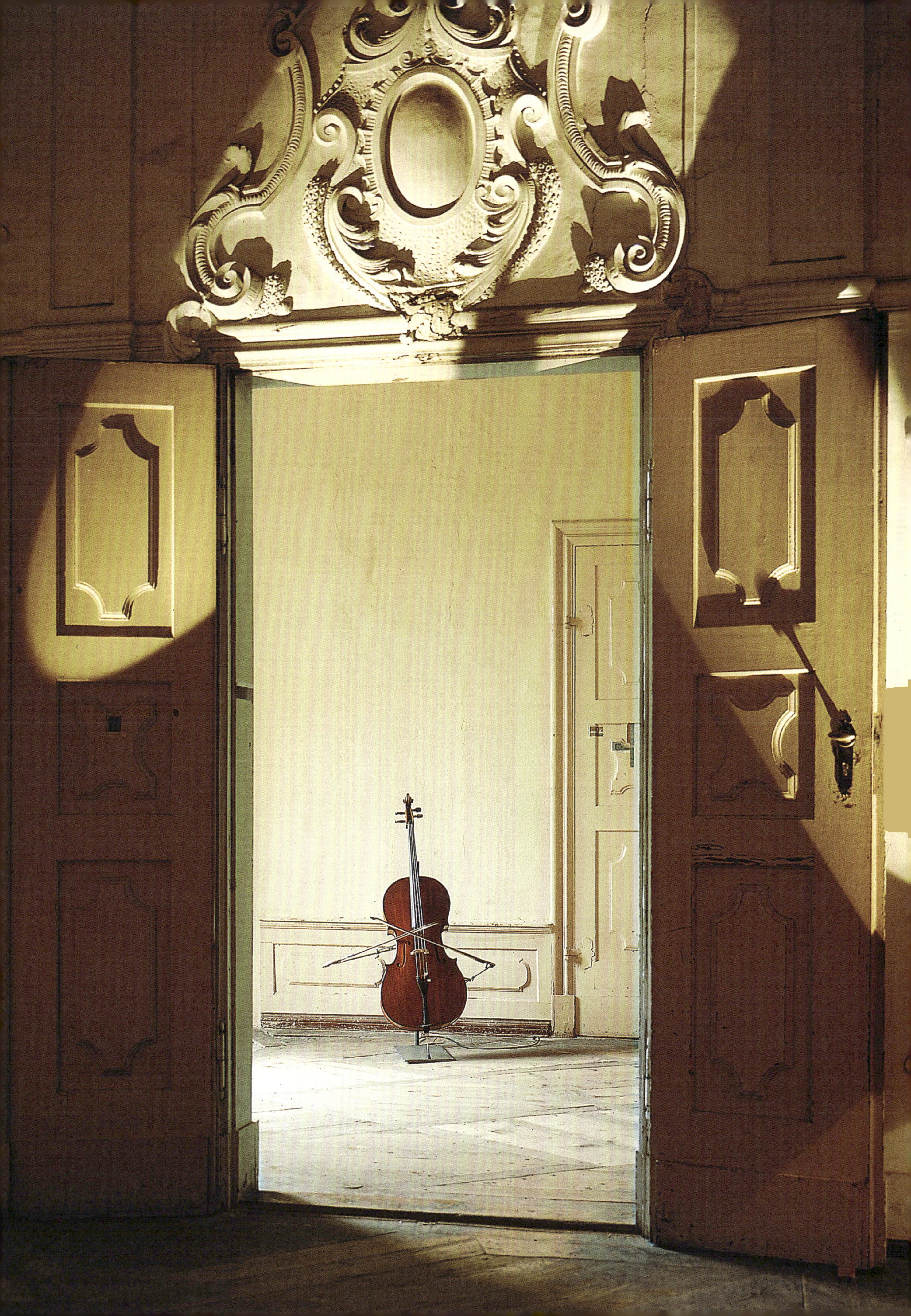

Edward J. Potokar

SILVER JUBILEE
1999

The son of a drummer of Latin, jazz and polka music, American industrial designer and musician Edward J. Potokar (Cleveland, 1958), has been making musical instruments, amplifiers, microphones and recording equipment since the 1970s. He loves the process of making instruments as much as playing them. According to him, "you get something unique, because one-of-a-kind instruments make one-of-a-kind sounds. I would rather hear something new than something good."

Known as "boxes," Potokar's musical instruments like *Silver Jubilee* and *Mopol* [exh.] are analog synthesizers based on an evolving lineage dating back to the late 1970s, with circuit boards either handmade or hard-wired, usually out of found or surplus parts. The components, not generally designed for audio, allow the instruments to produce unique sounds.

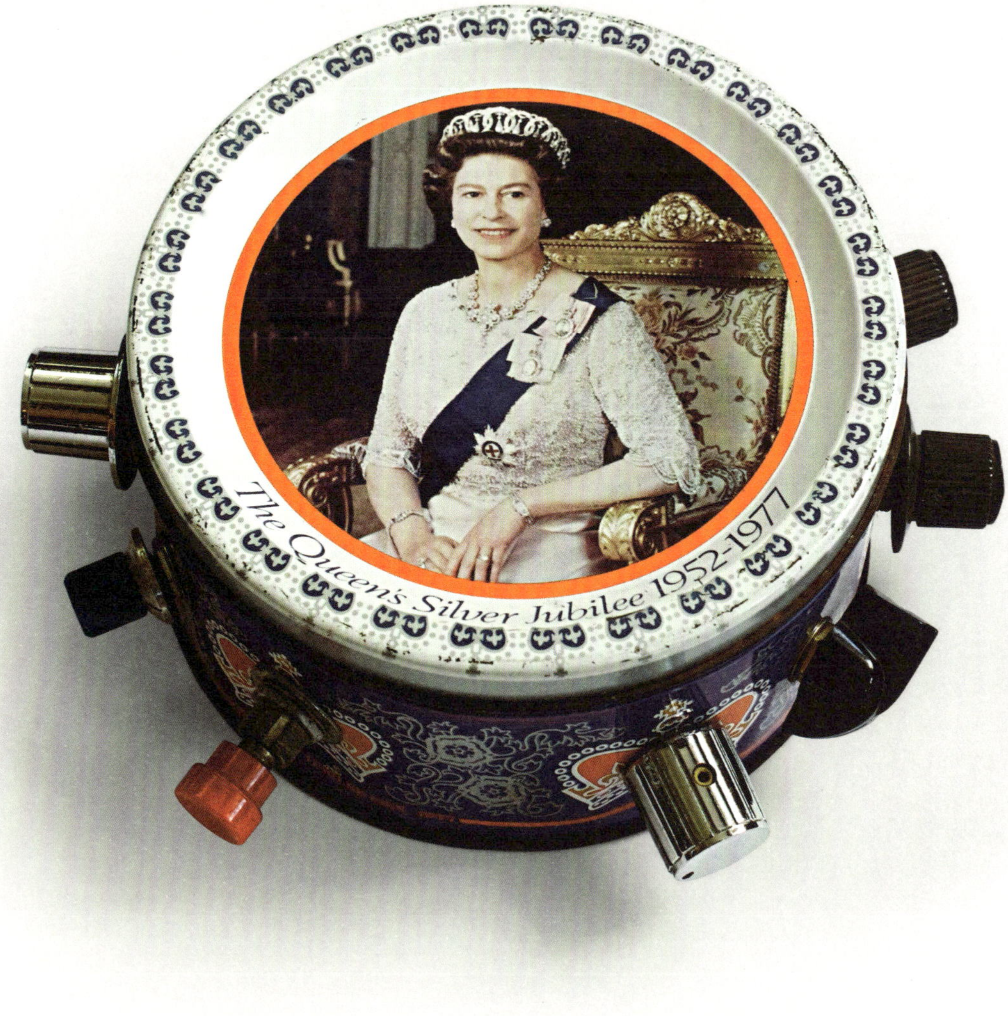
The Queen's Silver Jubilee 1952-1977

Neil Feather

THE VEGAS
2000

Sound mechanic Neil Feather (Sharon, 1956) has been creating radical and unusual musical instruments for thirty years, most of which can be described as mechanical, using strings or motors that are amplified via electro-magnetic or electro-acoustic pickups.

Feather creates several families of instruments grouped by the physical principles they explore. *The Vegas* represents three of these families since it is 3 instruments in 1: a former guitar, a roto-zither and the first vibro-wheel. The former guitar has 2 strings that are stretched against each other and controlled by a whammy bar so that one string will become tighter as the other loosens. The changing string tensions produce a broad range of timbres. The roto-zither has a cylinder with 8 strings that rotates past 2 electric guitar pickups, played with a solid metal ball. The vibro-wheel is a free-spinning flywheel mounted with small (1.5v DC) motors. The running motors create an audio signal as they move past the guitar pickup. Physical forces cause the motors to slow as the flywheel speeds up and vice versa, while the movement also creates a Doppler effect.

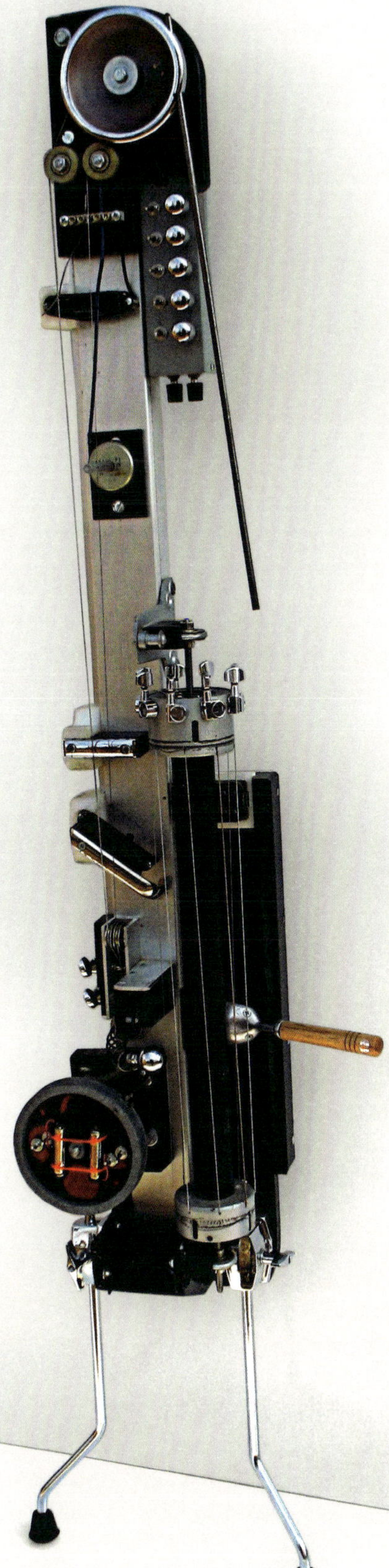

Christian Marclay

VERTEBRATE
2000

Christian Marclay (San Rafael, 1955) is an American artist, performer and musician. In the early 1970s he lived in Geneva, where he came into contact with Fluxus movements. By the end of the decade he moved to New York and started performing in a duo called The Bachelors, experimenting with turntables and vinyl records. Since then, Marclay has been working on the transformation of objects and other media—such as sound, video, photography and film—into new compositions.

Vertebrate is an altered acoustic guitar with its neck bent backwards, then cut and re-glued along the frets, *Prosthesis* (2000) [exh.] is an electric guitar molded from silicone rubber, with its neck curving to the floor as though from weakness or exhaustion, similar to a human body. Perception of these instruments creates a surreal displacement because they have all been transformed to become unplayable sculptures. As Marclay says, "Silence is the negative space that defines sound."*

* From Christian Marclay, *Arranged and Conducted*, exhibition catalogue (Zurich: Kunsthaus Zurich, 1997).

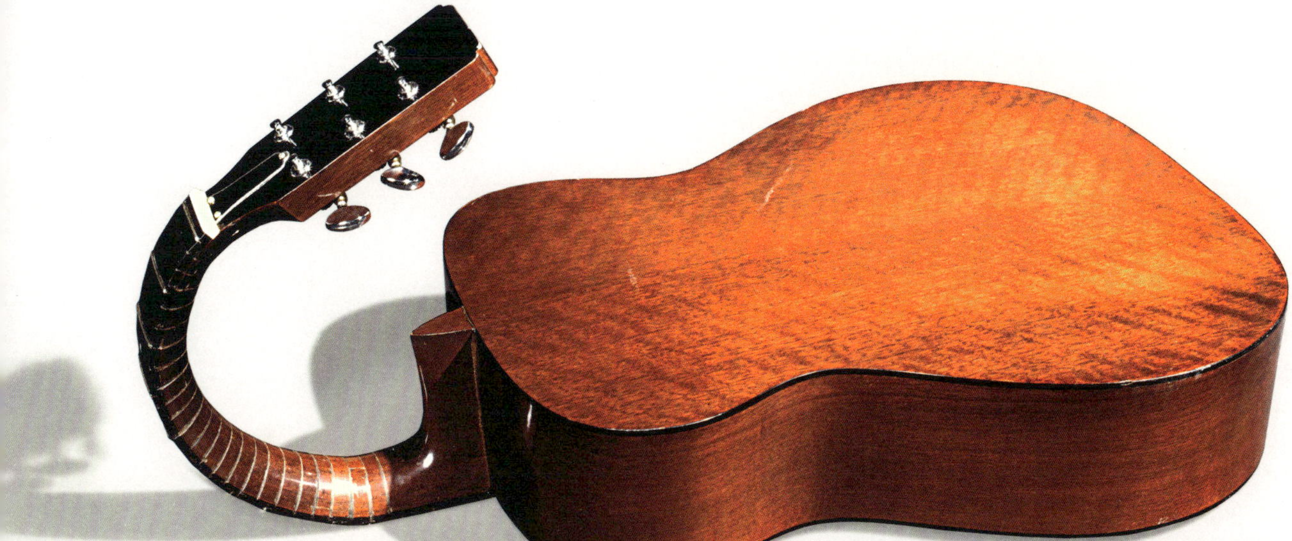

Tom Sachs

TOYAN'S JR.
2001

Tom Sachs (New York, 1966) appropriates objects commonly found in the public domain or commercial marketplace and transforms them into artworks that address consumer culture. Even as Sachs's work maintains an overt antagonism toward consumerism and globalization, it reveals an inherent idealism, championing transparency of production and homespun craftsmanship.

Toyan's Jr. is a massive, almost architectural speaker installation inspired by Jamaican sound systems, called boomboxes. "The wall of speakers is almost a physical representation of sound: the low-frequency speakers are big, and the high-frequency speakers are small, just like the wavelengths. It's literally a speaker, and it looks like an ear. In Jamaica in the 1960s and 1970s there were these competitions over sound systems, with rival sound operators, and they would have these sound wars that sometimes ended in violence." *

* Germano Celant, *Tom Sachs*, exhibition catalogue (Milan: Fondazione Prada, 2006), p. 381.

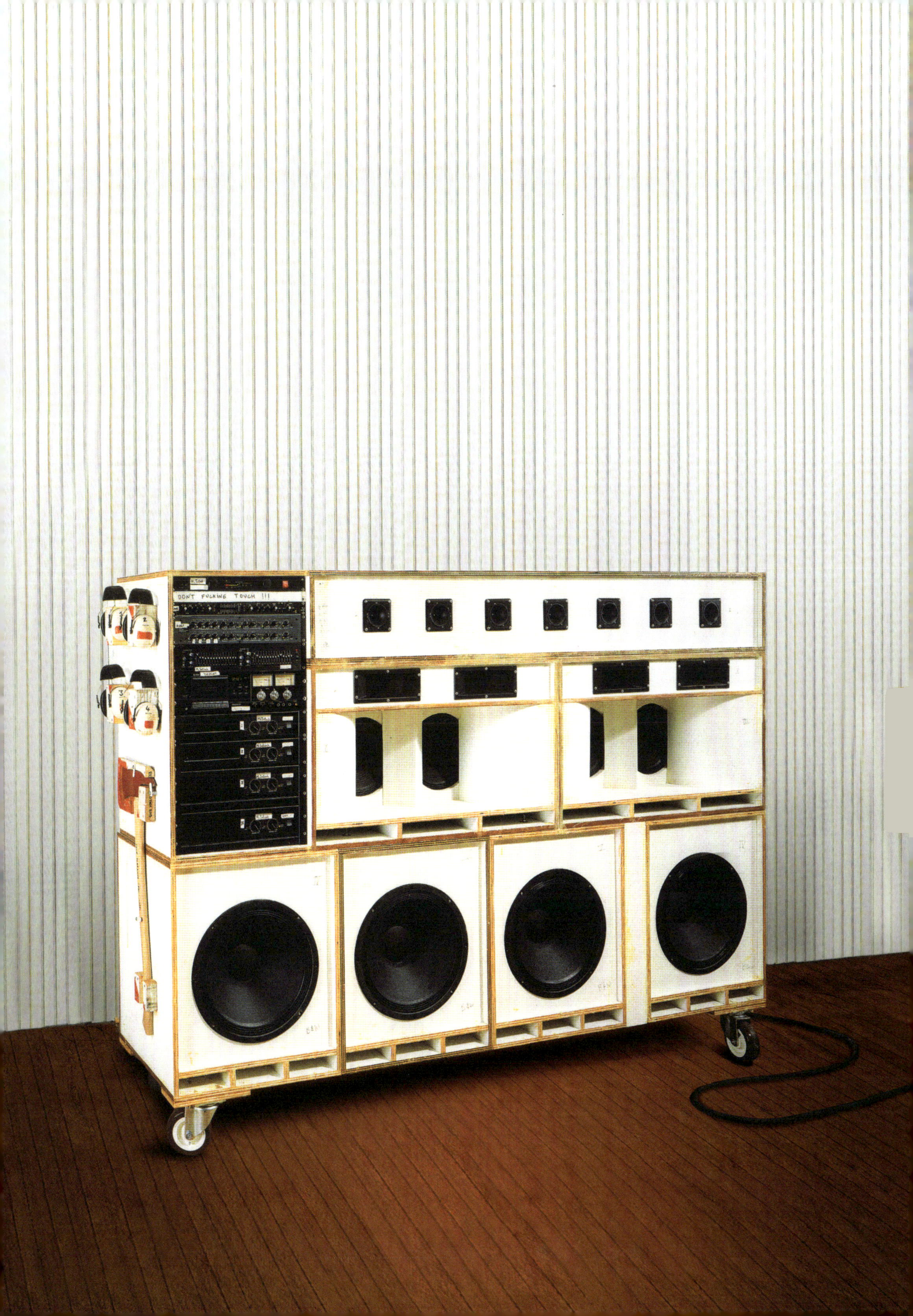
DON'T FUCKING TOUCH !!!

Terry Adkins

OMOHUNDRO
2002

American artist Terry Adkins (Washington, D.C., 1953 – New York, 2014) grew up with a solid family background in music. Assembling found objects, he realized sculptures as musical instruments playable in his "recitals" or exhibitions that combined visual art, performance, video and live music. *Omohundro* was part of a performance entitled *Black Beethoven*. Starting from anecdotes about the composer being black (some apocryphal genealogy from the 1940s traced him to the Moors), Adkins investigates whether if it is possible to tell someone's race and physical characteristics from his art. The sculpture is made of 2 sousaphone bells joined by braided metal tubing. A large nautilus shell is lodged deep inside one of the bells. As Adkins says, "This piece emphasizes Beethoven's orchestral vision as a composer and musician by virtue of the resonance and prior function of the materials themselves, suggesting the outward projection of sound as well as a pair of enormous ears capable of supra-sensory perception."

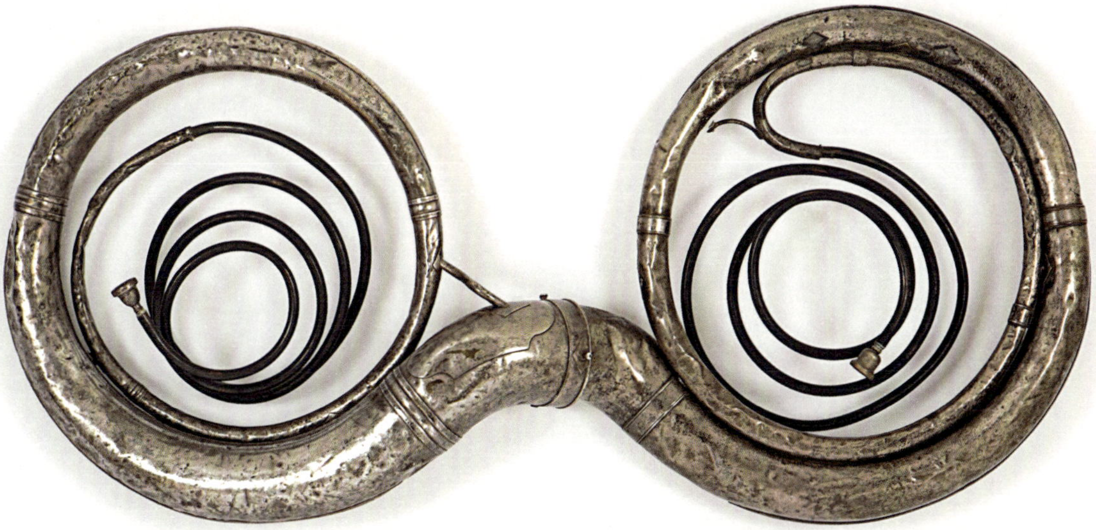

Andreas Slominski

BASS FÜR MÄUSE
2002

Ever since the beginning of his career, German artist Andreas Slominski (Meppen, 1959) has searched for a direct dialogue with his viewers. His works range from absurdly counter-intuitive actions to sculptures, that integrate everyday objects into an art context in order to visualize the process of artistic perception.

His most important series of artworks are the so-called “traps,” absurd-looking baits conceived to capture a specific animal, which combine an aesthetically pleasing object with a threatening instrument. In *Bass für Mäuse* (Bass for Mice), a musical instrument is converted into a device to capture and kill mice, turning our immediate empathy with the object onto to fear and a horror of potential harm.

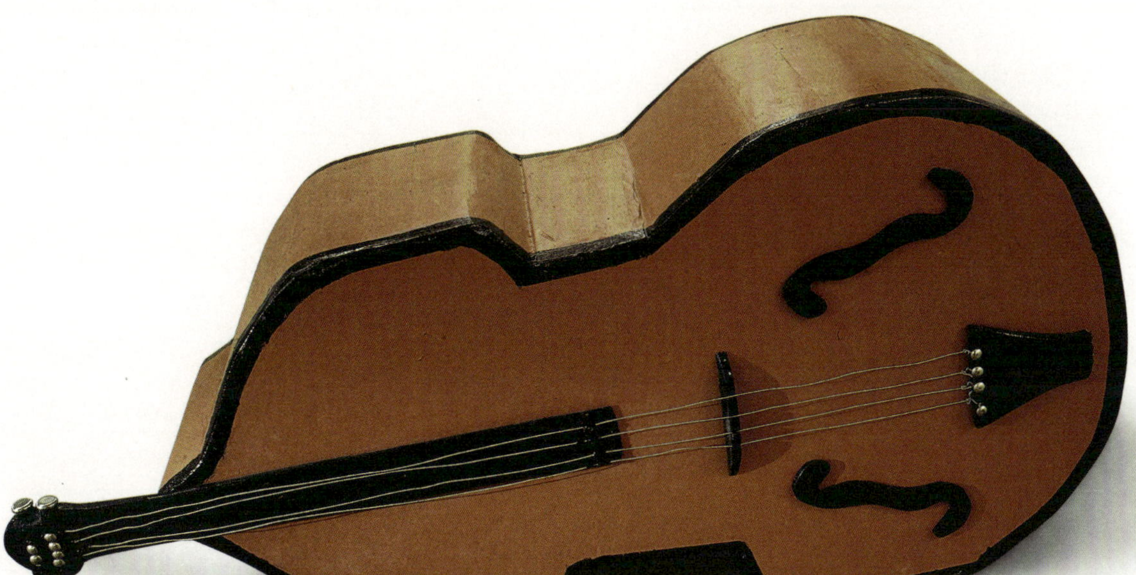

Maurizio Cattelan

SENZA TITOLO
2003

Maurizio Cattelan (Padova, 1960) is known for his interest in paradox, the sense of surprise, and of going beyond what is considered universally acceptable. This work is part of an artistic tradition that started with Dadaism, which transforms irreverence, provocation, black humor and nihilism into forms that illustrate the absurdity of contemporary society, the limits of the art system, and the uselessness of the figure of the artist.

Many of this Italian artist's works have the appearance of realistic sculptures with human and animal features. *Senza titolo* (Untitled) is a musical automaton inspired by Oskar Matzerath, the primary character in Günter Grass's book *The Tin Drum* (1959). It looks like a seated young boy who, activated by a hidden electronic-radio device, rhythmically strikes a drum. When the automaton is exhibited, it is placed in a precarious or unusual position, such as on a ledge or the roof of an exhibition venue.

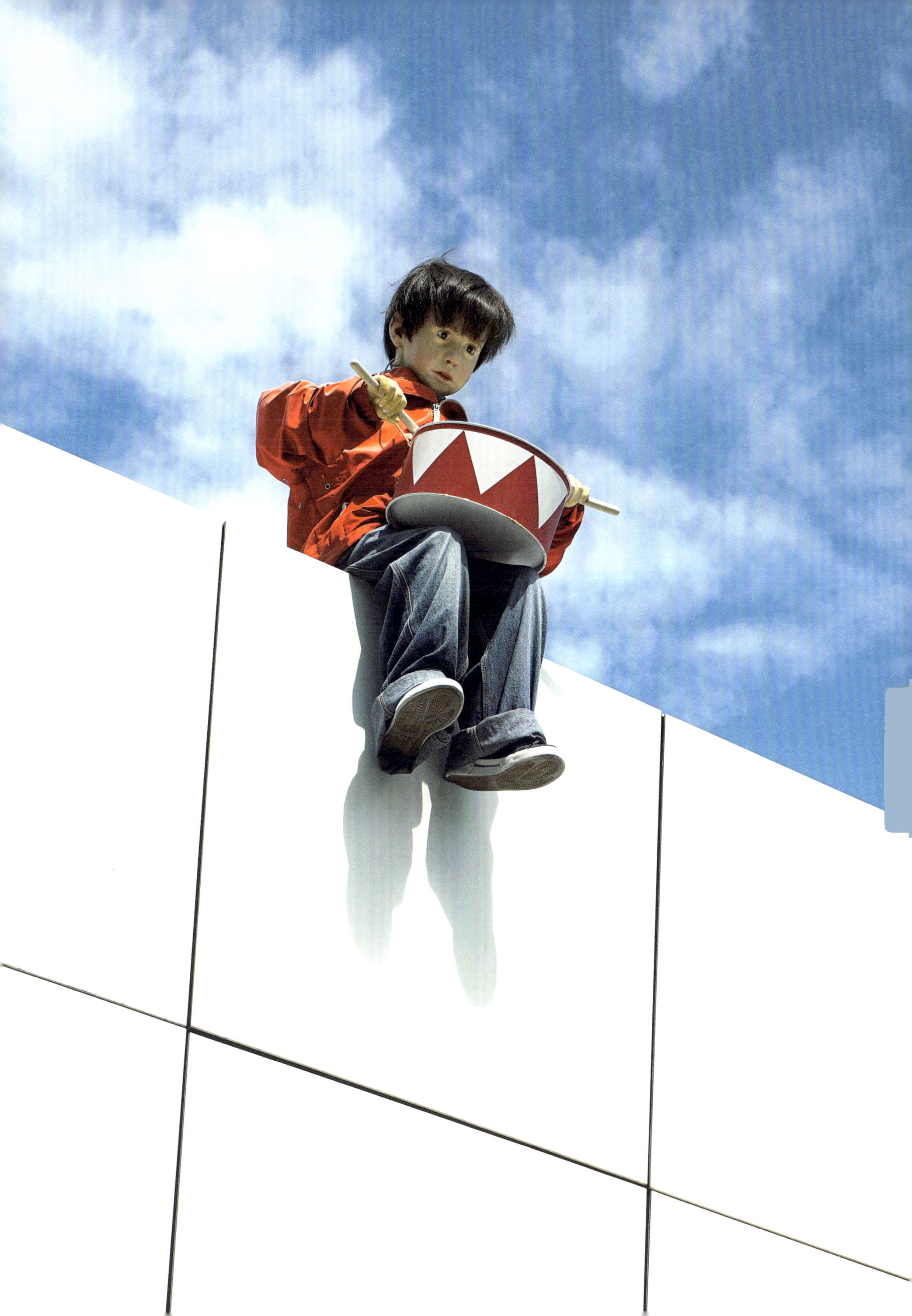

Ruth Ewan

A JUKEBOX OF PEOPLE TRYING TO CHANGE THE WORLD 2003—

Ruth Ewan (Aberdeen, 1980) is a London-based artist whose work takes many forms, including performance, installation and printed matter. Ewan explores overlooked areas of political and social history, reviving forgotten thoughts and ideas and highlighting their continued relevance today.

A Jukebox of People Trying to Change the World is a CD jukebox, which straddles digital and analogue technologies, containing a growing collection of songs addressing a spectrum of social issues, some directly political in motive, some vaguely utopian and some that chronicle specific historic events. The archive currently contains over 2,200 tracks, with no more than 2 by the same artist, ordered into over 70 categories such as feminism, land ownership, poverty, civil rights and ecology.

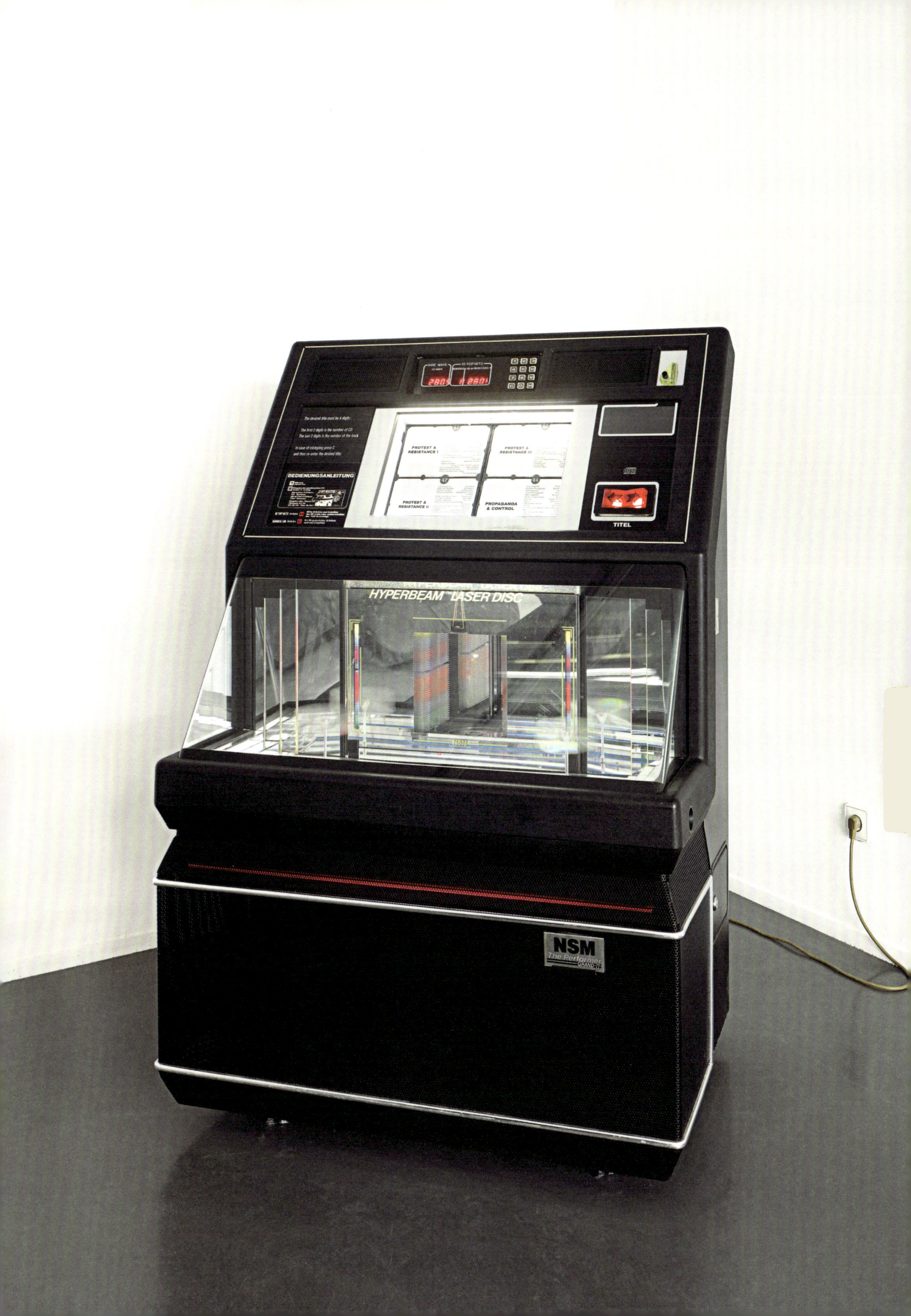
PROTEST &
RESISTANCE I
PROTEST &
RESISTANCE III
PROTEST &
RESISTANCE II
PROPAGANDA
& CONTROL
BEDIENUNGSANLEITUNG
TITEL
HYPERBEAM LASER DISC
NSM
The Performer

Loris Gréaud

CROSSFADING SUITCASE
2004

Loris Gréaud (Eaubonne, 1979) is a French artist working with film, sound and installation. His artistic research often relies on teamwork with scientists, graphic designers, video makers, engineers, composers and writers, whom he considers coauthors of his works rather than collaborators.

Crossfading Suitcase is the "portable" version of *Crossfading*, a sound installation/live performance also realized by the artist in 2004. The main difference is that *Crossfading Suitcase* has to be activated by the viewer. The artwork title refers to the sound-engineering process of gradually merging two or more different sound sources. In fact, it works on the basis of synchrotherapy: each ear perceives a different wave frequency that enables the synchronization of activity in both sides of the brain, inducing a state of "lucid dreaming." The body is asleep but the receiver is fully conscious, optimizing his or her brain functions, simultaneously experiencing emotions, thought control, analytical thinking, creativity, spatial intelligence, time and timelessness.

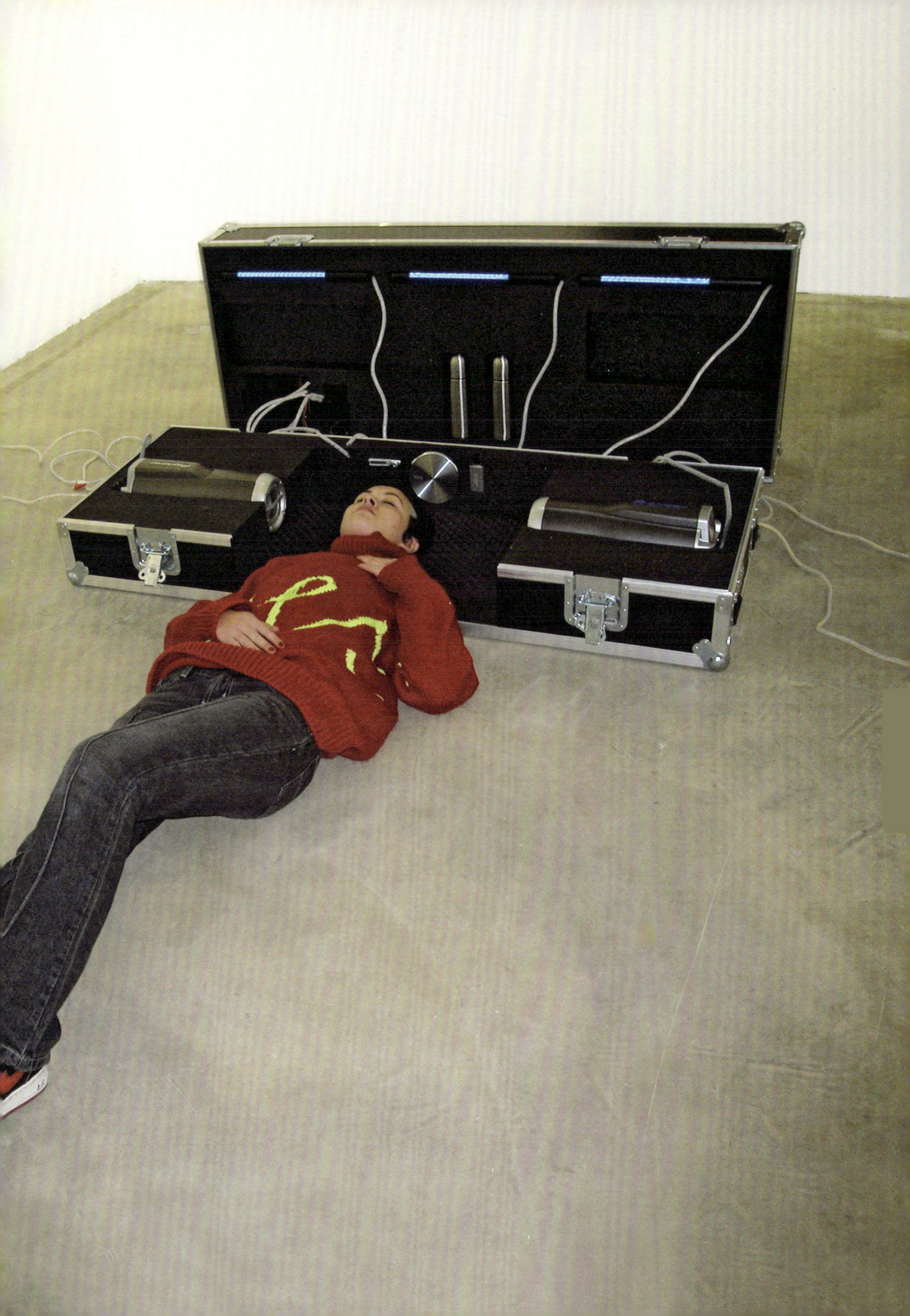

Brian Dewan

THE ROTARY ORDINATOR
2005

Dewanatrons are hand-crafted, experimental electronic instruments that can be played by a person or function on their own, built by cousins Brian (Boston, 1963) and Leon Dewan. Brian creates the sculptures, while Leon is in charge of the electronics. The machines are not digitally programmed and include analog sound elements that can be set by anyone manipulating the dials. Pitch, frequency, volume, on-off and other settings can be controlled through each of the dials. The systems are contained in wooden boxes or wall-hung constructions that also include speakers.

The Rotary Ordinator is a quartet of analog synthesizers that can be fired independently in cyclical patterns of varying lengths determined by dialing an integer from 1 to 16.

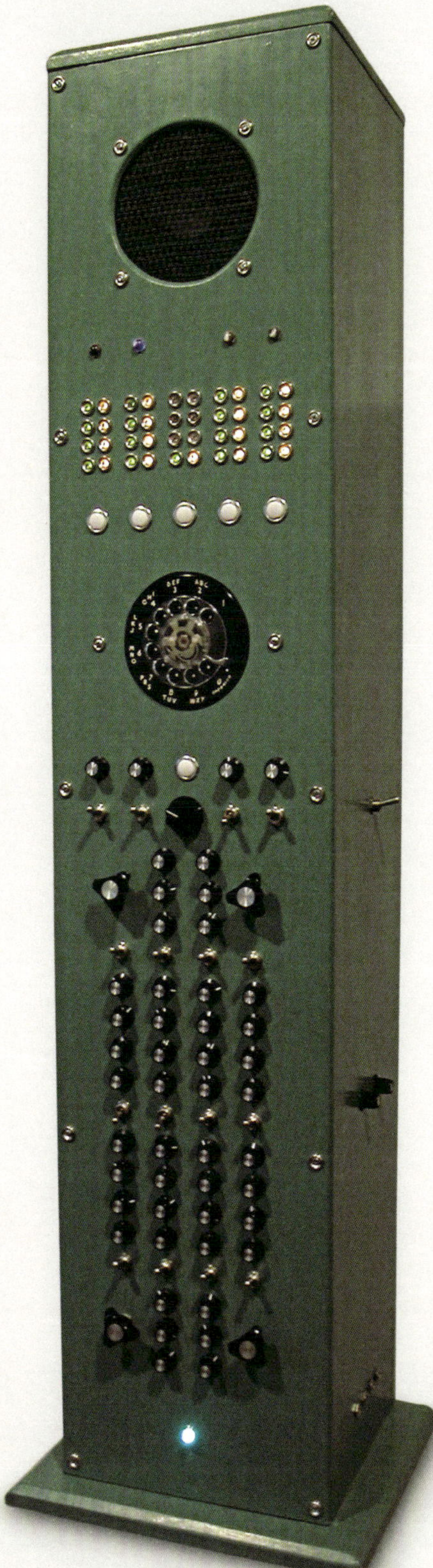

Subodh Gupta

JUTHA
2005

Subodh Gupta (Khagaul, 1964) is an Indian artist whose works include sculptures, installations, paintings, photos, performances and videos.

Jutha is composed of 3 wall-hung stainless steel sinks containing aluminum utensils. Hidden speakers are located at the base of the sinks and emit metallic sounds. As in many other works by the artist, *Jutha* incorporates objects from everyday life in India and transports them into the world of art, encouraging a reflection on the deep social changes associated with the country's recent and intense economic development. The choice of objects is strictly related to Gupta's life and reminiscences: "I am the idol thief. I steal from the drama of Hindu life. And from the kitchen—these pots, they are like stolen gods, smuggled out of the country. Hindu kitchens are as important as prayer rooms." *

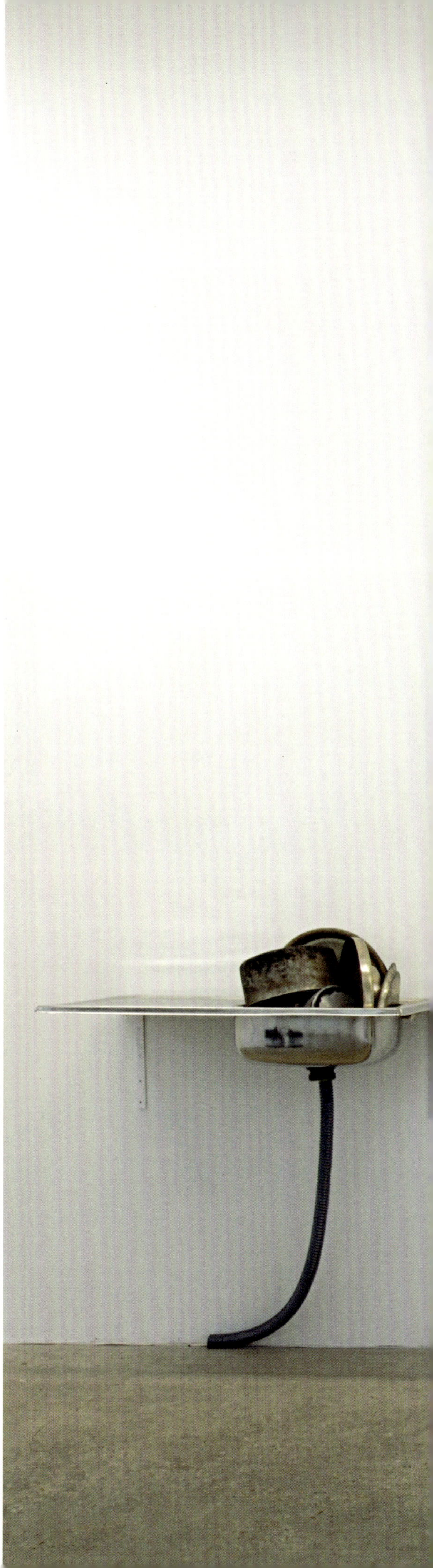

* Subodh Gupta, in Christopher Mooney, "Subodh Gupta: Idol Thief," *Art Review*, no. 17, December 17, 2007, p. 57.

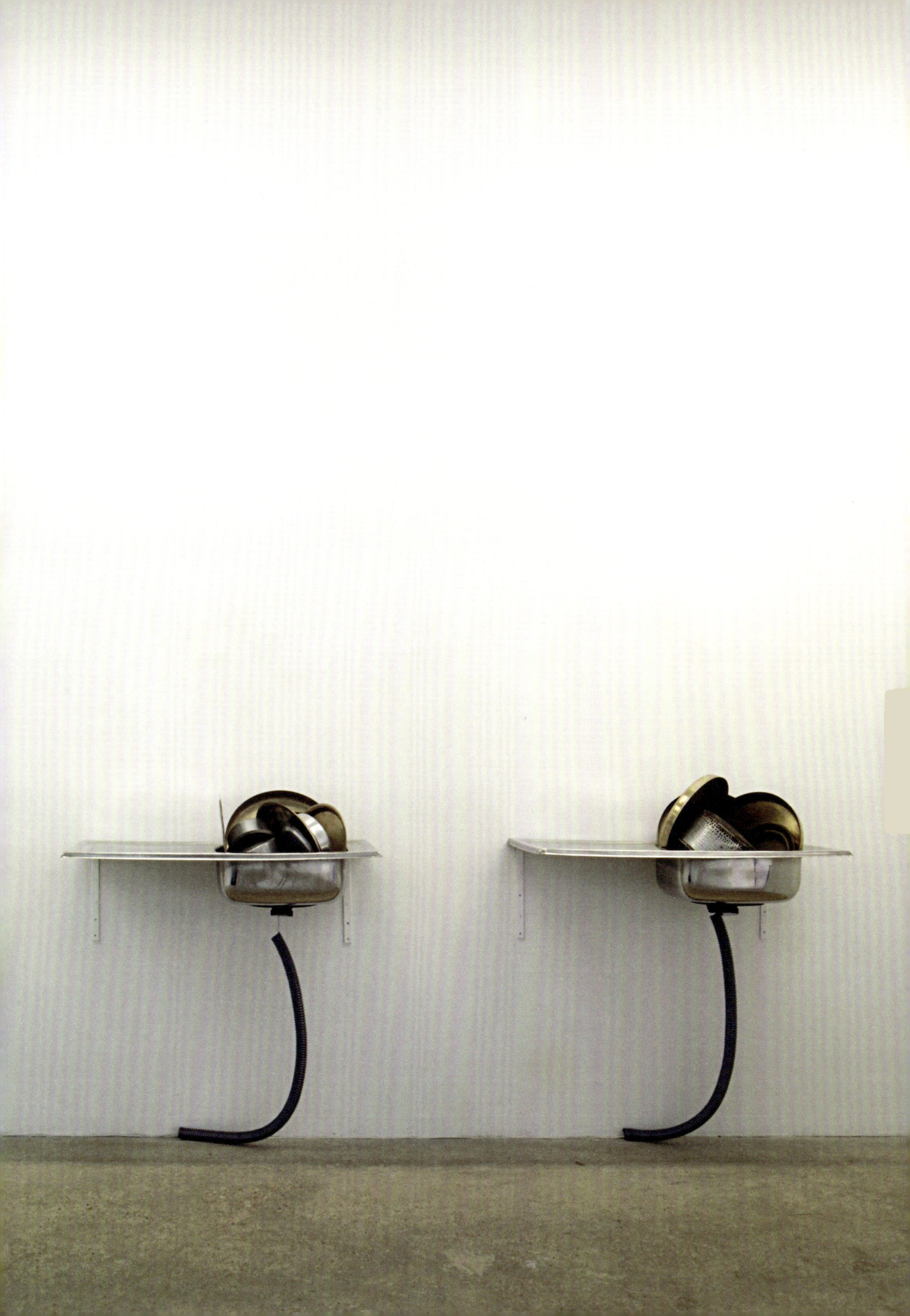

Claes Oldenburg and Coosje van Bruggen

SILENT METRONOME, 16 INCH, VERSION THREE 2005

An interest in everyday objects underlies the work of Claes Oldenburg (Stockholm, 1929), one of the most representative exponents of Pop Art. Since the start of his career in the early 1960s, Oldenburg has manipulated and decontextualized objects: he alters their scale, smashes them to pieces, and turns rigid objects into flaccid ones.

Musical instruments have featured strongly in his oeuvre since 1967, when he created the sculpture *Giant Soft Drum Set* in vinyl and polyurethane foam. The *Silent Metronome, 16 Inch, Version Three,* and *Study for Beached Lutes – Version Three* (2005) [exh.] made with Coosje van Bruggen (Groningen, 1942 – Los Angeles, 2009) focus attention on this category of object, are charged with references to the history of art and become metaphors for the voluptuous forms of the human body. Whereas the solidity of the metronome has been adapted to the canvas structure, the 2 lutes, as in 17th-century Dutch paintings, explore the eroticism inherent in the form of the musical instrument.

[See also p. 380]

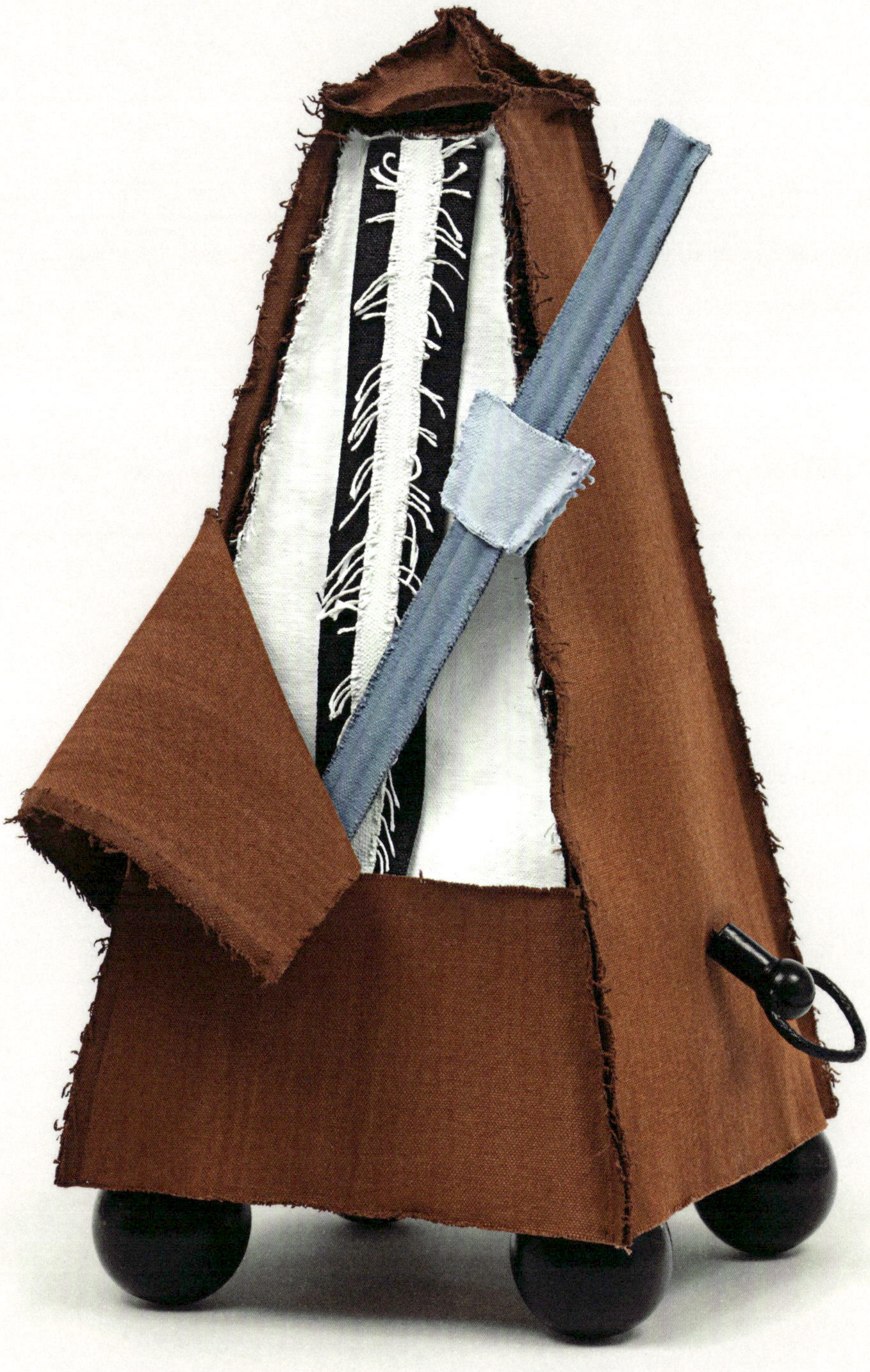

Claes Oldenburg
and Coosje van Bruggen

FRENCH HORNS, UNWOUND AND ENTWINED
2005

In 1992 Claes Oldenburg and Coosje van Bruggen bought a property in the countryside at Beaumont-sur-Dême, near Tours. Château de la Borde became their refuge and, in 2000, they installed a group of sculptures in the form of musical instruments—studies for large-scale works—in the 18th-century salon. The viola, hunting horn and clarinet were all removed from their customary function, broken down into pieces and recomposed to the extent that they became poor substitutes for themselves. Transformed, they became the primary elements in what was to become known as The Music Room. An enormous clarinet—*The Leaning Clarinet* [exh.]—was installed in the same room. It was conceived as a twisted and bent tree, its keys wilting and sagging like branches and leaves tossed by the wind. The room also hosted *French Horns: Unwound and Entwined*, which returns to the metaphor of the human figure and symbolizes a relationship between 2 people.

[See also p. 378]

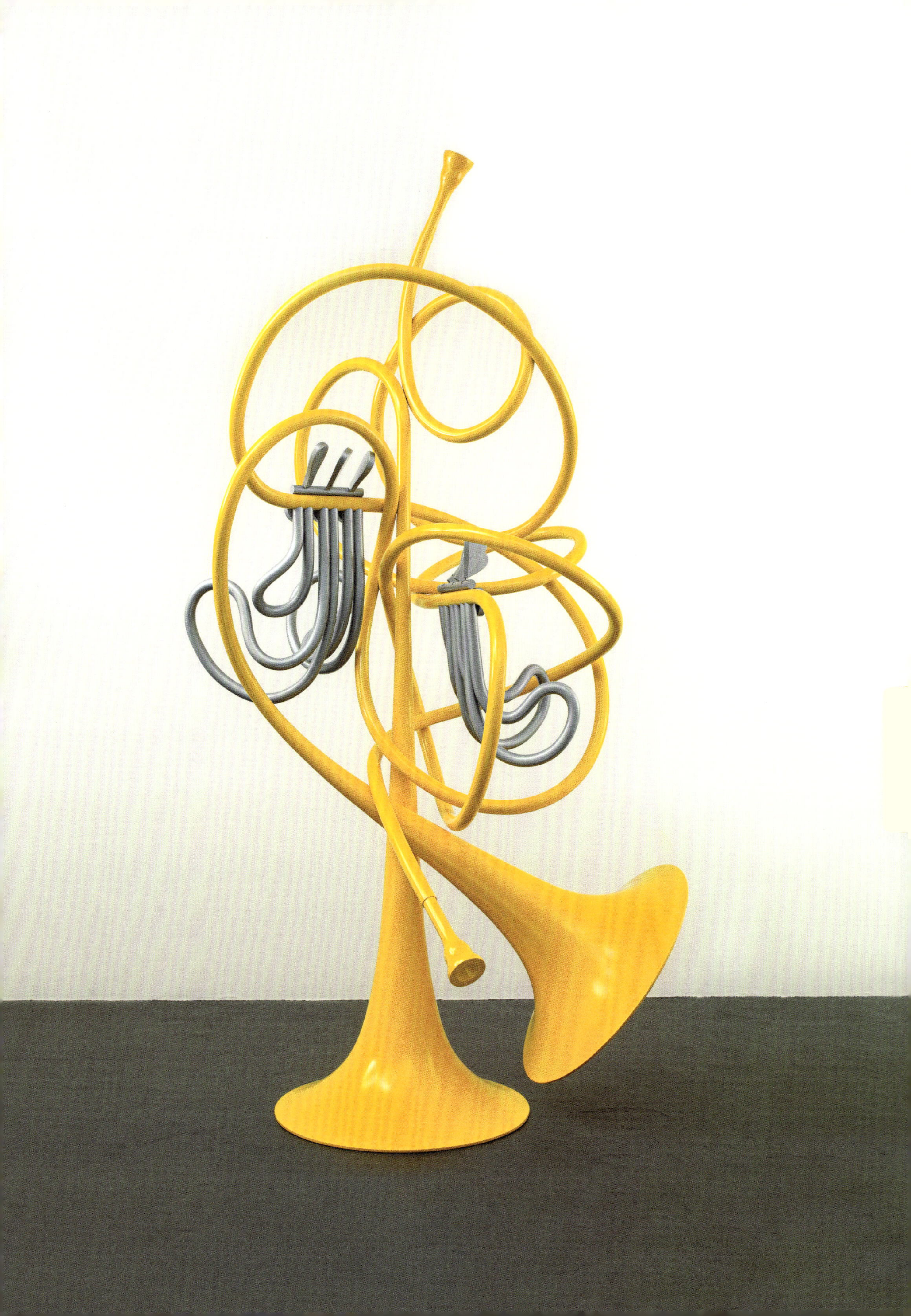

HOLLY ROGERS

TWISTED SYNAESTHESIA

MUSIC VIDEO AND THE VISUAL ARTS

The music video represents the direct convergence of music and image and marks a significant moment in audiovisual history. Although the sonic and the visual arts had been moving toward each other with great speed during the 20th century, the combination of music with the moving image enabled an unprecedented form of audiovisual synthesis.

Film is an example of such synthesis: music and image, moving through time together, can directly respond to one another to form a coherent audiovisual whole. However, in mainstream film, music is often scored during post-production, when film editing is well under way. In the golden age of Hollywood, symphonic composers, such as Max Steiner and Erich Korngold, were given only a number of weeks to score an entire film. Although this practice has changed in recent times, with many composers now involved in the filmmaking process from the outset, film narratives are still often governed by a literary, or visual, progression. Music video, a genre that was catapulted into the public sphere after the meteoric rise of MTV during the 1980s, offers an inversion of this process: in music videos, as Carol Vernallis explains in her seminal book *Experiencing Music Video*, image can underscore music.[1]

The use of the moving image to underscore music can be placed on a well-trodden path that goes back to the earliest recorded artistic endeavors. Opera has always been an audiovisual genre, but it was during the 19th century that Richard Wagner significantly theorized the notion that image and music could combine so closely as to form a new audiovisual art form (1849). Wagner not only wrote his own music and libretti, but also took control of stage and costume design to produce what he called *Gesamtkunstwerk*, or a total work of art. This idea suggested that all elements of theater—words, music and image—could come together to create a unified, and higher, form of art.

The notion of a total work of art has been very important to the development of a music video style. Many directors specialize in this tightly-knit, audiovisual genre, while others, including Spike Jonze and Michel Gondry, also make feature-length fiction films (it's interesting to note that their films are often intensely musical). Perhaps most interesting, however, are the visual artists who have engaged with the format. Music video provides artists not necessarily trained in music with the opportunity to create a highly sonic piece of work, to combine the visual arts with popular music culture, and to create images in accordance with a pre-existing musical structure.

Music video directed by visual artists is important as it signifies a history that extends from well before the birth of MTV, which launched on August 1, 1981, with the video for The Buggles song *Video Killed the Radio Star* (Russell Mulcahy). During the 1920s, several Berlin-based artists explored the ways in which sound and image could combine to produce filmed visual music that unfolded through time. Pioneer Hans Richter created intensely rhythmic films, such as *Rhythmus 21* (1921), by harmonizing abstract images with musical forms and structures to create what he referred to as silent, visual "instrumental themes."[2]

While Richter was exploring the ways in which music could be translated into silent, musical images, his contemporary Walter Ruttmann embarked on live forms of visual sound by commissioning new music to be performed during his films. For his animated film *Lichtspiel No 1* (1921), for instance, he asked his friend Max Butting to compose a string quartet (Ruttmann himself reportedly played the cello part) that would unfurl in closely synchronized rhythms with the abstract images to produce an artform that rested "somewhere in-between music and painting."[3]

But it is the colorful and rhythmic play of Oskar Fischinger's audiovisual Plexiglas animations that best illustrates the relationship between early visual music and music video. Fischinger is best known for his work on the "Toccata and Fugue" scene from Disney's *Fantasia* (1940), although the artist wasn't happy with the results and removed his name from the credits before the film was released.

Fischinger insisted that his images were not meant to be "translated music, because music doesn't need to be translated on the screen:" rather, they were meant to enter into a conversation with the music in order to create something totally new.[4] Fischinger was particularly interested in visualizing pre-existent music, and created a hand-drawn animated dialogue with Liszt's Second Hungarian Rhapsody in his 1938 film *An Optical Poem*, where music and image constantly dance around each other, but never descend into empty repetition; in *Motion Painting No 1* (1947), he gives us a complex and energetic visual response to Bach's Brandenburg Concerto No. 3.

The work of these early visual music artists had a great impact on later filmmakers, including Norman McLaren, the Whitney Brothers and Stan Brakhage. But another important line of influence that fed into the music video genre came from video art. In some ways, early video art was the precise opposite of the visual music school. Whereas the latter were artists seeking to sound their visual work, the people who acquired the first portable video equipment, when it became available in 1965, were often musicians interested in visualizing their sonic experiments.

In 1971, Korean-born Nam June Paik, who originally trained in composition, created his *TV Cello*, an instrument created from three television screens stacked one on top of the other to resemble the shape of a cello and strung with a single string (a later version had several strings). Conceived as a collaborative piece with avant-garde cellist Charlotte Moorman, the instrument was linked to a closed-circuit feed of the performance space (live images of the audience members and Moorman were interspersed with iconic images from the 20th century). As Moorman played the instrument, she produced what she called "TV Cello sounds" via electronic signals. These sounds were electronically linked to the images, which were warped and morphed out of recognition by the music.

Another important piece of visual music created by a video artist is Steina Vasulka's 1978 *Violin Power*. The Icelandic artist was trained in music and began her career as a violinist with the Prague Symphony Orchestra. This training profoundly influenced her video style, as she explains: "For me, it is the sound that leads me into the image. Every image has its own sound and in it I attempt to capture something flowing and living. I apply the same principle to art as to playing the violin: with the same attitude of continuous practice, the same concept of composition."[5]

This "concept of composition" can clearly be seen in *Matrix* (1970–72), a multi-screen, audiovisual work Vasulka co-created with her husband Woody, who explained that, at the time of its creation, he was "very much involved in changing the image in sound or controlling the sounds with images, and generating images from sounds and vice versa."[6] Like the *TV Cello*, *Violin Power* operates via an electronically-manipulated instrument (the Zeta-violin): when Vasulka played her modified violin, a closed-circuit feed of her own actions received different distortions according to pitch and bowing technique.

While evidence of these forms of video audiovisuality are traceable throughout the history of music video, the possible cross-over between the two genres was clearly articulated in the work of Laurie Anderson. Although Anderson trained in sculpture and often works with video, much of her work is highly musical: her performance-based art, for instance, has included her electric *Tape-bow Violin* (1977), an instrument that uses a bow fitted with a strip of recorded tape instead of horse hair, and a magnetic tape head on the violin where the bridge would normally be. During the 1980s, she graduated to a MIDI violin that allowed her to control synthesizers connected to the instrument, but, significantly, Anderson also released several songs, and directed the accompanying music videos. In 1981, for instance, she released *O Superman (For Massanet)*, a song loosely based on an aria from Massenet's 1885 opera *Le Cid*. The music video is closely synced to Anderson's vocoder-morphed voice: we see close-ups of her singing and gesticulating alongside images of the sun and the moon in what was to become a highly influential piece of sonic art.

Once music video became a well-established genre, video artists began to play on, and parody, its forms and conventions. Pipilotti Rist's single-channel video *I'm Not the Girl Who Misses Much* (1986), for example, is a parody of Beatles song *Happiness Is a Warm Gun*. The work can be read as a commentary on the (particularly visual) representation of women in popular music: as the Swiss artist delivers the song directly to the camera, her voice and image are fragmented through fast-forward and extreme slow motion, distortions that highlight the materiality of the video medium. Anne McGuire's nostalgic *I Am Crazy and You're Not Wrong* (1997) articulates a different relationship between technology and sound. Here, the San Francisco-based artist performs Kennedy-era lounge songs to a disinterested camera. Highlighting what Catherine Elwes has described as a "struggle for self-expression within the mimetic confines of the musical vernacular," the video style moves away from the real-time interactivity of *Violin Power* and *TV Cello* toward the pre-recorded style of music video.[7] Sam Taylor-Wood's *Killing Time* (1994) presents four people miming badly to arias from Richard Strauss's *Elektra*. Out of time with the recording and frequently distracted from their task, the miming characters emphasize the materiality of the medium and the strange conventions of music video's revisualization.

By the "birth" of the music video, then, the genre had enjoyed a long and rich history within the audio and visual arts, a lineage that has not gone unacknowledged by contemporary music video directors: for example, Michel Gondry's video for The Chemical Brothers' *Star Guitar* (2002) is a tribute to Ruttmann's 1927 film, *Berlin: Symphony of a Great City*; the musically-warped images of David Lynch's work pay loose stylistic homage to the earlier visual music artists; and Chris Milk's interactive music video for Arcade Fire's *The Wilderness Downtown* (2010) nods toward the interactive proclivities of early video artists.

The ability of music video to construct and disseminate an artist's image was a major factor in its rise to popularity. But the need to add a visual accompaniment to a song was not only fuelled by an image-driven pop industry, it also went a long way toward reintroducing the energy and physicality of live performance into song. Embodied within recorded music is an unavoidable absence: what we hear is an imprint of a recording session now over. Adding video to song provides the disembodied music with the illusion of a source, even if the source offered is every bit as disembodied as the recorded music it accompanies. While some critics have argued that grounding a song in image suggests that music alone is not enough—that visual support is required for its success—it could be argued that such support in fact empowers music. Rarely does a music video present a straightforward, unedited performance. Instead, music, liberated from its source, is able to construct an alternative world on its own terms. And the world depicted, to quote Vernallis again, can "become very strange."[8]

The creative potential of these "strange" worlds has been very appealing to those working in the visual arts. For some artists, the progression from image to the audiovisual can be seen as a natural, almost inevitable, phenomenon. After Warhol's involvement with The Velvet Underground and The Exploding Plastic Inevitable, for instance, his music video for The Cars' *Hello Again* (1984, in which he played the role of bartender) seemed like a very fluid transition. For others, the jump to visualized sound was more drastic, and, perhaps as a result, rather short-lived: American pop artist James Rizzi created the animated video for Tom Tom Club's *Pleasure of Love* (1983); video artist Doug Aitken turned his camera on Interpol to produce blue-and-white close-ups of the band for their *NYC* (2002) video; digital artist and painter Jeremy Blake created the slowly superimposing, abstract and static images for Beck's *Round the Bend* (2002); and artist, filmmaker and musician Martha Colburn produced the cut-out animation for Deerhoof's *Wrong Time Capsule* (2005) and Serj Tankian's *Lie, Lie, Lie* (2007). Despite representing only a brief foray into music video, all these examples demonstrate a clear extension of each artist's style into an audiovisual progression.

Some of the most successful collaborations between musicians and artists have arisen from within a shared aesthetic. Damien Hirst's 1995 video for Blur's *Country House*, for instance, is an excellent example of a mutual response to the Cool Britannia movement that was sweeping through England's music and art scenes during the 1990s. Best known for placing dead animals in formaldehyde and for his spot paintings, the Young British Artist met the nascent Britpop band while studying for an art degree at Goldsmith's College, London. Their subsequent collaboration on *Country House*, which bass player Alex James describes as "a clash between high and low art," seemed inevitable.[9]

The video's concept is simple yet effective: scenes of the characters running wildly around a board game ("Escape from the Ratrace") are interspersed with images of the protagonist (Keith Alan) strolling around his country house. Many conventional music video tropes are clearly present: singer Damon Albarn addresses the song's lyrics directly to camera; stop-motion filming and quick-fire editing creates a strong visual rhythm; and there are numerous instances of literal linguistic-visual play ("caught in a rat race" is accompanied by the suited protagonist dropping piles of money onto a board game covered in rats, for instance). But, throughout, Blur's Britpop vibe and Hirst's media-savvy sensationalism unite to form a completely integrated whole: there is a heightened lack of realism, the colorful sets are overly theatrical, and, most significantly, there are humorous references to British cultural phenomena, such as Benny Hill and Queen's famous video for *Bohemian Rhapsody* (Bruce Gowers, 1975). Here, image does not underscore music, but rather acts in partnership with it to form a *Gesamtkunstwerk* that feeds into a much larger cultural aesthetic that was raging through Britain at the time.

While Hirst created only one music video, other artists have worked with several musicians. Media artist and filmmaker Jim Blashfield directed videos for Michael Jackson, Talking Heads, Paul Simon and Joni Mitchell, among others. Similarly, American artist Bruce Conner, who worked with paint and photography as well as film and assemblage, created videos for Devo, Brian Eno and David Byrne. Perhaps the artist most devoted to the music video genre, however, was Derek Jarman. The experimental filmmaker directed the videos for numerous songs, including The Sex Pistol's *The Sex Pistols: The Sex Pistols Numbers One* (1976), The Smiths *Ask* (1986) and Suede's *The Next Life* (1993). But his longest collaboration was with the Pet Shop Boys, for whom he directed five music videos and designed their 1989 tour. These videos are incredibly filmic, and demonstrate a remarkable likeness to Jarman's experimental cinematic style. The alternative world of *It's a Sin* (1987) bathes band members Neil Tennant and Chris Lowe in *chiaroscuro* as they enact a symbolic narrative highly reminiscent of Jarman's *Caravaggio* (1986). And yet, because the image track highlights the rhythmic and semantic aspects of the completed song, the result is very different from the contemplative nature of Jarman's film. The images, edited to musical beats, form quick-fire sequences rarely encountered in the long, relatively unobtrusive shots of cinema, while the structure of the repetitive, brief and relatively abstract lyrics results in strange and disjointed visual montage sequences.

Artists have enjoyed a long and productive partnership with musicians, an alliance that has come in many different forms. But the interaction between media is not only the result of collaboration, it can also be found in the form of homage. Within the "strange" worlds of music video, references to the visual arts abound, and popular musicians have frequently played homage to well-known artworks. The video for Mine (2014) from Beyoncé's self-titled visual album, for instance, recreates Magritte's Les amants: Rihanna makes a clear reference to Andy Warhol's Marilyn Monroe's Lips (1962) in her Rude Boy video (Melina Matsoukas, 2009); while Nicki Minaj's video for The Boys (Colin Tilley, 2012) was inspired by the dotted world of Yayoi Kusama.

But it is Lady GaGa who has most consistently alluded to artwork as a means to fabricate the visual worlds for her songs. References to Warhol, Cindy Sherman and Botticelli's Birth of Venus are littered throughout her video for Applause (Inez & Vinoodh, 2013), while Nick Knight's images for Born This Way (2011) bore such stylistic reference to the work of French artist ORLAN that GaGa found herself facing a lawsuit for plagiarism.

It is clear that a two-way flow of influence between musicians and artists has resulted in an expressive and innovative synthesis that continues to expand the boundaries of audiovisual possibility. But it is also important to notice that the music video was not just the result of a televisual revolution in the 1980s, but rather has a long and fruitful history that extends back through both the musical and the visual arts.

1
Carol Vernallis, *Experiencing Music Video: Aesthetics and Cultural Context* (New York: Columbia University Press, 2004), p. 8.

2
Hans Richter quoted in "Early Colour Organs," paradise2012.com/visualMusic/musima/ (accessed February 5, 2014).

3
Walter Ruttmann quoted in Dieter Daniels, "Sound & Vision in Avantgarde & Mainstream," medienkunstnetz.de/themes/image-sound_relations/sound_vision/print (accessed February 5, 2014).

4
Oskar Fischinger quoted in William Moritz, *Optical Poetry: The Life and Work of Oskar Fischinger* (Bloomington, IN: Indiana University Press, 2004), p. 185.

5
Steina Vasulka, "Orka," vasulka.org/Steina/Steina_Orka/Orka.html (accessed February 5, 2014).

6
Steina Vasulka speaking in *Six Programs for Television*, "Matrix;" quoted in Yvonne Spielmann, *Video: The Reflexive Medium* (Cambridge, MA/London: The MIT Press, 2008), p. 99.

7
Catherine Elwes, *Video Art: A Guided Tour* (London: University of the Arts), p. 165.

8
Carol Vernallis, *op. cit.*, p. 7.

9
Alex James, quoted in Madeleine Baker, "Childhood Psychedelia," madeleine-baker.blogspot.co.uk/p/dissertation.htmlat (accessed February 5, 2014).

ANDREA LISSONI

THE IMPORTANCE OF UNPRE-DICTABILITY

UNEXPECTED DIALOGUES IN SOUND WORKS

The comparison of a work of art that is based on sound may be founded upon the basis of a mysterious form of dialogue. Whether it is established in bpm or Hz, in other words, beats per minute or the unit of measurement of frequency, this dialogue is established with the human body, its metrics and perceptual dynamics, or rather with the beat of the heart and with the aural tuning to middle A, at precisely 432Hz (according to some studies—or opinions—the frequency to which the buzzing of bees and the rustling of plants are tuned).

In considering this potential dialogue initiated by the artist, how much has the relationship between a work of sound art and the public changed, when the sounds, the ways in which they are produced, transmitted and consequently heard has been so radically transformed? Sound was revolutionized at the end of the 20th century as never before, opening up possibilities for manipulating sound that no one had conceived of until then, from the opportunity to sample from the widest range of sources, to the possibility to profoundly modify and alter sound through digital instruments. Sound has become digital. And like the image, its nature has been inevitably and definitively altered.

The realm of art quickly took stock of the phenomenon at hand, so much so that at the beginning of the new decade it was condensed into three essential exhibitions: "Sonic Boom – The Art of Sound," Hayward Gallery (2000), "Sonic Process – A New Geography of Sound," Macba/Centre Georges Pompidou (2003) and "Frequenzen (Hz), Audiovisuelle Raume," Schirn Kunsthalle, Frankfurt (2002). In a certain sense, all three dealt with ambiances and artistic forms tied to the analysis of an environment that was undeniably redefining itself. But they also bore witness to the way in which the (debatable) category of "sound art" was paradoxically rooting itself in an autonomous and impermeable semantic field rather than gliding along like a centrifugal and hybridizing practice. As a consequence, or rather as a collateral effect of the revolution underway, the 1990s saw an explosion of widely varied genres and sonorities over an extremely brief period of time, with an unprecedented general innovative momentum, before the archival fever for sources and the attitude of recovery became an obsession during the successive decade, slipping from revolution to derivation.

From a creative perspective (on the topic of fruition, we could depart on a wide tangent addressing the consumption of electronic dance/"artsy" tapestry, stigmatized by Mike Kelley in the work *Runway for Interactive DJ Event*, 2000), during that explosive moment the visual-artistic scene essentially embraced three distinct fronts:

– *noise*, ranging from a mixture of the historic tradition of futurist noise, bathed in improv-performance, to proto-punk, enmeshed in refined Californian electronica (generated and disseminated by *Destroy All Monsters* as well as by various formations that included collaboration from Mike Kelley), to radical Viennese electronica (centered around the Mego label).

– *electronica*, based on the rhythmic structures of frequencies (rather than notes), practiced by artists interested in describing landscapes and monotonous sound "tapestries" (or interpretations that would ultimately be diffused in public spaces. Examples include Carsten Nicolai's ∞, acoustic "graffiti" in the form of microstructures for documenta 10 in Kassel, 1997), or live musicians/artists who made spectators' bodies vibrate or resonate in live performances, sometimes accompanied by video images that were purely representational of the sound waves, thereby approaching a "minimalist" if not structural logic (Pan Sonic, Mika Vainio, the same Nicolai *aka* Alva Noto and Carl Michael Von Hausswolff, primarily a continental European phenomenon).

– *illbient*, the definition of which is disputed by DJ Spooky and DJ Olive, essentially deals with the assumption of an "ill" urban environment, sampled and re-incorporated into cinematic scenarios accompanied by dark rhythms derived from live dub and DJ sets (Paul D. Miller, *aka* DJ Spooky, was the prime exponent of this largely New York phenomenon, which had reverberations in Europe with the English Scanner, among others).

Within the artistic scene, however, these phenomena of absorption and regeneration of musical practices and genres primarily favor the consideration of live performance (with the sanctioned consecration of "A Fantastic World Superimposed on Reality: A Select History of Experimental Music," a festival curated by Mike Kelley for Performa at the Gramercy Theater in New York in 2009), or the more or less domestic diffusion of support media, rather than examining the potential mutation of the relationship between artwork and audience.

What would happen, therefore, if we hypothesized focusing our attention exclusively on the work and not on the performance? How much has the sound of the work changed if examined from an *ontological* perspective? And what about the public's perspective, its perceptual posture?

A series of rhetorical figures stands out against the background of an environment that, despite the retro-mania of the Zero years, which re-examined the recovery of obsolete media, supports and formats (radio, TV, Theremin, synthesizers, vinyl and cassette tapes), was certainly no longer electric, nor electronic, but inevitably digital.

Without a doubt, the metric obsession remains, partially inherited from historical musical Minimalism (*Music for Pure Waves, Bass Drums and Acoustic Pendulums*, 1980 by Alvin Lucier is a prime example).

The need to represent tempo in its regular and uniform subdivisions emerges as a central practice, as demonstrated in the majority of works by Carsten Nicolai or the installations by Martin Creed [338]. By all accounts, marking the tempo is as much a potentially authoritarian gesture as it is inevitably a political one, and it is not a random occurrence if the body—mechanical and mechanized, fragmented or robotic—appears in the work of artists that appear
1 quite distant from one another, such as Stephan von Huene (*Der Mann von Jüterbog*, 1995–96) [342] and Maurizio Cattelan (*Senza titolo*, 2003) [368]. All of this in light of the fact that in the transition from workshop to the center of art (and to the club), that which wandered around freely and in plain sight against a background of dark air was a specter working carefully to synchronize production regimens with those of consumption, and therefore of pleasure.

Traditional instruments, re-contextualized in the typical logic of readymades, are subjected to the same forms of manipulation that are simultaneously altering audio tracks: decomposed, disassembled and more or less ironically hybridized. The instrument, in fact, can be completely unknown yet incorporate a compositional logic ranging from cold and modernistic (as is often still the case with Carsten Nicolai), to vernacular (as in the works of Walter Kitundu [402, 404] or in the surprising sculpture *Burlesque* (2012) [430], by Janet Cardiff and George Bures Miller), or a combination of both (the sculptures and installations of Haroon Mirza [400, 434] constitute an impeccable example).

Or they may be clearly recognizable, the result of a *détournement*—in a tradition that began with Georges Braque and evolved to Ken Butler's hybrid instruments [298, 306], Christian Marclay's guitars [360], as well as instruments by Andreas Slominski [366], Yoshihiko Satoh and Pedro Reyes [432]; or to emblematic devices from the electronica era like *Mopol* (1999) [356] by Ed Potokar, or *Melody Gin* (2005) by Brian Dewan—, or potentially performable (*The Metastable Circuit* [2012] [426], by Tarek Atoui) or performative (like *A Solo in the Doldrums* [2009] [408] by Anri Sala, or *Keyboardchairs* [1997–2005] by Céleste Boursier-Mougenot) pieces, or pieces that are the consequence of having been performed (as in the long Fluxus tradition ranging from Raphael Montañez Ortiz to Arman [218, 344]).

Another recurring figure is the theme of equating a sound with an action/tempo in flux, regardless of whether the work is an assemblage of instruments or hybrid instruments (in the open tradition of Jean Tinguely's "anti-machines," [226] of which *Mindfall* [2004–07], a concert for electric motors in a container by Micol Assaël, may be a curious heir): the genesis of sound that develops over time is the result of a visible chain reaction (as in the electric/electronic works by Alberto Tadiello [422]).

Finally, there is that which may be defined as the mythology of diffusion and all the images associated with it. Therefore in the era of digital headphone listening, the sound-system—conceived as a combination of speakers and loudspeakers—becomes a monumental sculptor's figure. On one hand we have readymades that suggest a tricking of the senses (*Crossfading Suitcase* [2004] [372], by Loris Gréaud, which forces the listener to invert his posture, lying down on the ground with his head surrounded by the sound-system in order to listen to the music), and slips of the unconscious (the color white in *Toyan's Jr.* [2001] [362], which pushes in an "electro-domestic" direction, or the black of *Trojan* [2008], both by Tom Sachs, yet another "electro-domestic" dimension, which plays on Brancusi and Minimalism while still belonging to Jamaican dub and reggae culture). On the other hand, elusive perception consists of the ears deceiving the eyes: a generation of artists has given form to works that, through the classic audio "speaker," have disseminated audible micro-perceptions with an attitude that could be compared to Divisionism or Pointillism.

There are those who interact with new forms of landscape, more neural than mental, and in a certain sense radically abstract (Carsten Nicolai, Florian Hecker), and those, on the other hand, such as Marcellvs L (*Klavierwellen*, 2011), who have combined multiple modes of diffusion of the sound itself, distilling it—or rather sectioning it—in the landscape and re-projecting it to the audience in a processed form that is recognizable as an underlying reverberation.

Curiously, despite their penetration and perpetuation in the visual-artistic field during the 1990s and Zero years, the sonorities of these works of art seem to offer no immediate association with any musical tradition within *noise*, frequency-based digital electronica and *illbient*. It may be interesting, on the other hand, to point out the singular proliferation of "chamber tables," deriving apparently from the realm of chess (or from psychotherapy sessions), such as *Chess Piano* (2009) [410] by Guido van der Werve, or *Marble Sonic Table* (2011) [420] by Doug Aitken, which seem to have anticipated an entire culture of the practices of "digital" relationships and the behaviors of the screen media industry.

Since every taxonomy risks becoming sterile (if not dangerous), what conclusion might underlie recognition of these various figures?

From the perspective of its exhibition, sound is traditionally considered hostile: unrestrained, perfidious and difficult to control without paying the price of cumbersome objects dominating the visual presence. Nevertheless, as difficult as it may be to present, particularly in a collective exhibition, the work of sound art (or the work of art that involves sound), holds an inestimable value: it demands concentration —on the work, on its perception—in a spectrum that ranges from phenomenological approach to historical context. And it pushes toward meditation, toward concentration. It is like a familiar object, but one that cannot necessarily be contextualized. It invites a certain posture, which in turn reverberates within an auto-perceptive dynamic. It is subtly perverse, in that it requires a time not sustained by an image in motion. Yet it remains ontologically pure and free from any illustrative or "para-textual" necessity.

This trait, even though it is purely conceptual in nature, is the neo-avant-garde period's most specific inheritance. It is the most profound intersection in the dialogue between work and audience.

This is its most stimulating and perhaps most potentially volatile resource.

JONATHAN STERNE

THE MAGIC IN INSTRUMENTS

MUSIC TECHNOLOGIES AND COMMODITY FETISHISM

"Commodity fetishism? I love that idea!"

It is 2011. I am in northern California, riding in a car with Matt Ward, then the president of a company called Universal Audio, which specializes in reproducing old analog audio recording hardware, both as reissues and as software. We are discussing musicians' lust for old equipment, and the fierce arguments that populate internet discussion boards, such as the question of whether software models of hardware instruments and sound technologies will ever sound "as good as the real thing." I ask him why he thinks musicians are so invested in technologies like the ones his company produces. We go through the usual reasons: some musicians have experience with the old technology from working in studios and want to own it themselves without paying inflated prices for vintage equipment. Some lust after the equipment because they know it was used on their favorite recordings. And some are just learning the craft of sound recording, but purchase a well-known tool with the hope that it has a little bit of magic inside that will rub off on them.[1]

When I tell him scholars have a name for this last phenomenon, and explain the basics of commodity fetishism as a concept, he is amused. It explains so much, and yet, it cannot explain away the phenomenon. It's almost impossible to escape. Wendy Chun calls commodity fetishism a kind of false causality. She writes: "a fetish allows one to visualize what is unknown—to substitute images for causes. Fetishes allow the human mind both too much and not enough control by establishing a 'unified causal field' that encompasses both personal actions and physical events. Fetishes enable a semblance of control over future events—a possibility of influence, if not an airtight programmability—that itself relies on distorting real social relations into material givens."[2]
In other words, in a fetish, relations among people crystallize in things.

Universal Audio is in the business of making tools for musicians. But they are also in the business of commodity fetishism. One part of the their business involves building hardware copies of famous sound-processing devices used in the 1960s and 1970s. They also produce software replicas of analog audio devices. For their software, the graphical interface on the screen looks like a photo of the hardware, and the algorithms beneath the surface model every part of the circuits, down to the level of components. The user "grabs" images of knobs with the mouse to turn them. Even with the added flexibility afforded by software, Universal Audio shies away from introducing new capabilities, apart from maybe making a monaural unit into a stereo unit, or adding presets.

Though Universal Audio is exceptionally committed to a notion of faithfulness to originals, they are hardly alone. Companies like Native Instruments in Berlin got their start modeling old Hammond Organs and analog synthesizers in the software domain. Line6 in Los Angeles builds tools for guitarists, based on models of classic instruments, amplifiers and effects, but goes more for the sound than the old interfaces. If you buy a comprehensive package for music recording, composition or performance, it is likely to come with models of old instruments and signal processing devices as part of the bundle. Its own operating parameters are likely to be skeuomorphic as well, presenting the user with knobs and faders as if he or she were sitting before a giant mixing desk, providing at once a degree of legibility and an illusion of control. In her discussion of software interfaces, Wendy Chun argues that interfaces are "driven by a dream of individual control: of direct personal manipulation of the screen, and thus, by extension, of the system it indexes or represents. … Interfaces offer us an imaginary relationship to our hardware."[3] Between a mouse-click and an action on the screen lie countless digital instructions deliberately obscured from the end-user, along with the labor that went into making the computer, mouse and screen, and the elaborate standards and protocols that allow them to work with one another consistently, to work with other systems, and to appear seamless to our imaginary end-user. While Chun singles out software, this illusion of control is true of *any* interface, any technology for making sound or music.

Barry Blesser and Linda-Ruth Salter make the connection explicitly in their discussion of artificial reverberators (devices that add roomlike echo) that Blesser worked on in the 1970s: "once a spatial parameter is connected to a knob, button, or key, [from the perspective of the person operating it] a reverberator becomes effectively indistinguishable from a musical instrument, played in real time by a musician." Here, media collapse into instruments, or rather, the mediatic dimension of instruments is revealed directly.[4]

We are used to thinking of instrument design as interface design when it is a digital instrument, like a synthesizer or sampler. But all instruments have interfaces. The apparent immediacy of an acoustic instrument conceals just as much labor, craft and standardization as a software interface or digital signal processor. Alongside a set of pedals, a piano key operates a hammer, mediating and modulating the percussive dimensions of the instrument. The taut head of a tabla presents itself as a target for the player, with vastly different tones available depending on where and how fingers strike the surface. The fingers are meant to hit one end of the keys on an *mbira*, not the other, just as Western string musicians know the strings on members of the violin and guitar families are meant to be played between the bridge and the nut (though many have violated that rule to interesting effect). As with the software example, all of these mechanisms obfuscate all manner of labor and decisions. Piano-tuning is one of the oldest continuously-practiced professions in music. The pitch and temperament compromises built into a piano keyboard or cut into a guitar's fretboard come from sustained collective reflection on the differences between the physics of sound and the cultured ears of musicians and listeners. Members of the violin or brass family —or the drums in the "kit"—exist in a relation of complementarity with one another. Even acoustic instruments like acoustic guitars and basses are systems. The chain of physical causes and effects are more readily apparent to the untrained observer, but they are no less real: move the bridge even a little and you will probably have to adjust the neck.[5]

The decisions built into these instruments have real ramifications for musicians, from the finer points of technique to the injuries one can suffer from playing too much or incorrectly. On the same trip to northern California, I am speaking with Roger Linn in his living room, seated near a piano and not far from an electric guitar connected to a pedal and then a computer. Linn made his mark as an inventor of electronic instruments. After designing one of the first sampling drum machines—in other words, a drum machine that used the sounds of recorded drums rather than synthesizing its own drum sounds—he developed the concept for the MPC (MIDI Production Center), which became the most important instrument for rap and hip hop after the turntable. Linn has been hard at work trying to create a properly "expressive" digital instrument. By *expressive*, he means that it has the qualities of a stringed instrument, where even subtle gestures are mapped to variations in pitch, timbre or loudness. Gazing over at the piano and guitar, he lambastes them both from the standpoint of interface design. For both, the hand gestures change when the musician wants to change key. In other words, the skills are not transferrable. From the standpoint of modern interface theory, they are unnecessarily difficult to learn and lack sound ergonomics, yet in the previous breath he was lambasting digital instruments for their lack of expressivity, where musicians lacked sufficient control to produce sophisticated melody, harmony and timbral variation in real time.[6]

From the musician's standpoint, mastery of an instrument is mastery of an interface and the system to which it is connected, whether we are talking about a set of energy transfers built into assemblies of metal, bone and wood, or whether we are talking about electronics and plastic. Playing an instrument is a form of embodied knowledge, a knowledge won with hours and years of practice, a "second nature" as Pierre Bourdieu puts it.[7] That embodied knowledge mediates the standardization that went into the instrument's interface. When I pick up an electric bass guitar, my hand movements are completely unthought, even if they were once—long ago—a struggle. After years of playing, my body seems to conform to the instrument, even though it feels like the instrument is conforming to my body or my will. Phenomenally, it is not that different from typing out this sentence on a keyboard, where again my body has conformed to a standard—a standard originally developed for a system entirely different from the computer.[8]

From the outside, gestures like these can appear effortless to the casual observer, because the labor behind them is obscured in the design of the instrument and the skill of the musician. When observing skilled musicians perform (or persons skilled with any technology), we might say that they are *in instrumentality*. This is to say that they are in a relationship to the instrument where intention and action conform to one another, where certain prior actions and knowledges are relegated to a background status so that the musician may "sustain a certain direction," to use Sara Ahmed's terms.[9] But what happens when we confront instruments outside of this moment of performance? They can be experienced as tools of learning, or partial objects, but they may also be experienced as magical in their own right.

In her classic ethnography of a South African recording studio, Louise Meintjes tells a story of a singer named Joana mistaking a MIDI clock that synchronizes all the studio's different devices for a sound-container. "The MIDI clock does not actually house the sounds to which Joana refers. … For her, there is a whole sonic world packed into that sleek machine … It is a world to which Joana can point, but that she cannot enter herself. It is invisible but sensed to be of enormous proportion."[10] Again, we have a specific case that illustrates a general condition: when people do not have access to the inner world of an instrument—because of knowledge, experience, power differences, custom—they are more likely to attribute vast, complex inner worlds to those instruments.

In part, this is a deliberate design feature, and an old one at that. My recording teacher Mark Rubel called this dimension of instruments "psychocosmetic." The ornate scrolls and flourishes of the violin family; the branded headstocks of the guitar family; the shiny, smooth surface of a piano; the blinking LEDs of a drum machine—all of these allude to the magic within, the agency held inside the thing, just beyond a person's fingertips.

Writing of the recording studio, Meintjes captures the futurism that enrobes so many electronic instruments today. For her, the studio space is: "constructed and experienced as magical and as a fetish by music-makers who work within it. By typifying the space as magical, I mean that it is remote from the ordinary and that through the art of illusion and the capacity of the imagination, it seems to house a natural force … that when tapped produces compelling art. By thinking of the studio as fetish, I reify it into an object that can procure for those who have earned access to it the services of that force, or 'spirit,' lodged within it."[11]

Hanging on walls at music stores, leaned up against the chairs on a stage, assembled into racks, or organologically classified at museums, instruments can suggest some kind of spirit when separated from their moments of making or use. It is perhaps more appealing to believe that magic lies in instruments than labor in people, because of the ways in which music and musicianship are mystified and separated from everyday life, because of the appearance of effortlessness that attends so much good performance, and because of the distance most people will have from the lived experience of most instruments.

In the fantasy lives of musicians and artists, we can also see a connection between instrumentality and spirit or magic. A quick Google search of the phrase "it's almost like an instrument" yields a steady drone of artistic longings attached to equipment. Like Blesser's parameter-assigned-to-a-knob, magic manifests in machinery the moment that parameter control and efficient operation yield inspiration and hidden resources for art. One genre of comments comes from communities of musicians online who are discussing equipment, sometimes as users, other times as reviewers. A reviewer for a microphone writes, "The mic is a pleasure to sing through, it's like an instrument for singers like a guitar is to a guitarist."

A user of digital reverb software called ValhallaShimmer writes "I see Shimmer as a special effect—it has so much character it's almost like an instrument." Another musician, writing about a sequencer (which controls other instruments but doesn't make any sounds on its own) says "it's almost like an instrument in itself when you start doingthings like assign knobs A + B to control when things happen relative to other things."[12]

Artists, too, use this terminology. Writing of an interactive visual art program called *Aleph Null*, artist Jim Andrews writes "it takes practice to tease the really good stuff out of it. It's like an instrument that way." Note that the good stuff is teased out by the user, but resides in the thing itself. In an ArtIT interview with Janet Cardiff and George Bures Miller, an installation is described in the same way:

"ART iT: [Discussing *Experiment in F# Minor* (2013)] If you don't know where the sensors are, it seems natural to explore the piece through movement, to see whether you can control the intensity by waving your arms in one direction or another. The work sucks you into performative behavior. With the walks too, there's a mechanism of unconsciously entering a different zone of behavior.

Janet Cardiff: You see how unlimited it could be. It's like an instrument.

George Bures Miller: That's the problem for us. We're always discovering these things that could be unlimited.

Janet Cardiff: With *Pandemonium* (2005), for which we installed robotic percussive beaters in the cells of the Eastern State Penitentiary Museum in Philadelphia, we discovered that was like an instrument too. You could have made any piece of music with it. You could invite percussionists in and say, here's an instrument, what do you want to do with it? There's all these offshoots that would be great to follow through. Our problem is we have too many ideas and not enough time."[13]

"You see how unlimited it can be:" this turn of phrase marks the moment where description calls forth fetishism. The limitlessness—"you could have made any piece of music with it"—is the fantasy of that "natural force that produces good art" to which Meintjes gestures.

It is no accident that this attribution of magic usually comes at some distance from the moments of instrumentality—either the moment of observation from a distance where practical knowledge is not fully available, or the moment of reflection from a distance of time. In *Queer Phenomenology*, Sarah Ahmed writes about what is revealed when "technologies are no longer ready for action." An experience of being unable to use a thing—or at a distance from the moment of use—leads to attribution of properties to the thing itself. Writing of failure, using Heidegger's famous example of a hammer, she says that it: "might then lead to 'the object' *being attributed* with properties, qualities and values. In other words, what is at stake in moments of failure is not so much access to properties but attributions of properties, which become a matter of how we *approach* the object. … The moment of "non-use" is … the same moment in which objects may be judged insofar as they are inadequate to a task, the moment when we "blame the tool."[14]

It may also be the moment when we credit the tool, and find ourselves inadequate for it. Christopher Small's critique of Western music cultures is apropos here: that some people possess innate musical ability while others do not.[15] The deep feeling that an instrument brings magic or power to musicians, rather than they to it, is a residuum of this more general way of thinking.

The nearly 500 years of instruments catalogued in "Art or Sound" play both sides of this phenomenon. What better example of the power of commodity fetishism than an exhibit of instruments, set out for observation and experimentation, consecrated as artworks in themselves, separated from the contexts of musical practice and performance, and recontextualized as objects of solemn contemplation—as art set apart from life? At the same time, framed as works of art in the museum context, the instruments seem to comment on themselves: from the exquisite craft embodied in a 17th-century viola or guitar [50, 52], to the explicit reflections on the nature of modern instruments in Tore Honoré Bøe's acoustic laptops [436] or Yoshihiko Satoh and Manuel Rocha Iturbide's [418] deconstructions of the modern drum set.

When anyone can walk up to a consecrated object and interact with it, the idea that instruments are for some people and not others begins to erode; the invitations to touch, hear, engage or interact break down the separation of performer and audience. The collection itself also offers a commentary on instruments as objects with magic inside them. When experienced as a totality, the collection's sheer size and breadth testifies to the human labor and creativity sedimented in the instruments themselves, the dead labor that lurks behind the fetish.

Thanks to Carrie Rentschler

1
Matt Ward, author interview, February 17, 2012.

2
Wendy Hui Kyong Chun, *Programmed Visions: Software and Memory* (Cambridge, MA: The MIT Press, 2011), p. 50.

3
Ibid., pp. 62, 66.

4
Barry Blesser, Linda-Ruth Salter, *Spaces Speak, Are You Listening?: Experiencing Aural Architecture* (Cambridge, MA: The MIT Press, 2007), p. 91; Paul Théberge, *Any Sound You Can Imagine: Making Music/Consuming Technology* (Hanover, NH: Wesleyan University Press, 1997).

5
Myles Jackson, *Harmonious Triads: Physicists, Musicians and Instrument-Makers in Nineteenth-Century Germany* (Cambridge, MA: The MIT Press, 2006).

6
Roger Linn, author interview, February 13, 2012.

7
Pierre Bourdieu, *The Logic of Practice*, trans. by Richard Nice (Stanford, CA: Stanford University Press, 1990).

8
Sarah S. Lochlann Jain, *Injury: the Politics of Product Design and Safety Law in the United States* (Princeton, NJ: Princeton University Press, 2006).

9
Sara Ahmed, *Queer Phenomenology: Orientations, Objects, Others* (Durham, NC: Duke University Press, 2006), p. 31.

10
Louise Meintjes, *Sound of Africa!: Making Music Zulu in a South African Studio* (Durham, NC: Duke University Press, 2003), p. 89.

11
Ibid., pp. 73–74.

12
Pete Pell, reviewer comment on the Sterling Audio ST66 Large Diaphragm Tube Condensor Microphone, *Musician's Friend* sales website: musiciansfriend.com/tube-microphones/sterling-audio-st66-large-diaphragm-tube-condenser-microphone; Aniston, user comment on thread "ValhallaShimmer or EOS?" *KVRAudio* music software forum: kvraudio.com/forum/viewtopic.php?t=302426; Gosh, user comment on "Sequentix Cirklon" thread, *Muff Wiggler* modular synthesis forum: muffwiggler.com/forum/viewtopic.php?p=933082&sid=c75363c58507ef9674cb8b3bac00fc8c (all accessed March 2, 2014).

13
Jim Andrews artist profile on: rhizome.org: http://rhizome.org/profile/jimandrews/?page=2; *ARTiT* magazine interviews, "Janet Cardiff and George Bures Miller: Pt II," in art-it.asia/u/admin_ed_itv_e/k15N0UpRnFSy6i3caPT9 (both accessed March 2, 2014).

14
Sarah Ahmed, *op. cit.*, p. 49.

15
Christopher Small, *Music-Society-Education* (London: John Calder, 1977).

Haroon Mirza

LO-TECH PROPOSED
2006–14

4

Lo-Tech Proposed is a central work in the oeuvre of British artist Haroon Mirza (London, 1977), which explores the fusion between various sensorial stimuli through sound and light installations that modify perception by interacting with the surrounding space.

The work is composed of a series of elements placed on top of one another. A bucket stands on a long sheet of glass, which is set on a speaker cone placed on an amplifier. The bucket contains a mist maker and a waterproof LED set in a small quantity of water, while electric wires are wound around a metal rod in the upper part of the bucket. Drops of water produced by the mist maker cause the LED to flash. The flow of electricity is captured by the amplifier, which produces a sound that is completely random and therefore always different. *Lo-Tech Proposed*, originally made in 2006 and reassembled by the artist with the same pieces in 2014, is Mirza's first work to transform the amplification of electricity through a lighting system into an artistic technique.

[See also p. 434]

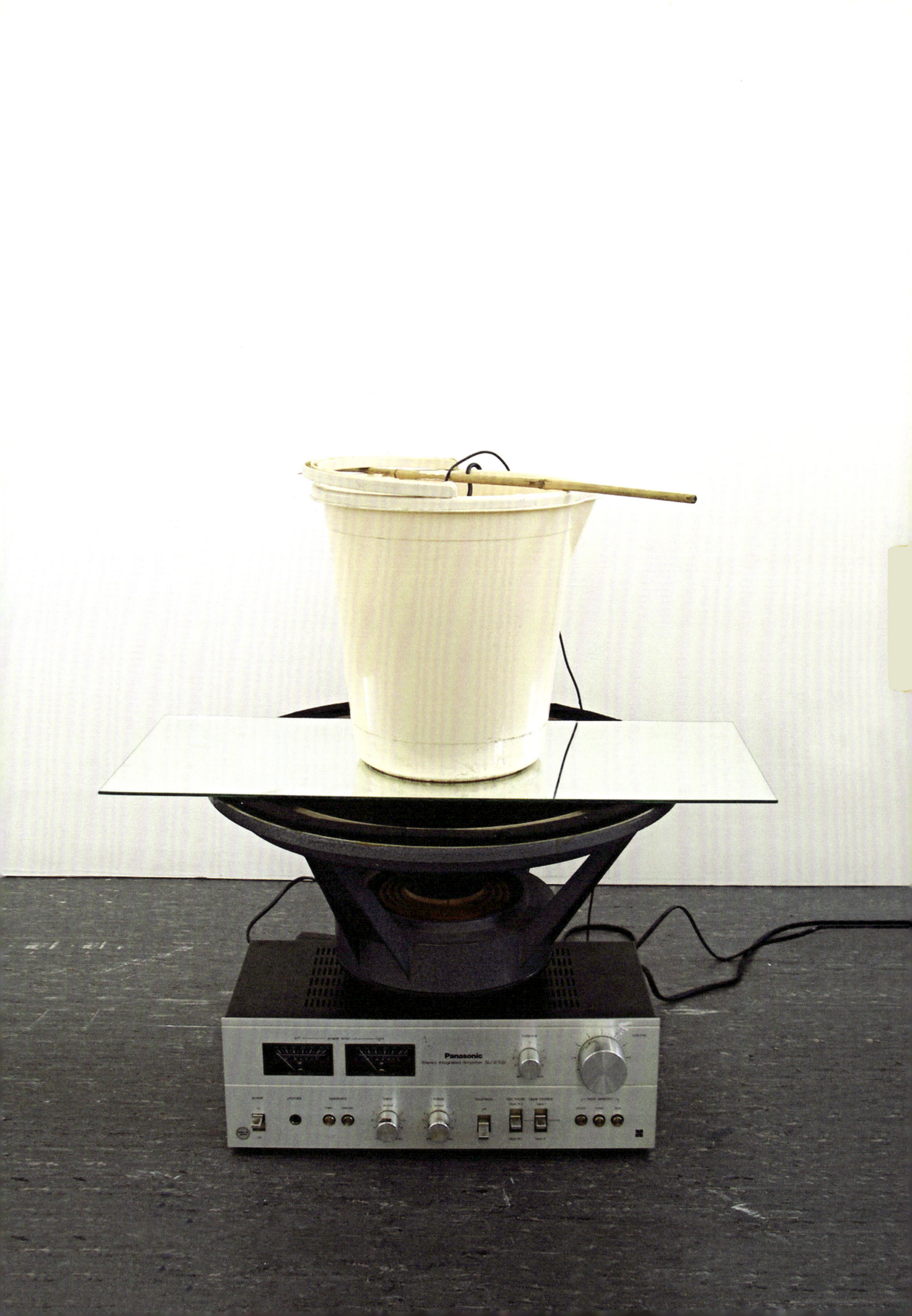
Panasonic

Walter Kitundu

EGRET PHONOHARP
2007

Walter Kitundu (Rochester, 1973) is an American sound artist and inventor of musical instruments who works at the boundary between live and recorded performance. Inspired by hip-hop, other modern musical forms and traditional Asian and African instruments, his phonoharps are hybrids of turntables and stringed instruments.

At once a sculptural art object and functional instrument, a phonoharp works by transmitting vibration in the strings through the body of the instrument into the stylus of a turntable. It can sound like a dulcimer, a bass guitar, a slide guitar, a percussive instrument or anything found in the vast library of available records.

[See also p. 404]

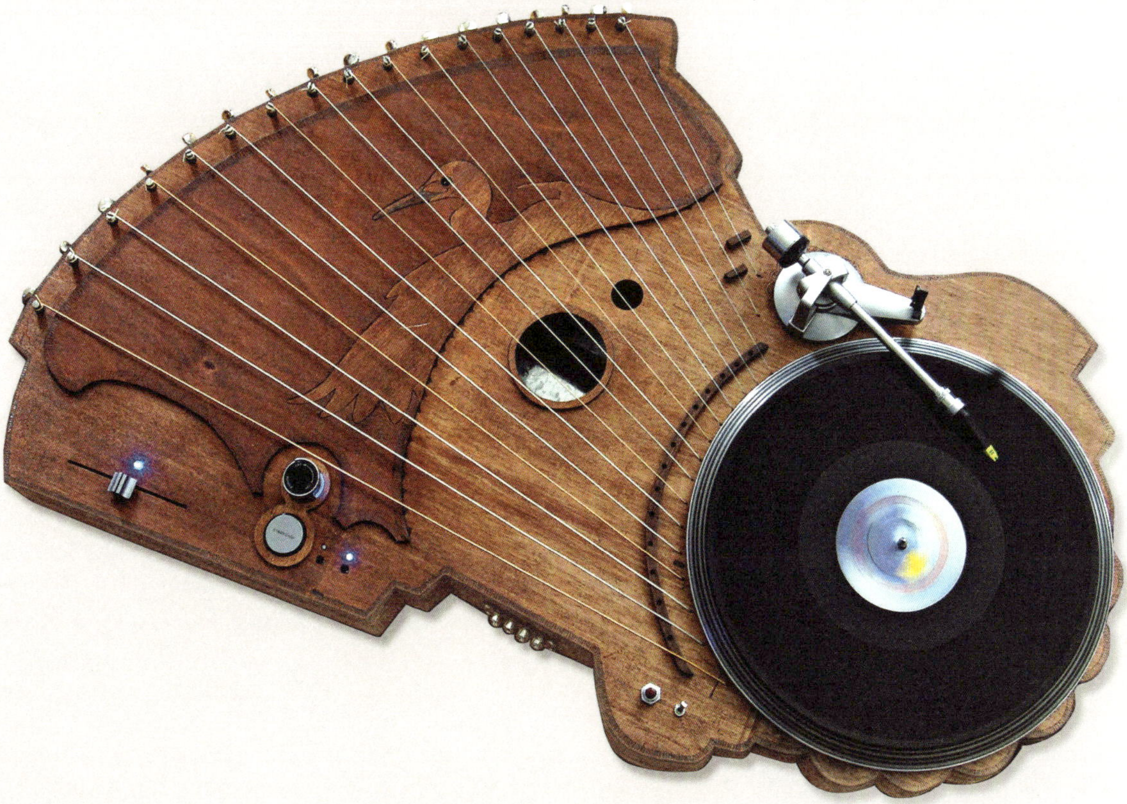

Walter Kitundu

BEGUÈNA MARIDHIA
2008

In 2008, American sound artist and instrument maker Walter Kitundu (Rochester, 1973) started a collaboration with Kronos Quartet, an ensemble based in San Francisco that combines a spirit of fearless exploration with a commitment to expanding the range and context of the string quartet. Kronos Quartet—which has collaborated with many of the world's most eclectic composers, such as Philip Glass and Steve Reich, and performers like Laurie Anderson, Tom Waits and David Bowie—asked Kitundu to construct 2 instruments: a Stroh cello [exh.] and the Beguèna Maridhia.

The Beguèna Maridhia was inspired by the Ethiopian beguèna and mimics its low, resonant, clattering sound. The Kronos Quartet commissioned it as one of 3 instruments used to perform a work by the great Ethiopian musician Alemu Aga.

[See also p. 402]

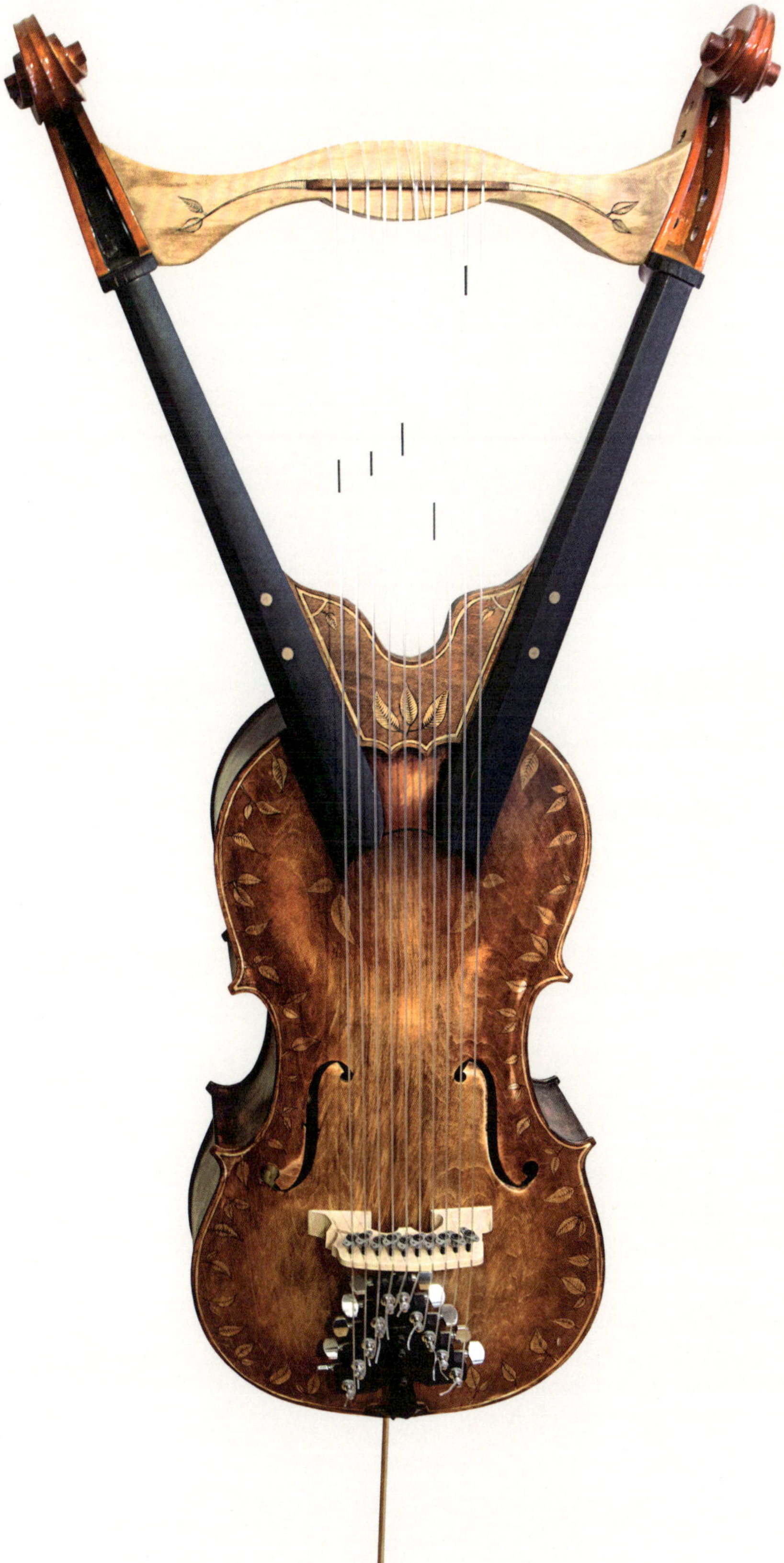

Thomas Demand

HELDENORGEL
2009

German artist Thomas Demand's (Munich, 1964) working method marries iconographic research, sculpture and photography. Using a found image (often taken from newspapers or magazines reporting crime news) as his starting point, Demand creates a life-size paper model which he then photographs and destroys. This work of reconstruction deceives our perception and holds our capacity for recognizing what we see in suspense.

The subject of *Heldenorgel* is the Heroes' Organ in Kufstein Fortress (Austria), which was built to commemorate Austrians and Germans who died in World War I. Based on a phonic design by Oskar Walcker in 1931, it is the largest outdoor organ in the world. Its soundboard is the natural environment formed by the surrounding mountains. With 4307 pipes, 46 registers and 4 keyboards, the organ is played every day for 10 minutes at the foot of the fortress, and can be heard throughout the river Inn valley between Austria and Bavaria.

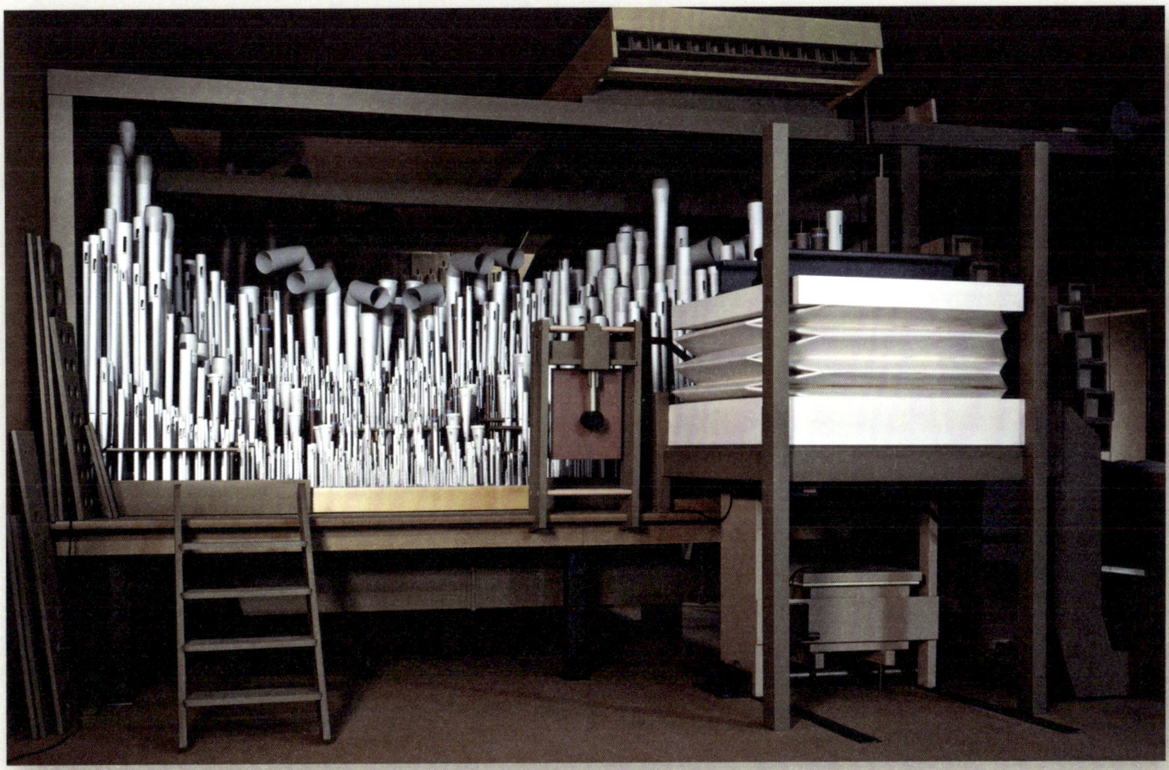

Anri Sala

A SOLO IN THE DOLDRUMS (BASED ON AN UNSEEN DANCE BY SIOBHAN DAVIES) 2009

In his works Albanian artist Anri Sala (Tirana, 1974) attempts to unite sounds and images to set correspondences between objects, creating physical and sound environments that give form to music.

A Solo in the Doldrums is a snare drum whose sticks are activated by vibrations caused by inaudible, low-frequency sounds. The sound signals are reproduced by a loudspeaker incorporated in the drum and are the result of the recording of a dance performed without an audience by dancer-choreographer Siobhan Davies. The sounds of her movements were captured by microphone and subsequently modified in a sound studio. The Doldrums—the equatorial region in the Atlantic Ocean mentioned in the work's title—used to be known for unpredictable weather conditions, in particular the absence of wind for long periods that made sailing through the region impossible. For a duration of 8 minutes and 50 seconds, *A Solo in the Doldrums* makes an unwitnessed dance both visible and audible.

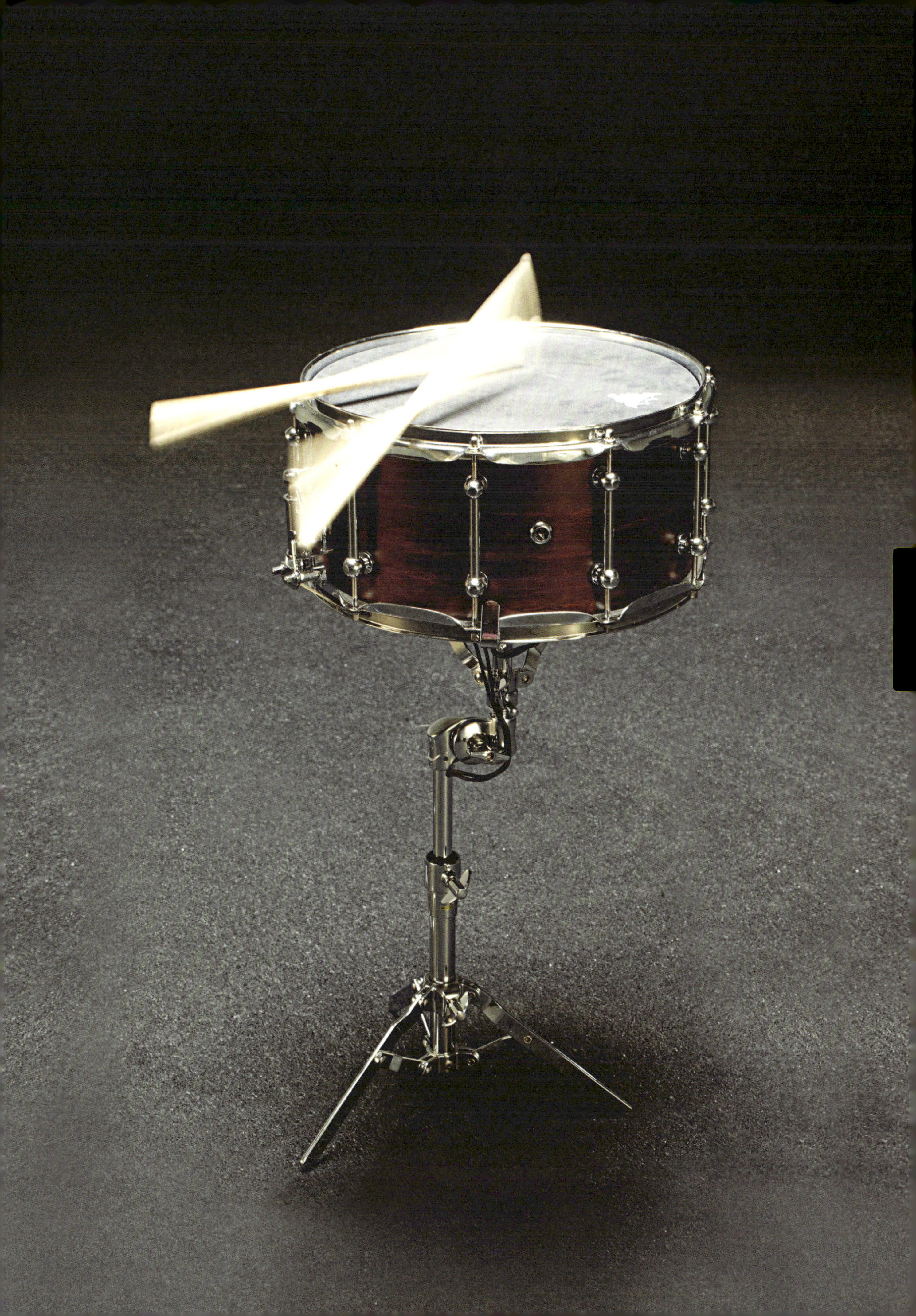

Guido van der Werve

CHESS PIANO
2009

Dutch artist Guido van der Werve (Papendrecht, 1977) studied industrial design, archaeology and music composition before starting his career at the turn of the century. His filmmaking is connected with music composition: he has performed the music in his films on several occasions with an orchestra.

Chess Piano is an instrument that merges piano mechanics with a traditional chessboard layout.
It allows 2 players seated on the 2 stools to interact through playing a chess match: the chessboard functions as a soundboard activated by moving the black and white pieces. Van der Werve composed *Nummer Elf: Chess Piano Concert in Three Movements* (2009) expressly for this instrument, to be performed together with violins, violas, cellos and bass.

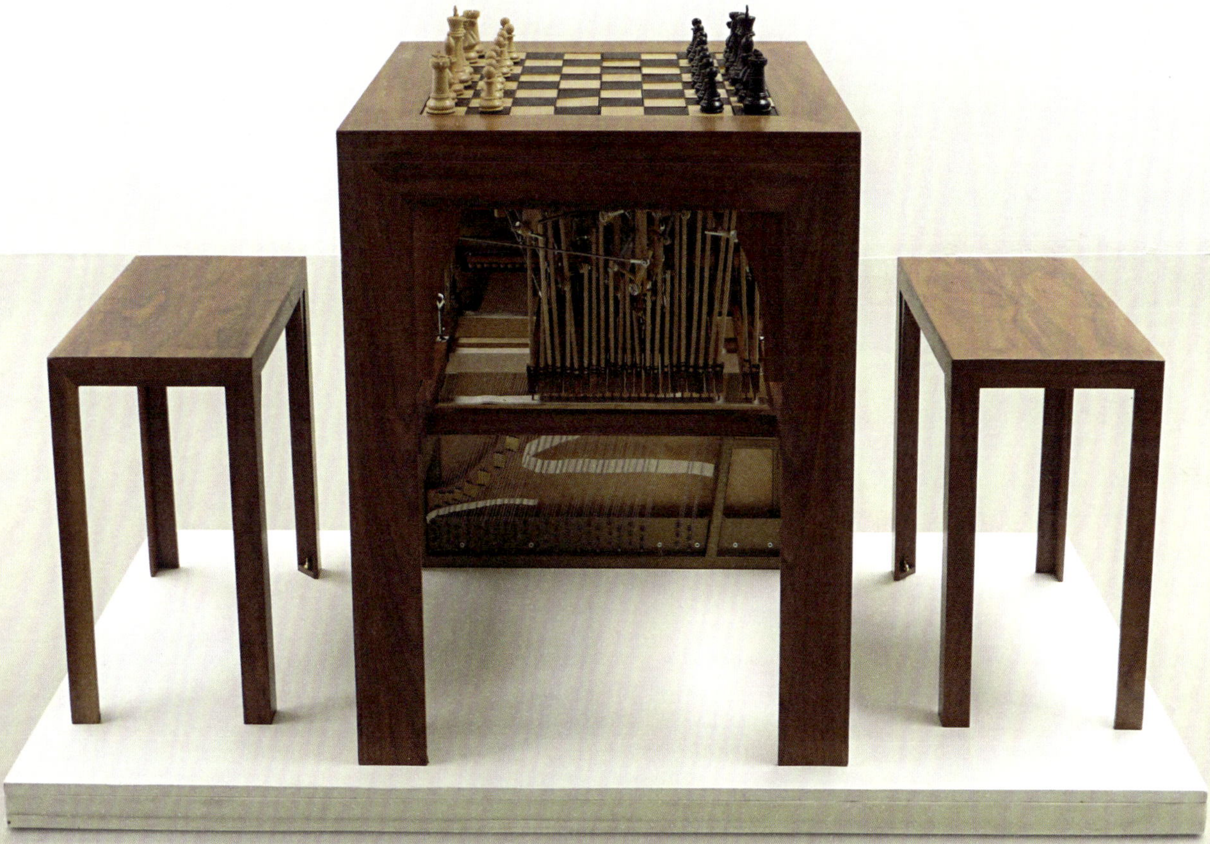

Athanasios Argianas

A DEMONSTRATION OF MANY VIEWS AS ONE 2009 (2014)

Greek artist Athanasios Argianas (Athens, 1976) develops his research between various media, creating objects, installations, performances and music. He makes objects that encode pieces of music, and designs objects to be used as structures for performances.

A Demonstration of Many Views as One started as a game between an object as a song, and a song as an object. Its first version began as a text and a sculpture. The text "I Was Swept Off Her Feet / I Was Swept Off Her Feet / She Was Swept Off Her Feet" formed the sculpture itself. Then the text became a song, written in the form of a canon (a cyclical piece of music) that was subsequently performed as a live work.

The work's numerous re-workings and translations have now resulted in a projection/sculpture where the physical object/projection surface mimics the structure of the music itself. It is a rotated triangle on 3 levels, much like the music is a singular melodic line, repeated thrice in its timeline, "rotated" through its duration, aiming to make an experience that oscillates between hearing the same song 3 times at different intervals, and the feeling of listening to a single harmonic entity.

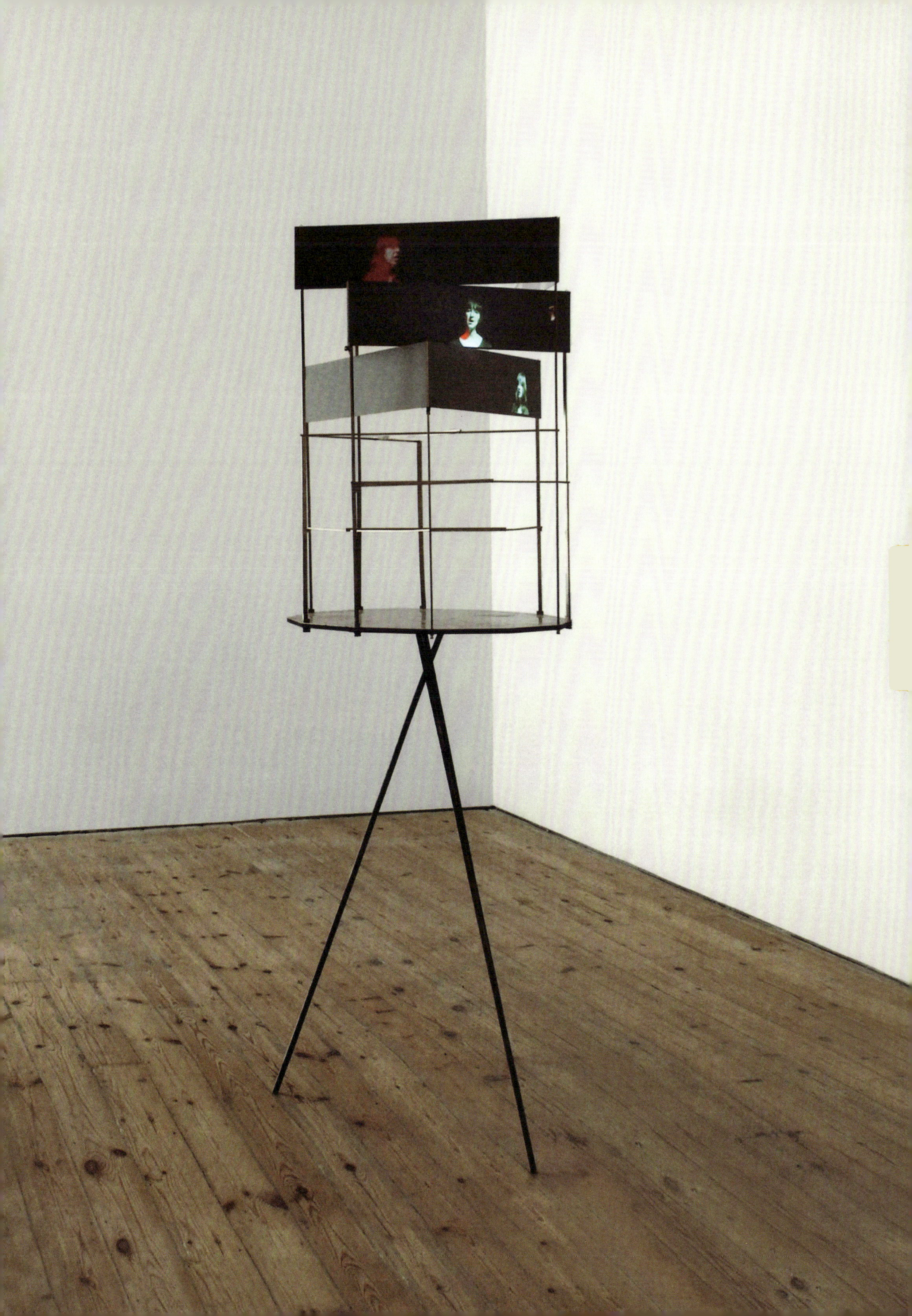

Nicolas Bernier and Martin Messier

LA CHAMBRE DES MACHINES
2010

La Chambre des Machines (The Mechanical Room) is an audio-visual performance by the Canadian duo formed by Nicolas Bernier (Ottawa, 1977)—an artist working with sound, performance, installation and video—and Martin Messier (Montreal, 1979), a composer, performer and video-maker. The artists' intention is to make inventive use of music, video and staging, blurring the lines between analog and electronic creation. Submerged in sound, the audience discovers the interaction between mechanical and synthetic sonorities. The sound machines made of gears and cranks used in the performance as musical instruments were purpose-built by Alexandre Landry and are manipulated to produce a sound construction at the crossroads of acoustics and electronics. Through specifically tailored programming, digital processing enlarges the machines' sound palette. The inspiration for a machine containing mysterious mechanisms (precisely what the computer is today), comes from the *intonarumori* [p. 160] created by Luigi Russolo at the beginning of 20th century.

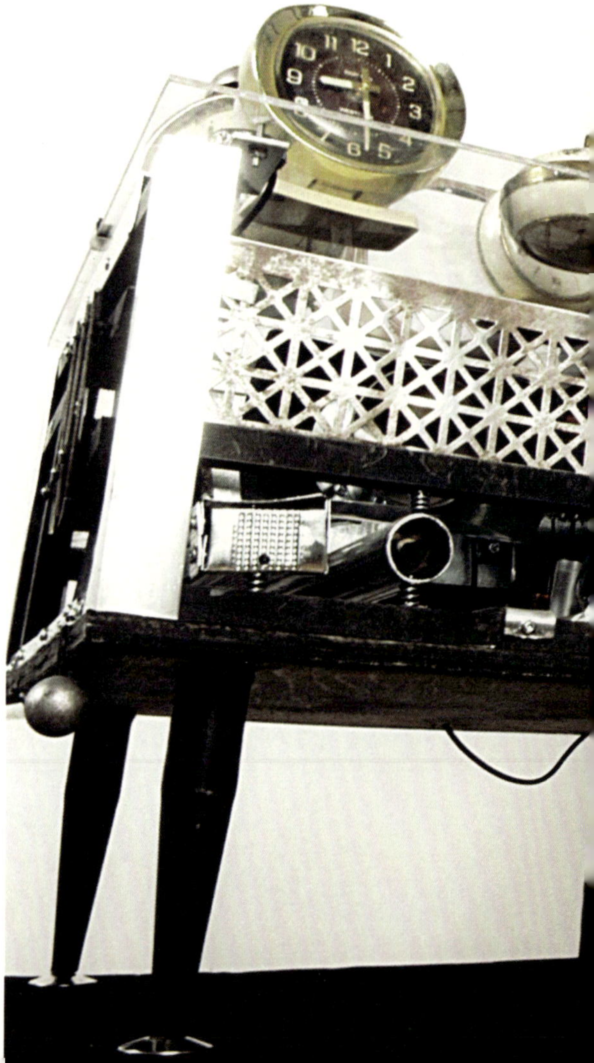

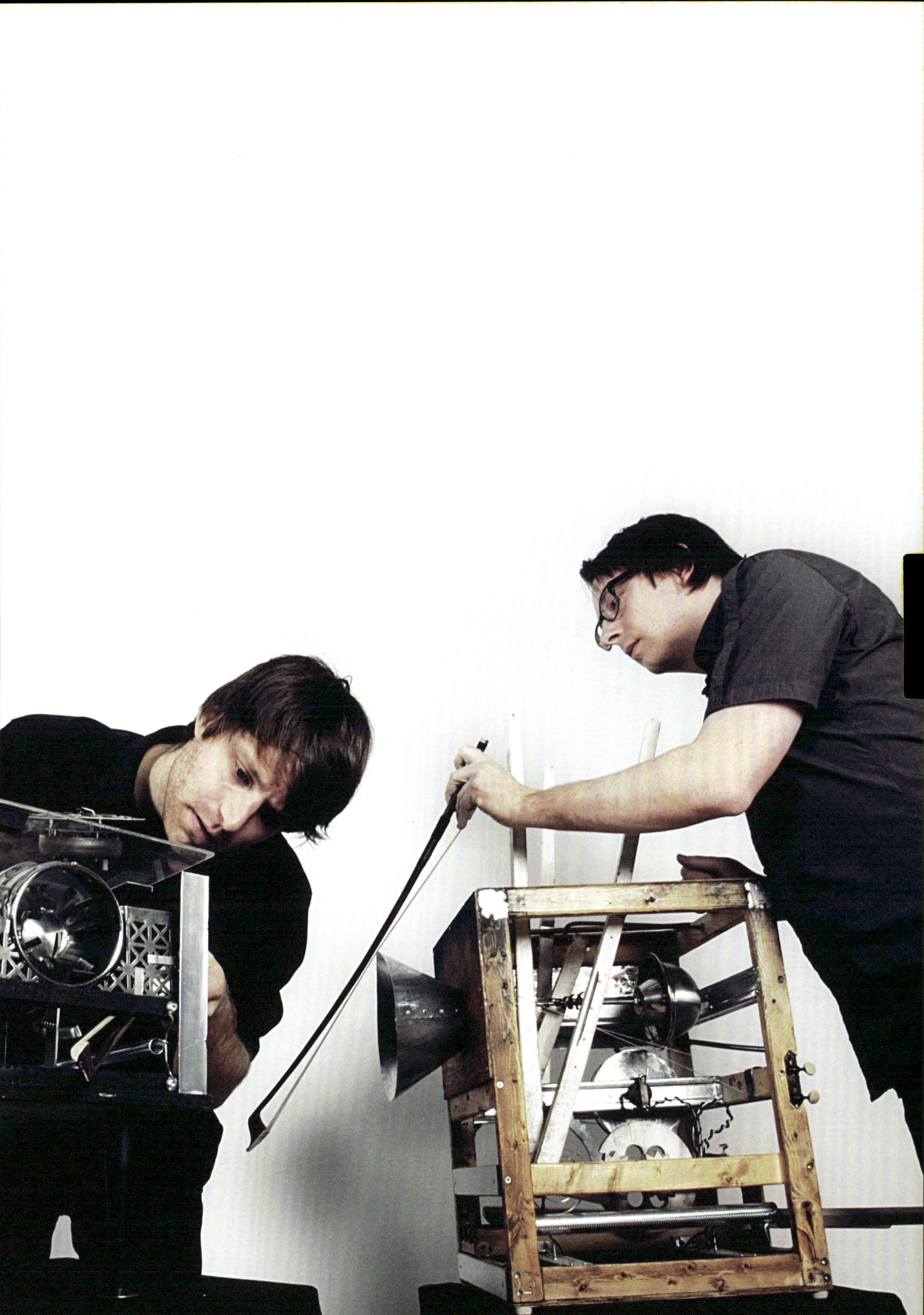

Martin Kersels

WARY (DICK)
2010

4

Martin Kersels (Los Angeles, 1960) is an American sculptor, performer and video artist. He began his career in 1984 as a member of the performing group Shrimps. After studying at University of California, Los Angeles, with Chris Burden and Paul McCarthy, he started work on a series of mechanical sculptures, including *MacArthur Park* (1996), in which the artist's body is represented by a set of painted wooden balls which a mechanism sets in motion to music from a loudspeaker, exploring his interest in the sonic and mechanical aspects of artworks.

Wary (Dick) and *Wary (Jane)* [exh.] are 2 real pairs of shoes on a wooden structure fitted with bells. Kersels claims that these are inspired by a nursery rhyme that "goes something like 'with rings on her fingers and bells on her toes, and she shall have music wherever she goes.'"

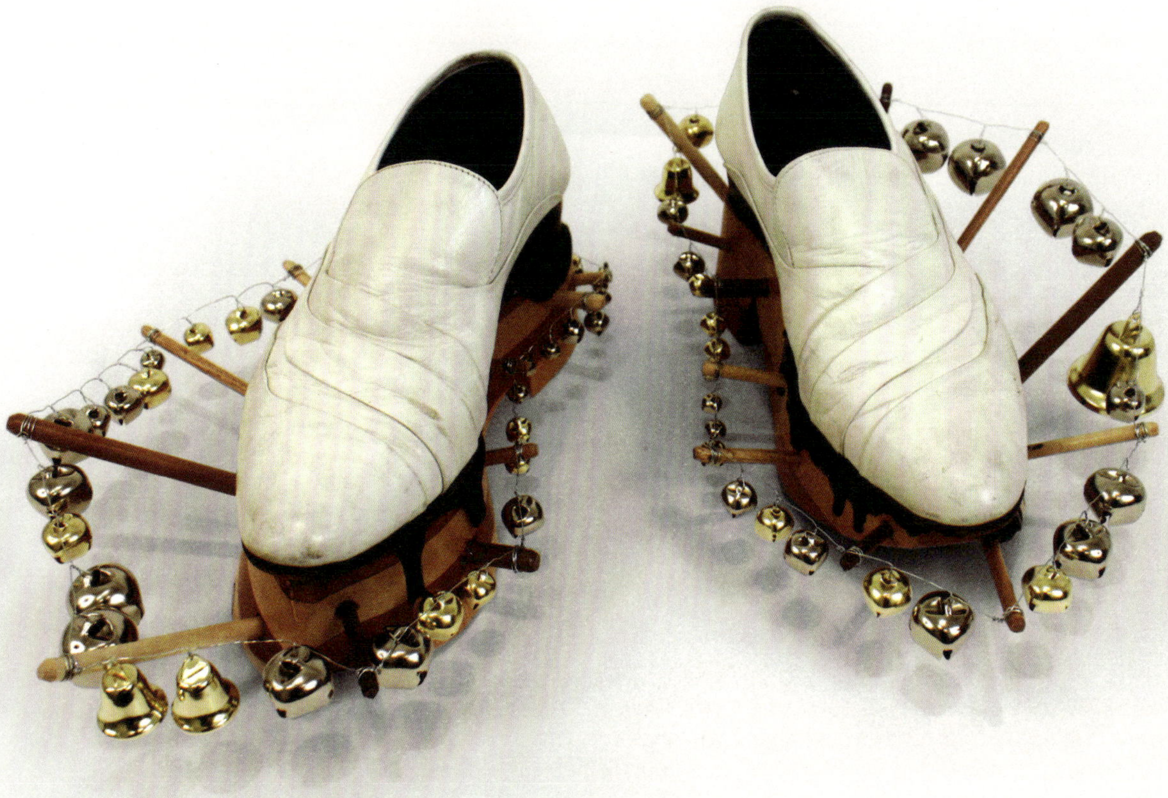

Manuel Rocha Iturbide

EL ETERNO RETORNO
2010

After studying music, Mexican composer and artist Manuel Rocha Iturbide (Mexico City, 1963) began exploring different creative paths beyond instrumental music, including electroacoustic music, sound art and curating practices.

In *El eterno retorno* (The Eternal Return) a toy train moves along a circular track placed on a drum. Endlessly turning in circles, the train produces a slight and continual noise amplified by 2 loudspeakers, thus creating a sound similar to that of a real train. The sound spreads throughout the installation space, generating a feedback loop that responds to the movement of the train, composing ever larger circles. "This eternal coming back of the train through the same starting point"—says the artist—"is not always the same; it changes in time, as in Friedrich Nietzsche's eternal recurrence philosophical theory."

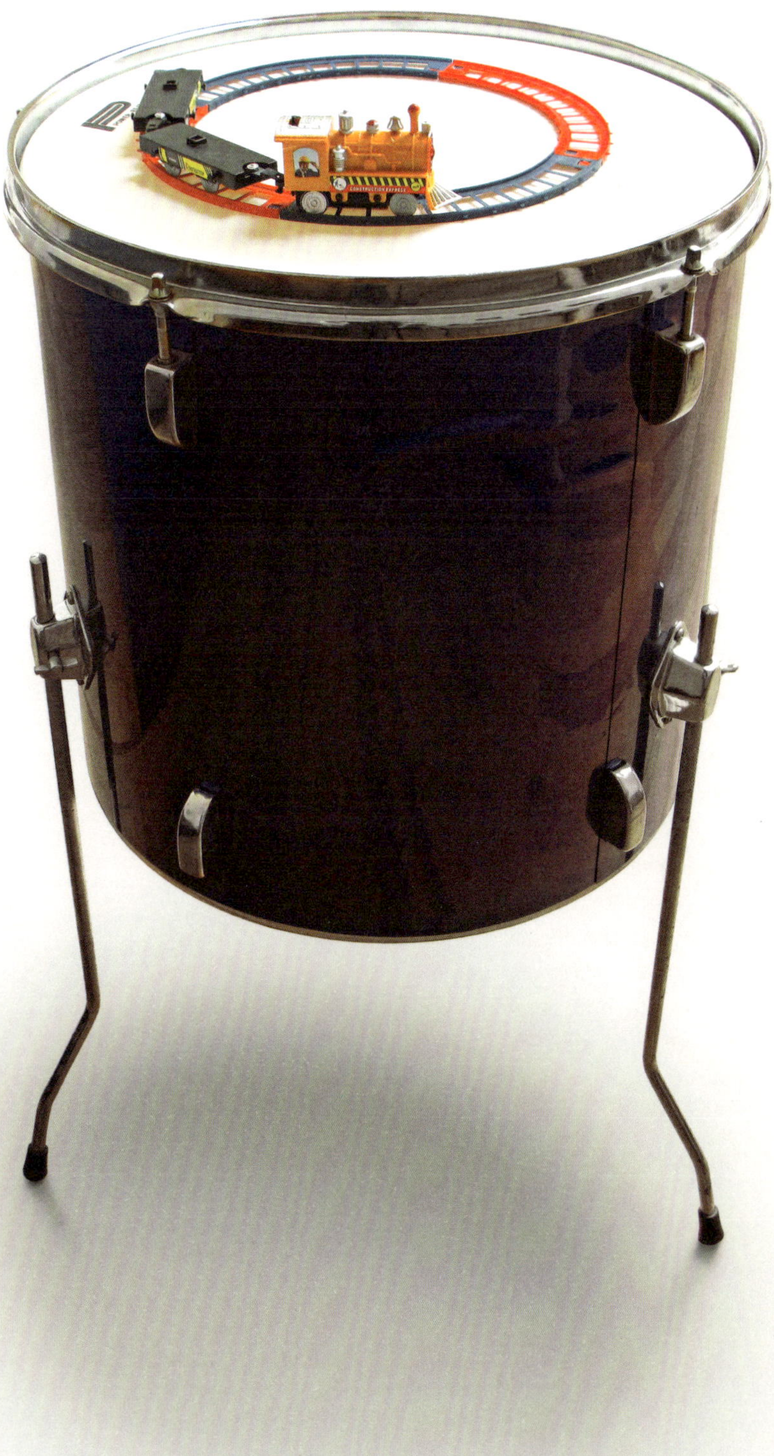

Doug Aitken

MARBLE SONIC TABLE
2011

American artist Doug Aitken (Redondo Beach, 1968) utilizes a wide array of media ranging from photography, print media, sculpture and architectural interventions to narrative films, sound, single and multi-channel video works and live performance. Aitken is conscious that sound and vision are equally important to the experience of his work: "The sound has to be intrinsic to the artwork—both specific and accurate at once." *

Marble Sonic Table is one in a series of polyphonic sculptural pieces that he created as a personal reflection on the limitations of language and as an attempt to transcend conversation in the creation of a social space. People sitting around the table can use drumsticks to bang on it: drumming chambers carved into the marble turn it into a living artwork where sound takes the place of dialogue.

* Daniel Birnbaum, "That's the Only Now I Get: Time, Space and Experience in the Work of Doug Aitken," in Daniel Birnbaum, Amanda Sharp, Jörg Heiser, *Doug Aitken* (London/New York: Phaidon, 2001), pp. 38–105: 75.

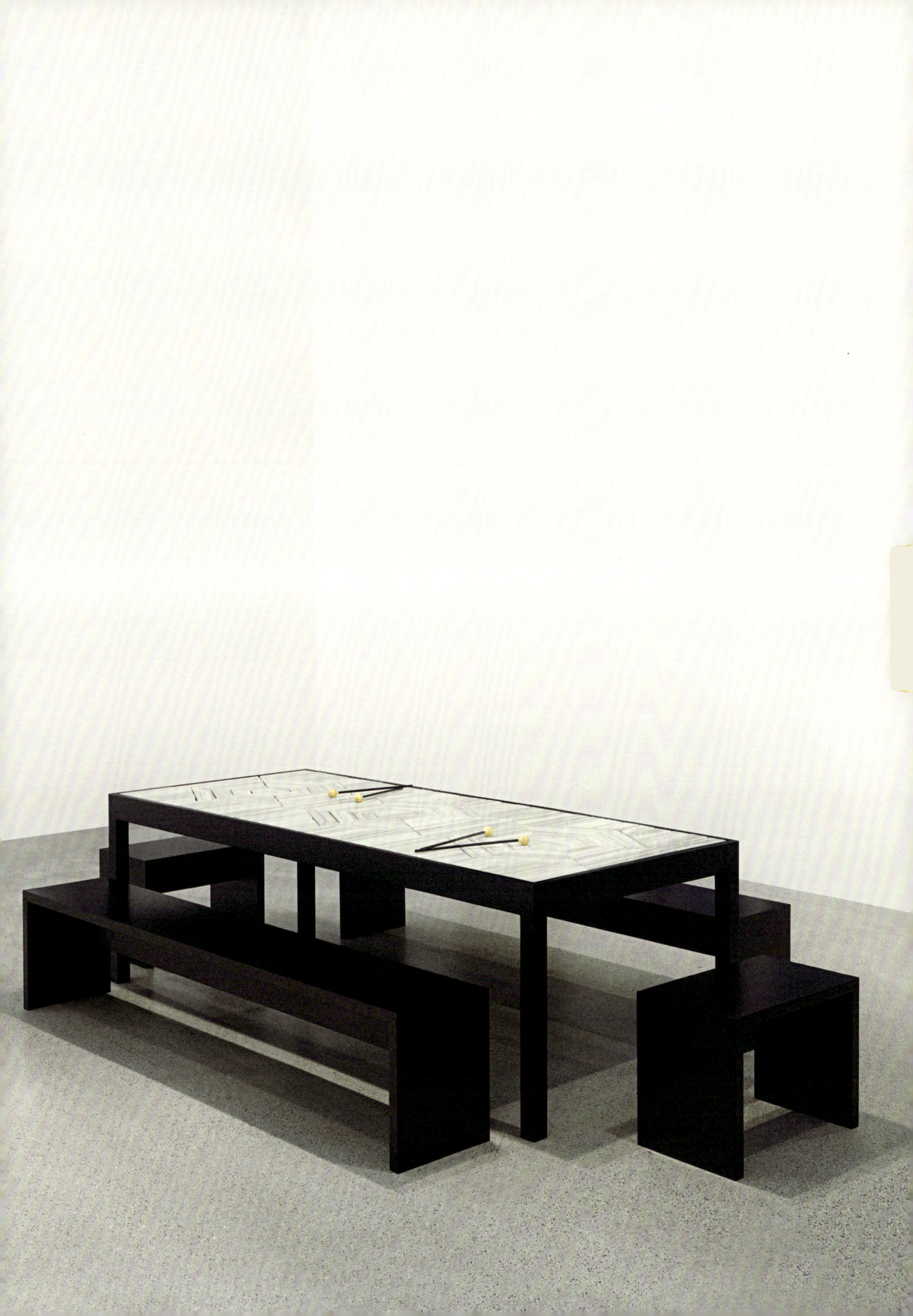

Alberto Tadiello

ELEKTRONSKAL
2011

With his carefully designed assemblages of electrical materials, Italian artist Alberto Tadiello (Montecchio Maggiore, 1983) creates complex machines and technological mechanisms that imitate the operations and forms of living organisms.

Elektronskal makes reference to a scientific term that identifies the structure of electrons around the nucleus of an atom. The work functions as an amplifier of sounds produced by an electric current. A perforated metal sheet connects a set of wires with 64 electric bells arranged to match the electronic configuration of an atom. The mechanical action of the motors inside the bells gently vibrates the external bell made from panels of medium-density fiberboard, creating an echo effect. The red color of the bells and wires emphasizes the dialectics between attraction and repulsion caused by the sound component of an object.

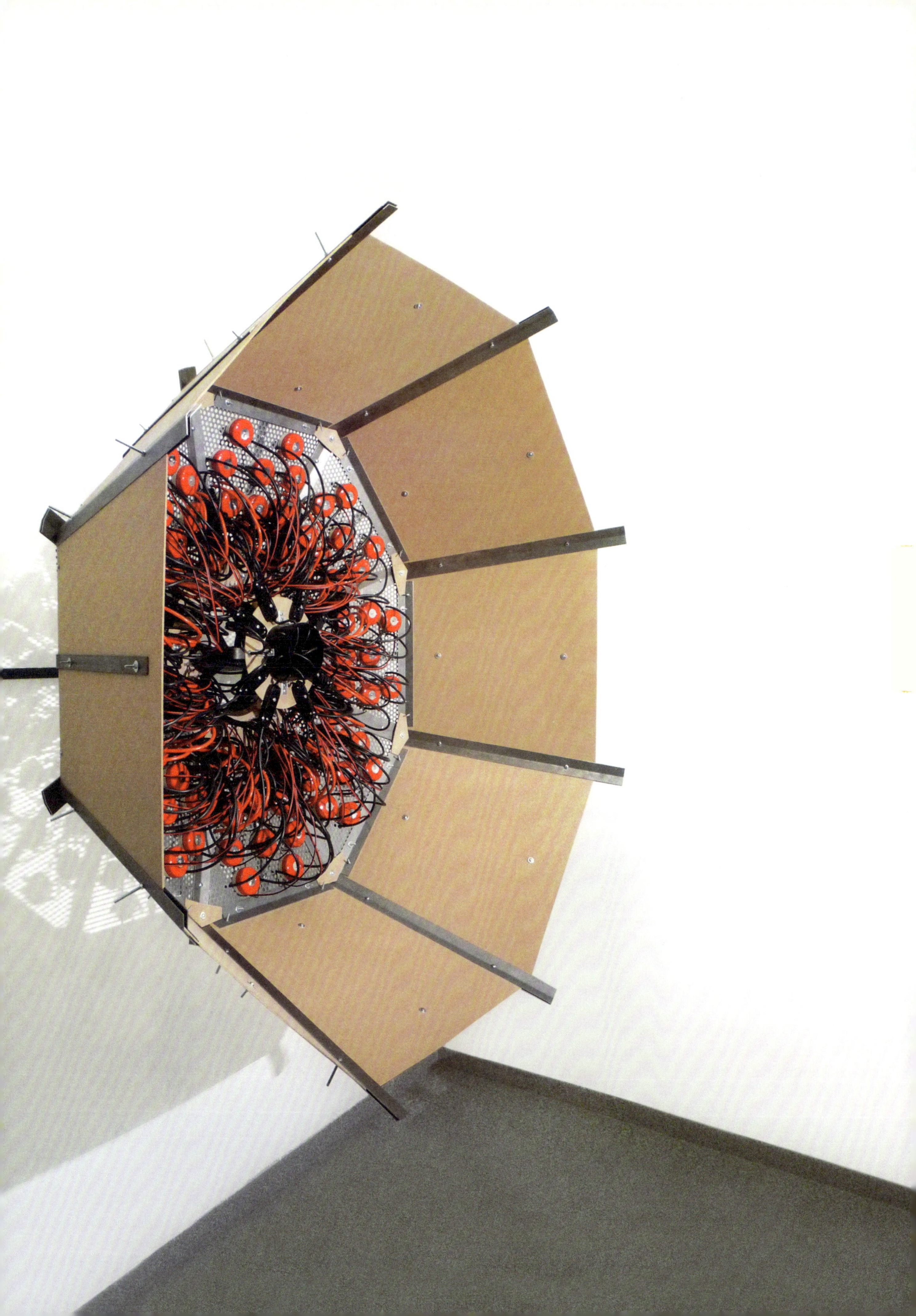

Riccardo Beretta

DONNERWETTER
2011–12

Italian artist Riccardo Beretta (Mariano Comense, 1982) creates choral works stemming from the interaction and pairing of different skills and techniques.

Donnerwetter (German for "Damn!") is a clavicytherium. The sound it produces is identical to that of a harpsichord, although it has more complex mechanics and its verticality means the sound is propagated differently. The best place to hear these sounds is right in front of it, where the player is the closest spectator. As the artist says, "the first level in this work is perceptive: the work operates as a platform for getting input and acts as a springboard. Smell and touch allow you to go beyond the sense of vision and to receive a physical impression. Another aspect is a primary reconstruction of your needs: things that we need, in order to live or to create relationships with others, and therefore to communicate."

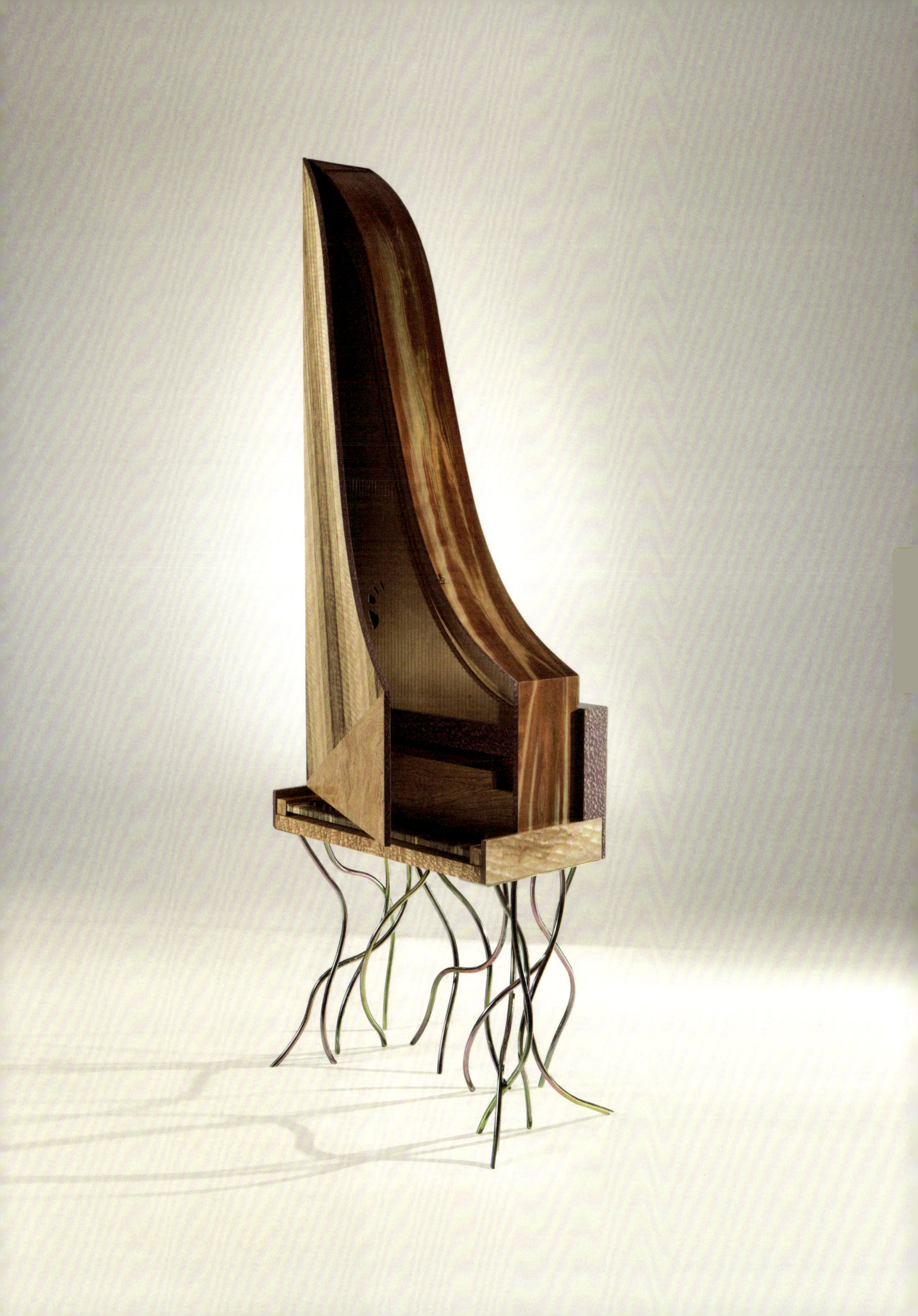

Tarek Atoui

THE METASTABLE CIRCUIT
2012

Lebanese artist Tarek Atoui (Beirut, 1980) studied contemporary and electronic music in Reims before moving to Paris. His abilities as a computer programmer characterize his sound-production techniques aimed at creating musical instruments.

The Metastable Circuit was commissioned for documenta 13 and used in performances during the exhibition in Kassel in 2012. It is composed of a set of 4 tables—4 sensor-based interrelated modules—with a computer station hidden inside one of them, and a selection of 250,000 sounds personally created and collected by the artist. Connected to the speakers, the instrument is played live by Atoui, who controls sound samples via the computer board, switches, buttons and knobs mounted on the surface of each table. Software conceived and written by Atoui is installed inside the computer station and runs the instrument in auto mode after the performance is finished.

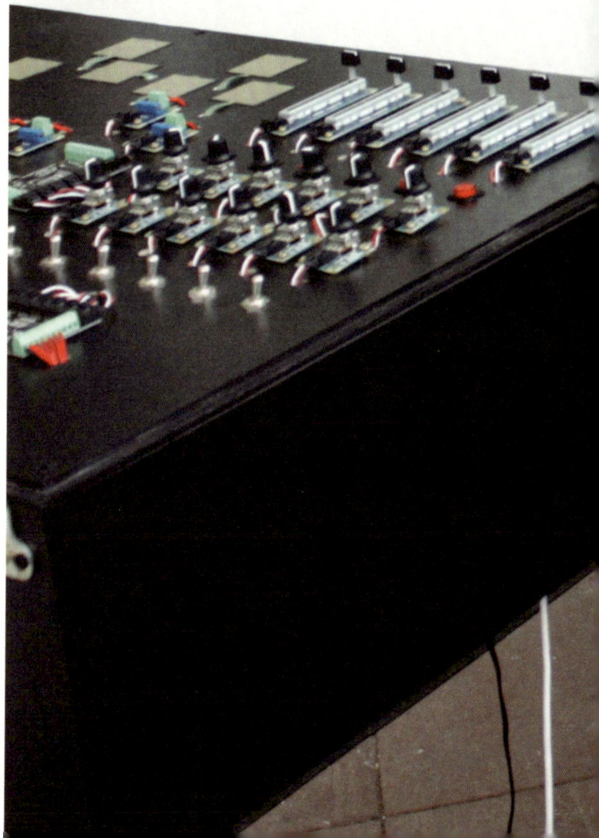

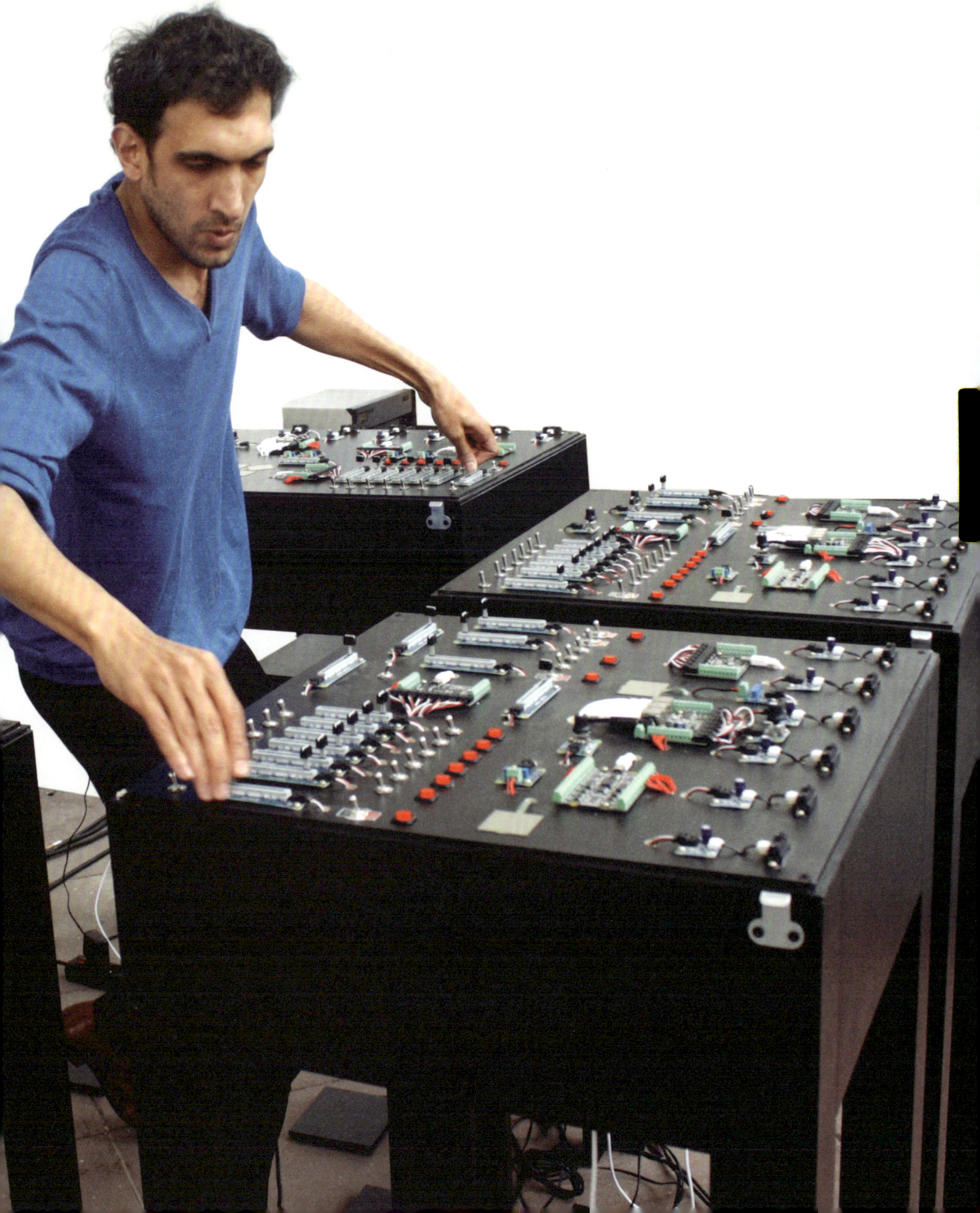

Céleste Boursier-Mougenot

U43
2012

French artist Céleste Boursier-Mougenot (Nice, 1961) started his career as a musician, working as a composer for the Pascal Rambert Theatre Company. In the 1990s Boursier-Mougenot started to create installations in places dedicated to contemporary art, such as *From Here to Ear* (1999), a sound installation at PS1, New York, made of sparrows and electronic guitars. The musical potential of everyday noises—such as the sounds of household appliances—synchronized with sophisticated electronic devices became element characteristic of his works, for example in *Keyboardchairs* (1997), hybrid objects which result from the coupling of a musical instrument and a chair.

The telephone in *U43*, mounted on a high black base, contains a wireless device synchronized with Google News, a web service that automatically aggregates information from online news sources. The telephone sounds whenever the news service is updated with an addition that matches the search query, and the piercing sound produced is that of a rotary dial telephone.

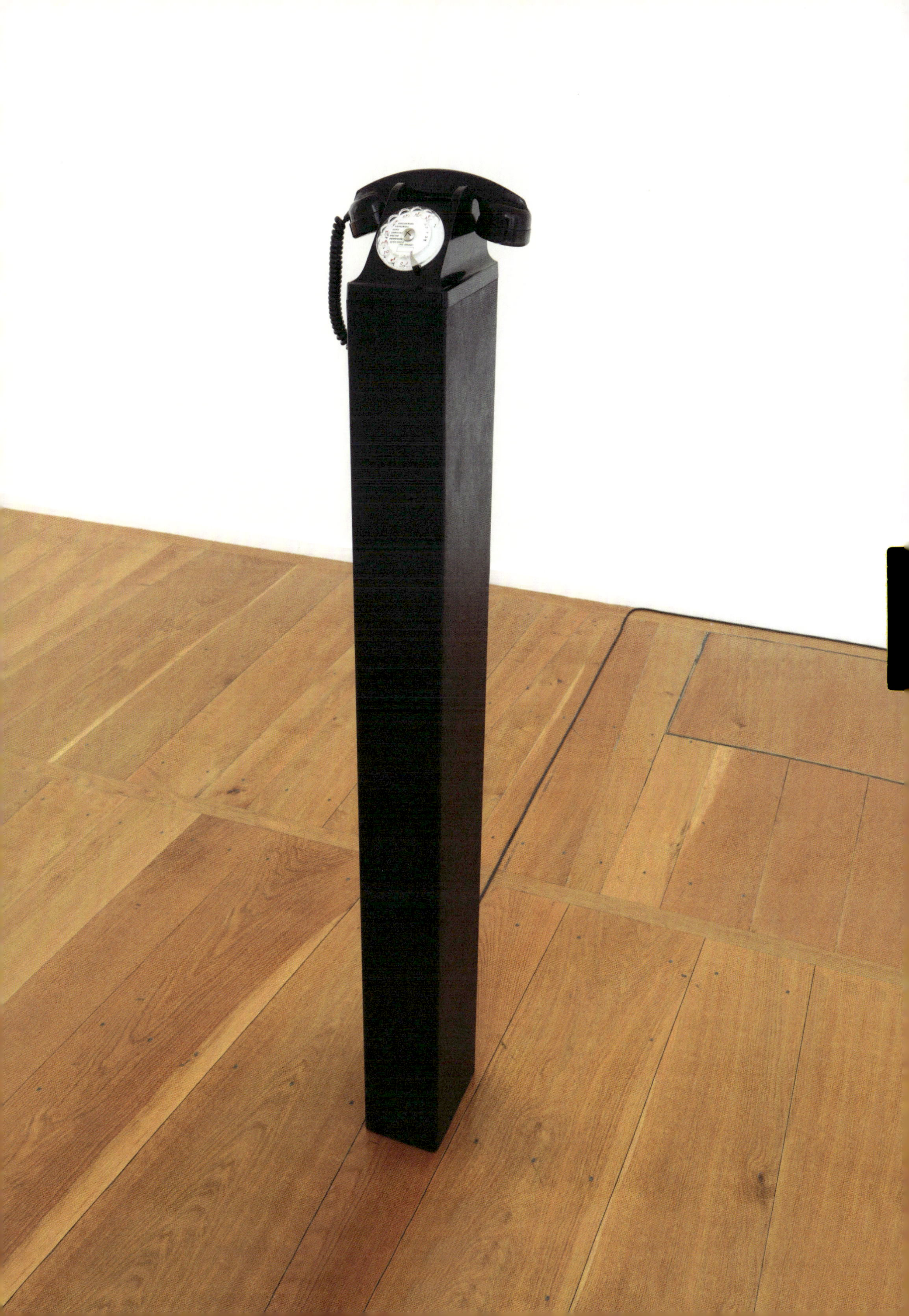

Janet Cardiff and George Bures Miller

BURLESQUE 2012

Canadian artists Janet Cardiff (Brussels, 1957) and George Bures Miller (Vegreville, 1969) formed as visual artists in Ontario and Alberta before moving to Berlin. Their first collaboration was a Super-8 film made in 1983, after which they began to include elements of sound in their sculptures. In *Burlesque* small sound speakers—composed together with a fan and little colored elements from a cabaret show—play music with words by Cardiff. Improvisation is an important part in performing and recording vocal audio tapes. The artists have investigated sound on a larger scale since the early 1990s, creating site-specific *Audiowalks*, interactive artworks where viewers are asked to touch, listen to, and often move through an environment: "It's not just an ear thing; it's a full-body thing. It affects you physically… It's also a way of bypassing the intellect and going straight to the core of our emotions, and that's really what we wanted to work with." *

* Interview with Catherine Crowston, in Catherine Crowston (ed.), *The Murder of Crows* (Ostfildern/Alberta, CA: Hatje Cantz Verlag/Art Gallery of Alberta, 2011), pp. 56, 58.

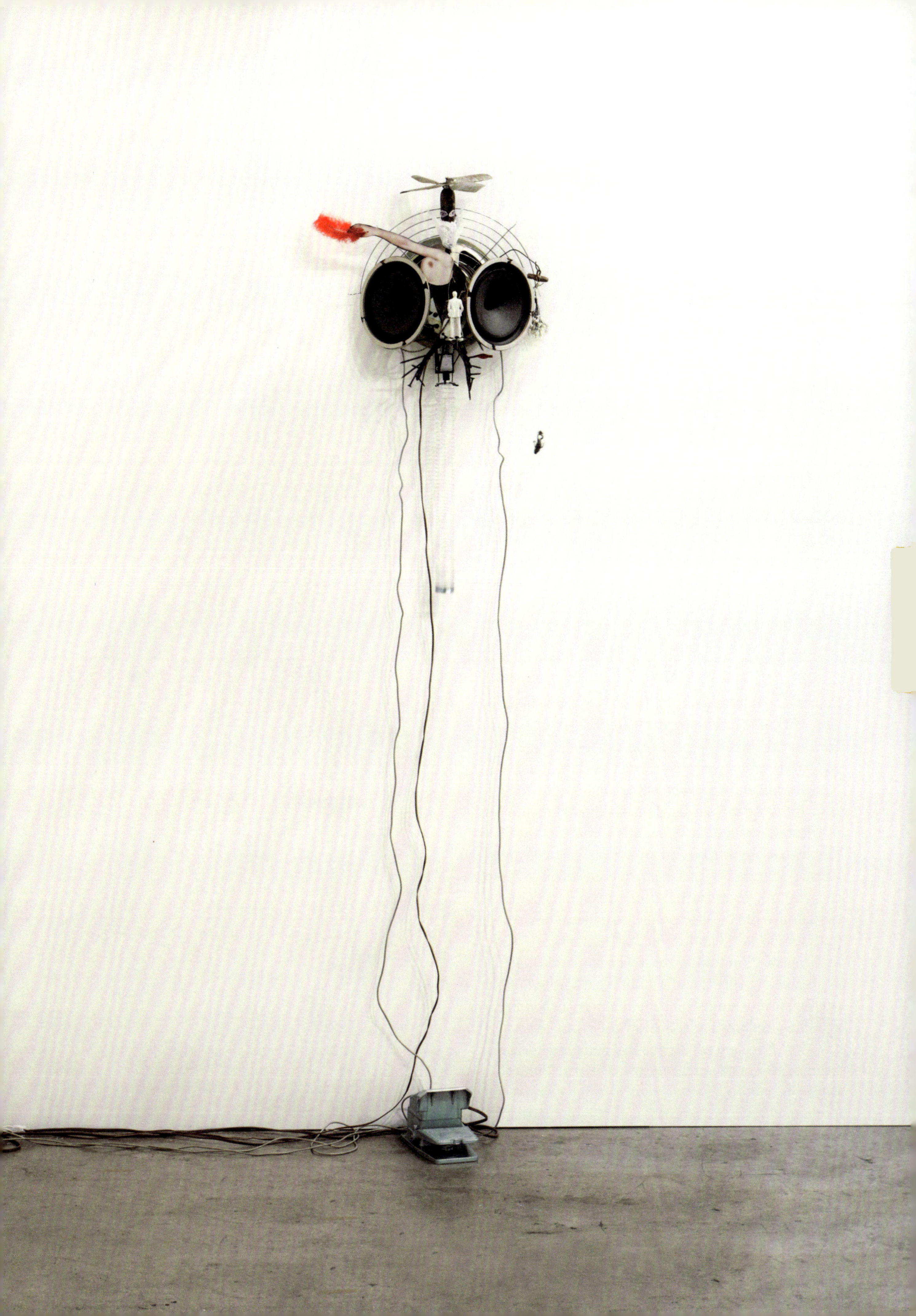

Pedro Reyes

IMAGINE (DOUBLE PSALTERY) 2012

Pedro Reyes (Mexico City, 1972) is a Mexican artist who works with sculpture, video and performance. He aims at increasing general awareness and collective action with reference to relevant social, educational and environmental themes.

Imagine is a series of musical instruments fabricated out of 6,700 dismantled weapons the Mexican Army seized from drug cartels. The artist worked together with a group of musicians to discover the auditory potential of each weapon in order to turn it into the most appropriate musical device (percussion, wind and string instruments). As the artist says, "I'm taking this piece of metal that represents our instinct for killing each other and turning it into a musical instrument, which is the most sophisticated form of communication on the planet." *

For a second series of 8 instruments derived from the remnants of weapons—*Disarm* (2013) [exh.]—, Reyes closely collaborated with a team of musicians and a media studio to create self-playing musical instruments, programmed and operated via computer.

* Video interview with the Artist in *Turning Weapons Into Instruments | Pedro Reyes 'Disarm'* (2013), youtube.com/watch?v=YwQp16D-TqQ

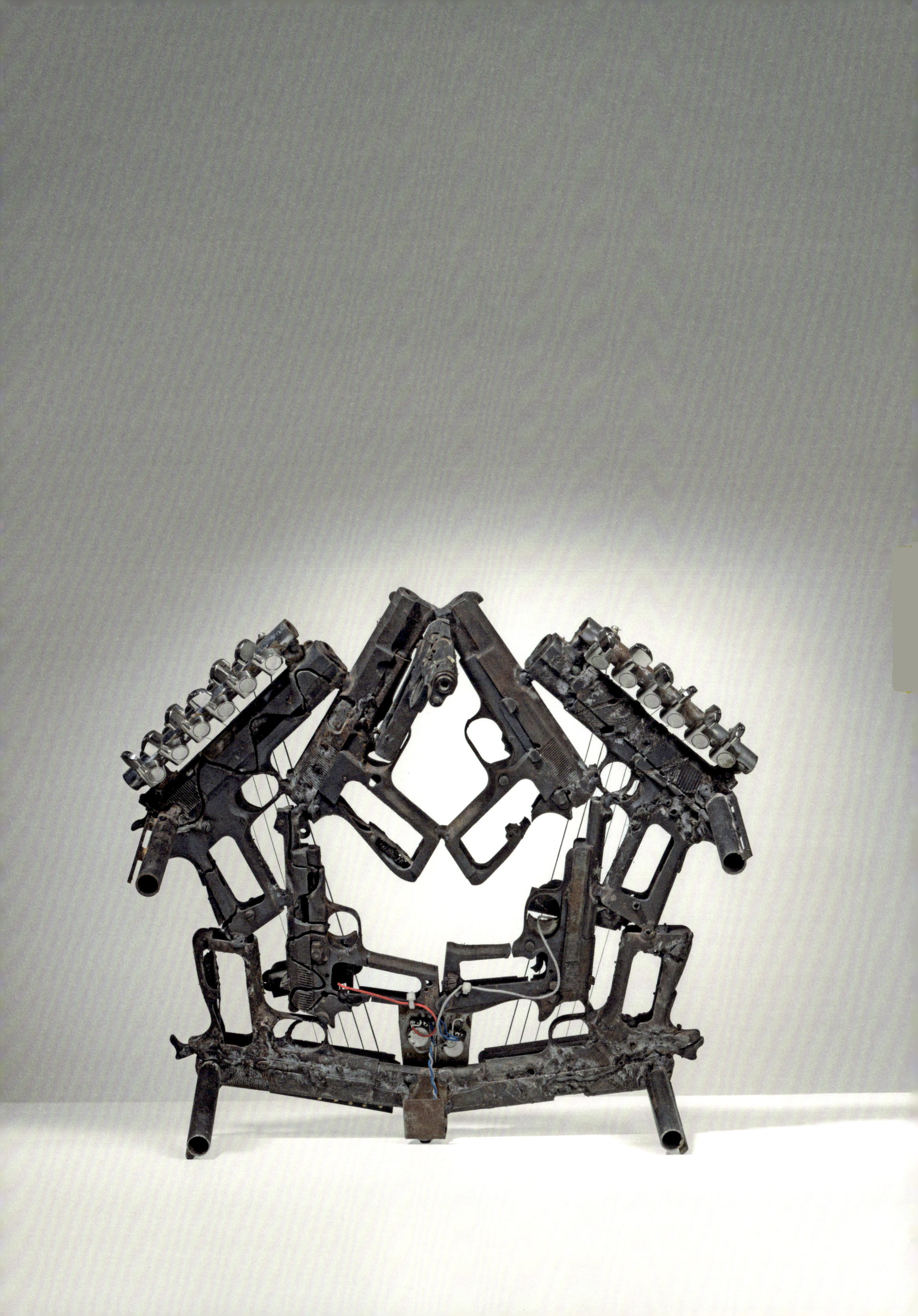

Haroon Mirza

STACK
2013

Created by British artist Haroon Mirza (London, 1977), *Stack* is the result of the assembly of electrical devices—lights, a radio, a keyboard and other materials—to generate the work's sound component. Electricity from the LEDs is amplified and transmitted in the form of impulses from a period radio. The loudspeaker underneath intercepts electricity that passes through the lights' wire and fades the sound. The alternating current powering the work interferes with the electric circuit of the keyboard, consequently producing an unpredictable series of notes that matches the rhythm of flashing lights.

The unconventional use of simple electrical materials and their combination in a creative manner are the result of a long preliminary research that forms the basis of Mirza's art.

[See also p. 400]

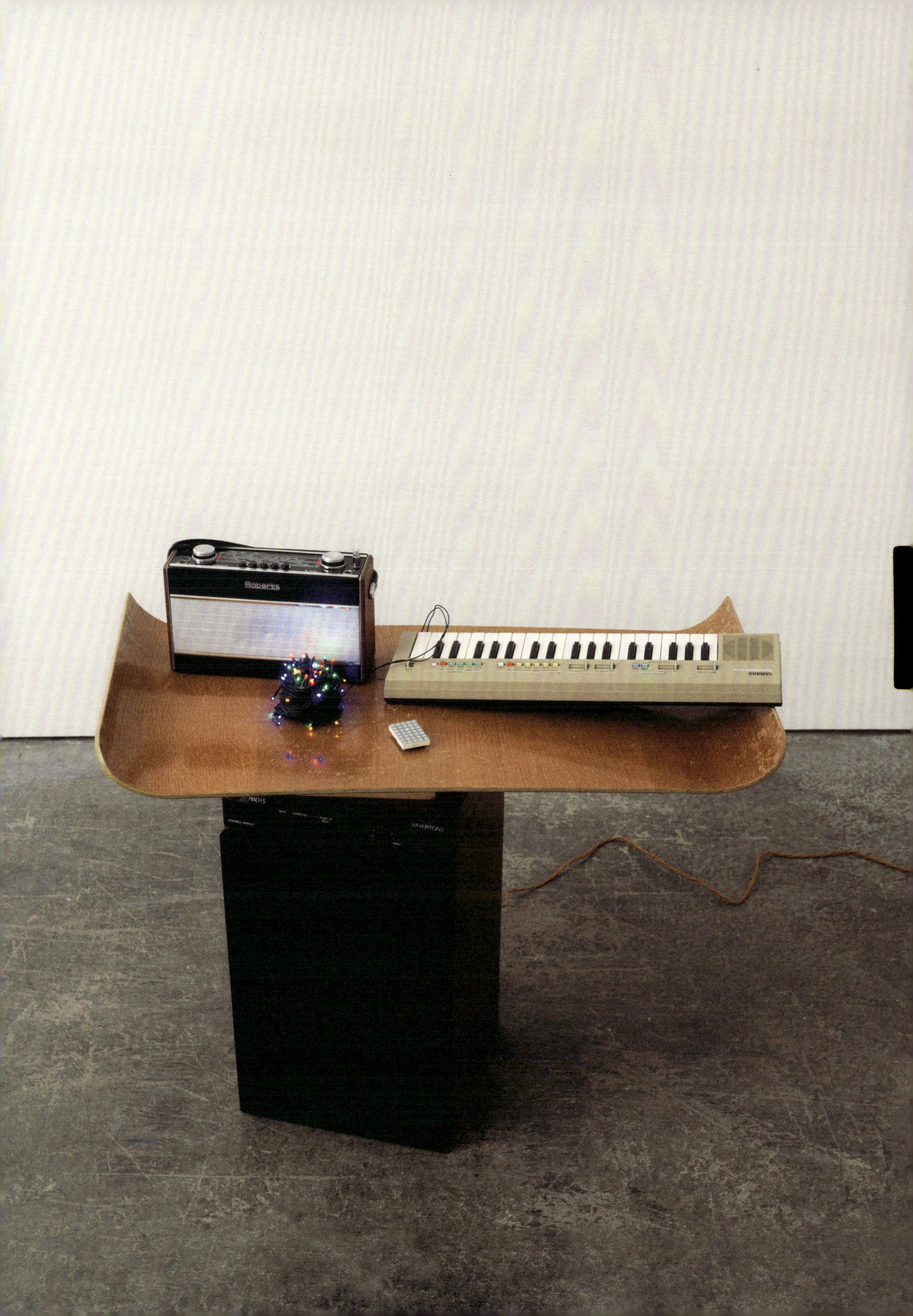
Roberts
YAMAHA
MACH 5
WHARFEDALE

Tore Honoré Bøe

CINCUENTA
2014

Since 2001 Norwegian artist Tore Honoré Bøe (Kristiansand, 1969) has produced a large number of instruments in the form of "acoustic laptops": small wooden boxes containing a large number of tiny objects. When these objects are touched, turned and pinched, the noises produced are amplified by small contact microphones, and the box itself functions as a sound resonator.

Cincuenta contains memories of the artist's former country, Norway, as well as a "global" perspective: drawings on the back of his father's wooden business card; the moose (pointedly plastic); a Danish coin; a bottle of wine hinting at Italy and his first travels there; bark from a birch tree; a runic —yet personal—symbol carved in stone; a chess piece recalling the game that lies behind all of artist's activities; a piece of electronic hardware the artist gave symbolically to a friend. The artist himself wonders: "Can you visualize the sound they make? Do the memories these non-sounding objects trigger affect the music I perform in any way? Will they affect the music you make, now that you know some of his stories? Can, at any point, sound / art / design / framing become entirely neutral, and is that something to strive for?"

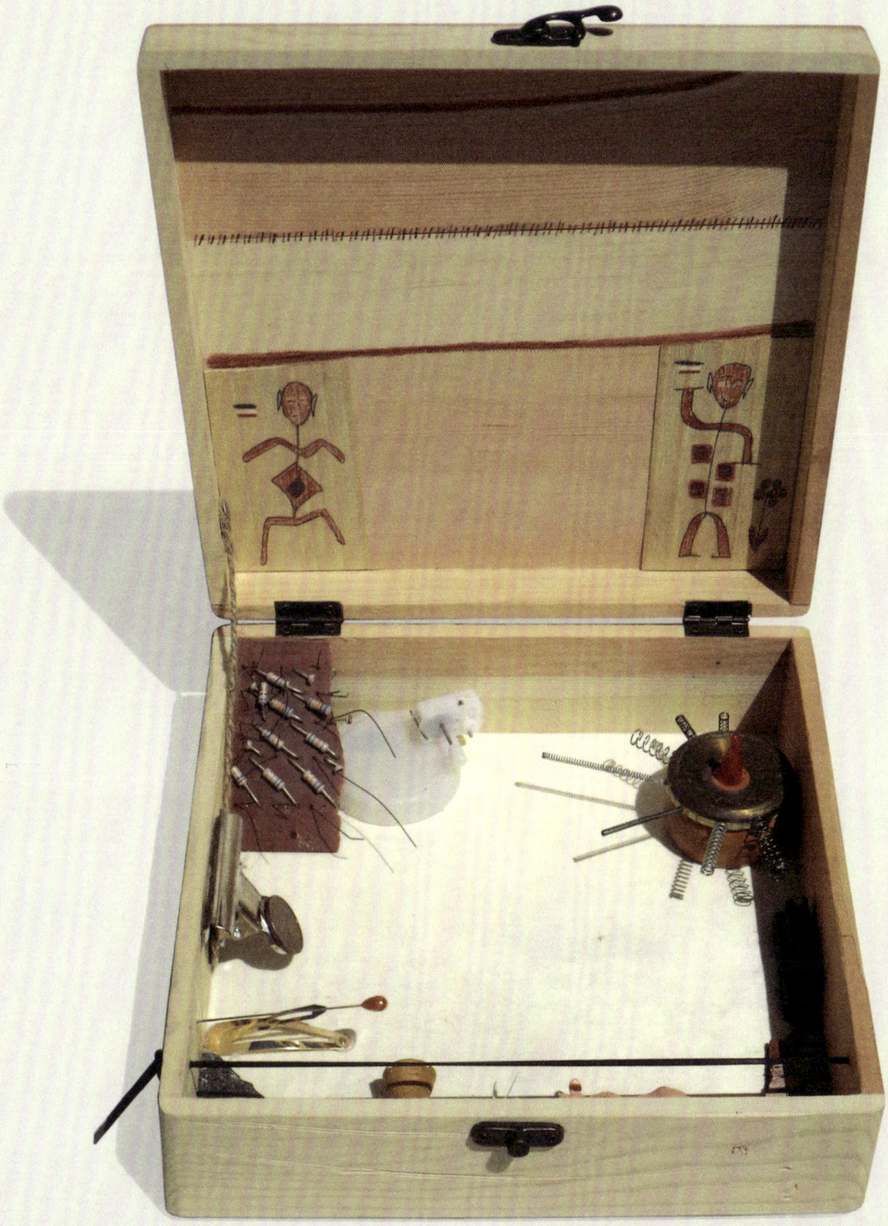

APPENDIX

LIST OF EXHIBITED AND ILLUSTRATED WORKS

NOTE TO THE READER

This section provides a complete chronological checklist of the works presented in the "Art or Sound" exhibition.
Each cited work includes:
— author's/maker's name when known
— execution date or production year; when in parenthesis, the year refers to a reconstruction or a replica
— media
— dimensions (height × width × depth)
— collection

All information is based on indications received by the lenders and confirmed wherever possible in documentation provided by artists, archives, museums, foundations, galleries and bibliographical references.
When illustrated in the book, works are linked to the related page, otherwise they are mentioned as exhibited, not illustrated (exh., not ill.).

Bartolomeo Veneto
WOMAN PLAYING A LUTE
1520
oil on panel
65 × 50 cm
Pinacoteca di Brera, Milan
p. 41

Nicola Giolfino
ALLEGORY OF MUSIC
1521–30
detached fresco
80 × 100 cm
Museo di Castelvecchio, Verona
p. 43

Unknown
LYRA-CISTER
16th–17th century
wood, paint, gilding, metal string and frets
160 × 36 × 4.5 cm
Collection of Historic Musical Instruments, Kunsthistorisches Museum, Vienna
p. 45

Unknown
TENOR CORNETT IN THE FORM OF A SERPENT WITH A DRAGON HEAD
17th century
paint, leather, wood, horn, ivory, gilding
11.5 × 68 × 36 cm
Musée de la musique – Cité de la musique, Paris
p. 47

Jean Dubois au Puy
CHIMING CLOCK WITH IRON CASE
beginning of 17th century
iron, paint
65 × 23 × 27 cm
Beyer Clock and Watch Museum, Zurich
p. 49

Michele Antonio Grandi
GUITAR
c. 1680
marble with marble-paste inlays
104 × 29 × 10 cm
Galleria Estense, Modena
p. 51

Giovan Battista Cassarini
CONTRALTO VIOLA
c. 1687
marble with marble-paste inlays
60.5 × 19.5 × 6 cm
Galleria Estense, Modena
p. 53

Onofrio Guarracino
RECTANGULAR SPINET
1692
spruce, cypress, walnut, tempera
20 × 148.5 × 43 cm
Museo Nazionale degli Strumenti Musicali, Rome
p. 55

Atelier Pons (attr.)
LYRE-GUITAR
late 18th century
wood, metal, gilding
87 × 36 × 11.5 cm
Musée de la musique – Cité de la musique, Paris
p. 57

Jean-Honoré Davrainville
BARREL ORGAN WITH FLUTES
c. 1770
metal, leather, paper, cardboard, wood, gut, gilding, thermometer
41.5 × 32.5 × 12.5 cm
Musée des Instruments de Musique, 4e Dép. des Musées royaux d'Art et d'Histoire, Brussels
p. 59

Pierre Jaquet-Droz
CLOCK WITH CARILLON AND AUTOMATON
1780
wood, bronze, gilding
40 × 33 × 62 cm
Museum für Musikautomaten, Seewen
p. 61

Pierre Jaquet-Droz and Henri Maillardet
SINGING BIRD CAGE WITH CLOCK
c. 1780
gilt bronze, enamel, feathers
55 × 29 × 29 cm
Beyer Clock and Watch Museum, Zurich
p. 63

Pierre Jaquet-Droz
SINGING BIRD CAGE WITH CLOCK
c. 1785
gilt bronze, enamel, painting, feathers, tin
51.5 × 30.5 × 30.5 cm
Private Collection
p. 64

Egor Kuznetsov
CARRIAGE WITH BARREL ORGAN
1785–1801
metal, wood, glass, velvet, painting
132 × 240 × 102 cm
The State Hermitage Museum, St. Petersburg
pp. 66–67

Pierre Jaquet-Droz
SINGING BIRD CAGE WITH CLOCK
1790s
gilt bronze, metal, enamel, feathers
75 × 35 × 35 cm
The State Hermitage Museum, St. Petersburg
p. 65

Unknown
CONCERTION
19th century
wood, metal, vellum
38.5 cm (height), 35 cm (width) closed
Museo Nazionale degli Strumenti Musicali, Rome
p. 69

Jean Arnold Antoine Tuerlinckx
BUCCIN
c. 1800
brass, paint

119 × 54.5 × 12.5 cm
Musée des Instruments de Musique,
4e Dép. des Musées royaux d'Art
et d'Histoire, Brussels
p. 71

Jean Arnold Antoine Tuerlinckx
BUCCIN
c. 1800
brass, paint
120.5 × 34 × 15 cm
Musée des Instruments de Musique,
4e Dép. des Musées royaux d'Art
et d'Histoire, Brussels
exh., not ill.

Unknown
CLOCK WITH FLUTES
c. 1800
wood
82 × 44 × 29 cm
Museum für Musikautomaten,
Sammlung Dr. h.c. H.
Weiss-Stauffacher, Seewen
p. 75

Unknown
SERPENT
c. 1800
wood, leather, brass
83.5 × 49.5 × 25.5 cm
Musée des Instruments de Musique,
4e Dép. des Musées royaux d'Art
et d'Histoire, Brussels
p. 77

François Antoine Sautermeister
BUCCIN
1809–30
brass, paint
103 × 60 × 18 cm
Musée des Instruments de Musique,
4e Dép. des Musées royaux d'Art
et d'Histoire, Brussels
p. 72

François Antoine Sautermeister
RUSSIAN BASSOON
1809–30
wood, leather, brass, paint
106 × 23 × 30 cm
Musée des Instruments de Musique,
4e Dép. des Musées royaux d'Art
et d'Histoire, Brussels
p. 103

Courtois neveu aîné
BUCCIN
1816–37
brass, paint
106.5 × 56 × 22 cm
Musée des Instruments de Musique,
4e Dép. des Musées royaux d'Art
et d'Histoire, Brussels
exh., not ill.

B. Schott's Söhne
BUCCIN
c. 1820
brass, paint
106 × 55.5 × 20.5 cm
Musée des Instruments de Musique,
4e Dép. des Musées royaux d'Art
et d'Histoire, Brussels
p. 73

Unknown
NAIL VIOLIN
early 19th century
wood, metal nails and strings
44.5 cm (height), 40.5 cm (diam.)
Musée de la musique –
Cité de la musique, Paris
p. 101

Ludwig Embach
SERPENT WITH KEYS
1820–44
wood, leather, brass, paint
63 × 56 × 42 cm
Musée des Instruments de Musique,
4e Dép. des Musées royaux d'Art
et d'Histoire, Brussels
p. 78

André Antoine Schmittschneider
HORN-SHAPED TRUMPET
c. 1823
copper
43.5 × 33.5 × 19 cm
Musée des Instruments de Musique,
4e Dép. des Musées royaux d'Art
et d'Histoire, Brussels
p. 105

Joseph Dupré
MILITARY SERPENT
c. 1825
wood, leather, metal
76.5 × 29 × 24 cm
Musée des Instruments de Musique,
4e Dép. des Musées royaux d'Art
et d'Histoire, Brussels
p. 79

Davrainville fils
BARREL ORGAN
WITH TRUMPETS
1828
brass, mahogany
81 × 61 × 40 cm
Musée des Instruments de Musique,
4e Dép. des Musées royaux d'Art
et d'Histoire, Brussels
p. 107

Unknown
DEMILUNE TRUMPET
first half of 19th century
brass
46.5 × 18.5 × 11.5 cm
Musée des Instruments de Musique,
4e Dép. des Musées royaux d'Art
et d'Histoire, Brussels
p. 109

Adolphe Sax
NATURAL TRUMPET
1866–84
brass
38 × 23.5 × 24 cm
Musée de la musique –
Cité de la musique, Paris
p. 111

Adolphe Sax
SIX-VALVE TRUMPET
c. 1875
brass
51 × 52 × 19.5 cm
Musée des Instruments de Musique,
4e Dép. des Musées royaux d'Art
et d'Histoire, Brussels
p. 113

Frédéric Kastner
PYROPHONE (GAS ORGAN)
c. 1876
wood, metal
225 × 130 × 110 cm
Musée historique de la ville de
Strasbourg, Strasbourg
p. 115

Charles-Abraham Bruguier
SINGING BIRD BOX WITH
BUILT-IN CLOCK
c. 1880
gilt bronze, enamel, feathers
4.5 × 9.5 × 6 cm
Beyer Clock and Watch Museum, Zurich
p. 117

Carlo Bugatti
TAMBURICORDE
c. 1885
wood, copper, vellum
instrument: 182 × 55 × 14 cm
stool: 45 × 94 × 44 cm
Museo Nazionale degli
Strumenti Musicali, Rome
p. 119

Roullet-Decamps
SNAKE CHARMER
AUTOMATON
1890
papier mâché with gutta-percha and
ivoirine, carpet, wood, metal
90 cm (height)
Nouveau Musée National
de Monaco, Monaco
p. 121

Paillard Vaucher & Co.
CYLINDER MUSICAL BOX
c. 1890
metal, wood
24.5 × 51 × 26 cm
Speelklok Museum, Utrecht
p. 123

Cocchi, Bacigalupo & Graffigna
BARREL ORGAN
1893
wood, fabric, metal, gilding
215 × 260 × 175 cm (cart included)
Speelklok Museum, Utrecht
p. 125

Victor-Charles Mahillon
EXPERIMENTAL TRUMPET
before 1900
copper
136 × 11.5 × 11.5 cm
Musée des Instruments de Musique,
4e Dép. des Musées royaux d'Art
et d'Histoire, Brussels
p. 127

François Millereau
TROMPE DE LORRAINE
c. 1900
copper
37.5 × 27 × 16.5 cm
Musée des Instruments de Musique,

4e Dép. des Musées royaux d'Art
et d'Histoire, Brussels
p. 129

Kalliope Musikwerke
MUSICAL DISC BOX
c. 1900
wood, glass, metal
114 × 80 × 45.5 cm
Speelklok Museum, Utrecht
p. 131

Gebrüder Wellershaus
FAIRGROUND ORGAN
early 20th century
carved painted wood, automata,
pipe organ, cymbal, bass drum,
snare drum, Gavioli system
235 × 250 × 130 cm
Fondazione Cassa di Risparmio, Bologna
p. 133

Arthur Howson
PHONOFIDDLE
1904–20
wood, brass
83 × 23 × 32.5 cm
Museum of Musical Instruments,
4th Department of the Royal Museums
of Art and History, Brussels
p. 135

Alexander Scriabin
and Alexander Moser
COLOR PIANO FOR THE PART
OF LUCE IN SCRIABIN'S
PROMETHEUS
1911 (2014)
exhibition copy of the model of the color
keyboard made for *Prometheus*
wood, light bulbs, plastic, electric wire
66 × 6 cm (board), 71 cm (diam.)
Scriabin State Museum, Moscow
p. 137

Luigi Russolo
INTONARUMORI
1913–14 (2002–09)
reconstructions by Pietro Verardo
p. 161

GRACIDATORE MEDIO (2002)
wood, steel, leather, cardboard
70 × 59 × 39 cm + trumpet
Collection of Pietro Verardo, Venice

ULULATORE ACUTO (2002)
wood, steel, leather, cardboard
60 × 40 × 36 cm + trumpet
Collection of Pietro Verardo, Venice

ULULATORE BASSO (2002)
wood, steel, leather, cardboard
140 × 58 × 45 cm + trumpet
Collection of Pietro Verardo, Venice

ULULATORE MEDIO (2002)
wood, steel, leather, cardboard
85 × 49 × 45 cm + trumpet
Collection of Pietro Verardo, Venice

FRUSCIATORE,
STROPICCIATORE (2006)
wood, steel, leather, cardboard
56 × 110 × 47 cm + trumpet
MART – Museo di arte
moderna e contemporanea di
Trento e Rovereto, Rovereto

RONZATORE (2006)
wood, steel, leather, cardboard
56 × 110 × 47 cm + trumpet
MART – Museo di arte
moderna e contemporanea di
Trento e Rovereto, Rovereto

ROMBATORE, FRUSCIATORE,
SIBILLATORE (2006–09)
wood, steel, leather, cardboard
110 × 52 × 42 cm + trumpet
Collection of Pietro Verardo, Venice

CREPITATORE ACUTO (2008)
wood, steel, leather, cardboard
62 × 48 × 42 cm + trumpet
Collection of Pietro Verardo, Venice

GRACIDATORE ACUTO (2008)
wood, steel, leather, cardboard
62 × 48 × 42 cm + trumpet
Collection of Pietro Verardo, Venice

GRACIDATORE BASSO (2008)
wood, steel, leather, cardboard
140 × 55 × 45 cm + trumpet
Collection of Pietro Verardo, Venice

RONZATORE SCOPPIATORE (2008)
wood, steel, leather, cardboard
70 × 59 × 39 cm + trumpet
Collection of Pietro Verardo, Venice

Giacomo Balla
CIAC CIAC
1914
painted wood, metal
60.5 cm (length), 10 cm (height)
Museo Nazionale degli
Strumenti Musicali, Rome
p. 163

Marcel Duchamp
WITH HIDDEN NOISE (À BRUIT SECRET)
1916 (1963)
ball of twine between 2 brass plates, joined
by 4 long screws, containing unknown object
11.5 (height), 13 cm (diam.)
Moderna Museet, Stockholm
p. 165

Theo van Doesburg
COMPOSITION IN GRAY (RAG TIME)
1919
oil on canvas
96.5 × 59 cm
Peggy Guggenheim Collection, Venice;
The Solomon R. Guggenheim Foundation,
New York
p. 167

Amelotti
ORCHESTRION ACCORDEO JAZZ
c. 1920
accordion, cymbal, bass drum,
snare drum, castanets, wooden cabinet,
pneumatic system, paper rolls
224 × 116 × 58 cm
Fondazione Cassa di Risparmio, Bologna
p. 169

Wladimir Baranoff-Rossiné
OPTOPHONIC PIANO
1920–23 (1971)
reconstruction by Jean Schifrine
wooden case, stereo cassette-player and
amplifier, projection device, screen
239 × 120 × 164 cm
Centre Pompidou, Paris. Gift of
Mme Wladimir Baranoff-Rossiné
and her son Eugène, 1972
p. 171

Man Ray
INDESTRUCTIBLE OBJECT
1923 (1965)
edition MAT/Galerie der Spiegel, Cologne
metronome, photograph
22 × 11 × 11 cm
Private Collection, Courtesy
Fondazione Marconi, Milan
p. 173

Man Ray
LE VIOLON D'INGRES
1924 (1980)
print of altered photograph
30 × 24 cm
Private Collection, Courtesy
Fondazione Marconi, Milan
p. 175

Man Ray
EMAK BAKIA
1926 (1970)
edition Studio Marconi
silver cello neck, horse-tail hairs
46.5 × 14 × 5 cm
Private Collection, Courtesy
Fondazione Marconi, Milan
p. 177

Henrik Neugeboren
SCULPTURAL REPRESENTATION
OF BARS 52-55 OF E-FLAT
MINOR FUGUE BY J.S. BACH.
PROPOSAL FOR A BACH MONUMENT
1928 (1966)
stainless steel on painted block board
76 × 80 × 80 cm
Bauhaus-Archiv, Berlin
p. 179

Georg Evans & Co.
STROH VIOLIN
1930
mahogany, maple, ebony, aluminum
59 × 30 × 19 cm
Musée des Instruments de Musique,
4e Dép. des Musées royaux d'Art
et d'Histoire, Brussels
p. 181

Columbia Graphophone Co.
DOUBLE HORN DISC GRAMOPHONE
c. 1930
wood, metal, enamel, spring
driver, hand crank system
40 × 21 × 40 cm
Fondazione Cassa di Risparmio,
Bologna
exh., not ill.

His Master's Voice
DOUBLE HORN DISC GRAMOPHONE
c. 1930

wood, metal, enamel, spring
driver, hand crank system
42 × 20 × 42 cm
Fondazione Cassa di Risparmio, Bologna
p. 183

Paillard
DISC GRAMOPHONE
c. 1930
wood, metal, enamel, spring
driver, hand crank system
50 × 24 × 37 cm
Fondazione Cassa di Risparmio, Bologna
exh., not ill.

Maurice Henry
HOMMAGE À PAGANINI
1936 (1968)
replica edition by Sergio Tosi, Milan
cloth, straw, wood
25 × 50 × 10 cm
Archivio Maurice Henry –
Collezione Peruz, Milan
p. 185

Alexander Calder
UNTITLED
c. 1940
sheet metal, iron, wire, paint
114 × 68.5 × 48 cm
Calder Foundation, New York
p. 187

Salvador Dalí
MÉTRONOME

1944
metronome, chalk on paper
22 × 16 × 12 cm
Private Collection. Courtesy of
Het Noordbrabants Museum,
Hertogenbosch
p. 189

Bernard Baschet
and François Baschet
FOLDING AND INFLATABLE
GUITAR
1952
plastic, metal, wood, paint
100 × 40 × 24 cm
Musée de la musique –
Cité de la musique, Paris
p. 191

Bernard Baschet
and François Baschet
CRISTAL
1952 (1980)
glass, metal, stainless steel, fiberglass
140 × 173 × 67 cm
Musée de la musique –
Cité de la musique, Paris
p. 193

John Cage
VARIATIONS I
1958
music score on paper and
transparent sheets
9 pages, 21.5 × 28 cm each
Music Division, The New York Public
Library for the Performing Arts,
Astor, Lenox and Tilden
Foundations, New York
p. 215

John Cage
WATER WALK
1959
music score on paper sheets
7 pages, 21.5 × 28 cm each
Music Division, The New York Public
Library for the Performing Arts, Astor,
Lenox and Tilden Foundations
p. 217

Arman
NBC RAGE
1961
broken bass violin on wood panel
200.5 × 157 × 25 cm
The Arman Marital Trust,
Corice Arman Trustee
p. 219

Robert Morris
BOX WITH THE SOUND
OF ITS OWN MAKING
1961
exhibition copy
walnut box, speaker, CD
25 × 25 × 25 cm
Collection of the Seattle Art Museum,
gift of Mr. and Mrs. Bagley Wright.
Courtesy of Castelli Gallery, New York
p. 221

Nam June Paik
URMUSIK
1961
wooden box, wire, various
materials and turnable tin can
52 × 92 × 68 cm
Moderner Kunst Stiftung Ludwig
Wien, Vienna; former Hahn Collection,
Cologne, acquired in 1978
p. 223

Edward Kienholz
THE CARNIVORE
1962
floor tiles, wood, metal, fur, glass,
adhesive tape, resin, light bulb,
electrical wiring
97 × 130 × 49 cm
Private Collection.
Courtesy Hauser & Wirth
p. 225

Jean Tinguely
RADIO WNYR NR. 15
1962
plexiglass, radio parts, cable,
red light bulb, electric motor
63.5 × 63.5 × 19.5 cm
Museum Tinguely, Basel
p. 227

Nam June Paik
LISTENING TO MUSIC
THROUGH THE MOUTH
1962–63
gramophone cabinet with old
running device, records
106 × 49 × 43 cm
Moderner Kunst Stiftung Ludwig
Wien, Vienna; former Hahn Collection,
Cologne, acquired in 1978
p. 229

Robert Rauschenberg
ORACLE
1962–65
galvanized sheet metal assembled
objects, water, audio system
5 elements
236 × 450 × 400 cm (overall)
Centre Georges Pompidou, Paris
Musée national d'art moderne. Centre
de création industrielle. Gift of M. and
Mme Pierre Schlumberger, 1976
pp. 230–31

Milan Knížák
DESTROYED MUSIC
1963 (1975)
altered vinyl record
25 cm (diam.)
The Museum of Modern Art, New York.
The Gilbert and Lila Silverman
Fluxus Collection Gift, 2008
p. 233

Milan Knížák
DESTROYED MUSIC
1963 (1980)
altered record player
41 × 33 × 17 cm (closed)
41 × 41 × 42 cm (open)
The Museum of Modern Art, New York.
The Gilbert and Lila Silverman
Fluxus Collection Gift, 2008
exh., not ill.

Joe Jones
BIRD CAGE
1964
bird cage, small guitar, plastic
birds, bells, mirror, electric motor,
propeller with wire, paint
36 × 57 × 23 cm
Moderner Kunst Stiftung Ludwig Wien,
Vienna; former Hahn Collection, Cologne,
acquired in 1978
p. 235

Günther Uecker
PIANO
1964
piano, nails, paint
138 × 150 × 80 cm
Collection Würth, Germany
p. 237

Tom Wesselmann
LITTLE STILL LIFE #18
1964
acrylic, collage and assemblage
on board with working radio
12.5 × 15 × 10.5 cm
Collection of Claire Wesselmann
p. 239

Richard Artschwager
PIANO
1965
Formica laminate on wood
81 × 122 × 48 cm
Collection of Barbara Bertozzi Castelli
p. 241

Roman Opalka
OPALKA 1965 / 1 – ∞
DÉTAIL 2450737 – 2473067
1965

acrylic on canvas, sound
196 × 135 cm
Marie-Madeleine G. Opalka and
Galleria Michela Rizzo, Venice
p. 243

Stephan von Huene
KALEIDOPHONIC DOG
1967
wood, leather, pneumatic system
190.5 × 61 × 56 cm
Los Angeles County Museum,
Gift of the Kleiner Foundation
p. 245

Stephan von Huene
WASHBOARD BAND
1967
wood, leather, washboard, cymbal,
gong, cowbell, harmonium, metal parts,
circuit, mechanics, pneumatics
2 elements
220 × 77 × 40 cm (overall)
Courtesy Private Collection, Düsseldorf
p. 247

Bruce Nauman
CONCRETE TAPE RECORDER PIECE
1968
tape recorder, plastic bag, concrete block
30.5 × 60.5 × 60.5 cm
Friedrich Christian Flick Collection
im Hamburger Bahnhof, Berlin
p. 249

Michelangelo Pistoletto
LE TROMBE DEL GIUDIZIO
1968
aluminum
3 elements: 200 cm (height),
100 cm (diam.) each
Cittadellarte – Fondazione Pistoletto,
Biella
pp. 250–51

Wolf Vostell
RADAR ALARM
1969
racing bicycle, mesh bag, sirens, TV
160 × 120 cm
The Wolf Vostell Estate
p. 253

George Maciunas
PIANO PIECE
1970
piano, nails
130 × 140 × 65 cm
Vehbi Koç Foundation Contemporary
Art Collection, Istanbul
p. 271

Akio Suzuki
ANALAPOS
1970
metal cylinders, steel coil
250 (height), 10 cm (diam.)
Courtesy of the Artist
p. 273

George Maciunas
JERALD ORDOVER. SPELL YOUR
NAME WITH THESE OBJECTS –
GREETINGS FROM GEORGE MACIUNAS
c. 1970
leatherette hinged box with velvet
lining containing: juniper, plastic egg,
rivet, asafetida, lens, domino, oregano,
reamer, dice, oats, vanilla, epoxy, rice
3 × 36 × 16.8 cm (closed)
Collection of the Zimmerli Art Museum
at Rutgers, New Brunswick
exh., not ill.

William T. Wiley
CANJO
1971
wooden stick, tin cans, rope, glass
67 × 25 × 11 cm
Collection of Sarah Hoe Sterling
p. 275

Jannis Kounellis
SENZA TITOLO
(DA INVENTARE SUL POSTO)
1972
oil on canvas, score of *La Pulcinella*
by Igor Stravinsky to be performed
by a ballerina and a violinist
270 × 300 cm
Rheingold Collection
pp. 278–79

Paul De Marinis
THE PYGMY GAMELAN
1973
electronic circuits, metal, glass,
wiring
2.5 × 18 × 10 cm
Courtesy of the Artist
p. 281

Dennis Oppenheim
THEME FOR A MAJOR HIT
1974
motor driven marionette, tape
recording, circular stage, spotlight
marionette: 76 × 25.5 cm
Collection Dennis Oppenheim Estate.
Courtesy Galleria Fumagalli, Milan
p. 283

Takis
SIGNALS
1974
iron
320 × 40 × 40 cm
Donation of Alexander Iolas-Permanent
collection of the Macedonian Museum of
Contemporary Art
p. 285

Yoshimasa Wada
EARTH HORN
1974
iron pipes, joints
145 × 43 × 37 cm
Fondazione Bonotto
p. 287

Bernhard Leitner
TON-LIEGE
1975 (2006)
wooden chair, 2 loudspeakers,
player with amplifier
60 × 210 × 68 cm
Collection of the Artist
p. 289

Joe Jones
UNTITLED
(MECHANICAL MANDOLIN)
1976
mandolin, electric motor, propeller with wire
71 × 75 × 20 cm
Collezione Peruz, Milan
exh., not ill.

William Anastasi
THE WORLD'S GREATEST MUSIC
1977
3 children's record players,
3 records (78-rpm)
15 × 96.5 × 25.5 cm
Collection of Alanna Heiss
p. 291

George Maciunas
JASMINKA BAN. SPELL YOUR NAME
WITH THESE OBJECTS
1977
plastic box with label containing objects
2 × 9.5 × 7 cm (closed)
Fondazione Bonotto
p. 293

Takako Saito
SOUND CHESS
c. 1977
printed board, plastic box-pieces
containing various objects, typewritten
set with pencil addictions
board: 50 × 51 cm; instruction sheet:
14 × 20 cm; 32 pieces: 4 × 3.5 × 3.5 cm each
The Museum of Modern Art, New York.
The Gilbert and Lila Silverman
Fluxus Collection Gift, 2008
p. 295

Laurie Anderson
HANDPHONE TABLE
1978
plywood table, 2 stools, photograph
variable dimensions;
table: 79.5 × 152.5 × 91 cm
Musée d'art contemporain de Lyon, Lyon
p. 297

Ken Butler
AXE VIOLIN
1978
axe, violin components, guitar
tuning pegs, contact microphone
61 × 18 × 12.5 cm (with case)
Collection of the Artist
p. 299

Takis
FLOWER
1978
iron
255 × 41 cm
Donation of Alexander Iolas-Permanent
collection of the Macedonian Museum of
Contemporary Art
exh., not ill.

Laurie Anderson
NUMBERS RUNNERS
1979
wood, telephone, audiotape, artificial grass
240 × 116 × 116 cm
Private Collection
p. 303

Ay-O
AY-O'S AND JOE'S SOUNDBOX
1979
wooden cube, various mechanical
and automatic sound sources
120 × 120 × 120 cm
Vehbi Koç Foundation Contemporary
Art Collection, Istanbul
p. 305

Ken Butler
FILM REEL PROTRACTOR
1979–80
film reel, metal components, matte cutter,
protractor arm, pocket watch, counter,
center finder, music box,
contact microphone
111.5 × 30.5 × 7.5 cm
Collection of the Artist
exh., not ill.

Ken Butler
BRIEFCASE GUITAR
1980
briefcase, guitar components
91.5 × 30.5 cm
Collection of the Artist
p. 300

Ken Butler
DOUBLE REEL
1981
film reel, audio reel, metal components,
tennis racket, shoe size guide,
motor cover, tuning pegs,
case handle, contact microphone
129.5 × 53.5 × 5 cm
Collection of the Artist
exh., not ill.

Ken Butler
POOL CUE ARCHERY BOW CELLO
1981
pool cue, archery bow, hockey stick,
walking stick, coat hanger, drumstick, cello
components, axe handle, contact microphone
162.5 × 51 × 10 cm
Collection of the Artist
exh., not ill.

Ken Butler
K-BOARD
1983
organ keyboard, electrical bulb
sockets, light bulbs, motorized bicycle
wheel, guitar picks, tripod, string
76 × 99 × 33 cm
Courtesy of the Artist, New York
p. 307

Eliseo Mattiacci
ECHI DI SUONI E CANI
CHE ABBAIANO
1983
cymbals in different types of bronze,
aluminum and steel, tape recorder,
2 loudspeakers
approx. 300 × 300 × 350 cm
Collection of the Artist
p. 309

Tom Wesselmann
BEDROOM BLONDE WITH TV
1984–93
oil on canvas on board, TV
106 × 152.5 × 66 cm
Collection of Claire Wesselmann
exh., not ill.

Joseph Beuys
KLAVIER OXYGEN
1985
piano, oxygen, cylinder, telephone,
blackboard on easel, bicycle inner tube
200 × 350 × 250 cm
Vehbi Koç Foundation Contemporary
Art Collection, Istanbul
p. 311

William T. Wiley
BANJO FOR J.B.
1985
wood, paint, plexiglass, metal, leather, rope
96.5 × 29 × 11.5 cm
Courtesy of the di Rosa Collection, Napa
p. 276

Ken Butler
VALVE BODY DOOR
HANDLE VIOLIN
1986
transmission valve body, refrigerator
door handle, violin components,
contact microphone
58.5 × 23 × 10 cm
Collection of the Artist
exh., not ill.

William T. Wiley
DEBILSLIED
1986
carved wood, metal
114.5 × 38 × 2.5 cm (variable)
Collection of Ethan Wiley,
Courtesy of Ochi Gallery, Ketchum
p. 277

Max Vandervorst
SAXOSOIR
1988
plastic watering can, ebony
tenor saxophone nozzle
33 × 60 × 20 cm
Courtesy of the Artist
p. 313

Ken Butler
HOCKEY GOLF GUITAR
1991
hockey stick, golf club, drumstick,
saw blades, palette knife, violin
chinrest, tuning pegs, cello strings,
guitar string, contact microphone
102 × 25.5 × 7.5 cm
Collection of the Artist
exh., not ill.

Ken Butler
WOOD COAT HANGERS GUITAR
1991
coat hanger, slide organizers,
guitar components ruler, popsicle
sticks, contact microphone
112 × 40.5 × 7.5 cm
Collection of the Artist
exh., not ill.

Walter Marchetti
MUSICA DA CAMERA N° 211
1991
rolls of toilet paper
variable dimensions
Courtesy of the Artist
pp. 314–15

Terry Allen
MONTEZUMA 6-CASTLE
1992
drawing in pastel, audio speakers,
plexiglass, shoe, cassette player,
cassette, power source, wires
variable dimensions
The John & Maxine Belger Family
Foundation in Kansas City
p. 317

Christian Marclay
CAGE
1993
bird cage, telephone, cable
69 × 69 × 69 cm
Museum für Kommunikation, Bern
exh., not ill.

Ken Butler
GUITAR CASE BASS
1994
guitar case, bass components
190.5 × 45.5 × 25.4 cm
Collection of the Artist
exh., not ill.

Martin Creed
WORK NO. 97
(A METRONOME WORKING
AT MODERATE SPEED)
1994
metronome, amplifier
variable dimensions
Courtesy of the Artist and Hauser & Wirth
pp. 338–39

Maywa Denki
GUITAR-LA
1994
electric guitars, pedal organ
253 × 235 × 190 cm
Collection of the Artist
p. 341

Stephan von Huene
DER MANN VON JÜTERBOG
1995–96
steel, wood, fabric, leather
250.5 × 91 × 56.5 cm
Staatliche Museen zu Berlin,
Nationalgalerie, Berlin. Purchased 1997 by
the Verein der Freunde der Nationalgalerie
p. 343

Arman
THE SPIRIT OF YAMAHA
1997
sliced piano, 2 Yamaha motorcycles
157.5 × 205.5 × 208.5 cm
The Arman Marital Trust,
Corice Arman Trustee
p. 345

Bart Hopkin
SAVART'S WHEEL
1997
wood, steel, plastic, polystyrene,
electric motor
wood element: 51 × 53 × 30 cm

cups: 11 × 11 × 9 cm
Collection of the Artist
p. 347

Marco Bagnoli
SONOVASORO
1997 (2000)
sound by Giuseppe Scali: *Sequenza 140100 "RanA" per Sonovasoro*
alabaster, sound diffuser
67 cm (height), 29 cm (diam.)
Collection of the Artist
p. 349

Carsten Nicolai
BAUSATZ NOTO ∞
1998
record player, vinyl records with locked grooves, mixing board, amplifier, loudspeaker
variable dimensions
Courtesy Galerie EIGEN + ART Leipzig/Berlin and The Pace Gallery
p. 351

Bart Hopkin
OVERHANGING
1999
wood, steel, audio electronics
main body: 15 × 15 × 17 cm; additional component 2 × 2 × 50 cm
Collection of the Artist
p. 353

Rebecca Horn
CONCERT FOR THE PATH OF THE BEES
1999
cello, motors, steel, electronic device
140 × 100 × 90 cm
Private Collection
p. 355

Edward J. Potokar
MOPOL
1999
aluminum, wood, electronics
43 × 40.5 × 15 cm
Courtesy of the Artist
exh., not ill.

Edward J. Potokar
SILVER JUBILEE
1999
tin, electronics
7.5 cm (height), 20.5 cm (diam.)
Phil Lenihan, Orphic Gallery, Roxbury
p. 357

Neil Feather
THE VEGAS
2000
metal, plastic, mechanical and electronic components
145 × 38 × 13 cm
Courtesy of the Artist
p. 359

Christian Marclay
PROSTHESIS
2000
silicone rubber, metal guitar stand
51.5 × 66 × 38.5 cm
Catharine and Jeffrey Soros
exh., not ill.

Christian Marclay
VERTEBRATE
2000
altered acoustic guitar
28 × 67 × 39.5 cm
Courtesy of the Artist and Paula Cooper Gallery, New York
p. 361

Tom Sachs
TOYAN'S JR.
2001
casters, ear protection, fire axe, hardware, plywood, resin, stereo components
187 × 261 × 59 cm
Gian Enzo Sperone, Switzerland
pp. 362–63

Terry Adkins
OMOHUNDRO
2002
brass, copper
152 × 74 × 20 cm
Courtesy the Estate of Terry Adkins and Salon 94, New York
p. 365

Ken Butler
BENT ARM VIOLIN
2002
mannequin arm, violin components
46 × 18 × 7.5 cm
Collection of Ruth Marten
p. 301

Andreas Slominski
BASS FÜR MÄUSE
2002
wood, iron, paint
48 × 100 × 27 cm
Prada Collection, Milan
p. 367

Maurizio Cattelan
SENZA TITOLO
2003
resin body, synthetic hair, clothes, electronic device, bronze drum
86.5 × 84 × 60 cm
The Rachofsky Collection
p. 369

Ruth Ewan
A JUKEBOX OF PEOPLE TRYING TO CHANGE THE WORLD
2003—
CD jukebox, listing of songs
140 × 100 × 70 cm
Courtesy of the Artist and Rob Tufnell, London
p. 371

Loris Gréaud
CROSSFADING SUITCASE
2004
made to measure suitcase, laser-cut high density foam, sub-woofers, CD reader, batteries, neons, mat
57.5 × 152 × 33.5 cm
FNAC, Paris
pp. 372–73

Brian Dewan
MELODY GIN
2005
wood, plastic, metal, fabric, electronics
42 × 24.5 × 21.5 cm
Courtesy of the Artist and Pierogi Gallery, Brooklyn, New York
exh., not ill.

Brian Dewan
THE ROTARY ORDINATOR
2005
wood, plastic, metal, fabric, electronics
100 × 33 × 29 cm
Courtesy of the Artist and Pierogi Gallery, Brooklyn, New York
p. 375

Subodh Gupta
JUTHA
2005
stainless steel sinks, brass, aluminum found utensils, plastic wire, sound
3 elements, 107 × 110 × 50 cm each
Private Collection, New York
pp. 376–77

Walter Kitundu
PHONOSARANGI
2005
wood, metal, string, record player
66 × 61 × 35.5 cm
Collection of Robin Sukhadia
exh., not ill.

Claes Oldenburg and Coosje van Bruggen
FRENCH HORNS, UNWOUND AND ENTWINED
2005
cast 3 of 3, edition of 3 + 1 AP
stainless steel and aluminum painted with polyurethane enamel
345.5 × 142 × 168 cm
Courtesy of the Artists and Pace Gallery
p. 381

Claes Oldenburg and Coosje van Bruggen
SILENT METRONOME, 16 INCH, VERSION THREE
2005
canvas, wood, cord, hardware; painted with acrylic and latex
38 × 21 × 20 cm
Courtesy of the Artists and Pace Gallery
p. 379

Claes Oldenburg and Coosje van Bruggen
STUDY FOR BEACHED LUTES – VERSION THREE
2005
muslin, wood, metal, polyurethane foam, cord; coated with resin and painted with latex
42 × 56 × 33.5 cm
Collection of the Artists
exh., not ill.

Claes Oldenburg and Coosje van Bruggen
TROMBONE BRIDGE
2005
pencil, colored pencil, pastel
96.5 × 127 cm
Collection of the Artists
exh., not ill.

Claes Oldenburg
and Coosje van Bruggen
LEANING CLARINET
2006
AP 1 of 1, edition of 3 + 1 AP
aluminum painted with
acrylic polyurethane
358 × 157.5 × 81.5 cm
Courtesy of the Artists and Pace Gallery
exh., not ill.

Haroon Mirza
LO-TECH PROPOSED
2006–14
plastic bucket, LED circuit, wood, mirror, speaker, amp, water, mist maker
65 × 66 × 46 cm
Collection of the Artist
p. 401

Walter Kitundu
EGRET PHONOHARP
2007
wood, metal, string, record player
66 × 122 × 23 cm
Collection of the Artist
p. 403

Walter Kitundu
BEGUÈNA MARIDHIA
2008
wood, metal, string, stain
66 × 122 × 40.5 cm
Kronos Quartet / Kronos
Performing Arts Association
p. 405

Thomas Demand
HELDENORGEL
2009
C-print/Diasec
240 × 380 cm
Prada Collection, Milan
p. 407

Rebecca Horn
SIGHS OF THE WINDS
2009
violin, stone, glass, pigment,
steel, electronic motor
69 × 77 × 38 cm
Private Collection
exh., not ill.

Walter Kitundu
STROH CELLO
2009
wood, aluminum, tin, mylar,
cello components
66 × 137 × 76 cm
Kronos Quartet /Kronos
Performing Arts Association
exh., not ill.

Anri Sala
A SOLO IN THE DOLDRUMS
(BASED ON AN UNSEEN DANCE
BY SIOBHAN DAVIES)
2009
brushbox natural satin finish snare drum with a built-in speaker, drumsticks, sound
75 × 56 × 41 cm
Friedrich Christian Flick Collection
p. 409

Guido van der Werve
CHESS PIANO
2009
soundboard, piano frame, piano mechanism, benches, ebony, walnut, maple, chess pieces
chess piano: 80 × 109 × 109 cm
benches: *c.* 50 × 30 × 50 cm
Courtesy of the Artists and
Luhring Augustine, New York
p. 411

Athanasios Argianas
A DEMONSTRATION OF MANY VIEWS
AS ONE
2009 (2014)
HD video projection with sound,
brass, steel
176 × 68 × 68 cm
Courtesy of the Artist and Aanant
& Zoo, Berlin, On Stellar Rays, NY
& Max Wigram Gallery, London
p. 413

Nicolas Bernier and Martin Messier
LA CHAMBRE DES MACHINES
2010
sound machines designed by Alexandre
Landry for the performance
La Chambre des Machines
mixed media, wood, plastic, metal
machine no. 1: 33 × 39.5 × 39.5 cm;
machine no. 2: 48.5 × 35.5 × 40.5 cm
Courtesy of the Artists, Montreal
pp. 414–15

Martin Kersels
WARY (DICK)
2010
shoes, wood, wire, bells, epoxy
16.5 × 51 × 46 cm
Courtesy of the Artist and Mitchell-
Innes & Nash, New York
p. 417

Martin Kersels
WARY (JANE)
2010
shoes, wood, wire, bells, epoxy
15 × 39.5 × 42 cm
Courtesy of the Artist and Mitchell-
Innes & Nash, New York
exh., not ill.

Manuel Rocha Iturbide
EL ETERNO RETORNO
2010
drum, speakers, small amplifier,
audio cable, toy train
80 × 60 × 60 cm
Courtesy of the Artist
p. 419

Doug Aitken
MARBLE SONIC TABLE
2011
onyx alabaster, wooden
table, mallets, benches
5 elements
table: 76 × 234.5 × 100 cm;
2 benches: 45 × 213.5 × 35.5 cm;
2 benches: 45 × 61 × 35.5 cm
Courtesy of the Artist; Galerie
Eva Presenhuber, Zurich and
303 Gallery, New York
p. 421

Alberto Tadiello
ELEKTRONSKAL
2011
metal structure, perforated metal sheet, electric bells, cables, MDF panels
160 × 160 × 100 cm
Courtesy T293, Naples – Rome
p. 423

Riccardo Beretta
DONNERWETTER
2011–12
clavicytherium; natural veneers,
Japanese ash, black limba; soundboard:
spruce; solid woods: lignum vitae,
Pau violeto, poplar, pine, maple,
mahogany; zinc-coated brass
246 × 86 × 49 cm (closed);
244 × 160 × 49 cm (open)
Private Collection, Switzerland
p. 425

Richard Artschwager
PIANO GRANDE
2012
formica laminate on wood, plexiglass
89 × 117 × 202 cm
Prada Collection, Milan
exh., not ill.

Tarek Atoui
THE METASTABLE CIRCUIT
2012
speakers, 4 sensor-based interrelated modules, 1 asynchronous digital system of 60 computer-based samplers and sequencers, and an indefinite number of sound libraries and samples
2 tables: 105 × 100 × 60 cm each;
2 tables: 105 × 60 × 60 cm each
Courtesy of the Artist and
Galerie Chantal Crousel
pp. 426–27

Céleste Boursier-Mougenot
U43
2012
wax, wood, Bakelite telephone – U43
model, internet connection, case
145 × 28 × 23 cm
Galerie Xippas, Paris
p. 429

Janet Cardiff
and George Bures Miller
BURLESQUE
2012
mixed media assemblage with audio
system
variable dimensions;
approx. 220 × 30 cm
Courtesy of the Artist and
Luhring Augustine, New York
p. 431

Pedro Reyes
IMAGINE (DOUBLE PSALTERY)
2012
recycled metal
20 × 50 × 50 cm
Fundación Alumnos47
p. 433

Tore Honoré Bøe
RECUERDOS
2013
wooden box, various objects
30 × 32 × 30 cm (open)
Courtesy of the Artist
exh., not ill.

Haroon Mirza
STACK
2013
fairy lights, radio, keyboard, LED display, speaker, wooden drawer, wooden shelf, LED circuit, amp, other electronic components
84 × 92 × 45 cm
Collection of Eric Ceputis and David W. Williams
p. 435

Tore Honoré Bøe
CINCUENTA
2014
wooden box, various objects
26 × 22 × 31 cm (open)
Courtesy of the Artist
p. 437

Pedro Reyes
DISARM (BASS V)
2014
metal
113 × 36 × 10 cm
Courtesy of the Artist and Lisson Gallery
exh., not ill.

Pedro Reyes
DISARM (CELLO)
2014
metal
163 × 57 × 25 cm
Courtesy of the Artist and Lisson Gallery
exh., not ill.

Pedro Reyes
DISARM (GUITAR)
2014
metal
approx. 86 × 30 × 10 cm
Courtesy of the Artist and Lisson Gallery
exh., not ill.

Pedro Reyes
DISARM (VIOLIN IX)
2014
metal
violin: 59 × 22 × 5.5 cm
bow: 75.5 × 4 × 3 cm
Courtesy of the Artist and Lisson Gallery
exh., not ill.

SELECTED BIBLIOGRAPHY

1840
Johann Wolfgang von Goethe, *Theory of Colours* (London: John Murray, 1840).

1866
Herman Von Helmholtz, *Handbuch der physiologischen Optik* (Leipzig: Leopold Voss, 1866).

1907
Charles Perregaux, *Les Jaquet-Droz et leurs automates* (Neuchâtel: Wolfrath & Sperlé, 1907).

1912
A. Wallace Rimington, *Colour-Music: The Art of Mobile Colour* (London: Hutchinson & Co., 1912).

1913
Wassily Kandinsky, *Klänge [Sounds]* (Munich: R. Piper & Co., 1913).

1915
Giacomo Balla, *Ricostruzione futurista dell'universo* (Milan: Direzione del Movimento futurista, 1915).

1916
Luigi Russolo, *L'Arte dei rumori* (Milan: Edizioni Futuriste di Poesia, 1916).

1918
"Manifest I of De Stijl," *De Stijl*, vol. 2, no. 1, November 1918, p. 4.

1923
Willard H. Wright, *The Future of Painting* (New York: B.W. Huebsch, 1923).

1927
Adrian Bernard Klein, *Colour – Music: The Art of Light* (London: Crosby Lockwood and Son, 1927).

1943
Alexander Calder, *A Propos of Measuring a Mobile*, manuscript (Archives of American Art, Smithsonian Institution, 1943).

1946
Mary Hallock Greenewalt, *Nourathar: The Fine Art of Light Color Playing* (Philadelphia, PA: Westbrook Publishing, 1946).

1947
László Moholy-Nagy, *Vision in Motion* (Chicago, IL: Paul Theobald, 1947).

Thomas Wilfred, "Light and the artist," *Journal of Aesthetics and Art Criticism*, vol. 5, no. 4, June 1947, pp. 247–55.

1948
Thomas Wilfred, "Composing in the Art of Lumia," *Journal of Aesthetics and Art Criticism*, vol. 7, no. 2, December 1948, pp. 79–93.

1949
Harry Partch, *Genesis of a Music* (Madison, WI: University of Wisconsin Press, 1949).

1950
Faber Birren, *Color Psychology and Color Therapy*: A Factual Study of the Influence of Color on Human Life (New York: McGraw Hill, 1950).

1951
Maitland Graves, *The Art of Color and Design* (New York: McGraw-Hill Book Company, 1951).

1952
Hans Richter, "Easel-Scroll-Film," *Magazine of Art*, February 1952, pp. 78–86.

1954
Rudolf Arnheim, *Art and Visual Perception: A Psychology of the Creative Eye* (Berkeley, CA: University of California Press, 1954).

1958
Eugenio Sansoni, "Automata," in *Enciclopedia universale dell'arte*, vol. 2 (Florence: Sansoni, 1958), pp. 246–48.

1961
John Cage, *Silence – Lectures and Writings* (Middletown, CT: Wesleyan University Press, 1961).

Johannes Itten, *The Art of Color: The Subjective Experience and Objective Rationale of Color* (New York: Reinhold, 1961).

1962
Various Authors, *Sonic Art*, exhibition catalogue (San Bernardino, CA: California State College, 1962).

1963
Josef Albers, *Interaction of Color* (New Haven, CT: Yale University Press, 1963).

George Brecht, John Cage, La Monte Young, Jackson Mac Low, George Maciunas *et al.* (ed.), *An Anthology of Chance Operations, Concept Art, Anti Art, Indeterminacy, Plans of Action, Diagrams, Music, Dance Constructions, Improvisation, Meaningless Work, Natural Disasters, Compositions, Mathematics, Essays, Poetry* (New York: George Maciunas & Jackson Mac Low, 1963).

Edward Ruscha, *Twentysix Gasoline Stations* (Los Angeles, CA: Edward Ruscha, 1963).

1964
Marina Repetto, *Nicola Giolfino: cronologia e vita* (Verona: Linotipia Veronese di Ghidini e Fiorini, 1964).

1965
John Cage, *Water Walk, for Solo Television Performer* (New York: C.F. Peters Corporation, 1965).

1966

Ralph T. Coe, *Sound Light Silence: Art That Performs*, exhibition catalogue (Kansas City, KS: Nelson-Atkins Gallery, 1966).

1967

Carl Dahlhaus, *Musikästhetik* (Cologne: Musikverlage Hans Gerig, 1967).

Marshall McLuhan, Quentin Fiore, *The Medium is the Message* (London: Penguin, 1967).

Frank Popper, *Naissance de l'art cinétique: L'image du mouvement dans les arts plastiques depuis 1860* (Paris: Gauthiers Villars, 1967).

Emanuel Winternitz, *Musical Instruments and their Symbolism in Western Art* (London: Faber & Faber, 1967).

1968

Luigi Gerli (ed.), *Museo di Strumenti Musicali Meccanici. Collezione Marino Marini* (Ravenna: Museo di Strumenti Musicali Meccanici/Musiani, 1968).

Karl Gunnar Pontus Hultén, *The Machine: As Seen at the End of the Mechanical Age*, exhibition catalogue (New York: The Museum of Modern Art, 1968).

Pierre Restany, *Nouveau Réalism* (Paris: Planète, 1968).

1969

Brent Berlin, Paul Kay, *Basic Color Terms: Their Universality and Evolution* (Berkeley, CA/London: The University of California Press, 1969).

Faber Birren, *Principles of Color* (Atglen, PA: Schiffer Publishing Ltd., 1969).

Guido Boursier, "Far scattare nella gente meccanismi di liberazione," *Sipario*, no. 276, Milan, April 1969.

John Cage, *Notations* (New York: Something Else Press, 1969).

Arturo Schwarz (ed.), *The Complete Works of Marcel Duchamp* (London: Thames & Hudson, 1969).

Herbert Alexander Simon, *The Sciences of the Artificial* (Cambridge, MA: The MIT Press, 1969).

LaMonte Young, Marian Zazeela, *Selected Writings* (Munich: Friedrich, 1969).

1970

László Moholy-Nagy, Richard Kostelanetz (ed.), *Moholy-Nagy: An Anthology* (New York: Da Capo Press, 1970).

Gene Youngblood, *Expanded Cinema* (New York: E.P. Dutton, 1970).

1971

David Cope, *New Directions in Music* (Dubuque, IA: William C. Brown & Co., 1971).

Jan van Loenen Martinet, Gerrit Kouwenaar, Dick Raaijmakers *et al.*, *Sound = Sight: Three Audio Visual Projects*, exhibition catalogue (Amsterdam: Stedelijk Museum Amsterdam, 1971).

Gordon Pask, "A comment, a case history and a plan," in Jasia Reichardt (ed.), *Cybernetics Art and Ideas* (London: Studio Vista, 1971), pp. 76–99.

1972

Jonathan Benthall, *Science and Technology in Art Today* (London: Thames & Hudson, 1972).

David Q. Bowers, *Encyclopedia of Automatic Musical Instruments* (Vestal, NY: The Vestal Press, 1972).

Jane Livingston, Marcia Tucker, *Bruce Nauman: Work from 1965 to 1972*, exhibition catalogue (Los Angeles, CA/New York: Co. Mus. A./Whitney, 1972).

Hans Maria Wingler, *Il Bauhaus. Weimar, Dessau, Berlino 1919–33* (Milan: Giangiacomo Feltrinelli Editore, 1972).

Wucius Wong, *Principles of Two-Dimensional Design* (New York: John Wiley & Sons, 1972).

1973

Ursula Block, Michael Glasmeier (eds.), *Broken Music – Artists' Recordworks* (Berlin: Berliner Kunstlerprogramm des DAAD/Gelbe Musik, 1973).

John Cage, Silence (London: Calder & Boyars, 1973).

Germano Celant, *Record as Artwork, 1959–73*, exhibition catalogue (London: Royal College of Art Gallery, 1973).

Lucy R. Lippard (ed.), Six Years: The Dematerialization of the Art Object from 1966 to 1972: a Cross-reference Book of Information on Some Esthetic Boundaries (New York/London: Praeger/ Studio Vista, 1973).

1974

Lawrence Alloway, *American Pop Art*, exhibition catalogue (New York: Whitney Museum of American Art/Collier Books, 1974).

Germano Celant, Jole De Sanna, Daniela Palazzoni *et al.*, *Nuovi Media: Film e Videotape*, exhibition catalogue (Milan: Centro Documentazione e Ricerca Jabik, 1974).

Frank J. Malina (ed.), *Kinetic Art: Theory and Practice. Selections from the Journal Leonardo* (New York: Dover Publications, 1974).

Michael Nyman, *Experimental Music: Cage and Beyond* (New York/London: Schirmer Books/Studio Vista, 1974).

Steve Reich, *Writings About Music* (Halifax/New York: Press of Nova Scotia College of Art and Design/New York University Press, 1974).

1976

Faber Birren, *Color Perception in Art* (New York/London: Van Nostrand Reinhold, 1976).

Mary Hillier, *Automata & Mechanical Toys: An Illustrated History* (London: Jupiter, 1976).

Robin MacOnie, *The Works of Karlheinz Stockhausen* (London/New York: Oxford University Press, 1976).

Robert Russett, Cecile Starr, *Experimental Animation: An Illustrated Anthology* (New York: Van Nostrand Reinhold Co., 1976).

1977

Jacques Attali, *Bruits – Essai sur l'économie politique de la musique* (Paris: Presses Universitaires de France, 1977).

Germano Celant, *Ambiente/Arte. Dal Futurismo alla Body Art* (Venice: Biennale, 1977).

Germano Celant, *Offmedia. Nuove tecniche artistiche: video disco libro* (Bari: Edizioni Dedalo, 1977).

Germano Celant, *The Record as Artwork: From Futurism to Conceptual Art*, exhibition catalogue (Fort Worth, TX: Fort Worth Art Museum, 1977).

Robert Murray Schafer (ed.), *European Sound Diary* (Burnaby: The Aesthetic Research Centre Publication, 1977).

Robert Murray Schafer, *The Tuning of the World* (New York: Alfred Knopf, 1977).

1978

Faber Birren, *Color & Human Response* (New York/London: Van Nostrand Reinhold, 1978).

W. Garner, "The Relationship between Colour and Music," *Leonardo*, vol. 11, no. 3 (Cambridge, MA: The MIT Press, 1978), pp. 225–26.

Roselee Goldberg, *Performance – Live Art 1909 to the Present* (New York: Harry N. Abrams, 1978).

Kim Levin, "Dennis Oppenheim: Post-Performance Works," *Arts Magazine*, September 1978, pp. 123–25.

1979

Paul Griffiths, *A Guide to Electronic Music* (London: Thames & Hudson, 1979).

Robert Smith, Robert Wilhite (eds.), *Sound: An Exhibition of Sound Sculpture, Instrument Building, and Acoustically Tuned Spaces*, exhibition catalogue (Los Angeles, CA: Los Angeles Institute of Contemporary Art, 1979).

Stephan von Wiese (ed.), *Günther Uecker: Schriften* (St.Gallen: Erker, 1979).

1980

Krome Barratt, *Logic & Design: in Art, Science and Mathematics* (New York/ London: Design Books/Godwin,1980).

Billy Klüver, *E.A.T. Bibliography 1965–1980* (New York: Experiments in Art and Technology, 1980).

Barry Rosen, Helene Winer (ed.), *A Sound Selection – Audio Works by Artists: A Traveling Exhibition*, exhibition catalogue (New York: Committee for the Visual Arts, 1980).

John Whitney, *Digital Harmony: On the Complementarity of Music and Visual Art* (Peterborough, NH: Byte Books, 1980).

1981

John Cage, *For the Birds* (Boston, MA: Marion Boyars, 1981).

Suzanne Delehanty, Dore Ashton, Germano Celant *et al.* (eds.), *Soundings*, exhibition catalogue (Purchase, NY: Neuberger Museum of Art, 1981).

Karl Gerstner, *The Spirit of Colors: The Art of Karl Gerstner* (Cambridge, MA: The MIT Press, 1981).

Robert Murray Schafer, *Sounds Unseen*, flexi disc with text (Vancouver/Stratford: Presentation House/The Gallery Stratford, 1981).

Henri Stierlin (ed.), *The Spirit of colors. The Art of Karl Gerstner: Nine Picture Chapters and Selected Essays* (Cambridge, MA/London: The MIT Press, 1981).

1982

Allan Doig, *Theo van Doesburg: Painting into Architecture, Theory into Practice* (Cambridge, UK: Cambridge University Press, 1982).

John G. Harnhardt (ed.), *Nam June Paik*, exhibition catalogue (New York: Whitney Musum of American Art, 1982).

Ilona Lindenberg, *Edward Kienholz. Roxys and Other Works aus der Sammlung Reinhard Onnasch*, exhibition catalogue (Bremen: Gesellschaft für aktuelle Kunst, 1982).

Kenneth C. Lindsay, Peter Vergo (eds.), *Wassily Kandinsky: Complete Writings on Art* (Boston, MA: G.K. Hall, 1982).

Franco Masotti, Roberto Masotti, Veniero Rizzardi *et al.* (eds.), *Sonorità Prospettiche – Suono Ambiente Immagine*, exhibition catalogue (Rimini: Comune di Rimini, 1982).

1983

Norman Andersen, Max Eastley (eds.), *Not Suitable for Framing: Arrangements For Sounds and Structure*, exhibition catalogue (Toronto: The Art Gallery at Harbourfront, 1983).

Germano Celant (ed.), *Jannis Kounellis* (Milan: Mazzotta, 1983).

Heidi von Gunden, *Music of Pauline Oliveros* (Washington D.C.: Scarecrow Press, 1983).

Stephan von Huene, *Stephan von Huene: Klangskulpturen. 15 Oktober bis 20 Novembre 1983*, exhibition catalogue (Baden-Baden: Staatliche Kunsthalle, 1983).

Janet Kardon (ed.), *Laurie Anderson Works From 1969 to 1983*, exhibition catalogue (Philadelphia, PA: Institute of Contemporary Art, University of Pennsylvania, 1983).

Ronald Pellegrino, *The Electronic Arts of Sound and Light* (New York: Van Nostrand Reinhold, 1983).

Man Ray, *Objets de mon affection* (Paris: Philippe Sers, 1983).

Various Authors, *Sound / Art*, exhibition catalogue (New York: Sound / Art Foundation, 1983).

1984

Laurie Anderson, *United States* (New York: HarperCollins, 1984).

Tom Darter, Greg Armbruster, *The Art of Electronic Music* (New York: William Morrow, 1984).

Barbara Haskell, *Blam: The Explosion of Pop, Minimalism and Performance 1958–1964*, exhibition catalogue (New York: Whitney Museum of American Art, 1984).

Peter S. Stevens, *Handbook of Regular Patterns: An Introduction to Symmetry in Two Dimensions* (Cambridge, MA: The MIT Press, 1984).

Frank Whitford, *Bauhaus* (London: Thames & Hudson, 1984).

1985

John Belton, Elisabeth Weis (eds.), *Film Sound – Theory and Practice* (New York: Columbia University Press, 1985).

J. Ronald Grele, *Envelopes of Sound – The Art of Oral History* (Chicago, IL: Precedent Publishing, 1985).

Thomas B. Holmes, *Electronic and Experimental Music* (New York: Charles Scribner's Sons, 1985).

Jolyon Laycock, *A Noise in Your Eye! An International Exhibition of Sound Sculpture*, exhibition catalogue (Bristol: Arnolfini Gallery, 1985).

Various Authors, *On the Wall/On the Air: Artists Make Noise*, exhibition catalogue (Boston, MA: The MIT Press/Hayden Corridor Gallery, 1985).

Various Authors, *Sound / Vision*, exhibition catalogue (Plymouth/Exeter, UK: Plymouth Arts Center/Spacex Gallery, 1985).

1986

Karl Gerstner, *Forms of Color: The Interaction of Visual Elements* (Cambridge, MA: The MIT Press, 1986).

Mary Jane Jacob (ed.), *Jannis Kounellis*, exhibition catalogue (Chicago, IL: Museum of Contemporary Art, 1986).

Friedrich A. Kittler, *Gramophone, Film, Typewriter* (Berlin: Brinkmann & Bose 1986).

Robert Murray Schafer, *The Thinking Ear* (Toronto: Arcana, 1986).

Various Authors, *Soundwave NYC: Audio Art Series*, exhibition catalogue (New York: City Gallery/Gerald Lindahl, Bill & Mary Buchen, 1986).

1987

Jonathan W. Bernard, *The Music of Edgard Varèse* (New Haven, CT/London: Yale University Press, 1987).

Ursula Block, Germano Celant, William Furlong *et al.*, *Sound Re Visited: An Anthology Concerning Sound in Art* (Amsterdam: Void Edition, 1987).

Tom DeWitt, "Visual Music: Searching for an Aesthetic," *Leonardo*, vol. 20, no. 2 (Cambridge, MA: The MIT Press, 1987), pp. 115–22.

Robert Dixon, *Mathographics* (Oxford, UK: Basil Blackwell, 1987).

Evan Eisenberg, *The Recording Angel – Explorations in Phonography* (New York: McGraw-Hill, 1987).

Mel Gordon, *Dada Performance* (New York: PAJ Publications, 1987).

H. Wiley Hitchcock (ed.), *The Phonograph and Our Musical Life*, monograph no. 14 (New York: I.s.A.M., 1987).

Richard Kostelanetz (ed.), *Conversing with Cage* (New York: Limelight Editions, 1987).

Wucius Wong, *Principles of Color Design* (New York: Van Nostrand Reinhold, 1987).

1988

Germano Celant, Claes Oldenburg, Coosje van Bruggen, *A Bottle of Notes and Some Voyages, Claes Oldenburg: Drawings, Sculptures and Large-scale Projects with Coosje van Bruggen*, exhibition catalogue (Sunderland/Leeds, UK: Northern Centre for Contemporary Art/Henry Moore Centre for the Study of Sculpture, 1988).

Anna Cirignola (ed.), *Il sesto senso della natura*, exhibition catalogue (Lecce: Comune di Lecce/Pari&Dispari, 1988).

John Hendricks, Robert Pincus-Witten, *Fluxus Codex* (Detroit, MI: Gilbert and Lila Silverman Fluxus Collection in association with HN Abrams, New York, 1988).

Kenneth Peacock, "Instruments to Perform Color-Music: Two Centuries of Technological Experimentation," *Leonardo*, vol. 21, no. 4 (Cambridge, MA: The MIT Press, 1988), pp. 397–406.

Wucius Wong, *Principles of Two-dimensional Form* (New York: Van Nostrand Reinhold, 1988).

1989

Enrico Crispolti (ed.), *Casa Balla e il Futurismo a Roma*, exhibition catalogue (Rome: Istituto Poligrafico e Zecca dello Stato, 1989).

Patricia Sloane, *The Visual Nature of Color* (New York: Design Press, 1989).

1990

Brian Evans, "Temporal Coherence with Digital Color," *Leonardo – Digital Image, Digital Cinema: SIGGRAPH '90 Art Show Catalog*, Supplemental Issue, vol. 3 (Cambridge, MA: The MIT Press, 1990), pp. 43–49.

Mary Lynn Kotz, *Rauschenberg: Art and Life* (New York: Abrams, 1990).

Marilyn S. Kushner, *Morgan Russell* (New York: Hudson Hills Press/Montclair Art Museum, 1990).

Juli Susin, Roberto Ohrt, *Man Ray: Assemblages*, exhibition catalogue (Paris: Galerie Marion Meyer, 1990).

1991

Guido Bizzi (ed.), *La collezione di strumenti musicali del Museo Teatrale alla Scala. Studio, restauro, restituzione* (Milan: Silvana Editoriale, 1991).

Maarten Franssen, "The Ocular Harpsichord of Louis-Bertrand Castel: The Science and Aesthetics of an Eighteenth-Century Cause Célèbre," *Tractrix Yearbook for the History of Science, Medicine, Technology and Mathematics*, no. 3, 1991, pp. 15–77.

Richard Kostelanetz (ed.), *John Cage: An Anthology* (New York: Da Capo Press, 1991).

Cathrine Pichler (ed.), *Sinneswerkzeug: Kunst als Instrument. Bernhard Leitner, Max Peintner, Alfons Schilling*, exhibition catalogue (Graz: Kulturhaus Graz/ Passagen Verlag, 1991).

Paola Elisabetta Simeoni, Roberta Tucci (eds.), *La Collezione degli Strumenti Musicali. Museo Nazionale delle Arti e Tradizioni Popolari* (Rome: Istituto Poligrafico e Zecca dello Stato, 1991).

1992

Rick Altman, *Sound Theory, Sound Practice* (New York: Routledge, 1992).

Douglas Kahn, Gregory Whitehead (eds.), *Wireless Imagination – Sound, Radio and the Avant-garde* (Cambridge, MA/London: The MIT Press, 1992).

Edward Lippman, *A History of Western Musical Aesthetic* (Lincoln, NE/London: University of Nebraska Press, 1992).

Michael Lloyd, Michael Desmond (eds.), *European and American Paintings and Sculptures 1870–1970 in the Australian National Gallery* (Canberra: Australian National Gallery, 1992).

Barton McLean, "Composition with Sound and Light," *Leonardo Music Journal*, vol. 2, no. 1 (Cambridge, MA: The MIT Press, 1992), pp. 13–18.

David Revill, *The Roaring Silence: John Cage, a Life* (New York: Arcade Publishing, 1992).

1993

Laurie Anderson, *Stories from the Nerve Bible: A Retrospective 1972–1992* (New York: Harper Perennial, 1993).

Jean-Yves Bosseur, *Sound and the Visual Arts* (Paris: Dis Voir, 1993).

John Gage, *Colour and Culture: Practice and Meaning from Antiquity to Abstraction* (London: Thames&Hudson, 1993).

Gregory Gargarian, *The Art of Design: Expressive Intelligence in Music*, MIT PhD Thesis, February 1993, web.media.mit.edu/~tristan/Classes/MAS.945/Papers/Main/Gargarian_Thesis.pdf

Janet Jenkins (ed.), *In the Spirit of Fluxus*, exhibition catalogue (Minneapolis: Walker Art Center, 1993).

Richard Kostelanetz (ed.), *Writings about Cage* (Michigan: The University of Michigan Press, 1993).

Robert Murray Schafer, *The Soundscape: Our Environment and the Tuning of the World* (New York: Destiny Books, 1993).

James Pritchett, *The Music of John Cage* (Cambridge, UK: Cambridge University Press, 1993).

Wucius Wong, *Principles of Form and Design* (London/New York: International Thomson/Van Nostrand Reinhold, 1993).

Various Authors, *Rebecca Horn*, exhibition catalogue (New York: Guggenheim Museum, 1993).

1994

Diana Augaitis, Dan Lander (eds.), *Radio Rethink: Art, Sound and Transmission* (Banff: Walter Phillips Gallery, 1994).

Luisa Cervelli (ed.), *La Galleria Armonica: Catalogo del Museo degli strumenti musicali di Roma* (Rome: Istituto Poligrafico e Zecca dello Stato, 1994).

Michel Chion, *Audio-Vision: Sound on Screen* (New York: Columbia University Press, 1994).

Fabrice Hergott (ed.), *Joseph Beuys*, exhibition catalogue (Paris: Éditions du Centre Pompidou, 1994).

Steven R. Holtzman, *Digital Mantras: The Languages of Abstract and Virtual Worlds* (Cambridge, MA: The MIT Press, 1994).

Thomas Kellein, *Fluxus*, exhibition catalogue (Basel: Kunsthalle Basel/ Edition Hansjörg Mayer, 1994).

Rosalind Krauss, Thomas Krens, David Anti *et al.*, *Robert Morris: The Mind/Body Problem*, exhibition catalogue (New York: The Guggenheim Museum, 1994).

Joseph Lanza, *Elevator Music: A Surreal History of Muzak, Easy-Listening, and Other Moodsong* (London: Quartet Books, 1994).

Scott McCloud, *Understanding Comics: The Invisible Art* (New York: Harper Perennial, 1994).

Joan Simon (ed.), *Bruce Nauman*, exhibition catalogue and catalogue raisonné (Minneapolis: Walker Art Center, 1994).

1995

Jean-François Augoyard, Henry Torgue, *À l'écoute de l'environnement: Repertoir des effets sonores* (Marseilles: Parenthèses, 1995).

Richard Buchanan, Victor Margolin (eds.), *Discovering Design: Explorations in Design Studies* (Chicago, IL: University of Chicago Press, 1995).

Michael Chanan, *Repeated Takes: A Short History of Recording and its Effects on Music* (London/New York: Verso, 1995).

Victor Margolin, Richard Buchanan (eds.), *The Idea of Design. A Design Issues Reader* (Cambridge, MA/London: The MIT Press, 1995).

Claes Oldenburg, Germano Celant (ed.), *Claes Oldenburg: An Anthology*, exhibition catalogue (New York: Guggenheim Museum Publications, 1995).

Robert Pepperell, *The Post-Human Condition* (Oxford, UK: Intellect Book, 1995).

Clifford A. Pickover (ed.), *The pattern book: Fractals, Art, and Nature* (Singapore/ River Edge: World Scientific, 1995).

Ian Stewart, *Nature's Numbers: The Unreal Reality Of Mathematics* (New York: Basic Books, 1995).

David Toop, *Ocean of Sound: Aether Talk, Ambient Sound and Imaginary Worlds* (London, Serpent's Tail, 1995).

Richard Torchia (ed.), *William Anastasi: A Retrospective, 1960–1995*, exhibition catalogue (Philadelphia, PA: Moore College of Art and Design, 1995)

Heidi E. Violand-Hobi (ed.), *Jean Tinguely: Life and Work* (Munich: Prestel, 1995).

1996

Rosetta Brooks, Monte Factor, Walter Hopps *et al.*, *Kienholz. A Retrospective*, exhibition catalogue (New York: Whitney Museum of Art, 1996).

Philip Dodd, Ian Christie (eds.), *Spellbound: Art and Film*, exhibition catalogue (London: Hayward Gallery/ British Film Institute Publishing, 1996).

Bart Hopkin, *Gravikords, Whirlies & Pyrophones: Experimental Musical Instruments*, book and CD (New York: Ellipsis Arts…, 1996).

Bart Hopkin, *Musical Instrument Design: Practical Information for Instrument Making* (Tucson: See Sharp Press, 1996).

Paola Igliori (ed.), *American Magus Harry Smith: A Modern Alchemist* (New York: Inanout Press, 1996).

Yasmin Kafai, Mitchel Resnick (eds.), *Constructionism in Practice: Designing, Thinking, and Learning in a Digital World* (Mahwah, NJ: Lawrence Erlbaum Associates, 1996).

Milan Knížák – Nový ráj. Výběr prací z let 1952–1995, exhibition catalogue (Prague: Galerie Mánes/ UPM, 1996).

Allen S. Weiss, "Experimental Sound & Radio," special issue of *The Drama Review*, vol. 40, no. 3, Fall 1996.

1997

David Batchelor, *Chromophobia: Ancient and Modern, and a Few Notable Exceptions*, exhibition catalogue (Leeds, UK: Centre for the Study of Sculpture/Henry Moore Institute, 1997).

Tobia Bezzola (ed.), *Arranged and Conducted by Christian Marclay*, exhibition catalogue (Zurich: Kunsthaus Zürich, 1997).

Joel Chadabe, *Electric Sound: The Past and Promise of Electronic Music* (Upper Saddle River, NJ/London: Prentice Hall Press, 1997).

Commotion: Martin Kersels, exhibition catalogue (Madison: Madison Art Center, 1997).

Edith Decker-Philips, *Paik Video* (Barrytown: Station Hill Arts, 1997).

Paul De Marinis, *Artist's Talk Delivered at International Conference on Auditory Displays*, Palo Alto, November 3, 1997, well.com/~demarini/icad.html

Paul De Marinis, interview by Shun-ichi Shiba. Originally published 1997 in Japanese in *ICC Journal*, well.com/user/demarini/shiba.html

Hajo Düchting, *Paul Klee: Painting Music* (Munich: Prestel, 1997).

C. L. Hardin, Luisa Maffi (eds.), *Color Categories in Thought and Language* (Cambridge, UK: Cambridge University Press, 1997).

Laura Pagnotta, *Bartolomeo Veneto: L'opera completa* (Florence: Centro Di, 1997).

Paul Théberge, *Any Sound You Can Imagine: Making Music/Consuming Technology* (Hanover, NH: Wesleyan University Press, University Press of New England, 1997).

Max Vandervorst, *Lutherie sauvage. Instruments de musique a inventer* (Paris: Éditions Alternatives, 1997).

Emmett Williams, Ann Noël (eds.), *Mr Fluxus: A Collective Portrait of George Maciunas 1931–1978* (London: Thames & Hudson, 1997).

1998

Bruce Andrews, "Praxis: A Political Economy of Noise and Informalism," in *Close Listening: Poetry and the Performed Word*, Charles Bernstein (New York: Oxford University Press, 1998).

Jadranka Bentini (ed.), *Sovrane passioni: le raccolte d'arte della Ducale galleria estense*, exhibition catalogue (Milan: Galleria Estense di Modena/Federico Motta Editore, 1998).

Nicolas Bourriaud, *Relational Aesthetics* (Paris: Les Presses du Réel, 1998).

Nicholas Cook, *Analysing Musical Multimedia* (Oxford, UK: Clarendon Press, 1998).

Nicholas Cook, *Music: A Very Short Introduction* (Oxford, UK: Oxford University Press, 1998).

Sydney Fels, Kazushi Nishimoto and Kenji Mase, "MusiKalscope: A Graphical Musical Instrument," *IEEE Multimedia*, July-September 1998, pp. 26–35.

Donald D. Hoffman, *Visual Intelligence: How We Create What We See* (New York: WW Norton & Company, 1998).

Bernhard Leitner, *Sound : Space* (Ostfildern: Cantz Verlag, 1998).

Christopher Phillips (ed.), *Voices*, exhibition catalogue (Rotterdam: Witte de With Publishers, 1998).

Akio Suzuki, *"A": Sound Works, Throwing and Following = Werfen und Folgen:* (Saarbrücken: Stadtgalerie, 1998).

1999

François Baschet, *Les Sculptures Sonores: The Sound Sculptures of Bernard and François Baschet* (Chelmsford: Soundworld, 1999).

Gabriele Bonomo (ed.), *Walter Marchetti*, exhibition catalogue (Vancouver/Verona: Morris and Helen Belkin Art Gallery/Archivio Francesco Conz, 1999).

John Bowers, *Introduction to Two-Dimensional Design: Understanding Form and Function* (New York: John Wiley & Sons, 1999).

Fred Collopy, Robert M. Fuhrer, David Jameson, "Visual Music in a Visual Programming Language," in *1999 IEEE Symposium on Visual Languages* (Tokyo: IEEE 1999), pp. 111–18.

John Gage, *Color and Meaning: Art, Science, and Symbolism* (Berkeley, CA: University of California Press, 1999).

Jürgen Geisenberger, *Joseph Beuys und die Musik* (Marburg: Tectum Verlag, 1999).

Douglas Kahn, *Noise, Water, Meat: A History of Sound in the Arts* (Cambridge, MA/London: The MIT Press, 1999).

Daniel Kurjakovic, Sebastian Lohse (eds.), *Other Rooms, Other Voices. Audio Works by Artists* (Zurich: Mermory/Cage Editions, 1999)

Ray Kurzweil, *The Age of Spiritual Machines – When Computers Exceed Human Intelligence* (London: Viking, 1999).

Brandon LaBelle, Steve Roden (eds.), *Site of Sound #2: Of Architecture and The Ear* (Los Angeles, CA: Errant Bodies Press/Smart Art Press, 1999).

Karin von Maur, *The Sound of Painting: Music in Modern Art* (Munich/London: Prestel, 1999).

Manuel Oliveira (ed.), *Lost in Sound*, exhibition catalogue (Santiago de Compostela: CGAC – Centro Gallego de Arte Contemporánea, 1999).

A.L. Rees, *A History of Experimental Film and Video: From Canonical Avant-Garde to Contemporary British Practice* (London: British Film Institute, 1999).

David Toop, *Exotica: Fabricated Soundscapes in a Real World* (London: Serpent's Tail, 1999).

2000

Fred Collopy, "Color, Form, and Motion: Dimensions of a Musical Art of Light," *Leonardo*, vol. 33, no. 5 (Cambridge, MA: The MIT Press, 2000), pp. 355–60.

Tia DeNora, *Music and Everyday Life* (Cambridge, UK/New York: Cambridge University Press, 2000).

Albert Glinsky, *Theremin: Ether Music and Espionage* (Urbana, IL: University of Illinois Press, 2000).

John G. Harnhardt, *The Worlds of Nam June Paik*, exhibition catalogue (New York: The Guggenheim Museum, 2000).

Els Hoek (ed.), *Theo van Doesburg: oeuvre catalogus* (Utrecht: Centraal Museum, 2000).

Lev Manovich, *The Language of New Media* (Cambridge, MA: The MIT Press, 2000).

Max Neuhaus, "Sound art?," liner for the CD *Volume: Bed of Sound* (New York: P.S. 1 Contemporary Art Center, July 2000), max-neuhaus.info/bibliography

Keith Potter, *Four Musical Minimalists: La Monte Young, Terry Riley, Steve Reich, Philip Glass* (Cambridge, UK: Cambridge University Press, 2000).

Mark J. Prendergast, *The Ambient Century. From Malher to Trance: The Evolution of Sound in the Electronic Age* (London: Bloomsbury, 2000).

Jacques Rancière, *Le Partage du sensible: Esthétique et politique* (Paris: La Fabrique, 2000).

Peter Shapiro, *Modulations: A History of Electronic Music: Throbbing Words on Sound* (New York: Caipirinha Productions, 2000).

Charles A. Stainback (ed.), *S.O.S.: Scenes of Sounds*, exhibition catalogue (Saratoga Springs, NY: The Frances Young Tang Teaching Museum and Art Gallery, 2000).

David Toop, *Sonic Boom: The Art of Sound*, exhibition catalogue (London: Hayward Gallery Publishing, 2000).

David Toop, *The Rap Attack 3: African Rap to New York Hip Hop*, 3rd revised, expanded, and updated edition [1st ed. 1994] (London: Serpent's Press, 2000).

Elisa Vaccarino (ed.), *Automi, marionette e ballerine nel Teatro d'avanguardia*, exhibition catalogue (Trento/Milan: Museo di Arte Moderna e Contemporanea di Trento e Rovereto/Skira, 2000).

2001

William Anastasi: A Retrospective, exhibition catalogue (Copenhagen: Nikolaj Contemporary Art Center, 2001).

Daniel Birnbaum, Amanda Sharp, Jörg Heiser, *Doug Aitken* (London/New York: Phaidon, 2001).

Germano Celant (ed.), *Dennis Oppenheim: Explorations* (Milan: Charta, 2001).

Sue Cramer (ed.), *Art > Music: Rock Pop Techno: An Exhibition Looking at Crossovers Between Visual Art and Music*, exhibition catalogue (Sydney: Museum of Contemporary Art, 2001).

Fred Collopy, Robert M. Fuhrer, "A Visual Programming Language for Expressing Visual Rhythms," *Journal of Visual Programming Languages*, no. 12, 2001, pp. 283–97.

Silke Feldhoff, *– zum Raum wird hier die Zeit. Günther Uecker Bühnenskulpturen und optische Partituren*, exhibition catalogue (Berlin: GH, 2001).

Celeb Kelly, *Sound* (Cambridge, MA: The MIT Press, 2001).

Manfred Kröplien (ed.), *Karl Gerstner: Review of 5 x 10 Years of Graphic Design etc.* (Ostfildern-Ruit: Hatje Cantz, 2001).

Anthony Hubermann (ed.), *Janet Cardiff: A Survey of Works, Including Collaborations with George Bures Miller* (New York: P.S.1 Contemporary Art Center, 2001).

Brandon LaBelle, Christof Migone (eds.), *Writing Aloud: The Sonics of Language* (Los Angeles/Downney: Errant Bodiees Press/Ground Fault Recordings, 2001).

Gloria Moure (ed.), *Jannis Kounellis: Works, Writings 1958–2000* (Barcelona: Éditiones Poligrafa, 2001).

Timothy D. Taylor, *Strange Sounds: Music, Technology, and Culture* (New York: Routledge, 2001).

Nabeel Zuberli, *Sound English: Transnational Popular Music* (Urbana, IL: Illinois University Press, 2001).

2002

Theodor W. Adorno, *Essays on Music* (Berkeley, CA/London: University of California Press, 2002).

Christine van Assche (ed.), *Sonic Process: A New Geography of Sound*, exhibition catalogue (Barcelona: Museu d'Art Contemporani de Barcelona/Actar, 2002).

Jacques Attali, *Noise: The Political Economy of Music* (Minneapolis: University of Minnesota Press, 2002).

Nicolas Bourriaud, *Postproduction: Culture as Screenplay* (New York: Lucas & Sternberg, 2002).

Hans-Joachim Braun, *Music and Technology in the Twentieth Century* (Baltimore/London: The John Hopkins University Press, 2002).

Christoph Brockhaus, Hubertus Gaßner, Christoph Heinrich (eds), *Stephan von Huene-Tune the World*, exhibition catalogue (Ostfildern: Hatje Cantz, 2002).

Marcel Duchamp, exhibition catalogue (Ostfildern: Hatje Cantz, 2002).

Hannah Higgins, *Fluxus Experience* (Berkeley, CA: University of California Press, 2002).

Margaret Livingstone, *Vision and Art: The Biology of Seeing* (New York: Harry N. Abrams Inc., 2002).

Trevor J. Pinch, *Analog Days: The Invention and Impact of the Moog Synthesizer* (Cambridge, MA: Harvard University Press, 2002).

Sandra Solimano, *The Fluxus Constellation*, exhibition catalogue (Genoa: Neos Edizioni, 2002).

2003

Roy Ascott, Edward A. Shanken (ed.), *Telematic Embrace: Visionary Theories of Art, Technology, and Consciousness* (Berkeley, CA/London: University of California Press, 2003).

Russel Ferguson (ed.), *Christian Marclay* (London: Thames & Hudson, 2003).

Christian Marclay, exhibition catalogue (Los Angeles, CA: Steidl/UCLA Hammer Museum, 2003).

Hatanaka Minoru (ed.), *Soundings Spaces – 9 Sound Installations*, exhibition catalogue (Tokyo: ICC – Intercommunication Center Publishing, 2003).

Michael O'Pray, *Avant-garde Film: Forms, Themes and Passions* (London: Wallflower, 2003).

Christiane Paul, *Digital Art* (London: Thames & Hudson, 2003).

Jonathan Sterne, *The Audible Past: Cultural Origins of Sound Reproduction* (Durham, NC: Duke University Press, 2003).

2004

Dawn Ades (ed.), *Dalí. The Centenary Retrospective*, exhibition catalogue (London: Thames & Hudson, 2004).

Germano Celant, *Arti & Architettura 1900/1968: Scultura, pittura, fotografia, design, cinema e architettura: un secolo di progetti creativi* (Genoa/Milan: Palazzo Ducale/Skira, 2004).

Christoph Cox, Daniel Warner, *Audio Culture: Readings in Modern Music* (London/New York: Continuum, 2004).

Jim Drobnick (ed.), *Aural Cultures* (Toronto: YYZ Books/Walter Philips Gallery, 2004).

Sophie Duplaix, Marcella Lista (eds.), *Sons & Lumieres: A History of Sound in the Art of the 20th Century*, exhibition catalogue (Paris: Centre Georges Pompidou, 2004).

Veit Erlmann (ed.), *Hearing Cultures: Essay on Sound, Listening and Modernity* (Oxford, UK: Berg, 2004).

Stephen Feeke, Jon Wood (eds.), *With Hidden Noise: Sculpture, Video and Ventriloquism*, exhibition catalogue (Leeds, UK: The Henry Moore Institute, 2004).

Mark Hansen, *New Philosophy for New Media* (Cambridge, MA: The MIT Press, 2004).

Thomas Jacobsen, "Kandinsky's Color-Form Correspondence and the Bauhaus Colors: An Empirical View," *Leonardo*, vol. 37, no. 2 (Cambridge, MA: The MIT Press, 2004), pp. 135–36.

Mark Katz, *Capturing Sound: How Technology Has Changed Music* (Berkeley, CA: University of California Press, 2004).

Igor Kipnis (ed.), *The Harpsichord and Clavichord: An Encyclopedia* (New York: Taylor & Francis Group, 2004).

Olivier Lussac, *Happening and fluxus: Polyexpressivité et pratique concrète des arts* (Paris: Editions L'Harmattan, 2004).

Walter Marchetti: Música visible, exhibition catalogue (Las Palmas: Centro Atlàntico de Arte Moderno, 2004).

Arthur W.J.G. Ord-Hume, *Automatic Pianos: A Collector's Guide to the Pianola, Barrel Piano, & Aeolian Orchestrelle* (Atglen: Shiffer Publishing, 2004).

Mark Rosenthal (ed.), *Joseph Beuys: Actions, Vitrines, Environments*, exhibition catalogue (Houston, TX: The Menil Collection/Tate Publishing, 2004).

David Toop, *Haunted Weather: Music, Silence and Memory* (London: Serpent's Tail, 2004).

Wolf Vostell, *Vostell y la música: Museo Vostell Malpartida* (Cáceres: Editora Regional de Extremadura, 2004).

2005

Marion Ackermann, Hans Werner Holzwarth (eds.), *Rebecca Horn: Moon Mirror. Site-specific installations, 1982–2003* (Ostfildern-Ruit: Hatje Cantz, 2005).

Bill Alves, "Digital Harmony of Sound and Light," *Computer Music Journal*, vol. 29, no. 4, Winter 2005, pp. 45–54.

Arman, *Arman: Conversations with Objects* (London: CV Publications, 2005).

Janet Cardiff, Mirjam Schaub (ed.), *Janet Cardiff: The Walk Book*, exhibition catalogue with CD (Vienna: Thyssen-Bornemisza Art Contemporary, 2005).

Computer Music Journal, vol. 29, no. 4 (Cambridge, MA: The MIT Press, Winter 2005).

Lina Dzuverovic Russel, Anne Hilde Neset (eds.), *Her Noise*, exhibition catalogue (London: Forma, 2005).

Jennifer A. González (ed.), *Christian Marclay* (New York: Phaidon 2005).

Janet Kraynak (ed.), *Please Pay Attention Please: Bruce Nauman's Words* (Cambridge, MA: The MIT Press, 2005).

Olivia Mattis, Kerry Brougher (eds.), *Visual Music: Synaesthesia in Art and Music Since 1900*, exhibition catalogue (London/New York: Thames & Hudson, 2005).

Thomas McEvilley, *The Triumph of Anti-Art: Conceptual and Performance Art in the Formation of Post-Modernism* (New York: McPherson and Co., 2005).

Pauline Oliveros, *Deep Listening: A Composer's Sound Practice* (New York: iUniverse, 2005).

2006

Jean-Francois Augoyard, *Sonic Experience: A Guide to Everyday Sounds* (Montréal: McGill Queen's University Press, 2006).

Germano Celant, *Tom Sachs*, exhibition catalogue (Fondazione Prada: Milan, 2006).

Valeria Dehò, Sound Zero. *Arte e musica dalla Pop alla Street Art*, exhibition catalogue (Merano/Bologna: Merano arte edificio Cassa di Risparmio/Damiani, 2006).

Ida Giannelli, Marcella Beccaria (eds.), *Claes Oldenburg. Coosje van Bruggen. Scultura per caso*, exhibition catalogue (Milan: Skira, 2006).

Gli strumenti musicali della collezione dei Musei Civici Veneziani (Venice: Musei Civici Veneziani d'Arte e di Storia/Marsilio Editori, 2006).

Mark Godfrey, Hans Ulrich Obrist, Liam Gillik, *Anri Sala* (London/New York: Phaidon, 2006).

J. J. Haspels (ed.), *Royal Music Machines*, exhibition catalogue (Zutphen: Walburg Pers, 2006).

Brandon LaBelle, *Background Noise – Perspectives on Sound Art* (London: Continuum International, 2006).

Max Vandervorst, *Nouvelle lutheries sauvages* (Paris: Alternatives, 2006).

2007

Les Back, *The Art of Listening* (Oxford/New York: Berg Publishers, 2007).

François Baschet, *Les mémoires sonores* (Paris: L'Harmattan, 2007).

Luca Beatrice (ed.), *Sound & Vision* (Bologna: Damiani, 2007).

Germano Celant, Gianfranco Maraniello (eds.), *Vertigo. Il secolo di arte off-media dal Futurismo al Web*, exhibition catalogue (Milan: Skira, 2007).

Fred Collopy, "Visual Music as a Performing Art," *Offscreen*, vol. 11, nos. 8–9, August-September 2007.

Sophie Diplaix (ed.), *Art contemporain. La Collection du Centre Pompidou, Musée national d'art moderne* (Paris: Centre Pompidou, 2007).

Double Vision: The Poetic Focus of Claes Oldenburg and Coosje van Bruggen, exhibition catalogue (London: Waddington Galleries, 2007).

Franca Faletti, Renato Meucci, Gabriele Rossi Rognoni (eds.), *Marvels of Sound and Beauty: Italian Baroque Musical Instruments*, exhibition catalogue (Florence: Giunti, 2007).

Danièle Freuensohn, *Bernard Baschet: Chercheur et sculpteur de sons* (Paris: L'Harmattan, 2007).

Christoph Grunenberg, Jonathab Harris (eds.), *Summer of Love: Psychedelic Art, Social Crisis and Counterculture in the 1960s*, exhibition catalogue (Liverpool, UK: Tate Liverpool/Liverpool University Press, 2007).

Paul Hegarty, *Noise/Music: A History* (London/New York: Bloomsbury Academic, 2007).

Bart Hopkin, Phil Dadson, Slap Tubes *et al.*, *Positive Aerophones: The greatest music-making idea that most people never heard of* (Point Reyes Station: Experimental Musical Instruments, 2007).

Thomas Kellein, *The Dream of Fluxus: George Maciunas. An Artist's Biography* (London: Edition Hansjörg Mayer/Thames & Hudson, 2007).

Udo Kittelmann, Mario Kramer (eds.), *Andreas Slominski*, exhibition catalogue (Frankfurt/Cologne: MMK-Museum für Moderne Kunst/Walther König, 2007).

Alan Licht, *Sound Art – Beyond Music, Between Categories*, exhibition catalogue (New York: Rizzoli, 2007).

Constance M. Lewallen, *A Rose Has No Teeth. Bruce Nauman in the 1960s*, exhibition catalogue (University of California Press, 2007).

Alberto Manodori Sagredo, *Italia Musicale: Strumenti e musicisti nell'iconografia fotografica soprattutto quotidiana 1860–2000* (Bologna: Bononia University Press, 2007).

Katalin Mollek Burmeister (ed.), *Marco Bagnoli* (Pistoia: Gli Ori, 2007).

Christopher Mooney, "Subodh Gupta: Idol Thief," *Art Review*, no. 17, December 17, 2007.

Arthur W.J.G. Ord-Hume, *Automatic Organs: A Guide Orchestrions, Barrel Organs, Fairgrounds, Dancehall & Street Organs Including Organettes* (Atgen: Shiffer Publishing, 2007).

Alex Ross, *The Rest is Noise. Listening to the Twentieth Century* (London: Fourth Estate, 2007).

Petra Stegmann (ed.), *Fluxus East: Fluxus Networks in Central Eastern Europe*, exhibition catalogue (Berlin: Künstlerhaus Bethanien, 2007).

Ghislaine Wood, *Surreal Things: Surrealism and Design*, exhibition catalogue (London: V&A Publications, 2007).

Yumi Yamaguchi, *Warriors of Art: A Guide to Contemporary Japanese Artists* (Tokyo: Kodansha international Ltd, 2007).

2008

Karin Bijsterveld, *Mechanical Sound: Technology, Culture and Public Problems of Noise in the Twentieth Century* (Cambridge, MA/London: The MIT Press, 2008).

Reinhard Braun, "Radio Amidst Technological Ideologies," in Heidi Grundmann, Elisabeth Zimmermann, Reinhard Braun *et al.* (eds.), *Re-Inventing Radio: Aspects of Radio as Art* (Frankfurt am Main: Revolver, 2008), pp. 233–46.

Daniela Cascella, *Scultori di suono* (Tuttle Edizioni: Camucia, 2008).

Subodh Gupta, *Subodh Gupta: Gandhi's Three Monkeys*, exhibition catalogue (New York: Jack Shainman Gallery, 2008).

Branden W. Joseph, *Beyond the Dream Syndicate: Tony Conrad and the Arts after Cage* (New York: Zone Book, 2008).

Daniel Makagon, Mark Neumann, *Recording Culture: Audio Documentary and the Ethnographic Experience* (Thousand Oaks, CA: Sage Publications, 2008).

Christian Marclay (ed.), *Ensemble*, exhibition catalogue (Philadelphia, Institute of Contemporary Art, University of Pennsylvania, 2008).

Martha Mockus, *Sounding Out: Pauline Oliveros and Lesbian Musicality* (New York/London: Routledge, 2008).

Theresa Sauer, *Notation 21* (Brooklyn, NY: Mark Batty Publisher, 2008).

Thomas Trummer, *Voice & Void* (Ridgefield: The Aldrich Museum, 2008).

Jonathan Watkins, Michael Archer, Yukie Kamiya, *Martin Creed*, exhibition catalogue (Birmingham: Ikon Gallery, 2008).

2009

John Luther Adams, *The Place Where You Go to Listen: In Search of an Ecology of Music* (Middletown, CT: Wesleyan University Press, 2009).

Berry Bergdoll, Leah Dickerman, *Bauhaus. Workshop for Modernity, 1919–1933*, exhibition catalogue (The Museum of Modern Art, New York, 2009).

Nicolas Bourriaud, *The Radicant* (New York: Lucas & Sternberg, 2009).

Fred Collopy, "Playing (With) Color," *GLIMPSE journal | the art + science of seeing*, a. II, vol. 2, no. 3, Autumn 2009, pp. 62–67.

Bruno Corà, Pietro Bellasi (eds.), *Corpo, Automi, Robot: Tra Arte, Scienza e Tecnologia*, exhibition catalogue (Lugano/Milan: Museo d'Arte Città di Lugano/Mazzotta, 2009).

Dietrich Diederichsen (ed.), *Shere – Stein – Papier. Pop-Musik als Gegenstand Bildender Kunst*, exhibition catalogue (Graz: Kunsthaus Graz/Walther König, 2009).

Francis Dyson, *Sounding New Media: Immersion and Embodiment in the Arts and Culture* (Berkeley, CA: University of California Press, 2009).

Gladys C. Fabre, Doris Wintgens Hötte, *Van Doesburg and the International Avant-Garde: Constructing a New World*, exhibition catalogue (London: Tate, 2009).

Ben Geraerts (ed.), *Nummer 10. Works 2003–2009: Guido van der Werve* (Rotterdam: Veenman Publishers, 2009).

Loris Gréaud, *Trajectories and Destinations. Vol. 1* (Berlin: Sternberg Press, 2009).

Caleb Kelly, *Cracked Media: The Sound of Malfunction* (Cambridge, MA/London: The MIT Press, 2009).

Seth Kim-Cohen, *In the Blink of an Ear: Toward a Non-Cochlear Sonic Art* (New York/London: Continuum, 2009).

Udo Kittelmann, *Thomas Demand Nationalgalerie*, exhibition catalogue (Berlin: Neue Nationalgalerie/Steidl Mack, 2009).

Joann Moser, *What's It All Mean: William T. Wiley in Retrospect*, exhibition catalogue (Berkeley, CA: University of California Press, 2009).

Susanne Neuburger (ed.), *Nam June Paik. Exposition of Music: Electronic television. Revisited*, exhibition catalogue (Wien: Museum Moderner Kunst Stiftung Ludwig/Walther König, 2009).

Tara Rodgers, *Pink Noises: Women on Electronic Music and Sound* (Durham, NC: Duke University Press Books, 2009).

Gabriele Rossi Rognoni (ed.), *Strumenti Musicali. Guida alle collezioni medicee e lorenesi* (Florence: Galleria dell'Accademia/Giunti, 2009).

Guido Van Der Werve: Minor Pieces, exhibition catalogue (Sheffield, UK: Site Gallery, 2009).

2010

Terry Allen, *Terry Allen* (Austin: University of Texas Press, 2010).

Mónica Artigas, "La Chambre des Machines experimenta amb el so industrial com a cloenda del festival de música avançada," *Publico*, June 21, 2010.

Manuela Ammer, Cosima Rainer, Stella Rollig *et al.* (eds.), *SEE THIS SOUND: Promises in Sound and Vision*, exhibition catalogue (Cologne: Walther König, 2010).

Carlos Basualdo (ed.), *Michelangelo Pistoletto: From One to Many, 1956–1974*, exhibition catalogue (Philadelphia, PA: Publishing Department Philadelphia Museum of Art, 2010).

Jean-Michel Bouhours (ed.), *Arman*, exhibition catalogue (Paris: Centre Pompidou, 2010).

Karin Bucher Trantow, Peter Pakesch, Andres Pardey *et al.*, *Roboter-träume*, exhibition catalogue (Basel/Graz: Museum Tinguely/Kunsthaus Graz – Kehrer, 2010).

José Manuel Costa (ed.), *ARTe SONoro*, exhibition catalogue (Madrid: La Casa Encendida, 2010).

Paul De Marinis, *Buried in Noise* (Berlin: Kehrer, 2010).

Joanna Demers, *Listening Through the Noise. The Aesthetics of Experimental Electronic Music* (New York: Oxford University Press, 2010).

Daniel Genthmann, *Klangmaschinen zwischen Experiment und Medientechnik* (Bielefeld: Transcript Verlag, 2010).

Brandon LaBelle, *Acoustic Territories: Sound Culture and Everyday Life* (New York/London: Continuum, 2010).

Carsten Nicolai, *Moiré index* (Berlin: Gestalten, 2010).

Julia Robinson, *New Realisms, 1957–1962: Object Strategies between Readymade and Spectacle*, exhibition catalogue (Madrid/Cambridge, MA: Museo Nacional Centro de Arte Reina Sofia, The MIT Press, 2010).

Franklin Sirmans (ed.), *Maurizio Cattelan: Is There Life Before Death?*, exhibition catalogue (Huston, TX: The Menil Collection/Yale University Press, 2010).

David Toop, *Sinister Resonance: The Mediumship of the Listener* (New York/London: Continuum, 2010).

Pietro Verardo, Riccardo Brazzale (eds.), *La Distruzione della Quadratura: Luigi Russolo, Francesco Balilla Pratella e l'avventura della musica futurista*, exhibition catalogue (Vicenza: Comune di Vicenza – Assessorato alla cultura, 2010).

Peter Vergo, *The Music of Painting: Music, Modernism and the Visual Arts from the Romantics to John Cage* (London/New York: Phaidon, 2010).

Salomé Voegelin, *Listening to Noise and Silence. Towards a Philosophy of Sound Art* (New York/London: Continuum, 2010).

2011

Larry Austin, Douglas Kahn (eds.), *Source: Music of the Avant-garde, 1966–1973* (Berkeley, CA/London: University of California Press, 2011).

Jacquelynn Baas (ed.), *Fluxus and the Essential Questions of Life*, exhibition catalogue (Hanover, NH/Chicago, IL/London: Hood Museum of Art/Dartmouth College/University of Chicago Press, 2011).

Catherine Crowsten (ed.), *Janet Cardiff and George Bures Miller: The murder of crows*, exhibition catalogue (Alberta/Ostfildern: Art Gallery of Alberta/ Hatje Cantz, 2011).

Jeremy Grimshaw, *Draw a Straight Line and Follow It: The Music and Mysticism of La Monte Young* (New York: Oxford University Press, 2011).

Rosalind Krauss, *Under Blue Cup* (Cambridge, MA/London: The MIT Press, 2011).

Pavle Levi, *Cinema by Other Means* (New York: Oxford University Press, 2011).

Christian Marclay, *Cyanotypes* (Zurich: JRP|Ringier, 2011).

Ingrid Pfeiffer, Max Hollein (eds.), *Surreal Objects. Three Dimensional Works from Dalí to Man Ray*, exhibition catalogue (Ostfildern: Hatje Cantz, 2011).

Edi Rama, Anri Sala, *Anri Sala/Inversion*, exhibition catalogue (Montréal: Musée d'art contemporain de Montréal, 2011).

Tom Sachs: Work 2008-2011, exhibition catalogue (New York: Sperone Westwater, 2011).

Christopher Scoates, *Bullet Proof… I Wish I Was: The Lighting and Stage Design of Andi Watson* (San Francisco: Chronicle Books, 2011).

Nancy Spector (ed.), *Maurizio Cattelan: All*, exhibition catalogue (New York: Guggenheim Museum, 2011).

2012

Heather Anderson (ed.), *Sounding Selves*, exhibition catalogue (Halifax: Dalhousie Art Gallery, 2012).

Richard Artschwager, exhibition catalogue (Rome: Gagosian Gallery, 2012).

Céleste Boursier-Mougenot, exhibition catalogue (Amherst, MA: UMass – University Museum of Contemporary Art, 2012).

Daniela Cascella, *En Abîme: Listening, Reading, Writing: An Archival Fiction* (Winchester: Zero Books, 2012).

Luciano Chessa, *Luigi Russolo, Futurist. Noise, Visual Arts and the Occult* (Berkeley, CA: University of California Press 2012).

David Crowley, Daniel Muzyczuk (eds.), *Sounding the Body Electric: Experiments in Art and Music in Eastern Europe*, exhibition catalogue (Lodz: Muzeum Sztuki Publisher, 2012).

Ruth Ewan, exhibition catalogue (Dundee/Copenhagen: Dundee Contemporary Arts/Kunsthal Charlottenborg, 2012).

Patrick Feaster, *Picture of Sound. One Thousand Years of Educed Audio: 980–1980* (Atlanta, GA: Dust-to-Digital, 2012).

Michael Goddard, Benjamin Halligan, Paul Hegarty (eds.), *Reverberations: The Philosophy, Aesthetics and Politics of Noise* (London/New York: Continuum, 2012).

Achim Hochdörfer, *Claes Oldenburg: The Sixties*, exhibition catalogue (Munich: Prestel, 2012).

Jeffrey Kastner, *Nature* (London: Whitechapel Editions, 2012).

Petra Kipphoff von Huene, Marvin Altner, *Stephan von Huene. Die Gespaltene Zunge. Split Tongue: Texte und Interviews* (Chicago, IL: Chicago University Press, 2012).

Sabrina van der Ley, *I Wish I Was a Song – Music in Contemporary Art*, exhibition catalogue (Oslo: Nasjonalmuseet for kunst, arkitektur og design, 2012).

Christine Macel (ed.), *Anri Sala*, exhibition catalogue (Paris: Centre Pompidou, 2012).

Christoph Metzger, Ulrich Eller (eds.), *Vol. 01 Drawings, Sounds & Ambiences*, exhibition catalogue (Berlin: Kehrer, 2012).

Christof Migone, *Sonic Somatic: Performances and the Unsound Body* (Los Angeles, CA: Errant Bodies Press, 2012).

Ellen Pearlman, *Nothing & Everything: The Influence of Buddhism on the American Avant-Garde. 1942–1962* (Berkeley, CA: Evolver Edistions, 2012).

Jonathan Sterne (ed.), *The Sound Studies Reader* (London/New York: Routledge, 2012).

Marco Tagliaferro, "Riccardo Beretta," in *Artforum*, October 2012.

2013

Ilaria Bonacossa (ed.), *Alberto Tadiello: High Gospel* (Milan: Mousse Publishing, 2013).

Germano Celant (ed.), *Eliseo Mattiacci* (Milan: Skira, 2013).

Emanuel Dimas de Melo Pimenta (ed.), *One Hundred Years with John Cage* (CreateSpace Independent Publishing Platform, 2013).

Fernando Frances (ed.), *Subodh Gupta: The Imaginary Order Of Things*, exhibition catalogue (Malaga: Centro de Arte Contemporáneo, 2013).

Giacomo Fronzi, *Electrosound. Storia ed estetica della musica elettronica* (Turin: EDT, 2013).

Loren Hansi Momodu, Andi-Asmita Rangari (eds.), *Word. Sound. Power.*, exhibition booklet (London: Tate Modern/ Tate Design Studio, 2013).

David Hendy, *Noise. A Human History of Sound & Listening* (London: Profile Books, 2013).

Kati Ilves, Ragne Nukk (eds.), *Out of Sync: Looking Back at the History of Sound Art*, exhibition catalogue (Tallin, EE: Eesti Kunstimuuseum, 2013).

Douglas Kahn, *Earth Sound Earth Signal: Energies and Earth Magnitude in the Arts* (Berkeley, CA: University of California Press, 2013).

Barbara London (ed.), *Soundings: A Contemporary Score*, exhibition catalogue (New York: The Museum of Modern Art, 2013).

Karen Marta (ed.), *Doug Aitken: 100YRS* (New York: Rizzoli, 2013).

Sylvia Martin, Sabine Röder, Bernward Wember (eds.), *Vibrating Pictures Noisy Sculptures 1958–63 – A Homage to Paul Wember*, exhibition catalogue (Krefeld, DE: Kunstmuseen Krefeld/Museum Haus Lange Publisher, 2013).

Haroon Mirza, *A User's Manual* (London: Spike Island, 2013).

Carsten Nicolai, *Syn chron* (Berlin: Gestalten, 2013).

Hans Ulrich Obrist, *A Brief History of New Music* (Zurich/Dijon: JRP|Ringier/ Les presses du reel, 2013).

Didier Ottinger (ed.), *Dictionnaire de l'Object Surréaliste* (Paris: Centre Pompidou/Éditions Gallimard, 2013).

Pedro Reyes, *Sanatorium Operations Manual* (Geneva: Head – Genève/ Ridinghouse, 2013).

Holly Rogers, *Sounding the Gallery: Video and the Rise of Art-Music* (Oxford, UK/ New York: Oxford University Press, 2013).

2014

Torsten Blume, Christian Hiller (eds.), *Human – Space – Machine: Stage Experiments at the Bauhaus*, exhibition catalogue (Dessau: Stiftung Bauhaus Dessau/Spector Books, 2014).

Subodh Gupta, *Everything is inside* (London/New Delhi: Penguin, 2014).

Tim Shephard, Anne Leonard (eds.), *The Routledge Companion to Music and Visual Culture* (New York/London: Routledge, 2014).

Wendy Vogel, Rachel Ward (eds.), *Index A to Z: Art, Design, Fashion, Film, and Music in the Indie Era* (New York: Rizzoli, 2014).

Peter Weibel, *Sound Art – Sound as a Medium of Art*, exhibition catalogue (Karlsruhe: ZKM Publisher, forthcoming).

ART OR SOUND. DAL PLURI-LINGUISTICO AL MULTISENSORIALE

Germano Celant

L'ABBRACCIO DEI SENSI

Il Museo è diventato il luogo dell'astinenza sensoriale: un territorio ottico-centrico che permette un viaggio conoscitivo che avviene solo attraverso lo sguardo, in cui l'impiego di sensi non visuali è rimosso. Tale focalità univoca, basata su molteplici restrizioni sensoriali – quelle dell'olfatto, dell'udito, del gusto e del tatto –, inizia a essere messa in discussione al fine di trovare sistemi capaci di sollecitare gli altri sensi. Questa svolta nell'approccio sensoriale, seppur ancora rara e limitata, sta portando a una concezione espositiva multisensoriale dell'arte, dove la dittatura del vedere lascia posto a una democrazia che include l'odorare, l'ascoltare, il gustare e il toccare. L'intento è di arricchire i modi dell'esporre includendo elementi e materie che non richiedano solo una risposta connessa al guardare, ma anche relativa al sentire con altri organi del corpo: dalle mani alla bocca, dal naso all'orecchio. Un simile procedere aspira a realizzare un'interazione diretta fra il visitatore e le opere, stabilendo un rapporto esperienziale che coinvolge tutta la persona. La possibilità di ascoltare, toccare, gustare e odorare comporta un coinvolgimento viscerale e sensuale; coinvolgimento che si amplifica con modalità divergenti e molteplici che immergono il visitatore in un universo più emotivo e sensibile, atto ad arricchire la lettura e la percezione estetiche del manufatto artistico esposto.

La pratica espositiva sensoriale può inoltre servire a rompere gli schemi normativi del mostrare e del guardare l'arte, oggi affidata a sistemi restrittivi che demonizzano la partecipazione e lo scambio corporeo fra cose e persone. Nella relazione che nei musei si instaura tra visitatore e manufatto artistico manca la concezione orientale che prevede quella potenziale dialettica per cui è la statua della divinità a guardare il suo fruitore, il quale risponde con offerte e regali, candele e doni, fiori e incensi. Nello spazio intermedio fra essi, scorre un flusso di energia, per cui – da un polo all'altro – l'arte innesca la persona e l'individuo attiva la scultura. Si stabilisce un rapporto di reciprocità e l'aspetto cerimoniale del trovarsi dinanzi a una rappresentazione sottende un rituale di sollecitazione esperienziale costante e infinita.

Diversamente dalla cultura orientale, gli allestimenti e le esposizioni si affidano quasi totalmente al primato del vedere e dell'osservare, del camminare e del percorrere, quasi che la fonte della conoscenza dovesse ruotare solo sull'osservazione dell'immagine, accompagnata a volte dalla scrittura, poiché entrambe – per tradizione antica e filosofica – appaiono più affidabili sul piano del raziocinio e del pensiero. Le ragioni storiche di questo lessico privilegiato della trasmissione della conoscenza, affidato all'osservare e al leggere, sono certamente rintracciabili nella consuetudine di favorire il pensiero a svantaggio dei sensi, seppur questi abbiano stretti legami e significative connessioni con i processi dell'intelletto e la coscienza del reale.

Di fatto il sistema del Museo e delle sue tecniche espositive mostra grandi difficoltà nell'includere le sensazioni e gli elementi naturali, sia che essi riguardino gli esseri umani o gli animali, la terra o il cielo. Esclude e confina il fuoco e l'acqua, la pioggia e la neve; si spaventa dinanzi alla presenza di qualsiasi forma di fauna e di flora – dal pappagallo al cavallo, dal cactus all'albero – quasi per evitare qualsiasi contatto con gli elementi organici e vitali. Perpetua la condanna aristotelica che identifica nel linguaggio – visuale e scritto – la distinzione tra essere umano e animale, da cui scaturisce la negazione di qualsiasi partecipazione istintiva, considerata negativa e inferiore, imprecisa e casuale, impura e disordinata. È un rinnegare le tentazioni degli altri sensi che fa parte di quell'induzione alla purezza professata dall'idealismo culturale occidentale. In esso si tende ad abolire ogni piacere fisico e carnale in quanto fonte di godimento corporeo antitetico all'ipersensibilità intellettuale e spirituale.

Per sottrarsi a questo processo di riduzione della partecipazione sensoriale, gli spazi museali non possono più essere pensati come territori clinici, asettici e sterili, in cui dedicarsi alla conservazione dei "corpi" imbalsamati e congelati della cultura visiva. È necessario sincronizzarli con la vita, con i suoi profumi e i suoi olezzi, i suoi rumori e i suoi suoni, le sue durezze e le sue mollezze e con qualsivoglia sensazione gustativa. Da ospedale o da casa del "defunto", il Museo va trasformato in un luogo vitale,[1] affinché si avvicini al mondo reale – pieno di stimoli sensoriali – e si allontani dal *diktat nevrotico* del non toccare, non ascoltare, non gustare e non odorare.

Tuttavia, questi ingredienti *invisibili* e *volatili* continuano a essere assenti nei sistemi del comunicare e del mostrare l'arte. Anzi, qualsiasi interesse per queste dimensioni del percepire viene guardato con sospetto e osteggiato, quasi che l'educazione, intesa come "cura culturale", non possa prevedere nelle sue "farmacie" la messa in circolazione della stimolazione fisica e sensoriale. Eppure l'odoroso, il rumoroso, il caldo, l'appiccicoso, il putrido, l'organico, il marcio sono un memento del vivere, come del morire. È forse per questo che negli ultimi decenni, dagli anni '60, la produzione artistica si è aperta al processo non solo del linguaggio cromatico e industriale – Pop Art e Minimalismo –, ma ha iniziato anche a includere elementi naturali e corporei che possono fornire nuove informazioni alla conoscenza. L'intento è stato di approntare una condizione interattiva dove l'inclusione di piante, di animali, di persone, di processi chimici e minerali, per il loro carattere attivo e mobile, possa indurre a considerare la reciprocità della situazione performativa in cui l'osservatore è elemento attivo di uno scambio vitale e dinamico.

Con le ricerche di Land Art, Arte Povera e Body Art si è sviluppata una *concezione olistica* dell'essere umano, in cui i piaceri intellettuali e spirituali non escludono quelli materiali e sensoriali che sono determinati dal tutto naturale. Questo modo di fare arte ha investito l'osservatore-visitatore ponendolo in una situazione circolare, non frontale, così da focalizzare la propria attenzione non su un punto bensì sull'intera situazione ambientale. È stimolato a lasciarsi coinvolgere, riverberando per tramite del proprio corpo l'oggetto presentato e mostrato. È un cambio di senso percettivo basato sul fatto di produrre suoni, rumori, odori e sensazioni che sono entità invisibili ma attive, e arrivano a occupare uno spazio che, non essendo unidirezionale, è avvolgente e sferico.

Un simile approccio a 360 gradi mette in discussione la centralità dello sguardo e la percezione si trasforma in una costellazione di ricezioni sensorie la cui localizzazione si manifesta in uno spazio omogeneo. Questa modalità del *comprendere* – nel senso letterale di prendere insieme – fa parte della serie di emancipazioni dai vincoli e dalle limitazioni territoriali nell'arte. Un percorso iniziato nel XX secolo con la messa in crisi, da parte dei futurismi italiani e russi e delle avanguardie storiche, della superiorità mitico-metafisica delle arti maggiori sulle minori, dell'arte e della scultura su fotografia, letteratura, musica, teatro e cinema. Si ipotizza allora un orientamento teso a non avvantaggiare un linguaggio sull'altro, per approdare a un'esteticizzazione plurilinguistica del reale.

Si apre così un territorio immenso, inesplorato e in divenire, che arriva a comprendere dichiarazioni e manifesti sul toccare sensuale e sul sentire sessuale. L'arte lascia posto alle arti che si dimostrano interessate a un intenso e reciproco scambio. Emergono zone inesplorate nelle quali è possibile far circolare insieme pittura e danza, fotografia e teatro, architettura e cinema, poesia e grafica, design e scultura... I limiti tendono a dissolversi e le arti viaggiano da un mondo visuale-plastico a un altro. L'itinerario, intorno al 1968, si adegua ai rivolgimenti originati dalle lacerazioni generazionali alla disperata ricerca di un'identità emancipata da un passato in cui tutto era coniugato secondo coordinate rigide e univoche, ideali e conservatrici. Lo strappo dovuto all'immissione di un'incandescenza linguistica connessa agli organismi naturali rende evanescenti, come abbiamo visto, i recinti per fare emergere energie sopite. Si sviluppa, accanto al vedere, l'attivazione del toccare, sentire, odorare e gustare, così da rituffare la creazione nel crogiolo ardente dei sensi. La vertigine creata da questi due momenti storici confluisce sul piano espositivo in progetti onnicomprensivi con eventi che vanno – sul piano delle esposizioni – da "When Attitudes Become Form", curata da Harald Szeemann alla Kunsthalle Bern nel 1969, alla serie "Paris-New York", "Paris-Berlin", "Paris-Moscow", curata da Pontus Hultén al Centre Pompidou di Parigi fra il 1977 e il 1979. Un coagulo di materiali e di espressioni – dallo storico al vivente –, che propone un pensiero artistico in relazione con il vissuto e le vicende del tempo, dall'architettura e dagli spazi urbani e culturali in cui si sono collocate o si collocano. È in questo pulsare espositivo – rappreso in un insieme di oggetti e di documenti, di manufatti e di media –, che l'asettico sguardo si intreccia all'eruzione dei materiali e dei dati. Le forme e i volumi precipitano in un contenitore, il Museo, che ritrova un contatto con la vita presente e passata e si scopre materiale attivo. A questo contribuiscono gli strumenti mediatici così come le comunicazioni sensoriali proposte dalle avanguardie di ieri e di oggi, che in molti casi ricorrono agli aspetti performativi e mutevoli delle cose, artificiali e naturali.

Negli anni '80 la mondializzazione scompagina l'asse artistico Occidente-Oriente, aprendo le porte a un pensare "globale" capace di declinare situazioni espressive e comunicative, estetiche ed etnico-religiose differenti. Dopo secoli di neutralizzazione e rimozione delle territorialità linguistiche "altre", il mondo artistico si apre alle interfacce identitarie non dominanti del femminile e dell'etnico. L'impostazione monolitica e univoca del suo sistema entra in cortocircuito ed emergono, o s'impongono, le produzioni locali; proliferano allora le comunità con interessi post-patriarcali e valori postcoloniali, da "Women Artists: 1550-1950", curata da Ann Sutherland Harris e Linda Nochlin (Los Angeles, 1976) a "Magiciens de la terre", curata da Jean-Hubert Martin (Parigi, 1989). Si pone il problema di una ricostruzione di una visione realmente cosmopolita – aperta a tutti i generi –, che tenda a ridefinire il termine "universale".

Questa ridotta mappatura dei diversi strappi attitudinali verso l'arte non aspira a definire le grandi trasformazioni avvenute, dal XX secolo a oggi, nell'universo del mostrare e dell'esperire il manufatto artistico, ma indica piuttosto passaggi fondanti che hanno spostato il baricentro dalla gerarchia tra tecniche, generi ed etnie a una democratizzazione di equivalenze tra i linguaggi estetici e hanno affiancato all'ordine unipolare dell'asse Occidente-Oriente una circolarità pluridirezionale. Sono riformulazioni del metodo precostituito di assumere l'arte per farla uscire da un dualismo e farla entrare in una dinamica aperta ed espansiva. Immessa in una struttura veramente cosmopolita, formata da tratti referenziali circolari e globali, la società dell'arte intesa come molteplicità di linguaggi e di culture arriva a nutrirsi di differenziazioni reticolari nuove e imprevedibili che colpiscono il cuore di una filosofia del mostrare gerarchica e vetero-occidentale.

Tuttavia molte sacche di resistenza si oppongono a queste aperture post-mediatiche e postcoloniali. Tra queste il passaggio a una dimensione multisensoriale che non è riconducibile soltanto al paradigma del vedere ma a tutto il sistema sensorio. Si continua a mantenere una subordinazione gerarchica per cui il vedere predomina, delimitando l'attenzione al di fuori e non dentro l'esperire. Il confrontarsi con i sensi assenti può dunque essere un altro progetto di apertura e di sconfinamento che porta a un'ulteriore manipolazione dei diversi rapporti possibili fra corpi e oggetti. Un mondo da apprendere attraverso l'ascolto e il tocco è più dinamico, complesso e indeterminato. Rispetto all'occhio, che stabilisce un territorio esplicito e amministrato, gli altri sensi sono impliciti, quindi più profondi e permeanti. Ciò porta a sviluppare nuove sensazioni delle cose e del proprio corpo, che intensificano la coscienza di sé e del reale. Il riverbero creato dall'udire e dal toccare nella persona la pone nella condizione di risperimentarsi. Allargando la relazione con il mondo, attraverso la presenza di strumenti sensoriali, si ottiene un'intensificazione in cui gli oggetti e i manufatti artistici si arricchiscono di aspetti complementari – suono, odore, gusto e tatto –, creando un paesaggio alternativo.

ART OR SOUND

Che cosa comporta, sul piano dei processi conoscitivi, il possibile mutamento di un paradigma multisensoriale? Si tratta innanzitutto di non cedere alla tentazione di dissimulare il dominio del visivo, tenendo presente l'esperienza dei sensi "scartati". Una condizione paradossale che di recente è stata sottoposta a critiche, teoriche e operative, in nome della globalità esperienziale. Oggi si tende a parlare e a professare una presenza simultanea e una co-appartenenza per far leva su una mescolanza di odore + suono + tatto + gusto + manufatto artistico in cui il pluralismo dell'esperienza è un tratto costitutivo dell'interpretazione ambientale e culturale nel mondo delle arti. Per superare la visione unilaterale del processo percettivo bisogna produrre un'interazione dove la parzialità venga ampliata e superata, comprimendo in uno stesso spazio esperienze sensoriali diverse. È quanto è successo negli ultimi decenni nell'attenzione posta alle relazioni tra suono + arte,[2] mentre molto rimane da fare per la composizione circolare e interagente tra gusto + tatto + odore + arte.[3] La coabitazione tra queste polarità e pluralità, in cui si incrociano soggetti ancora "in conflitto", può creare un luogo dell'esperire mobile e imprevedibile. La dinamica trans-sensoriale produce un flusso che risucchia nel suo vortice ogni unilateralità così da sollecitare un'altra differenza, basata sulla sollecitazione di tutti i sensi. Tale stereoscopia sensoriale si può esprimere oggi in senso sia congiuntivo sia disgiuntivo per cui, in una visione museale ed espositiva, gli elementi possono convivere come distinti e indistinti, in un gioco di specchi tra immagini tattili, sonore, visive e olfattive. Nella saldatura o nella messa in circuito di tali elementi, intimamente intrecciati, si tende a far sparire qualsiasi supremazia e a far nascere una costellazione esperienziale unitaria e totale.

La mostra "Art or Sound" vuole fare emergere tanto la contrapposizione quanto l'associazione tra le due polarità. Sottolinea sia il rispetto per l'identità del singolo linguaggio – arte e suono – sia la possibilità che si intreccino a formare una somma – arte + suono –, che convergono nel mettere in mostra oggetti nei quali si attuano i reciproci influssi e flussi. Inoltre, una volta affermatene la fusione e la dialettica, cerca di evidenziare un percorso storico, offrendo una raccolta, dal Rinascimento alla contemporaneità, di manufatti sonori, così da definire un'esperienza mnemonica, visiva e uditiva durante il percorso espositivo. Questo passaggio da ieri a oggi elude la riduzione al presente di una contaminazione pittorica e scultorea tra oggetti e musica al fine di cercarne le radici nel tempo.

Negli spazi di Ca' Corner della Regina a Venezia la questione dei confini tende a essere rimossa così da affermare una metodologia che dà forma alle due facce di una medaglia. L'ipotesi è di verificare e di documentare come sin dal XVI-XVII secolo, attraverso strumenti come l'orologio o l'automa, e in alcune collezioni pubbliche, tra cui il British Museum e l'Hermitage, si attui una tendenza a uniformare le zone sensoriali. È quindi un discorso sull'interdipendenza e sullo scambio, ma anche sull'autonomia che concerne la memoria e il ritorno di un esperire che nell'epoca vittoriana era indiviso, tanto che nella frequentazione delle collezioni e dei relativi musei erano contemplati il toccare, l'ascoltare, il gustare, l'odorare e il guardare.[4]

L'esercizio del suono anima l'intero corpo, echeggia in esso e gli conferisce una dimensione dinamica. Qualcosa che proviene dal suo esterno e non è prodotta dall'interno, pur arrivando a esso amplificata e modificata. Il risultato di un agente o di un produttore di suoni e di rumori – oltre che di silenzi –, che si afferma come sostanza avvolgente da ascoltare secondo una durata e una direzione. Il suono crea un rapporto interlocutorio che ha effetti su chi ascolta, intenzionalmente e casualmente. Il suono occupa uno spazio e torna in un circuito di feedback sonoro. S'irradia in un ambiente e si espande in esso. Si amplifica o si riduce rispetto all'assenza o alla presenza di ostacoli. Le informazioni acustiche e sensoriali sollecitano una coscienza amplificata delle qualità timbriche dei luoghi e dei corpi, quindi del mondo. La storia degli oggetti che sviluppano dimensione, presenza e colore mediante il suono – e quindi animano gli spazi – data nel tempo; quella che intreccia i sensi e l'arte per arrivare a costruire un nucleo autonomo, formato dalla somma suono + arte e arte + suono, si può far risalire ai primi tentativi di missaggio tra le arti, finalizzato a unire l'umano e il disumano, il naturale e l'artificiale, l'organico e l'industriale. Un'osmosi che s'inizia a praticare[5] già nel XIII secolo nel mondo arabo e poi si perfeziona nel Medioevo e nel Rinascimento per arrivare al suo apice nel XVII secolo, quando si pensò di aggiungere una dimensione motoria e sonora agli oggetti artistici – dagli orologi agli automi, come le creazioni dei Jacquet-Droz (il musicista, lo scrittore, il disegnatore) o il suonatore di flauto di Jacques de Vaucanson – che superava, ampliandola, l'informazione sta tica e visuale del tempo. È un primo passo verso la connessione e la reciprocità tra entità sensoriali in cui l'invisibile materialità del suono illumina l'esperienza personale rispetto alla distanza e alla posizione dell'oggetto. Al tempo stesso esso offre una memoria altra rispetto all'immagine visiva, che è temporanea, frammentata e contingente.

La creazione di un'equivalenza tra orecchio e occhio, visibile e invisibile, limitato e illimitato stabilisce un'eco di reciprocità che, in molti casi, produce connessioni tra stanza e stanza, oggetto vicino e oggetto lontano, anche se la posizione del visitatore ricettore non si trova dinanzi all'oggetto. Un effetto d'interazione rilevante che oggi è definito come "disturbo", ma che è invece un procedere multimodale e multisensoriale dell'arte ma anche della circostanza contingente che la accoglie. È la messa in discussione dell'autoritarismo dello sguardo, che continua anche nella pubblicazione – catalogo o libro – concernente l'evento espositivo, dove i rumori o i suoni delle cose, per non parlare del tatto, del gusto e dell'odore, sono del tutto assenti, evocati mediante mere riproduzioni.

La scelta di iniziare il percorso di "Art or Sound" con oggetti ibridi in cui la funzione si riflette in un altrove fantastico, che è intreccio tra i diversi sensi, il suono e l'immagine, è indice di un'articolazione estetica e iconografica che non intende fissarsi in una struttura rigida, per cui le cose si costruiscono mediante metamorfosi, amplificazioni e riduzioni. L'intento è di liberare istanze eterodosse e marginali per cui gli oggetti, dagli automi meccanici agli strumenti musicali, arrivino a evocare il meraviglioso e il sorprendente, seppur trasmettendo il suono in maniera tradizionale e classica. In tal modo l'oggetto può entrare in un complesso gioco di slittamenti: un violino o una chitarra diventano una scultura di marmo, come nel caso dell'opera di Michele Antonio Grandi [50], mentre un cornetto [46], un buccin [70] o un fagotto russo [102] assumono le sembianze di un serpente con la testa di un drago. Altrove una carrozza si tramuta in organo [66], mentre una gabbietta o una scatola [62] si trasformano in carillon, quasi si nutrissero della collisione fortuita che segna la successiva esperienza surreale. La scelta dell'iconografia zoomorfa potrebbe essere legata a un desiderio di "realismo" pittorico e scultoreo, tipico dell'Ottocento, per cui ci si aspettava che anche lo strumento musicale rendesse manifesta la sua presenza nella natura così come nella vita delle cose. In tale modo riusciva a legarsi al tipo di musica borghese o popolare, liturgica o laica. Tendeva ad apparire solenne o fantastico a seconda del contesto sociale in cui operava – dalla carrozza all'organo di strada e da fiera [132]. Era il risultato di un'iconografia che rappresentava l'antico – come nel caso del teatrino con automi –, oppure riprendeva le scoperte dell'industria e creava composizioni sonore mediante l'uso di gas, come nel caso del pirofono [114] di Frédéric Kastner, o di meccanismi che solennizzano le nuove tecniche industriali: il violino Stroh [180] inventato da J. M. Augustus Stroh, il *phonofiddle* [134] di Arthur Howson o la scatola musicale a cilindro [122] di Paillard Vaucher & Co.

La scelta di esporre simili oggetti diviene pertanto un metodo per evidenziare un'energia visiva e sonora, meccanica e materica che non è riconducibile a un'univoca catalogazione. Non si vuole impedire solo di vederli, ma ripensare a essi come emettitori di suono che hanno subito modificazioni estetiche e tecniche anche in relazione all'emergere della cultura industriale, che le avvicina sempre più a una scultura o a un oggetto artistico. Quasi aspirare a creare un effetto d'instabilità percettiva e plastica delle sostanze visibili e invisibili, materiali e immateriali, che possono essere sottoposte a un interscambio di proprietà estetiche e di effetti performativi. Un *perpetuum mobile* dove il manufatto artistico o il suono, come tutte le altre possibili espansioni sensoriali, come l'odore, il tatto, il gusto, siano incluse, approdando a un "altro-da-sé" multisensoriale. Ne sono un esempio storico gli strumenti o le sculture che aspirano a una sinestesia in cui l'orecchio e l'occhio "ascoltano" – come la rappresentazione plastica di una fuga di Johann Sebastian Bach [178] concepita da Henrik Neugeboren – perché in esse si attua l'irruzione di un sentire che mescola il dipinto alla musica, lo strumento musicale alla scultura, così da smontare e disfare i contorni tracciati dall'univocità del vedere e del percepire.

Una polisemia illimitata che inizia ad affermarsi con le avanguardie storiche del Novecento, dal Cubismo al Futurismo, in cui il principio di metamorfosi genera e distrugge le forme, da Giacomo Balla [162] a Pablo Picasso, e plasma l'oggetto per connotarlo plasticamente fuori della sua funzione. Lo stesso avviene nell'ambito musicale, dove l'immissione di strumenti "non musicali", come negli intonarumori [160] con cui l'artista Luigi Russolo porta a un incontro con il rumore, suggerisce la rottura o il limite della musica, mentre nel caso di Duchamp la sorpresa è l'*hidden noise* [164], il "rumore nascosto" che annuncia l'attenzione al silenzio di John Cage. L'inizio del XX secolo si annuncia all'insegna della contrapposizione tra rumore e musica,[6] anche in relazione alla consapevolezza di uno scontro tra silenziosa campagna e frastornante città, tra suoni della civiltà contadina e rumori di macchine, trame e treni. È altresì la conseguenza di uno sviluppo della storia della musica, da Claude Debussy a Igor Stravinskij, da Erik Satie a Edgard Varèse, da Henry Cowell a Karlheinz Stockhausen, in cui i lavori musicali nascevano da un collage o da un montaggio di note, armonie e rapsodie intrecciate a suoni provenienti dalla vita quotidiana. Insiemi caotici e aleatori che sul piano della pittura corrispondono a una dinamica dei colori come in Vasilij Kandinskij o a una strutturazione come in Theo van Doesburg [166]. Sono cromatismi sonori – concitati o calmi – che definiscono valori e significati contrapposti, dal disordine all'ordine.

La questione dell'intreccio tra arte e suono attraversa tutta la vicenda artistica dal XVII secolo a oggi come aspirazione a tematizzare uno spazio o un territorio sensoriale non rispondente alla tradizione occidentale e alle sue schematiche coordinate. Segna tutto il percorso della modernità e le relazioni sinestetiche tra diversi linguaggi espressivi della comunicazione, al fine di trovare un altro ordine che non sia convenzionale. "Art or Sound" entra inoltre nei territori della storia e della memoria per evidenziare che l'aspirazione multisensoriale esiste da secoli come indagine metodologica che corrisponde a una mappatura diversificata di suoni e di effetti musicali, così che in simultanea i risultati influenzino sia la vista che l'udito.

Pertanto, piuttosto che affermare la coincidenza o la somma "art=sound" o "art+sound", che può rimandare a un cannibalismo linguistico dove il tutto si fonde, si è mantenuta l'indipendenza "art or sound" che non alimenta un antagonismo ma piuttosto una fluttuazione e un transito libero: uno spazio dell'erranza e dello scambio che serve a far fluire l'arte nel suono e il suono nell'arte, ma anche a distinguerne le trame e i percorsi.

La scommessa dell'esposizione alla Fondazione Prada a Venezia è la designazione di un *teatro di cose* e di eventi in cui la connotazione – È arte o suono? È *sound art* o artigianato musicale? – tende a decadere. A volte esistono la coincidenza e l'attrazione, altre il rifiuto e il rigetto e ciò serve ad aprire un interstizio tra i modelli di decifrazione. Il processo enigmatico condotto da Man Ray [172, 174, 176] a Salvador Dalí [188], dove a contare è la pluralità mobile dell'oggetto, che si fa molteplice e sfuggente, vive di accostamenti improvvisi e inattese apparenze. Non molto dissimili dal violino a chiodi [100] e dal tamburicorde di Carlo Bugatti [118], sono le creazioni di Edward Kienholz [224], dei fratelli Baschet [190, 192], di Christian Marclay [360] e di Terry Adkins [364], che si avvicinano agli innesti e ai sovvertimenti tipici delle sperimentazioni cubiste, dadaiste e surrealiste.

L'aspirazione a una sensorialità totale che ricerca una sensibilità sonora delle cose, costruite o trovate, si configura con la scoperta linguistica che al termine della seconda guerra mondiale porta alla definizione di uno spazio autonomo delle arti. Una necessità a far funzionare la comunicazione multisensoriale fuori da intenzioni funzionali al conscio o all'inconscio, al sociale o all'individuale, all'astratto o al reale, così che il manufatto artistico non si riduca a una ragione "altra" ma sia la manifestazione, quasi tautologica, dello stesso linguaggio espressivo. Le distinzioni del fuori e del dentro cadono, così che l'oggetto esprima varie "proiezioni" che fanno convergere su esso parabole distinte di significati e di logiche.

La scultura ritrova un valore a essa estraneo come il suono in Alexander Calder [186], dove i frammenti in metallo dipinto che compongono il *mobile* riportano una numerazione, quasi fossero segni di uno spartito. Arrivano a produrre suoni – come nel caso di Takis [284] – quando le parti, mosse dall'aria, vengono a contatto o si scontrano. In altri casi lo strumento musicale si ripiega al suo interno quasi biografico – quello del mondo di Paganini, come nel violino bendato [184] di Maurice Henry o trasforma – con Ay-o [304] – il box sonoro degli orologi e dei carillon automatici in un contenitore silenzioso e onirico che rimanda a contenuti tanto voluttuosi e sognati quanto sconosciuti e angosciosi. È un confondere l'esperienza per abbandonarsi al fluttuare dei sensi, privati di colpo della sicurezza delle discriminazioni ottiche rispetto a quelle fisico-sensoriali. Dal 1945 gli artisti, segnati dalla tragedia delle guerre, sono consapevoli del fallimento storico di un comprendere unico o unificato, ne denunciano il formalismo e attraverso l'Action Painting e l'Espressionismo Astratto affermano il *regno dell'informe*.[7] Tutto si sbriciola e si decompone, le ricerche esprimono la loro fragilità dinanzi ai drammi della storia. Le forme esplodono con la sensazione che sarà difficile ricomporle. È un momento straziante in cui ogni linguaggio si spossessa di sé, per abbandonarsi alla gestualità e all'azione che corrode e divora ogni narrazione o racconto, se non quello di una vacuità esistenziale, che dà spazio solo al vuoto o al silenzio. Saltano tutte le coordinate di riferimento, che sono sospese perché non è pensabile tracciarne un'identità sicura.

Tuttavia in questo periodo di vitalità essiccata, l'arte e la musica, il cinema e la letteratura producono aggregati in cui si cristallizzano gli estremi del vivere, dal rumore al silenzio, dalla realtà all'astrazione, e si rimettono in circolazione le radici prime del dipingere e dello scrivere, del documentare e del narrare. Tale situazione effervescente porta alla corrosione totale di tutti i limiti linguistici. Sottoponendo i materiali a una strategia frantumatrice, immergendoli in una sorta di crogiolo dell'informe, le arti dimostrano la volontà di "rifarsi", rifondando in sé tutte le loro polarità, con la conseguenza di riattivare e di mettere sullo stesso piano tutti gli strumenti dell'esprimersi e dell'esporre.

IL MANUFATTO SONORO

Dagli anni '60, oltre all'equiparazione dei media condotta da Andy Warhol, decadono le distinzioni fra pavimento, parete e soffitto; si scardinano le linee di demarcazione tra linguaggi e sensi; mentre il territorio del visibile arriva a includere ogni emissione di corpi e oggetti, crollano le differenze tra basso e alto, maggiore e minore, pieno e vuoto, significante e insignificante: tutto si investe di senso. Accanto ai rumori industriali che hanno segnato gli inizi del secolo – dalla fabbrica ai mezzi di trasporto, dal tram al treno, dall'automobile alla motocicletta –, la percezione dei suoni si amplia con le nuove tecnologie di comunicazione del suono, dalla radio al grammofono [182], per cui i concerti e i brani musicali cominciano a contemplare registrazioni e trasmissioni radiofoniche. Il paesaggio sonoro dal secondo dopoguerra si dilata a comprendere i rumori di guerra e i fragori degli eventi naturali, come le eruzioni o le tempeste. Ingloba le partiture filmiche e teatrali, per cui la commistione dei suoni diventa totale. È a John Cage [214, 216] che si deve questa equiparazione tra alto e basso musicale, tra opera e Muzak, tra concerto e jukebox, tra *silent music* e radio.

Non meraviglia pertanto che gli artisti New Dada e Pop, da Robert Rauschenberg [230] a Tom Wesselmann [238], includano nelle loro sculture e nelle loro pitture i programmi radiofonici, tanto che li definiscono "still life", e Jasper Johns inserisca in un suo dipinto un orologio a molla ricaricabile. A contare è la percezione del mondo, così come accade, attraverso l'udito, ma anche mediante la cassa di risonanza del proprio corpo in *Listening to Music Through the Mouth* [228] di Nam June Paik. In altri casi l'influenza dell'introversione, che arriva a diffondersi con la conoscenza della cultura Zen, porta alla ricerca del proprio suono interno, che nel caso di una scultura o di una pittura risiede nelle sue parti attive, come in *Radio WNYR Nr. 15* [226] di Jean Tinguely. È il concetto dell'illuminazione individuale a informare tali combinazioni di arte e suono, sia in senso attivo sia passivo, che in Milan Knížák [232] si traduce in una equivalenza tra vita e morte, tra suono e silenzio, tra registratore funzionante e registratore distrutto.

Tutti questi insiemi inscindibili, spesso in movimento, che incorporano entità diverse, trovate o costruite, e non si fissano su una struttura rigida – così da mettere in campo opposizioni tradizionali buio e luce, autonomo ed eteronomo, meccanico e organico – trovano precursori negli anni '20 negli intonarumori di Russolo, nel "color piano" [136] di Alexander Scriabin o nel pianoforte optofonico [170] di Wladimir Baranoff-Rossiné. Essi aprono la strada ai pianoforti modificati o preparati e alle accumulazioni di giradischi di Cage o agli insiemi sonori e le apparecchiature musicali di Paik, e trovano seguito nei *combine* sonori di Edward Kienholz e di Wolf Vostell [252]. Meccanismi e strumenti che presentano forti analogie con gli orologi e con gli automi storici che aprono la fase dei "motori" artistici e sonori, con cui modificare e potenziare gli effetti del vedere e del sentire. Con essi l'arte del vedere e del suonare diventa una commistione, un marchingegno che, una volta innescato, dirama entità in cui soggettivo e oggettivo non possono essere considerati più antinomici, permettendo così una continuità e un'integrazione tra costruito e trovato, scultura e readymade, rumore e silenzio. In tale senso si afferma sempre più sulla scena il linguaggio dell'arte e del suono come protagonista del suo auto-svelamento iconico; Richard Artschwager [240], Arman [218, 344], Günther Uecker [236], Walter Marchetti [314], Claes Oldenburg e Coosje van Bruggen [378, 380] sembrano rinunciare alla componente sonora a favore dell'impatto cromatico, scultoreo e ambientale, ma formano e compongono un silenzio visivo e plastico.

I neodadaisti europei e americani, specialmente nell'area Fluxus – da George Maciunas [270, 292] a Joe Jones [234] – stabiliscono una simbiosi tra corpo e manufatto sonoro e macchina musicale. Permettono di trasformare l'oggetto in una protesi che mette in connessione l'artificiale e l'umano, facendo uscire entrambi da una dimensione unica e fissa. Tale passaggio chiude il cerchio della reciproca mutazione tra arte e suono, si afferma come evoluzione che porta alla nascita di un essere-macchina o di una macchina-essere che, quale oggetto o quale figura, è capace di un'autonomia del suonare e dell'immaginare, del ballare e dell'ascoltare, come in Robert Morris [220] e Bruce Nauman [248], in Joseph Beuys [310] e Stephan von Huene [244, 246]: altri esempi di una sonorità interiore e nascosta che trova una relazione con l'esperienza fisica nella camera anecoica di Cage, descritta in *Silence* (1961). La simbiosi e l'osmosi creano un transito continuo per cui lo strumento sonoro e la scultura con o senza suono, entrambe legate al fare umano, sollecitano un immaginario con effetti sul guardare e sul sentire, che sembrano richiamare le forze dinamiche definite in *Les machines célibataires* (1954) da Michel Carrouges.

Sul piano della macchina che turba la sensibilità e crea inquietudine, a causa di una serie di slittamenti di significato e d'innesti nelle sue valenze espressive, si colloca la dimensione instabile e aleatoria dell'oggetto che, fatto entrare in unwa successione di aspettative circa materia e struttura, arriva a produrre un "altro" fantastico e meravigliante. Tale amplificazione visuale e funzionale avviene spesso, com'è tradizione dal Cubismo al Dadaismo e al Surrealismo, con il recupero del residuo. Questo è assunto e riutilizzato per uno scopo estraneo, come nel caso degli strumenti musicali di Ken Butler [298, 306], dove i prodotti della quotidianità, dall'ascia alla valigia, si tramutano in chitarre. È un'attitudine che ricorre anche in William T. Wiley [274] e in Pedro Reyes [432], per cui le cose, dalle lattine di metallo alla armi dei narcos, subiscono un rimaneggiamento pla stico così da produrre un esito inatteso che le iscrive in un apparato musicale, mentre i risultati e gli effetti strumentali rimangono tradizionali. E se gli androidi settecenteschi miravano a imitare il vivente con l'intenzione di far vivere ed evolvere i meccanismi per trasformarli in figure umane – aspirazione che continua con i robot o le marionette di Dennis Oppenheim [282] e di Maurizio Cattelan [368] –, avviene oggi anche il contrario: l'ambito passaggio dall'artificiale all'organico, dall'arte e dalla musica alla gestualità corporale.

Una specularità che evidenzia la componente vitale dell'arte, che in Jannis Kounellis si manifesta con l'associazione tra elemento visivo – un brano dello spartito per il balletto *Pulcinella* di Igor Stravinskij – e la sua esecuzione ripetuta fino alla stanchezza, e al relativo abbandono da parte degli esecutori, un violinista e una ballerina [278]. L'associazione simbiotica tra oggetto e soggetto, per cui diventano due organismi che tendono a scambiarsi i ruoli, è riproposta nel dipinto che si traduce in materia vocale e sonora in Roman Opalka [242] e in Terry Allen [316] o nello strumento musicale che Thomas Demand trasforma in fotografia [406]: quasi un'evoluzione bio-tecnologica del suono in arte.

Un'osmosi tra astrazione e performance, tra racconto musicale e ambiente, tra oggetto e ricezione sonora che nei lavori di Laurie Anderson [296, 302], Takako Saito [294], Céleste Boursier-Mougenot [428], richiede la partecipazione del pubblico. In altri casi i manufatti sonori che rimandano ad armonie orientali e classiche occidentali, così come a musiche etiopi e brasiliane, alla country music o all'hip-hop, al gracidio delle rane e a scampanii rituali sono attivabili mediante interruttori, timer o connessioni elettriche – Bernhard Leitner [288], Rebecca Horn [354], William Anastasi [290], Eliseo Mattiacci [308], Marco Bagnoli [348], Manuel Rocha Iturbide [418], Janet Cardiff & George Bures Miller [430], Carsten Nicolai [350], Anri Sala [408] –, così da estroflettersi e occupare lo spazio e il tempo con emissioni di suono che raccontano della natura o delle vicende romantiche della persona o degli attraversamenti corporei degli oggetti e dei luoghi. Protesi pittoriche e scultoree che producono informazioni e narrazioni musicali e sonore, e costituiscono una nuova architettura del percepire e del comprendere multisensoriale, i cui significati appartengono alla storia autobiografica dell'artista.

Dagli anni '90 i sistemi di comunicazione, artistica e non, continuano a evolversi e a compiere enormi progressi sull'uso dell'informatica e dell'elettronica, della scienza logica e dell'intelligenza artificiale. Si ampliano attraverso strumenti tecnologici come il computer e la rete, che con i suoi flussi influenza l'impianto e la funzionalità, l'estetica e la logica degli oggetti e delle macchine. Sul piano dei manufatti ibridi, che intrecciano suono earte, le nuove creazioni sonore si avvicinano sempre più ai cyborg. Si fondono con il corpo umano, ne diventano protesi tecnologiche, seppur ancora primitive come succede per Yoshimasa Wada [286] e Martin Kersels [416], oppure ritornano a copiare le sembianze umane. In altri casi l'uso di meccanismi sembra avere cambiato il modo di eseguire e di produrre un suono e mutato l'apparenza dello strumento. Quasi che la configurazione e l'iconografia dell'oggetto strumentale non fossero più importanti, quindi da perdere di vista, e contassero solo le emissioni sonore: ecco gli oggetti astratto-tecnici e puramente meccanici di Paul De Marinis [280], Tarek Atoui [426], Martin Creed [338], Brian Dewan [374], Ed Potokar [356], Bart Hopkin [346, 352], Haroon Mirza [400, 434] e Alberto Tadiello [422].

In altri casi, invece, l'evoluzione meccanica è rifiutata in nome della memoria del passato in cui gli strumenti di emissione sonora erano ancora a dimensione umana; si ritorna così alla scala ottocentesca del già citato pirofono di Kastner, o dell'organo di strada di Cocchi, Bacigalupo & Graffigna [124], con le riproposte contemporanee del *boombox* di Tom Sachs [362] o del jukebox di Ruth Ewan [370] che si identificano con la cultura musicale dei rapper o delle canzoni politiche. Altrove le costruzioni si fanno complesse e si integrano nell'architettura. Non si distanziano dall'ambiente quotidiano, ma si integrano in esso, ne propongono un doppio, che può assumere diverse connotazioni linguistiche, dall'arredo all'utilizzo funzionale – come in Guido van der Werve [410], Doug Aitken [420], Mirza e Subodh Gupta [376] –, oppure creano un cortocircuito che, come in Athanasios Argianas [412], utilizza l'elemento architettonico per dare dimensione inedita al suono e alle immagini in relazione alla sua collocazione urbana, interna o esterna.

L'intensità del procedere intorno al manufatto sonoro potrebbe rimanere circoscritta alla polarità arte-suono, ma non bisogna dimenticare la forza attiva dell'esecutore o del performer che rappresenta l'altra sostanza del passaggio dall'una all'altro. Il protocollo dell'autoproduzione, via carica manuale o accensione elettrica, resta fondamentale per comprendere il senso di un generare autonomo, ma esiste pure la fusione tra le due polarità, quando queste sono attivate da una persona o da un essere umano. Qui la fissità si traduce in un percorso mobile, fisico e spirituale, organico e artificiale, che rappresenta una visione d'uso e di riflesso sulla cosa. L'esecutore la prende, la espropria e le dà un'altra vita. Fa uscire il suono e l'oggetto dalla loro specificità e attiva un altro status per entrambi. Esalta quindi la virtualità come nel caso di Kounellis o di Atoui, Anderson e Michelangelo Pistoletto [250], Akio Suzuki [272] e Maywa Denki [340]. Alimenta il potenziale del medium musicale e sonoro, sia esso quadro o strumento, e gli attribuisce una vita vera, lo fa "incarnare" in una corporeità come in Loris Gréaud [372], Tore Honoré Bøe [436] e Neil Feather [358]. Il coinvolgimento del corpo ne rinnova l'immaginario e lo traduce in un gesto reale, quello del concerto. Qui la musica non è tuttavia spazio da palcoscenico ma profondità di una rigenerazione delle cose, trovate, composte e assemblate, che a volte passa attraverso l'alienazione dei gesti – ancora Kounellis –, oppure si esprime liberamente nell'incandescenza dell'esecuzione: Butler o Walter Kitundu [402, 404], Riccardo Beretta [424] o Reyes, Max Vandervorst [312] o Nicolas Bernier e Martin Messier [414]. In tal caso la dimensione muta e immobile è sottoposta a una trasmutazione e il silenzio iconico dà spazio a un suono oggettuale. Ne scaturiscono rumori e bagliori musicali che rivelano come il manufatto artistico sia di solito avvolto in una trance che lo rende statico e paralizzato, mentre esso può essere anche organico e dinamico. È nell'attesa del gesto umano, con tutte le sue complicazioni sensoriali, che esso rivive, risuona e odora dei suoi umori. Il suo pulsare genera una circolarità e una gestazione senza fine. La ricerca artistica si conferma allora irradiante in tutte le direzioni, una cerniera tra l'alto e il basso, tra lo spirito e la carne, tra il silenzio e il rumore, tra il profumo e la puzza, tra il visibile e l'invisibile, tra la vita e la morte. Il fine è di optare per un labirinto dei sensi, per una geografia dell'informe e dell'indefinito esperienziale così da addentrarsi in una cavità multisensoriale dove covare una nuova nascita.

1
Nina Levent, Alvaro Pascual-Leone (a cura di), *The Multisensory Museum*, Rowan & Littlefield, Plymouth 2014.

2
La bibliografia sui rapporti tra arte e suono è molto ampia. Recenti contributi tra storia e presente comprendono sia mostre che pubblicazioni.
Fra le mostre più significative, segnaliamo: "Sons & Lumières. Histoire du son dans l'art du XXe siècle", Centre Pompidou, Parigi, 22 settembre 2004-3 gennaio 2005, catalogo della mostra a cura di Sophie Duplaix e Marcella Lista, Centre Georges Pompidou, Parigi 2004; "Visual Music: Synaesthesia in Art and Music Since 1900", Museum of Contemporary Art, Los Angeles (CA), 13 febbraio-22 maggio 2005 e Hirshhorn Museum and Sculpture Garden, Smithsonian Institution, Washington (D.C.), 23 giugno-11 settembre 2005, catalogo della mostra a cura di Olivia Mattis e Ker ry Brougher, Thames & Hudson, London/New York 2005; "SEE THIS SOUND. Premises in Sound and Vision", Lentos Kunstmuseum, Linz, 28 agosto 2009-10 gennaio 2010, catalogo della mostra a cura di Manuela Ammer, Cosima Rainer, Stella Rollig *et al.*, Walther König Publisher, Köln 2010; "Sound Art. Sound as a Medium of Art", ZKM | Media Museum, Karlsruhe, 17 marzo 2012-6 gennaio 2013, catalogo della mostra a cura di Peter Weibel, ZKM Publisher, Karlsruhe – in corso di pubblicazione); "Soundings: A Contemporary Score", Museum of Modern Art, New York, 10 agosto-3 novembre 2013, catalogo della mostra a cura di Barbara London, The Museum of Modern Art, New York 2013. Fra le pubblicazioni più recenti, sono degne di nota: David Toop, *Haunted Weather: Music, Silence and Memory*, Serpent's Tail, London 2004; Alan Licht, *Sound Art: Beyond Music, Between Categories*, Rizzoli, New York 2007; Caleb Kelly, *Cracked Media: The Sound of Malfunction*, The MIT Press, Cambridge (MA)/London 2009; Seth Kim-Cohen, *In the Blink of an Ear: Toward a Non-Cochlear Sonic Art*, Continuum, New York/London 2009; Salomé Voegelin, *Listening to Noise and Silence: Toward a Philosophy of Sound Art*, Continuum, New York/London 2010; Patrick Feaster, *Picture of Sound: One Thousand Years of Educed Audio: 980-1980*, Dust-to-Digital, Atlanta (GA) 2012; Jonathan Sterne (a cura di), *The Sound Studies Reader*, Routledge, London/New York 2012; Douglas Kahn, *Earth Sound Earth Signal: Energies and Earth Magnitude in the Arts*, University of California Press, Berkeley (CA) 2013.

3
Solo alcuni musei, dal Victoria & Albert di Londra all'Art Gallery of Ontario, Toronto, hanno tentato di istituire una relazione tra arte e odore, mentre intrecci sensoriali, tra profumi e oggetti estetici, sono stati presentati in mostre come "Five Holes: reminiSCENT" curata da Jim Drobnick e Paul Couillard, Toronto 2003; e "Odor Limits", curata da Jim Drobnick e Jennifer Fisher, Philadelphia (PA) 2005. Per una bibliografia vedi Jim Drobnick, "Reverie, Assault and Evaporating Presences: Olfactory Dimensions in Contemporary Art", *Parachute*, n. 89, Montreal 1998, pp. 10-19; Anna Barbara, Anthony Perliss, *Architetture invisibili: L'esperienza dei luoghi attraverso gli odori*, Skira, Milano 2006; Patrizia di Bello, Gabriel Koureas (a cura di), *Art History and the Senses: 1830 to the Present*, Ashgate, London 2010; Peter De Cupere, "The Use of Smell in Art, an 'Olfactology' Art Research", lezione tenuta alla College Art Association Conference, New York, 16 febbraio 2013. L'implicazione del tatto è spesso confinata a copie di manufatti o alla manipolazione di materiali non elaborati ma grezzi. Qui è la preziosità dell'oggetto e la sua rarità storica a impedire l'esplorazione sensoriale. Tuttavia l'arte contemporanea dovrebbe aprirsi a tale avvicinamento corporale, per offrire esperienze non soltanto visuali.
Per una bibliografia recente: Helen J. Chatterjee (a cura di), *Touch in Museums: Policy and Practice in Object Handling*, Berg, Oxford 2008; Sandra H. Dudley (a cura di), *Museum Materialities: Objects, Engagements, Interpretations*, Routledge, New York 2010.

4
Constance Classen (a cura di), *The Book of Touch*, Berg, Oxford-New York 2005.

5
Gian Paolo Ceserani, Gli *automi. Storia e mito*, Laterza, Bari 1983; Mario G. Losano, *Storie di automi*, Einaudi, Torino 1990 e Id., *Automi d'Oriente*, Medusa, Milano 2003; Carlo Sini, *L'uomo, la macchina, l'automa*, Bollati Boringhieri, Torino 2009; Bruno Corà, Pietro Bellasi (a cura di), *Corpo, Automi, Robot: Tra Arte, Scienza e Tecnologia*, catalogo della mostra, Museo d'Arte Città di Lugano/Mazzotta, Lugano/Milano 2009; Roberto Peverelli (a cura di), *Gli automi sono tra noi*, Medusa, Milano 2013.

6
Stefano Pivato, *Il secolo del rumore*, Il Mulino, Bologna 2001; Kyle Gann, *Il silenzio non esiste*, trad. di Melinda Mele, Isbn Edizioni, Milano [2010] 2012.

7
Yve-Alain Bois, Rosalind Krauss, *L'informe: istruzioni per l'uso*, trad. di Elio Grazioli, Bruno Mondadori, Milano [1997] 2003.

MERAVIGLIOSI INGANNI. LA BELLEZZA VISIVA E MUSICALE DAL RINASCIMENTO AL XVIII SECOLO

Deirdre Loughridge

Si suppone che un suono eccelso abbia origine da oggetti esteticamente sublimi. Era questo l'assunto che guidava musicisti e fabbricanti di strumenti nel periodo compreso tra il XVI e il XVIII secolo. I frutti di questo pensiero sono apprezzabili non soltanto nei meravigliosi strumenti del periodo conservati nei musei, come il clavicembalo attribuito a Vitus de Transuntinus (1560), presso il Musikinstrumenten-Museum di Berlino, che palesa le proprie duplici finalità estetiche nell'iscrizione posta sopra alla tastiera: "Rendo lieti in un tempo gli occhi e 'l core". Il principio è evidente anche nelle forme standard degli strumenti musicali che quell'epoca ci ha tramandato. I fisici sono tuttora all'opera per cercare di comprendere in che modo le linee e le proporzioni del violino – una forma che apparve in Italia a metà del XVI secolo e che raggiunse la perfezione con gli Amati, i Guarneri, gli Stradivari – contribuiscono alla sua resa qualitativa del suono. Facendo ricorso a laser scanner tridimensionali per tracciare una mappatura dei violini e dei loro schemi vibrazionali, il progetto Strad3D ha sottoposto ad analisi alcuni degli esemplari più quotati al mondo, scoprendo che la collocazione e le dimensioni delle F e delle C, per esempio, sono essenziali per determinare la flessibilità della parte superiore del violino, e di conseguenza il modo in cui esso diffonde il suono. Come spiega Samuel Zygmuntowicz, "una forma raffinata è indice di un elevato grado di sofisticatezza dal punto di vista progettuale; … cambiare uno qualsiasi di quegli elementi avrà delle ripercussioni sul suono".[1]

Se è vero che la relazione sinestetica tra oggetto artistico e strumento acustico rappresentava una condizione sostanziale per la cultura musicale nel periodo fra il Rinascimento e l'inizio del XIX secolo, è altrettanto certo che si verificarono casi straordinari – esempi che misero in discussione o suggerirono nuove forme di correlazione tra bellezza visiva e acustica, duratura ed effimera. In effetti, questi secoli possono essere letti come un periodo di transizione, durante il quale l'interesse congiunto per la bellezza visiva e musicale andò via via scomparendo, per lasciare il passo all'ipotesi che queste arti sensoriali fossero separate e indipendenti tra loro. È possibile ripercorrere le tappe di questo processo, prendendo in esame i casi straordinari che hanno messo alla prova l'assunto della correlazione sopra citata: casi come quelli degli strumenti visivamente eccelsi, ma acusticamente inadeguati, di Michele Antonio Grandi [50] o Giovanni Battista Cassarini [52]; della forma perfetta, ma inarrivabile, congetturata da filosofi come Athanasius Kircher; della tastiera a colori di Louis-Bertrand Castel creata per una ricezione ambivalente; e degli automi musicali del XVIII secolo, nei quali la bellezza non era incarnata solo dallo strumento musicale in sé, ma anche dalla sua associazione a una figura umana in movimento.

La bellezza degli oggetti non scaturisce soltanto dalla loro forma, ma anche dai materiali con i quali sono realizzati. Se è dunque vero che la forma tipica degli strumenti come il violino, la chitarra e il liuto nacque da una duplice attenzione nei confronti del suono e della struttura, essi poterono in seguito essere perfezionati grazie all'uso di materiali preziosi. Oro, marmo e gemme arricchivano lo splendore visivo degli strumenti di cui erano materia costitutiva o che adornavano, rendendoli particolarmente ambiti presso ricchi collezionisti. Ma in casi come questi, spesso la bellezza visiva surclassava le qualità acustiche, il che si traduceva in strumenti musicali impraticabili se non "impossibili". È il caso, per esempio, della chitarra in marmo attribuita a Michele Antonio Grandi, realizzata intorno al 1687 per Casa d'Este. Costruita con straordinaria precisione sul modello di una chitarra barocca, è dotata di un cavigliere e di un tallone del manico per l'attacco di cinque ordini di corde tutte doppie, di tasti realizzati con estrema accuratezza, nonché di due anelli ricavati nel marmo per il fissaggio di una tracolla. Ma lo strumento è sprovvisto di buca, al posto della quale vi è una decorazione circolare sul marmo. Non solo, l'altezza del capotasto sul piano della tastiera è insufficiente per elevare le corde al di sopra delle traversine metalliche, e la prima e l'ultima corda sono oltremodo vicine al bordo della tastiera. La conclusione a cui è giunto Gabriele Rossi-Rognoni, curatore della sezione di strumenti musicali presso il Royal College of Music di Londra, è che "sembra che questo strumento sia un meraviglioso 'inganno' destinato esclusivamente agli occhi, capace di riprodurre con straordinaria accuratezza i dettagli costruttivi della chitarra senza, tuttavia, essere (o essere mai stato) in condizione di suonare".[2]

Laddove gli artigiani, lavorando su commissione delle case reali, si trovavano talvolta costretti a sacrificare la resa acustica e la suonabilità, a favore dell'aspetto estetico, i filosofi concepivano i progetti di origine musicale basandoli su modelli matematici, facendo apparire come per magia strumenti speculativi che non avrebbero mai soddisfatto le aspettative. All'inizio dell'era moderna, molti intellettuali ritenevano che la spirale avesse un'affinità speciale con il suono. Questa struttura ha un ruolo dominante nel *Tempio della musica* (1618) del filosofo inglese Robert Fludd, nel quale sono raffigurati i movimenti a spirale dell'aria, che si pensava fossero prodotti dal suono. Era convinzione altrettanto comune che la spirale fosse un amplificatore acustico, principio fondato sul fatto che l'orecchio è dotato di una struttura spiraliforme – la coclea – all'epoca ritenuta strettamente correlata all'incremento del volume dei suoni. Sulla base di questo ragionamento, l'eclettico gesuita Athanasius Kircher illustrò uno strumento chiamato "tubo cocleato"; egli lo riteneva in grado di amplificare in modo esponenziale il suono della voce che vi passava attraverso. Riprendendo il bizzarro progetto di Kircher, Filippo Bonanni, nel suo trattato sugli strumenti musicali intitolato *Gabinetto armonico pieno d'istromenti. Indicati e spiegati* (1722), osservò che difficilmente ci si poteva aspettare di far funzionare un tubo cocleato, in quanto attribuire al tubo la forma di una spirale perfetta avrebbe richiesto un talento smisurato.

Ci fu un'altra tendenza speculativa che venne messa alla prova con maggiore facilità: si tratta dell'analogia tra colore e tono. Isaac Newton rese popolare questa correlazione nel suo trattato *Opticks*, pubblicato nel 1704. Osservando i colori che comparivano quando la luce del sole attraversava un prisma, egli suggerì che la loro ampiezza relativa potesse corrispondere alla lunghezza relativa di una corda che produceva il suono delle sette note della scala diatonica. Sulla base della stessa analogia, il filosofo francese Louis-Bertrand Castel avanzò l'ipotesi di creare una nuova arte dei colori temporale: una musica per gli occhi. Benché non avesse ricevuto una formazione artigianale, l'interesse suscitato dalla sua proposta lo spinse a costruire un clavicembalo oculare, nel quale la percussione dei tasti attivava visivamente stoffe di seta o vetri colorati. "Il principale vantaggio di questo nuovo clavicembalo", scriveva Castel, "è quello di dare ai colori … una certa vivacità e leggerezza, che su una tela immobile e inanimata non riescono mai a raggiungere".[3] La bellezza visiva, anziché presentarsi come statica e solida, poteva ora assumere le qualità effimere e dinamiche della musica.

Tuttavia, nell'applicazione pratica, in molti trovavano il clavicembalo di Castel deludente. Il professore tedesco Johann Gottlob Krüger era d'accordo con il principio sul quale si basava lo strumento, sostenendo che "quando la nostra mente giudica come piacevoli o sgradevoli le cose che passano davanti agli occhi, soggiace alle stesse leggi e alle stesse regole che sono prese in considerazione quando si tratta di distinguere la musica armonica da quella discordante".[4] Ma lo studioso identificò un problema a livello concettuale: lo strumento di Castel catturava la dimensione sequenziale della musica, ma non il suo aspetto di simultaneità: una melodia di colori priva di armonia. Così Krüger propose una tastiera che avrebbe prodotto "accordi a colori": le sue luci colorate sarebbero apparse sotto forma di cerchi concentrici, in cui i toni più bassi erano più ampi e i toni più alti erano sovrapposti al loro interno.

Nel contempo, altri trovarono priva di fondamento l'idea stessa di una musica per gli occhi. Come osservò Jean-Jacques Rousseau, "rendere una pluralità di suoni in modo da farli udire simultaneamente o sviluppare i colori l'uno dopo l'altro è mutarne l'essenza: è mettere l'occhio al posto dell'orecchio e l'orecchio al posto dell'occhio".[5] Per Rousseau e per un numero crescente di intellettuali, la bellezza visiva e quella musicale non erano paragonabili: la prima si fondava sul principio della simultaneità a livello spaziale, la seconda su quello della sequenzialità temporale. Poiché, secondo il suo punto di vista, le due arti erano profondamente diverse nella loro stessa essenza, era folle pensare di mettere in atto una correlazione fra di loro.

Un'analoga inconciliabilità riguardava la correlazione tra il suono della musica

e la vista dei musicisti intenti a suonare – una correlazione che gli artigiani produttori di automi musicali del XVIII secolo esplorarono nei loro oggetti d'arte. Come osservò Johann Carl Enslen, un costruttore di automi musicali, a proposito di uno dei suoi esemplari, "non ho risparmiato lavoro, fatica, né denaro per fabbricare un teatro musicale meccanico nel quale la musica, la pittura e la scultura si combinano tra loro, dando origine agli oggetti protagonisti, che si presentano sotto forma di automi dalle sembianze vitali o di figure artificiali in movimento… Gli spettatori sono piacevolmente divertiti e impressionati".[6] Incarnando l'unione dell'arte meccanica, visiva e musicale, gli automi musicali rispecchiavano gli ideali di come dovesse apparire o risuonare una bella performance musicale. Gli insegnanti di musica del XVIII secolo consigliavano sempre ai propri studenti di evitare le espressioni affettate, poiché tali smorfie del volto – benché naturale conseguenza dello sforzo intenso – erano incompatibili con gli ideali di grazia. Il clavicembalista francese François Couperin suggeriva di collocare uno specchio sul leggio del clavicembalo, in modo tale da potersi osservare e correggere da soli, e insegnava che "bisogna mostrarsi a proprio agio con questo clavicembalo, evitando di mantenere lo sguardo fisso troppo a lungo su un oggetto o di assumere un'espressione vuota".[7] I fabbricanti di automi musicali presero alla lettera questo consiglio, dando vita a marchingegni che simulavano l'immagine aggraziata richiesta agli interpreti. Così facendo, questi artigiani andarono oltre il loro compito di realizzare strumenti dalle qualità estetiche eccezionali, in quanto resero la figura umana parte integrante dell'oggetto d'arte.

Se è vero che il clavicembalo oculare aveva dato l'occasione a Rousseau di teorizzare la differenza tra arti visive e musicali, dalla figura dell'interprete egli imparò anche che la bellezza visiva e la bellezza musicale non procedevano di pari passo. Durante un viaggio a Venezia, nel 1743, il filosofo entrò in una chiesa nella quale udì dei mottetti cantati da alcune ragazze nascoste dietro a delle grate. "Non conosco nulla di così voluttuoso, di così commovente come quella musica", scrisse. "Ciò che mi angustiava erano quelle maledette grate, che lasciavano filtrare soltanto i suoni, e mi nascondevano gli angeli di cui essi erano degni". Così cercò di incontrare di persona le fanciulle che producevano quella musica sublime. Ma quando ebbe modo di fare la loro conoscenza, scoprì che quelle voci non provenivano affatto da visi angelici, anzi le ragazze avevano volti "orribili", una di loro era guercia e un' altra era "sfigurata dal vaiolo".[8] Un paio di ragazze gli erano sembrate "discrete", ma apprese che non si trattava delle angeliche soliste, bensì di semplici componenti del coro. Il racconto servì dunque per mettere in luce la divergenza tra bellezza musicale e visiva – una divergenza che colse di sorpresa gli ascoltatori della metà del XVIII secolo e servì da spartiacque tra le arti visive e sonore.

Bartolomeo Cristofori, fabbricante di strumenti per il Gran Duca di Toscana, inventò il pianoforte agli inizi del XVIII secolo. Eleanor Selfridge-Field Eldridge sostiene che Cristofori abbia concepito il pianoforte non per rispondere a bisogni o interessi musicali, ma piuttosto per cimentarsi in una sorta di esperimento scientifico – una applicazione delle leggi della forza e del moto appena scoperte da Newton.[9] In altri termini, il pianoforte nacque da un interesse per l'oggetto d'arte – per le sue forme e proprietà materiali – che era in misura quanto meno equivalente all'interesse nei suoi confronti in quanto oggetto musicale.

Durante la metà del XVIII secolo, tuttavia, gran parte della sperimentazione nell'ambito della progettazione di strumenti musicali focalizzò la propria attenzione sulla qualità del tono e sulla suonabilità, riservando sempre meno spazio all'interesse per la bellezza visiva. L'armonica a bicchieri e i suoi diversi esemplari possono essere considerati parte di questa tendenza: strumento celebre per i suoi toni celestiali, era altresì apprezzato perché consentiva all'ascoltatore di percepire "l'effetto ma non la causa" – un suono puro, privo di qualsiasi traccia della fonte materiale.[10] Questo ideale emerse dalle fratture che si erano via via create tra bellezza musicale e visiva nel periodo compreso tra il XVI e il XVIII secolo, quando gli artigiani e i filosofi erano alla ricerca, rivelatasi illusoria, degli svariati modi che potessero unificarle. In conclusione, queste divergenze hanno contribuito a rendere ancora più ampio il divario tra l'oggetto d'arte e l'oggetto sonoro, divario che gli artisti e gli spettatori hanno continuato a esplorare fino ai giorni nostri.

1
"Strad3D: Video Tour", strad3d.org/demo/videoTour2.html (ultimo accesso: 7 marzo 2014).

2
Franca Falletti, Renato Meucci, Gabriele Rossi Rognoni (a cura di), *Meraviglie sonore. Strumenti musicali del Barocco italiano*, catalogo della mostra, Giunti, Firenze 2007, p. 178.

3
Louis-Bertrand Castel, "Clavecin pour les yeux avec l'art de peindre les sons et toutes sortes de pièces de musique", *Mercure de France*, Paris novembre 1725, p. 2573; trad it. di G. Giordano in Frank Popper, *L'arte cinetica. L'immagine del movimento nelle arti plastiche dopo il 1860*, Einaudi, Torino 1970, pp. 195-96.

4
Johann Gottlob Krüger, "De novo musices, quo oculi delectantur, genere", in *Miscellanea Berolinensia ad incrementum scientiarum, ex scriptis Societatis regia Scientiarum exhibitis edita*, vol. 7, Johann Gottfried Michaelis, Berlin 1743, pp. 345-357: 352; trad. nostra.

5
Jean-Jacques Rousseau, Paolo Bora (a cura di), *Saggio sull'origine delle lingue*, Einaudi, Torino 1989, pp. 92-93.

6
Wiener Zeitung, 7 dicembre 1791, p. 3131.

7
François Couperin, *L'Art de toucher le Clavecin*, Sieur Foucaut, Paris 1716; trad. nostra.

8
Jean-Jacques Rousseau, *Le confessioni, Garzanti*, Milano 1988, pp. 324-25.

9
Eleanor Selfridge-Field, "The Invention of the Fortepiano as Intellectual History", in *Early Music*, vol. 33, n. 1, Oxford University Press, Oxford febbraio 2005, pp. 81-94.

10
Anne Ford, *Instructions for Playing on the Musical Glasses*, London 1761; trad. nostra. Sulla tematica della ricerca del tono ideale nel XVIII secolo, si veda Emily I. Dolan, "E. T. A. Hoffmann and the Ethereal Technologies of 'Nature Music'", in *Eighteenth-Century Music*, vol. 5, n. 1, Cambridge University Press, Cambridge 2008, pp. 7-26.

SUONO, MUSICA, TEMPO E MOVIMENTO. ANTICHE MACCHINE MUSICALI

Christoph E. Hänggi

Scienziati, tecnici e meccanici dell'antichità greco-romana e dell'area araba, quali Ctesibio d'Alessandria, Filone di Bisanzio, Erone d'Alessandria, i fratelli Muhammad, Ahmed e Hassan Mûsà, nei loro scritti e trattati scientifici descrivevano spesso organi, suonatori di flauto, uccelli canori o suonerie azionate dall'acqua o dal vento. In Europa, durante il Medioevo, la conoscenza di organi meccanici e altri automi musicali andò via via disperdendosi, e di quegli oggetti oggi non resta traccia.

SUONO E TEMPO

I più antichi strumenti musicali che hanno superato la prova del tempo sono i carillon azionati da rulli dentati. Il suono delle campane accompagnava la vita quotidiana delle persone, oltre a simboleggiare le autorità ecclesiastiche e le comunità urbane medievali. Nel corso del XII e XIII secolo gli orologi delle torri vennero muniti di suonerie di campane, da cui si svilupparono nel XIV secolo i carillon meccanici. Si pensa che il carillon più antico sia quello installato intorno al 1310 nell'orologio della cattedrale di Beauvais. Fra i primi carillon vi fu anche quello realizzato tra il 1352 e il 1354 per un analogo orologio nella cattedrale di Strasburgo. Aritmetica, geometria, astronomia e musica – l'antico *quadrivium* – fornivano i fondamenti per la costruzione di questi monumentali orologi astronomici, ai quali venivano talvolta aggiunti non solo i carillon, ma anche altri effetti musicali, come il suono di cembali, il canto del gallo, il suono dell'organo, e così via. I carillon automatici messi in funzione con rulli dentati e quelli con azionamento manuale si diffusero rapidamente nella Francia del nord e nei paesi dell'attuale Benelux, dove ancora oggi circa 400 carillon risuonano ogni giorno in molte cattedrali e chiese minori, nonché in municipi ed edifici istituzionali. Questo tradizionale strumento si ritrova, sempre con il nome di carillon, anche nell'area tedescofona fino a Lubecca, Praga o Berna.

TEMPO E MUSICA

Fra il tardo Medioevo e i primi del Rinascimento l'avvento dell'orologeria diede impulso allo sviluppo di strumenti musicali meccanici. Il complicato meccanismo degli orologi poteva essere contemporaneamente usato come azionamento per gli automi musicali. Agli inizi del XIV secolo gli orologi fecero il loro ingresso nei palazzi dei nobili, nei monasteri del clero e nelle abitazioni della ricca borghesia. I meccanismi e i carillon divennero via via più piccoli, e furono realizzati orologi da casa con carillon dalle forme più disparate.

Nel XVI secolo si svilupparono gli organi meccanici detti *Flötenuhren* (orologi a flauti, ossia orologi meccanici dotati di registri di flauto), che nelle stanze di principi e sovrani scandivano in musica le ore del giorno. Questi strumenti musicali meccanici – con canne aperte o coperte – suonano per mezzo di un rullo dentato su cui erano predisposte melodie appositamente composte o arrangiate. A intervalli regolari, il meccanismo a orologeria asservito innesca la rotazione del rullo.

In questa fase del tardo Rinascimento il centro dell'orologeria europea era Augusta, fiorente città della Germania meridionale, che insieme a Norimberga primeggiava anche nella costruzione degli automi. Oltre ai carillon e agli organi meccanici già menzionati, i maestri di Augusta e Norimberga approntavano orologi da tavolo e armadi con orologi corredandoli di strumenti a corda meccanici – come spinetta, virginale e clavicembalo – e congegni acustici azionati da ruoteo rulli dentati inseriti in preziosi *secrétaires*. Tra i costruttori di automi, i capiscuola furono Samuel Bidermann, Achilles e Veit Langenbucher, Matthäus Rungell o Hans Schlottheim. Stimatissimi musicisti quali Hans Leo Hassler o Christian Erbach composero la musica per i rulli di questi strumenti.

Nell'epoca del Barocco e del Rococò gli armadi con orologio e altri mobili si trasformarono e i contenuti musicali diventarono più elaborati. La possibilità di cambiare i rulli dentati consentì di aumentare il numero di canne e quindi di ampliare il repertorio. Negli *Harfenuhren* (orologi ad arpa) e nei *Flötenuhren* furono inseriti strumenti di dimensioni ridotte. I *Flötenuhren* furono molto in voga nella seconda metà del XVIII secolo e per tutto il XIX secolo. Le casse artisticamente lavorate indicano che erano destinati agli ambienti nobiliari. Ciò nonostante questi strumenti, originariamente ideati come regali per imperatori e sovrani, si diffusero largamente anche tra la ricca borghesia, nei gabinetti artistici aperti al pubblico e nelle sale da ballo. A volte i registri di flauto venivano svincolati dagli orologi, come nel caso delle serinette, che imitavano il cinguettio degli uccelli.

Nel XVIII secolo, nell'area anglofona il centro di produzione dei *Flötenuhren* era Londra. Fu qui che, intorno al 1718, Henry Bridges costruì un orologio denominato *The Microcosm* e che Charles Clay, John Pyke e suo figlio George, James Cox, James Newton ed Eardley Norton montarono i registri di flauto negli orologi. Nella Svizzera del XVIII secolo erano attivi Pierre Jaquet-Droz [60, 62], suo figlio Henri-Louis e Jean-Frédéric Leschot, mentre a Parigi vi erano Julien Leroy, suo figlio Pierre, e Jacques de Vaucanson. Il famoso suonatore di flauto di Vaucanson conteneva un meccanismo musicale analogo, mentre Jaquet-Droz installò registri di flauto nei suoi automi e orologi. Anche nei Paesi Bassi e in Svezia, tuttavia, vi erano dei centri specializzati che annoveravano orologiai di grande perizia, fra i quali Cornelius Engeringh, Dietrich Nicolas Winkel e Per Strand. I *Flötenuhren* venivano fabbricati anche in diverse città tedesche, come Hannover, Gottinga o Neuwied am Rhein, dove David Roentgen e il suo meccanico Peter Kinzing costruivano orologi a pendolo con registri di flauto e strumenti a corda di grande qualità. A Berlino, in particolare, la produzione di *Flötenuhren* era più che fiorente, anche grazie all'iniziativa di Federico il Grande che, con l'intento di pungolare l'orgoglio degli orologiai tedeschi, fece arrivare in città famiglie di orologiai svizzeri originari di Neuchâtel e Ginevra, oltre a orologiai ugonotti dalla Francia. I più famosi fabbricanti di *Flötenuhren* berlinesi erano Christian Ernst Kleemeyer e Christian Möllinger.

Rinomati compositori, tra cui Georg Friedrich Händel, Franz Benda, Wilhelm Friedemann Bach e suo fratello minore Carl Philipp Emanuel, composero delle opere espressamente per questi strumenti. Le composizioni di Händel si trovano su uno strumento di Charles Clay, un orologiaio specializzato in raffinati orologi a pendolo con campane e registri di flauto, vissuto a Londra nel primo decennio del XVIII secolo. Anche un *Flötenuhr* di Charles Pyke contiene un pezzo di questo compositore, che operò a Londra dalla fine del 1709.

Durante il Classicismo viennese, la capitale austriaca divenne il secondo centro di fabbricazione di *Flötenuhren* dopo Berlino. Verso il 1800 in città si contava almeno una ventina di fabbricanti di *Flötenuhren*, tra i quali Joseph Primitivus Niemecz che, essendo allievo e amico di Franz Joseph Haydn, fu da questi incoraggiato a realizzare diversi *Flötenuhren*. Niemecz potrebbe essere anche l'artefice del *Flötenuhr* sito nel gabinetto artistico del conte Josef Deym von Stržitéž a Vienna, sul quale fu eseguita la *Fantasia per rullo d'organo* di Wolfgang Amadeus Mozart. Fra i fabbricanti di *Flötenuhren* vanno inoltre menzionati Franz Aegidius Arzt, Eduard Heinrichssohn, Franz Erbs, Johann Nepomuk Mälzel (l'inventore del metronomo), Johann Georg Strasser, Leonhard Mälzel, Johann Adam Hoyer, Georg Anton Reinlein e suo figlio Rudolf, Johann Christian Heinrich, Joseph Franz Erbs e Joseph Arzt.[1]

Nella seconda metà del XVIII secolo furono trasposti per *Flötenuhren* brani musicali alla moda, come arie d'opera o pezzi strumentali famosi. Ma anche Haydn, Antonio Salieri, Franz Anton Hoffmeister, Mozart, Joseph Ignaz Pleyel, Luigi Cherubini o Ludwig van Beethoven scrissero composizioni espressamente destinate a questi strumenti meccanici. Avendo operato a Vienna per almeno un certo periodo, tutti questi compositori non poterono ignorare il generale entusiasmo per i *Flötenuhren*, al punto da comporre brani anche per questi strumenti.

I numerosi e brevi componimenti di Haydn per gli automi musicali comprendono rielaborazioni di sinfonie, quartetti o pezzi per pianoforte del compositore stesso, ma anche molti pezzi specifici per il*Flötenuhr*. E ancora oggi esse figurano tra le più famose composizioni del genere, oltre a rispecchiare in modo straordinario le tendenze sociali e culturali del tempo, al pari della musica militare stilizzata o dei caffè all'ultima moda. Nondimeno, questi brani possono testimoniare semplicemente l'inoperosità di un ricco annoiato che passa il tempo trastullandosi con la banale musica degli automi mentre l'orologio segna le ore. Haydn compose tre serie di pezzi con dedicatari differenti. La prima serie di 16 brevi composizioni fu composta nel 1772 per il battesimo della figlia del maestro di cappella Florian Leopold Gassmann a Vienna, ed è dedicata alla moglie dell'amico. Al 1792 risale il secondo gruppo, composto per la casa regnante del Liechtenstein. Infine, la terza serie venne dedicata nel 1793 al principe Nikolaus Esterhazy, in procinto di partire per Londra. Haydn utilizzerà alcuni brani della prima serie anche negli anni 1792-93, quindi i pezzi per *Flötenuhr* sono complessivamente 32. Conservatisi fino a oggi, i tre strumenti per i quali questi pezzi furono scritti ci consentono di trarre importanti conclusioni sul modo di fare musica dell'epoca.

Le tre opere per *Flötenuhr* di Mozart furono composte per la Wunderkammer, la "camera delle meraviglie" del conte Deym von Stržitéž. In uno scenario tipico per la fine del XVIII secolo, nel 1770 il conte Deym tornò dall'esilio per debuttare nella Vienna mondana con il nome di Joseph Müller. La sua "Müllersche Galerie" aveva velleità ludiche simili a quelle di un museo delle cere. Parte della galleria era una Wunderkammer, che dal 1790 ospitò anche una sala dedicata alla memoria del feldmaresciallo Gideon Ernst Freiherr von Laudon. Vi erano conservati diversi orologi musicali, tra cui alcuni realizzati in collaborazione con l'orologiaio Niemecz. La Fantasia KV 594 fu composta da Mozart nel dicembre 1790 su commissione del conte Deym, che chiese un componimento funebre da dedicare al feldmaresciallo defunto e da predisporre per il suo *Flötenuhr* di maggiori dimensioni; fu eseguita per la prima volta il 23 marzo 1791. Questo strumento disponeva sia del registro di flauto che del registro di canne a lingua ed era più grande di un normale *Flötenuhr*; sembra che nella sua melodia programmatica in memoria dell'alto militare, Mozart ne avesse ristretto notevolmente lo spettro musicale. Anche le altre opere per *Flötenuhr* di Mozart furono composte su richiesta del conte Deym. Il 3 marzo 1791 Mozart completò la Fantasia KV 608, che fu eseguita per la prima volta dopo il trasferimento della Wunderkammer in un altro quartiere di Vienna e dopo la trasformazione del mausoleo in elisio per il feldmaresciallo Laudon e l'imperatore Giuseppe II, deceduto nel febbraio del 1790. L'Andante KV 616, infine, fu l'ultima opera della serie ed era destinata a un *Flötenuhr* più piccolo.

Sempre al XVIII secolo risalgono le opere composte da Ludwig van Beethoven per strumenti musicali meccanici. All'epoca giovane compositore, Beethoven era fortemente influenzato da Haydn e Mozart. Inoltre, egli era anche amico del conte Deym ed è ragionevole supporre che sia stato lo stesso Deym a incoraggiare le sue composizioni viennesi del 1799.

Nel XIX secolo i *Flötenuhren* furono sostituiti da orologi con meccanismi musicali moderni. La fine del XVIII secolo segnò la nascita di scatole musicali svizzere e quindi di una nuova tipologia di produzione del suono mediante pettine musicale e lamelle di acciaio. La nuova tecnica fu introdotta anche negli orologi da comò e a parete del XIX secolo, soppiantando i più complicati e dispendiosi carillon e i *Flötenuhren*. Le arie per opere più comuni e i pezzi strumentali più famosi vennero arrangiati per la riproduzione meccanica delle scatole musicali, dal momento che per questi nuovi strumenti non venivano più composti brani appositi, poiché vi erano arrangiatori specializzati che adattavano i pezzi musicali più in voga.

MUSICA E MOVIMENTO

Sincronizzare la musica con il movimento era la sfida dei maestri costruttori di automi, quali Jacques de Vaucanson o Pierre Jaquet-Droz, due pionieri della storia degli automi musicali. Ancora oggi, lo stesso obiettivo si manifesta variamente nell'arte contemporanea. Una delle sfide dell'orologeria era aggiungere agli antichi orologi delle torri campanarie o agli orologi a parete gotici delle figure moventi – dette *jacquemarts* –, che sembravano suonare le campane o battere sulle incudini.

Poco prima dell'inizio del XVII secolo, Leonardo da Vinci lavorò a macchine in grado di imitare i movimenti umani.[2] Come Cartesio qualche anno dopo, Leonardo studiò i corpi di animali e persone e li confrontò con macchine e automi. Costruttori tecnicamente abili iniziarono inoltre a realizzare figure moventi suscitando l'entusiasmo di scienziati e gente comune. Il costruttore di automi François-Joseph Camus, per esempio, progettò una carrozza con cavalli e figure moventi che presentò alla corte di Ludovico XIV. Nel secolo dei Lumi, l'ingegnere francese Jacques de Vaucanson o gli orologiai svizzeri Pierre Jaquet-Droz e Henri Maillardet ammaliarono il pubblico con i loro automi.

Il suonatore di flauto traverso di Vaucanson costituisce il primo collegamento riuscito di un automa che imita i movimenti umani con un meccanismo musicale meccanico. Vaucanson, originario di Grenoble, si dedicò a questo primo automa dal 1733, completandolo quattro anni dopo. Presentò il suo suonatore di flauto a un'esposizione a Parigi, riscuotendo grande successo. Verso la fine del 1738 i visitatori, però, scemarono e Vaucanson decise di aggiungere altri due automi: un suonatore di *galoubet* (simile a un piccolo flauto) con una mano e tamburello con l'altra e un'anatra meccanica.

L'orologiaio svizzero Pierre Jaquet-Droz del cantone di Neuchâtel, dopo gli studi di filosofia a Basilea, si dedicò all'orologeria e alla costruzione di automi, raggiungendo un ottimo livello. Attraverso la manifattura e le filiali, Jaquet-Droz recapitava le sue opere a buona parte della nobiltà francese e inglese e dell'alta borghesia a Ginevra, Parigi e Londra. Viaggiando su una carrozza appositamente costruita, nel 1758 si recò a Madrid alla corte di Ferdinando VI, e il sovrano acquistò sei orologi a pendolo di particolare pregio.

Tra il 1768 e il 1774 Jaquet-Droz, insieme al figlio Henri-Louis e a Jean-Frédéric Leschot, costruì degli automi la cui fama perdura ancora oggi: lo scrittore, il disegnatore e la musicista. Nel 1774 li presentò – riscuotendo uno strepitoso successo – a La Chaux-de-Fonds, nel cantone di Neuchâtel, nella regione del Giura. I tre automi erano costituiti da 2000 a 6000 pezzi: veri e propri capolavori del loro genere. Visto che in molti si recavano a La Chaux-de-Fonds per ammirare gli automi, Jaquet-Droz decise di portarli in tournée. A Parigi, la prima tappa, le qualità meccaniche degli automi furono accolte con entusiasmo. Le tappe successive furono le corti reali di Bruxelles, Londra, Mosca e Madrid. Poco prima dello scoppio della Rivoluzione francese, gli automi di Jaquet-Droz furono venduti in Spagna, dove superarono indenni i tumulti rivoluzionari. Dal 1812 ripresero a viaggiare in Europa e nel mondo intero, giungendo fino in Cina, India e Giappone. Nel corso del XIX secolo cambiarono spesso proprietario, anche se nella mente del pubblico si impresse un unico nome: Jaquet-Droz. Jaquet-Droz e le sue creazioni (orologi, tabacchiere e uccelli canori) divennero famosi in tutto il mondo. I tre automi fecero ritorno in Svizzera agli inizi del XX secolo.[3]

A tutt'oggi non è stato possibile rintracciare un quarto automa costruito da Jaquet-Droz, denominato Grotta, raffigurante un idilliaco paesaggio con rocce, capanne, mulino, cascata, ruscello, sorgente, pecore, uccelli e figure umane. Nella sua descrizione si parla di un pastore che, come il suonatore di flauto di Vaucanson, era in grado di suonare meccanicamente, e di uccelli il cui funzionamento parrebbe analogo a quello degli uccelli canori montati sulle scatole di tabacco da fiuto. Oggi non vi è più traccia di quest'opera, e anche se ne esiste un'illustrazione non è certo che sia mai stata realizzata.[4]

Oltre a Vaucanson e Jaquet-Droz, alla fine del XVIII secolo vi erano altri costruttori di automi, quali Friedrich von Knaus, che per la corte imperiale asburgica di Vienna aveva approntato uno scrittore; l'abate Mical con le sue teste parlanti e un suonatore di flauto; o Peter Kinzing e David Roentgen, che nel 1784 crearono una suonatrice di cembali per la corte francese a Versailles. Anche l'orologiaio svizzero Henri Maillardet, contemporaneo di Jaquet-Droz e con il quale ebbe un'intensa collaborazione [62], costruì insieme ai fratelli Jacques Rodolphe e Jean-David degli automi, alcuni dei quali sono giunti sino a noi. L'automa di uno scrittore fu realizzato come regalo per l'imperatore cinese da parte del sovrano inglese Giorgio III. Un altro automa era in grado di scrivere e disegnare, mentre sono noti anche diversi prestigiatori. Infine, va citato anche l'automa trombettista di Friedrich Kaufmann, realizzato a Dresda tra il 1810 e il 1812.[5]

Gli automi costruiti nel XVIII secolo erano pezzi unici che impressionavano enormemente il pubblico e furono utilizzati anche per studi di medicina. Oltre a dedicarsi all'orologeria e alla costruzione di automi, in molti – fra i quali Jacquet-Droz e il suo socio Leschot – iniziarono a produrre anche protesi, in grado talvolta di eseguire movimenti meccanici come i piegamenti del ginocchio. I risultati raggiunti dai costruttori di automi erano sbalorditivi per l'epoca, ma lo sono ancora oggi. Non solo la musica del flauto o del piano, la scrittura o il disegno venivano riprodotti automaticamente, ma anche la fisionomia e la respirazione umane venivano previste nelle fasi di progettazione e costruzione, dando così origine a esseri antropomorfi o zoomorfi che suscitavano negli osservatori sensazioni contrastanti di entusiasmo e paura.

In un certo senso, il XIX secolo determinò una sorta di "industrializzazione" della produzione di automi. Le fasi costruttive furono semplificate, mentre alcuni soggetti – suonatore di flauto, pifferaio, suonatore di tamburo o disegnatore – furono realizzati inserie, sebbene non si possa parlare ancora di prodotti di massa. Con i suoi numerosi laboratori, la Parigi del XIX secolo era il principale centro di produzione di automi.

1
Si veda Helmut Kowar, *Die Wiener Flötenuhr: "Sie spielt besser als das Orchester im Kärntnertor"*, Technisches Museum Wien, Vienna 2001, e dello stesso autore, "P. Primitivus Niemecz (1750-1806): seine musikalischen Spieluhren und Maschinen", in *Das mechanische Musikinstrument*, n. 115, Gesellschaft für Selbstspielende Musikinstrumente e.V., Essen dicembre 2012, pp. 7-13.

2
Claudio Giorgione, "Leonardo da Vinci e le macchine: forme del corpo e automatismi", in Bruno Corà, Pietro Bellasi (a cura di), *Corpo, Automi, Robot*, Edizioni Mazzotta, Milano 2009, pp. 65-69.

3
Si veda anche *Automates & Merveilles, Les Jaquet-Droz et Leschot*, catalogo della mostra, Musée d'Art et d'Histoire de Neuchâtel, Editions Alphil, Neuchâtel 2012.

4
Charles Perregaux, François-Louis Perrot, *Les Jaquet-Droz et Leschot*, Attinger Frères, Neuchâtel 1916.

5
Si veda Rebecca Wolf, *Friedrich Kaufmanns Trompeterautomat. Ein musikalisches Experiment um 1810*, nella collana Beihefte zum Archiv für Musikwissenschaft, vol. 68, Franz Steiner Verlag, Stuttgart 2011.

I GRAFICI DEL SUONO. UNA PANORAMICA DEI MANUFATTI SONORI

Patrick Feaster

Il suono in sé è evanescente, sia come movimento vibratorio sia come percezione sensoriale, ma si può dire che, in modi diversi, certi manufatti permanenti incarnino il suono in una struttura materiale.

In primo luogo c'è lo *strumento*: un oggetto che viene adoperato per generare suoni a piacimento sfruttandone la capacità di produrre una specifica vibrazione acustica, sia essa intrinseca o estrinseca (come nel caso degli altoparlanti elettrici). Così, si può dire che una corda incarna il suono che produce se pizzicata o strofinata, o un paio di noci di cocco incarnano lo scalpitio degli zoccoli di un cavallo dietro le quinte di un palcoscenico. In secondo luogo c'è la *partitura*: una scrittura che fornisce istruzioni a un suonatore umano affinché la traduca in suoni, anche se forse non esclusivamente in suoni. Si potrebbe dire che queste scritture incarnano un poema che qualcuno dovrà declamare, o un monologo che qualcuno dovrà recitare, o una composizione musicale che qualcuno dovrà eseguire. In terza istanza c'è il *programma*: un oggetto che, quando è eseguito da un apposito strumento (un *automatofono*), produce una sequenza di suoni senza bisogno di ulteriori interventi umani. È in questo senso che il cilindro perforato di un organetto a rullo può incarnare una composizione musicale, o un disco da grammofono può incarnare l'esecuzione musicale o le parole che vi sono registrate. E poi c'è un *surrogato del suono*: la forma visiva o tattile che rappresenta il suono in una modalità sensoriale diversa dall'ascolto, e funge da alternativa o da integrazione al suono stesso: le forme d'onda sullo schermo di un computer, per esempio, o il BOOM! in un fumetto. Gli oggetti di questa categoria incarnano i suoni nel senso che ne rappresentano una traduzione intersemiotica: "significano" i suoni anche in assenza di suoni effettivi.

Nella classificazione che propongo, ciascuna delle quattro categorie implica un rapporto specifico tra manufatti e suoni, ma le categorie non si escludono a vicenda e continuano a compenetrarsi e a pervadersi. Può accadere che un unico manufatto permanente rientri in più d'una di queste categorie, o sia libero di migrare dall'una all'altra. In questo saggio vorrei esplorare un particolare approccio alla materializzazione del suono, un approccio che si ascrive alla tradizione occidentale e trascende le distinzioni tra strumenti, partiture, programmi e surrogati del suono: ovvero, la rappresentazione del suono in forma di *grafico*, in cui il significato si posiziona lungo l'ascissa e l'ordinata di un sistema di coordinate cartesiano.

Certe metafore grafiche per la rappresentazione del suono oggi sono talmente familiari che tendiamo a darle per scontate e a ritenerle "naturali", ma in realtà sono prodotti squisitamente culturali. La notazione musicale del Medioevo europeo ne racchiudeva due. La prima: il posizionamento spaziale delle note lungo l'asse che corre da sinistra a destra fungeva da metafora per il passare del tempo, seguendo le convenzioni della scrittura alfabetica latina. La seconda: il "su" o il "giù" fungeva da metafora per una maggiore o minore rapidità delle vibrazioni. Per consuetudine, oggi usiamo la stessa metafora quando parliamo della distinzione tra note più "alte" e più "basse", o di un'intonazione che "sale" o "scende". Nell'antichità classica, invece, si usava la metafora dell'acutezza, per cui le altezze delle note variavano da "acute" a "gravi" (da *ὀξεῖα* a *βαρεῖα*; da *acutus* a *gravis*). Non è chiaro se sia stata la nuova metafora grafica dell'altezza nella notazione musicale del Medioevo a dar vita alla metafora linguistica o viceversa, ma l'accoppiata dei due sviluppi ha promosso uno specifico approccio culturale che esprime l'altezza di una nota in termini spaziali, e vige ancora ai giorni nostri.

La notazione musicale del Medioevo aveva quindi già adottato spontaneamente una logica di tipo grafico secoli prima di Cartesio, disponendo il tempo lungo l'asse delle ascisse e la frequenza lungo l'asse delle ordinate. Un altro formato grafico che assegna agli assi x e y questi stessi parametri è lo spettrogramma del suono prodotto dai Bell Laboratories negli anni Quaranta, strumento ideato per rendere visivamente intelligibili i segnali acustici, facendoli prima passare attraverso una serie di stretti filtri analogici che isolavano la frequenza e poi visualizzando i dati – disposti dalla parte "alta" alla parte "bassa" dello spettro – come funzioni del tempo. Oggi, la visione "spettrale" dei programmi informatici per l'elaborazione di suoni svolge essenzialmente la stessa operazione. Mentre la notazione musicale del Medioevo era intesa come partitura per un'esecuzione, gli spettrogrammi in origine erano concepiti come *surrogati del suono*; infatti, il primo scopo del progetto era permettere ai non udenti di usare i telefoni "come mezzi per l'ascolto visivo".[1] Ma gli spettrogrammi sono utilizzati anche come *programmi* per gli automatofoni. Un esempio precoce di questo utilizzo è il Pattern Playback messo a punto dagli Haskins Laboratories nel 1950, che esponeva alla luce modulata spettrogrammi trasparenti o riflettenti affinché fossero tradotti in suoni; oggi i software per la sintesi additiva di solito producono gli stessi risultati in modalità digitale. La dimensione visiva degli spettrogrammi spesso mantiene la sua importanza anche in questi casi: così, il valore del Pattern Playback deriva dal fatto che permette ai ricercatori di manipolare le rappresentazioni grafiche del suono e poi ascoltarne i risultati, mentre i software per la sintesi additiva permettono di nascondere immagini spettrografiche nelle registrazioni sonore, come ha fatto Aphex Twin quando ha inserito l'immagine di una spirale nel brano *Windowlicker* (1999) e una fotografia del suo volto nell'"equazione" incisa sul lato B. Il formato spettrografico si usa anche per costruire voci di sintesi con immagini grafiche (una delle funzioni dichiarate di programmi come Photosounder e MetaSynth), un utilizzo che lo ascrive all'ambito degli *strumenti*. Perciò i grafici che rappresentano il tempo e la frequenza attraversano tutte le quattro categorie della mia classificazione, e i grafici creati per fini diversi sono almeno parzialmente interscambiabili: ho scoperto che i software per la sintesi additiva si possono anche utilizzare per tradurre direttamente in suoni le scansioni digitali di notazioni musicali del Medioevo, e che i risultati presentano un'apprezzabile somiglianza con certe esecuzioni vocali degli stessi brani eseguite in età moderna sulla base di un'accurata documentazione.[2] La limitata compatibilità della notazione medievale con le tecnologie foniche del XX secolo evidenzia una sorprendente persistenza a lungo termine delle convenzioni culturali per la rappresentazione del suono.

Certo sono esistite anche tendenze compensative. La notazione musicale d'Occidente spesso ha integrato la "pura" rappresentazione grafica del tempo e della frequenza con altre convenzioni comprensibili, come l'acutezza e la gravità e le corrispondenze arbitrarie tra la forma di una nota e la sua durata (semibreve, minima, semiminima). Ma nella prima era moderna, certi sviluppi tecnici e culturali hanno favorito un ritorno alla logica "pura" del grafico.

Primo: la preparazione manuale di *programmi* per automatofoni – come gli organetti a rullo, le serinette, gli orologi musicali e i carillon – introduceva nuove possibilità di iscrizione dei suoni e implicava requisiti diversi da quelli della partitura leggibile da un suonatore umano. Per certi aspetti gli automatofoni potevano superare le capacità degli esecutori umani, per esempio suonando trilli di una rapidità impossibile, incarnati magari da punti di spillo o graffette lungo un asse del tempo che correva lungo la circonferenza del cilindro. Tuttavia, i programmi dovevano avere una "completezza" non richiesta nelle normali partiture, perché non potevano contare sulla discrezionalità di un suonatore immerso in una tradizione esecutiva coerente. Così, Marie Dominique Joseph Engramelle – importante teorico e sperimentatore della programmazione di automatofoni nel XVIII secolo – ha profuso un notevole impegno nel tentativo di codificare i "silenzi dell'articolazione" che a suo avviso i suonatori di strumenti a tastiera inserivano abitualmente tra le note, ma che non erano indicati nella notazione convenzionale.[3] L'approccio grafico, invece, si prestava a questa meticolosa determinatezza, proprio come i copioni predisposti in un'analoga forma grafica si prestavano all'esecuzione automatica di spettacoli spesso associati agli automatofoni; si pensi per esempio alle ruote dentate che controllavano i movimenti corporei degli androidi musicali e degli uccelli canterini meccanici di Jaquet-Droz [60, 62].

E al Settecento europeo risale anche l'invenzione di un apparecchio che permetteva di *registrare* esecuzioni estemporanee sugli strumenti a tastiera. Alla base del suo funzionamento vi era un meccanismo grazie al quale, ogni qualvolta il suonatore premeva un tasto, una puntina entrava in contatto con un foglio di carta mobile e lasciava un segno, la cui lunghezza corrispondeva al tempo per cui era stato premuto. John Creed, che a quanto è dato di sapere è il primo europeo ad aver sviluppato un progetto di questo tipo, ne ha spiegato i vantaggi evidenziandone l'elevato grado di precisione: "Con questa macchina è più facile distinguere il grave dal presto, l'allegro dall'adagio, rispetto al modo ordinario di indicarli con *Adagio, Allegro, Grave, Presto* e così via, perché con queste parole sappiamo solo che in linea generale l'esecuzione deve essere lenta o veloce, ma non sappiamo in quale grado, perché questo è lasciato all'immaginazione dell'esecutore; ma qui, invece, io so esattamente quante note devo suonare in un secondo; viceversa, so esattamente quante note sono contenute in un centimetro di rotolo... Mentre nel modo ordinario di scrivere la musica non esistono caratteri per gli abbellimenti, o quanto meno quelli esistenti non denotano il tempo e la maniera dell'esecuzione, qui invece sono delineate tutte le minime particelle del suono che compongono matematicamente anche gli abbellimenti più fuggevoli".[4]

Entrambi questi progetti – la programmazione di automatofoni e la registrazione di improvvisazioni alla tastiera – aspiravano a travalicare i limiti della notazione ordinaria, vuoi per fissare gli elementi di un'esecuzione dal vivo, vuoi per sostituirli. Ed entrambi ricorrevano a incarnazioni in forma grafica che intrattenevano rapporti diretti e causali con i suoni che codificavano, sia quand'era la forma grafica a produrre i suoni, sia quand'era la produzione di suoni a creare la forma grafica. L'approccio grafico liberava la scrittura musicale dai vincoli significativi della convenzione notazionale, perché in teoria un grafico poteva accogliere *qualunque* grado dell'altezza, *qualunque* grado del tempo e *qualunque* interazione tra questi due parametri, a qualunque livello di complessità.

Il modello grafico adottato nel XVIII secolo per programmare gli organetti a rullo e per registrare le esecuzioni su tastiera è stato poi ereditato dalla pianola meccanica, e a sua volta il modello del "rullo per pianola" sarà ripreso da strumenti digitali come i sequenziatori MIDI. Così i compositori continuano a cimentarsi con questo modello su base regolare, e alcuni trovano che il suo peculiare carattere grafico sia interessante di per sé. Jonathan Pitkin ritiene che il modello del "rullo per pianola" usato per programmare il Disklavier della Yamaha – un pianoforte digitale – offre una "rappresentazione visiva" delle sue composizioni per pianoforte che è "più fedele" rispetto a quella offerta dalla notazione convenzionale. Parte dell'attrattiva di questo strumento è data dal fatto che "preme da sé i suoi tasti come per magia", e Pitkin ha voluto esplorare questo aspetto visivo nelle esecuzioni. "Spesso mi trovo a 'comporre' le forme e i disegni che voglio vedere apparire sulla tastiera del Disklavier." La notazione convenzionale occulterebbe queste forme e questi disegni, mentre il modello del "rullo per pianola" li mostra chiaramente.[5]

A partire dal XVII secolo si è delineato un approccio più diretto all'incarnazione del suono al livello della singola vibrazione, che ha finito per dar vita alla forma d'onda sonora che trascrive graficamente il tempo rispetto all'ampiezza. Un obiettivo di questo approccio era creare strumenti di *produzione* del suono. Nella ruota di Savart, le sequenze di denti equidistanti percuotevano dei pezzi di carta o altri oggetti facendoli vibrare; l'origine di questo strumento, riattualizzato di recente da Bart Hopkin [346], risale agli anni Settanta del Seicento, quando Robert Hooke lo ha messo a punto per dimostrare che in teoria era possibile "creare tutti i toni musicali con i colpi di un martelletto.[6] In altri casi, sequenze di fori equidistanti venivano passate davanti a un getto d'aria o d'acqua, come nella sirena di Cagniard-Latour; ma in ciascuno dei casi era la velocità di rotazione a determinare l'altezza del suono. Negli anni '10 del Novecento, Luigi Russolo ha inserito in alcuni dei suoi intonarumori [160] delle ruote dentate che funzionavano in base allo stesso principio: "Così alcuni rumori ottenuti con un movimento rotativo possono offrire un'intera scala cromatica ascendente o discendente, se si aumenta o diminuisce la velocità del movimento".[7]

Nel frattempo, l'incorporazione del suono al livello delle singole vibrazioni era usata anche per creare *surrogati del suono* visibili. A tal fine, in certi casi si tracciavano a piacere semplici impulsi o complesse forme d'onda in base a elaborati principi matematici. In altri casi si registravano automaticamente le vibrazioni sonore, per esempio attaccando una puntina a un diapason in modo che fosse quest'ultimo a trascrivere su una superficie mobile le proprie oscillazioni. All'inizio, la rappresentazione grafica delle vibrazioni sonore come funzioni del tempo era intesa come una forma d'arte prettamente *visiva*. La testimonianza più lampante è data dalle ambizioni che Édouard-Léon Scott de Martinville nutriva per il suo fonoautografo, brevettato nel 1857: un'invenzione che sostituiva il diapason dell'esempio precedente con una membrana simile a un timpano, destinata a captare e a trascrivere i suoni dell'ambiente circostante. Scott pensava che le linee ondulate tracciate dalla sua invenzione avessero un'attrattiva di per sé – a tal punto che propose di usarle "per produrre disegni industriali per ricami, filigrane, gioielli, paralumi"[8] – ma a entusiasmarlo era soprattutto la prospettiva di combinare l'estetica visiva e sonora su una base scientifica e oggettiva. Uno dei testi che amava recitare era un passaggio dell'*Otello* di Jean-François Ducis, che declamava ripetutamente nel suo fonoautografo per studiare le corrispondenze tra le intonazioni espressive e la "stenografia naturale" della voce. Il codice delle forme d'onda si rivelava completamente diverso da quello del grafico per la rappresentazione del tempo e della frequenza (anche se si potevano convertire matematicamente i due codici con una trasformata di Fourier): un aumento dell'altezza di un suono produceva ondulazioni più fitte, mentre un aumento del volume produceva ondulazioni di maggiore altezza. Ma dopo aver studiato queste correlazioni, Scott elaborò una visualizzazione ibrida del testo dell'*Otello* sovrapponendo gli elementi della "naturale stenografia" tracciata dal suo fonoautografo alla scrittura alfabetica convenzionale, nella speranza che in futuro le persone avrebbero imparato a "leggere" le grandi esecuzioni musicali e oratorie in questa modalità.

Nel 1877 il fonografo di Thomas Edison introduceva la riproduzione nel campo della registrazione automatica delle vibrazioni sonore, reinventando quest'ultima per trasformarla in uno strumento capace di fissare i suoni in modo che fossero ripetuti *come suoni*. Ora le forme d'onda registrate diventavano programmi per automatofoni, più che surrogati del suono, e pertanto inducevano i legali a chiedersi se questi prodotti dovessero intendersi come scritture soggette al diritto d'autore o come macchine soggette al brevetto.[9] Molti ritengono che il fonografo di Edison si basi su una tecnologia intrinsecamente documentaria, che sia un mezzo per *copiare* i suoni in modo più o meno trasparente, e alcuni teorici hanno voluto distinguere i fonogrammi analogici da quelli digitali in base alla "autenticità" percepita della loro testimonianza: si pensa che i suoni, nel mondo analogico, lascino dietro di sé delle effettive tracce fisiche.[10] Ma di per sé i modelli analogici di registrazione del suono non hanno nulla di intrinsecamente documentario – sono semplicemente un diverso tipo di grafico – e fin dall'inizio il fonografo è stato adibito a usi prettamente creativi. Nel 1878 lo stesso Edison si è accostato alla sintesi del suono, sostituendo il consueto lamierino con sequenze di fori praticati con estrema precisione, al fine di generare un "segnale acustico per vari scopi",[11] e dunque inserendo la sua invenzione nella più antica tradizione di oggetti congegnati per produrre suoni, come le ruote dentate e le sirene. Negli anni seguenti, i fonogrammi tracciati a mano hanno conquistato un notevole interesse tra gli artisti d'avanguardia; in particolare László Moholy-Nagy immaginava i solchi del grammofono che venivano "incisi manualmente nella piastra di cera, senza l'intervento esterno di strumenti meccanici, e poi producevano effetti sonori che avrebbero comportato – senza nuovi strumenti e senza un'orchestra – un'innovazione fondamentale nella produzione del suono."[12] A partire dal 1929-30, qualcosa di simile al sogno di Moholy-Nagy è stato realizzato da Arkady Avraamov, Rudolf Pfenninger, Oskar Fischinger e altri con la creazione manuale di colonne sonore ottiche su pellicola.[13] I suoni creati con questa tecnica avevano un rapporto diretto e indessicale con le immagini, poiché erano prodotti in base alla quantità di luce che passava attraverso ogni punto della pellicola; ma in molti casi la qualità artistica delle immagini assumeva forme irrilevanti per la produzione del suono, perciò rivelavano un interesse a coltivare sia la dimensione visiva, sia la dimensione acustica della nuova arte, invece di subordinare l'una all'altra.

Nelle composizioni prodotte con intenti sperimentali, la determinatezza notazionale raggiunge esiti estremi, quando il compositore specifica i parametri del lavoro musicale fino al livello della singola forma d'onda. Ma le arti visive e sonore si sono incontrate anche all'estremo opposto, con la creazione di partiture musicali innovative sul piano grafico che restano volutamente indeterminate, privilegiando nuove esperienze rispetto a quelle statiche e prevedibili, e dunque perseguendo i valori espressi da John Cage [214, 216]: "La musica che preferisco è quella che non ho ancora sentito. Non ascolto la musica che scrivo; scrivo per ascoltare musica che non ho ancora sentito."[14] Un eccellente esempio del tipo di partitura che Cage aveva in mente è *December 1952* di Earle Brown: una composizione rappresentata da una pagina di linee e rettangoli sparsi, da "eseguire in qualunque direzione, da qualunque punto nello spazio definito per qualunque estensione di tempo e… da una qualunque delle quattro posizioni rotatorie di ogni sequenza."[15] In questo caso, come negli altri che abbiamo esaminato, l'approccio grafico intendeva liberare la musica dai vincoli delle convenzioni, ma qui lo scopo specifico era di fissare la composizione in modi meno rigidi del solito: non per facilitare la realizzazione automatica della musica bensì per opporvi la massima resistenza. E non vi è nulla di contraddittorio in questo. I grafici permettono simultaneamente l'iperdeterminatezza e l'ipodeterminatezza della forma compositiva perché incarnano il suono in dati puri e semplici, svincolati dalla simbologia arbitraria della tradizione; e i loro significati sono riferiti ad assi che possono essere designati e ridesignati a volontà, con un approccio più inclusivo o più esclusivo, più rigoroso o più libero. Ed è proprio per queste sue potenzialità multiformi che a mio avviso il grafico sonoro e i suoi utilizzi offrono una finestra importante per osservare la storia dell'immaginazione sonica.

1
Ralph K. Potter, George A. Kopp, Harriet C. Green, *Visible Speech*, D. Van Nostrand, New York 1947, p. ix.

2
Cfr. Patrick Feaster, *Pictures of Sound: One Thousand Years of Educed Audio: 980-1980*, traccia 25k, Dust-To-Digital, Atlanta (GA) 2012; e l'esecuzione di *Rex caeli domine*, in J. Peter Burkholder, Claude V. Palisca, *Norton Recorded Anthology of Western Music*, vol. 1, W. W. Norton & Company Inc., New York 2009.

3
Dom François Bedos de Celles, *L'art du Facteur d'Orgues*, vol. 4, L. F. Delatour, Paris 1778, pp. 599-600.

4
"A Letter from Mr. John Freke", in *Philosophical Transactions (1683-1775)*, n. 483, Royal Society, London marzo-maggio 1747, pp. 445-450: 450.

5
Jonathan Pitkin, "Con Spirito", in Theresa Sauer (a cura di), *Notations 21*, Mark Batty, New York 2009, p. 177.

6
Henry W. Robinson, Walter Adams (a cura di), *The Diary of Robert Hooke*, Taylor & Francis, London 1935, p. 211.

7
Luigi Russolo, *L'arte dei rumori*, Edizioni Futuriste di Poesia, Milano 1916, p. 14.

8
Patrick Feaster (a cura di), *The Phonautographic Manuscripts of Édouard-Léon Scott de Martinville*, FirstSounds.org, Bloomington (IN) 2010, p. 14.

9
Si veda E. Fulton Brylawski, Abe Goldman (a cura di), *Legislative History of the 1909 Copyright Act*, 6 voll., F. B. Rothman, South Hackensack (NJ) 1976.

10
Si veda in particolare Eric W. Rothenbuhler, John Durham Peters, "Defining Phonography: An Experiment in Theory", in *Musical Quarterly*, vol. 81, n. 2, Oxford University Press, Oxford 1997, pp. 242-64.

11
Reese Jenkins, Robert A. Rosenberg, Paul Israel et al. (a cura di), *The Papers of Thomas A. Edison. The Making of an Inventor, February 1847-June 1873*, vol. 1, Johns Hopkins University Press, Baltimore (MD) 1989, pp. 4-48.

12
László Moholy-Nagy, "Production-Reproduction", in Krisztina Passuth, *Moholy-Nagy*, Thames & Hudson, New York 1985, pp. 289-90.

13
Andrey Smirnov, "Sound out of Paper", asmir.theremin.ru/gsound1.htm (consultato: 20 agosto 2010; Thomas Y. Levin, "'Tones from out of Nowhere': Rudolf Pfenninger and the Archaeology of Synthetic Sound", *The Grey Room*, n. 12, The MIT Press, Cambridge (MA) estate 2003, pp. 32-79.

14
John Cage, "An Autobiographical Statement", newalbion.com/artists/cagej/autobiog.html (ultimo accesso: marzo 2014).

15
Earle Brown, *Folio and 4 Systems*, Associated Music Publishers Inc., New York 1961.

L'ARCAICO ALTROVE DEI CIELI. TERMODINAMICA ED ELETTRICITÀ, PANTEISMO E VUOTO

John Tresch

"Si potrebbe chiamare baccano, fracasso, fragore, martellamento, sommo strepito, divino mugghiare, ma rumore è una parola semplice che designa altrettanto male quella cosa che non ha nome".
Roberto Bolaño, *2666*[1]

LA CATTURA DEI SENSI

"Art or Sound" offre al nostro sguardo oggetti che possiamo ascoltare: strumenti che producono, trasformano o racchiudono un suono. Tendiamo a restare disorientati per le mutevoli posizioni assegnate al visibile e all'udibile nel corso dei secoli documentati in mostra, e per la quantità di strumenti in cui il visibile e l'udibile si fondono, si scambiano, si giustappongono e si separano.

La scultura musicale cattura e converte. È qualcosa che imbriglia, trattiene, trasforma e impone una direzione nuova. Se uno strumento cattura l'aria, il movimento e la vibrazione di un suono, le sculture musicali catturano anche l'occhio, trattengono lo sguardo, lo ripagano con un'armonia, una complessità, un invito a indugiare, una trappola: due secoli di gabbie.

I legni e gli strumenti a corda ricevono, plasmano e amplificano il respiro e il movimento (e il talento e l'emozione) del musicista: l'azione arriva dall'esterno, dal corpo umano che suona. Negli oggetti semoventi, o automi, il procedimento è inverso. Nel corso del XIX secolo, la differenza tra uno *strumento* e una *macchina* è un quesito ricorrente, anche se molti negano che esista una distinzione: uno strumento suonato con abilità sembra suonare da sé, mentre le valvole e i pistoni di vari strumenti, tra cui i legni e gli ottoni costruiti dal XIX secolo in poi, sono macchine semplici integrate in un unico congegno.[2]

La forma visiva degli oggetti musicali, così come i suoni che emettono, solleva interrogativi sulle fonti dell'azione: chi detiene il controllo e attraverso quali mezzi? Gli automi musicali ci inducono in particolare a chiederci: chi ha caricato la molla, chi ha alimentato la fornace, chi ha progettato i circuiti e azionato l'interruttore? Tutti questi interrogativi aleggiano sull'arpa eolica, uno strumento a corda – dal funzionamento semplice ma splendidamente elaborato – che è suonato dal vento.

Nel poema *The Eolian Harp* (1795), Samuel Taylor Coleridge prende spunto dall'arpa eolia per insinuare una provocazione metafisica.[3] Al tramonto, lo scrittore e la moglie siedono circondati dai fiori, ascoltando la "lieve, fluttuante malia" di "quell'umile liuto": "una luce nel suono, un potere sonoro nella luce". Il poeta è colto dal pensiero di un'unica sostanza che unisce l'universo, e si azzarda a chiedersi: "e se tutta la natura animata altro non fosse che tante arpe organiche di forme diverse", accarezzate e vivificate da una "brezza intellettuale" che è al contempo Dio e anima? Poi quella finestra speculativa si chiude: sua moglie lo guarda con "mite rimprovero" e gli ricorda di "affiancarsi umilmente al [suo] Dio".[4] Con deferenza il poeta abbandona le sue fantasie, ma è difficile non sospettare che la vera perla della poesia, il cuore pulsante e la fonte della sua fascinazione, sia proprio quella trasgressiva rêverie panteista posizionata al centro, incastonata tra le innocenti convenzioni del dualismo cristiano e domestico.

La poesia di Coleridge istituisce un legame di lungo corso tra strumenti innovativi, sinestesia e tentazione metafisica. Dietro il richiamo della sinestesia, che fonde i sensi, si schiudono possibilità cosmologiche che fondono la materia, la mente e le fonti ineffabili di ogni attività. L'ebbrezza degli strumenti sinestetici apre le porte del panteismo. Pur essendo oggetto di dibattiti perenni, nell'Ottocento queste congetture acquistano una particolare pregnanza storica grazie a una serie di assemblaggi tecnologici che scatenano forze inedite.[5]

CIMBALO, CILINDRO, CIGNO

I musicologi amano dire che all'inizio del XIX secolo gli europei sentirono due tipi di suoni prodotti dall'uomo che erano più forti di qualunque suono mai udito prima: le esplosioni delle guerre napoleoniche e le sinfonie di Beethoven. Che sia vero o meno, l'aneddoto mette in luce una rottura ma anche un legame, terrificante e sublime, tra le nuove tecnologie della distruzione e le nuove tecnologie della creatività. Molti hanno voluto istituire questa disarmante equazione tra il dispiegamento strategico dell'artiglieria e l'assemblaggio tecnologico dell'orchestra, con i compositori e i direttori d'orchestra posizionati sul campo come generali, gli strumenti schierati come soldati, le sinfonie condotte come campagne militari. L'*Eroica* (1802-04) di Ludwig van Beethoven è ispirata a Napoleone; mentre la *La vittoria di Wellington* (1813) è composta non solo per commemorare la battaglia di Waterloo ma anche per testare una nuova, gigantesca macchina orchestrale, con bombardamenti di tamburi, squilli di fanfare, rimbombi di percussioni a molle e fragore di cimbali.[6]

Il panarmonico per cui Beethoven compone *La vittoria di Wellington* è progettato dallo stesso Johann Nepomuk Mälzel che ha inventato il metronomo, ha elaborato una laringe artificiale e viaggia con un automa che gioca a scacchi. Se nel Settecento gli automi erano intesi come emblemi della padronanza di sé della borghesia, agli inizi dell'Ottocento molti uomini considerano queste macchine dei rivali invadenti: nella loro mente si insinua il sospetto che forze impersonali stiano prendendo il sopravvento sull'individuo.[7] Per di più, sentono tuonare in lontananza il fragore degli strumenti e delle armi prodotti in serie, e presto sentiranno il cigolio e lo sferragliare delle macchine a vapore che azioneranno i mulini, le presse e le locomotive.

A queste soverchianti esperienze acustiche si affiancano le sconcertanti esperienze visive offerte dai diorami, che spesso raffigurano scene di battaglie e vedute vertiginose da torri e montagne, e da tutta la cultura visiva che nei diorami trova un punto di riferimento: dagli enormi e travolgenti dipinti a soggetto storico ai fondali vividi e mobili del teatro e dell'opera. Il viaggio in treno offre una "veduta panoramica" in cui i dettagli in primo piano svaniscono in uno sfondo che si muove adagio, ma anche scossoni disorientanti per il violento sobbalzare del treno.[8] Battaglie, diorami, macchine a vapore e orchestre comprendono suoni, visioni e scossoni troppo potenti, troppo vasti o dinamici da cogliere tutti in una volta: minacciano di sopraffare, divorare, inghiottire il corpo e la mente di chi li esperisce, di sgretolare i confini che tengono separati gli esseri umani dal loro ambiente.

Di recente, Helmut Müller-Sievers ha rinquadrato la scienza, la tecnologia e le arti del XIX secolo utilizzando come categoria d'analisi un'unica forma geometrica: il cilindro e le azioni che rende possibili. Tra le necessità imposte dal motore a vapore non c'è solo la capacità di trattenere il calore e l'energia – che James Watt perfeziona introducendo un secondo cilindro o condensatore da affiancare al pistone principale – ma anche di tradurre il moto lineare in moto circolare e viceversa.[9] Essendo la forma geometrica di motori, contenitori e utensili (compresi gli strumenti usati per registrare i processi fisiologici), il cilindro assicura sia l'ordine prestabilito, sia il progresso richiesto dai consumi e dall'attività creatività. Il cilindro è anche la forma elementare (seppure rielaborata) di nuove integrazioni che si aggiungono all'arsenale audiovisivo: i tubi di ottone allungati e curvati da Adolphe Sax [110, 112] e Ludwig Embach [76], gli amplificatori vocali usati dal coro dei demoni in *Robert le diable* (1831) di Giacomo Meyerbeer, i binocoli da teatro e gli stessi diorami. La forma cilindrica offre una logica per comprendere la struttura dei sensi e il loro svolgersi nel corso del tempo: il dispiegarsi del diorama corrisponde alla sequenza del viaggio in treno, al meccanismo del carillon e della pianola, allo srotolarsi della partitura di una sinfonia.

Negli studi della fisiologia dei sensi si profila un interessante movimento biunivoco: i sensi vengono separati per poi essere riuniti.[10] Sul versante della separazione, Johannes Müller e Charles Bell ritengono che i nervi che trasportano le informazioni dei segnali acustici e visivi sono distinti, e ciascuno di essi utilizza specifiche energie nervose. Félix Savart e Jean-Baptiste Biot usano la strumentazione della fisica sperimentale laplaciana per captare le costanti specifiche dei fenomeni ottici e acustici, costruendo repliche meccaniche sia degli occhi, sia degli orecchi. Con il supporto di François Arago – divulgatore scientifico e collaboratore di Alexander von Humboldt nell'elaborazione della scienza cosmica ed estetica dei cieli e della terra – gli scienziati svolgono esperimenti per misurare la velocità del suono e della luce. Per misurare il suono, l'operatore spara una palla di cannone nel momento stesso in cui viene alzato un segnale visivo: gli osservatori, posizionati a una distanza predefinita, misurano il tempo impiegato per udire l'esplosione dopo aver visto il segnale. Per misurare la velocità della luce, invece, Arago impartisce istruzioni a Léon Foucault e Hippolyte Fizeau per costruire un sistema di specchi rotanti che, con la loro rapida rotazione, fanno viaggiare un raggio di luce per una lunga distanza in uno spazio delimitato; questo meccanismo rotante è mutuato da uno strumento utilizzato per regolare il flusso d'aria nelle canne dell'organo di una chiesa.

Sull'altro versante, emergono invece profonde analogie che sembrano unire il suono alla luce e ad altri fluidi "imponderabili" studiati dai fisici: alcuni fisici ipotizzano che tutti questi fenomeni siano sottoponibili a identici procedimenti matematici, e altri insistono per trovare prove dell'unità fisica sottesa a tutti questi fenomeni, poiché li intendono come diverse vibrazioni in un etere pervasivo o come varianti di un'unica forza o energia primaria. Ad avallare e tradurre queste teorie sono le arti di ispirazione romantica e fantastica, in cui l'unità dei sensi suggerisce una più profonda unità metafisica: Johann Wolfgang von Goethe propone un'analisi dei colori basata su risonanze emotive e musicali che si accorda a questa natura spinoziana; e Charles Baudelaire annuncia programmaticamente le sue *Corrispondenze* (1857). Così, l'equivalenza e la conversione tra i sensi diventa un obiettivo delle arti e un topos della critica: dipingere con i suoni, la musica dei colori.

La commistione dei sensi suggerisce una commistione di essenze; non di rado, proporre la sinestesia è il primo passo per asserire un'unità mistica tra la mente e la natura, secondo un monismo o panteismo di ispirazione vitalista. Entrambe le proposte rispondono agli sconvolgimenti che hanno alterato la distribuzione dei dati sensibili, alla nuova sensibilità tecnologica in cui i confini analitici sono tracciati in modo implacabile ma anche implacabilmente valicati.[11] La sinestesia collima sia con il panteismo filosofico, sia con gli sviluppi tecnici e scientifici del primo Ottocento: macchine a vapore, reti elettriche, mezzi di comunicazione di massa.

Questa concordanza genera anche nuove visioni sociali. I seguaci di Charles Fourier propongono analogie tra i sensi e le varie combinazioni che la "tastiera delle personalità" consente di esprimere; grazie a un'armoniosa organizzazione del lavoro appariranno nuove lune, nuovi continenti, nuovi corpi e creature. Il sansimonismo, una religione inventata dagli ingegneri per la diffusione dell'industria, combina le conversioni del vapore, dell'elettricità, del lavoro manuale e dei sensi. L'equivalenza del calore e del lavoro nella macchina a vapore ricalca, a loro avviso, l'equivalenza cosmica tra mente e materia, intesi come due aspetti di Dio. Perciò, la termodinamica, il panteismo e la sinestesia procedono mano nella mano.

Nel 1832 un sansimoniano denuncia la Parigi contemporanea come una "grande danza satanica" dominata dalla dissonanza visiva e acustica: gli scienziati del Quartiere Latino sono faccia a faccia con gli "animali ululanti" del Jardin des Plantes; i trovatelli che strepitano nell'orfanotrofio disturbano gli astronomi dell'Osservatorio. Nella "Città Nuova dei sansimoniani", invece, il lavoro manuale, il sapere, il commercio e gli svaghi vengono dapprima separati e poi riuniti su un unico piano, distribuito su tutta la città, che assume la forma di un uomo. La scienza occupa la parte sinistra, dove si erge la gigantesca piramide di un'università accanto a lucenti edifici che "scalano il cielo con cristalli di luce"; l'industria occupa la parte destra, con "camini cilindrici che spalancano fauci piene di fiamme" al "ritmo dei colpi di martelli e asce."

Tra le gambe dell'uomo sorgono edifici "consacrati all'estasi della mente e al delirio dei sensi", tra cui "teatri di prosa e d'opera con il loro apparato di strumenti, costumi e scenografie." Separare le funzioni del pensiero, dell'azione e della sfera emotiva è il prodromo della loro armoniosa fusione. Alla testa di questo "nuovo colosso", sull'Île de la Cité, sorge un nuovo tempio che ha la forma di una donna sovradimensionata. Ai suoi fianchi si aprono scalinate a spirale, come ghirlande e cinture ornamentali; la musica sgorga da un organo posto all'altezza del suo seno; dietro di lei, lo strascico del suo abito si drappeggia in un anfiteatro. Questo feticcio sinestetico e altamente tecnologico converte tutte le forze della natura in un melodioso e straordinario strumento per unire l'umanità, i suoi prodotti e le multiformi energie di una natura divinizzata.[12]

Le visione panteista dei sansimoniani, la loro idea di una città più efficiente e armoniosa grazie alla tecnologia, è una fonte cruciale per la trasformazione di Parigi operata da Georges Eugène Haussmann negli anni '50 e '60 dell'Ottocento. Ma il nuovo assetto voluto da Napoleone III non sembra lasciare spazio ad armonie più intime e pacate. Come Victor Hugo, Baudelaire si sente un esule in questo paesaggio urbano così mutato, tra le impalcature, il clangore degli attrezzi, il tramestio di strade e marciapiedi. Nel poema *Il cigno* (1861) Baudelaire plasma un legame allegorico tra sé e un cigno fuggito da un serraglio, che vede vagare disperato e morente vicino alle Tuileries: "nella foresta in cui il [suo] spirito si esilia"[13] sente risuonare le note di un corno, che richiama alla sua mente altre persone derelitte, lontane da casa. La simbolica fusione tra suono, vista e senso in questo cigno/segno implica l'espulsione del poeta dal suo ambiente. Il senso e il significato sono enfatizzati, mentre la materia scivola via, consegnata alla "fatuità moderna".[14]

La poetica di Baudelaire, la cattura fonetica dell'eterno nel fuggevole e dell'eroismo specifico della modernità, risuonerà a lungo nelle arti. Ma con un curioso paradosso: negli anni '30 e '40 dell'Ottocento, le visioni di un'unità dei sensi, delle azioni e della tecnica sono condivise da ingegneri, scienziati, poeti e artisti, il fronte che Saint-Simon battezza "avanguardia". Tuttavia, nella seconda metà del secolo, molte aspirazioni mistiche dei riformatori trovano attuazione in conquiste meramente materiali (spesso fondate sullo sfruttamento). Per reazione, gli artisti ridefiniscono se stessi e il loro territorio, scegliendo in molti casi una fuga – anche se condannata al fallimento – nel regno della memoria, della sensibilità, dell'emozione e dell'immaginazione. E Baudelaire è la luce del faro di questo naufragio.

LA MATERIA DEL VUOTO

La *Gesamtkunstwerk* (opera d'arte totale) di Wagner è la fusione di musica, poesia e teatro. I suoi seguaci parigini recepiscono la sua opera sinestetica in termini spiritualisti: non solo come esperienza sensoriale, emotiva e intellettuale ma anche, e soprattutto, come spettacolo che trascende i supporti materiali. L'arte diventa un medium ideale che consente di travalicare la materia. Wagner incoraggia questa smaterializzazione nascondendo l'orchestra: i leitmotiv musicali e visivi ora aleggiano sopra gli strumenti. Dopo il 1850, le dottrine panteiste che nella prima metà del secolo proclamavano l'unità di spirito e materia vengono spesso sublimate in una nuova forma di dualismo, che risuona nello spiritismo, nello spiritualismo e nelle ricerche sulla "quarta dimensione" dell'etere, una sostanza mercuriale e incorporea che si ritiene agisca tanto nelle sedute spiritiche quanto nei tubi a scarica gassosa, precursori delle odierne luci fluorescenti e al neon.[15]

Nondimeno, per produrre effetti trascendenti servono supporti materiali; l'arte spiritualista di fine Ottocento è inscindibile dalle sue produzioni materiali. Si dice che Wagner sia un cultore del magnetismo animale, quasi volesse gareggiare con il rivale Hector Berlioz, che medita di dirigere un'orchestra via telegrafo.[16] Alle reti idrauliche e alle reti dei trasporti a vapore si affiancano le reti di comunicazione elettrica: le linee del telegrafo corrono accanto alle ferrovie, e sotto o davanti ai piroscafi. E viene lanciato un nuovo equipaggiamento sinestetico. Dopo il caleidoscopio manuale (1815) di David Brewster, arriva il pirofono (1873) [114] di Frédéric Kastner, con i suoi cilindri che si aprono a fiamme di gas danzanti, seguito dall'organo a colori (1893) di Alexander Wallace Rimington. Il compositore russo Alexander Scriabin inserisce le variopinte luci elettriche del "color piano" [136] nelle sue sinfonie audiovisive.[17] Anche la psicologia sperimentale, con il suo vasto apparato di cronometri, sirene e luci lampeggianti, insegue la sinestesia: Théodule Ribot in *La psychologie des sentiments* (1896) e Frederic W. H. Myers in *Human Personality and Its Survival of Bodily Death* (1903) cercano equivalenze ricorrenti tra suoni specifici (vocali, note, timbri), colori e forme, partendo da esperimenti condotti su pazienti sinestetici, compreso Scriabin, uno dei soggetti studiati da Myers.[18]

Artisti, compositori e poeti operano sulla stessa lunghezza d'onda di queste ricerche. Paul Valéry dirà in seguito che gli artisti degli anni '80 dell'Ottocento erano intenti a sviluppare "una dottrina dell'arte derivata dalle tesi allora in voga nell'ambito della psicofisica. Lo studio della sensibilità con i metodi della fisica, le ricerche sulla corrispondenza (ipotetica) tra le sensazioni, le analisi energetiche del ritmo" incidono a fondo sulla poesia e la pittura.[19]

Sottoporre i sensi a macchinari elaborati – per analizzarne il funzionamento o sollecitare nuove esperienze – può avere effetti edificanti, ma anche deleteri. Jean des Esseintes, l'antieroe di *Controcorrente* (1884)[20] di Joris K. Huysmans, si immerge nelle sperimentazioni estetiche dei suoi contemporanei d'avanguardia e nell'era delle macchine. Dalla residenza trasformata in un piroscafo, all'immaginario viaggio a Londra senza lasciare la stazione del treno, alla tartaruga incastonata di pietre preziose, il romanzo segue la sua disperata ricerca di nuove sensazioni. Il punto culminante è l'organo che emette profumi, un congegno wagneriano per comporre sinfonie di fragranze; dopo una suggestiva esecuzione, des Esseintes collassa.

Stéphane Mallarmé, uno degli idoli di des Esseintes, riconosce le seduzioni di Wagner ma si ferma "a metà del sacro monte", prima di raggiungere "l'ossessione troppo lucida di questa vetta minacciosa di assoluto" che il culto dell'arte sta proclamando.[21] Per Mallarmé, la folla trascinata dall'opera d'arte totale, dalle mostre o dai grandi magazzini rivela delle continuità con i riti religiosi del passato. Ma gli adepti del culto dell'arte credono con troppa convinzione e con troppa avventatezza di poter raggiungere il vago obiettivo di una perfetta comunione attraverso la transustanziazione tecnica. Mallarmé ammonisce: "non fate l'errore di alcuni predicatori andando a stingere, in non so quale diluizione, colore, elettricità e popolo, il blu arcaico dei cieli".[22] Con l'illuminazione sbagliata, o una musica troppo forte, si ottiene in cambio solo un cielo contraffatto.

Mallarmé, invece, avanza una concezione che intende la poesia come musica primordiale, un'arte che accenna indirettamente al vuoto indicibile che sta alla radice del pensiero e delle cose. Il linguaggio poetico indica ciò che non può essere detto, mediante la sua intrinseca incapacità di custodire ciò che rivela. La disposizione delle parole come figure statiche di una ballerina può così "evocare, in un vago tratteggio, l'oggetto taciuto, usando parole allusive, mai dirette, che si riducono a un equivalente del silenzio."[23] Al posto delle fantasie troppo letterali degli spettacoli ottocenteschi illuminati dal colore dell'elettricità, Mallarmé chiede luci più soffuse e pazienti, che illuminino con cautela le orme di un vuoto.

Il linguaggio disposto sulla pagina diventa silenzio rarefatto. Per Marcel Duchamp, devoto di Mallarmé, il gesto diventa una modalità dell'inerzia; le tracce materiali diventano una specie di vuoto. Le avanguardie si inoltrano in questi dirupi esplorando l'automatismo, i limiti specifici del medium, gli enigmi autoriflessivi che non offrono soluzione [164].

Due secoli di segnali e rumori interconnessi, di distruzione coreografata dalla corrente elettrica, separano le blasfeme supposizioni di Coleridge sull'arpa suonata dal vento dall'era contemporanea. Nelle sculture di parole e luce create da Bruce Nauman forse vediamo e sentiamo ancora la pienezza del panteismo che pulsava nella sinestesia romantica, ma la vediamo implodere nel suo gemello cosmico: la silente materia oscura che sottende alla "luce nel suono". In uno degli enigmatici circuiti di Nauman, il neon sussurra nei tubi giallo limone, arancio e fucsia; le parole si accendono, si spengono e ripetono: VIOLINI... VIOLENZA... SILENZIO.[24]

Questo saggio è dedicato a Robert Michael Brain

1
Roberto Bolaño, *2666*, trad. di Ilide Carmignani, Adelphi Edizioni, Milano 2001, p. 795.

2
Helmut Müller-Sievers, *The Cylinder: Kinematics of the Nineteenth Century*, University of California Press, Berkeley (CA) 2012; John Tresch, Emily I. Dolan, "Toward a New Organology: Instruments of Music and Science", in *Osiris*, "Music, Sound, and the Laboratory from 1750-1980", vol. 28, n. 1, The University of Chicago Press, Chicago (IL) gennaio 2013, pp. 278-98.

3
Thomas L. Hankins, Robert J. Silverman, Willem Hackmann. *Instruments and the Imagination*, Princeton University Press, Princeton (NJ) 1995.

4
Samuel Taylor Coleridge, *The Eolian Harp*, vv. 13, 21, 29, 45-46, 48, 50, 53, in Ernest H. Coleridge (a cura di), *The Complete Poetical Works of Samuel Taylor Coleridge*, vol. I, Clarendon Press, Oxford 1912, pp. 100-02; trad. nostra.

5
Douglas Kahn, *Earth Sound, Earth Signal: Energies and Magnitudes in the Arts*, University of California Press, Berkeley (CA) 2013.

6
Emily I. Dolan, "The Origins of the Orchestra Machine", *Current Musicology*, n. 76, 2003, pp. 7-23; Nicholas Mathew "History under Erasure: 'Wellingtons Sieg,' the Congress of Vienna, and the Ruination of Beethoven's Heroic Style", *The Musical Quarterly*, vol. 89, n. 1, Oxford University Press, Oxford primavera 2006, pp. 17-61.

7
Adelheid Voskuhl, *Androids in the Enlightenment: Mechanics, Artisans, and Cultures of the Self*, University of Chicago Press, Chicago (IL) 2013.

8
Wolfgang Schivelbusch, *Geschichte der Eisenbahnreise: Zur Industrialisierung von Raum und Zeit im 19.Jahrhundert*, Hanser, München/Wien 1977; trad. inglese Anselm Hollo, *The Railway Journey: Trains and Travel in the 19th Century*, Urizen Books, New York 1979, pp. 57, 137; trad. nostra.

9
Helmut Müller-Sievers, *op. cit.*, pp. 23-40.

10
Jonathan Sterne, *The Audible Past: Cultural Origins of Sound Reproduction*, Duke University Press, Durham (NC) 2003; Jonathan Crary, *Techniques of the Observer: On Vision and Modernity in the 19th Century*, The MIT Press, Cambridge (MA) 1992; Mara Mills, John Tresch, "Introduction: Audio/ Visual", *Grey Room*, vol. 43, The MIT Press, Cambridge (MA) primavera 2011, pp. 6-15.

11
Jacques Rancière, *Le Partage du sensible: Esthétique et politique*, La Fabrique-Éditions, Paris 2000; Todd H. Weir (a cura di), *Monism: Science, Philosophy, Religion, and the History of a Worldview*, Palgrave Macmillan, New York 2012.

12
Charles Duveyrier, "La ville nouvelle, ou le Paris des Saint-Simoniens", in C. Ladvocat (a cura di), *Paris, ou le livre des cent-et-un*, vol. 8, Ladvocat, Paris 1832, pp. 315-44.

13
Verso 49 da *Il cigno* (*Le Cygne*). Questa poesia è la quarta nella sezione intitolata "Tableaux Parisiens" (Quadri Parigini), aggiunta alla seconda edizione de *I fiori del male*. Prima edizione *Les fleurs du mal*, Poulet-Malassis et de Broise, Paris 1857, seconda edizione 1861. [N.d.R.]

14
Charles Baudelaire, "Exposition universelle – 1855 – Beaux-arts", in Id., *Œuvres complètes*, vol. II, Michel Lévy frères, Paris 1868, pp. 211-244: 219; trad. nostra.

15
Falk Müller, "Johann Wilhelm Hittorf and the Material Culture of Nineteenth-Century Gas Discharge Research", *British Journal for the History of Science*, vol. 44, n. 2, Cambridge University Press, Cambridge giugno 2011; Linda Dalrymple Henderson, "Etherial Bride and Mechanical Bachelors: Science and Allegory in Marcel Duchamp's 'Large Glass'", *Configurations*, vol. 4, n. 1, Johns Hopkins University Press, Baltimore (MD) inverno 1996, pp. 91-120.

16
Alison Winter, *Mesmerized: Powers of Mind in Victorian Britain*, University of Chicago Press, Chicago (IL) 2000.

17
Edizione originale *À rebours*, G. Charpentier et Cie, Paris 1884. Pubblicato in Italia in diverse edizioni, anche con i titoli *Controcorrente, Contro corrente e A ritroso*. [N.d.R.]

18
Kerry Brougher, Judith Zilczer, Jeremy Strick, Ari Wiseman, Olivia Mattis, *Visual Music: Synaesthesia in Art and Music Since 1900*, Thames & Hudson, New York 2005.

19
Roy Porter, "Synaesthesia", in Adam Lowe, Simon Schaffer (a cura di), *N01se*, catalogo della mostra, Kettle's Yard, Cambridge 1999.

20
Paul Valéry citato in Robert Michael Brain "Genealogy of 'ZANG TUMB TUMB': Experimental Phonetics, Vers Libre, and Modernist Sound Art", *Grey Room*, vol. 43, The MIT Press, Cambridge (MA), primavera 2011, pp. 88-117.

21
Stéphane Mallarmé, "Richard Wagner, rêverie d'un poëte français", *Revue wagnérienne*, n. 7, Paris agosto 1885, pp. 199-200; trad. nostra. Jacques Rancière, *Mallarmé o la politica della sirena*, trad. di Alessandro Serra, Clueb, Bologna 2000.

22
Stéphane Mallarmé, *Divagations*, Eugène Fasquelle Éditeur, Paris 1897, p. 306, trad. nostra.

23
Stéphane Mallarmé, *Divagations*, cit., p. 326, trad. nostra.

24
Bruce Nauman, *VIOLINS VIOLENCE SILENCE*, 1981-82, Tate and National Galleries of Scotland.

VEDERE NON È SENTIRE. SINESTESIA, ANESTESIA E ARTE AUDIOVISIVA

Christoph Cox

DALLA *GESAMTKUNSTWERK* ALLA SINESTESIA

Noi esseri umani siamo dotati di cinque sensi, cinque canali distinti attraverso i quali recepiamo informazioni sull'ambiente in continuo mutamento. Nella comunicazione tra interno ed esterno, quattro canali sensoriali partono da particolari orifizi – occhi, orecchi, naso e bocca – grazie ai quali il nostro corpo si apre al mondo. L'altro senso, il tatto, è distribuito su tutta la superficie dell'epidermide, una membrana avvolgente che registra anche gli stimoli più lievi.

La suddivisione delle arti rispecchia in gran parte questa distribuzione e frammentazione sensoriale, anche con la tradizionale gerarchia che separa i due sensi "nobili" (la vista e l'udito) dai loro omologhi più volgari (il gusto, il tatto e l'olfatto). L'arte rivolta al primo senso trova spazio nel museo o nella galleria; l'arte rivolta al secondo si esprime nella sala da concerto. Ciascuna ha la sua storia, che solo di rado si sovrappone alla storia dell'altra. E se è vero che esistono forme d'arte miste – per esempio il teatro e il cinema – parlare di "arti visive" e distinguerle dalla musica è scontato quanto distinguere le mele dalle arance.

Eppure, nonostante la frammentazione sensoriale, la dispersione fisiologica e la gerarchia estetica, noi esperiamo il mondo come unità. Non ci accorgiamo di coordinare e tradurre continuamente flussi eterogenei di dati sensoriali. Il mondo sembra offrirsi alla nostra esperienza nella sua integrità e completezza. Nel riconoscere il misterioso emergere dell'unità dalla pluralità, Aristotele postulò una sorta di sesto senso – un *sensus communis* o senso comune – deputato a coordinare i cinque flussi sensoriali e a garantirne la concordanza.[1] L'ipotesi di Aristotele ha profondamente influenzato i suoi successori, tanto che alcune varianti di quel postulato continuano a permeare le attuali ricerche scientifiche.[2]

Ma allora, se esperiamo il mondo come unità sensoriale, le arti non dovrebbero affermare la loro unione? Sulla questione il modernismo era profondamente diviso. Uno dei suoi teorici più autorevoli e influenti, Clement Greenberg, sosteneva che il tratto distintivo della modernità risiede precisamente nella separazione delle arti e nello sviluppo autonomo di ciascuna. In particolare, Greenberg si preoccupava di emendare le arti visive da ogni sensazione extravisiva (quelle tattili, per esempio) per renderle "puramente ottiche".[3] Fin dagli albori, però, figure di spicco dell'arte moderna hanno assunto una posizione alternativa, sposando la causa dell'unità sensoriale ed estetica. A lanciare l'appello nel 1849 fu il compositore e drammaturgo Richard Wagner con *L'opera d'arte dell'avvenire*, in cui affermava che solo una vera sintesi delle arti – una *Gesamtkunstwerk* o "opera d'arte totale" – può realizzare il vero scopo dell'arte. Wagner scriveva: "Ogni facoltà umana è limitata, ma tali facoltà riunite, in accordo tra loro, si soccorrono a vicenda; in altri termini le facoltà umane, *amandosi* reciprocamente, costituiscono la facoltà universale umana che è senza limiti, che basta a se stessa. Parimenti tutte le facoltà *artistiche* dell'uomo hanno i loro limiti naturali perché l'uomo non ha *un senso solo*, ma *dei sensi* in generale. Ogni facoltà deriva solo da un determinato senso e ha i propri limiti nei limiti del senso che le dà origine. I limiti dei sensi particolari sono anche punti di contatto tra loro, i punti in cui si confondono e si comprendono. Allo stesso modo le facoltà da loro derivate si toccano e si comprendono: i loro limiti si aboliscono nel loro accordo. Ma solo ciò che si ama si può comprendere, e amare significa riconoscere altri e, nello stesso tempo, riconoscere se stessi. Il riconoscere mediante l'amore è la libertà; la libertà delle facoltà umane è *la facoltà universale*. Solo l'arte corrisponde a questa facoltà universale dell'uomo e per conseguenza è *libera*, mentre un *genere* d'arte non è libero perché deriva da una sola facoltà umana. Danza, musica e poesia, prese separatamente, sono ognuna limitata a se stessa; se si oppongono ai propri limiti, ognuna si sente schiava, a meno che, giunta agli estremi, non tenda la mano all'altro genere d'arte corrispondente con un amore tutto pieno di riconoscimento. Basta l'azione del tendersi la mano perché siano poste al di sopra di quel limite; l'assorbimento assoluto, il dono assoluto all'arte sorella, cioè il dono assoluto di se stessa oltre il limite che le divide, fa crollare il limite stesso; e quando tutti i limiti saranno in tal modo eliminati, spariranno i generi d'arte, che anch'essi altro non sono che limiti, e non resterà che l'*Arte*, l'arte comune, universale, illimitata."[4]

Nei decenni successivi al proclama di Wagner, l'invocazione di un'unità dei sensi nella sfera artistica veniva rafforzata dall'interesse scientifico per la sinestesia, il fenomeno neurologico in virtù del quale la stimolazione di una modalità sensoriale induce una sensazione in un'altra; per esempio si vedono suoni o si odono profumi. Questo fenomeno affascinava filosofi e scienziati da secoli, per lo meno dal 1690, quando John Locke riportava il caso di un cieco che sosteneva di intendere il rosso scarlatto come il suono di una tromba.[5] Ma negli ultimi decenni dell'Ottocento la sinestesia acquisì un ruolo centrale nella ricerca neurologica.

Negli stessi anni il tema accendeva l'interesse di pittori, compositori e poeti attratti dalle possibilità immaginative e dalle risonanze spirituali dell'unità sensoriale. Nel celebre poema *Corrispondenze* (1857), Charles Baudelaire scriveva di "odori [...] dolci come òboi, verdi come giardini."[6] Il suo compatriota simbolista Arthur Rimbaud tracciava equivalenze tra vocali, colori, consistenze e odori: così, la lettera "A" evocava il "nero corsetto villoso delle mosche lucenti / che ronzano intorno a fetori crudeli"; la "E" "candori di vapori e di tende" e così via.[7] Quando gli artisti cominciavano ad abbandonare la figurazione per avvicinarsi all'astrazione, spesso cercavano ispirazione nella musica, concepita come la forma d'arte più elevata e più libera dai vincoli della rappresentazione. "Tutte le arti aspirano costantemente alla condizione della musica", dichiarava il critico Walter Pater nel 1877, un'osservazione che sarebbe riecheggiata nella pittura di molti pionieri dell'astrattismo. Per *Impressione III (Concerto)* [*Impression III (Konzert)*], un dipinto proto-astratto del 1911, Vasilij Kandinskij si ispirava direttamente alle emozioni suscitate dalle prime composizioni atonali di Arnold Schönberg, che il pittore aveva ascoltato durante un concerto a Monaco. Altri artisti, come Paul Klee, Marsden Hartley e František Kupka, si ispiravano a Bach, tentando di elaborare un contrappunto visivo simile alla struttura ritmica della fuga.

La passione per gli intrecci sinestetici dilagata nel secondo Ottocento catturò anche la fantasia degli inventori, molti dei quali si misero all'opera per costruire macchine che potessero collegare direttamente il suono e il colore. Nei tardi anni '60 del secolo, il chimico e musicista alsaziano Frédéric Kastner elaborò uno strumento chiamato *Pyrophone* [114], in cui una piccola tastiera riusciva a produrre suoni e luci colorate infiammando getti di gas che illuminavano 13 tubi di cristallo emergenti da una console. Due decadi e mezzo dopo, l'inglese Alexander Rimington, inventore e insegnante d'arte, costruì uno strumento che fu il primo a definire "organo a colori", uno strumento che non produceva suono ma utilizzava una comune tastiera d'organo per illuminare 14 lampade ad arco che assumevano diverse tonalità e intensità di colore.

Le invenzioni di Kastner e Rimington scatenarono una fascinazione dilagante per la musica visiva, stimolando artisti e inventori a costruire numerose varianti dell'organo a colori. Nei primi decenni del Novecento, le avanguardie artistiche furono attratte da queste invenzioni, adottate non come strumenti musicali ma come meccanismi capaci di animare le forme astratte attraverso una sorta di proiezione cinematica. Il russo Wladimir

Baranoff-Rossiné, pittore futurista, elaborò il suo *piano optophonique* (pianoforte optofonico, 1920-23) [170], un congegno complesso azionato da una tastiera che produceva suoni proiettando una luce attraverso una serie di specchi, filtri, lenti e dischi dipinti a mano. Al Bauhaus di Weimar, Ludwig Hirschfeld-Mack costruì una macchina che traduceva la musica in proiezioni di forme mobili e luci colorate. Thomas Wilfred, artista danese-americano, aggiornò la concezione dell'organo a colori con il suo *Clavilux* (1922), utilizzando un banco di comandi e tasti scorrevoli per far ruotare lampadine e specchi che proiettavano eterei ciuffi di colori danzanti simili a fiamme.

Accanto a queste proiezioni di musica visiva, gli artisti esploravano la possibilità di creare un astrattismo su pellicola ispirato alla musica. Sull'onda delle invenzioni di Baranoff-Rossiné e Hirschfeld-Mack, il "cinema assoluto" di Hans Richter e Viking Eggeling mirava ad animare i rapporti ritmici e armonici tipici della pittura non figurativa arricchendoli con la dimensione temporale tipica della musica. I rettangoli pulsanti che si espandono, si contraggono e si sovrappongono in *Rhythmus 21* (1921) di Richter evocano l'esperienza musicale del volume, dell'altezza delle note, dell'armonia e, soprattutto, del ritmo. In *Symphonie Diagonale* (1924) di Eggeling, una serie di forme simili a pettini si alternano e si susseguono come accordi, mentre le linee curve e spezzate si sviluppano sullo schermo come melodie silenziose. Connessioni ancora più dirette tra immagini e suoni cinematici furono esplorate da Oskar Fischinger, creatore di "ornamenti" visivi – motivi seghettati, stelle, punti e onde – che potevano fungere simultaneamente da contenuto e da colonna sonora di un film, poiché nei suoi lavori la colonna sonora è impressa sulla celluloide come forma d'onda visiva, e viene letta da uno strumento ottico che la converte in un segnale acustico.

L'interesse per una connessione diretta tra suono e immagine ricorre continuamente nell'arte moderna e contemporanea, ed è stato riproposto nei primi anni '70 da Guy Sherwin con i suoi filmati ottici e sonori, da Steina e Woody Vasulka con le loro sperimentazioni video, da Alvin Lucier con *The Queen of the South* (1972), e in tempi più recenti da artisti come Carsten Nicolai [350], Billy Roisz, Haroon Mirza [400, 434] e altri.

SUONO E IMMAGINE

Con l'avvento del film sonoro nei tardi anni '20, il sogno di un'arte sinestetica sembrava finalmente realizzato. Ma questo sviluppo non trovò unanimi consensi tra i cineasti e gli artisti. Nel 1928, tre dei più illustri esponenti del cinema sovietico, Sergej Eisenstein, Vsevolod Pudovkin e Grigorij Aleksandrov, pubblicarono una dichiarazione sul sonoro in cui esprimevano il timore che il suono sincronizzato venisse usato per rafforzare l'illusione cinematografica: "Usare il suono in questo modo", notavano, "distruggerà la cultura del montaggio." I tre registi favorivano l'asincronia tra suono e immagine, un rapporto contrappuntistico tra i due elementi che avrebbe resistito alla subordinazione del suono all'immagine e avrebbe incoraggiato una tensione capace di contrastare l'illusione naturalistica.[8]

Qualche anno dopo, Walter Ruttmann raccolse l'appello in favore di un rapporto non sincronico tra suono e immagine e lo portò alle estreme conseguenze. Animato da interessi affini a quelli di Richter, Eggeling e Fischinger, Ruttmann invertì le sperimentazioni dei compagni per creare "musica visiva". Una sua pellicola del 1930, *Wochenende* (*Weekend*), consiste unicamente in una colonna sonora senza immagini: un collage di 11 minuti di rumori concreti – martelli, registratori di cassa, sirene, voci, musica accidentale e così via – che lasciavano immaginare allo spettatore i loro corrispettivi visivi. Questa pellicola si può considerare complementare alla silenziosa musica visiva di *Rhythmus 21* o di *Symphonie Diagonale*; ma il predominio del visivo sul sonoro impedisce questa complementarietà e fa sì che l'assenza dell'elemento visivo sia esperita non come invito alla sinestesia ma come mancanza, come una sconnessione tra suono e immagine.

Nell'arte moderna e contemporanea, la resistenza a fondere suono e immagine è diffusa quanto l'aspirazione all'unità sinestetica. In parte questa resistenza si fonda sul sospetto che qualunque convergenza dei sensi manterrà in vigore la gerarchia che subordina tutte le modalità sensoriali alla vista. Ma nasce anche dal rifiuto di abrogare le differenze irriducibili tra i sensi e le ricche tensioni estetiche che queste differenze sanno generare.

Questo atteggiamento prevale in tutta la storia dell'arte del XX e del XXI secolo, dai readymade di Duchamp alle declinazioni contemporanee dell'arte sonora. Mentre Kandinskij, Klee e molti altri pionieri della modernità inseguivano il sogno della *sinestesia*, Duchamp inseguiva l'*anestesia*. La scelta dei readymade, notava, "era fondata su una reazione di indifferenza visiva, unita a una totale assenza di buono o cattivo gusto… di fatto una completa anestesia."[9] Infatti, in tutta la sua attività, Duchamp si è opposto all'idea che l'arte dovesse essere principalmente "retinica", spingendola invece verso la dimensione verbale e concettuale. In diverse sue opere, la critica all'arte retinica si è avvalsa dell'elemento sonoro. Prendiamo per esempio *With Hidden Noise – À bruit secret* (Con rumore segreto, 1916) [164], un "readymade assistito" composto da un gomitolo di spago racchiuso tra due lastre di ottone fissate con dei bulloni, che al suo interno contiene un oggetto misterioso inserito dal mecenate di Duchamp, Walter Arensberg. Anche se in sostanza è un oggetto o un giocattolo che produce suono, quando assurge al rango di opera d'arte, di oggetto inerte esposto in un museo, perde il suo carattere sonoro, di cui resta traccia solo nel titolo, e la sua origine sonica è ora "nascosta" o "segreta".

Un'intera progenie di opere ispirate da Duchamp sembra voler separare il suono dall'immagine. *Das Schweigen* (Il silenzio, 1973) di Joseph Beuys, è un oggetto scultoreo che si compone semplicemente di 5 bobine del film di Bergman *Il silenzio* (*Tystnaden*, 1963); la pellicola di celluloide è rivestita di rame e zinco e sigillata in dischi galvanizzati. Eliminando sia il suono che l'immagine del film di Bergman, la pila di bobine di Beuys attira l'attenzione sulla pellicola non come insieme di "immagini parlanti" ma come materiale muto. Christian Marclay porta avanti questa offensiva con *The Sound of Silence* (Il suono del silenzio), una fotografia del disco di Simon & Garfunkel del 1964 *The Sounds of Silence*. L'opera di Marclay rimanda a Beuys e a Duchamp ma anche a René Magritte, in particolare al dipinto del 1929 *La trahison des images* (Il tradimento delle immagini), che sotto l'immagine stilizzata di una pipa afferma perentoriamente *Ceci n'est pas une pipe* (Questa non è una pipa). Il dipinto di Magritte sottolinea la sconnessione tra parola, immagine e oggetto. E la fotografia di Marclay estende questa sconnessione al suono, notando, come Beuys, che questa immagine muta, questo oggetto afono, cattura quel silenzio a cui la canzone di Simon & Garfunkel e il film di Bergman potevano solo paradossalmente alludere.

La fotografia di Marclay si inscrive nella linea dell'arte concettuale classica che, spesso ispirandosi a Duchamp, metteva in discussione lo status dell'opera d'arte sia in qualità di immagine, sia in qualità di oggetto. La ricerca di un'arte non retinica ha portato molti artisti concettuali a muoversi in direzione del concetto puro e del linguaggio inteso come sua manifestazione fisica. Ma ha avvicinato diversi artisti concettuali – tra cui Robert Barry, Christine Kozlov, Robert Morris e Bruce Nauman – a un suono inteso come materiale effimero, invisibile e al contempo poderosamente fisico. *Box with the Sound of Its Own Making* (Scatola con i rumori della sua creazione, 1961) [220], un'opera di Robert Morris che riveste un valore pionieristico sia per l'arte concettuale, sia per l'arte sonora, è esattamente ciò che il titolo descrive: un cubo di legno accompagnato da una registrazione di tre ore che documenta il processo della sua costruzione. L'anodino oggetto visivo non è nulla senza la sua colonna sonora, che esprime tutta l'attrattiva dell'opera e offre una critica implicita alla reificazione estetica: l'arte non è un risultato bensì un processo, immortalato con maggiore efficacia nella *dimensione* temporale del sonoro che nell'inerzia dell'oggetto visivo.

Diverse opere di Nauman del 1968 e del 1969 indagano le possibilità del suono per destabilizzare lo status visivo e oggettuale dell'arte. In *Six Sound Problems for Konrad Fischer* (Sei problemi sonori per Konrad Fischer, 1968), l'elemento visivo è un semplice registratore a bobine, con un riccio del nastro che si protende nello spazio e si avvolge intorno a una matita attaccata a una sedia. I sei nastri che accompagnano l'opera registrano i suoni di semplici attività svolte da Nauman alla Galerie Konrad Fischer di Düsseldorf: l'artista cammina, fa rimbalzare una palla, suona il violino o compie simultaneamente alcune di queste azioni. Nauman aveva già registrato queste attività su video; qui le presenta unicamente come suoni, come "*problemi* sonori", perché, ancora una volta, l'audio da solo bandisce la veridicità documentaria del video, che registra più chiaramente le fonti e le cause degli eventi. Questa capacità di separare il suono dalla sua fonte è un tratto distintivo di quello che il compositore Pierre Schaeffer ha definito ascolto "acusmatico", una forma di attenzione acustica che la registrazione

sonora consente di sperimentare in quanto ci permette di ascoltare il suono in assenza delle sue fonti visive. La musica concreta di Schaeffer – opere sonore composte interamente da registrazioni di suoni musicali e documentari – indagava le possibilità acusmatiche del suono registrato. *Concrete Tape Recorder Piece* (Registratore di cemento, 1968) [248] di Nauman è al contempo un accenno duchampiano a Schaeffer, un omaggio a *With Hidden Noise* e una risposta a *Box with the Sound of Its Own Making*. L'opera, un blocco di cemento da circa 240 chili da cui sporge un cavo elettrico, è descritta dall'artista come un "registratore con un nastro su cui è registrato un urlo, avvolto in un sacchetto di plastica e inserito al centro di una colata di cemento." Anche in questo caso il rumore è nascosto, silenzioso, o meglio *ridotto al silenzio*, perché l'opera di Nauman è più sinistra di quella di Duchamp o quella di Morris. Il suo suono è sepolto, ripetuto all'infinito ma inudibile, seppellito dalla forma scultorea che lo nega. Così, l'opera si può forse intendere come un omologo del silenzioso *Urlo* (1893) di Edvard Munch, ma più sconcertante, poiché alla sua fonte sonica è negata ogni espressione.

Da un lato l'aspirazione, dall'altro l'opposizione all'unità sensoriale: queste due tendenze (e le tensioni tra le due polarità) guidano l'arte moderna e contemporanea da oltre un secolo e mezzo. Da Wagner e Kandinskij in poi, gli artisti hanno inseguito il sogno della sinestesia, l'unificazione delle arti che avrebbe potuto superare la frammentazione e permettere all'esperienza estetica di affermare l'unità dell'esperienza esistenziale: quello che Aristotele chiamava senso comune, che Wagner celebrava come la "facoltà universale degli uomini" e Brian Massumi ha definito di recente la "amodale confusione esperienziale", la "virtuale interezza multidimensionale della sensazione" che caratterizza le nostre ordinarie vite sensoriali.[10] Oggi questa spinta verso l'unità artistica e sensoriale trae nuova linfa dal fatto che l'arte diventa sempre più multimediale, intermediale o postmediale, e che le tecnologie digitali facilitano la trasposizione dei dati da un medium all'altro, per esempio la conversione di un'immagine in un suono, o viceversa. Ma la tesi della fusione estetica e sensoriale ha incontrato una vigorosa antitesi, articolata nel modo più enfatico da Clement Greenberg e Michael Fried, i quali patrocinavano un gruppo di artisti che a loro avviso spingevano nella direzione opposta: verso l'attenta e rigorosa indagine di media specifici e di specifiche modalità sensoriali.[11] Se lo schieramento modernista sembrava aver perso la battaglia contro la produzione ibrida e multimediale che domina l'arte contemporanea, di recente la voce autorevole di Rosalind Krauss ha rinnovato l'appello di Greenberg e Fried in favore della specificità del medium, affermando che "l'abbandono di un medium specifico decreta la morte dell'arte seria."[12] Gran parte dell'arte sonora emersa negli ultimi due decenni si è allineata implicitamente con queste posizioni critiche. Senza rifiutare la produzione multimediale e multimodale, l'arte sonora ci ricorda con forza che il vedere e l'udire non si equivalgono, e mira a contrastare le ambizioni imperialistiche dell'elemento visivo. Nell'ultimo secolo e mezzo, nessuna opera ha aderito appieno a una sola di queste due tendenze. Ma alcune delle opere più interessanti hanno raccolto la provocazione lanciata dalle due polarità, opponendo resistenza sia all'assimilazione, sia alla segregazione dei sensi, e operando in quello spazio gravido di tensione che si estende tra l'una e l'altra.

1
Si veda Aristotele, *De Anima* III, 1-2.

2
Si veda Richard Cytowic, *Synesthesia: A Union of the Senses*, The MIT Press, Cambridge (MA) 2002, p. 75.

3
Si veda Clement Greenberg, "Modernist Painting", in John O'Brian (a cura di), *The Collected Essays and Criticism, Volume 4: Modernism with a Vengeance, 1957-1969*, University of Chicago Press, Chicago (IL) 1993.

4
Richard Wagner, *L'opera d'arte dell'avvenire*, trad. di Alfio Cozzi, Rizzoli, Milano 1983, pp. 143-144.

5
John Locke, Roger Woolhouse (a cura di), An *Essay Concerning Human Understanding*, Libro IV, Penguin, London 1997, §11.

6
Charles Baudelaire, "Corrispondenze" in *I fiori del male*, trad. di Gesualdo Bufalino, Mondadori, Milano 1989, pp. 18-19.

7
Arthur Rimbaud, "Vocali" in *Poesie*, trad. di Dario Bellezza, Garzanti, Milano 1983, pp. 30-31.

8
Sergej Eisenstein, Vsevolod Pudovkin, Grigorij Aleksandrov, "Statement [on Sound]", in Gerald Mast, Marshall Cohen, Leo Braudy (a cura di), *Film Theory and Criticism: Introductory Readings*, IV ed., Oxford University Press, Oxford 1992, pp. 317-19.

9
Marcel Duchamp, "Apropos of 'Readymades'", in Michael Sanouillet, Elmer Peterson (a cura di), *Salt Seller: The Writings of Marcel Duchamp*, Oxford University Press, Oxford 1973, p. 141.

10
Si veda Brian Massumi, *Parables for the Virtual: Movement, Affect, Sensation*, Duke University Press, Durham (NC) 2002, p. 169, e Id., *Semblance and Event: Activist Philosophy and the Occurrent Arts*, The MIT Press, Cambridge (MA) 2011, p. 88.

11
La classica difesa della specificità del medium espressa da Greenberg si trova in "Modernist Painting", in John O'Brian (a cura di), *op. cit.*, pp. 85-93. Le influenti argomentazioni di Fried contro la mescolanza dei media si trovano in "Art and Objecthood", in Michael Fried, *Art and Objecthood: Essays and Reviews*, University of Chicago Press, Chicago (IL) 1998, pp. 148-72.

12
Rosalind Krauss, *Perpetual Inventory*, The MIT Press, Cambridge (MA) 2010, p. xiii.

LA RIVELAZIONE DEI RUMORI NASCOSTI. STRUMENTI MUSICALI E DISPOSITIVI SONORI IMPROBABILI NELLA LETTERATURA DEL PRIMO NOVECENTO

Douglas Kahn

Lo scrittore francese Raymond Roussel è uno dei più importanti creatori di strumenti musicali del Novecento, e il fatto che i suoi strumenti e i suoi brani siano costituiti solo da parole non deve far passare in secondo piano le loro caratteristiche sonore, la semplicità costruttiva e il significato organologico. Un estroso orchestrion alimentato dall'espansione e dalla contrazione di un metallo particolarmente sensibile al calore da poco scoperto, il bexio; 12 cocchi dotati di ruote sonore che eseguono un repertorio di arie popolari; un lungo verme che distilla gocce di acqua pesante sulle corde di una cetra producendo una melodia familiare come una sorta di pianola biologica; due fratelli giocolieri che si lanciano tra loro 12 palle, colpendole con dei tamburelli; un uomo senza una gamba che suona un flauto ricavato intagliando la sua stessa tibia; un uomo che canta quattro brani simultaneamente da diverse parti della sua enorme bocca; una banda musicale composta da un solo artista privo di gambe e monco delle braccia; un bambino che sa imitare con incredibile verosimiglianza i suoni di un treno che corre a tutta velocità, i gridi degli animali domestici, lo stridere di una sega su una pietra da taglio, il salto improvviso di un turacciolo da champagne, il gorgogliare di un liquido versato da una bottiglia, fanfare del corno da caccia, un assolo di violino, un canto lamentoso di un violoncello; per finire, un pelo di roditore che quando viene piegato produce due note.

Sono queste le suggestioni sonore contenute in *Impressioni d'Africa*, scritto da Roussel nel 1910[1]: definirlo un romanzo può essere fuorviante, in quanto la sua struttura narrativa ricorda più le musiche da processione dei film di Federico Fellini, che accompagnano per esempio la sfilata di abiti ecclesiastici in *Roma* (1972) o alcune scene isolate di *Fellini Satyricon* (1969), visibili attraverso porte lasciate opportunamente aperte, con la sola differenza che ciò che viene mostrato qui è più vario, minuzioso sino all'ossessione, bizzarro, come se si assistesse al sogno febbricitante di un orologiaio in cui un fumoso Jules Verne si stempera nei tratti di William Burroughs. Il libro racconta di una nave che naufraga sulla costa atlantica, vicino all'Equatore, un'invenzione costruita sulla realtà del colonialismo francese in Africa. I passeggeri si rivelano essere un gruppo dal talento fuori dal comune, e mentre attendono di essere liberati dall'Imperatore Talù, organizzano un gala che si sovrappone allo spettacolo offerto dai riti, dalle rappresentazioni, dai decreti, dalle pubbliche torture e dalle esecuzioni dello stesso Talù.

Insieme alle fantasmagoriche scene tratte da un altro suo importante romanzo, *Locus Solus* (1914),[2] *Impressioni d'Africa* fa di Roussel un prolifico capomastro di quelli che definisco "strumenti concettuali", sebbene non pochi tra quelli che egli immaginò, più che improbabili, siano impossibili, come l'oggetto che mi accingo a descrivere,[3] uno strumento che potrebbe essere facilmente realizzato se non si pensasse alla crudeltà come a un ostacolo. Roussel è a buon diritto al centro di ogni discussione sugli strumenti musicali e sull'arte, anche a causa della sua influenza su figure chiave quali Marcel Duchamp, Guillaume Apollinaire, André Breton e molti altri.[4]

Più concretamente, con Francis Picabia, Duchamp e Apollinaire egli partecipò a una rappresentazione teatrale di *Impressioni d'Africa;* più tardi, Duchamp avrebbe affermato che Roussel e il suo libro erano "responsabili" nientemeno che della sua opera più importante, *The Large Glass* (*Il grande vetro*, 1915-23). *Impressioni d'Africa* si basa sul racconto di Apollinaire *Il Re luna* (1916), in cui il narratore vaga nelle viscere di una montagna fino a imbattersi nell'eponimo personaggio che dispone di una speciale tastiera, i cui tasti sono collegati ad altrettanti microfoni sparsi per il mondo.

I microfoni perfetti che il re ha a sua disposizione sono sintonizzati in modo da trasmettere nel sotterraneo i rumori più lontani della vita terrestre. Ogni tasto aziona un microfono regolato su questa o quella distanza. Ora sono i rumori di un paesaggio giapponese che s'avvertono. Il vento soffia tra gli alberi, e deve esservi un villaggio, perché si sentono le risate delle serve, la pialla di un falegname e il getto ghiacciato delle cascate.[5] E poi, la fatica socialista e i geyser della Nuova Zelanda, e così via con i suoni dei più svariati luoghi del mondo. Apollinaire descrisse uno strumento musicale se non impossibile quanto meno improbabile per l'epoca, ma che un secolo dopo non sarebbe stato difficile realizzare e suonare.

Vorrei concentrarmi ora su uno strumento che fa la sua comparsa nelle primissime pagine di *Impressioni d'Africa*, talvolta definito "intreccio musicale", sottolineando come lo sfarzo e la magnificenza della decorazione formale di un'uniforme militare possa essere accompagnata da una sua musica specifica. Tale denominazione non permette tuttavia di comprendere in che modo è generato il suono, ovvero dal polmone di una persona ancora in vita. Inoltre, molto prima di scorgere gli intrecci o di comprendere altri aspetti dello strumento nel suo complesso, si avvertono gli echi di un suono che può essere definito un "respiro musicale". All'infuori di esso, tutto è nascosto in una piccola struttura scura simile a una casetta, con quattro pareti nere, prive di finestre o porte, a eccezione di una piccola uscita, anch'essa coperta. Roussel è solito introdurre una scena inusuale, che non spiega; lascia, al contrario, che i suoi meccanismi e comportamenti interni, se non addirittura i significati, rimangano tali fino a che la scena non si ripete una seconda volta. E questo è ciò che intendo fare nel descrivere questo strumento: lascerò che per ora i suoi dettagli rimangano nascosti e li rivelerò solo in seguito.

Gli strumenti di Roussel erano il prodotto di un'inverosimile abilità, a sua volta figlia di un benessere ostentato e di una personalità ossessiva. Quali altri soggetti maniacali di incredibile ricchezza erano mai stati altrettanto estrosi? Egli era guidato da un'aspettativa di genialità universalmente riconosciuta, alimentata da uno stato di diritto di matrice aristocratica. Invece di mostrarsi incline agli svaghi, egli si impose un'arbitraria quanto ardua disciplina di scrittura. Una genialità eccentrica che fu infine riconosciuta da artisti e intellettuali, ma non da chi aveva determinato la popolarità dei personaggi che egli più ammirava: Victor Hugo o Jules Verne.

Roussel fu spinto sin da giovane allo studio della musica dalla madre, ma fu ammesso al Conservatorio solo a sedici anni, dopo ben tre tentativi. Raggiunse ben presto i più alti livelli di tale istituzione, tanto che uno dei suoi insegnanti lo soprannominò "il pianista dalle cento mani". La signora Roussel non badò a spese per le sue lezioni, né per qualunque altra attività si svolgesse nella casa di rue de Chaillot, dove un ascensore rivestito di boiserie di legno esotico conduceva al piano al quale Raymond viveva; ogni figlio, come riporta il suo biografo François Caradec, viveva a un piano diverso e aveva le sue stanze private, nonché domestici, cavalli e carrozza propri, alloggiati nell'edificio della corte interna.[6] La famiglia aveva un palco all'opera; la musica, in casa, era ovunque: il salotto della madre era frequentato da musicisti e Roussel presenziava regolarmente a concerti, spettacoli di musica popolare e teatro musicale, sebbene nessuna di tali eccezionali esperienze possa spiegare la fertilità della sua immaginazione musicale.

La sua carriera letteraria ebbe inizio quando il verso creato per essere cantato prese il sopravvento sulla musica; i suoi risultati possono essere in parte considerati come l'applicazione alla scrittura delle complesse regole che governano determinati tipi di composizioni musicali. Il carattere ossessivo e la "mania per le regole", come egli stesso ebbe a definirla, gli permisero di sviluppare una tecnica che ne imprigionava la scrittura e al tempo stesso rompeva gli argini della sua immaginazione. La sua tecnica consisteva nel prendere una frase e cambiare una lettera, una sola, ma in grado di mutare il significato di una parola. Tale cambiamento si sarebbe poi fatto rumorosamente strada nella frase, con calembour e altre forme di trasformazione, creando una seconda frase, molto diversa dalla prima. Egli inseriva quindi una delle due all'inizio di una storia, e l'altra alla fine. Un'unica leggerissima variazione, quindi, per produrre un sostanziale *non sequitur*, che a sua volta richiede la risolutezza di un'esposizione complessa, simile a quella di una fuga musicale.

Possiamo paragonare tale involuzione a un nodo nel legno di un albero. L'inserimento di una piccola imperfezione dà origine a una realtà alternativa. In natura la maggior parte delle mutazioni non sopravvive, ma i nodi esistono in quanto espressione caotica dei loro ospiti, in quanto deviazione della venatura da un percorso lineare verso una turbolenza di cui percepiamo la bellezza. Per Roussel il gioco di parole generava un altro gioco di parole. L'effetto trasformistico di una lettera porta con sé cambiamenti fonici nella frase, che a loro volta generano parvenze che si rivelano meramente esteriori, azioni e comportamenti che in seguito si fanno da parte a favore della descrizione di un meccanismo. Si tratta di una forma di ribellione alla codificazione che invece di entrare in collisione con una macchina che fornisce informazioni, ne crea una del tutto nuova. Il frutto più importante di tale meccanismo fu *Impressioni d'Africa*.

Lo strumento letterario di Roussel prefigurò nel corso del secolo tecniche meccanicistiche simili, tra cui il metodo dodecafonico, le tecniche e i giochi dadaisti e surrealisti, i diversi modi di John Cage di creare libertà e indeterminatezza, i cut-up di William Burroughs e molto altro, tra cui la costrizione produttiva delle prime forme di arte digitale.[7] Le intenzioni alla base della creazione di dispositivi così arbitrari furono le più svariate, in primis la volontà di prendere le distanze dalle normali finalità e consuetudini dell'espressione, come spiega magistralmente Michel Leiris: "Il fatto di ridurre se stessi a schiavi volontari di una regola difficile e complicata, sulla quale è necessario concentrare la propria attenzione ignorando ciò che ci circonda, produce una liberazione dell'inconscio, che è più reale di quanto non avvenga nel puro e semplice abbandono della scrittura automatica [surrealista]. Forse, concentrare la propria attenzione su qualcosa di futile rimuove le inibizioni in modo più efficace di quanto non riescano a fare la passività o il tentativo di liberare la mente."[8]

La spinta alla tecnica può essere paragonata anche alla ricerca anagrammatica di Ferdinand de Saussure nel verso latino, che sovverte la frase, senza tuttavia ottenere un risultato in tutto equivalente all'originale. Roussel rivelò il suo segreto solo nella sua opera postuma *Come ho scritto alcuni miei libri* (1935).[9] Dalla sua tomba – o in altre parole, dalla sua cripta – fornì la chiave per "decriptare" i suoi scritti, rivelando che cosa nascondevano i loro più remoti meccanismi. Lo fece senza però raggiungere la popolarità che pensava di meritare, e rivelandolo destinò le sue opere al medesimo esame ossessivo con il quale erano state scritte, dedicandole a tutti i futuri de Saussure.

Pur avendo l'intenzione di scrivere per un ampio pubblico, Roussel finì per unirsi a quel gruppo di illustri scrittori e artisti le cui opere furono realizzate innanzitutto per essere studiate: *Finnegans Wake* (1939) di James Joyce o *The Large Glass* di Marcel Duchamp sono tra le prime a venire in mente. Tali opere, che non furono concepite né come rivelazione al grande pubblico né come ricerca ermetica, rivelano la loro natura sin dall'inizio ma, a differenza dei rompicapo, che possono avere una soluzione, studiarle non fa che liberarne l'enorme complessità. Non vi sono mete da raggiungere né luoghi sacri nascosti da profanare, come invece dovrebbe essere. Svelare un enigma è un atto violento in quanto ne distrugge i misteri, la forza generativa. È meglio eludere la distruzione cambiandone la forma proprio sotto il nostro naso, creando nuovi scorci di conoscenza proprio sotto la nostra gnosi.

Tanto Joyce quanto Duchamp costruirono gran parte delle loro opere ermetiche trasponendo i giochi di parole in codici e crittogrammi. I giochi di parole godono di una cattiva reputazione in quanto terribilmente ovvi, inconsapevoli o semplicemente ostici. Ma anche quando danno il peggio di sé, testimoniano di un'economia di tempo e di spazio. Il senso non è mai uno solo; ve n'è infatti sempre un altro, umoristico, nascosto, come se la lingua fosse divisa a metà e ciascuna delle due parti parlasse per sé, in modo non dissimile dal cantante Ludovic dall'enorme bocca in *Impressioni d'Africa*, che si allena a "frazionare labbra e lingua in settori indipendenti l'uno dall'altro, e senza fatica poteva articolare nello stesso tempo molteplici parti sovrapposte, diverse nella melodia e nelle parole..."[10] Più una bocca o una parola riesce a dire o a cantare simultaneamente, meglio è.

"Wheatstone's magic lyer" è un gioco di parole musicale creato da James Joyce in *Finnegans Wake*, che ruota intorno alle parole "*lyre*" (lira) e "*liar*" (bugiardo). Joyce fu probabilmente il più grande mago delle parole mai vissuto; basta chiedere agli studiosi che hanno dedicato la loro vita a quest'opera, dove ogni vocabolo sembra celato dietro un produttivo conflitto con se stesso, quasi come un "harpsidischord" (gravecembalo).[11] Joyce definisce Sir Charles Wheatstone, scienziato e inventore vittoriano, un bugiardo, per l'inganno perpetrato dal suo insolito strumento musicale, la "lira incantata": una lira sospesa al soffitto che all'apparenza sembra produrre musica da sola, e non unicamente musica da lira, mentre il suono proviene da una camera al piano superiore, da strumenti le cui vibrazioni sono trasmesse attraverso il cavo di sospensione. Un inganno, in effetti, al pari del *Turco meccanico* di Wolfgang von Kempelen che, pur sembrando un automa giocatore di scacchi, in realtà al suo interno celava un uomo in carne e ossa; è quindi comprensibile che in seguito Wheatstone abbia sviluppato macchine per la crittografia e i messaggi in codice. Wheatstone aveva coniato un altro termine per la sua lira incantata, e l'aveva chiamata "acucriptofono", ovvero "ascolto di suoni nascosti" (da *acu*, ascoltare; *cripto*, nascosto; e *fono* da voce o suono).

With Hidden Noise – À bruit secret (Con rumore segreto, 1916) [164] di Marcel Duchamp è un gomitolo di spago posto tra due lastre metalliche fissate da quattro viti lunghe e strette, che fanno anche da base all'intera opera. All'interno, vi è un oggetto non identificato che, quando viene scosso, fa rumore. Fu Walter Arensberg, amico di Duchamp, a inserire tale oggetto nell'opera e non disse a nessuno di che cosa si trattasse, neppure a Duchamp. Entrambe le lastre recano iscrizioni enigmatiche, tre righe di tre parole ciascuna, con intervalli che denotano l'assenza di lettere nelle parole. Sostituire le lettere permette di creare abbastanza facilmente delle parole sia in inglese sia in francese, ma il loro significato può rivelarsi problematico. Duchamp affermò che le iscrizioni erano come lettere mancanti su un'insegna al neon difettosa, che la rendono priva di significato. Non può quindi esservi altra soluzione se non cercare di identificare la fonte del suono o il significato delle iscrizioni, entrambi destinati a rimanere nascosti. Lo storico dell'arte Seymour Howard ideò un gioco di parole su chi questi giochi li crea, definendo la tendenza di Duchamp un *hidden naos*, un tempio nascosto, un luogo sacro dove risiede un'autorità invisibile.[12] L'interpretazione, a questo punto, si fa labirintica, nutrendosi del piacere dello schema, come la vita si nutre di ciò che è vivo: mentre il *noise/naos* (rumore/tempio) nasconde, come la cripta nella criptografia, dall'aldilà il defunto esprime una "mania per le regole".

Tra i readymade di Duchamp, *With Hidden Noise* è spesso accostato a *Fresh Widow* (1920), una finestra di dimensioni ridotte con otto riquadri di vetro coperti da pelle nera recante l'iscrizione "FRESH WIDOW COPYRIGHT ROSE Selavy 1920." Il titolo è un gioco di parole tra *fresh widow*, "fresca vedova", e *French window*, letteralmente "finestra francese", completato da uno degli pseudonimi di Duchamp, *Eros, c'est la vie*. L'opera lascia ampio spazio alle interpretazioni più svariate, ma di fatto tutte riconducono all'atto di nascondere e mascherare, alla decifrazione di iscrizioni più o meno oscure. *With Hidden Noise* e *Fresh Widow* sono entrambe riconducibili a un oggetto e strumento musicale introdotto e descritto nelle prime pagine di *Impressioni d'Africa* di Raymond Roussel. Come nell'opera *With Hidden Noise* esiste un rumore nascosto, ma in questo caso si tratta di un rumore che sarà rivelato.

Torniamo per un momento a *Impressioni d'Africa*, nella nazione fittizia di un'Africa equatoriale francese reale: ecco una passeggiata che contorna una piazza e, nelle vicinanze, una fila di statue e accanto a esse una loggetta. Si tratta di una costruzione piccola e bizzarra: priva di porta, ha solo quattro pareti di spessa tela nera, che ricordano i riquadri di vetro annerito di *Fresh Widow*. Il tetto, che sembra di paglia, è fatto in realtà con le pagine ingiallite di un romanzo di Sir Walter Scott. L'interno sarebbe immerso nell'oscurità, se non fosse per una *judas window* ("finestra di Giuda"), un'apertura che ha la funzione di uno spioncino che, come implica il termine, rappresenta un luogo di sorveglianza, tradimento, un nascondiglio e una fuga clandestina. Essendo la costruzione priva di una porta e di vere e proprie finestre, rappresenta infatti il solo modo per uscirne. Anch'esso è però coperto dalle pagine del romanzo, mascherando la possibilità di origliare dall'interno; non essendo completamente scure, le pagine che coprono l'apertura (così osserva il narratore) diffondono nella costruzione una luce tenue e giallastra. Ciò che non viene celato è un suono proveniente dalla casa: "Una specie di accordo che ricordava, sia pure molto attutito, il timbro degli strumenti in ottone usciva a intervalli regolari dal centro della loggetta, dando la sensazione precisa di un respiro musicale".[13]

L'assenza di finestre e la scelta di nascondere lo spioncino ricordano l'estasi di Roussel al tempo in cui la sua scrittura s'impadronì di lui. "Ero circondato da raggi di luce, tanto da voler chiudere le tende per timore che i raggi abbaglianti emanati dalla mia penna potessero sfuggire da una seppur minuscola fessura; volevo ritirare lo schermo e d'improvviso illuminare il mondo."[14] I teli neri della piccola costruzione erano come tende tirate che non lasciavano sfuggire, o piuttosto, entrare la luce. Neppure un puntino luminoso brillava nella stanza; da questa sorta di *camera obscura*, con le finestre completamente chiuse, trapelava solo un suono. La presenza dello spioncino di quello spazio oscurato potrebbe far pensare alla finestrella che si trovava sul coperchio della bara della madre di Roussel. Niente di particolarmente strano se si pensa che la donna, durante un lungo viaggio nell'Oceano indiano, aveva portato con sé un feretro, casomai le fosse accaduto l'irreparabile e il suo corpo imbalsamato avesse avuto bisogno di far degnamente ritorno a casa. Roussel era con lei durante quel viaggio ed era stato proprio lui a far mettere la finestrella per poter vedere il volto della madre sino all'ultimo. Peccato che *Impressioni d'Africa* fosse stato scritto prima del viaggio con il figlio o della sua dipartita: la cripta seguiva la criptografia.

La fonte di quel suono rimane un mistero per molte pagine, sospesa tra il suono segreto e la sua rivelazione, proprio come la frase originaria e la sua variazione che completa la storia. Nel frattempo, il crudele imperatore Talù si avvicina alla piazza con il suo nutrito seguito, mentre gli abitanti della città si radunano in attesa, pronti ad assistere a uno spettacolo fatto da una serie di eventi fuori dall'ordinario, torture ed esecuzioni. Un altro momento musicale è quello della danza di dieci mogli di Talù (il quale ne ha molte di più) dopo che esse hanno ingerito una sostanza misteriosa che si svelerà solo attraverso i suoi effetti, animandone il canto e la danza; una danza che ha inizio con "lente evoluzioni accompagnate da movimenti agili e ondulanti"[15] che in un crescendo si fanno delirio frenetico e contorto, mentre il canto irrompe come un rumoroso coro di rutti sonori, resi ancora più intensi dalla gara che si scatena per fragore e sonorità. La rappresentazione raggiunge un assordante punto culminante prima di concludersi, accompagnata da un accordo finale prolungato che si spegne gradualmente, fino al silenzio. E tuttavia si continua a percepire il respiro musicale.

Eppure nulla rimane nascosto a lungo nel romanzo di Roussel. Lo spioncino si apre e si palesa una giovane donna, Louise Montalescot, in divisa da ufficiale francese. Sulla spalla destra, a indicarne lo status nell'esercito, cordelline dorate, intrecciate, dal cui nodo pendono dei piccoli puntali metallici cavi, che ospitano al loro interno delle lamelle vibranti. Sono questi piccoli elementi sonori a emettere quel respiro musicale che si sente fuoriuscire dalla loggetta. "Era di lì che partiva l'accordo sommesso che fino allora avevamo sentito uscire dalla loggetta; era prodotto dallo stesso respiro della giovane donna grazie a un canale artificiale che univa la base del polmone con l'insieme delle cordelline, sistemate apposta per nascondere tubi flessibili, liberi e sonori … A ogni contrazione del polmone una parte dell'aria espirata passava dai molteplici condotti e, mettendo in moto le lamelle, provocava un'armoniosa vibrazione."[16] Quando Louise Montalescot vide il cadavere di uno dei giustiziati "pianse nervosamente, col petto scosso da orrendi singulti, che rendevano più netti e più acuti gli accordi delle cordelline".[17] Tecnicamente, si tratta di uno strumento a metà strada tra la cornamusa e la voce. Il polmone equivale a un otre o a una vescica, come quella della cornamusa, e analoga è la distribuzione dei suoni che genera, attraverso svariati apparati sonori; tuttavia, al pari di un serbatoio, si svuota molto velocemente in mancanza di un effetto espressivo diretto, come accade per la voce. Questo strumento mostra una certa affinità con il coro delle donne che ruttano, in quanto il loro suono supera la capacità dei polmoni di controllare la riserva d'aria, utilizzando una vescica più adatta alle cornamuse: lo stomaco. Il tubo che collega il polmone di Louise Montalescot con la rappresentazione pubblica di una musica celebrativa potrebbe essere il cordone ombelicale. Un cordone collegato a quello del respiro sonoro, che nell'autobiografia di Roussel rappresenta il rapporto di dipendenza e devozione nei confronti della madre e della sua influenza in ambito musicale; allo stesso modo la successiva tossicodipendenza della donna riecheggia nella misteriosa sostanza ingerita dalle mogli di Talù, che ne anima lo strano canto e la danza.

Il piacere di queste interpretazioni risiede in uno schema che, pur percepibile nel linguaggio, può essere esistito davvero oppure no. In letteratura, fortunatamente, non si tratta di una distinzione importante, né dovrebbe esserlo nella progettazione e nella costruzione degli strumenti musicali. In quanto appena detto non vi è una correlazione percepibile tra James Joyce, la lira incantata di Charles Wheatstone e Raymond Roussel, a parte la comune propensione per i giochi di parole e i suoni musicali nascosti. Marcel Duchamp, per contro, è impresso in modo indelebile nello spettro di Roussel, ma se esso sia un fantasma che infesta una determinata opera o un rumore nascosto, resta occultato…

In *Impressioni d'Africa* Roussel ci ha generosamente edotti su che cosa creasse il suono nascosto del respiro musicale. Ciò era parte della struttura narrativa di una serie di racconti rivelatori che esisteva al posto di una storia più lunga, identificabile come un romanzo. Forse alla fine non ne conosceremo il motivo, ma attraverso il processo autorivelatore del romanzo, una duplice azione di presentazione e una spiegazione conclusiva, se non altro avremo una nozione più chiara del *come*, la pura parvenza superficiale del *come*, rispetto ai rumori del *perché*. Ciò che rimane nascosto è lo slittamento di quella lettera che ha creato questo strumento. E se anche poi fosse stato rivelato, ne sapremmo forse di più?

1
Edizione originale *Impressions d'Afrique*, Lemerre, Paris 1910; *Impressioni d'Africa*, trad. di Laura Lovisetti Fuà, Rizzoli, Milano 1964.

2
Edizione originale *Locus Solus*, Lemerre, Paris 1914; in Raymond Roussel, *Locus Solus, seguito da Come ho scritto alcuni miei libri*, trad. di Paola Dècina Lombardi, Einaudi, Torino 1975.

3
Douglas Kahn, "Conceptual Instruments", *Experimental Musical Instruments*, vol. 6, n. 3, ottobre 1990, pp. 17-20.

4
Insieme ad Alfred Jarry e alla sua Patafisica, Roussel fu anche l'ispiratore di Oulipo (*Ouvroir de littérature potentielle*, ovvero "officina di letteratura potenziale"), un gruppo di scrittori che applicava regole costruttive attraverso le quali liberare scrittura, e del romanziere Alain Robbe-Grillet, il quale propendeva per una "superficie" in cui ciò che si vede è ciò che si ha, piuttosto che per una realtà superficiale che al di sotto lasciasse intravedere ulteriori tematiche.

5
Guillaume Apollinaire, *Re luna*, trad. di Marisa Vallini, Milano, Rosa e Ballo 1946.

6
François Caradec, *Raymond Roussel*, trad. inglese di Ian Monk, Atlas Press, London 2001, p. 26.

7
Oliver Grau (a cura di), "Between a Bach and a Bard Place: Productive Constraint in Early Computer Arts", *MediaArtsHistory*, The MIT Press, Cambridge (MA) 2007.

8
Michel Leiris, *Roussel l'Ingénu* (1987) citato in Caradec, *op. cit.*, p. 64.

9
Edizione originale *Comment j'ai écrit certains de mes livres*, Lemerre, Paris 1914; pubblicato in Italia in Raymond Roussel, *Locus Solus, seguito da Come ho scritto alcuni miei libri*, cit., pp. 263-85.

10
Raymond Roussel, *Impressioni d'Africa*, cit., pp. 56-57.

11
James Joyce, *Finnegans Wake*, Penguin Classics, London 1999, p. 13.

12
Seymour Howard, "Hidden Naos: Duchamp Labyrinths", *Artibus et Historiae*, vol. 15, n. 29, 1994, pp. 153-80.

13
Raymond Roussel, *Impressioni d'Africa*, cit., p. 13.

14
Citato in Pierre Janet, "The Psychological Characteristics of Ecstasy", in Alastair Brotchie *et al.* (a cura di), *Raymond Roussel: Life, Death and Works*, Atlas Press, London 1987, pp. 40-41; trad. nostra.

15
Raymond Roussel, *Impressioni d'Africa*, cit., pp. 19-20.

16
Ibid., p. 23.

17
Ibid.

UN'ORCHESTRA METAFISICA. UNA RICERCA SULLA RICOSTRUZIONE DEGLI INTONARUMORI

Luciano Chessa

Nel 1913 il pittore italiano Luigi Russolo firma il Manifesto futurista *L'arte dei rumori* con l'intento di cambiare radicalmente la musica da concerto dell'epoca, obbligandola a venire a patti con un mondo sonoro contemporaneo fatto di rumori essenzialmente urbani.

Abbandonando la pittura e dedicandosi anima e corpo alla costruzione di nuovi strumenti musicali, alla composizione e alla direzione d'orchestra, Russolo espone concretamente e in modo quanto più efficace la propria visione rivoluzionaria di come avrebbe dovuto essere la musica del XX secolo.

La rivoluzione sonora di Russolo, inizialmente irrisa e accolta con riluttanza, andò assumendo sempre maggiore importanza, mentre la figura di Russolo, soggetto all'attenzione dei musicologi almeno dagli anni '70 del Novecento in poi, gode oggi di non pochi estimatori.

Tuttavia, quando nel 2001 iniziai a lavorare alla mia tesi di dottorato sull'estetica di Russolo,[1] scoprii con mia grande sorpresa che la storia della musica era ancora carente di un considerevole sforzo ermeneutico, tanto nell'ambito dell'arte dei rumori, quanto in quello degli intonarumori [160] che Russolo realizzò a dimostrazione delle sue tesi.

Le opere compilative pubblicate a metà degli anni '70 da Giovanni Lista e Gianfranco Maffina,[2] pionieri della materia, che attingevano in gran parte a documenti conservati e catalogati da Maria Zanovello, vedova dell'artista, non sono state tuttavia seguite da alcuna interpretazione sistematica del lavoro effettivamente compiuto da Russolo sul suono. E sebbene nella letteratura minore il nome di Russolo ricorra spesso, gran parte degli scritti sulle sue attività musicali sono considerati aneddotici e il significato di queste stesse attività non è ancora stato studiato con la dovuta attenzione.

Una situazione tanto più sorprendente se si pensa alla mole delle ricerche riguardanti il Russolo pittore, ancor più agli occhi di chi, come noi, ritiene che il contributo di Russolo in ambito musicale sia stato ben più importante che non nella pittura.

Detto ciò, l'importanza della formazione di Russolo come pittore nel contesto dei suoi interessi filosofici e, ovviamente, in quello dei suoi esperimenti sonori, non dovrebbe essere sottovalutata. La chiave per comprendere il suo lavoro sulla musica risiede infatti nella comprensione di ciò che mosse, o piuttosto animò, la sua pittura.

La mia ricerca ha avuto inizio dall'analisi del processo creativo di Russolo alla luce delle prime proposte estetiche del Futurismo, al quale inizialmente aderì in qualità di pittore. Riesaminando le numerose fonti secondarie sull'arte futurista, la mia attenzione fu catturata da alcune ricerche che rivelavano l'influsso di teorie occulte all'interno del movimento.

Su tutte, un rivelatore articolo di Germano Celant intitolato *Futurismo Esoterico*[3] che, insieme ad alcuni brani tratti dal volume *Il futurismo: la fusione della vita nell'arte* di Maurizio Calvesi, del 1967,[4] è alla base di tutte le mie scoperte successive.

Grazie a tali letture potei comprendere che sotto una superficie apparentemente modernista, i futuristi s'interessavano alle arti occulte. Seguendo il filo dei miei pensieri, mi resi conto innanzitutto di quanto l'opera di Russolo fosse influenzata dall'arte e dalla tecnologia rinascimentali, nell'armonioso connubio rappresentato dalla figura proto-futurista di Leonardo da Vinci.

Mi fu quindi chiaro che l'intera opera di Russolo aveva avuto origine da un reale interesse per l'occulto, un interesse che egli mostrò durante tutta la sua vita, dalle primissime tele (*Autoritratto con teschi* del 1908 o *Autoritratto con doppio eterico* del 1910-11) ai dipinti, più tardivi, del suo periodo "classico-moderno" degli anni '40 del Novecento.

Una prima revisione degli scritti e dell'opera di Russolo ha portato alla luce il legame tra musica e pittura che egli aveva stabilito attraverso la sinestesia, un concetto che Russolo e i futuristi investivano di un significato occulto, in quanto campo di possibili correlazioni arcane tra le diverse sfere sensoriali.

La sua nozione di sinestesia derivava da una branca specifica delle arti occulte in voga verso la fine dell'Ottocento, la Teosofia. Russolo e i suoi amici futuristi avevano verosimilmente fatto propria tale dottrina, sedotti dal fascino del pensiero francese di fine secolo, in particolare del Simbolismo, movimento artistico che aveva ispirato il Futurismo sin dalle sue prime mosse.[5] Il primo Futurismo attinse al Simbolismo e di conseguenza all'occulto, attraverso la Teosofia, e la parola chiave fu sinestesia.

Se si considerano i primissimi membri del Futurismo, a cominciare dal fondatore del movimento Filippo Tommaso Marinetti, si trattò di artisti e intellettuali che, in modo o nell'altro, prima di abbracciare il Futurismo, avevano tutti guardato con interesse al Simbolismo. Le poesie che Marinetti scrisse in francese s'ispirano senza dubbio alla poesia simbolista, così come le prime opere di Umberto Boccioni o Russolo presentano tematiche comuni alla pittura di quel movimento artistico.

Ma in che modo l'interesse per l'occulto influì sull'arte dei rumori di Russolo? Ritengo che l'arte dei rumori possa essere letta come un processo di evocazione degli spiriti articolato in due fasi: nella prima il rumore diventa spirituale, nella seconda sono gli spiriti a materializzarsi. Credo che Russolo abbia prima rappresentato il processo, nel 1911-12, nel grande dipinto intitolato *La musica*, per poi metterlo in pratica un anno più tardi, elaborando la teoria dell'arte dei rumori e costruendo l'intonarumori.[6]

Un semplice sguardo a *La musica* dovrebbe essere sufficiente per capire che si è di fronte a un'opera altamente simbolica, impregnata di dottrina teosofica, in particolare della teoria eminentemente sinestetica delle *forme sonore* introdotta per la prima volta nel 1904 da Charles Webster Leadbeater[7] come un ampliamento della più nota teoria delle *forme di pensiero*, che egli elaborò insieme ad Annie Besant.

A un livello ermeneutico più profondo, il dipinto mostra un performer/medium quasi in stato di trance, in grado pertanto di evocare gli spiriti e produrre materializzazioni di *forme sonore* e *di pensiero*. Quanto alle maschere, che nell'opera si muovono senza sosta mostrando una gamma completa di espressioni, ritengo siano materializzazioni figurative di *forme di pensiero*, al pari di un concetto chiave dei primi dipinti futuristi, ovvero gli "stati d'animo". Il sinuoso nastro blu che occupa una parte significativa del quadro è, a mio avviso, una sua personale rappresentazione della *forma sonora* appropriata relativa a una fonte sonora che Russolo avrebbe definito *enarmonica*, per esempio una fonte che si muove incessantemente nello spazio-tono in glissando, una fonte incorrotta, di cui non si percepisce alcuna suddivisione in intervalli.

L'arte del rumore, che Russolo formulò un anno dopo, e l'orchestra intonarumori che costruì per realizzarla concretamente possono essere anch'essi letti in questo stesso ambito di riferimento. La propensione di Russolo per l'occulto ha fatto sì che considerassi l'intonarumori non solo uno strumento musicale di intelligente fattura, bensì uno strumento filosofico, parte quindi di un'illustre tradizione il cui lignaggio risale a Pitagora e ai suoi esperimenti sul monocorde.

L'intonarumori, anch'esso essenzialmente monocorde, genera dapprima rumore, una sorta di materia grezza che, nell'interfaccia rappresentata da una leva, diventa spirituale nell'attimo in cui è suonata per la prima volta, quindi creata, al di là di qualunque suddivisione tonale, nel fluido spazio "enarmonico". Producendo dei glissando, l'intonarumori prova che l'altezza di un suono è una quantità infinitamente divisibile e che, pertanto, lo spazio-tono è continuo. Russolo definì questa caratteristica dell'intonarumori "continuità dinamica".[8]

Sebbene la parola "intonarumori" possa essere usata tanto al singolare quanto al plurale, Russolo non pensava a un singolo strumento musicale, a un modulo sonoro individuale; egli intendeva, al contrario, realizzare un'orchestra che, nel suo complesso, potesse sintetizzare la totalità dei suoni disponibili nel mondo naturale e in quello meccanico.

Dopo averne introdotto la tassonomia nel manifesto *L'arte dei rumori*, nel 1913 Russolo si dedicò alla costruzione degli strumenti in grado di produrli. Nelle intenzioni dell'artista, l'orchestra intornarumori doveva ricreare il mondo attraverso i suoi diversi suoni: dapprima come caos simultaneo, quindi – grazie all'energia e alla capacità del direttore/medium – come unità sostanziale. Quando il direttore/medium avesse raggiunto il culmine dell'ispirazione suprema, gli spiriti dei morti sarebbero stati evocati, risvegliati e si sarebbero infine materializzati.

Il principio della materializzazione/risveglio, onnipresente nella dottrina teosofica, è alla base dell'opera di Marinetti e di altri futuristi, tra cui Ardengo Soffici, Boccioni e, naturalmente, Russolo. E per quanto la mia precedente lettura riguardo i primi intenti dell'arte dei rumori possa sembrare inverosimile, essa è avvalorata dalla fonte futurista coeva Paolo Buzzi, il quale spesso spiegava l'arte dei rumori di Russolo proprio come un'operazione dell'occulto, il cui fine era l'evocazione degli spiriti: innanzitutto in un breve testo del 1919 intitolato "Russolo ferito"; quindi nel poema *Russolo*, scritto probabilmente nel 1919 ma pubblicato nel 1922 in una raccolta intitolata *Poema dei Quarant'anni*; infine, nel romanzo *La cavalcata delle vertigini*[9] del 1924.

Ampliando la lettura dell'arte dei rumori in quanto procedimento di evocazione, come espresso nei testi del 1919, *La cavalcata delle vertigini* fa vestire a Russolo i panni del protagonista principale, sotto il nome di Marzio, e dedica un intero capitolo a Russolo/Marzio intento a dirigere un'orchestra/evocare gli spiriti.

Seppure indirettamente e non letteralmente, la comprensione approfondita dell'arte dei rumori in quanto ambiziosa operazione metafisica è stato il principio fondamentale che ha guidato la mia ricostruzione dell'orchestra intonarumori, una ricostruzione resa possibile dall'incarico conferitomi da RoseLee Goldberg di Performa, la Biennale d'Arte di New York.[10]

È stato infatti per Performa 2009 che ho completato la prima ricostruzione moderna mai tentata della prima orchestra intonarumori di Russolo, oggi andata perduta.

Per la mia ricostruzione presi a modello l'orchestra costituita da 16 strumenti intonarumori che fu presentata l'11 agosto 1913 a un concerto per la stampa presso la Casa Rosa di Marinetti, il quartier generale milanese del movimento futurista, e di cui si persero le tracce nel corso della Seconda guerra mondiale, durante la ritirata tedesca.

Quando iniziai a lavorarvi, vi erano già stati alcuni progetti a riguardo, primo fra tutti quello di Pietro Verardo e Mario Abate, commissionato dalla Biennale di Venezia nel 1977. Verardo era stato il primo a ricostruire alcuni degli intonarumori di Russolo, attenendosi alle informazioni tratte dai brevetti allora disponibili. Naturalmente, pensavo di dover produrre qualcosa di diverso da ciò che esisteva già: ero alla ricerca di un'identità, nonché di un concetto unificatore per il mio progetto di ricostruzione.

Non ebbi subito l'intenzione di ricostruire interamente la prima orchestra e quando iniziai a pensarci, mi fu chiaro che ciò che desideravo davvero era costruire una serie di strumenti che mi permettessero di suonare il brano di Russolo tratto da *Risveglio di una città* – un brano del 1913 apparso per la prima volta sul periodico futurista *Lacerba* nel 1914 e che non veniva eseguito dal vivo da quasi un centinaio d'anni – riproducendo il più possibile i suoni originari.[11]

All'epoca in cui decisi di dedicarmi alla mia opera di ricostruzione, nessuna orchestra intonarumori avrebbe potuto eseguire tale brano dal vivo, e poiché sapevo esattamente di volere riprodurre l'intonarumori proprio a tal fine, la partitura di *Risveglio di una città* divenne la mia guida.

Quando Performa mi contattò, nel 2008, mi occupavo di Russolo da circa sette anni. Fino a quel momento, il mio lavoro era stato soprattutto teorico, in quanto mi interessavo alla sua estetica per la mia tesi di dottorato, interamente dedicata all'arte dei rumori.[12]

Tra le informazioni che avevo raccolto a sostegno delle mie argomentazioni c'erano i brevetti e i disegni a essi relativi riguardanti l'intonarumori. Poiché Russolo non aveva brevettato i singoli strumenti, ma solo il meccanismo base, quel materiale mi fu utile solo fino a un certo punto, pur fornendomi comunque moltissime informazioni. Ancor più importante, grazie alla ricerca a cui ho appena fatto cenno, sentivo che sarei riuscito a capire ciò che Russolo aveva in mente quando decise di costruire l'intonarumori.[13]

Nel dicembre 2008 contattai il fabbricante di strumenti musicali Keith Cary, amico di amici che vivevano in California, e gli chiesi di aiutarmi a realizzare il primo prototipo del progetto. Si trattava di un ululatore medio, che fra gli strumenti intonarumori di Russolo è il più intuitivo da costruire, in quanto è quello più direttamente correlato ai disegni del brevetto.

Dopo avere realizzato il prototipo, dovetti affrontare una questione particolarmente spinosa: come ricostruire gli altri 15 strumenti, molti dei quali sapevo essere notevolmente diversi dal meccanismo base che Russolo aveva brevettato. Iniziai quindi a raccogliere qualunque elemento mi fosse possibile. Oltre ai brevetti, c'erano gli scritti di Russolo: articoli, libri, grafici che riportavano le estensioni degli strumenti, scritti di artisti futuristi e simpatizzanti del movimento, riviste ecc.

Un aiuto prezioso mi giunse dalle fotografie che riproducevano l'esterno degli strumenti. Probabilmente le più utili furono due famose immagini scattate nel luglio 1913 presso lo studio di Russolo a Milano, in cui l'artista e il suo assistente Ugo Piatti sono immortalati fra 16 strumenti.

Queste due fotografie furono estremamente importanti per la mia ricostruzione. Raffigurano infatti l'intera orchestra di 16 intonarumori che avevo intenzione di ricostruire, quella presentata al concerto per la stampa nell'agosto 1913. Essendo ancora in fase di realizzazione e sparsi per la sala, se ne possono osservare dettagli utili, non più visibili in fotografie successive: la posizione delle manovelle rispetto alle leve, quella dei coni, le attrezzature utilizzate ancora su un tavolo, e molto altro.

Altrettanto utili si rivelarono le immagini delle rappresentazioni dal vivo, alcune scattate ai concerti londinesi del giugno 1914, altre a quelli parigini del 1921, le quali mostrano tutte l'involucro esterno dell'intonarumori. Solo una, presumibilmente del 1914, rivela l'interno di uno strumento, nella fattispecie il ronzatore, ed è superfluo dire che si rivelò decisiva.

Lo strumento fotografato era dotato di un motore elettrico. Sappiamo che la prima orchestra intonarumori di Russolo era formata da 16 strumenti, 10 dei quali azionati a mano mediante manovelle senza l'ausilio dell'elettricità, mentre 6 erano elettrici. Volendo fare delle distinzioni, quindi, la prima può essere questa: strumenti elettrici o azionati a mano.

Avere un'idea dell'aspetto di entrambe le tipologie era molto importante per ricostruirle: dal brevetto potevo capire più o meno come fosse uno strumento a manovella; dalla fotografia del ronzatore, invece, avere un'idea di come fosse internamente uno strumento elettrico, una conoscenza che potevo applicare ad altri strumenti analoghi.

Sfortunatamente non vi sono immagini che riproducano internamente altri strumenti. Una simile circostanza avrebbe facilitato enormemente la ricostruzione dell'intera orchestra e io ne avrei potuto completare la replica molto prima di quanto realmente feci.

Tra le informazioni raccolte durante tale processo ve ne erano alcune correlate a ricostruzioni precedenti, tutte parziali, la prima delle quali – realizzata, come ho già avuto occasione di menzionare, per la Biennale di Venezia del 1977 – si rivelò senza dubbio la più utile.

Pietro Verardo aveva lavorato sin dall'inizio basandosi su documenti originali e aveva dedicato decenni a perfezionare gli strumenti, selezionando con cura i materiali e così via. Nell'estate del 2009 ebbi la possibilità di visitarne lo studio, con il compositore e personaggio radiofonico Robert Worby e una troupe della BBC. C'erano ben 9 strumenti e potei suonarne qualcuno.

Le altre ricostruzioni di cui ero a conoscenza (una in Giappone, un'altra in Olanda ecc.) avevano ambizioni molto più limitate e comprendevano pochi strumenti a manovella molto simili a quello raffigurato nel brevetto. Tuttavia, tutte le informazioni mi furono d'aiuto.

Se è vero che le mie conoscenze riguardo l'estetica di Russolo hanno alimentato la mia opera di ricostruzione, è ugualmente vero che la responsabilità progettuale di tale opera e la possibilità di potermi concentrare intensamente sui dettagli costruttivi più insignificanti hanno finito a loro volta per influire sul giudizio concernente la sua estetica.[14]

L'eclettica personalità di Luigi Russolo, che oscillava senza sosta tra la sfera teorica e quella pratica, ne sarebbe lieta.

Vorrei ringraziare Todd Lerew per avermi aiutato nella revisione di questo saggio

1
Luciano Chessa, *Luigi Russolo and the Occult*, tesi di dottorato, University of California, Davis (CA) 2004. La mia tesi è stata in seguito pubblicata con il titolo *Luigi Russolo, Futurist. Noise, Visual Arts, and the Occult*, University of California Press, Berkeley (CA) 2012.

2
Giovanni Lista, "Russolo, peinture et bruitisme", in Luigi Russolo, *L'art des bruits*, l'Age d'Homme, Losanna 1975; Gianfranco Maffina, *Luigi Russolo e l'Arte dei rumori: Con tutti gli scritti musicali*, Martano, Torino 1978.

3
Germano Celant, "Futurismo esoterico", *il verri 15*, nn. 33-34, ottobre 1970.

4
Maurizio Calvesi, *Il futurismo: la fusione della vita nell'arte*, Fratelli Fabbri Editori, Milano 1967.

5
Si può infatti affermare che l'atto con cui nasce il Futurismo, il manifesto di Filippo Tommaso Marinetti pubblicato da *Le Figaro*, non prendeva a modello tanto il *Manifesto del Partito Comunista*, quanto piuttosto il *Manifesto del Simbolismo* di Jean Moréas, anch'esso pubblicato da *Le Figaro* nel 1886.

6
Per una dettagliata spiegazione dell'intero procedimento, si veda Luciano Chessa, *Luigi Russolo, Futurist. Noise, Visual Arts, and the Occult*, cit., capp. 7, 8, 11.

7
Charles Webster Leadbeater, "L'aspect occulte de la musique", in *Le Lotus Bleu: la revue théosophique*, vol. 15, n. 10, dicembre 1904.

8
Luigi Russolo, *L'arte dei rumori*, Edizioni Futuriste di Poesia, Milano 1913.

9
Si veda Luciano Chessa, *Luigi Russolo, Futurist. Noise, Visual Arts, and the Occult*, cit., pp. 209-18.

10
Per la Biennale del 2009, la curatrice RoseLee Goldberg scelse i manifesti futuristi come tema unificatore degli oltre 100 eventi che ne costituivano l'imponente programma. Fu grazie al suggerimento di un amico, Tom Welsh, che Esa Nickle, Direttrice Generale di Performa, mi contattò nel 2008 chiedendomi di ricostruire l'intonarumori di Russolo.

11
Luigi Russolo, "Grafia enarmonica per gl'intonarumori futuristi", *Lacerba*, 1 marzo 1914.

12
Si veda nota 1.

13
Il 2008 fu per me l'anno perfetto per passare a un approccio più partecipativo e spostare i miei interessi dall'estetica alla costruzione degli strumenti. Nel 2004 avevo depositato la mia tesi presso l'UC Davis e per un po' il mio lavoro aveva preso altre strade (per esempio, avevo scritto sul compositore Giacinto Scelsi). Nel 2008 – circa quattro anni dopo avere archiviato la mia tesi – si presentò un buon momento per tornare a occuparmi di Russolo.

14
Dopo questa esperienza, ripresi a dedicarmi alla bozza di *Luigi Russolo, Futurist. Noise, Visual Arts, and the Occult* con rinnovato entusiasmo e nuove conoscenze. Un'esperienza che mi ha guidato fino alla conclusione del libro nel 2012.

COSÌ BRUTALMENTE FORZATA. LA SVOLTA DEL DOPOGUERRA DALLA MUSICA AL SUONO

Rob Young

Si potrebbe affermare che la trasformazione della musica in "suono" sia cominciata su una pista da sci. Nei primi mesi del 1948, il produttore Pierre Schaeffer, che aveva lavorato per Radio France nel corso della Seconda guerra mondiale, provava un senso di disillusione per lo stato della musica e rifletteva con una certa irrequietezza sulla stagnazione delle forme musicali tradizionali. Mentre si trovava sulle Alpi Francesi, aggrappato a uno skilift, ebbe la percezione di come le sue orecchie si sintonizzassero con i cigolii e gli scricchiolii metallici del macchinario, resi ancora più stranianti dall'acustica ovattata inusuale creata dalla neve farinosa. Con ogni probabilità fu il primo essere umano a riconoscere che quei rumori concreti potessero, in quanto tali, suscitare un piacere estetico e anche essere percepiti, se non come una composizione, quanto meno come una forma di esperienza musicale.

Quell'escursione gli aprì le orecchie a qualcosa di nuovo, tanto che, nel giro di poche settimane, Schaeffer fece sigillare una stanza del suo posto di lavoro e ne ricavò una sala insonorizzata rudimentale, nella quale intraprese una serie di esperimenti con giradischi, vinili con il solco chiuso, nastri magnetici ed effetti di eco/riverbero. Negli anni seguenti i risultati di questo lavoro furono proposti come opere musicali originali, che vennero trasmesse alla radio francese o rappresentate "acusmaticamente" (in forma anonima per mezzo di sistemi audio); tra queste figurano gli *Études de bruits* (1948), *Les paroles dégelées* (1952) e la *Symphonie pour un homme seul* (1952, in collaborazione con Pierre Henry).

Il nuovo modo di lavorare di Schaeffer si configurava in maniera necessariamente diversa rispetto a quella del compositore musicale tradizionale. Invece di mettersi seduto alla scrivania per comporre combinazioni armoniche e melodiche su partitura, la sua ricerca di nuove sonorità gli richiedeva di impegnarsi fisicamente con apparecchiature come giradischi e magnetofoni, o con dispositivi da lui stesso progettati. Avventuratosi nella ricerca dell'*objet sonore* – l'oggetto sonoro – ispezionò tutto il quartier generale di Radio France, assemblando una collezione di cianfrusaglie, strumenti a percussione, oggetti di metallo e un vecchio pianoforte; in seguito si spinse fino a rovistare tra canne di organi abbandonati in una chiesa bombardata, o a dirigersi alla Gare des Batignolles con un camion che conteneva un impianto di registrazione portatile e dei microfoni, con i quali riprodusse i rumori delle locomotive ferme alle banchine che avrebbero dato origine a uno dei primi capolavori di musica elettronica: l'*Étude aux chemins de fer* (1948).

Per comprendere a fondo il cambiamento paradigmatico che ha caratterizzato la musica elettronica durante la prima decade successiva alla Prima guerra mondiale, è necessario ripercorrere lo stesso viaggio intrapreso dai compositori nel corso di quegli anni e abbandonare il concetto di musica come struttura armonica, melodica e ritmica. Nell'introduzione a una delle prime storie della musica elettronica, l'autore Paul Griffiths dichiara: "Gran parte della nostra musica fa uso unicamente di una ristretta selezione all'interno di una smisurata serie di suoni percepibili dal nostro orecchio, ... la gamma di strumenti a disposizione del compositore, benché relativamente variegata nella moderna orchestra sinfonica, ne restringe ancora l'immaginazione entro modalità piuttosto specifiche".[1]

Lo sviluppo della musica elettronica in tutte le sue forme fu a tutti gli effetti un tacito assalto ai limiti di un mezzo musicale che, da secoli, era rimasto più o meno costituito dagli stessi elementi. Le caratteristiche principali della musica elettronica, in ogni sua forma, sono la combinazione di frequenze, microtoni, rumore, interferenze, suoni concreti e trovati, così come la manipolazione e distorsione del timbro strumentale. Pierre Schaeffer era però insoddisfatto anche della strada che la musica contemporanea aveva intrapreso fino a quel momento, dominata dall'atonalità e dalla serialità di cui all'inizio del secolo era stata antesignana la Seconda Scuola di Vienna di Arnold Schönberg, Alban Berg e Anton von Webern. "Dopo la guerra", scrive egli stesso, "nel periodo tra il 1945 e il 1948, avevamo respinto l'invasione tedesca, ma non avevamo contrastato l'invasione della musica dodecafonica austriaca. Ci eravamo liberati politicamente, ma la musica continuava a subire l'occupazione di una potenza straniera: la musica della Scuola di Vienna".[2] Schaeffer e il movimento della musica concreta da lui ispirato, quindi, operavano come una corrente di resistenza volta a sabotare l'asse teutonico che aveva conquistato la musica moderna nella prima metà del XX secolo.

Schaeffer incominciò a rimissare su vinile e su nastro magnetico brevi frammenti audio catturati *ad anello* e ripetuti all'infinito, facendoli girare in *backwards*, variando la loro velocità e inventando sistemi di playback tridimensionali. Fino ad allora la musica non aveva avuto qualità plastiche: i compositori concepivano la musica nella loro mente, poi annotavano su carta dei simboli che corrispondevano alle frequenze stabili delle note musicali, da utilizzare nell'esecuzione. D'altro canto, la musica concreta non poteva definirsi preesistente, poiché era il prodotto dei suoi componenti registrati. "Fintanto che il significato sarà preponderante, e costituirà l'aspetto centrale, avremo a che fare con la letteratura e non con la musica", scrive Schaeffer. "Ma come è possibile dimenticarsi del significato e isolare il fenomeno suono come unità percettiva?"[3]

Il 15 maggio 1948 Schaeffer annuncia: "Ho coniato il termine *Musique Concrète* per indicare questa dedizione a comporre con materiali sonori 'preesistenti' sperimentali, al fine di mettere in luce la nostra dipendenza, non più da suoni astratti precostituiti, ma da frammenti di suono che esistono nella realtà e che sono considerati oggetti sonori distinti e compiuti..."[4] L'*objet sonore* (oggetto sonoro) – un frammento su nastro magnetico o su *shellac* (gommalacca) – gli consentiva di analizzare e trasformare i suoni, esaltandone le qualità come mai prima era stato possibile; Schaeffer parlava di vivisezionare alcuni atomi del suono per fondere elementi audio non uditi.

A questo concetto della violenza della musica elettronica allude un brano del 1964 per voce e sintetizzatore del compositore americano Milton Babbitt. *Philomel* si ispira al mito greco di Filomela, una principessa ateniese rapita e "così brutalmente forzata"[5] dal re Tereo, suo cognato, che poi le recide la lingua affinché non possa rivelare la notizia dello stupro. In seguito, insieme alla sorella Procne (moglie di Tereo), Filomela si vendica dell'uomo dandogli in pasto il proprio figlioletto, e si trasforma infine in un usignolo, l'uccello celebre per la bellezza del suo canto. Nella *Philomel* di Babbitt è presente l'eco di una voce solista femminile riprodotta su nastro e distorta dal sintetizzatore; in questo modo viene rappresentata in forma mitologica la "forza bruta" della musica elettronica rispetto a quella che la precede. Facendo eco alle paure contemporanee, secondo le quali la musica basata su nastro e su strumentazione elettronica avrebbe finito per sterminare l'esperienza musicale derivante dalla consueta partecipazione ai concerti, la voce – ossia lo strumento musicale più originale ed essenziale – viene prima "divorata" dal sintetizzatore, quindi subisce una trasformazione e infine spicca il volo assumendo una forma completamente nuova. Si tratta di un'opera rilevante che, mescolando procedure seriali con la voce solista dal vivo, il riverbero su nastro e il processo di sintesi, mette a confronto e supera molte delle problematiche della musica contemporanea dell'epoca.

Una delle tecniche più impressionanti elaborate da Schaeffer negli anni '40 e nei primi anni '50 prevedeva una sorta di taglio della linguetta. Facendo uso di un solco in vinile o di un nastro magnetico per isolare un frammento di suono – come il rintocco di una campana o di una nota di pianoforte – Schaeffer riusciva a insinuarsi nella profondità microscopica della struttura atomica dell'oggetto sonoro e a tagliarne i primi millisecondi, noti anche come "transitorio d'attacco". Questa tecnica trasformava un rumore percussivo in una tessitura dolcemente ossessionante che, anziché manifestarsi in modo esplosivo, dava l'impressione di affiorare dal silenzio. Come accadeva a Filomela nell'antica leggenda dopo aver subito la mutilazione della lingua, i suoni trascendevano le forze reazionarie per trasformarsi e volare svincolati dalle loro origini fisiche.

In questa nuova realtà, il "compositore" doveva essere qualcosa di più di un soggetto seduto alla scrivania a tracciare note su un manoscritto; a lui era inevitabilmente richiesto un impegno molto diverso, spesso fisico, nei confronti della concretezza del suono. Il compositore americano Harry Partch, che finì per progettare e costruire il proprio strumentario in grado di dare vita a un universo di suoni tutto suo, una volta parlò di se stesso come di "un musicista-filosofo affascinato dalla carpenteria".[6] Il compositore non poteva più essere colui che affidava la produzione musicale a terzi (solisti, ensemble, orchestra, direttore d'orchestra), ma anzi, come capitò a La Monte Young, David Tudor, Morton Subotnick, Pierre Henry, Alvin Lucier, Pauline Oliveros e a molti altri, non di rado si ritrovava costretto a esibirsi in prima persona. Lo stesso Schaeffer fu impegnato in varie creazioni, dal cosiddetto *pupitre d'espace* (banco dello spazio) – un sistema di "aloni" elettromagnetici, controllato dai gesti dell'esecutore, che diffondeva i suoni in uno spazio performativo –, al *fonogeno* (generatore di suono), che offriva la possibilità di variare il tono e la velocità del nastro, e al morfofono, una primitiva *effects box* le cui testine multiple producevano un rudimentale effetto di *delay* e di riverbero. Un critico ne descrisse i risultati come qualcosa di simile alla "scoperta di un continente di suoni vergine come l'isola di Robinson Crusoe".

Benché nel corso degli anni '50 e '60 Schaeffer si fosse allontanato dalla *performance* per dedicarsi a un approccio più teorico, i compositori e i musicisti della sua "scuola", riuniti intorno al GRM (Groupe de Recherches Musicales, fondato nel 1958), accolsero le sue metodologie e le convertirono per i propri scopi. Uno di questi era Luc Ferrari (1929-2005), il quale sosteneva che la registrazione di un'ora di tramonto lungo il litorale francese fosse musica (in *Presque rien no 1 – Le lever du jour au bord de la mer*, 1967-70), precorrendo la tendenza *ambient* e la pletora di *field recordists* attivi ai giorni nostri. Un'altra esponente, Eliane Radigue, nel 2011 è stata celebrata a Londra con una grande retrospettiva. Radigue, oggi più che ottantenne, aveva utilizzato i sintetizzatori già alla fine degli anni '60, sfruttandone la malleabilità e le frequenze estreme come elementi significanti dell'"eterno presente" che andava cercando attraverso il buddhismo. Tuttora attiva come compositrice, Radigue si dedica allo studio di analoghe proprietà trascendentali in strumenti acustici come il violoncello e l'arpa. L'universo di suoni è probabilmente assai diverso dai primi "studi sui rumori" di Schaeffer, ma la compenetrazione di forma e materia e la ricerca di una realtà sonora che va oltre i confini della "musica" convenzionale sono sempre vivi.

Le potenzialità della musica elettronica sembravano invocare nuovi ambiti per la ricezione. Adattata con scarso successo allo spazio formale dell'auditorium (ancora oggi il pubblico e i critici guardano con scetticismo ai concerti che diffondono i suoni in playback o utilizzano il computer), l'habitat naturale della musica elettronica era rappresentato dalla radio, dal *long-playing* o dall'installazione. Un'altra pietra miliare di quel periodo fu il Padiglione Philips, costruito in occasione dell'Expo di Bruxelles del 1958. La titolarità dell'opera, commissionata dall'azienda elettronica olandese all'avanguardia nello sviluppo di sistemi hi-fi e di apparecchi per la registrazione, fu attribuita al grande architetto francese Le Corbusier, ma l'edificio venne di fatto progettato sulla base di uno schizzo di Corbusier da uno dei suoi assistenti dell'epoca, Iannis Xenakis. Di origine greca, il trentenne architetto offrì il proprio contributo anche in qualità di compositore: tra i suoi interessi vi era infatti l'applicazione dei principi matematici e architettonici al processo di composizione musicale.

Secondo la concezione di Le Corbusier, il padiglione, il cui aspetto era simile a un oscuro "solido platonico" sottoposto a elevati livelli di torsione, doveva funzionare come un sistema digestivo, con il pubblico che entrava attraverso la bocca, procedeva subendo bombardamenti di suoni, colori, luci e immagini all'interno dello stomaco, e veniva poi espulso dal foro di uscita dopo un'esperienza multisensoriale che durava sette minuti. *Concret PH* di Xenakis fu all'epoca, e rimane tutt'oggi, uno degli esempi più astratti di composizione musicale, un continuum di proto-rumori realizzato a partire dalla registrazione del suono prodotto dalla combustione del carbone, tagliato in frammenti della durata di un secondo e ricuciti insieme per formare una tessitura granulosa che accompagnava i visitatori all'ingresso e all'uscita dell'edificio. Una volta entrati nella struttura, si veniva bersagliati da un'opera di Edgard Varèse dal titolo *Poème électronique*, un collage fatto di rumori percussivi, nastri tagliati e manipolati e trasformazioni soniche. Varèse era stato uno dei principali compositori d'avanguardia ben prima della Seconda guerra mondiale, una figura profetica le cui idee erano sempre troppo avanti rispetto alla tecnologia disponibile. Già nel 1916 aveva assunto uno sguardo decisamente futurista nei confronti delle potenzialità del suono, come manifestato in questa dichiarazione a un giornalista: "Mi rifiuto di sottomettermi a suoni che sono già stati uditi. Sono alla ricerca di nuovi strumenti tecnologici che si prestino a tutte le espressioni del pensiero e che siano in grado di mantenere il passo con quel pensiero".[7] Per la fine degli anni '50, quando ebbe compiuto i settant'anni di età, le apparecchiature stavano iniziando a tenere il passo con il suo radicalismo. Diffuso per mezzo di una rete di altoparlanti incorporati nei muri, *Poème électronique* sfruttava tutta l'ampiezza dello spazio architettonico, grazie a un sistema di commutatori telefonici che controllavano le onde sonore in tre dimensioni.

Sempre in quegli anni, a Colonia, altri artisti erano alla ricerca di vari tipi di apparecchiature musicali che potessero procedere di pari passo con la velocità del pensiero umano. La Westdeutscher Rundfunk (WDR), diretta da Herbert Eimert, aveva intrapreso quello che era stato definito "il primo passo nel controllo musicale della natura",[8] facendo ricorso a metodi sensibilmente diversi rispetto a quelli di Pierre Schaeffer e dei suoi seguaci. Anziché trasformare suoni autentici, registrati alla maniera della musica concreta, i tedeschi generavano toni elettronici grazie all'ausilio di oscillatori, modulatori ad anello e rudimentali sintetizzatori. Un concerto per radio del 1951 venne descritto come un "universo sonoro della musica";[9] in tempi recenti l'espressione è diventata un cliché giornalistico, ma all'epoca il suo impiego era indice di un desiderio di creare uno spazio ermetico per nuove forme musicali evolutive. Come Varèse, Eimert era affascinato dalle possibilità di reinventare interamente il linguaggio della musica attraverso la tecnologia. Invocava una "rivoluzione sotterranea dell'ascolto"[10] e sosteneva che la musica classica, se congelata nel tempo sotto forma di registrazione al grammofono, sarebbe potuta "essere il surrogato mentre la composizione elettronica la musica vera".[11]

Tra gli allievi-assistenti di Eimert vi era Karlheinz Stockhausen, che sarebbe diventato la figura più rappresentativa della corrente musicale contemporanea tedesca postbellica. Se composizioni su larga scala come *Gruppen* (1955-57), per tre orchestre, rivelavano un iniziale interesse per il superamento dei confini della musica classica canonica, egli si spinse presto verso forme di arrangiamento non tradizionali. È il caso di *Gesang der Jünglinge* (1955-56) per coro maschile modulato ad anello, del gong amplificato di *Mikrophonie*

I (1964), e di pezzi elettroacustici concettuali imponenti come *Telemusik* (1966) e *Hymnen* (1966-67). Di portata potenzialmente epica, le opere di Stockhausen acquisirono dimensioni più vaste negli anni '60 e '70. Il 1970 lo vide impegnato nell'allestimento di un erede del Padiglione Philips, il Padiglione tedesco a Osaka Expo 70, un auditorium sferico attrezzato con sette altoparlanti disposti ad anelli concentrici nel quale fu trasmessa una selezione delle sue opere. *Hymnen* apriva uno spazio acustico immaginario in cui tutti i Paesi del mondo erano rappresentati da versioni distorte dei loro inni nazionali, segnalandosi come la prima composizione a proporsi attraverso una diffusione su scala globale. *Telemusik* giocava invece con idee telematiche di chiusura e distanza, utilizzando come materie prime campioni di musica etnica. La musica oltrepassava i confini per estendersi a un pubblico planetario, in un gesto cosmico che Stockhausen avrebbe in seguito reso ancora più universale dichiarando che la sua musica aveva origine nel lontano sistema stellare di Sirio.

Altrove, la musica si stava espandendo come una nana bianca. Già nel 1948, John Cage aveva proposto per la prima volta uno spazio in cui tutti i suoni avevano ragion d'essere: il famigerato e per nulla silenzioso brano *4'33"*. Le sue composizioni per nastro, come *Williams Mix* (1953) e *Fontana Mix* (1958), non avevano a che fare tanto con i suoni realmente prodotti, quanto piuttosto con la possibilità di offrire agli interpreti una scelta democratica su come e quali elementi avrebbero prediletto per presentare la partitura. L'utilizzo da parte di Cage di operazioni casuali, suoni indeterminati e partiture su fogli trasparenti intercambiabili sottraeva il controllo del lavoro musicale al compositore e affidava gran parte del potere decisionale (o forse tutto) alle facoltà dell'interprete, o degli interpreti, o addirittura al caso [214, 216].

Nel 1969, la risposta di Cage all'era spaziale fu *HPSCHD*, un happening multimediale che combinava suoni generati al computer (in collaborazione con Lejaren Hiller) con materiale visivo del programma spaziale della NASA. Come avveniva nel padiglione di Xenakis/Varèse, *HPSCHD*, che fu presentato in anteprima quando mancava un mese all'allunaggio dell'Apollo, immergeva lo spettatore in suoni e immagini, con il timbro insistente del clavicembalo barocco che veniva accompagnato da nastri di suoni generati al computer, mentre immagini del pianeta terra provenienti dallo spazio davano vita a uno scenario ovattato e sospeso. L'installazione, che durava cinque ore, fu visitata in quel periodo da circa 60.000 persone.

Il ricorso a tempi così dilatati stava diventando una pratica sempre più comune nelle forme elastiche dell'arte della fine degli anni '60 e la musica non faceva eccezione. I film di Andy Warhol e Michael Snow documentavano il dispiegarsi di eventi in tempo reale in tutta la loro banalità, senza filtri, creando condizioni in cui il passare del tempo diventava impercettibile. Effetti analoghi erano presenti nei primi brani musicali di La Monte Young, già seguace del movimento Fluxus. Il suo complesso, Theater Of Eternal Music, si ispirava ai lunghi suoni della musica classica indiana e asiatica sia come elemento strutturale, sia facendo ricorso a bordoni e frequenze provenienti da strumenti ad arco, canti, *tambura* e oggetti riverberanti. Tra gli ospiti del gruppo figuravano John Cale, che continuò a influenzare il mondo del rock d'avanguardia con i Velvet Underground (una band nata sotto gli auspici della Warhol's Factory), e il violinista Tony Conrad, che proseguì il suo percorso esplorando i sistemi di ottimizzazione del suono e la *just intonation*. A sua volta, Young, insieme alla moglie Marian Zazeela, diede vita a un altro degli spazi fisici più influenti della *sound art* contemporanea, la Dream House di Manhattan, a New York. Qui i due realizzarono una installazione permanente all'interno di una serie di stanze in costante mutamento, che contenevano suoni di bordoni riverberanti ed effetti luce a colori trasformati con la tecnica del *morphing*, il cui scopo era di produrre un effetto ipnotico e meditativo. Young e Zazeela continuarono a vivere secondo il proprio ritmo circadiano – sfasato rispetto alle 24 ore dell'orologio – suffragato da una musica correlata a forze universali avulse dalla realtà umana.

Tra gli artisti del suono che hanno fatto dello spazio il proprio oggetto di ricerca figura Alvin Lucier, il cui brano *I Am Sitting in a Room* (1970) sfrutta i riverberi accumulatisi intorno a una voce fatta oggetto di molteplici registrazioni su nastro, poi riproposte all'interno di una stanza, producendo sonorità provenienti da uno spazio unico. *Public Supply I* (1966) di Max Neuhaus utilizzava la rete telefonica come nodo di accesso per creare un missaggio in tempo reale dei contributi provenienti dall'etere. Attraverso un sistema di microfoni, *Sound-Joined Rooms* (1980) di Maryanne Amacher metteva in comunicazione spazi chiusi scollegati e posti a una certa distanza tra loro. Infine *Electrical Walks*, opera intrapresa da Christina Kubisch sul finire degli anni '70, dà allo spettatore la possibilità di vagare per la città indossando cuffie senza fili che amplificano gli impulsi elettromagnetici, rivelando una dimensione elettroacustica nascosta delle abitazioni umane. Un processo che aveva avuto inizio con le peregrinazioni urbane di Pierre Schaeffer si è evoluto fino a invitare l'ascoltatore a ricoprire un analogo ruolo di flâneur.

Tutti questi esempi servono a dimostrare che, se è vero che la lingua dell'armonia musicale è stata con ogni probabilità oggetto di mutilazioni dopo la guerra, essa ha continuato, malgrado ciò, a librarsi in volo ad ali spiegate.

1
Paul Griffiths, *A Guide to Electronic Music*, Thames & Hudson, London 1979, p. 713.

2
Pierre Schaeffer, *In Search of a Concrete Music*, University of California Press, Oakland (CA) 2012.

3
Ibid.

4
Ibid.

5
Dal poema *La terra desolata* (1922) di T.S. Eliot: "...la metamorfosi di Filomela, dal re barbaro / così brutalmente forzata; eppure l'usignolo / riempiva tutto il deserto con inviolabile voce / e ancora lei gemeva...", in T.S. Eliot, *La terra desolata. Quattro quartetti*, trad. e cura di Angelo Tonelli, Universale economica Feltrinelli, Milano 2004, p. 39.

6
Harry Partch, *Genesis of a Music*, Da Capo Press, Cambridge (MA) 1947.

7
Malcolm MacDonald, *Varèse: Astronomer in Sound*, Kahn & Averill, London 2002.

8
Paul Griffiths, *op. cit.*

9
Ibid.

10
Ibid.

11
Ibid.

IL SILENZIO COME SCULTURA MUSICALE. JOHN CAGE E GLI STRUMENTI DI 4'33"

Eric de Visscher

Ai confini dell'arte e del suono, là dove musica e silenzio, spazio e tempo, strumenti e concetti non cessano di incrociarsi, il percorso del compositore John Cage appare fondamentale. Tra le sue opere, a detta dell'autore stesso, la *silent piece* (composizione silenziosa) *4'33"* occupa un posto centrale. Fu pensata ed elaborata entro un periodo di quattro anni (1948-52), cosa che la colloca esattamente a metà del XX secolo, anni sui quali si irradia a tutto tondo, affondando le sue radici nelle avanguardie del suo inizio ed estendendo la sua influenza sugli oltre cinquant'anni che la seguiranno. Ma qual è il significato attuale di questa opera silenziosa, e quale rapporto intrattiene con lo strumento o la tecnologia?

Nato nel 1912, Cage sviluppa il suo percorso creativo a partire dagli inizi degli anni '30, che si possono considerare come il suo periodo di formazione. In quegli anni compie un viaggio in Europa passando per Parigi e la Germania, dove visita la scuola del Bauhaus. Al suo ritorno a Los Angeles, si inserisce presto negli ambienti dell'avanguardia, popolati principalmente da musicisti e artisti visivi tedeschi emigrati negli Stati Uniti. In questo contesto artistico e intellettuale incontrerà personalità diverse tra loro quali Galka Scheyer, Pauline Schindler, Oskar Fischinger e chiaramente Arnold Schönberg, mentore temuto e riverito al tempo stesso. Su consiglio di Henry Cowell – compositore, curatore del *New Music Quarterly* e docente alla New School of Social Research (presso la quale Cage avrebbe insegnato vent'anni più tardi) – Cage si trasferisce per un anno a New York, dove studia con il pianista Adolphe Weiss, allo scopo di prepararsi a studiare con Schönberg, il grande maestro. Durante il periodo newyorkese, Cage incontra diverse figure eminenti della musica moderna quali Edgard Varèse e Virgil Thomson. Cowell, insieme a Schönberg, è l'elemento strutturante di un pensiero in fase di gestazione: dal primo Cage deriverà l'immensa apertura nei confronti della musica più sperimentale dell'epoca (le opere per percussione di Varèse, Carlos Chavez, Amadeo Roldán e molti altri, le prime sperimentazioni con i suoni elettronici...); dal secondo la consapevolezza del bisogno di rigore e strutturazione per questi universi necessariamente caotici.[1]

Dalle prime opere "seriali" ai pezzi per percussione, da cui deriverà in seguito l'invenzione del *Prepared piano* (piano preparato), il pensiero musicale di Cage impiegherà una buona decina di anni a formarsi, dalla *Sonata per clarinetto* (1933) ai diversi *Constructions (in Metal)* o *Imaginary Landscape*, la cui presentazione in concerto al Museum of Modern Art di New York nel febbraio del 1943 costituisce il momento culminante. In parallelo Cage conduce una intensa attività di scrittura sulla propria musica, i cui esiti più importanti sono i testi "The Future of Music: Credo" (1937) e "For More New Sounds" (1942). Questi costituiscono al tempo stesso la sintesi del fermento creativo che scuote gli Stati Uniti nel protrarsi delle avanguardie europee e un programma di lavoro per il compositore, che gli permette di realizzare la definizione varèsiana della musica: il suono organizzato. La musica deve dunque aprirsi a tutti i suoni, includendo tutti i rumori ma anche il silenzio, considerato come parte integrante del *continuum* sonoro; per organizzare tutto questo è necessaria una struttura, la quale non potrà che basarsi sulle durate e non sulle altezze, poiché solo il tempo è comune al suono e al silenzio.

Allo scopo di ottenere "un maggior numero di suoni", Cage, così come altri compositori, si interessa alla percussione facendola chiaramente sconfinare dal campo ben definito dello "strumento di percussione": se Varèse aveva già inserito nella strumentazione di *Ionisation* (1929-31) sirene dei pompieri e incudini, Cage si spinge oltre, aggiungendo parti di automobili e altri *objets trouvés*. Per lui, tuttavia, le percussioni sono solo una tappa di questo processo di apertura: "Credo che l'uso del rumore per fare musica proseguirà e crescerà finché non avremo raggiunto una musica prodotta attraverso l'ausilio di strumenti elettrici, che renderanno disponibili per scopi musicali qualunque minimo suono si possa udire. Per la produzione sintetica della musica si esploreranno dispositivi fotoelettrici, pellicole e mezzi meccanici."[2]

In "For More New Sounds", Cage esplicita la sua ricerca di nuovi mezzi tecnici e strumenti di composizione. Riferendosi allo sviluppo in corso di nuove macchine, tanto in contesti radiofonici quanto in quello dell'industria del cinema, al tempo nascente, Cage rivendica la possibilità che i compositori se ne impadroniscano: "Molti musicisti, compreso chi scrive, hanno sognato compatte scatole tecnologiche, dall'interno delle quali tutti i suoni percettibili, rumore compreso, fossero pronti a fuoriuscire su comando del compositore. Quelle scatole sono ancora in qualche punto del futuro. Per il momento la scelta è tra aspettare e lamentare che non siano disponibili subito per scopi sperimentali e musicali, o continuare a lavorare, cercando di costruirle con qualunque 'ascia e secchio' siano disponibili o realizzabili."[3] La frustrazione che traspare in queste parole fa seguito al fallimento delle iniziative messe in atto da Cage presso diverse compagnie di ricerca e istituzioni per tentare di allestire uno studio di musica sperimentale. Anche la collaborazione con il Columbia Workshop di Chicago, che diede luogo nel 1942 alla realizzazione della composizione radiofonica *The City Wears a Slouch Hat*, si rivelò deludente, tanto i mezzi messi a sua disposizione non erano all'altezza delle sue ambizioni. Per quanto riguarda "ascia e secchio", questi sono precisamente gli strumenti base che il musicista utilizza nelle sue prime opere elettroniche quali *Imaginary Landscape No. 1* (1939), per la quale ricorre a 3 fonografi a velocità variabile provenienti dallo studio radiofonico della Cornish School di Seattle, dove al tempo insegnava, su cui suona dischi destinati a test acustici. Le variazioni di tono erano ottenute modificando la velocità del piatto, mentre il ritmo era prodotto sollevando la puntina. Nelle due opere successive della stessa serie (*Imaginary Landscape No. 2* e *Imaginary Landscape No. 3*, entrambe del 1942), Cage continua a perfezionare la sua tecnica "elettroacustica"; nella *No. 2*, amplifica una bobina di filo collegandola al braccio del fonografo, mentre per la *No. 3* aggiunge ai dispositivi esistenti un oscillatore, la registrazione di un generatore di frequenza, un buzzer, una marimbula amplificata. Nelle tre opere, questi suoni nuovi si trovano sempre mescolati a strumenti a percussione, così come accade in una composizione dello stesso anno, *Credo in US*, che includeva trasmissioni radiofoniche dal vivo.

La necessità di una articolazione ritmica che consenta di strutturare questi suoni e "non suoni" diviene evidente dagli anni '40 e ossessionerà Cage fino al 1952: poiché sarà precisamente l'evoluzione del suo pensiero sul silenzio durante questo periodo a modificare il suo rapporto con la tecnologia e la sua concezione della forma, passando in particolare per l'accettazione dell'intervento del caso nella composizione. Questa evoluzione è nota: viene introdotta per la prima volta in un testo del 1948 ma scoperto molto più tardi, "A Composer's Confessions", nel quale Cage afferma di voler scrivere un'opera fatta di "silenzio interrotto" per venderla alla Muzak Co.:[4] "Durerà da 3 a 4 minuti e mezzo – secondo le lunghezze standard della musica preregistrata – e si intitolerà *Silent Prayer*."[5] È sostanzialmente un'opera di reazione, che combatte contro l'invasione del suono nell'ambiente, mutuando gli strumenti propri della musica commerciale, e nella fattispecie i successi commerciali di breve durata. Cage si ferma tuttavia a una concezione chiusa del silenzio, assimilata all'assenza di suono. Questa visione prevale anche quando parla della struttura che può accogliere tanto il suono – tutti i suoni – quanto i "non suoni", i silenzi. La sua posizione si evolve tuttavia sul volgere degli anni '50, sotto l'influenza simultanea del buddhismo zen ("tutto è un'eco del nulla"), dei *White Paintings* di Robert Rauschenberg e dei suoni dolci della musica di Morton Feldman.

La dimensione spaziale, e perciò plastica, del silenzio si afferma quando Cage scrive che "Il silenzio circonda molti dei suoni, così che essi esistono nello spazio senza ostacolarsi a vicenda e tuttavia penetrandosi a vicenda…".[6] La famosa esperienza della camera anecoica dell'università di Harvard aggiunge un altro mattone all'edificio, permettendo a Cage di asserire che il silenzio assoluto non esiste, poiché il nostro corpo produce continuamente suoni. Il silenzio diventa allora l'insieme di tutti i suoni non intenzionali e si esprime sotto forma di un atteggiamento di ascolto e di apertura, senz'altro eccessivamente positivo, a tutti i suoni provenienti dallo spazio circostante. Questo conduce anche alla creazione dell'opera silenziosa *4'33"*, presentata nel 1952 nel corso di un recital per pianoforte di David Tudor alla Maverick Concert Hall di Woodstock.

Questa evoluzione da un lato rivela uno spostamento che parte da una visione polemica presente in *Silence Prayer* (l'assenza di suoni come alternativa all'inquinamento sonoro) per arrivare a una concezione apparentemente più filosofica e individualista (l'ascolto e l'attenzione alla non-intenzionalità); dall'altro traduce un costante, anzi crescente, ricorso alla tecnologia per comunicare questo concetto. La camera anecoica e il piano acustico, entrambi muti, nel servire all'interpretazione di *4'33"*, materializzano concretamente quello che può sembrare un dispositivo puramente concettuale. Anche la tecnologia elettrica, propria degli *Imaginary Landscape*, viene utilizzata a partire dal 1951 per avvicinarsi al silenzio: nel quarto di questa serie di *Landscapes*, scritto per un gruppo di 12 radio azionate da 24 esecutori, senza alcun altro strumento, la partitura indica una struttura di durate e di frequenze per ciascuna radio. Leggenda vuole che al momento in cui l'opera fu eseguita, il 10 maggio 1951 a un'ora assai tarda, molte delle stazioni radio locali non stessero più trasmettendo e che, per questo motivo, la composizione somigliasse a una lunga sequenza di silenzi e di disturbi elettrici. Le radio funsero così da rivelatori di suoni inudibili o inauditi, una funzione che Cage avrebbe assegnato a tutta la tecnologia utilizzata in seguito.

Come ha ben dimostrato Douglas Kahn,[7] una delle conseguenze logiche di *4'33"* è che la tecnologia non servirà più esclusivamente a produrre nuovi suoni: è qui per dare suono al silenzio o, più precisamente, all'inudibile, quello che non possiamo sentire con le orecchie e che tuttavia è sempre presente. Il microfono e l'altoparlante costituiscono gli strumenti indispensabili di questo dispositivo, diffondendo nello spazio, in tempo reale, questi suoni (o silenzi) amplificati. Dal 1952, le versioni successive della *silent piece* comprenderanno sempre questa dimensione di amplificazione, fortemente influenzate dalle idee liberatorie di Marshall McLuhan sul decentramento, la de-gerarchizzazione e l'anarchismo portati dalla tecnologia. Questo processo di ingrandimento e di frantumazione della materia sonora appare, certo, in altre composizioni dello stesso periodo, e nella fattispecie nella serie delle *Variations* [214], nella *Cartridge Music*, e anche in quelle che ricorrevano a supporti fissi come il nastro magnetico (*Williams Mix*, *Rozart Mix*) o il disco (*33 1/3*, per un minimo di 12 giradischi e un gran numero di dischi azionati dagli spettatori).

La serie delle *silent pieces* permette tuttavia una comprensione particolarmente interessante del rapporto che il pensiero di Cage intrattiene con la tecnologia e le sue implicazioni politico-sociali. Perché è dieci anni più tardi, nel 1962, che Cage ritorna su questo momento saliente proponendo *0'00" (4'33" No. 2)*, una composizione radicalmente diversa, un "assolo che chiunque può suonare in qualunque modo"; la partitura consta unicamente di una frase principale e di qualche precisazione: "In una situazione di massima amplificazione, si metta in scena un'azione disciplinata". Si passa dunque da una prescrizione comune, quella dello spartito che indica al musicista ciò che deve suonare – o non suonare – a una performatività molto più esplicita e direttiva. E Cage proseguirà sull'onda di questo suo slancio con uno *0'00" No. 2* che comparve in *Song Books* (1970): "Su un'area da gioco (per esempio un tavolo o una scacchiera) attrezzata con microfoni a contatto (preferibilmente a quattro canali, con altoparlanti attorno al pubblico, al volume massimo senza ritorno, si giochi con una (per esempio a scacchi, a domino) o più persone (per esempio a Scarabeo o a bridge)". Anche in questo caso c'è stata la massima amplificazione, anche se questa volta in forma di azione sociale (che implica una pluralità di partecipanti); un'azione che è musicale perché amplificata e necessariamente simile a un gioco. Perché se la musica non è un gioco, vero è che si dice "giocare" uno strumento musicale[8] e che certe regole del gioco possono comunque essere utilizzate per comporre musica – come ha dimostrato Marcel Duchamp. D'altra parte, una delle interpretazioni più famose di *0'00" n. 2* fu realizzata a Toronto nel 1968 – cioè prima della pubblicazione della partitura – con il titolo *Reunion*; durante l'intera serata Marcel e Teeny Duchamp, così come Cage, giocarono a scacchi in pubblico su una scacchiera, realizzata da Lowell Cross, che trasformava il movimento dei pezzi in suoni diffusi dagli altoparlanti nell'auditorium.

Poco più di vent'anni dopo, nel 1989, Cage interpreta un'ultima opera silenziosa, che intitola *One*^{3} *= 4'33" (0'00") +* 𝄞. I riferimenti nel titolo sono tanto numerosi quanto misteriosi. *One*^{3} fa parte delle *Number Pieces*, opere tarde di Cage in cui la cifra in lettere indica il numero di musicisti richiesti per eseguire l'opera, mentre l'esponente indica il numero progressivo di opere della stessa serie. Il titolo cita poi le due opere silenziose principali, seguite da una chiave di Sol, riferimento a un'osservazione fatta a Cage dalla compositrice russa Sofija Gubajdulina, che non amava cronometrare il tempo nelle opere musicali. Il titolo indica perciò tre diversi tipi di tempo: un tempo misurato cronometricamente, il tempo zero che in qualche modo nega il concetto di durata e di flusso e infine un tempo musicale che prende forma dal movimento dei suoni… L'unica esecuzione dell'opera ebbe luogo a Tokyo; Cage si sedette in sala, mentre sulla scena alcuni microfoni portavano la risonanza dell'ambiente a un livello prossimo al ritorno senza tuttavia raggiungerlo. Dopo un tempo non determinato dallo spartito (pare che fossero trascorsi più di 12 minuti), Cage si alzò, andò sul palco e fermò l'amplificazione…[9]

Cosa pensare di questa evoluzione dell'opera silenziosa nell'arco di quasi quarant'anni? Molti elementi sono comuni alle diverse versioni: l'accettazione dei suoni non desiderati, una forma di teatralità che supera i confini di un'interpretazione "musicale" per collocarsi in altre situazioni sociali (come il gioco), infine il ricorso a un elemento tecnico (lo strumento musicale o il dispositivo elettroacustico) concepito come una sorta di protesi del corpo dell'interprete. Questa tecnologia, considerata in senso lato, secondo Cage è necessaria per distinguere l'interpretazione dell'opera silenziosa in una data situazione (durante un concerto, per esempio) dalla semplice possibilità dell'ascolto del silenzio in qualunque momento, in una situazione di attenzione auditiva e di apertura a tutto quello che può accadere. Per Cage l'opera silenziosa esiste sempre: "[…] in natura, e in ogni momento. 'Percepibile' significa che c'è amplificazione. È un'opera su un'opera, come tutte le mie opere indeterminate! […] Ecco come lavoro, con l'aiuto della tecnologia."[10] La tecnologia rivela i suoni che non possiamo sentire in sua assenza, ma contribuisce anche alla messa in scena dell'opera, dando vita a un evento nuovo rispetto all'esperienza del quotidiano. In qualche modo, la tecnologia crea l'opera, ma non prende il sopravvento. È proprio questa la conclusione a cui arriva Cage alla fine della sua vita, quando si rende conto che esiste un vero potenziale pericolo causato da un cattivo impiego della tecnica: "Adesso si tratta di un silenzio tecnologico […] adesso il silenzio comprende la tecnologia. In un certo senso non è necessariamente… una buona cosa", afferma nel corso di un'intervista a Joan Retallack.[11]

Passare da *4'33"* a *One*^{3} significa dunque per Cage che il silenzio degli anni '90 non è lo stesso di quello del 1952, poiché il nostro ambiente si è necessariamente modificato, in particolare sotto l'influenza della crescente presenza degli strumenti tecnici. Se non può più, come in *Silent Prayer*, imputare alla Muzak l'onnipresenza del suono nell'ambiente, è proprio la tecnologia, un tempo considerata liberatoria, che sembra porre un problema, in quanto invadente od opprimente.

Un'aggiunta manoscritta sullo spartito delle *Sculptures Musicales*, l'elaborazione allestita da Cage nel 1989 a partire da un appunto scritto da Marcel Duchamp per la sua *La Boîte verte* del 1934: *Sculpture musicale. Sons durant et partant de différent points et formant une sculpture sonore qui dure* (Scultura musicale. Suoni duraturi che partono da diversi punti a formare una scultura sonora che dura) sembra confermare questa diagnosi. La partitura dell'opera realizzata per un balletto di Merce Cunningham, presentata unicamente sotto forma di testo, prescrive la realizzazione di una serie di blocchi di almeno tre suoni, provenienti da fonti spaziali differenti, separati da silenzi aperti ai suoni esterni, tra cui quelli dei ballerini... La parola "silenzio", scritta tra virgolette, indica lo sguardo ambiguo che Cage le rivolge (è un "non suono"? Oppure è un suono riempito di rumori che l'autore non può controllare?), mentre una delle bozze stampate della partitura riporta alla fine del testo la nota, scritta di pugno del compositore: *The dark side of silence* (Il lato oscuro del silenzio).

Questo lato nascosto del silenzio rivela un certo disincanto da parte di Cage, che rimanda a una visione più pessimistica del silenzio, simile a quella del poeta Edmond Jabès, che l'associa all'"ascolto di una nave che affonda". Probabilmente era giunto anche lui all'idea che la camera anecoica, che tante possibilità aveva schiuso rivelando la permanenza del suono, potesse anche significare una privazione di senso, l'obbligo al silenzio e quindi il contrario della libertà.

Cage aveva presagito questa forma di rassegnazione già nel 1980, quando, attraverso un mesostico compilato a partire dalla parola SILENCE scritta in verticale, ammette l'ineluttabilità del concetto, la necessità di adattare il nostro spirito a una forma di anarchia e di lasciare che il "caos" sia.[12]

Se la parola "anarchia" percorre tutto il pensiero politico di Cage dal 1960, quello di "caos" è un concetto cui ricorre più di rado. Cage vi si riferisce parlando della teoria matematica omonima, che cerca di descrivere la complessità di certe situazioni, mescolando determinismo e incertezza, ciò che avvicina "caos" a "caso" (possibilità), il metodo di composizione impiegato da Cage dal 1951. Ma si potrebbe pensare anche al *Chaos* della mitologia greca, quello stato di confusione violenta che darà vita al mondo. In entrambi i casi ci troviamo di fronte a una percezione della complessità, non scevra di pericoli ma portatrice di futuro. Ed è proprio la stessa complessità della nostra relazione con il mondo a caratterizzare la visione del silenzio di Cage. Lungi dal negare la possibilità dell'anarchia positiva, la presa in considerazione di un aspetto nascosto del silenzio la arricchisce. Il potenziale pericolo di una onnipresenza della tecnologia non impedisce che essa possa giocare un ruolo estremamente liberatorio nel nostro processo di esperienza del mondo. Ed è questa ambivalenza, questa accettazione del complesso a rendere il pensiero di Cage sempre attuale. Oggigiorno ognuno di noi può ascoltare il silenzio, ma ormai può anche interpretare *4'33"* mediante un'applicazione per iPhone e comunicare la propria esperienza agli altri usando questa nuova tecnologia.[13] Chiunque può isolarsi dal mondo circostante infilandosi gli auricolari nelle orecchie, ma può anche costruire il suo proprio "paesaggio immaginario" mescolando musiche registrate e suoni esterni. Pensare silenzio oggi significa dunque costruire la nostra relazione con il mondo attraverso l'ascolto, e distinguerla da altri modi di cogliere la realtà. Se usiamo le orecchie come ricettacoli ora aperti ora chiusi, forse il silenzio può offrirci anche la possibilità di pensare il futuro, come scrive la scrittrice giapponese Yoko Ogawa: "Se si vuole conoscere l'avvenire, bisogna tapparsi le orecchie. Allora si potrà sentire la propria storia in lontananza, nel futuro"[14].

1
Su questi anni di formazione, si veda Thomas S. Hines, "'Then Not Yet Cage': The Los Angeles Years, 1912-1938", in Marjorie Perloff, Charles Junkerman (a cura di), *John Cage. Composed in America*, University of Chicago Press, Chicago (IL) 1994.

2
John Cage, "The Future of Music: Credo", in Id., *Silence*, Wesleyan University Press, Middletown (CO) 1961, pp. 3-6: 3-4; trad. nostra.

3
John Cage, "For More New Sounds", in Richard Kostelanetz (a cura di), *John Cage*, Praeger, New York 1970, pp. 64-66: 65; trad. nostra.

4
La parola Muzak, ispirata dal successo ottenuto dal nome Kodak, è stata inventata da George Owen Squier, un ingegnere che negli anni '20 aveva brevettato un sistema di diffusione della musica mediante cavi elettrici. Questo sistema a pagamento fu però presto superato dalle trasmissioni via radio, tant'è che Squier decise di proporre la sua musica per la diffusione in spazi pubblici, centri commerciali e aziende. Basati sul concetto di "Stimulus Progression", i programmi proposti dalla Muzak puntavano a condizionare l'ascoltatore (lavoratore o consumatore) offrendo trasmissioni diversificate a seconda del momento della giornata, a volte stimolanti, altre rilassanti. Fortemente criticata per questo suo aspetto manipolatore, l'idea alla base della Muzak ha comunque suscitato l'interesse di musicisti del calibro di John Cage e Brian Eno; l'espressione *Musique d'Ameublement* (musica da arredamento) di Erik Satie può essere considerata un'antesignana involontaria di questa idea.

5
A Composer's Confessions fu pubblicato nel 1991-92, contemporaneamente nelle riviste *Musicworks*, n. 52, Toronto, e *MusikTexte*, nn. 40-41, Köln, su iniziativa di chi scrive e di Gisela Gronemeyer, curatrice di *MusikTexte*, e comparve successivamente in Richard Kostelanetz (a cura di), *John Cage Writer*, Limelight Editions, New York 1993, p. 65.

6
John Cage, "Julliard Lecture", in John Cage, *A Year From Monday*, Wesleyan University Press, Middletown (CO) 1967, p. 100.

7
Douglas Kahn, *Noise Water Meat. A History of Sound in the Arts*, The MIT Press, Cambridge (MA) 1999, p. 192.

8
Per comprendere questa affermazione, è importante ricordare che, diversamente da quanto succede in italiano, "suonare" e "giocare" si dice nello stesso modo in molte lingue: in inglese (*to play*), in francese (*jouer*), in tedesco (*spielen*)... [N.d.T.]

9
Sulle diverse versioni della *silent piece*, comprendenti anche delle versioni per la televisione e il cinema (*WGBH-TV*, *One[11]* ecc.), così come le sue trasposizioni contemporanee, si veda Dieter Daniels, Inke Arns (a cura di), *Sounds Like Silence. John Cage.4'33". Silence Today*, catalogo della mostra, Spector Books, Leipzig 2012.

10
John Cage (conversazione con Daniel Charles), *For the Birds*, Marion Boyars, London 1981, p. 210.

11
John Cage (conversazione con Joan Retallack), *Musicage. Cage Muses on Words, Art and Music*, Wesleyan University Press, Middletown (CO) 1996, p. 77.

12
John Cage, *X – Writings '79-'82*, Wesleyan University Press, Middletown (CO) 1983, p. 117.

13
L'applicazione *4'33"* è disponibile per iPhone e iPad; si veda il sito johncage.org

14
Yoko Ogawa, *Chinmoku Hakubutsukan*, Chikuma Shobo, Tokyo 2000, citazione tradotta dall'edizione francese *Le Musée du Silence*, Actes Sud, 2003.

MUSICA SPERIMENTALE E PERFORMANCE. DAVID TUDOR, JOHN CAGE E MERCE CUNNINGHAM

Geeta Dayal

Oggi, reperire e ascoltare musica sembra essere più facile che mai. Pochi e semplici clic, e si può scaricare o ascoltare in streaming l'intera discografia di un artista. Difficile è, casomai, riuscire a orientarsi in un'offerta sconfinata e resistere a questa travolgente valanga digitale. L'ampia disponibilità di creazioni rare – tra cui sconosciute registrazioni dal vivo di musica elettronica, *out-takes* e *B-sides* di free jazz – rende ancora più difficile scegliere tra generi musicali che forse non sono mai stati ritenuti all'altezza per essere registrati sul lato A di un vinile. Uno dei problemi è che in passato gran parte della musica sperimentale e d'avanguardia era accompagnata innanzitutto da un forte elemento visivo, che oggi si tende a dimenticare. Qualche volta si faceva ricorso alla danza, al teatro e alle performance. In altri casi, dietro ai performer scorrevano le immagini di un film, oppure la musica stessa era una colonna sonora strettamente correlata alla pellicola. O ancora si usavano costumi stravaganti, sculture o dipinti. Una cosa è certa: in origine la musica non era intesa come una lunga lista di mp3. In questo senso, laddove vi sono dei filmati disponibili, YouTube si avvicina al concetto, anche se si tratta di un'esperienza ancora alquanto incompleta.

Nel suo affascinante libro *Records Ruin the Landscape*, il musicista e professore David Grubbs spiega come molte forme emergenti di musica d'avanguardia e sperimentale degli anni '60 non fossero per nulla adatte a essere conservate e ascoltate sotto forma di registrazioni. A ragion veduta, Grubbs concentra gran parte del libro su John Cage, ricordando il difficile rapporto di quest'ultimo con le registrazioni, per disquisire di libera improvvisazione, minimalismo, degli happening e delle performance degli artisti di Fluxus, di musica elettronica dal vivo, installazioni artistiche sonore e di molti altri generi che allora fiorivano sulla scena artistica. "Il complesso rapporto di Cage con il suono registrato divenne una pietra miliare dell'epoca ed è un atteggiamento che si riflette in tutti quei generi musicali", scrive Grubbs. Gli artisti di Fluxus creavano partiture che richiedevano di essere rappresentate nel corso di una performance, i minimalisti opere che potevano durare ore oppure, come nel caso di La Monte Young, settimane, mesi o persino anni. La musica elettronica live e la libera improvvisazione erano un proclama contro la trasferibilità e la ripetibilità della musica registrata. Gli artisti visivi che lavoravano con il suono creavano performance e installazioni che enfatizzavano le qualità scultoree della produzione sonora".[1]

Alcuni anni fa, mentre cercavo di venire a capo della voluminosa produzione di Cage ascoltandone i CD, mi resi conto di avere perduto la visione d'insieme. Pur rimanendo impressionata da gran parte di ciò che ascoltavo nelle registrazioni e leggevo nei libri, solo assistere a una serie completa di rappresentazioni dal vivo riuscì a fare nascere in me un profondo apprezzamento per il suo lavoro. In una piccola sala, con i musicisti giusti, l'opera di Cage divenne ai miei occhi tridimensionale e, in alcuni casi, le dimensioni erano sette o addirittura nove, tanto le immagini erano vivide. Sinceramente commossa da quelle performance, provavo per l'opera di Cage sentimenti intensi, viscerali ed emotivi, non solo teorici e intellettuali. Fu un'esperienza che mi colpì nel profondo. Ricordo perfettamente quando corsi fuori dalla sala coprendomi le orecchie durante un'interpretazione di *Cartridge Music*, così intensa e assordante che le cuffie non poterono fare altro se non lasciare che la crescente e soffocante marea del feedback acustico prendesse forma. Dopo che il mio udito ebbe ritrovato l'orientamento, tornai, sempre di corsa, in sala, impaziente di continuare l'ascolto. Oltre alla straordinaria esperienza sonora, l'opera teatrale si rivelò misurata ma coinvolgente, con artisti alle prese con molle Slinky, strumenti di vario genere e testine di amplificazione.

Quello stesso elemento visivo, ricco e affascinante, mancava, a mio avviso, anche in Harry Partch, le cui splendide opere attingono in gran parte agli anni '60. Quando si ascoltano le sue registrazioni, non si può fare a meno di pensare che vi è qualcosa di grande che ci sfugge. Per approfondire le proprie conoscenze sulla musica di Partch bisogna essere in grado di comprenderne gli strumenti musicali, bizzarri e unici al mondo, tra cui il Chromelodeon, costruito nel 1945. Inoltre Partch compose gran parte dei suoi lavori più importanti perché fossero rappresentati, e questo, quando ci si limita ad ascoltare le sue registrazioni audio, è un altro pezzo mancante del puzzle. *Delusion of the Fury*, una delle sue opere maggiori, andò in scena per la prima volta nel 1969, a Los Angeles: prevedeva mimi e ballerini, miti e folklore dell'Africa occidentale ed elementi del teatro giapponese Nō. Le note a corredo della registrazione del 1971 per CBS specificano che sulla scena ciascun musicista indossava "un enorme paio di pantaloni", "un capo simile a un poncho", "un trucco estroso" e "un copricapo fantastico".[2]

Le stesse note si riferivano agli strumenti come fossero sculture, opere d'arte nate per essere godute in quanto parte della *Gestalt* dell'intera produzione. Persino i nomi degli strumenti utilizzati in *Delusion* erano pervasi da un potente mistero: il Chromelodeon, con le sue colorate chiavi dipinte a rappresentare una scala di 43 tonalità, la Diamond Marimba, le Cloud Chamber Bowls. "Gli strumenti di Partch non servono unicamente a rappresentare qualcosa, ma sono essi stessi una rappresentazione", scrive Eugene Paul nelle note dell'album. "Oltre a contribuire a diffondere la sua musica, insieme a cantanti e ballerini, sono opere scultoree imponenti che entrano nella struttura scenica, non solo in *Delusion*, ma anche in tutte le composizioni di Partch che sono state messe in scena. Veicolo della sua creatività, intorno a queste creazioni l'artista ha strutturato un suo specifico sistema di notazione".

Una delle prime opere di musica elettronica d'avanguardia che ho ascoltato, *Poème électronique*, del grande compositore Edgard Varèse, era su CD. Su uno stereo decente, la resa era bellissima – tutti gli 8 minuti di registrazione – tuttavia non era quell'incredibile esperienza multimediale di cui avevo letto nei libri. *Poème électronique* fu rappresentata all'Esposizione Universale di Bruxelles del 1958, nel Padiglione Philips, dove quest'eccezionale esperienza sonora – diffusa da innumerevoli amplificatori – si mescolava a innovazioni nell'ambito dell'architettura, della cinematografia, delle luci e delle proiezioni. Nello scenario architettonico unico di quello straordinario padiglione, la cui realizzazione era stata affidata proprio al grande Le Corbusier (all'epoca in viaggio – presumibilmente nella città indiana di Chandigarh – tanto che fu il compositore greco Iannis Xenakis, suo collaboratore, a coordinare il progetto), contemporaneamente alla musica si proiettavano film di Le Corbusier. A intervalli regolari, risuonavano anche le note di *Concrete PH*, una composizione di Xenakis. Personaggio poliedrico, Xenakis era matematico, ingegnere, architetto, compositore e artista: personificazione della contaminazione tra arte visiva e musica elettronica. I suoi numerosi progetti architettonici, le partiture grafiche e i diagrammi erano opere d'arte di per sé.

Cartridge Music (1960) è una delle rare opere di Cage che la registrazione non penalizza, come nota Grubbs nel suo libro. Una rappresentazione dal vivo, se da un lato sottrae parte dell'intenso mistero della registrazione – si possono infatti collegare i diversi oggetti sul palco ai suoni percepiti in quel momento – dall'altro offre una ricca informazione visiva. Il partner di Cage, Merce Cunningham, ha fatto ancora di più utilizzando *Cartridge Music* come colonna sonora di un balletto da lui coreografato nel 1973, intitolato *Changing Steps*, di cui

Charles Atlas disegnò i variopinti costumi, oggi conservati al Walker Art Center di Minneapolis. Tra i costumi di Atlas vi erano "felpe di cotone con zip (alcune dotate di tasche) in un'ampia gamma di colori, tra cui panna, verde, blu e marrone".[3] Sempre presso il Walker Art Center sono conservati molti dei coloratissimi costumi che Robert Rauschenberg creò per la compagnia di Merce Cunningham dagli anni '60 in poi.

La collaborazione tra Rauschenberg, Cage e Cunningham era di lunga data e coinvolgeva altri personaggi della cerchia di Cage, tra i quali David Tudor, il cui lavoro realizzato con Cunningham e Rauschenberg nel corso degli anni '60, aveva dimostrato quanto potesse essere profondamente interdisciplinare il nascente genere della musica elettronica. Molte delle opere di Tudor – tra cui la famosa *Rainforest* (1968) – sono esempi dirompenti di come danza e scultura possono essere inserite senza soluzione di continuità in un genere all'avanguardia come quello della musica elettronica.

All'inizio della sua carriera Tudor si impose come pianista di talento, e il suo nome era spesso accostato al più famoso Cage, il quale più di una volta affermò di avere composto le sue opere pensando a Tudor.

Dopo avere studiato pianoforte dal 1944 al 1950 con Stefan e Irma Wolpe, all'inizio degli anni '50 Tudor iniziò una lunga e leggendaria collaborazione con Cage, che segnò la nascita di un'amicizia per la vita. E se è vero che Cage scrisse molte delle sue composizioni per Tudor, è altrettanto vero che a renderle possibili furono la straordinaria bravura di quest'ultimo, e la sua capacità unica di interpretare le partiture più complesse.

Negli anni '60 Tudor si reinventò, trasformandosi da irrinunciabile interprete di composizioni altrui a compositore di musica elettronica dal vivo, creando installazioni e paesaggi sonori, e lavorando a lungo con la Merce Cunningham Dance Company e artisti del calibro di Rauschenberg e Jasper Johns.

Nel 1961 fu presentata a Parigi una curiosa opera performativa, intitolata *Homage to David Tudor*, interessante da analizzare pensando a Tudor e ai suoi molteplici interessi e nella quale Rauschenberg fa ricorso al talento del compositore, di Jasper Johns, Jean Tinguely e Niki de Saint Phalle. Davanti al pubblico, Rauschenberg dipinge un'opera intitolata *First Time Painting*, di cui tuttavia si vede solo il retro della tela; il pubblico è quindi all'oscuro di ciò che l'artista sta realizzando. Lo storico dell'arte Jeffrey Saletnik scrive che "Tudor eseguì la composizione di Cage *Variations II* (1961), Johns realizzò un bersaglio fatto di fiori, de Saint Phalle uno dei suoi *shooting paintings*, mentre Tinguely diede il suo contributo con una scultura automatizzata".[4] Rauschenberg smise di dipingere al suono della sveglia e il dipinto fu rimosso frettolosamente dalla scena.

Nel 1964, Rauschenberg e Tudor misero in scena un evento presso il Moderna Museet di Stoccolma, che faceva parte della serie *5 New York Nights*. L'opera musicale che Tudor creò per l'occasione era spontanea e strettamente correlata al luogo; il curioso titolo, *Fluorescent Sound*, si riferiva letteralmente al sistema di luci fluorescenti che era stato creato per l'occasione. L'opera, considerata la prima composizione di Tudor, non è tuttavia inclusa nel recente cofanetto edito dalla New World Records, peraltro piuttosto completo, che racchiude le opere di musica elettronica composte dall'artista, poiché a quanto pare non fu mai registrata. Il contributo di Rauschenberg era intitolato *Elgin Tie*; a quanto riportano le fotografie e i racconti di chi vi assistette, egli fu lentamente calato dal soffitto in un barile d'acqua, dove trattenne il respiro finché poté, per poi essere nuovamente sollevato. Per l'evento venne portata in scena anche una mucca.

Competere con le idee di Rauschenberg e con le sue performance istrioniche non era per nulla facile. "Camminavo per il museo e intanto pensavo, 'che cosa farò?'", confessò in seguito Tudor in un'intervista del 1988 al critico e giornalista Teddy Hultberg. "A un certo punto notai che c'erano, dovevano esserci quasi ... un migliaio di lampadine fluorescenti. Un giorno, mentre ero in sala, qualcuno accese le lampadine ma non sapevano quali accendere, e improvvisamente ecco una musica meravigliosa. Pensai, ok, perché non mettere qualche microfono a contatto accanto alle lampadine, per vedere se il suono era davvero percepibile. ... Così presi una lunga scala e sistemai dei microfoni a contatto, operazione che richiese quasi tre giorni interi. Fortunatamente fui aiutato dal fatto che c'erano, credo, 75 interruttori a cui erano collegate 3 lampadine ciascuno. Fu un lavoro lungo, ma venne benissimo. L'unico problema fu quando Bob Rauschenberg fece il suo ingresso dall'entrata laterale del museo con una mucca e mi chiese di fermare la musica... (ride) Fu la mia prima composizione".[5]

A giudicare dall'esigua documentazione esistente, la serata suscitò grande stupore: un riuscito connubio tra arte performativa, luci, intrattenimento e una musica unica, strana e interessante. E una mucca.

Meglio documentata è *Bandoneon! (A Combine)*, una performance realizzata sotto la guida di Tudor nel 1966, che faceva parte della serie delle leggendarie *9 Evenings* all'Armory di New York. Il termine *combine* (combinazione) fa ovviamente riferimento ai *combine paintings* di Rauschenberg. L'utilizzo del termine da parte di Tudor è interessante ed è indicativo del suo legame con l'artista, quantomeno ne sottolinea la reciproca ammirazione e comprensione.

Rainforest, rappresentata per la prima volta nel 1968, è l'opera più famosa di Tudor. Ne esistono varie versioni, contraddistinte da numeri diversi, ma in tutte per creare dei suoni fu utilizzato un sistema di trasduttori collegati ad alcuni oggetti. Nonostante l'abbondanza di circuiti elettronici e gli oggetti utilizzati, *Rainforest* non è mai sembrata del tutto una foresta; al contrario, il suono, incredibilmente naturale, ricordava il cinguettio degli uccelli nel verde fogliame degli alberi.

Nella filosofia di Tudor, ogni altoparlante od oggetto in grado di produrre una risonanza può essere considerato uno strumento a sé, dotato di personalità propria. "Ho pensato a lungo che i diffusori fossero una cosa oscena, ma in ogni caso è una falsa nozione: non esistono", affermò lo stesso Tudor nel 1972. "È stato fatto un tentativo di creare un oggetto che riproduca il suono. Ma per me, il diffusore, è una membrana che vibra, in grado di creare dei suoni proprio attraverso questa vibrazione. Così, un giorno, immerso nei miei pensieri, immaginai che sarebbe stato bello potere avere a disposizione un'intera foresta, o un'orchestra di diffusori, ciascuno con un suono un po' diverso. Il che significa che avrebbero dovuto esprimere proprio il loro suono, più che riprodurre un suono partorito dalla mente di qualcuno ... un altoparlante è uno strumento".[6]

La musica elettronica di Tudor era già bella di per sé, ma sul palcoscenico creava un'atmosfera magica. A fare da scenografia c'era *Silver Clouds* (1966), un'opera di Andy Wharol minimalista e luminosa. La coreografia di Cunningham era meravigliosamente intensa ed essenziale e i costumi dei ballerini, color carne, strappati ad arte e di grande impatto, erano stati disegnati da Jasper Johns.

Johns ne parlò molti anni dopo ad Alastair Macaulay durante un'intervista al *New York Times*. "Avevo chiesto ad Andy [Warhol] di disegnare dei costumi che si intonassero ai cuscini, ma il suo unico suggerimento fu che i ballerini dovessero danzare nudi, un'idea che però non piaceva a Merce. Merce mi mostrò una sua vecchia calzamaglia lacera e strappata e io la presi a modello".[7]

Rainforest – grazie a Cunningham e ai suoi ballerini, a Johns, a Warhol, e alla colonna sonora all'avanguardia creata da Tudor – era divenuto qualcosa di più delle sue singole componenti. Guardare un filmato d'archivio della versione del 1968 è un'esperienza che ti trafigge, come un'opera d'arte viva di irresistibile bellezza.

Le successive versioni di *Rainforest* prodotte negli anni '70 erano più installazioni sonore interattive e scultoree, da cui era stata eliminata la danza. Un'esperienza coinvolgente, seppure in modo diverso.

"Nel 1973 realizzai *Rainforest IV* in cui gli oggetti che producevano suoni erano molto grandi e avevano una propria dimensione spaziale", spiegò Tudor a

Hultberg nel 1988. “Intendo dire che il loro suono si diffondeva a livello locale nel luogo in cui erano appesi ed erano inoltre completati da un sistema di diffusione sonora ... il che creava, credo, un'atmosfera piuttosto piacevole e armoniosa, perché ovunque ci si muoveva, nella stanza, si percepiva l'eco dei suoni ascoltati in altri punti dello spazio... è importante che ognuno crei la propria scultura, decida come programmarla e la rappresenti”.[8]

Era proprio al concetto di creazione della propria scultura che pensavo quando scelsi una registrazione di *Rainforest IV*, e lasciai che il vinile suonasse mentre mi dedicavo ad altro. Su uno stereo normale non potevo vivere pienamente l'esperienza sognata da Tudor di “un'orchestra di altoparlanti”, né vedere la poetica produzione di danza di Merce Cunningham del 1968. Non c'erano le nuvole argentee di Andy Warhol né i costumi lacerati ad arte di Jasper Johns, né installazioni sonore scultoree. Con un po' di sforzo, potevo vedere tutto questo con gli occhi della mia immaginazione, in un sogno sonoro.

1
David Grubbs, *Records Ruin the Landscape: John Cage, The Sixties and Sound Recording*, Duke University Press, Durham (NC) 2014.

2
Eugene Paul, note del libretto per il CD di Harry Partch, *Delusion of the Fury*, CBS recording, New York 1971.

3
Costumi per *Changing Steps*, collezioni del Walker Art Center, Minneapolis, MN. Si veda walkerart.org/collections/artworks/costumes-for-changing-steps

4
Jeffrey Saletnik, *Trophy IV (For John Cage)*, Rauschenberg Research Project, luglio 2013. San Francisco Museum of Modern Art, sfmoma.org/explore/collection/artwork/25848/essay/trophy_iv_for_john_cage

5
Teddy Hultberg, “I smile when the sound is singing through the space”, intervista a David Tudor, Dusseldorf, 17-18 maggio 1988; davidtudor.org/Articles/hultberg.html

6
David Tudor, “From Piano to Electronics”, *Music and Musicians*, n. 20, London agosto 1972.

7
Alastair Macaulay, “Cunningham and Johns: Rare Glimpses Into a Collaboration”, *The New York Times*”, New York 7 gennaio 2013.

8
Teddy Hultberg, intervista a David Tudor, cit.

RUMORE OTTICO. IL SUONO DELLA SCULTURA NEGLI ANNI '60

Jo Applin

Negli anni '60, la preferenza accordata dal modernismo alla ricezione visiva rispetto ad altre risposte sensoriali all'opera d'arte – presupposto dominante fin dal XIX secolo – ha subito numerosi attacchi. Negli stessi anni in cui autorevoli critici modernisti come Clement Greenberg e Michael Fried insistevano sull'aspetto puramente ottico della pittura modernista – il suo rivolgersi “solo all'occhio” – indicandolo come il suo tratto distintivo e determinante, altri tipi di esperienze ambientali ed esistenziali, altri tipi di incontri e stimoli esterni cominciavano nel lavoro di diversi pittori, scultori, musicisti, danzatori, registi e scrittori a filtrare nella produzione e nella ricezione dell'arte.[1] Una quantità di note dissonanti e interventi corporei eccessivi, rumorosi e talvolta violenti ha segnato il panorama artistico degli anni '60. Per esempio, Dick Higgins istruiva i performer a “Urlare! Urlare! Urlare!” per la sua *Danger Music 17* (1961), mentre Bruce Nauman, per *Concrete Tape Recorder Piece* (Registratore di cemento, 1968) [248], registrava su una cassetta una donna urlante che poi metteva a tacere chiudendola in un blocco di cemento. L'unica prova che l'opera di Nauman contenesse del suono era il filo elettrico, ormai superfluo, che ancora sbucava da un lato. Così, l'opera di Nauman portava in primo piano il ruolo del medium, l'irriducibile supporto materiale per la registrazione del suono, ma ne metteva in luce anche i limiti e i fallimenti. Altri artisti associati a movimenti internazionali come Fluxus, Zero e il Nouveau Réalisme sferravano attacchi al feticismo dei beni di consumo dilagato nel dopoguerra con la distruzione di oggetti quotidiani e familiari, accanendosi non di rado sugli strumenti musicali per scagliarsi contro i valori elitari incarnati dallo strumento orchestrale. Per alcuni artisti la musica era diventata un motivo contemporaneo attraverso il quale sviluppare una riflessione sul contesto storico in cui operavano, mentre altri, più che a distruggere, miravano a riformare la concezione dell'arte sonora fino a esiti dalla portata radicale: il famigerato arrangiamento “silenzioso” di John Cage, *4'33"* (1952), in cui gli unici suoni erano quelli ambientali prodotti nella sala da concerto, come colpi di tosse, pagine voltate, vaghi fruscii e respiri, assumeva un valore esemplare ed evidenziava l'interesse di molti artisti degli anni '60 per il lavoro con il suono.

C'è qualcosa di definitivo nell'atto di distruggere, aggredire, zittire o rompere uno strumento musicale. La netta cesura tra il suono di uno strumento che viene suonato e il suono che uno strumento produce quando viene aggredito e quindi ridotto al silenzio veniva esplorata da artisti inclini a sovvertire il predominio del visivo su altri canali espressivi della produzione artistica. Gesti iconoclasti come questi rappresentavano anche l'ostentazione pubblica di un rifiuto: rifiuto di ascoltare, di imparare, di accettare, ripetere o riprodurre i suoni di un altro musicista, compositore o strumento. Nel 1964, in *Piano* [236], Günther Uecker piantava centinaia di chiodi appuntiti nella superficie di un pianoforte verticale, un'azione riecheggiata formalmente da *Piano Piece* [270] di George Maciunas nel 1970, anche se in quest'ultimo caso Maciunas ha invertito il procedimento, collocando i chiodi in verticale, uno su ogni singolo tasto, come fossero surrogati di dita che evocano e al contempo ostacolano il tocco. In una vena analoga, le opere di Arman della serie *Colères* (dal 1961) [218] in cui l'artista distruggeva strumenti musicali per poi montarli e incorniciarli come assemblaggi scultorei, mettevano in atto un'aggressiva riduzione al silenzio che presenta marcati parallelismi anche con *Concrete Tape Recorder Piece* di Nauman. Nel frattempo *Piano Activities*, la partitura di Philip Corner eseguita al festival Fluxus di Wiesbaden nel 1962, coinvolgeva un gruppo di artisti Fluxus che collettivamente riducevano in frantumi un pianoforte. Ciascuna di queste opere si affidava alla partecipazione fisica o all'intervento di un altro corpo, sia che dovesse suonare, urlare, attaccare, o restare muto. E allora, più che ridotto al silenzio, il suono veniva ricreato, attingendo a una dissonante varietà di rumori ambientali che esistono oltre la partitura scritta per accogliere il mondo esterno.

L'ARTE DEL RUMORE

Operazioni come queste, che implicano un registro sonico sia sul versante della produzione, sia sul versante della ricezione, stimolano riflessioni sulle conseguenze derivanti dalla volontà di proporre un linguaggio che sostituisce l'ascolto alle più convenzionali risposte visive all'opera d'arte. Che cosa significa “ascoltare” invece di “guardare” un'opera d'arte, comprese quelle che non producono suoni percettibili? Pongo questa domanda per valutare come si potrebbe tracciare una mappatura del tessuto connettivo concettuale e formale che lega un'opera d'arte all'altra attraverso il tempo e la storia, non tanto scorgendo analogie e differenze visive ma prestando attenzione al registro acustico, per stabilire se sia legittimo affermare che certe opere “suonano” come altre. Come possono certe sculture essere recepite sia a livello visivo, sia a livello acustico, e cosa accade quando questi due ambiti sensoriali si confondono e s'intrecciano?

Affrontare questa domanda richiede di considerare non solo il ruolo del suono nella scultura a livello della rappresentazione o dell'intenzione (anche se sono aspetti importanti) ma anche le modalità collaborative ed effimere con le quali il suono è prodotto, proiettato, dislocato o variamente evocato dall'oggetto scultoreo.

Nel suo *The Life of the Mind: Thinking* Hannah Arendt notava come sovente "si è pensato al pensiero in termini di *visione*", un postulato che ha finito per scalzare l'antica metafora del narratore "di cui si ascoltano le storie" (Arendt ricorda il motivo del "poeta cieco"). "Ci si potrebbe chiedere", continuava Arendt, "perché l'ascolto non sia diventato la metafora guida del pensiero".[2] Alcune opere tridimensionali degli anni '60 tentano di assegnare all'ascolto, e a volte all'inabilità ad ascoltare, un ruolo centrale sia nell'attività intellettuale, sia nella pratica fattiva.

Due assemblaggi dell'artista americano Robert Rauschenberg, *Broadcast* (1959) e *Oracle* (1962-65) [230], incorporano apparecchi radiofonici. Sono opere d'arte che richiedono all'osservatore di essere ascoltate, oltre che guardate.[3] *Broadcast* è un *combine painting* montato a parete e accoglie sulla sua superficie un caotico assortimento di giornali e altri materiali stampati. Sotto la superficie si celano 3 radio, di cui sono visibili solo 2 manopole: una per regolare il volume, l'altra per cambiare stazione. Le radio sono impostate per far sì che, in qualunque modo l'osservatore regoli la frequenza, non saranno mai sintonizzate sullo stesso canale, producendo una cacofonia di suoni sovrapposti e interferenze spettrali che sembra emanare dalla superficie stessa dell'opera.

Oracle, di dimensioni più ambiziose, è un'installazione tridimensionale che richiama gli intonarumori [160] del futurista Luigi Russolo, autore del manifesto *L'arte dei rumori* (1913). La sua attinenza con le macchine di Russolo non risiede nelle analogie formali ma piuttosto nel tentativo di fondere i rumori e la cacofonia della vita quotidiana con l'opera d'arte. *Oracle* include 5 radio funzionanti, nascoste all'interno di una serie di *objets trouvés*, fra i quali una serie di condotti d'areazione, il telaio in legno di una finestra, il relitto metallico di una portiera sventrata sul tavolo di una macchina da scrivere e una varietà di oggetti assemblati come cestini di fili metallici e componenti elettronici. Le radio nascoste sono azionate da alcuni motori a velocità variabile che, come in *Broadcast*, producono rumori dissonanti costituiti da flussi e aperture che si accordano al cacofonico impatto visivo dell'assemblaggio. *Broadcast* e *Oracle* non solo evocano le proprietà materiali ed emotive del paesaggio mediatico contemporaneo, ma sventano l'idea che la loro interpretazione possa esaurirsi unicamente nella contemplazione visiva. In questo caso il pensiero – sia per Rauschenberg, sia per l'osservatore di queste opere – richiede l'ascolto.

RUMORE OTTICO

I suoni prodotti dai *Combines* di Rauschenberg sono deliberatamente confusi e i confini fisici dell'opera sono volutamente indeterminati. Il suono deborda. Questi oggetti sono, in ogni senso, ambientali: la risposta sensoriale che sollecitano in chi si aggira intorno all'opera è multipla, come quando si cammina in una strada trafficata di città, in quanto attingono e incorporano il mondo esterno (per Russolo, l'orecchio umano doveva rapidamente adattarsi all'assalto dei nuovi suoni prodotti dalla metropoli). Ed è proprio questo aspetto dell'opera di Rauschenberg che Leo Steinberg rilevava nel celebre saggio "Other Criteria" (1968-71). Steinberg sosteneva che l'opera di Rauschenberg produce "rumore ottico", offrendo un potente correttivo all'insistenza della critica modernista sul rivolgersi delle opere d'arte unicamente all'occhio dell'osservatore. Per Steinberg, è avvenuto uno spostamento nella "direzione psichica" dell'opera, che riorienta quest'ultima in modo che non sia più "l'omologo di un'esperienza visiva" ma un "piano" orizzontale che si presenta come propagazione laterale e matrice di informazioni.[4]

Per Steinberg questo passaggio era particolarmente visibile nell'uso, da parte di Rauschenberg, di una macchina da stampa piana per le sue trasposizioni, in cui immagini, testo, carta e giornali venivano stampati e sovrapposti l'uno sull'altro in una posizione orizzontale per produrre un palinsesto di informazioni o di "rumori" visivi. Steinberg notava che la sovrapposizione delle immagini era l'equivalente della modalità di trasmissione del suono, se non che, invece di produrre una comunicazione chiara, i frammenti di informazioni stranamente assortiti "interferivano" l'uno con l'altro, espandendosi a creare una sorta di interferenza visiva. Assomigliano a "scarti e detriti della comunicazione; come trasmissioni via radio con interferenze, rumori e significati sulla stessa lunghezza d'onda, e visivamente sulla stessa superficie piana."[5] I commenti di Steinberg si riferivano unicamente alle opere bidimensionali di Rauschenberg, ma le sue argomentazioni si possono anche estendere per descrivere nuove tipologie di oggetti tridimensionali che in modo analogo incorporano aspetti di "rumore ottico".

RUMORE NASCOSTO

Un altro tipo di "rumore ottico" era prodotto da due sculture realizzate in punti diversi della traiettoria modernista che ha attraversato il XX secolo, rispettivamente nel 1916 e nel 1961, le quali si prestano a una descrizione che mette in luce non tanto "l'ascolto" da parte dell'osservatore quanto il modo in cui i due oggetti si "ascoltano" a vicenda. La prima è *With Hidden Noise – À bruit secret* (Con rumore segreto, 1916) [164] di Marcel Duchamp e il secondo è l'omaggio a Duchamp di Robert Morris, *Box with the Sound of Its Own Making* (Scatola con i rumori della sua creazione, 1961) [220]. Come gli assemblaggi sonori di Rauschenberg, entrambe le opere rilasciano suoni c informazioni nel corso del tempo e si rinviano, l'una con l'altra, un'eco formale e acustica (che si propaga anche alla scatola di cemento di Nauman). In entrambe le opere vi è un elemento temporale che, come un metronomo che ticchetta, oscilla avanti e indietro tra "l'oggetto" finito e il suono di precedenti fasi di elaborazione (un tintinnio nel caso di Duchamp; i suoni dello studio nel caso di Morris).

Questo modello di scambio suggerisce l'istituzione di un rapporto di parentela che non si basa su una discendenza patriarcale verticale, in cui un oggetto porta inesorabilmente al successivo (poniamo da Higgins a Nauman, da Russolo a Rauschenberg, o da Duchamp a Morris). Piuttosto, questi rapporti si possono intendere come modelli alternativi di parentela, articolati mediante un registro acustico in cui gli artisti e le opere precedenti restano ad "ascoltare" gli eco e le risonanze degli altri. Come Michel Foucault sosteneva che la "genealogia" non dovrebbe essere considerata solo in termini di "continuità ininterrotta" e lineare da una cosa all'altra, ma come un "assemblaggio instabile di faglie, fessure e strati eterogenei che minacciano il fragile erede da dentro o da sotto", così il rapporto tra le opere d'arte non è sempre determinato da mezzi formali puramente visivi; un ricordo sonoro, come un odore che si ricorda vagamente, può produrre altri tipi di richiami che non attengono unicamente al regime visivo.[6]

Potremmo infatti immaginare, sviluppando le argomentazioni di Steinberg, un modello di affiliazione e affinità che non si allinea alla marcia in avanti della cronologia, ma al campo laterale e diacronico delle pratiche in virtù del quale tutti gli artisti lavorano contemporaneamente.[7] Per esempio, Nauman nel 1967 ha realizzato un orecchio utilizzando una corda annodata e della cera modellata. L'oggetto, intitolato *Knot an Ear*[8], era parte di una serie di opere scultoree in omaggio all'opera di Henry Moore e H.C. Westermann, due artisti rispetto ai quali Nauman tentava di collocarsi e ai quali era legato da un rapporto teso e contraddittorio. Ricorrendo alla sineddoche dell'orecchio, Nauman si rappresenta mentre presta "ascolto" alle opere dei suoi pari, mentre il titolo, con il suo gioco di parole, evoca le possibilità del linguaggio. Nondimeno, *Knot an Ear* richiama anche le frustrazioni, la sordità e il carattere non lineare di questo modello di parentela. In un diverso contesto, queste "eredità fragili" si potrebbero definire "influenze fantasma", mutuando l'espressione di Nicolas Abraham e Maria Torok per descrivere gli effetti ritardati di antichi incontri, ricordi, visioni e suoni che si ripercuotono sulle generazioni successive. Questi effetti sono racchiusi, o "criptati", l'uno nell'altro; come il "rumore nascosto" nell'opera di Duchamp, o il ricordo della musica quando si osservano i pianoforti rotti di Uecker o di Maciunas, queste opere si parlano a vicenda guardando in avanti e voltandosi indietro, a volte di lato, con esiti spesso inattesi.

L'ASCOLTO DELLA STORIA

Nel suo saggio del 1933 intitolato "The Lamp", Walter Benjamin descriveva il lavoro dello storico in termini uditivi, spiegando che il suo approccio lo portava ad appoggiare l'orecchio all'involucro del XIX secolo.[9] Evocando i vari modi in cui lo storico acquisisce il materiale da una varietà di fonti, che non sono necessariamente le più ovvie e manifeste, Benjamin intendeva il compito dello storico come un'attenta ricerca per scorgere tracce appena distinguibili, per captare il suono a stento udibile di personaggi, episodi e motivi quasi dimenticati. Il modello "uditivo", secondo il quale le opere d'arte non richiedono solo l'osservazione ma anche "l'ascolto", si potrebbe accostare a modi di pensare alla storia in termini acustici, ricorrendo a quella che in un diverso contesto T.S. Eliot ha notoriamente definito "immaginazione uditiva", che è un requisito della poesia e che, scriveva Eliot, "agisce attraverso i significati, non di certo i significati nel senso ordinario del termine, e fonde insieme i vecchi, obliterati e triti con i più attuali, nuovi e sorprendenti".[10]

"Ascoltare" un'opera d'arte suggerisce un modo di prestare attenzione ad aspetti dell'oggetto che non sono articolati visivamente ma risuonano attraverso segnali meno nitidi, forse effimeri o comunque veicolati diversamente. A volte questi segnali possono risuonare con chiarezza stentorea, ma in altri momenti gli effetti possono essere tenui e opachi e assumere il tenore di una vibrazione, un crepitio, un ronzio o persino un urlo silenzioso e ossessivo. Il suono può essere discordante, persino disturbante, un rumore violento che segnala una spaccatura o un momento di rottura. E negli anni '60, appunto, c'è stato un momento in cui il suono, per diversi artisti che creavano oggetti tridimensionali, è diventato sia un materiale artistico, sia un concetto che una metafora. Ascoltare opere d'arte era dunque una richiesta esplicita tanto quant'era un problema storico. Nel valutare un tempo in cui il visivo e l'uditivo si sono combinati in modi così inattesi e spiazzanti, potremmo dunque richiamarci ad Hannah Arendt e all'idea che l'ascolto diventi una "metafora guida" per riflettere su come questi artisti hanno usato il suono e la sua ricezione per sovvertire l'inveterato predominio del "visivo" nell'ambito della modernità in generale e nel discorso modernista in particolare.

1
Si veda Michael Fried, "Three American Painters: Kenneth Noland, Jules Olitski and Frank Stella", in Id., *Art and Objecthood: Essays and Review*, University of Chicago Press, Chicago (IL) 1998, pp. 213-65: 224; Clement Greenberg, "Modernist Painting" in John O'Brian (a cura di), *Clement Greenberg: the Collected Essays and Criticism*, vol. 4, Chicago University Press, Chicago (IL) 1993, pp. 85-93. Per due diverse prospettive sull'elemento visivo nella modernità, si veda Jonathan Crary, *Techniques of the Observer: On Vision and Modernity in the Nineteenth Century*, The MIT Press, Cambridge (MA) 1992; Martin Jay *Downcast Eyes: The Denigration of Vision*, University of California Press, Berkeley (CA) 1994.

2
Hannah Arendt, "Metaphor and the Ineffable", in *The Life of the Mind: Thinking*, Harcourt Brace, New York 1981, pp. 110-29: 110-11; trad. nostra. Si veda anche Angela Frattarola, "Developing an Ear for the Modernist Novel: Virginia Woolf, Dorothy Richardson and James Joyce", *Journal of Modern Literature*, vol. 33, no. 1, Indiana University Press, Bloomington (IN) autunno 2009, pp. 132-53: 132.

3
Brandon Joseph, *Random Order: Robert Rauschenberg and the Neo-Avant-Garde*, The MIT Press, Cambridge (MA) 2007, p. 187.

4
Leo Steinberg, "Other Criteria" [1968/1971] in Id., *Other Criteria: Confrontations with Twentieth Century Art*, Oxford University Press, New York 1972), pp. 55-93: 84.

5
Leo Steinberg, "Other Criteria", in *op. cit.*, p. 88.

6
Michel Foucault, "Nietzsche, History, Genealogy" [1971] in Donald F. Bouchard (a cura di), Donald F. Bouchard, Sherry Simon (trad.), *Language, Counter-Memory, Practice*, Blackwell Publishing, Oxford 1977, p. 146; trad. nostra. Ovviamente, l'intimo compenetrarsi dei sensi è il motivo d'apertura della *Recherche*. Cfr: Marcel Proust, "Dalla parte di Swann", in Id., *Alla ricerca del tempo perduto*, vol. 1, trad. di Giovanni Raboni, Mondadori, Milano 2005.

7
Per una discussione sui modelli di parentela alternativi e non lineari tra artisti e opere che esplorano un registro acustico legato all'"ascolto" e al "fraintendimento" si veda il quinto capitolo del mio *Eccentric Objects: Rethinking Sculpture in 1960s America*, Yale University Press, New Haven (CT)/London 2012.

8
Il titolo *Knot an Ear* letteralmente tradotto potrebbe significare "Annodare un orecchio", ma pronunciandolo si dà vita al gioco di parole sotteso *Not an Ear*: ossia "Non è un orecchio". [N.d.R.]

9
Walter Benjamin, "The Lamp" [1933], trad. di Rodney Livingstone, in Michael W. Jennings, Howard Eiland, Gary Smith (a cura di), *Walter Benjamin: Selected Writings, vol. 2, part 2, 1931-1934*, The Belknap Press/Harvard University Press, Cambridge (MA)/London 1999, p. 692.

10
T.S. Eliot, "Matthew Arnold", in *The Use of Poetry and the Use of Criticism* Faber and Faber, London 1933, pp. 118-19; trad. nostra. Anche Jonathan Sterne cita questo celebre passaggio nel suo saggio "Sonic Imaginations", in Jonathan Sterne (a cura di), *Sound Studies Reader*, Routledge, New York 2012, pp. 1-19: 5-6.

IL RUMORE DELLA SUPERFICIE. RENDERE L'ARTE UDIBILE NEL XX SECOLO

Alan Licht

Per quanto il suono possa sembrare etereo, la sua esistenza è fondata sull'interazione della materia. Il suono è il risultato acustico del contatto fisico; vale per l'acqua che gocciola dal rubinetto al lavandino, come per il segnale elettrico che esce dall'altoparlante di un sistema di amplificazione. Anche lo strumento musicale è basato sul tatto; è costruito allo scopo specifico di generare suoni facendo vibrare una o più delle sue parti. E il più delle volte è destinato a produrre note di altezza definita che attendono di essere attivate dalla mano del suonatore.

Il suono è l'indicatore di una presenza. Anche oggetti intrinsecamente afoni si possono *far* suonare toccandoli. Il suono è una parte fondamentale dell'identità, non solo per gli animali che lo emettono, come una balena che canta o una mosca che ronza, ma anche per gli oggetti motorizzati (un'automobile, un tosaerba) e gli elementi della natura (pioggia, vento e fuoco). Uno strumento musicale è l'estensione dell'innata capacità umana di produrre suoni. Nel caso degli strumenti a fiato o a corda, lo strumento musicale diventa una sorta di sostituto della voce umana, mentre gli strumenti a percussione richiamano piuttosto il battito cardiaco e il movimento del corpo. Ogni strumentista usa la musica come un secondo linguaggio, una forma di comunicazione e di espressione che, pur essendo non verbale (a esclusione del canto), è comunque soggetta all'equivalente di un sistema grammaticale.

Le arti visive sono fondamentalmente un'estensione della vista, non del linguaggio né dell'udito. Prive di una componente sonora, sono mute. All'osservatore è vietato toccare l'opera: il contatto è prettamente visivo, non tattile. E la condizione generale in cui l'opera si osserva e si contempla è il silenzio. Solo nel XX secolo gli artisti cominciano a concepire opere d'arte visiva dotate di un suono. Nel 1916 Marcel Duchamp concepisce *With Hidden Noise – À bruit secret* (Con rumore segreto) [164], un gomitolo di spago che racchiude un oggetto misterioso destinato a emettere un suono quando viene agitato. Negli anni '50, artisti cinetici come Jean Tinguely [226], Takis [284], Len Lye e Nicholas Schöffer creano sculture semoventi che producono un suono quando le singole parti entrano in contatto. Negli anni '60, due artisti dell'era concettuale giustappongono gli oggetti ai loro suoni: nel 1961 Robert Morris realizza *Box with the Sound of Its Own Making* (Scatola con i rumori della sua creazione) [220] inserendo in una scatola una registrazione di tre ore dei suoni prodotti nella costruzione della scatola stessa; mentre William Anastasi [290] crea una serie di "oggetti sonori", con cui rivisita la concezione del readymade incorporando gli effetti sonori che l'oggetto produce, come nel caso di un termosifone accompagnato da un nastro sul quale è registrato il suo clangore. In *Musical Chairs* (1983), Bruce Nauman appende al soffitto due sbarre d'acciaio e una sedia e le fa entrare in rotta di collisione con l'intento specifico di "fare rumore, fare musica". Più di recente, nel 2002, Stephen Vitiello attacca dei microsensori alle sculture di Donald Judd a Marfa per scoprirne le vibrazioni, solitamente inudibili. In tutte queste operazioni, gli artisti danno voce all'opera d'arte, le conferiscono un'individualità sonica che trascende il piano della visione.

Harry Bertoia e i fratelli Baschet sono i pionieri della creazione di sculture concepite espressamente come opere da esporre e al contempo come strumenti da suonare. Dopo aver udito la collisione tra due pezzi di filo metallico mentre realizzava una scultura, Bertoia comincia a creare "sculture tonali" composte da asticelle sottili saldate a una base che, se attivate, improvvisano suoni vibranti simili ai rintocchi di una campana. I suoni prodotti dalle sculture variano a seconda delle dimensioni e del tipo di materiale, e possono assumere un respiro quasi orchestrale quando le opere sono azionate collettivamente. Molte di queste sculture sono ora conservate in sedi museali o esposte in spazi aperti come installazioni permanenti. Anche i fratelli Baschet creano sculture sonore [190, 192] usando aste di vetro o di metallo combinate ad altri oggetti e materiali (vetro, plastica, corde di pianoforte, palloni), che interagiscono quando la scultura è fatta vibrare con le dita bagnate. Per la composizione e l'esecuzione delle musiche, i Baschet collaborano con Jacques e Yvonne Lasry, e le loro creazioni saranno utilizzate da altri compositori, come Toru Takemitsu e Jean Michel Jarre. Le sculture dei fratelli Baschet verranno poi esposte in musei di tutto il mondo, a cominciare dal Museum of Modern Art di New York nel 1965.

In parallelo a questi sviluppi nel mondo dell'arte e all'avanzata di un'estetica dell'astrazione, all'alba del modernismo emerge un interesse nel ricalibrare la definizione del concetto di musica e di ciò che può essere considerato uno strumento musicale. Le tendenze d'avanguardia nella musica classica procedono a infrangere i tabù, con il prevalere sempre più insistito dell'atonalità, della dissonanza e di ritmi più complessi. Ma a formulare idee veramente radicali per una nuova musica sono i futuristi, che esortano a portare la dissonanza alle sue estreme conseguenze incorporando nella musica i rumori dell'era dominata dalle macchine, come proclama Luigi Russolo nel Manifesto L'arte dei rumori (1913). La poesia sonora del futurista Marinetti e dei dadaisti Hugo Ball e Raoul Hausmann scompone il linguaggio in suoni vocalici, mantenendo un'idea di struttura ma sostituendo le parole con suoni insensati e gutturali, e dunque suggerendo che il rumore potrebbe sostituire le scale di note nella musica strumentale (che, come ho già rilevato, è intrinsecamente un sistema linguistico). E così Kurt Schwitters compone *Die Sonate in Urlauten* (1922-32), un poema sonoro in termini musicali (scherzo, cadenza e così via) per suoni vocali ad altezza indeterminata. Nei saggi sul Neoplasticismo del 1921 e del 1922, Piet Mondrian invoca un "nuovo ordine di suoni e non suoni (rumore determinato)", e László Moholy-Nagy, nel saggio del 1923 dal titolo *New Form in Music: Potentialities of the Phonograph*, pensa al fonografo come a un potenziale produttore di musica, invece di un semplice riproduttore del suono.

Alcuni compositori accolgono queste istanze: Edgar Varèse concepisce una musica intesa come "organizzazione di suoni". George Antheil compone il suo *Ballet Méchanique* (1927) per un eccentrico assortimento di pianole, eliche di aeroplani, xilofoni, ventilatori elettrici, grancasse, una sirena e un tam tam. John Cage usa piatti di giradischi in *Imaginary Landscape No. 1* (1939). Ma anche gli strumenti classici vengono rivisitati: nelle sue composizioni, Harry Cowell comincia a pizzicare manualmente le corde del pianoforte, mentre Cage, suo ex allievo, "prepara" un pianoforte inserendo dadi e bulloni tra le corde, un intervento che espande le possibilità cromatiche del pianoforte e ne enfatizza la qualità strutturale di strumento da percussione. Negli anni '40, compositori di musica concreta come Pierre Schaeffer e Karlheinz Stockhausen impiegano il nastro magnetico e lo studio di registrazione come strumenti da utilizzare di per sé, manipolando i suoni ambientali e strumentali registrati sul nastro per comporre paesaggi sonori stranianti attraverso la rielaborazione elettronica. *Cartridge Music* (1960) di Cage trasporta l'amplificazione e il missaggio di suoni non strumentali (di suppellettili e altri ammennicoli equipaggiati con testine da fonografo o microfoni da contatto) nel contesto di un concerto dal vivo. L'ormai celebre visita del 1951 a una camera anecoica aveva già sancito, nella mente di Cage, che il silenzio non esiste. Se *ogni cosa* ha un suono, allora *qualunque cosa* può essere arruolata al servizio della composizione musicale.

Nello spirito della musica d'avanguardia vengono fabbricati nuovi strumenti. Negli anni '60 David Tudor, illustre interprete di musica sperimentale per pianoforte, comincia a costruirsi strumenti elettronici che utilizza per comporre e per suonare. Il compositore Harry Partch costruisce speciali strumenti per suonare la sua musica microtonale, invece della musica a temperamento equabile prevalsa in Occidente fin dal XIV secolo (basata su 12 note per ottava; mentre gli strumenti di Partch possono avere fino a 43 note per ottava). Oltre ad "adattare" viole e chitarre, Partch inventa personali varianti dell'organo (il Chromelodeon), della marimba (il Boo, una marimba di bambù; la Diamond Marimba e la Mazda Marimba, composta da 24 lampadine accordate) e creazioni inedite (Spoils of War, una pila di ferrivecchi da utilizzare come percussioni, tra cui diversi bossoli di artiglieria; Kithara, che somiglia a un'arpa ma si suona come una chitarra orizzontale; Cloud Chamber Bowls, gong di vetro appesi a un telaio ligneo). Nei primi anni '80 Glenn Branca mette a punto le sue insolite chitarre (la chitarra armonica, la chitarra per mazzuoli) da suonare in orizzontale con l'ausilio di mazzuoli e composte da assi di legno equipaggiate di corde e ponticelli, anch'esse nate per conformarsi a uno speciale sistema di accordatura, più ampio del sistema convenzionale e basato su serie armoniche.

Anche tra gli artisti visivi vi è chi sperimenta la costruzione di strumenti musicali, principalmente per i propri fini artistici. Propugnando la ricerca di una musica del rumore, Russolo arriva a elaborare gli intonarumori, casse armoniche in legno munite di un diffusore che si attivano con una manovella e sono provviste di una leva per regolare l'emissione del suono; ne costruisce 27 tipi, ciascuno destinato a produrre suoni specifici, tra cui l'Ululatore, il Rombatore, il Crepitatore. A distanza di tempo, negli anni '60, nel clima neo-dada proprio del Fluxus nascono le macchine musicali [234] di Joe Jones (strumenti meccanici che suonano automaticamente – tra cui chitarra, violino, tamburo, vibrafono, gong e così via – e talvolta si uniscono a comporre la Mechanical Fluxorchestra) e nel 1971 Nam June Paik propone il suo *TV Cello*, suonato con perizia da Charlotte Moorman. Nei tardi anni '60 lo scultore Robert Rutman presenta i suoi *Steel Cello* e *Bow Chime* (entrambi realizzati in metallo e suonati con un archetto), e poco dopo Akio Suzuki inventa il suo *Analapos* (1970) [272], ideato per studiare il riverbero dell'eco e composto da due cilindri metallici e una molla a spirale. Più di recente compaiono gli *Hybrid instruments* (strumenti ibridi) di Ken Butler [298], strumenti a corda il cui corpo è composto da oggetti di recupero (per esempio un violino ricavato da un telefono a disco, un monopattino che diventa una chitarra). La musica che questi artisti producono con i loro strumenti tende a essere per lo più una dimostrazione delle potenzialità di questi ultimi. *The Pygmy Gamelan* (1973) [280] di Paul De Marinis, un oggetto che suona cinque note in risposta alle sollecitazioni del campo elettromagnetico in cui viene collocato, è concepito sia come un'opera scultorea destinata alla visione, sia come qualcosa che *l'osservatore* può suonare (e a un certo punto De Marinis pensa addirittura di metterlo in commercio come una sorta di radio alternativa). Anche *Handphone Table* (1978) [296] di Laurie Anderson, che esplora la combinazione tra un tavolo e l'ossatura umana come conduttore del suono, si offre all'esperienza del visitatore, più che a una performance dell'artista: il tavolo è collegato a suoni registrati su nastro che si possono udire/sentire appoggiando i gomiti sul tavolo e posando le mani sulle orecchie.

Talvolta gli artisti si avvalgono di strumenti convenzionali per piegarli a fini diversi dalla loro funzione originaria. Per l'installazione *Playing the Building* (2008), David Byrne preleva un organo preesistente, lo posiziona nello spazio di una vecchia architettura, attacca lunghi cavi al posto delle canne e li collega a vari punti dell'ambiente: condutture, impianti, travi e altri elementi strutturali. I visitatori sono invitati a premere i tasti, che attivano rumori stridenti, sibili flautati e profondi rimbombi (distribuiti in base ai registri alti, medi e bassi della tastiera, accennando così alle consuete potenzialità dello strumento). L'operazione dell'artista trasforma l'intera stanza in una scultura sonora, ma al contempo trasforma l'organo in un campionatore non elettronico. È un'installazione, ma priva di suono continuo. Per *I Play the Drums with Frequency* (2007), Manuel Rocha Iturbide dissemina i componenti di una batteria in vari punti di una stanza, dotando ciascun componente di un amplificatore. Un computer invia onde sinusoidali agli amplificatori, che fanno vibrare le membrane dei tamburi e i cimbali, creando un ambiente sonoro ad alta complessità timbrica che è latente in una batteria ma di norma, quando la si suona con le bacchette, non emerge appieno. Nel video *Guitar Drag* (2000), Christian Marclay collega una chitarra elettrica – una Fender Stratocaster integra – a un amplificatore collocato sul retro di un pick-up, e la attacca al veicolo con una corda legata al manico. Il camioncino si trascina dietro la chitarra attraverso i campi e lungo le strade del Texas. Poiché tutte le corde si rompono dopo pochi minuti, di fatto i magneti della chitarra diventano microfoni da contatto posati su pezzi di legno e metallo che sfregano contro la terra e l'asfalto; ora è il corpo della chitarra a vibrare, non le corde, e sono le asperità del terreno a determinare le variazioni del tessuto sonoro. In quest'opera ispirata alla morte di James Byrd Jr., ucciso in Texas con questa raccapricciante modalità, Marclay cita anche la performance in cui Nam June Paik trascina dietro di sé un violino attaccato a una corda, così come richiama le chitarre fracassate sul palco da chitarristi rock come Pete Townshend e Jimi Hendrix. Il fatto che le chitarre di Townshend, Hendrix e Marclay continuino a emettere suoni anche se rese "insuonabili" segnala le conseguenze accidentali dell'amplificazione e rivela fino a che punto il ripensamento dei parametri musicali promosso dal modernismo ha inciso sia sugli strumenti, sia sugli idiomi musicali nell'ultimo mezzo secolo.

Nei primi anni '60, durante la sua formazione artistica, Townshend sente la conferenza di Gustav Metzger sull'autodistruzione nell'arte, mentre un altro giovane inglese, Keith Rowe, pittore e autore di improvvisazioni musicali allora impegnato negli studi d'arte, è forse il primo a stendere una chitarra convenzionale su un tavolo per suonarla. Seguendo le orme di Jackson Pollock, che aveva mostrato cosa si poteva creare lasciando gocciolare il colore su una tela distesa a terra, e tenendo a mente il pianoforte preparato di Cage, Rowe lascia cadere vari oggetti sulla chitarra supina, attacca microfoni da contatto al corpo dello strumento e non lo suona mai in maniera tradizionale (non lo accorda nemmeno). Rowe è tra i primi a praticare quella che verrà definita "improvvisazione libera", una tendenza minore ma influente che attraversa gli anni '60 e '70, animata da musicisti che intendono creare una musica atonale, aleatoria, eseguita in tempo reale senza alcuna partitura prestabilita, sotto l'influsso dell'improvvisazione jazz e delle partiture aperte della musica contemporanea. Poiché l'enfasi cade sul tessuto sonoro invece che sull'armonia, l'idea di tecniche sperimentali che ne amplino le possibilità (come quelle di Rowe) è accolta da improvvisatori musicali che devono continuamente trovare modi per sorprendere se stessi, i loro compagni d'improvvisazione e il pubblico. Anche se con un approccio meno radicale rispetto a Rowe e al suo rifiuto di adottare il metodo tradizionale per suonare il suo strumento, percussionisti come Han Bennink, Tony Oxley e Jamie Muir modificano pesantemente le loro percussioni con vari oggetti in legno e in metallo e persino strumenti elettronici (nel caso di Oxley) per ottenere una gamma sonora estremamente estesa (Bennink e Muir erano anche artisti visivi). Altri improvvisatori costruiscono da sé i loro strumenti elettrici o acustici, come Paul Burwell, Max Eastley, Hans Reichel e Hugh Davies.

Tuttavia, a precorrere la nascita di un'improvvisazione libera tra i musicisti di orientamento sperimentale sono gli esperimenti sonori di un artista visivo, Jean Dubuffet, che nel 1960 realizza sfrenate registrazioni cacofoniche insieme ad Asger Jorn. I due artisti aggrediscono il pianoforte, il violino, il violoncello, la tromba, il magnetofono, il flauto sahariano, la chitarra e il tamburello: strumenti che non hanno mai suonato. Anche la Artists Jazz Band e la Nihilist Spasm Band, entrambe formatesi in Canada negli anni

’60, sono composte da artisti visivi privi di una formazione musicale che intendono l’improvvisazione libera come un modo per comporre musica (diversi membri della Spasm Band si costruiscono i propri strumenti, in particolare chitarre elettriche e un kazoo elettrico). Eliminando buona parte dei sistemi convenzionali per suonare musica, e preferendo un approccio “improprio” agli strumenti, l’improvvisazione libera è aperta tanto ai profani quanto ai musicisti esperti. Marclay, anche lui privo di una formazione musicale, si unisce allo schieramento internazionale degli improvvisatori ma usa come strumenti dei dischi e piatti di giradischi, e soprattutto crea “dischi riciclati” incollando frammenti di dischi rotti che, oltre a essere utilizzati nelle sue performance, vengono anche esposti e venduti come opere d’arte.

Se con Marclay, Bertoia e i fratelli Baschet sfumano i confini tra scultura e strumento, con Bill Fontana, che parla di “scultura sonora” per le sue installazioni, si delinea una prospettiva ulteriore. Fontana fa largo uso delle registrazioni sul campo per percepire il suono di un dato luogo da ogni posizione possibile, proprio come accade quando ci si muove intorno a una scultura. Fontana scrive: “Nello sviluppare la definizione scultorea della mia opera ho usato diverse strategie per superare le qualità effimere del suono, che mi paiono in netto contrasto con il senso di certezza fisica e di permanenza che di norma attengono alla scultura e all’architettura”.[1] Cita per esempio un suo lavoro del 1976, *Kirribilli Wharf*, una registrazione effettuata al mare in Australia: “L’azione percussiva delle onde a Kirribilli Wharf aveva in sé continuità e permanenza. Questo nastro a otto canali non registrava un unico momento, come nel caso di un’eclisse totale, bensì un brano di un processo sonoro perpetuo”.[2]

L’approccio di Fontana prevede di ascoltare innanzitutto, e poi di scolpire. Nell’arte sonora, che in Fontana trova uno dei suoi massimi esponenti, l’ascolto è la semente; l’opera finita si esprime in un rapporto da ascoltatore ad ascoltatore. La musica composta, invece, prevede un rapporto tra suonatore e ascoltatore. A Kirribilli Wharf il suono è già presente, autonomo; esiste prima, durante e dopo la registrazione di Fontana. Non è “suonato” da Fontana, ed è udibile da chiunque nella sua versione originale. Buona parte dell’arte sonora nasce dal desiderio di dare al suono una continuità, simile alla continuità quasi sistematica di paesaggi sonori ambientali come quelli che esplora Fontana. La relativa regolarità dell’azione nei dintorni del molo permette un ascolto profondo, immersivo, diverso dall’ascolto che segue l’arco narrativo di una composizione. La musica in sé è episodica, destinata a durare per un determinato segmento temporale e poi concludersi. Viene suonato uno strumento che ritorna al silenzio non appena lo si posa. Quando un artigiano crea uno strumento musicale, crea un oggetto d’uso: dev’essere durevole e bello, ma non ornamentale. Quando gli artisti creano oggetti destinati all’ascolto, producono l’incarnazione stessa del paradosso tra la finalità dell’opera d’arte e la trascendenza del suono.

1
Bill Fontana, Resoundings, resoundings.org
2
Ibid.

SOUND ART? ARTI VISIVE E SUONO AL PASSAGGIO FRA ANNI ’70 E ’80

Simone Menegoi

Gli anni a cavallo del 1980 segnano un passaggio significativo nella storia dei rapporti fra arti visive e suono nel XX secolo. Nell’arco di un numero relativamente ridotto di anni, una serie di mostre tematiche in Europa e negli Stati Uniti presentarono come un fenomeno unitario il lavoro degli artisti che, nei due decenni precedenti, avevano cominciato a usare il suono in congiunzione ad altri media tipici delle arti visive, o a farne addirittura un uso esclusivo, accordando però alla dimensione spaziale del suono un’importanza almeno altrettanto importante di quella temporale. Era la nascita ufficiale, a livello istituzionale, del fenomeno che proprio in quegli anni prese il nome di “sound art”, e che continua a essere oggetto di mostre e dibattiti critici e che pone, oggi più di allora, problemi di definizione, di *collocazione*, di presentazione all’interno delle istituzioni.

Se si scorre una cronologia delle mostre tematiche dedicate ai rapporti fra arte e suono dagli anni ’60 in poi,[1] il fenomeno che abbiamo segnalato appare abbastanza evidente. L’inizio degli anni ’60 è marcato da una fioritura di eventi e mostre legati al movimento Fluxus, grande laboratorio di contaminazioni fra le discipline, seguita, fra gli anni ’60 e ’70, da una relativa penuria di eventi. (Fra le non molte mostre spicca “Record as Artwork, 1959-73” curata da Germano Celant alla Royal College of Art Gallery di Londra nel 1973). L’interesse per le relazioni arte-suono a livello espositivo riemerge in modo evidente nella seconda metà degli anni ’70 e culmina, tra la fine del decennio e l’inizio del successivo, in un proliferare di mostre piccole e grandi, alcune delle quali riconosciute già all’epoca come pietre di paragone. In Europa si segnala “Für Augen und Ohren. Von der Spieluhr zum akustischen Environment – Objekte, Installationen, Performances” (Akademie der Künste, Berlino, 1980, a cura di René Block), grande rassegna di oggetti/strumenti/macchine per produrre suono, dagli automi musicali del XVII-XVIII secolo ai cacofonici vinili manipolati del Fluxus Milan Knížák [232]. Di dimensioni più ridotte, e assai meno ricordata, ma con un indubbio primato cronologico sulla mostra di Berlino, è un’altra esposizione tedesca, “Sehen um zu hören: Objekte und Konzerte zur visuellen Musik der 60er Jahre” (Städtische Kunsthalle, Düsseldorf, 1975, a cura di Inge Baecker), che in Fluxus trovava il suo fulcro. Negli Stati Uniti si succedono, in un crescendo di dimensioni e ambizione, le mostre: "Audio Works" (Artists Space, New York, 1978, a cura di Helene Winer), che a sua volta includeva opere su nastro magnetico provenienti da una mostra di poco precedente ("Narrative Themes / Audio Works", Los Angeles Institute for Contemporary Art, 1977, a cura di Robert Smith), di artisti concettuali quali Barbara Kruger, John Baldessarri, Robert Barry, Guy de Cointet, Bernard Venet, Lawrence Weiner; "Sound. An Exhibition of Sound Sculpture, Instrument Building and Acoustically Tuned Spaces" (Los Angeles Institute of Contemporary Art e in seguito, in versione ampliata, PS1, New York, 1979, a cura di Robert Smith e Bob Wilhite); e "Soundings" (Neuberger Museum, SUNY at Purchase, New York, 1981, a cura di Suzanne Delehanty), presentata come "la prima rassegna sull'uso del suono, della musica e dei fenomeni acustici nell'arte visiva americana dal 1900 a oggi".[2]

Le rassegne più vaste e significative, come "Für Augen und Ohren" e "Soundings", che si lanciavano nella titanica impresa di dar conto di un secolo (e anche più, poiché la mostra berlinese includeva oggetti che datavano fino a quattro secoli prima) di relazioni arte-suono avevano un terreno comune con quelle più ristrette e mirate: le opere degli anni '60 e '70 che si accampavano in una terra di nessuno fra suono e visione. Un produzione ormai abbastanza consistente, elaborata e varia da fare massa critica; tanto più che i suoi autori, provenienti tanto dalle arti visive e dalla performance quanto dalla musica di ricerca, avevano iniziato a sviluppare delle poetiche e una terminologia specifica per definire il loro lavoro. Se si considera il fenomeno dalla visuale di oggi, appare abbastanza chiaro che è l'emersione di questa nuova area di ricerca nella prassi e nella teoria degli artisti che motiva, in maniera più o meno dichiarata, le mostre che abbiamo elencato (e molte altre degli anni successivi). Anche nel caso delle grandi esposizioni dal taglio storico-artistico, uno dei moventi principali dei curatori era la volontà di dare a un fenomeno artistico recente (e stimolante) un adeguato retroterra nella cultura del XX secolo. In questo, senso, il testo con cui Suzanne Delehanty presentava "Soundings" nel catalogo della mostra eponima è esemplare. Il lavoro di artisti emersi negli anni '60 e '70 che impiegavano il suono, da solo o insieme ad altri media, nell'ambito dell'arte visiva – da Nam June Paik [222, 228] a Max Neuhaus, da Harry Bertoia a Bruce Nauman [248] – e di musicisti sperimentali le cui opere, in termini di strumentazione, formato e ricezione, avevano abbandonato già da anni le sale da concerto – da Takehisa Kosugi a Alvin Lucier, da John Cage a David Tudor – erano inquadrati in una prospettiva storica e culturale che abbracciava due secoli, il XIX e il XX, e che faceva riferimento a due filoni di ricerca principali. Il primo riguardava le innovazioni tecnologiche – in questo caso, in particolare, quelle che permettono la registrazione e la trasmissione del suono, o che associano suono e immagine, come il cinema e il video – e il suo impatto sulle arti, sia a livello dei nuovi mezzi espressivi che mettevano a disposizione, sia delle nuove poetiche che contribuivano e definire. Il secondo filone di ricerca riguardava i tentativi di sintesi fra le arti: l'utopia, ricorrente nella cultura del periodo fra XIX e XX secolo, di una *Gesamtkunstwerk* capace di risalire all'unità originaria che le arti avevano, per esempio, nel dramma greco antico; le ricerche sulle corrispondenze sinestetiche fra suono e colore del Simbolismo e della pittura astratta delle origini; le esperienze fra segno grafico, senso e suono della poesia di Stéphane Mallarmé, di Guillaume Apollinaire, dei futuristi. Nel complesso, un'argomentazione critica solida ma tradizionale, preoccupata meno di definire la nuova tendenza in atto che di darle una genealogia culturale. (Un merito, quest'ultimo, non indifferente, beninteso). L'impostazione del testo rifletteva fedelmente quella della mostra, che ricapitolava effettivamente, attraverso opere, oggetti e documenti, le esperienze della prima metà del XX secolo, per poi spingersi alle creazioni degli artisti citati.[3]

Per converso, un'altra grande mostra americana di quegli anni sui rapporti fra arte e suono, "Sound" (1979), si concentrava proprio sulla produzione artistica recente, schierando oltre quaranta artisti emersi soprattutto negli anni '70. Anche i testi in catalogo si concentravano sulle ultime tendenze. Il saggio più importante, a firma di Peter Frank[4] delineava essenzialmente tre aree di ricerca presenti in mostra: la prima (rappresentata da Connie Beckley, Charlemagne Palestine, Peter Gordon, Laurie Anderson [296, 302], il duo Christina Kubisch e Fabrizio Plessi) è strettamente legata alla dimensione del tempo, e annovera come componente più o meno costante la presenza fisica dell'artista stesso, impegnato a produrre suoni con la propria voce o con uno strumento. La seconda è quella che si colloca fra la scultura e la creazione di strumenti musicali, in cui Frank individua due sottoinsiemi: *sculptural instruments*, creati per lo più da musicisti per superare i limiti della strumentazione ereditata dall'Ottocento, e dotati di proprietà scultoree (oltre a veterani come Harry Partch e i fratelli Baschet [190, 192], Susan Rawcliffe, Bob Bates, Emmet Chapman); e *sounding sculptures*, create per lo più da artisti visivi con l'intento di esplorare una dimensione ulteriore, dinamica e temporale, della scultura (fra gli altri, Bob Wilhite, Jim Pomeroy, Tom Recchion, Stephan von Huene [244, 246]). Infine, la terza area di ricerca, quella più radicalmente "intermediale", difficile da concepire perfino come somma di discipline diverse in quanto sembra costituire un ambito di ricerca a sé stante, è quella degli *acoustically tuned spaces*, come recita il sottotitolo della mostra, o delle *sonic installations*, come le chiama Frank. Opere ambientali in cui l'accento cade meno sul tempo in cui il suono si svolge e più sullo spazio che occupa, in un'accezione fenomenologica del loro rapporto. (I nomi, in questo caso, sono quelli di Michael Brewster, Max Neuhaus, Liz Phillips, Bill Fontana, Marvin Torffield, Karen Wolff e William Klingsbury). Al termine del testo, a mo' di ricapitolazione e per sottolineare la complessità e la varietà dell'area artistica di cui sta cercando di tracciare i confini, Frank inserisce un piccolo elenco delle opere più singolari, apparentemente irriducibili l'una all'altra, che ha evocato nel testo: "Una partitura musicale che somiglia a Mondrian. Istruzioni di una sola parola per brani da concerto. Tintinnii che pervadono un ambiente. Gruppi rock e country impegnati in performance artistiche. Musicisti che suonano delle sculture. Un artista che suona il flauto attraverso una maschera antigas. Un'opera frammentaria composta con le parole di uno scialbo romanzo vittoriano".[5] Per quanto il catalogo sia effettivamente eterogeneo, avrebbe potuto esserlo molto di più. Per esempio, avrebbe potuto includere gli *art tapes* di artisti concettuali che costituivano il nerbo della mostra "Narrative Themes / Audio Works", oppure le singolari audio-performance che in quegli anni trasmetteva "Close Radio", una trasmissione condotta da John Duncan e Neil Goldstein sulle frequenze di KPFK di Los Angeles.

Le parole sono importanti. La nominazione è un atto di creazione, o quantomeno di appropriazione. E così, lo sforzo dei critici e dei curatori di quegli anni non era solo quello di circoscrivere l'area comune alla musica (o al suono) e alle arti visive, ma anche quello di trovare le parole per farlo. Alcune le abbiamo già lette: *sculptural instruments* e *sounding sculptures*, *sonic installations* e *acoustically tuned spaces*, alle quali va aggiunta *musical art* – la definizione complessiva che Peter Frank dà del fenomeno arte/suono nel 1979.[6] Dalle due mostre tedesche ricordate all'inizio, "Sehen um zu hören" e "Für Augen und Ohren" provengono definizioni come, rispettivamente, *visuelle Musik* – una sintesi che calzava in modo particolare a determinate ricerche Fluxus, che costituivano infatti il nerbo della mostra di Düsseldorf del 1975 – e *akustischen Environment*. In una recensione di "Soundings", un altro critico di nome Frank, Robert, utilizzava l'espressione destinata ad avere più fortuna nei decenni successivi, tanto da diventare, in ambito anglosassone "la" definizione di un determinato ambito di ricerca: *sound art*. L'espressione non si trova nel testo di Suzanne Delehanty o nel titolo della mostra; Robert Frank la impiega perché era nell'aria da alcuni anni. Di solito se ne attribuisce la paternità a William Hellermann, un artista newyorkese (non presente in "Sounds", ma citato nel catalogo della mostra) che alla fine degli anni '70 aveva creato la SoundArt Foundation, e che nel 1983 curò allo Sculpture Center di New York un'altra mostra significativa del periodo, chiamata appunto "Sound/Art", che includeva, fra gli altri, artisti eminentemente visivi/sonori già presenti in "Sounds", accanto ad altri piuttosto affermati che non venivano identificati in maniera primaria con il suono ma ne facevano comunque uso, come Vito Acconci, Hannah Wilke, Keith Sonnier.

Non è certo un caso che Hellermann fosse un artista. Anche uno dei due curatori di "Sounds", Bob Wilhite, lo era. È dallo sforzo degli artisti di trovare un'identità nuova per ricerche che, all'epoca, erano nuove, che nasce il fenomeno di cui le mostre fra gli anni '70 e '80 cercavano di dar conto da angolature diverse e con diversa ampiezza. È un altro degli artisti che abbiamo già incontrato, Max Neuhaus, ad aver utilizzato per primo la definizione di *sound installation* per descrivere un ciclo di opere, iniziato nel 1967, che cercavano di liberare il suono dalla logica spaziale della sala da concerto, e da quella temporale del brano musicale con un inizio e una fine determinati.

A circa 35 anni di distanza da quelle ricerche artistiche e dalle mostre che le documentavano, qual è il bilancio del passaggio fra gli anni '70 e '80? E quale inquadramento teorico si può darne? All'epoca, la comprensione e la definizione di determinate esperienze oscillava fra una cornice teorica nata più di dieci anni prima in ambito Fluxus da Dick Higgins e ormai familiare, quella di "intermedia", e il tentativo di creare una terminologia e delle classificazioni nuove e specifiche, non sempre perfettamente a fuoco; fra una visione di lungo termine che si ricollegava alle ricerche sinestetiche fra XIX e XX secolo e agli esperimenti delle avanguardie storiche, e a una mappatura, a tratti troppo minuziosa, a tratti lacunosa, delle vicende in atto. Oggi, con il vantaggio di una visione retrospettiva, può essere suggestivo tentare altre strade. Per esempio, si può provare a ripensare le esperienze artistiche fra suono e visione, e lo sforzo di definizione (e auto-definizione) che comportavano, nei termini di uno fra i più interessanti filosofi dell'estetica emersi a livello mondiale nell'ultimo decennio, Jacques Rancière; in particolare, nei termini in cui Rancière rilegge l'intera vicenda della modernità artistica, dal XVIII secolo agli anni '50 del XX secolo, in *Aisthesis. Scènes du régime esthétique de l'art* (2011). Nell'introduzione ai quattordici capitoli del testo, dedicati ognuno a un'opera e alla sua ricezione presso i contemporanei, Rancière precisa che la sua potrebbe essere considerata addirittura una "contro-storia" della modernità artistica, tanto si oppone ai criteri correnti con cui di solito quest'ultima viene identificata e descritta. "Le storie e le filosofie della modernità artistica che fanno testo", scrive il filosofo francese, "la identificano nella conquista, da parte di ciascuna arte, della propria autonomia, la quale si esprime attraverso opere esemplari, che creano una cesura nel corso della storia separandosi al tempo stesso dall'arte del passato e dalle forme 'estetizzate' della vita prosaica. Quindici anni di lavoro mi hanno condotto a conclusioni esattamente opposte: il movimento proprio del regime estetico, quello che ha sostenuto i sogni di novità artistica e di fusione fra l'arte e la vita riuniti nell'idea di modernità, tende a cancellare le specificità delle arti e a rendere incerti i confini che le separano, sia l'una dall'altra, sia dalla vita ordinaria. Le opere non creano una cesura se non prestandosi a condensare i tratti di regimi di percezione e di pensiero che preesistono loro e si sono formati altrove".[7] È certamente affascinante l'idea di provare a leggere l'emergere della (necessariamente fra virgolette) "sound art" come un *case study* – escluso dalla trattazione di Rancière per i limiti cronologici che il filosofo si è imposto – della logica essenziale delle arti nella modernità, ovvero la loro tendenza ad abbracciare, da un lato i mezzi e i fini delle altre arti, e dall'altro aspetti del reale fino a quel momento esclusi dalla sfera delle arti stesse, e appartenenti piuttosto alla quotidianità prosaica. Sul primo aspetto non è neppure il caso di insistere: la tendenza di cui abbiamo parlato finora, qualunque nome le si voglia attribuire, nasce costitutivamente ibrida, in debito tanto con le arti della visione e dello spazio, quanto con quelle dell'udito e del tempo. Per quanto riguarda il secondo aspetto che segnala Rancière, la tendenza ad annettersi aspetti del reale fino a quel momento esclusi dalla sfera artistica, le ricerche arte/suono non sono meno esemplari. Già Cage, lasciando vergine la partitura di *4'33"* (1952), aveva focalizzato l'attenzione degli spettatori sulla trama di suoni che ci accompagna ovunque, anche nella presunta neutralità acustica della sala da concerto, l'equivalente del *white cube* delle arti visive. Max Neuhaus, in una serie di eventi acustici intitolati *LISTEN*, iniziati nel 1966 e proseguiti, in forme diverse, fino al 1978, aveva chiuso il cerchio: i suoi "concerti" consistevano nel condurre gli spettatori fuori dal luogo deputato all'ascolto, che fosse il teatro o il museo, e nel farli passeggiare per le strade della loro città, soffermandosi in luoghi che l'artista riteneva di particolare interesse acustico – e si trattava di luoghi in cui il suono era talvolta del tutto ordinario, talvolta, come nel caso del rumore del traffico, addirittura spiacevole secondo il senso comune. È solo un esempio fra i molti. Una delle poste in gioco nelle esperienze visivo-sonore fra gli anni '70 e '80 è senz'altro la conquista di una dimensione acustica integrale, capace di annettersi, e di usare come materiale, qualunque suono, qualunque spazio, qualunque interazione fra il primo e il secondo.

Un altro criterio di lettura della modernità elaborato da Rancière appare stimolante in rapporto alle ricerche arte/suono, e segnala una via di ricerca che, in alcuni casi, è stata già imboccata. In *Aisthesis*, il filosofo non parla solo di opere (di pittura, scultura, cinema, teatro, fotografia) ritenute unanimemente "maggiori". Altrettanto spesso, si dedica a opere ed eventi "minori", in qualche caso addirittura oscuri – pantomime di clown, mostre di arti decorative – seguendo un'intuizione affascinante; che quelle opere ed eventi abbiano contribuito a creare il "regime percettivo", il sottofondo di idee e intuizioni sulle arti e sul reale, da cui sono sorte le opere "maggiori". Applicata alla sound art, l'intuizione di Rancière consiglia di esaminare non solo ciò che accadeva in musei, gallerie e festival, ma anche intorno a loro; nel cinema, per esempio, in cui, con il lavoro di Walter Murch per il film di Francis Ford Coppola *La conversazione* (1974), nasceva la definizione di "sound designer" (e, nel caso di Murch, si trattava di un sound designer che applicava le tecniche dell'avanguardia musicale contemporanea al tessuto sonoro del film); oppure cogliendo, per esempio, l'emergere di una coscienza ecologica applicata al suono, di cui il libro di R. Murray Schafer *The Tuning of the World*[8] (1977) è il primo manifesto. Si tratta, come dicevo, di ricerche già avviate: entrambe le informazioni, per esempio, sono citate in un utile profilo sintetico della sound art redatto da uno dei più attenti studiosi del campo (e artista sonoro lui stesso), Alan Licht.[9]

Una considerazione finale. Già al principio degli anni '80, l'ambito delle ricerche fra arte e suono appariva ai critici più informati tanto complesso ed eterogeneo da far sorgere dei dubbi sulla possibilità (e sull'opportunità) di darne un'immagine unitaria, tanto in una mostra quanto in un testo.[10] A distanza di più di tre decenni, il dubbio è già, a nostro parere, una ragionevole certezza. Retrospettivamente, le differenze saltano agli occhi in modo ancora più evidente. Differenze fra coloro che impiegavano il suono in modo occasionale, e altri che ne avevano fatto il fulcro del proprio lavoro. Differenze fra artisti che cercavano nel suono un'estensione del proprio lavoro visivo, ed erano a casa propria in musei e gallerie, e fra musicisti d'avanguardia la cui ricerca, a causa del conservatorismo delle istituzioni musicali, finiva per trovar posto in quelle artistiche per mancanza di alternative. Differenze fra coloro si concentravano sulle proprietà spaziali, plastiche del suono in rapporto all'architettura, e coloro che, come gli artisti concettuali, lo concepivano come puro veicolo di informazioni verbali. Verso il 1980, queste differenze potevano apparire tutto sommato trascurabili rispetto a un elemento unificante – l'uso del suono nelle arti visive – ancora abbastanza recente, e poco censito a livello istituzionale, da rendere accettabile un approccio unitario. Con il tempo, le differenze sono diventate divergenze, tanto a livello teorico quanto nella pratica degli artisti e dei curatori; contemporaneamente, come per altri media (il video, in primo luogo), l'uso del suono si è diffuso in modo tanto capillare nel lavoro degli artisti contemporanei da perdere parecchia della sua efficacia come collante tematico. Nel 2004, dando eco alla ripresa di interesse per le sperimentazioni visivo/sonore che si era manifestata fra la fine degli anni '90 e l'inizio del nuovo secolo, la rivista *Artforum* organizzò una tavola rotonda via internet fra alcuni artisti, curatori, critici e filosofi particolarmente interessati all'area in questione, o direttamente attivi in essa: Cristoph Cox, Carl Michael Von Hausswolff, Anthony Huberman, Branden W. Joseph, Steve Roden, Marina Rosenfeld, David Toop, Stephen Vitiello.[11] Leggendo i loro scambi, la prima cosa che colpisce è la divergenza babelica delle opinioni perfino su questioni basilari per circoscrivere l'ambito della discussione. Neuhaus, di cui Cox cita uno scritto intitolato polemicamente *Sound Art?*[12] (il punto di domanda mette in questione la legittimità stessa del termine), sostiene che la maggior parte di ciò che viene rubricato sotto l'etichetta in questione è semplicemente musica contemporanea, e conclude che è fuorviante definirlo in altro modo e collocarlo nelle istituzioni per l'arte. Von Hausswolff concorda sulla premessa, ma al contrario di Neuhaus si rallegra che musei e gallerie abbiano confini disciplinari abbastanza ampi da permettere alla nuova musica di trovar posto al loro interno. Christoph Cox propone di cercare il discrimine fra musica e ricerche di sound art nel modo di affrontare la dimensione del tempo, lineare e dotato di un principio e una fine nel caso della musica, tendenzialmente aperto nel caso della sound art; ma la sua proposta, ragionevole in via di principio, è accolta tiepidamente dagli altri partecipanti al dibattito, in particolare dagli artisti, tipicamente restii a rinunciare a delle possibilità espressive per ragioni teoriche. E così via. La discussione offre una rappresentazione plastica del fatto che raggruppare gli artisti sulla sola base del medium "suono", senza specifiche ulteriori (e specifiche tanto sofisticate quanto è vasta la diffusione del medium) è ormai da tempo, più che difficile, insensato. Del resto, come ricordava ironicamente Douglas Kahn alcuni anni fa, "più che *sound artists*, molti artisti che da tempo usano il suono preferirebbero essere chiamati semplicemente *artisti*".[13]

1
Quella che abbiamo consultato, incompleta ma comunque di grande utilità, è stata compilata e messa on line (all'indirizzo joy.nujus.net/SAET) da due musicisti/artisti e studiosi degli incroci fra arti visive e suono, il francese Jerôme Joy e l'americano Thom Holmes. Al momento della consultazione (28 marzo 2014) la cronologia copre il periodo dal 1964 al 2004, cercando, soprattutto per i primi quindici anni, di utilizzare un criterio inclusivo piuttosto che esclusivo, registrando cioè qualunque "collective public presentation of sound and musical works (which overlap the concert framework such as installations, performances, etc., i.e. sound and music events which use another public configuration than the presentation on a stage, and which haven't a beginning nor an end) under a topic, a theme or a specific question and problem expressed by the title of the event" ["presentazione pubblica di opere sonore e musicali (che travalicano la cornice del concerto, come installazioni, performance e così via, ovvero eventi musicali e sonori che usano una configurazione pubblica diversa dalla presentazione su un palcoscenico, e che non hanno un inizio né una fine) all'insegna di un argomento, un tema o una questione e un problema specifico espresso dal titolo dell'evento"; trad. nostra].

2
Come riportato in "Soundings at SUNY", la recensione della mostra redatta da Robert Frank su *Art Journal*, vol. 42, n. 1, primavera 1982, pp. 58-62; trad. nostra.

3
Suzanne Delehanty, "Soundings", in Suzanne Delehanty, Dore Ashton, Germano Celant, Lucy Fischer (a cura di), *Soundings*, catalogo della mostra, Neuberger Museum, SUNY at Purchase, New York 1981, pp. 21-38.

4
Peter Frank, "The Sound and the Theory. A Short History of the Art-Music Interface", in Robert Smith, Robert Wilhite (a cura di), *Sound. An Exhibition of Sound Sculpture, Instrument Building and Acoustically Tuned Spaces*, catalogo della mostra, Los Angeles Institute of Contemporary Art, Los Angeles, e Project Studio 1, New York 1979, pp. 9-16.

5
"A musical score that looks like Mondrian. One-word instructions for concert pieces. A room full of clicking. Rock and country bands in art performances. Musicians playing sculpture. An artist playing flute through a gas mask. A fragmentary opera gleaned from the words of a sodden Victorian novel", in *Sound. An Exhibition of Sound Sculpture, Instrument Building and Acoustically Tuned Spaces*, cit., p. 16.

6
"One result of this open acceptance in the art world of new musical and fused musical-artistic forms is the emergence of musical art of all kinds" ["Un risultato di questa accoglienza, nel mondo dell'arte, di nuove forme musicali e forme ibride artistico-musicali è l'emergere di un'arte musicale declinata in una molteplicità di generi"; trad. nostra], in *Sound. An Exhibition of Sound Sculpture, Instrument Building and Acoustically Tuned Spaces*, cit., p. 12.

7
Jacques Rancière, *Aisthesis. Scènes du regime esthétique de l'art*, Éditions Galilée, Parigi, 2011, p. 13; trad. dell'autore. Il bersaglio polemico del filosofo non è difficile da indovinare, e viene chiamato in causa esplicitamente nella pagina successiva, con parole di durezza non comune: è Clement Greenberg, primo e più illustre promotore del "monumento di quel modernismo retrospettivo che, mancando della capacità di fondare un'arte di qualche importanza, riuscirà tuttavia a imporre la leggenda dorata delle avanguardie e a riscrivere a suo vantaggio la storia degli sconvolgimenti artistici di un secolo"; trad. dell'autore.

8
Raymond Murray Schafer, *The Tuning of the World*, McLelland and Stewart/Knopf, Toronto/New York 1977.

9
Alan Licht, "Sound Art: Origins, Development and Ambiguities", in *Organised Sound. An International Journal of Music Technology*, Cambridge University Press, vol. 14, n. 1, aprile 2009, p. 8.

10
Nella già citata recensione a "Soundings", Robert Frank scriveva: "*Soundings* is only the most recent, and most formally museological, in a series of sound art shows of all shapes, sizes, forms and focuses. This series is likely to continue, although future sound art exhibitions here are likely to concentrate on specific aspects of the polymedium (sound-producing object, say, or the visual representation of music)" ["*Soundings* è solo la più recente, e la più formalmente museologica, di una serie di mostre sull'arte sonora che hanno assunto varie configurazioni, dimensioni, forme e tagli tematici. È probabile che questa serie continuerà, ma è verosimile che le future mostre di arte sonora si concentreranno su aspetti specifici dell'opera multimediale (l'oggetto che produce suono, poniamo, o la rappresentazione visiva della musica)"; trad. nostra].

11
"Audio Files. Sound Art Now: An Online Symposium", 2004. Consultabile all'indirizzo artforum.com/symposium/id=6682 (ultimo accesso: 27 marzo 2014).

12
Max Neuhaus, "Sound art?", note di copertina per il CD audio *Volume: Bed of Sound*, P.S. 1 Contemporary Art Center, New York, luglio 2000. Ora consultabile su max-neuhaus.info/bibliography (ultimo accesso: 28 marzo 2014).

13
"Many artists who have been using sound for a long time would rather be called artists than sound artists". (Il corsivo è mio). Douglas Kahn, "The Arts of Sound Art and Music", *The Iowa Review Web*, vol. 8, n. 1, febbraio/marzo 2006, iowareview.uiowa.edu/TIRW/TIRW_Archive/feb06/kahn.html (ultimo accesso: 27 marzo 2014).

STRUMENTI DELLA NON-ESISTENZA (ATTRAVERSO I QUALI CIELO E TERRA CERCANO UNA RICONCILIAZIONE)

David Toop

Nella sezione "Sound (Acoustics)" di *Science & Civilisation in China* di Joseph Needham, Kenneth Robinson descrive uno strano strumento. È una costruzione ad ambiente unico di fatto composta da tre stanze concentriche, come scatole cinesi. Le pareti sono intonacate in modo da coprire ogni fessura e le porte impediscono l'ingresso di correnti d'aria dall'esterno. Poi ancora una tenda di seta arancione protegge la parte centrale dell'ambiente, dove vengono montati su supporti lignei degli accordatori, ciascuno dei quali è disposto in corrispondenza di un punto cardinale. Le estremità superiori dei tubi vengono riempite di cenere di canna. Quando la preparazione è terminata, si osserva ciò che accade agli accordatori in base al calendario. Come scrisse Tshai Yung, esperto di astronomia, acustica e musica del periodo Han: "Quando arriva l'emanazione (*chii*) di un (determinato) mese, le ceneri (dello specifico accordatore) volano via e il tubo si libera."[1]

Ammesso che nel XXI secolo una macchina cosmologica come questa possa rientrare approssimativamente nelle forme di "arte sonora", viene da chiedersi se le disposizioni di ambienti e i loro contenuti esoterici si possano intendere come strumenti musicali. La risposta è no, certo che no. Eppure questo strumento d'osservazione, di "attesa del *chii*", fonda la sua esistenza sull'uso di tubi di bambù per determinare l'accordatura in relazione al mondo fisico, ciclico, proporzionato, e al *chii*, la misteriosa e ineffabile emanazione che nel suo sorgere dalla terra verso gli antenati celesti, e poi dal cielo ancestrale di nuovo sulla terra, può dare vita a tutti i fenomeni. Perciò i tubi di bambù erano strumenti divinatori, "tubi sonori" usati da maestri di musica sciamanica per predire il futuro e influenzare il corso degli aventi. "Ogni uomo ha in corpo il suo *chii*", scrive Robinson. "Il divinatore usa il suo per emettere un'interferenza nel mondo esterno quando soffia nel suo tubo sonoro. Un *chii* allora, 'per una sorta di misteriosa risonanza', si ripercuoterà su un altro *chii*, proprio come uno strumento musicale influisce sull'altro con cui è accordato."[2]

E quindi, nell'antico pensiero cinese il *chii* (*chi*, *ch'i* o *qi*, nelle traslitterazioni più diffuse ai giorni nostri) e il suono si possono intendere come emanazioni strettamente correlate, così come queste azioni divinatorie si possono ricondurre a una più ampia categoria di voci spiritiche o ancestrali che "parlano" attraverso distorsori vocali, flauti e altri strumenti esoterici, e dunque a una forma di comunicazione extraumana che un tempo era ampiamente praticata in diverse regioni della Papua Nuova Guinea, dell'Africa centrale e dell'Africa occidentale. Anche se nel 1930, quando Henry Balfour raccoglieva materiale in Nigeria per il suo *Ritual and Secular Uses of Vibrating Membranes as Voice-Disguisers* (Rituali e usi secolari delle membrane vibranti che alterano la voce), questi "misteri", come li definiva Balfour, "avevano già perso molto del loro antico significato e del loro potere terrorizzante",[3] tale principio rimane costante in tutta la storia universale degli strumenti di produzione del suono.

Le proprietà del suono, fisiche e dimostrabili eppure suadentemente intangibili, possono essere modificate, proiettate, trasformate e vivificate attraverso oggetti che attengono alla prosaica materialità terrena. Per quanto belli possano risultare ai nostri sensi, rimangono pur sempre prosaici. Per quanto si possa attribuir loro un valore feticistico, monetario o una qualità antropomorfa, gli strumenti musicali sono il prodotto di scarto del suono. Rappresentando il suono attraverso la materia in una forma statica da percepire con lo sguardo, costituiscono il tradimento della musica, specialmente nella prospettiva radicale del taoismo, secondo il quale "la musica più grande ha le note più ineffabili." Nell'*Ulisse* (1922) di James Joyce, Leopold Bloom medita sul suono e sui limiti del concetto di strumento musicale: "un filo d'erba, conchiglia di mani, poi soffiare".[4] Uno strumento simile a questo è illustrato nel volume di Jaap Kunst *Hindu-Javanese Musical Instruments*,[5] dov'è descritto come un aerofono composto da uno stelo di riso, uno strumento che non ha memoria; appena è privo d'aria, labbra, mani a coppetta, torna subito afono, torna a essere semplicemente quel che era prima di aver emesso il suono, che Kunst definisce "un suono potente, simile a quello di una cennamella"; torna a essere un infimo ciuffo di materia organica. Nel museo del flusso di coscienza di Bloom ci sono un pettine e un foglio di carta, un corno da caccia, una zampogna, un fischietto da poliziotto, il bestiario della strumentazione – "Ottoni asini raglianti con proboscidi erette... Pianoforte a mezzacoda aperto coccodrillo la musica ha mascelle" –, i pertugi del corpo femminile da suonare come un flauto, la musica da camera di Molly Bloom che urina nel vaso da notte: "È un genere di musica cui ho spesso pensato quando lei. Acustica si chiama. Tintinnabolio. I vasi vuoti fan più rumore. Perché l'acustica, la risonanza cambia secondo che il peso dell'acqua è uguale alla legge di gravità dell'acqua... Perle. Gocce. Pioggia. Plim plim plem plem plom plom. Sssssss."

Nel tentativo di trascrivere i suoni, Joyce usa un linguaggio semplicistico, un registro volgare quanto l'azione stessa, eppure in quel tentativo di ascoltare un'eco riecheggia un'acquaforte di Rembrandt, *Uomo e donna che urinano* del 1631: entrambi contemplano in modo spassionato (anche se in entrambi si capta un accenno di interesse sessuale) una storia non scritta di quella musica abietta, infima, che è forse una delle fonti originarie della musica (sebbene mai menzionata in quelle ricostruzioni congetturali che immaginano i primi esseri umani intenti ad ascoltare gli uccelli e a imitarne il canto).

ORGANO(LOGIA) SENZA CORPO

"Comporre significa: costruire uno strumento", scrisse Helmut Lachenmann.[6] Viceversa: costruire uno strumento significa comporre (se l'atto del comporre s'intende in senso ampio, ovvero radunare materiali).

Uno strumento può mettere in discussione la fissità dei confini tra tutto ciò che è inteso come umano e tutto ciò che non lo è. Molti strumenti musicali si ispirano a forme animali. Sono canali, serbatoi, superfici, camere da cui il suono si proietta nel mondo attraverso varie forme di violenza – frizione, espulsione del fiato, percosse, pizzichi – come fossero animali vuoti le cui grida si possono tramutare nelle forme del desiderio umano. Nei sogni, nelle allucinazioni, nella macchina generatrice delle fantasie di Raymond Roussel, nella simbologia alchemica e nei dipinti, troviamo combinazioni che infrangono i confini tra umano, animale e celeste, proponendo un ripensamento del giardino dell'Eden. Pensiamo alla creatura ibrida raffigurata in un trattato alchemico del XIV secolo, l'*Aurora Consurgens*: il suo strumento è un'aragosta suonata con quello che potrebbe essere un serpente, e dunque produce un'armonia che scaturisce dal caos primordiale della Materia Prima; o al coro di angeli e all'asino che cantano al cielo nella *Natività* di Piero della Francesca; o al *Il riposo durante la Fuga in Egitto* del Caravaggio, in cui un angelo suona il violino, mentre Giuseppe si presta a tenere lo spartito del *Cantico dei Cantici* nella versione musicata da Noël Baulduin, strofinando un piede sull'altro per restare sveglio, mentre la Madonna e il Bambino dormono, il muso dell'asino accostato allo spartito in un'intima vicinanza tra aliti sacri e profani, il suono dell'erba e delle foglie, la frizione dei sassi sotto i piedi, una melodia: "Quanto sei bella, amata mia, quanto sei bella."

Questa musica silenziosa è quanto di più vicino all'ideale di uno strumento che si è smaterializzato per concedere appieno alla musica di esprimere la sua essenza celeste. La musica senza suono è un ideale taoista. Una nota non farebbe altro che sciupare la precedente o la successiva; lo strumento più esaltato, il *ch'in* (che Robert Hans van Gulik traduce con "liuto")[7] può ridurre gli strumenti terreni al silenzio, con il suo suono celeste udibile solo agli esseri soprannaturali: "Ora, quando si suona il liuto [ch'in], il suono degli strumenti di metallo e pietra si estingue, e il soffio in quelli di zucca o bambù cessa. Wang Pao interrompe il suo canto, Ti Ya perde il suo gusto." Nel silenzio che ora risuona, un silenzio più sottile desta gli immortali, una coppia di fenici e le fate celesti.

In cerca di una rigenerazione (o del cielo), la musica disconosce i suoi strumenti, affidando il suo mandato alle forze della natura o agli automatismi, ipotizzando (contrariamente al corpo libero senza organi proposto da Antonin Artaud) una (libera) organologia senza corpo. Bo Diddley, cantante di Rhythm and Blues nato nel Mississippi, creatore di chitarre che per forma e materiali – rettangolari, affusolate, rivestite di pelliccia rosa – celebravano la natura di non-chitarra dell'oggetto chitarra, identificava il catalizzatore di suoni in uno strumento senza corpo: "C'era questo tipo, Sandman, che portava un'asse, un sacchetto di sabbia e una scopa di saggina, sai, di quelle per spazzare. Metteva l'asse a terra, ci buttava sopra la sabbia e poi la spargeva, e poi cantava e spazzava, ed era come una specie di passo strisciato, ma veniva chiamato *sandin'*. Tirava fuori un ritmo, come un ballerino di tiptap, e mica tutti sono capaci di farlo. Quando finiva, ramazzava la sabbia e se la rimetteva nel sacchetto."[8]

Questa modulazione tra il terreno (letteralmente) e l'immateriale rinvia ai mandala di sabbia dei lama tibetani, che dopo giorni e giorni di elaborazione vengono ritualmente distrutti e affidati alle acque, così come rinvia alle *dōtaku*, le campane di bronzo giapponesi esposte al British Museum. Prodotte nell'era Yayoi tra il 500 a.C. e il 250 d.C., venivano sepolte ai margini dei terreni agricoli. Decorate con immagini di animali ritenuti predatori degli insetti che avrebbero distrutto i raccolti e poi sottratte ritualmente alla vista e alla possibilità di attivazione, queste campane suggeriscono un profondo rapporto tra le invisibili e dunque misteriose energie del suono, gli inosservabili processi della natura e l'imponderabile operato degli dèi. L'oggetto sparisce; la sua presenza e il suo suono effettivo non sono necessari affinché si attui la magia.

Questi strumenti del suono e queste azioni a distanza si collocano in equilibrio precario ai confini della concezione contemporanea dello strumento musicale. Si avvicinano a quella zona extraumana che va oltre la cultura, all'involontario e all'incontrollabile, alla meteorologia, alla bioacustica, alle cavernose riverberazioni del suono e agli inesplicabili rumori sottomarini e sotterranei, alle concezioni sovrannaturali dei fenomeni acustici. Sembrano sorgere da profondità inaccessibili, spesso trattati come esseri viventi, emblematici di una vita oltremondana e dotati di un linguaggio tutto loro. Gli enormi garamut documentati da Ragnar Johnson nel 1976 in Papua Nuova Guinea erano decorati con figure ancestrali dai lunghi nasi, scolpite come protuberanze alle estremità della fessura dello strumento. Ricavati da un unico tronco, questi gong in legno erano usati sia per inviare messaggi a lunga distanza, sia per accompagnare i flauti sacri i cui suoni erano ritenuti le voci degli spiriti. "In passato", scriveva Johnson, "quando un garamut era stato scolpito e portato al villaggio, gli abitanti si vestivano a festa e tenevano una cerimonia, in cui decoravano il garamut e gli tributavano offerte di sagù e latte di cocco."[9]

I gong in legno intagliato avevano un ruolo centrale nella comunità Bamum in Camerun,[10] in cui erano utilizzati per inviare segnali dalla sede della corte reale in momenti di crisi. Quando i colonizzatori tedeschi raggiunsero il regno Bamum nel 1910, scoprirono otto di questi spettacolari strumenti, ciascuno intagliato a partire da una fessura centrale, lasciati a decomporsi sul terreno del palazzo come giganti caduti da un passato mitico. Due funzionari dell'amministrazione coloniale francese che erano di stanza nella regione di Bamilike notarono che questi gong venivano intagliati in occasione dell'insediamento di un sovrano; quando il sovrano moriva, venivano abbandonati nella piazza del mercato e lasciati a marcire, subendo lo stesso processo di decomposizione del corpo del re. Un gong di Bamilike fotografato nel 1911 era scolpito a forma di bufalo, più grande del vero; sul corpo erano scolpite le figure di un camaleonte, una lucertola, un rospo ed era riprodotto il doppio gong in ferro usato dalle società segrete. Consacrato con un'offerta sacrificale, un gong come questo poteva essere usato dalle società segrete interamente maschili per chiamare a raccolta i membri in occasione di un funerale. Il suono di questi gong strappati alla terra dava voce ai defunti e all'oltremondo sotterraneo, ed è con questi ignoti strumenti che dalle profondità ter restri risuonano le lingue ctonie e i confusi *rambling undergroans* (gemiti sotterranei) di *Finnegans Wake*.

Mentre la musica della diaspora europea si sottraeva alle leggi dell'armonia elegante per diventare un'opera di ascolto, i sacrifici assumevano un ruolo centrale nelle cerimonie musicali del XX secolo. Immergersi in un abisso ben più oscuro e profondo del formalismo della metrica equivaleva in un certo senso a confrontarsi con la morte. Nel 1966 l'artista Raphael Montañez Ortiz ha partecipato al "Destruction in Art Symposium" di Londra, dove ha distrutto un pianoforte. Certo non è stato il primo né sarà l'ultimo artista ad assassinare quest'oggetto così carico di valori simbolici, ma il suo gesto cadeva in un momento in cui "molti andavano in cerca di un quadro di riferimento estetico extraeuropeo, e molti nativi e africani americani cercavano un'identità esterna a quelle esistenti di matrice coloniale". Per Ortiz, si trattava di colmare il divario tra la sua formazione cattolica e i suoi antenati Yaqui, "che ai processi sacrificali attribuivano un valore sciamanico di riconciliazione".[11]

I temi dell'occultamento, del sacrificio e dell'organo senza corpo ricorrono con insistenza nell'arco del XX e del XXI secolo. Nel 1964 Hugh Davies, improvvisatore, compositore, inventore e studioso, ha suonato nell'ensemble che ha registrato *Mikrophonie I* di Karlheinz Stockhausen, in cui l'attivazione della superficie di un grande tam tam con una varietà di utensili (molti dei quali di uso domestico) è amplificata dai microfoni.[12] A loro volta, questi suoni amplificati sono trasformati da altri operatori che usano filtri e diffusori. Stockhausen aveva messo a punto la tecnica di attivazione e di amplificazione del tam tam insieme al suo tecnico, Jaap Spek. In questo caso lo "strumento" non è posizionato nello spazio. Il suo corpo non si trova da nessuna parte nell'ecosistema di rapporti che intercorrono tra la violenza delle azioni (sfregamenti, scricchiolii e brontolii sono alcuni dei termini cui ricorre Stockhausen per descrivere i suoni prodotti), gli umili oggetti usati per l'attivazione, i microfoni, la rielaborazione elettronica e la diffusione del suono e la superficie piatta di un tam tam con le sue segrete potenzialità. Nella prefazione alla partitura, Stockhausen descriveva il processo come una sorta di auscultazione; come un medico che indaga con uno stetoscopio, lui era capace di rivelare al mondo quei suoni nascosti all'interno del metallo.

Tornato a Londra con il bagaglio di conoscenze acquisite da Stockhausen (di cui era assistente) e da Jaap Spek, Hugh Davies ha cominciato a sviluppare una pratica basata sull'amplificazione di suoni minimi o nascosti usando microfoni da contatto. I suoni provenivano da umili oggetti recuperati in cucina, nel credenzino del bagno, in garage o nel capanno degli attrezzi: molle, affettauova, un guanto di gomma, una cartolina in 3D, un pelaverdure e così via. Questi oggetti erano raccolti in assemblaggi chiamati *Shozyg*, postazioni di lavoro multifunzionali, tattili e peculiari che richiedevano intimità nel tocco e sensibilità nell'ascolto; perseguendo una finalità al contempo scultorea e microa custica, esprimevano appieno le loro potenzialità segrete nell'improvvisazione (dal 1968 al 1971 Davies ha suonato nella Music Improvisation Company con Evan Parker, Derek Bailey, Jamie Muir, Christine Jeffrey, poi con molti altri improvvisatori fino alla morte, avvenuta nel 2005).

Negli anni '70 Davies faceva parte di un gruppo non organizzato di inventori che riflettevano su questi interrogativi: che cos'è uno strumento musicale? quali sono i suoi confini? cosa potrebbe diventare la musica che produce? Anna (ora Annea) Lockwood, collega di Davies, ha pubblicato e diffuso 15 copie del suo *Sound-Hat* attraverso la "rivista-disco" *OU*, diretta da Henri Chopin.[13] Le istruzioni scritte da Chopin per questo semplice cappello da cui pendevano fili di bambù e conchiglie invitavano a indossarlo e a voltare la testa "lentamente da destra a sinistra, da nord a sud e da est a ovest", come in un gesto divinatorio per prevedere il futuro attraverso il suono e la sua corrispondenza ai punti cardinali. "Pensate che Anna, l'artefice del cappello", scriveva Chopin, "intende abolire il culto del compositore chiedendo a voi – a ciascuno di voi – di comporre la propria musica."

Se queste proposte delineavano in qualche misura una nuova visione degli strumenti musicali (quanto meno gli strumenti della diaspora europea), questa visione non era più basata sulla musica o sulla composizione. Persino l'azione umana veniva sacrificata. Artisti come Max Eastley costruivano strumenti azionati da agenti atmosferici, dal vento o dall'acqua, o da motori che funzionavano in base alla propria complessità. L'attivazione degli strumenti non era più generata dalla violenza umana ma dall'imprevedibile intervento della natura o dal fugace stridore di un cavo metallico. Lo strumento diventava incorporeo, disincarnato, sede di un'emanazione, non dissimile dalla misteriosa stanza cinese in cui erano disposti gli accordatori pieni di cenere. Essere dentro agli "strumenti" di Haroon Mirza [400, 434], Tetsuya Umeda,[14] Samson Young e Rie Nakajima,[15] – per menzionare quattro artisti che hanno accolto la proposta dell'organo(logia) senza corpo – significa sentirsi una parte dinamica di campi sonori dotati di una volontà autonoma che opera in territori dai confini instabili. Come porgendo uno specchio alla vita del XXI secolo, i loro materiali sono umili e (de)composti, un caos primordiale di disparità; da modeste, fragili esistenze, si librano domande sul cielo e sulla terra, si librano altri suoni.

1
Joseph Needham (con Kenneth Robinson), *Science & Civilisation in China*, vol. IV, n. 1, Cambridge University Press, Cambridge 1962, pp. 187-88.

2
Ibid., p. 138.

3
Henry Balfour, "Ritual and Secular Uses of Vibrating Membranes as Voice Disguisers", in *Journal of the Royal Anthropological Institute of Great Britain and Ireland*, vol. 78, n. 1/2, London 1948, p. 47.

4
Edizione originale *Ulysses*, Sylvia Beach, Paris 1922; *Ulisse*, trad. di Giulio de Angelis, Mondadori, Milano 2000. I passaggi sono tratti dell'episodio delle Sirene.

5
Jaap Kunst, *Hindu-Javanese Musical Instruments*, Nijhoff, The Hague 1968, p. 29 (e tavola illustrata, tratta da Curt Sachs, fig. 103).

6
Helmut Lachenmann, "Über das Komponieren, 1986", in Joseph Häusler (a cura di) *Musik als existentielle Erfahrung*, Breitkopf & Härtel, Wiesbaden 1996, p. 73.

7
Robert Hans van Gulik, *Hsi K'ang and his Poetical Essay on the Lute*, Sophia University, Tokio 1968, p. 118.

8
George R. White, *Bo Diddley: Living Legend*, Castle Communications, Chessington 1995, p. 27.

9
Ragnar Johnson, *Sacred Flute Music from New Guinea, Madang*, 33 giri, Quartz Publications, London 1977.

10
Christraud M. Geary, "Slit Gongs in the Cameroon Grassfields: Sights and Sounds of Beauty and Power", in Marie-Thérèse Brincard, Sounding Forms: African Musical Instruments, American Federation of Arts, New York 1989, pp. 63-71.

11
Raphael Montañez Ortiz, "Art Attack At Tate Britain", *The Guardian*, 28 settembre 2013, gu.com/p/3j4tb

12
Hugh Davies, youtube.com/watch?v=wPT9A0I-sGgs

13
Annea Lockwood, "Sound-Hat", *OU, Revue de Poésie Evolutive*, n. 36/37, 1970.

14
Tetsuya Umeda, youtube.com/watch?v=TYpta-Kc-LMo

15
Rie Nakajima, rienakajima.com/_work/_installations.html/carpet.html

TIPOLOGIE E PEZZI UNICI. LA FUNZIONE SOCIALE ED ESTETICA DEGLI STRUMENTI MUSICALI

Bart Hopkin

In inglese è d'uso una convenzione linguistica alquanto originale: in frasi come "Laura suona il violino" si parla di un oggetto chiamato "il violino". La stranezza risiede nell'uso dell'articolo determinativo "il". A quale violino ci si riferisce? Non sarebbe più corretto dire che "suona *un* violino"?[1]

Tale consuetudine riflette la tendenza a suddividere gli strumenti musicali in tipologie. Ci sono il violino, il fagotto e, ancora, il waterphone, la tromba marina, lo snaggleodeon (d'accordo, lo ammetto, quest'ultimo strumento è frutto della mia immaginazione) e molti altri ancora. Ciascuno di essi rappresenta una categoria tassonomica definita da caratteristiche ben determinate; il violino, per esempio, fa parte del gruppo degli strumenti musicali a quattro corde, suonati di solito con un archetto, tutti di una certa forma e dimensione… e così via: fondamentale in questa categorizzazione è l'esistenza di una cultura locale del violino. Ciò comprende un ampio repertorio di opere composte nella più classica tradizione europea, cui si aggiungono consuetudini popolari, etniche o folcloriche, e uno stuolo di violinisti ben preparati. E volendo essere ancora più precisi, include un insieme di idee generiche su come dovrebbe essere il suono del violino e per quale tipo di musica sia più indicato. Un violino specifico, per esempio quello suonato dalla nostra amica sopra, è un singolo strumento rappresentativo della categoria. Mentre "quel" particolare violino ha inevitabilmente caratteristiche specifiche, quando se ne parla in generale esso deve rientrare nella descrizione standard tanto da poter essere considerato tale.

Il motivo per cui gli strumenti musicali sono concepiti come "tipologie" – una caratteristica, questa, che si riflette nel linguaggio – è che in genere tale opinione riproduce fedelmente una situazione già in essere. Gran parte degli strumenti musicali ricrea tipologie di strumenti preesistenti, con poche variazioni individuali di scarsa importanza. Ovviamente non deve essere per forza così. Se si decide di costruire uno strumento musicale, si è liberi di fare ciò che si vuole, compreso un pezzo unico non riconducibile ad alcuna classificazione e che non rappresenti nulla che si sia già visto. Ma bisogna riconoscere che, se gli strumenti musicali sono raggruppati in categorie, è per una buona ragione.

Quanto appena detto – il gruppo di artisti di talento, il repertorio esistente e (il che è più importante di quanto non possa sembrare) un'opinione condivisa circa le possibilità sonore dello strumento in questione – si è dimostrato d'immenso valore nel mondo della creazione musicale. Quest'accumulazione di capitale culturale nei confronti di determinate tipologie di strumenti ha creato infatti le giuste condizioni per la composizione di molta della musica più bella che il mondo abbia mai ascoltato.

Uno dei fattori in gioco nella tipizzazione degli strumenti musicali potrebbe risiedere nella natura del progresso umano, come si evince dal loro sviluppo nel corso della storia. Essi evidenziano infatti uno schema evolutivo sorprendentemente parallelo a quello dell'evoluzione biologica. Tanto nell'ambito degli strumenti musicali quanto in quello della biologia, infatti, con il trascorrere del tempo si sviluppano gradualmente tipi e organismi diversi, in una trasformazione simile alla selezione naturale che, in un'incessante opera di perfezionamento, assicura la sopravvivenza di ciò che dimostra di possedere le caratteristiche più adatte. Certo è possibile, come si è già detto, che un giorno dalla creatività di alcuni singoli individui nasca una nuova stirpe di strumenti, ben diversa da quella che l'ha preceduta; in alcuni casi ciò accade davvero, sebbene non sia così frequente. Rientra però maggiormente nella norma uno sviluppo evoluzionistico in cui si ampliano e si perfezionano idee ereditate da chi ci ha preceduto. Ecco perché viviamo in un mondo con un numero apparentemente finito di "specie" di strumenti musicali dallo sviluppo sostanzialmente armonico.

Non meno interessante è l'analisi del ruolo che l'individuo ha in tale processo. In tutta la storia dell'uomo e nella maggior parte delle culture, lo sviluppo materiale, tanto nell'ambito degli strumenti musicali quanto in altri campi, non è stato solo graduale ed evolutivo, bensì anche collettivo e anonimo. L'idea che un individuo possa apportare innovazione e rivendicare un riconoscimento nel corso della storia è stata celebrata in misura minore di quanto non accada oggi nel mondo occidentale. È tuttavia necessaria una precisazione: in un certo senso, la visione di un passato tradizionalista e collettivista potrebbe riflettere una prospettiva superficiale e potenzialmente imprecisa, propria dell'uomo contemporaneo. Non sapendo a chi attribuire talune innovazioni specifiche, è facile sposare tesi che si richiamano a concetti vaghi come "tradizione" e "cultura", che si stemperano in una sorta di primitivismo sentimentale. Tutto ciò potrebbe risultare irritante per chi, in un determinato momento di un passato non codificato, ha effettivamente apportato un'innovazione di valore e gradirebbe che questo venisse almeno in parte riconosciuto, ma sfortunatamente non è più tra noi per poterlo esprimere. Detto questo, la tensione tra individuale e collettivo è un dato di fatto nella vita umana e, a quanto pare, sia nel campo della liuteria sia in altri ambiti, la celebrazione dell'inventore o dell'artista negli ultimi secoli è stata maggiore in Occidente di quanto non lo sia stata in altre epoche e luoghi.

Pertanto, se oggi qualcuno mi domandasse chi, tra i costruttori di strumenti musicali, ammiro di più, invece di scegliere un nome della Cremona secentesca o uno tra i più creativi artisti della nostra epoca, forse getterei lo sguardo sul gruppo di chi non ha mai ottenuto un riconoscimento e risponderei "Chi ha ideato la tastiera della m'bira! Quella persona era un vero genio!". O, se me lo domandassero domani, potrei dire "Chiunque abbia pensato di inserire l'anima all'interno del violino sotto una corda ben precisa! Che idea intelligente!" Naturalmente non ho idea dei loro nomi e queste tecnologie potrebbero essere il risultato non tanto di una ricerca individuale, quanto di processi collettivi ed evolutivi.

In "Art or Sound" si possono ammirare opere di individui di particolare creatività, i quali hanno in qualche modo superato l'idea che gli strumenti musicali e i dispositivi sonori in genere debbano essere realizzati in base a modelli preesistenti. Come si può vedere, non appena ci si allontana da un'esigenza di standardizzazione, ecco che si aprono orizzonti interessanti. Passando all'osservazione dei processi chiave della costruzione sonora, davanti ai nostri occhi si dischiudono ampie possibilità esplorative, non solo nell'ambito del suono, ma anche in relazione ai modi in cui tale "lavoro sonoro" fa nascere una forma visiva, alle leggi fisiche che danno origine a entrambi, e allo studio di come l'"arte sonora" sia in grado di riflettere prospettive concettuali più ampie.

Ne sono un primo esempio i musicisti e gli artisti che a lungo sono stati affascinati dal concetto della sinestesia, ovvero l'idea che le percezioni di uno dei nostri sensi possano essere trasmesse a un altro senso. La testimonianza più interessante riguarda la comunicazione trasversale tra musica e arte visiva. Possono i toni, i timbri e i modelli musicali tradursi in schemi di forme e colori?

Tra i primi esperimenti sinestestici troviamo il "color piano" (1911) [136] di Alexander Scriabin, la cui tastiera era collegata a lampadine elettriche, ideato come parte integrante della partitura per il poema sinfonico *Prometheus*, e costruito da Alexander Moser per la prima rappresentazione, nel 1910. Il dispositivo chiamato *Farblichtmusiken* di Mack & Schwerdtfeger è un altro dei primissimi sforzi di traduzione del suono musicale in schemi di luce. Più recentemente, la *Sculptural Representation of Bars 52-55 of E-Flat Minor Fugue by J.S. Bach, Proposal for a Bach Monument* (1928) [178] di Henrik Neugeboren è un tentativo meno letterale e più poetico di ottenere un risultato analogo.

Spostandosi nel regno dell'olfatto, *The Smell Organ* (l'organo olfattivo), descritto sin nei minimi dettagli ma a quanto pare mai realizzato dal chimico francese Septimus Piesse negli anni '20 del Novecento, aveva una tastiera studiata per diffondere aromi e profumi, proprio come la musica riempie l'aria di una stanza. Il progetto comprendeva una consolle d'organo di impressionanti dimensioni e partiture in notazione musicale. Uno dei disegni giunti sino a noi raffigura un virtuoso della tastiera di altera bellezza, intento a suonare di fronte a un pubblico rapito dalla sua musica, in una grande sala da concerto.

Elemento centrale di molti sforzi trasversali è il problema della schematizzazione: dato un evento o schema musicale, come decidere in quale tipo di schema visivo (od olfattivo) tradurre (o schematizzare) la musica? Tra gli addetti ai lavori vi è chi ha postulato modelli specifici e presumibilmente universali, e sembra suggerire per esempio che la nota Do è in qualche modo inerente al colore blu, mentre il Sol al rosso. Altri hanno ragionato in termini di associazioni personali o soggettive tra suoni musicali, colori e forme visibili. Su tutt'altro piano, alcuni artisti hanno trovato il modo di tradurre i fenomeni visivi in modelli sonori, per esempio trasformando lo schema di uno skyline urbano o di una fila di uccelli appollaiati lungo un cavo telefonico in composizioni musicali. Ma sono stati esplorati anche altri tipi di mappatura, per esempio tracciando uno schema dello spettro luminoso dei composti chimici e correlandolo ai timbri sonori corrispondenti o trasponendo i segnali radio provenienti dallo spazio (sotto forma di radiazioni elettromagnetiche a frequenze che l'udito umano non riesce a percepire) in segnali audio.

La schematizzazione è inoltre al centro di un'area di studio oggi di particolare importanza: esiste la possibilità di tradurre la gestualità del corpo umano in modelli sonori? Per comprendere il profondo significato di una simile ricerca, torniamo per un attimo con la memoria agli inizi della commercializzazione degli strumenti musicali elettronici (si pensi a organi e sintetizzatori degli anni '60 e delle decadi successive; il marchio Casio è di solito tra i primi a venire in mente). A quell'epoca la cosa più naturale parve essere quella di collegare qualunque innovazione nella produzione di suoni elettronici a una tastiera da pianoforte. Negli anni successivi si è fatta sempre più strada l'idea che potessero esistere altre modalità potenzialmente interessanti per l'uomo di interagire con tali dispositivi. Man mano che la tecnologia del suono digitale si andava arricchendo in potenzialità e flessibilità operative, è stato possibile affrontare la questione dell'interfaccia sonora, domandandosi in che modo si possano schematizzare i movimenti umani in una produzione musicale. Associando i tasti bianchi e neri a determinate note, la tastiera diventa l'interfaccia di una schematizzazione, ma le possibilità sono infinite. Dispositivi per controllare i passi di danza? Piattaforme interattive per più suonatori? Sensori per leggere e digitare altri tipi di movimento in uno spazio tridimensionale? Il solo limite posto all'immaginazione di chi si occupa di progettazione è quello dettato dalla necessità di definire in che modo interagire con lo strumento per ottenere un prodotto udibile. Don Buchla, un pioniere del settore, ha progettato dispositivi in grado non solo di riconoscere un'ampia e fantasiosa varietà di movimenti, bensì di offrire caratteristiche di estrema flessibilità e controllo nella schematizzazione dei gesti in suoni, manipolazioni sonore e interazioni.

Ma superiamo i costrutti di artisti e ingegneri, per giungere alla fonte dove tutto ha origine: la natura e le sue espressioni. In anni recenti, non pochi artisti sonori sono rimasti affascinati dalla fisica e dalla fisicità del suono, e hanno cercato, dopo averne compreso le caratteristiche, di renderle visibili, mettendone in luce la bellezza o l'interessante singolarità. Per descrivere il tentativo di rendere visivamente riconoscibile la vibrazione sonora facendo ricorso a mezzi fisici e acustici, è stato coniato un termine – cimatica – suggerito dal fisico e scienziato svizzero Hans Jenny (1904-72) il quale, rifacendosi al fisico e musicista tedesco Ernst Chladni (1756-1827), creò strutture visive affascinanti cospargendo di sabbia e altre sostanze granulari alcune lastre e membrane, e sottoponendole a frequenze diverse. Ispirandosi a sua volta alla teoria cimatica, il fotografo tedesco Alexander Lauterwasser ha realizzato impressionanti opere visive utilizzando dei recipienti pieni d'acqua che, agitata, assumeva forme di artistica bellezza per via delle vibrazioni sonore.

I pirofoni [114], in cui i suoni sono prodotti dal calore del fuoco in un tubo, rappresentano un primo tentativo di suono che si rende visibile e di una fisica che si rende udibile. I primi pirofoni, ma anche alcuni di quelli contemporanei, furono progettati per essere utilizzati quasi come degli organi e produrre musica convenzionale. Sperimentazioni più recenti, come quella dell'artista francese Michel Moglia, hanno assunto la forma di selvagge e possenti dimostrazioni delle forze pirotecniche in gioco, tanto a livello visivo quanto sonoro.

Anche Alvin Lucier si è addentrato, con forza visionaria, nella fisicità del suono. Nella sua *Music for Pure Waves*, pendoli leggeri rimbalzano su membrane di tamburo mossi da onde sonore generate elettronicamente, creando strutture visibili in movimento, in risposta ai fattori acustici e meccanici in gioco. Lo stesso vale per *Modell zur Visualisierung* (2001) di Carsten Nicolai, dove le strutture sonore assumono forme contorte in un fascio elettronico visibile.

Tra i molti altri artisti contemporanei che hanno esplorato con le loro opere la fisica a esse sottesa, possiamo citare Ellen Fullman e i suoi strumenti a corde lunghe, un'installazione composta da decine di metri di fili che non vibrano in senso trasversale come avviene nel caso delle corde musicali, bensì in senso longitudinale. Ciò comporta, a sua volta, intriganti implicazioni fisiche che si traducono non solo in una musica esotica e piacevole, ma anche in una sorta di coreografia naturale: la principale gestualità sonora consiste nel camminare lungo le corde suonandole con le mani ricoperte di pece greca, mentre la scelta dei percorsi assume la valenza di una partitura musicale, rendendo la composizione visibile. Allo stesso modo, diventa possibile osservare anche la fisica sottesa alla teoria della scala musicale, evidente nella lunghezza delle corde, nel posizionamento di pesi sonori lungo di esse e nelle distanze percorse.

Si fanno strada, inoltre, inedite riflessioni sulla struttura fisica degli strumenti, che si uniscono a quelle sulle strutture sociali e individuali della produzione musicale. Invece di considerare gli strumenti principalmente come mezzi indispensabili a un concerto, gli artisti esplorano nozioni alternative su come il suono possa avere una funzione individuale o sociale. Ricordo un rapido commento di Sharon Rowell, ideatrice di un flauto globulare a tripla camera, ricercato tanto nell'aspetto quanto nella produzione sonora. Non fu un'osservazione rivoluzionaria, ma mi rimase impressa nella mente: disse che i suoi strumenti erano concepiti innanzitutto come oggetti personali, privati, adatti a una meditazione solitaria e rivelò di suonarli da sola tra le dune di una spiaggia vicino al luogo in cui viveva. Sebbene tali strumenti siano spesso usati per arricchire i concerti di sontuosi effetti di movimento, l'immagine di Sharon sola tra le dune è di quelle che non si dimenticano.

D'altro canto, gli strumenti sonori possono anche fungere da mediatori nell'interazione tra coppie o gruppi di persone. Se ne è già parlato brevemente nell'ambito dei sistemi informatici che controllano input provenienti da soggetti diversi, generando nuovo materiale musicale attraverso questa interconnessione. Le configurazioni sonore in cui varie persone interagiscono con lunghe corde multiple e congiunte, che piegate e tirate producono variazioni nella tensione e nei prodotti sonori, sono un ulteriore esempio, più fisico, di tale interazione. Alcuni strumenti *multiplayer* assumono la forma di giochi, come il tavolo da ping pong del gruppo francese Les Phones la cui superficie è formata da barre metalliche, posizionate l'una accanto all'altra, che al rimbalzare casuale della pallina producono suoni diversi. Per un'interazione più intima, basta pensare a un unico strumento a vento suonato da due musicisti, in cui il suono varia al variare della pressione del loro respiro, che all'interno dello strumento si mescola e si confonde. Come accade in *Due Capi* (1994) di Oliver DiCicco, dove un terzo musicista muove le valvole.

All'estremità opposta dello spettro sociale, si può pensare a strumenti e dispositivi sonori suonati da... nessuno o, se non altro, non da mani umane. A questo proposito vanno citati gli automi, un'idea mai accantonata che ha vissuto un revival grazie ai recenti progressi nella robotica computerizzata. Tra le molte opere che si avvalgono di questa tecnologia musicale, si possono menzionare gli impressionanti automi di Compressor Head, una band rock robotica, o del gruppo EMMI (Expressive Machines Musical Instruments, costituita da tre compositori-costruttori), la cui produzione spazia in stili musicali più astratti. All'interesse per la robotica computerizzata fa da contrappunto quello per la meccanica vecchio stile, esemplificato dalle costruzioni a bassa tecnologia del francese Pierre Bastien o dell'australiano Ernie Althoff.

La musica è di per sé un potente creatore di messaggi, in quanto forgia naturalmente l'opera di chi costruisce strumenti. Ma se si impara a non correlare la costruzione del suono a una categoria specifica, ecco che si aprono nuove prospettive che permettono di riflettere sul suono puro indipendentemente dalle considerazioni musicali. Un importante concetto della moderna arte sonora suggerisce che tutti i suoni possono essere ascoltati come musica: basta imparare ad ascoltare con le "orecchie ben aperte". Per quanto mi riguarda, il modo in cui le persone rispondono alla musica o ad altri tipi di suono può essere molto diverso, e tentare di fondere questi due aspetti è essenzialmente improduttivo. Ciò lascia ampio spazio a un apprezzamento acustico del suono che prescinde da parametri musicali, tant'è che molti artisti contemporanei ritengono che l'idea di perseguire un suono fine a se stesso sia maggiormente liberatoria. Per la purissima bellezza astratta visiva e sonora, tra i tanti nomi rappresentativi che si potrebbero citare, vi è il designer italo-americano Harry Bertoia (1915-78), con le sue sculture *Sonambient*: dei boschetti – perché ne realizzò più d'uno – di tondini metallici montati verticalmente su basi di legno o altro materiale, in qualche caso dotati di pesi sulla parte superiore. L'oscillazione prodotta da un tocco anche lieve li fa scontrare tra loro, creando un'immagine luccicante e lievemente ondeggiante, accompagnata da suoni sostenuti, movimentati e fruscianti, come se i sospiri della brezza potessero essere di metallo.

All'interno di una mostra che pone particolare enfasi sulle caratteristiche visive, gli strumenti musicali e gli altri dispositivi sonori mostrano potenzialità leggermente diverse rispetto ad altri generi di arte creativa. Basti pensare alla questione dell'astratto in relazione all'aspetto rappresentativo/relazionale/programmatico. Apparentemente, gli strumenti musicali non sono astratti, nel senso che guardandoli non li si può considerare unicamente come forme astratte. In qualche modo, coinvolgono chi li osserva raccontando una storia visiva, in particolare i pezzi unici che non sono riconducibili ad alcuna categoria. Ma questa è un'altra storia, che ha a che vedere con la loro funzione, il loro scopo, ciò che essi possono fare e il modo in cui possono farlo, o ancora perché hanno una determinata forma e come tali caratteristiche specifiche derivino dall'obiettivo prefissato. In un certo qual modo si tratta di un processo liberatorio, in quanto il messaggio, pur essendo accattivante, non è gravato da implicazioni sociali, politiche o estetiche; parla solo di come le cose fanno ciò che devono fare.

Allo stesso tempo, gli strumenti musicali sono diversi da altri dispositivi, per esempio da quelli in grado di produrre belle immagini, e gli strumenti qui esposti in quanto oggetti artistici sono essi stessi destinati a creare arte (seppur di tipo auditivo). Le decisioni progettuali si basano sia sul prodotto finale sia sul modo in cui è visto e percepito l'oggetto in sé. Si tratta di un processo profondamente umano, poiché il suo intento principale è raggiungere un'intima interazione fisica tra lo strumento e colui che lo suonerà.

E per quanto possa sembrare una verità lapalissiana, negli strumenti musicali la forma è conseguente alla funzione. In questo caso, le forme si rivelano belle o, se di bellezza non convenzionale, di certo affascinanti per gli usi insoliti e suggestivi.

1
Tale convenzione – ovvero l'utilizzo dell'articolo determinativo in una frase semplice, come "Laura suona il violino" – è comune a molte lingue neolatine, tra cui anche l'italiano. È interessante notare come invece non lo sia nella maggior parte delle lingue germaniche (in tedesco, per esempio, si dice "Sie Geige spielt").

OGGETTI ETNOGRAFICI ANIMATI. STRUMENTI MUSICALI DAL QUOTIDIANO ALLO STRAORDINARIO

Noel Lobley

"L'arte dell'oggetto etnografico", secondo Barbara Kirshenblatt-Gimblett, "è un'arte di asportazione, distacco, estrazione".[1] Quando gli strumenti musicali vengono selezionati, estrapolati dai territori e dalle culture da cui e per cui producono suoni e musica, quali pratiche e concezioni si distaccano e restano sepolte? E quali invece si manifestano e si rivelano? In quale modo l'esperienza di un oggetto e delle sue risonanze locali ci aiuta a identificare l'arte della produzione di suoni?

In "Art or Sound" vedo e immagino strumenti ibridi, gamelan in miniatura e arpofoni; pistole, schermi televisivi e bombolette d'ossigeno inseriti negli strumenti; scatole sonore idiofone, solchi muti e chitarre che si sciolgono. Quando pensiamo a una mostra che esplora i rapporti tra strumenti musicali, suoni e oggetti d'arte contemporanea, può essere opportuno guardare oltre le collezioni, le gallerie e le esposizioni di strumenti musicali prodotti in Occidente negli ultimi tre secoli per elaborare una concezione più ampia di ciò che un oggetto sonoro può o non può essere. L'idea stessa di "strumento musicale", se considerata in una prospettiva transculturale, ci chiede di localizzare il concetto di arte e artisticità musicale in luoghi diversi, e talvolta nascosti, dentro e oltre gli oggetti materiali.

LA COLLEZIONE E L'ESPOSIZIONE DI STRUMENTI MUSICALI

Nelle raccolte museali di strumenti etnografici, di solito gli strumenti spaziano dal quotidiano allo straordinario. Per esempio, il Pitt Rivers Museum dell'Università di Oxford conserva una gamma variegata di strumenti che vanno dai kazoo per bambini, usati come semplici giocattoli, ai classici qin cinesi (simili a cetre), che con il loro corpo e le loro corde stimolano la contemplazione filosofica e cosmologica. Chi assembla una raccolta etnografica è incoraggiato ad adottare un approccio che vada al di là della propria concezione estetica della musica e dell'arte, e a includere tra gli strumenti musicali "qualunque oggetto, naturale o artificiale, a prescindere da quanto sia semplice, che sia impiegato allo scopo di produrre suono (sia esso 'musicale' in senso estetico o no)". Henry Balfour, erudito poliedrico e primo curatore del Pitt Rivers Museum di Oxford, diede nel 1929 questo consiglio agli antropologi impegnati a raccogliere strumenti musicali; consiglio che oggi sembra preveggente se consideriamo che, nella riflessione contemporanea sugli oggetti musicali, sonici e artistici, le distinzioni tra suono e musica si dissolvono e si riarticolano continuamente. Balfour era interessato a tutto ciò che riguardava l'etnografia, alle arti decorative, agli utensili e alle armi non meno che alla musica. Come molti curatori, riconosceva l'importanza di partire dall'oggetto per valutarne e studiarne le proprietà materiali e visive, e poi indagarne le più ampie risonanze e valenze culturali.

Che cosa impariamo dall'esposizione di oggetti sonici in teche di vetro o negli spazi museali, specialmente quando molti strumenti musicali sono personificati e concepiti come essenze vitali dell'atto creativo in sé? Gli oggetti etnografici, ci ricorda Kirshenblatt-Gimblett, "diventano etnografici in virtù di un processo di definizione, segmentazione, distacco e asportazione attuato dagli etnografi" con il risultato che "come una rovina, il frammento etnografico è informato a una poetica del distacco."[2] Il processo in cui si chiede ai visitatori di osservare attentamente gli oggetti musicali quando il loro valore spesso trascende l'aspetto visivo è parte di questo distacco. Ed è proprio quel processo che permette a questi oggetti di entrare e attraversare gli ambiti della curiosità, del campionario e dell'oggetto d'arte, "anche se non necessariamente in questo ordine".[3] Il processo di assemblaggio, esposizione e studio di oggetti che esemplificano quel repertorio di materiali che è la raccolta etnografica di strumenti musicali e delle registrazioni dei suoni che producono – talvolta una massa indifferenziata di oggetti e documenti – indaga queste intersezioni tra strumenti musicali, oggetti sonori e oggetti d'arte.

Gli strumenti musicali e gli oggetti sonori africani continuano a essere ampiamente rappresentati nei musei etnografici, nelle esposizioni dedicate alla musica e nelle mostre che più estesamente riguardano l'arte africana tradizionale e contemporanea. Le modalità di presentazione di questi oggetti sono estremamente diversificate e sfaccettate. Per esempio, la mostra itinerante "Dogon Now: Masks in Motion" organizzata dal Museum for African Art di New York indaga le rappresentazioni musicali e in maschera in un allestimento teatrale che riunisce diversi manufatti dei Dogon: sculture antiche e recenti, dipinti e disegni, tessuti e abiti, maschere e costumi;[4] l'angolano Victor Gama, compositore e costruttore di strumenti musicali, esplora le potenzialità di strumenti interattivi che fondono i principi musicali delle tradizioni africane e orientali con il minimalismo e la musica contemporanea più avanzata;[5] mentre l'artista beninese Calixte Dakpogan ha inserito frantumi di dischi africani a 45 giri tra i rottami di un'automobile per creare la figura antropomorfa che ha intitolato *Perroquet (Parrot)* (pappagallo, 2005).[6] Tutti questi esempi indagano espressamente il rapporto tra comunicazione sonica e visiva nell'ambito dell'oggetto sonoro. "Sounding Forms: African Musical Instruments", una mostra itinerante presentata in quattro sedi negli Stati Uniti e a Parigi, esponeva strumenti africani come oggetti d'arte per evidenziare il "silenzio degli studi sul tema dello strumento musicale africano inteso come scultura".[7] Probabilmente era il primo tentativo concertato di colmare questa vistosa lacuna nella cornice di una mostra occidentale.

CONTESTUALIZZARE LE CONCEZIONI DEGLI STRUMENTI MUSICALI

Nel riconoscere che le culture di tutto il mondo hanno elaborato specifiche concezioni degli strumenti musicali, sia a livello formale sia a livello informale, l'etnomusicologa e organologa Margaret Kartomi afferma che "classificare gruppi di oggetti come gli strumenti musicali è un principio fondamentale del pensiero umano".[8] Si riferisce a studi di casi specifici condotti in diverse comunità di varie parti del mondo per indagare le proprietà degli strumenti e il loro rapporto con i popoli e la loro arte. "Diversamente dalla concezione greca ed ellenistica, secondo la quale gli strumenti musicali sono "oggetti" inanimati, scrive Kartomi, "i popoli dell'Africa occidentale, come i Dan della Costa d'Avorio e della vicina Liberia (dove si chiamano Gio) o i Kpelle liberiani, aderiscono a una concezione antropocentrica degli strumenti."[9] "Per i Dan o i Gio", continua, "gli esseri umani non detengono la proprietà esclusiva degli strumenti che suonano, e dunque nemmeno della musica che producono", perché "si ritiene che gli strumenti, per la maggior parte, appartenessero in origine a entità non umane come spiriti, maschere, stregoni o animali."[10]

L'etnomusicologo ghanese J.H. Kwabena Nketia ha analizzato la dimensione estetica di numerosi strumenti musicali per esplorare la relazione tra oggetti, potere, linguaggio e modalità di comunicazione visiva. Nketia rileva che alcuni strumenti sono venerati come simboli religiosi del potere politico. In alcune comunità Akan del Ghana, spiega Nketia, la struttura gerarchica dell'autorità è regolata dai tamburi; quando assume il potere, il capo presta giuramento sui tamburi e "in passato si poteva destituire un capo, o ridurne il potere, sottraendogli con la forza i suoi tamburi."[11] Nketia ritiene che si possa individuare una "tendenza a considerare gli strumenti musicali come 'entità' dotate di un'intrinseca vitalità (analoga a quella di una persona immobile e silenziosa), che si manifesta quando vengono suonati."[12] Riporta a titolo d'esempio l'iya ilu degli Yoruba nigeriani, un tamburo costruito con il legno di un albero che cresce ai margini di una strada molto battuta perché si crede che, grazie al passaggio delle persone, l'albero abbia acquisito la capacità di "parlare" assorbendo i suoni dell'ambiente circostante. Viceversa in Tanzania, quando i tamburi muti vengono appesi fuori dalle case dei Wagogo per simboleggiare che in famiglia si sta svolgendo una cerimonia di iniziazione, è chiaro che gli strumenti hanno assimilato anche l'arte della comunicazione silenziosa.

Molto spesso gli strumenti musicali sono anche personificati come corpi e oggetti animati. Marie-Thérèse Brincard osserva che "per alcuni africani, [gli strumenti] sono amplificazioni o estensioni delle vibrazioni del corpo e dell'anima del suonatore."[13] Per le comunità Bakongo dello Zaire meridionale, dell'Angola e della Repubblica del Congo in Africa centrale, Robert Farris Thompson spiega che "la musica è un'entità vivente, presente con la sua vitalità quanto un corpo umano in movimento" e afferma che "i Bakongo credono che gli strumenti musicali siano esseri viventi."[14]

Allo stesso modo, anche la simbologia e la rappresentazione visiva possono essere estremamente potenti. Nella sua analisi di alcune varietà di arpe nell'Africa centrale – molte delle quali si pensa risalgano a tradizioni antiche di almeno 5000 anni – Sue Carole DeVale sostiene che l'elevato prestigio tradizionalmente attribuito all'arpa di solito si esprime con una simbologia complessa e una decorazione molto elaborata. Nel Gabon, per esempio, lo ngombi dei Bwiti ha una cassa di risonanza che simboleggia lo stomaco femminile, mentre il collo simboleggia il maschio e il corpo di solito è dipinto in rosso e bianco per simboleggiare il sangue e il seme; inoltre, otto corde dello strumento sono considerate femminili e altre otto maschili. Un elaborato sistema di rituali ruota intorno allo strumento, che in sé è concepito come il potere della notte, perciò non deve essere esposto alla luce del sole. Così, terminate le cerimonie notturne, lo ngombi dev'essere accompagnato da specifici canti di congedo per poi essere depositato in un luogo buio, al riparo dalla luce del giorno.[15]

LA CONCEZIONE DELLA MBIRA IN ZIMBABWE

Paul Berliner, nella sua fondamentale analisi della musica e delle tradizioni legate alla mbira (lamellafono) tra gli Shona dello Zimbabwe, dipinge un quadro molto ricco delle concezioni fluide che ruotano intorno agli strumenti musicali.[16] Esaminando alcuni miti delle origini e gli usi di diverse varianti di mbira, Berliner osserva che, se alcuni strumenti possono essere poco più che giocattoli, altri acquistano uno status molto elevato, e spesso vengono riservati solo a specifici suonatori e cerimonie ancestrali. L'associazione della mbira con il mondo degli antenati spesso è definita la "legge della mbira" e "alcuni musicisti ritengono che la 'legge della mbira' li obblighi a riconoscere formalmente gli antenati come proprietari dei loro strumenti e a prendersi cura di questi ultimi con il dovuto rispetto: nella cultura degli Shona, il furto di una mbira era tradizionalmente considerato un reato molto grave."[17]

Berliner dimostra che spesso al rapporto tra lo strumento e il suonatore è attribuito un valore supremo e, infatti, "certi suonatori suggeriscono... che lo strumento ha una forte volontà di suo" e che di fatto ti chiami.[18] Alcuni suonatori di mbira si riferiscono al loro strumento con personificazioni, sia al livello dell'intero strumento, sia al livello dei singoli tasti. Oggi molti suonatori spiegano il rapporto tra i diversi tasti in termini di altezza delle note, ma i musicisti più radicati nella tradizione, come Mubayiwa Bandambira, un rispettato anziano all'epoca in cui Berliner svolgeva le sue ricerche, nei tardi anni '70, hanno un sistema molto più dettagliato per descrivere i rapporti familiari sia tra i registri della tastiera, sia tra i singoli tasti. I tre registri della tastiera di una mbira dzavadzimu (la "mbira degli antenati") – registro basso, medio e alto – sono indicati rispettivamente come le "voci dei vecchi uomini", le "voci dei giovani uomini" e le "voci delle donne".[19] Bandambira, inoltre, espone nel dettaglio il sistema complesso e articolato con cui nomina e personifica ciascuno dei 22 tasti della sua mbira dzavadzimu, in base a classificazioni che "si possono considerare in gran parte come nomi qualificativi, nomi che esaltano ogni tasto per la sua qualità musicale e per l'effetto che produce negli ascoltatori."[20] Bandambira spiega che le voci acute dei tasti *Benzi* (persona matta) sono quelle che ti spaventano o ti fanno sentire irrequieto e agitato; i tasti *Duri* (mortaio) forniscono suoni ritmici come quando si pesta il miglio; i tasti *Nhiura* (grande tamburo) aprono i polmoni e fanno venire voglia di muovere il corpo; e i tasti *Shumba* (leone) sono animali in cui vivono spiriti potenti.

Alle pareti della International Library of African Music di Grahamstown, in Sudafrica, sono appesi centinaia di mbira, e decine di questi strumenti sono esposti in teche di vetro nel cortile del Pitt Rivers Museum a Oxford; ma qual è il legame che resta tra l'allestimento di questi oggetti muti e le personificazioni e le animazioni dei rapporti originari tra musicisti, materiali e spiriti? Un simile interrogativo è analizzato in pittura dall'artista e musicista zimbabwese Moses Bikishoni. La mbira di solito si suona dentro una zucca svuotata o in un contenitore in vetroresina a forma di zucca (*deze*) profilato da tappi metallici di bottiglia, per amplificare il suono e aggiungere ritmi, estensioni e timbri più complessi. Nel corso del tempo, e nelle mani di un suonatore esperto, i cicli melodici reiterati possono evolversi in tessuti sonori dall'intreccio fantasioso, che rivelano polifonici mondi sonori di acque che scorrono e zampillano, di bambini che cantano e ridono in mezzo a grandi strutture che cigolano e scricchiolano. I dipinti di Bikishoni, come *Gwariva – Inside the Gourd* (2014) fondono Cubismo e Impressionismo con visioni idiosincratiche di spirali simili a ossa per rappresentare gli intricati rapporti che corrono tra gli strumenti mbira, il suono e la comunicazione con un mondo spiritico ancestrale.[21]

GLI STRUMENTI MUSICALI DEI CACCIATORI-RACCOGLITORI

Ogni qualvolta ci adagiamo su una classificazione, un paradigma o un quadro di riferimenti, possiamo sempre cercare raffronti culturali e stimoli critici nell'antropologia. L'antropologia promuove il relativismo culturale, chiedendoci di mettere in discussione il predominio o la naturale validità di qualunque sistema o idea, sottoponendo a un esame critico tutto ciò che potrebbe considerarsi normativo.

Le comunità seminomadi dei pigmei Baka, cacciatori-raccoglitori che vivono nella Repubblica Centrafricana e in alcune aree del Congo settentrionale, hanno creato un sistema musicale di straordinaria ricchezza, basato su un ingegnoso rapporto tra strumentazione, corpo e ambiente. La strumentazione musicale per le cerimonie nella foresta pluviale, come i *boyobis*, è tanto ampia da includere la volta della foresta – che offre un'acustica simile a quella di una cattedrale – utensili ordinari come taniche, vecchi barattoli e ciotole rovesciate, e l'illusorio spettacolo visivo e sonoro di spiriti mascherati che danzano, per creare reti splendidamente complesse di rapporti intrecciati polifonici e poliritmici.

I Baka sono intimamente consapevoli delle proprietà soniche dei materiali e degli oggetti del loro ambiente. Nella foresta, i giovani maschi a volte scelgono, flettono e poi legano un giovane albero per formare un arco che poi fissano a una fossa scavata sottoterra che funge da cassa armonica. L'estemporaneo "arco della terra" (*ngbindi*) è poi pizzicato e bloccato con pollici e indici per creare un potente e fragoroso giro di basso che, a un attento ascolto, può evocare i suoni della musica dubstep, frutto di una pesante elaborazione elettronica. Quando i Baka si spostano verso un altro accampamento, lo strumento viene semplicemente abbandonato. E allora i giovani creano sorprendenti ritmi incrociati percuotendo con palmi, pugni e bastoni i loro corpi e le radici respiratorie d'albero appositamente selezionate. Di solito gli alberi sono scelti come strumenti per il modo in cui lo schiaffo, l'urto e il tonfo delle mani sul legno e sul corpo umano riecheggiano, risuonano e riverberano nell'ambiente circostante.

In altri casi, anche un fiume si può trasformare in uno strumento musicale. Quando si bagnano nei fiumi, le donne e i bambini Baka possono dedicare del tempo alla musica, mettendo le mani a coppetta in varie forme specifiche per generare ritmi e melodie percuotendo l'incresparsi dell'acqua e schizzando ritmi incrociati in un gioioso divertimento.

L'arte di uno strumento musicale può anche andare tragicamente perduta. Lo mbyo è un flauto a quattro fori che fino a tempi relativamente recenti si sentiva spesso suonare tra le comunità Baka di Yandoumbé, vicino alla città di Bayanga, nell'area sudoccidentale della Repubblica Centrafricana. Suonatori esperti come Mindumi, Mobila e Contreboeuf potevano suonare variazioni fino all'alba, ascoltando i suoni pulsanti degli insetti della foresta, accompagnando il morbido scrosciare della pioggia notturna, esplorando e intrecciando le loro note con i suoni sospesi che risuonavano nella volta della foresta. Offrendo una colonna sonora per la comunità che dormiva nell'accampamento, questa musica era letteralmente suonata per entrare nei sogni delle persone. Quando è morto l'ultimo suonatore esperto, nessun bambino ha preso in mano lo strumento e l'ultimo mbyo conosciuto era conservato da Louis Sarno, ricercatore e documentatore di musica che viveva con i Baka a Yandoumbé. Nel marzo 2013, durante un colpo di stato nella Repubblica Centrafricana, i ribelli Séléka hanno saccheggiato le case di Yandoumbé. Così l'ultimo mbyo, costruito con una pianta locale e tanto fragile da rompersi se maneggiato con incuria, si è frantumato sotto gli scarponi delle milizie.[22]

CONCLUSIONE: L'ARTE DI VEDERE GLI OGGETTI SONORI

La lente etnografica con la quale gli strumenti musicali vengono selezionati ed esposti spesso privilegia la dimensione visiva ed estetica degli oggetti, chiedendoci di osservarli da vicino, invece di ascoltare l'arte dei rapporti sonori. Esaminando le varie concezioni degli strumenti musicali nell'ambito di diverse civiltà africane, si scopre che gli oggetti sonori potevano essere proprietà degli dèi o degli spiriti; si scoprono reti di relazioni tra esseri umani, oggetti e ambienti; si scoprono materiali naturali effimeri che vengono abbandonati dopo l'uso sulle rotte dei cacciatori; o che i simboli visivi del potere e del rango vengono usati con maestria da alcuni individui, sono severamente vietati a certi gruppi sociali e vengono addirittura distrutti da altri.

1 Barbara Kirshenblatt-Gimblett, "Objects of Ethnography", in Ivan Karp, Steven Levine (a cura di), *Exhibiting Cultures: the Poetics and Politics of Museum Display*, Smithsonian Institution Press, Washington (DC)/London 1991, pp. 386-443: 388.

2 *Ibid.*, p. 387.

3 *Ibid.*, p. 392.

4 africanart.org

5 victorgama.org

6 caacart.com

7 Marie-Thérèse Brincard (a cura di), *Sounding Forms: African Musical Instruments*, The American Federation of the Arts, New York 1989, p. 15.

8 Margaret Kartomi, *On Concepts and Classifications of Musical Instruments*, The University of Chicago Press, Chicago (IL)/ London 1990, p. 3.

9 *Ibid.*, p. 241.

10 *Ibid.*, p. 243.

11 J.H. Kwabena Nketia, "The Aesthetic Dimensions of African Musical Instruments" in Marie-Thérèse Brincard (a cura di), *op. cit.*, pp. 21-29: 22.

12 *Ibid.*, p. 23.

13 Marie-Thérèse Brincard (a cura di), *op. cit.*, p. 17.

14 Robert Farris Thompson, "Body and Voice: Kongo Figurative Musical Instruments", in Marie-Thérèse Brincard (a cura di), *op. cit.*, pp. 39-45: 39.

15 Sue Carole DeVale, "African Harps: Construction, Decoration, and Sound", in Marie-Thérèse Brincard (a cura di), *op. cit.*, pp. 53-61.

16 Paul F. Berliner, *The Soul of Mbira: Music and Traditions of the Shona People of Zimbabwe*, University of Chicago Press, Chicago (IL)/ London 1978, reprint 1993.

17 *Ibid.*, p. 236.

18 *Ibid.*, p. 128.

19 *Ibid.*, p. 56.

20 *Ibid.*, p. 58.

21 mosesbikishoni.com

22 Questa sezione si basa sulle ricerche condotte nel ricco archivio di musica Baka composto dalle registrazioni Louis Sarno e curato dal Pitt Rivers Museum di Oxford. Per ulteriori approfondimenti si veda: web.prm.ox.ac.uk/ reel2real/index.php/collections-sarno.html10

SINESTESIA DISTORTA. IL VIDEO MUSICALE E LE ARTI VISIVE

Holly Rogers

Il video musicale rappresenta la diretta convergenza tra musica e immagine e contraddistingue un momento significativo nella storia dell'audiovisivo. Benché nel corso del XX secolo le arti visive e sonore si stessero già muovendo a grande velocità l'una verso l'altra, la combinazione della musica con l'immagine in movimento generò una forma di sintesi audiovisiva senza precedenti.

Il film ne è un esempio: la musica e l'immagine, procedendo insieme nel tempo, sono in grado di compensarsi reciprocamente per dare origine a un prodotto audiovisivo unitario coerente.

Nel cinema convenzionale la musica viene però composta nella fase di post-produzione, quando il montaggio del film è già a uno stadio avanzato. Nell'epoca d'oro di Hollywood, per esempio, compositori di musica sinfonica come Max Steiner ed Erich Korngold avevano a disposizione solo qualche settimana per scrivere la musica di un intero film. Sebbene questa pratica sia cambiata in tempi recenti e molti compositori siano coinvolti nel processo di realizzazione del film sin dall'inizio, le narrazioni filmiche spesso si reggono ancora su una progressione sequenziale letteraria o visiva. Il videoclip, genere catapultato nella sfera pubblica dopo la rapidissima ascesa di MTV negli anni '80, dall'altro lato offre un esempio di inversione di questo processo: nei video musicali, come spiega Carol Vernallis nel suo autorevole *Experiencing Music Video*, l'immagine è in grado di enfatizzare la musica.[1]

L'uso dell'immagine in movimento per dare enfasi alla componente musicale può essere collocato all'interno di un percorso chiaramente definito, che risale alle primissime imprese artistiche documentate. L'opera lirica è da sempre un genere audiovisivo, ma è stato nel corso del XIX secolo che Richard Wagner ha formulato una teoria significativa intorno al concetto che immagine e musica potessero combinarsi in modo così stretto da dare vita a una nuova forma artistica audiovisiva (1849). Wagner non si occupava soltanto di scrivere la propria musica e i propri libretti, ma assumeva anche il controllo del palcoscenico e dei costumi per concepire ciò che definiva *Gesamtkunstwerk*, un'opera d'arte totale. Questa idea suggeriva che tutti gli elementi del teatro – parole, musica e immagine – potessero unirsi per creare una forma artistica superiore unificata.

Il concetto di "opera d'arte totale" è stato di grande importanza ai fini dello sviluppo di uno stile del video musicale. Sono numerosi i registi che dedicano la propria attività esclusivamente a questo preciso genere audiovisivo, mentre altri, tra i quali Spike Jonze e Michel Gondry, si occupano anche di lungometraggi (è interessante notare come i loro film siano spesso dotati di grande intensità musicale). Forse ancora più degni di attenzione sono però gli artisti visivi che si sono cimentati con questo format: il videoclip offre ad artisti che non hanno necessariamente una formazione musicale l'opportunità di creare un lavoro a elevato impatto sonoro, di combinare le arti visive con la cultura musicale popolare e di dare vita a immagini correlate a una struttura musicale preesistente.

Il video musicale diretto dagli artisti visivi è di grande interesse in quanto è parte di una storia che risale a un periodo ben precedente la nascita di MTV, la quale lanciò il nuovo canale televisivo il 1° agosto 1981 con il video *Video Killed the Radio Star* dei The Buggles, diretto da Russell Mulcahy. Durante gli anni '20, per esempio, un nutrito numero di artisti berlinesi esplorò i modi in cui suono e immagine potevano combinarsi per creare una musica visiva su pellicola che si sviluppasse lungo una certa durata. Il pioniere Hans Richter realizzò cortometraggi di grande intensità ritmica, come nel caso di *Rhythmus 21* (1921), predisponendo immagini astratte conformi ai modelli e alle strutture musicali, al fine di creare quelle che definiva "colonne sonore strumentali" visive mute.[2]

Mentre Richter stava esplorando i modi per tradurre la musica in immagini musicali mute, il suo contemporaneo Walter Ruttmann si imbarcò nell'impresa di creare forme vive di suono visivo, commissionando la realizzazione di nuove musiche destinate a essere eseguite durante i suoi film. Per il film d'animazione *Lichtspiel Opus 1* (sempre del 1921), per esempio, chiese al suo amico Max Butting di comporre un quartetto d'archi (pare che lo stesso Ruttmann suonò la parte del violoncello) che avrebbe dovuto far combinare i ritmi serrati e sincronizzati con le immagini astratte, al fine di produrre una forma artistica che poggiasse le fondamenta "da qualche parte a metà tra la musica e la pittura".[3]

Ma è il gioco variopinto e ritmico delle animazioni visive in plexiglas di Oskar Fischinger che illustra al meglio la relazione tra le prime forme di musica visiva e il video musicale. Fischinger è celebre per aver disegnato la sequenza "Toccata e fuga" per *Fantasia* (1940) della Disney, anche se alla fine, insoddisfatto del risultato, l'artista fece cancellare la sua firma prima dell'uscita del film. Fischinger insisteva sul fatto che le sue immagini non fossero concepite per essere "musica tradotta, poiché la musica non ha bisogno di essere tradotta sullo schermo"; piuttosto, erano pensate per dialogare con la musica allo scopo di creare qualcosa di completamente nuovo.[4] Fischinger era particolarmente interessato alla visualizzazione di una musica già esistente: nel film *Poema ottico* (1938), nel quale disegnò a mano un dialogo animato della *Rapsodia ungherese n. 2* di Liszt, musica e immagine volteggiano incessantemente l'una intorno all'altra, senza mai sconfinare nella vacua ripetizione; in *Motion Painting No. 1* (1947), l'artista ci offre una complessa ed energica risposta visiva al *Concerto brandeburghese n. 3* di Bach.

Il lavoro di questi antesignani della *visual art* musicale, tra i quali figurano Norman McLaren, i fratelli John e James Whitney e Stan Brakhage, ebbe un notevole impatto sui successivi registi cinematografici. Ma un'altra importante corrente d'influenza che confluì nel genere del video musicale giunse dalla video arte. Per certi aspetti le prime forme di *video art* costituivano l'esatto opposto rispetto alle tendenze della *visual music*. Mentre i promotori di quest'ultima erano artisti che cercavano di attribuire un'espressione sonora alle proprie opere visive, coloro che acquistarono la prima attrezzatura video portatile, quando questa fu resa disponibile nel 1965, furono spesso musicisti interessati a rappresentare visivamente i propri esperimenti sonori.

Nel 1971, per esempio, l'artista di origine coreana Nam June Paik, che inizialmente aveva studiato per diventare compositore, creò il suo *TV Cello*, uno strumento costituito da tre schermi televisivi impilati uno sopra l'altro che richiamavano la forma di un violoncello monocorde (una versione successiva fu realizzata con più corde). Concepito in collaborazione con la violoncellista d'avanguardia Charlotte Moorman, lo strumento era collegato allo spazio in cui si svolgeva la performance per mezzo di un circuito chiuso (immagini dal vivo del pubblico e di Charlotte Moorman venivano così trasmesse inframmezzate da immagini emblematiche del XX secolo). Quando Moorman suonava lo strumento, produceva mediante segnali elettronici quelli che chiamava "suoni del videocello". Tali suoni erano collegati elettronicamente alle immagini, che dalla musica venivano distorte e trasformate in irriconoscibili.

Un altro importante esempio di *visual music* creata da un video artist è *Violin Power* (1978) di Steina Vasulka. La formazione musicale di questa artista islandese, che ha iniziato la sua carriera come violinista con l'Orchestra Sinfonica di Praga, ha avuto una profonda influenza sullo stile dei suoi video. Come lei stessa spiega: "È il suono che mi guida dentro all'immagine. Ogni immagine ha il proprio suono e, in esso, io cerco di catturare qualcosa di fluido e vitale. Applico all'arte e al violino il medesimo principio: la stessa inclinazione al continuo esercizio, lo stesso concetto di composizione".[5]

Questo "concetto di composizione" emerge chiaramente in *Matrix* (1970-72), un'opera audiovisiva costruita con diversi monitor e realizzata assieme al marito Woody. Come spiega egli stesso, al tempo della produzione era "molto impegnato a trasformare l'immagine in suono o a controllare i suoni con le immagini, così come a generare immagini dai suoni e viceversa".[6] Analogamente al *TV Cello*, *Violin Power* funzionava mediante uno strumento manipolato elettronicamente (il violino Zeta): quando Vasulka suonava il suo violino modificato, un alimentatore a circuito chiuso sincronizzato con i movimenti dell'artista trasmetteva immagini distorte a seconda del tono e dell'archeggio.

Se è vero che queste forme di audiovisivo sono rintracciabili in tutta la storia del video musicale, la possibilità di un incrocio tra i due generi si manifesta in modo evidente nel lavoro di Laurie Anderson. Benché Anderson abbia una formazione da scultrice e lavori spesso con il video, molte delle sue opere sono estremamente musicali: la sua *performance art* si avvale per esempio del *Tape-bow Violin* (1977), uno strumento dotato di un nastro magnetico inciso al posto dei crini dell'archetto e di una testina da registratore nella posizione normalmente occupata dal ponticello. Negli anni '80 la musicista passa a un violino midi che le consente di controllare dei sintetizzatori collegati allo strumento. Ma Anderson è particolarmente apprezzata anche per i suoi numerosi brani cantati e per avere diretto i videoclip che li accompagnano: nel 1981, per esempio, esce *O Superman (For Massanet)*, brano liberamente ispirato all'opera *Le Cid* di Massenet (1885). Il videoclip è perfettamente sincronizzato con la voce dell'artista grazie all'uso del filtro *vocoder* e mostra alcuni suoi primi piani mentre canta e gesticola, accompagnata da immagini del sole e della luna, in uno di quelli che sarebbe diventato un modello molto influente di *sonic art*.

Quando il video musicale diventa un genere consolidato, i video artist incominciano a sfruttare e a parodiare le sue forme e le sue convenzioni. Il video a canale singolo *I'm Not the Girl Who Misses Much* (1986) di Pipilotti Rist, per esempio, è una parodia della canzone *Happiness is a Warm Gun* dei Beatles. L'opera può essere interpretata come un commento sul modo in cui le donne sono rappresentate (visivamente, in particolare) nella musica popolare. Nel momento in cui l'artista svizzera esegue il brano, rivolgendosi direttamente alla macchina da presa, la sua voce e la sua immagine si presentano frammentate grazie agli espedienti del fast-forward e dello slow-motion estremo, offrendo allo spettatore distorsioni che mettono in evidenza la concretezza del mezzo video. La nostalgica *I Am Crazy and You're Not Wrong* (1997) di Anne McGuire esprime una diversa relazione tra suono e tecnologia. In questo caso l'artista, originaria di San Francisco, esegue musica lounge dell'era Kennedy davanti all'occhio disinteressato della cinepresa. Mettendo in luce quella che Catherine Elwes ha descritto come una "lotta per l'espressione di sé entro i confini mimetici del gergo musicale", lo stile del video si allontana dal modello di interattività in tempo reale caratteristico di *Violin Power* e *TV Cello* per accostarsi allo stile pre-registrato del videoclip.[7] *Killing Time* (1994) di Sam Taylor-Wood mostra quattro persone mentre eseguono una pessima pantomima delle arie di *Elektra* di Richard Strauss. Fuori tempo con la registrazione e spesso distratti dal loro compito, i mimi enfatizzano la concretezza del mezzo e le bizzarre convenzioni della rivisualizzazione del video musicale.

Quando "nasce" il video musicale, il genere ha pertanto già alle spalle una lunga e ricca storia nell'ambito delle arti sonore e visive, un'eredità che non è certo passata inosservata ai registi di videoclip contemporanei: il video *Star Guitar* (2002), realizzato da Michel Gondry per i Chemical Brothers, per esempio, è un tributo al film Berlino – Sinfonia di una grande città (1927) di Walter Ruttmann; lo stile musicale distorto che accompagna le immagini dei lavori di David Lynch rende liberamente omaggio ai primi visual artist musicali; il filmato interattivo di *The Wilderness Downtown* (2010), realizzato da Chris Milk per gli Arcade Fire, allude alle tendenze interattive dei primi video artista.

La capacità del video musicale di costruire e disseminare l'immagine di un artista ha costituito un fattore rilevante per la crescita della sua popolarità. Ma non è certo stata soltanto un'industria pop orientata all'immagine ad alimentare il bisogno di associare un accompagnamento visivo a una canzone; senza dubbio ha avuto un ruolo fondamentale la reintroduzione nella canzone dell'energia e della fisicità della performance dal vivo. L'essenza della musica riprodotta è inevitabilmente immateriale: ciò che sentiamo è la traccia di una sessione incisa ormai conclusa. Aggiungere un video a un pezzo cantato conferisce all'astrattezza della musica l'illusione che abbia una genesi concreta, benché la fonte sia di per sé astratta quanto la musica registrata che l'accompagna. All'osservazione mossa da alcuni critici, secondo cui basare una canzone sull'immagine suggerisce l'idea che la musica, da sola, non sia autosufficiente – che per il suo successo sia necessario il supporto visivo –, si potrebbe obiettare che tale supporto, di fatto, rinvigorisce la musica stessa. Di rado un video musicale presenta una performance genuina, priva di revisioni. Al contrario, la musica, svincolata da questa sua sorgente, è in grado di per sé di costruire un mondo alternativo: e il mondo rappresentato, per citare di nuovo Vernallis, può "diventare molto strano".[8]

Il potenziale creativo di questi "strani" mondi ha costituito una forte attrattiva per coloro che lavorano nell'ambito delle arti visive. Per alcuni artisti la progressione dall'immagine all'audiovisivo può essere vista come un processo naturale, quasi inevitabile. Dopo la collaborazione di Andy Warhol con i Velvet Underground e The Exploding Plastic Inevitable, per esempio, il videoclip di *Hello Again* realizzato dall'artista per i Cars (1984), nel quale interpreta il ruolo del barista, apparve come un'evoluzione naturale. Per altri, il salto verso il suono visualizzato è stato più drastico e, forse per questo motivo, ha avuto vita piuttosto breve: l'artista pop americano James Rizzi creò un video d'animazione per il pezzo *Pleasure of Love* dei Tom Tom Club; il video artist Doug Aitken puntò la macchina da presa sugli Interpol per produrre una serie di primi piani in bianco e blu della band durante la realizzazione del loro video *NYC* (2002); il digital artist e pittore Jeremy Blake, nel video *Round the Bend* (2002) di Beck, diede vita a immagini statiche e astratte che si sovrappongono lentamente; l'artista, regista e musicista Martha Colburn produsse un collage di immagini animate per il video *Wrong Time Capsule* (2005) dei Deerhoof e per *Lie, Lie, Lie* (2007) di Serj Tankian. Benché si tratti solo di brevi incursioni nell'ambito del video musicale, tutti questi esempi quantomeno illustrano come lo stile di ciascun artista si sia progressivamente esteso nel territorio dell'audiovisivo.

Alcune delle collaborazioni di maggior successo tra musicisti ed artisti sono nate da una condivisione di carattere estetico. Il video *Country House*, realizzato per i Blur da Damien Hirst nel 1995, è un esempio eccellente di risposta congiunta al movimento Cool Britannia, che si stava diffondendo sulla scena musicale e artistica inglese negli anni '90. Più celebre per i suoi animali in formaldeide e per i suoi *spot paintings*, il giovane Young British Artist incontra la nascente band Britpop mentre frequenta la facoltà di Arte al Goldsmith's College di Londra. La loro successiva collaborazione per *Country House*, che il bassista Alex James descrive come "una collisione tra arte superiore ed inferiore", sembrerà inevitabile.[9]

Il concept del videoclip è semplice ma efficace: inquadrature dei personaggi che corrono selvaggiamente intorno a un gioco da tavolo (*Escape from the Ratrace*)[10] si alternano ad immagini del protagonista (Keith Alan) che passeggia accanto alla sua casa di campagna. È evidente la presenza di numerosi tropi convenzionali dei video musicali: il vocalist Damon Albarn esegue il brano rivolgendosi direttamente all'obbiettivo; le riprese a passo uno e il montaggio accelerato creano un ritmo incalzante a livello visivo; numerosi sono anche gli esempi di giochi linguistico-visivi nel senso letterale del termine (l'espressione *caught in a rat race* è associata all'immagine del protagonista in abito elegante che fa cadere mazzette di denaro su un gioco da tavolo ricoperto di topi). Dall'inizio alla fine del video, l'atmosfera Britpop dei Blur e il sensazionalismo del maestro della comunicazione Hirst si fondono per dare vita ad un corpo completamente integrato: scopriamo infatti una progressiva assenza di realismo; i variopinti scenari appaiono oltremodo teatrali e, aspetto di maggior rilievo, ritroviamo riferimenti umoristici a fenomeni culturali britannici, come Benny Hill e il popolare video dei Queen *Bohemian Rhapsody* (Bruce Gowers, 1975). In questo caso l'immagine non ha la funzione di potenziare la musica, bensì agisce di concerto con essa per formare una *Gesamtkunstwerk* che foraggia un'estetica culturale più ampia, all'epoca in voga in Gran Bretagna.

Se è vero che Hirst ha dato il suo contributo a un solo videoclip, altri artisti hanno collaborato con diversi musicisti. Il media artist e cineasta Jim Blashfield, per esempio, ha diretto video per Michael Jackson, i Talking Heads, Paul Simon e Joni Mitchell. Analogamente, l'artista americano Bruce Conner, che si è occupato di pittura e fotografia, così come di cinema e montaggio, ha prodotto video per i Devo, Brian Eno e David Byrne. Quasi certa-
11 mente però, l'artista più devoto al genere del videoclip è stato Derek Jarman, regista sperimentale che ha diretto tra gli altri *The Sex Pistols: The Sex Pistols Numbers One* (1976) dei Sex Pistol, *Ask* (1986) degli Smiths e *The Next Life* (1993) degli Suede. La sua collaborazione più duratura è stata tuttavia quella con i Pet Shop Boys, per i quali ha realizzato cinque videoclip e ideato il tour del 1989. Questi video hanno un'incredibile valenza filmica e dimostrano un'evidente analogia con lo stile cinematografico sperimentale di Jarman. Il mondo alternativo di *It's a Sin* (1987), per esempio, immerge i due componenti della band Neil Tennant e Chris Lowe nel chiaroscuro mentre inscenano una narrazione simbolica che ricorda molto il *Caravaggio* (1986) di Jarman. Eppure, poiché la colonna visiva esalta gli aspetti ritmici e semantici della canzone, l'effetto che ne risulta è assai diverso rispetto alla natura contemplativa del film di Jarman. Le immagini, che seguono un montaggio basato sulla ritmica musicale, formano sequenze accelerate difficilmente riscontrabili nelle lunghe e, per certi versi, discrete riprese cinematografiche; di contro, la struttura del testo, ripetitivo, breve e relativamente astratto, si manifesta in sequenze caratterizzate da un montaggio visivo straniante e disconnesso.

Gli artisti hanno pertanto goduto di una lunga e produttiva collaborazione con i musicisti, un'alleanza che si è manifestata in svariati modi, anche se l'interazione tra media non è soltanto frutto della cooperazione: essa si palesa anche sotto forma di tributo. All'interno degli "strani" mondi del video musicale, i riferimenti alle arti visive abbondano e musicisti famosi hanno spesso reso omaggio ad opere d'arte molto popolari. Il video del brano *Mine* (2013) di Beyoncé, tratto dal visual album che porta il suo stesso nome, ripropone *Gli amanti* (1928) di René Magritte; nel video *Rude Boy* (prodotto da Melina Matsoukas, 2009) Rihanna fa riferimenti espliciti a *Marilyn Monroe's Lips* (1962) di Warhol; il video *The Boys* (realizzato da Colin Tilley, 2012) di Nicki Minaj allude invece al mondo a pois di Yayoi Kusama. Ma l'artista che più si è ispirata alle opere d'arte come mezzo per costruire i mondi visivi delle proprie canzoni è Lady Gaga. Per esempio, il video *Applause* (Inez & Vinoodh, 2013) è disseminato di allusioni a Warhol, Cindy Sherman e alla *Nascita di Venere* di Botticelli, mentre *Born This Way* (2011), realizzato da Nick Knights, contiene così tanti riferimenti stilistici all'opera dell'artista francese Orlan che Lady Gaga si è ritrovata addirittura con un'accusa di plagio.

È dunque evidente che un flusso bidirezionale tra musicisti e artisti abbia dato vita a una sintesi espressiva e innovativa che continua ad estendersi oltre i confini delle possibilità dell'audiovisivo. Allo stesso modo, è importante prendere atto del fatto che il videoclip non è stato soltanto il risultato di una rivoluzione televisiva avvenuta negli anni '80, bensì, piuttosto, il frutto di un lungo e proficuo processo evolutivo che ha caratterizzato la storia delle arti visive e musicali.

1
Carol Vernallis, *Experiencing Music Video: Aesthetics and Cultural Context*, Columbia University Press, New York 2004, p. 8.

2
Hans Richter, citato in "Early Colour Organs", paradise2012.com/visualMusic/musima (ultimo accesso: 5 febbraio 2014).

3
Walter Ruttmann, citato in Dieter Daniels, "Sound & Vision in Avantgarde & Mainstream", 2004, medienkunstnetz.de/themes/image-sound_relations/sound_vision (ultimo accesso: 5 febbraio 2014).

4
Oskar Fischinger, citato in William Moritz, *Optical Poetry: The Life and Work of Oskar Fischinger*, Indiana University Press, Bloomington (IN) 2004, p. 185.

5
Steina Vasulka, "Orka", vasulka.org/Steina/Steina_Orka/Orka.html (ultimo accesso: 5 febbraio 2014).

6
Steina Vasulka speaking in *Six Programs for Television: Matrix*; citata in Yvonne Spielmann, *Video: The Reflexive Medium*, The MIT Press, Cambridge (MA)/London 2008, p. 99.

7
Catherine Elwes, *Video Art: A Guided Tour*, University of the Arts, London 2005, p. 165.

8
Carol Vernallis, *op. cit.*, p. 7.

9
Alex James, citato in Madeleine Baker, "Childhood Psychedelia", madeleine-baker.blogspot.co.uk/p/dissertation.html (ultimo accesso: 5 febbraio 2014).

10
Escape from the ratrace, letteralmente "fuga dalla corsa del topo". *The Ratrace* è un'espressione idiomatica anglosassone che designa – rifacendosi alla corsa dei topi da laboratorio che corrono all'infinito all'interno del labirinto della loro gabbia cercando un'uscita che non potranno mai trovare – alla corsa al successo: una corsa senza fine in un'inutile ricerca di qualcosa che è irraggiungibile, destinata a un esito autodistruttivo. [N.d.T.]

L'IMPORTANZA DI NON POTER PREVEDERE. GLI INSOSPETTABILI DIALOGHI DELLE OPERE SONORE

Andrea Lissoni

Il confronto con un'opera d'arte basata sul suono si determina anche sulla base di una misteriosa forma di dialogo. Che si fondi sui bpm o sugli Hz, cioè sui battiti al minuto o sull'unità di misura della frequenza, quel dialogo si instaura con il corpo umano e con le sue metriche e le sue dinamiche percettive, cioè con il battito cardiaco o con l'accordatura aurea, il La centrale intonato a 432Hz (su cui secondo alcuni studi – o opinioni – sarebbe sintonizzato il ronzio emesso dalle api, o il fruscio delle piante).

Quanto è cambiata la relazione fra l'opera sonora e il pubblico fondata sul potenziale dialogo innescato dall'artista quando i suoni, il modo di produrli, di diffonderli e, di conseguenza, di ascoltarli si è radicalmente trasformato? In effetti, alla fine del XX secolo, il suono ha incontrato una rivoluzione senza precedenti, aprendosi a possibilità di manipolazione fino ad allora impensabili, sia per le opportunità di campionamento delle fonti più disparate, sia per le possibilità di profonda modificazione e alterazione che gli strumenti digitali mettevano a disposizione. Il suono, in sostanza, è divenuto digitale. E, come l'immagine, la sua natura è inevitabilmente e definitivamente cambiata.

Il campo dell'arte ha prontamente registrato il fenomeno in corso, tant'è vero che agli inizi del nuovo decennio si sono concentrate tre essenziali mostre: "Sonic Boom-The Art of Sound", Hayward Gallery (2000), "Sonic Process-A New Geography of Sound", Macba/Centre Georges Pompidou (2003) e "Frequenzen (Hz), Audiovisuelle Raume", Schirn Kunsthalle, Frankfurt (2002). In un certo senso, tutte si occupavano di ambiente e di forme artistiche legate all'analisi di un ambiente che si stava senz'altro ridefinendo. Ma testimoniavano anche come la (opinabile) categoria di *sound art* stesse radicandosi paradossalmente in un campo semantico autonomo e impermeabile più che slittare come una pratica centrifuga e ibridante.

Come conseguenza, o piuttosto come effetto collaterale della rivoluzione in corso, il decennio '90 ha visto esplodere generi e sonorità fra loro molto differenti lungo un arco temporale molto ridotto, con una generale spinta innovativa senza precedenti, prima che la febbre archivistica per le radici, e le attitudini di recupero del decennio successivo diventassero un'ossessione, scivolando dalla rivelazione alla derivazione.

Comunque sia, in quel momento dirompente, dal punto di vista della creazione (sulla fruizione si potrebbe aprire una lunga parentesi sul consumo di elettronica dance/tappezzeria "arty" stigmatizzato da Mike Kelley nell'opera *Runway for Interactive DJ Event*, 2000), la scena artistico-visiva ha accolto, sostanzialmente, tre distinti fronti:

– il *noise*, in un'accezione che va dall'impasto sgorgato dalla tradizione storica del *rumore* futurista bagnato nella scena impro-performativa, poi proto-punk innestata nell'elettronica colta californiana (generata e disseminata da *Destroy All Monsters* e dalle varie formazioni di cui Mike Kelley ha fatto parte o con cui ha collaborato) fino all'elettronica radicale Viennese (coagulatasi intorno alla *label* Mego).

– l'elettronica basata su strutture ritmiche di frequenze (invece che di note) praticata da artisti interessati a disegnare paesaggi e "tappeti" sonori monotonali (o loro interpretazioni, eventualmente diffusi negli spazi pubblici, come ha fatto Carsten Nicolai con ∞, "graffiti" sonori in forma di microstrutture per documenta X a Kassel, 1997), o musicisti/artisti attivi in live che facevano vibrare o risuonare il corpo dello spettatore in live eventualmente accompagnati da immagini video di pura rappresentazione delle onde sonore e vicini ad una logica "minimalista", se non strutturale (Pan Sonic, Mika Vainio, lo stesso Nicolai come Alva Noto, Carl Michael Von Haswolff, un fenomeno prevalentemente europeo continentale).

– l'*illbient*, definizione disputata da DJ Spooky e DJ Olive, che in sintesi si confronta con il presupposto di un ambiente urbano "malato", campionato e re-intarsiato in scenari cinematici abitati da ritmiche scure di derivazione dub in live e dj set (Paul Miller *aka* DJ Spooky ne è stato l'esponente di punta, per un fenomeno soprattutto newyorkese che ha conosciuto riverberi in Europa per esempio con l'inglese Scanner).

Questi fenomeni di assorbimento e di rigenerazione nella scena artistica di pratiche e generi musicali, tendono però a far considerare soprattutto la dimensione performativa e live (con la consacrazione sancita da "A Fantastic World Superimposed on Reality: A Select History of Experimental Music", il festival curato da Mike Kelley per Performa al Gramercy Theater a New York nel 2009), o la diffusione più o meno domestica di supporti, piuttosto di interrogare la potenziale mutazione della relazione fra opera e pubblico.

Cosa accadrebbe quindi se ipotizzassimo di concentrare la nostra attenzione esclusivamente sull'opera e non sulla performance? Quanto è cambiato il suono dell'opera, presa dal punto di vista *ontologico*? E quanto lo sguardo, la postura percettiva del pubblico?

Una serie di figure retoriche si stagliano sullo sfondo di un ambiente che, nonostante la retromania degli anni Zero che ha riguardato anche il recupero di media, supporti e formati obsoleti (radio, tv, theremin, sintetizzatori, vinile e cassette), non era certamente più elettrico, né elettronico ma inevitabilmente digitale.

Senz'altro l'ossessione metrica, parzialmente ereditata dal minimalismo musicale storico (*Music for Pure Wawes, Bass Drums and Acoustic Pendulums*, 1980 di Alvin Lucier ne è un ottimo esempio), permane. La necessità di rappresentare il tempo nelle sue partiture regolari e uniformi, si consolida come una pratica centrale, come dimostrano la maggior parte delle opere di Carsten Nicolai [350] o le installazioni di Martin Creed [338]. Del resto, scandire il tempo è un gesto tanto potenzialmente autoritario quanto inevitabilmente politico, e non è un caso se il corpo meccanico e meccanizzato, frammentato o da parata, compare nell'opera di artisti apparentemente fra loro lontanissimi come Stephan von Huene (*Der Mann von Jüterbog*, 1995-96) [342], o Maurizio Cattelan (*Senza titolo*, 2003) [368]. Il tutto considerando che, nella transizione dalla fabbrica al centro d'arte (e, altrettanto – dettaglio non trascurabile – al club), quel che vagava comunque libero e ben percepibile nello sfondo dell'aria buia, era lo spettro che si occupava accuratamente di sincronizzare i regimi della produzione con quelli del consumo, e quindi del piacere.

Gli strumenti tradizionali, ricontestualizzati nella logica tipica del readymade, subiscono le forme di manipolazione che stanno subendo contemporaneamente le tracce sonore stesse: decomposti, disassemblati, ibridati, più o meno ironicamente: lo strumento infatti, può essere completamente ignoto e incorporare una logica compositiva ora algida e modernista (è spesso ancora il caso di Carsten Nicolai), ora vernacolare (come nelle opere di Walter Kitundu [402, 404] o nella sorprendente scultura *Burlesque* (2012) [430], di Janet Cardiff e George Bures Miller), o una combinazione fra le due (che sculture e installazioni di Haroon Mirza [400, 434] rappresentano in modo impeccabile). Oppure ben riconoscibile, come risultato di un *détournement* (in una tradizione che partendo da Georges Braque va agli strumenti ibridi di Ken Butler [298, 306], alle chitarre di Christian Marclay [360], agli strumenti di Andreas Slominski [366], Yoshihiko Satoh e Pedro Reyes [432], o agli apparecchi-emblema dell'era elettronica come *Mopol* (1999) [356] di Ed Potokar o *Melody Gin* (2005) di Brian Dewan, eventualmente performabile (*The Metastable Circuit*, 2012 [426] di Tarek Atoui), performativo (come *A Solo in the Doldrums*, 2009 [408] di Anri Sala o *Keyboardchairs*, 1997-2005 di Céléste Boursier-Mougenot), o tale perché conseguente all'essere stato performato (come nella lunga tradizione Fluxus che va da Raphael Montañez Ortiz ad Arman [218, 344]).

Un'altra figura ricorrente è la tematizzazione dell'equivalenza suono in azione/tempo in divenire, a prescindere che l'opera sia un assemblaggio di strumenti o di oggetti ibridi (nella tradizione aperta dalle "antimacchine" di Jean Tinguely [226], di cui *Mindfall* (2004-07) di Micol Assaël, un concerto per motori elettrici in un container, potrebbe essere un curioso erede): la genesi del suono che si articola nel tempo è la conseguenza di una reazione a catena visibile (come nelle opere elettrico/elettroniche di Alberto Tadiello [422]).

Infine c'è quello che si potrebbe definire la mitologia della diffusione, e tutti gli immaginari che vi sono associati. E, quindi, nemmeno troppo paradossalmente, nell'epoca dell'ascolto digitale in cuffia, il *sound-system* – inteso come combo di forme di casse e altoparlanti – diventa una figura scultorea monumentale. Da una parte abbiamo i readymade che suggeriscono tradimenti percettivi (*Crossfading Suitcase*, 2004 [372] di Loris Gréaud, che obbliga a invertire la propria postura di ascolto musicale, sdraiandosi a terra con la testa fra il sound-system) e slittamenti di immaginario (il colore bianco in *Toyan's Jr.*, 2001 [362] che spinge verso una direzione elettrodomestica o il nero di *Trojan*, 2008, entrambi di Tom Sachs, ancora una dimensione "elettrodomestica" che gioca con Brancusi e il minimalismo pur appartenendo alla cultura dub e reggae giamaicana). Dall'altra, la percezione ingannevole consiste in un tradimento del regime sonoro nei confronti di quello visivo: una generazione di artisti ha dato infatti forma a opere che, attraverso la classica "cassa" audio, hanno disseminato micro percezioni uditive con un atteggiamento che si potrebbe confrontare con il divisionismo, o con il puntinismo. Chi interagendo con nuove forme di paesaggio, più neuronali che mentali e in un certo senso radicalmente astratti (Carsten Nicolai, Florian Hecker), chi invece, come Marcellvs L (*Klavierwellen*, 2011), che ha combinato più modi di diffusione del suono stesso, distillandolo – cioè sezionandolo – nel paesaggio, e rilanciandolo al pubblico processato e riconoscibile come un riverbero di fondo.

Ma, nonostante la penetrazione e la perpetuazione in ambito artistico-visivo nel periodo anni '90/anni '00, dei fronti *noise*, elettronico/digitale basato su frequenze e *illbient*, curiosamente le sonorità delle opere d'arte del periodo sembrano non prevedere un'immediata associazione a una tradizione musicale. Può essere interessante rilevare piuttosto la singolare proliferazione di "tavoli da camera", di apparente derivazione scacchistica (o da seduta parap sicologica) come il *Chess Piano* (2009) [410] di Guido van der Werve, o il *Marble Sonic Table* (2011) [420] di Doug Aitken, che sembra aver anticipato un'intera cultura delle pratiche di relazione e comportamentali "digitali" dell'industria dei media schermici.

Dal momento che ogni tassonomia rischia di essere sterile (se non pericolosa), quale potrebbe essere la conclusione a margine di questa ricognizione per figure?

Dal punto di vista della sua esposizione, il suono è tradizionalmente considerato ostile: incontenibile, infido, difficile da gestire se non rischiando di pagare il prezzo di presenze visive ingombranti.

Tuttavia, per quanto difficilmente esponibile, specie in mostre collettive, l'opera sonora (o che prevede suono), contiene un valore inestimabile: obbliga alla concentrazione sull'opera, sulla sua percezione, in uno spettro che va dall'approccio fenomenologico a quello storico. E spinge al raccoglimento. Alla concentrazione. Al confronto con qualcosa di noto, ma non necessariamente collocabile. Invita a una postura, che si riverbera su una dinamica di auto-percezione. È sottilmente perversa, in quanto richiede un tempo non sorretto dall'immagine-movimento.

Ma rimane in sé, ontologicamente pura e sganciata da ogni necessità illustrativa o paratestuale.

E questo tratto, ancorché di natura puramente concettuale, è la più specifica eredità dell'età delle neoavanguardie. È naturalmente il punto di incontro più profondo nel dialogo fra opera e pubblico, ed è la sua risorsa più stimolante, se non potenzialmente dirompente.

LA MAGIA DEGLI STRUMENTI. TECNOLOGIE MUSICALI E FETICISMO DELL'OGGETTO DI CONSUMO

Jonathan Sterne

"Feticismo dell'oggetto di consumo? Bella idea!"

Siamo nel 2011. Mi trovo nel nord della California e sto viaggiando in auto con Matt Ward, all'epoca presidente della Universal Audio, un'azienda specializzata nel produrre copie di vecchi apparecchi analogici per la registrazione del suono, creandone delle repliche o riproducendoli in forma digitale. Stiamo commentando la passione dei musicisti per le vecchie apparecchiature e le animate discussioni che popolano i forum dei siti internet, in cui ci si domanda per esempio se i software che emulano vecchi hardware saranno mai capaci di eguagliare il suono di "quelli veri". Gli chiedo perché, a suo avviso, i musicisti sono così interessati a tecnologie come quelle che propone la sua azienda. Passiamo in rassegna le motivazioni più comuni: alcuni musicisti sono abituati a utilizzare le vecchie tecnologie perché hanno lavorato in studi di registrazione e vogliono averne di proprie senza pagare prezzi esorbitanti per le attrezzature vintage. Altri stanno invece muovendo i primi passi nell'arte della registrazione del suono, eppure preferiscono acquistare un'attrezzatura collaudata sperando di poter assimilare quel pizzico di magia che pensano contenga.[1]

Quando accenno al fatto che gli studiosi hanno un nome per quest'ultimo fenomeno, ed espongo i fondamenti del feticismo dell'oggetto di consumo, Ward appare divertito. L'idea spiega buona parte del fenomeno di cui discutiamo; sebbene non lo esaurisca ancora, è quasi impossibile evitare di confrontarsi con essa. Wendy Chun definisce il feticismo dell'oggetto di consumo una falsa causalità. Scrive: "un feticcio permette di visualizzare l'ignoto, di sostituire le cause con le immagini. I feticci offrono alla mente umana un controllo eccessivo e al contempo insufficiente, istituendo un 'campo causale unificato' che comprende sia le azioni personali, sia gli eventi fisici. I feticci offrono una parvenza di controllo sugli eventi futuri – la possibilità di influenzarli, se non di programmarne ogni aspetto – che si fonda sul travisamento dei veri rapporti sociali di cui vengono investiti gli oggetti materiali.[2] In altre parole, nel feticcio i rapporti tra le persone si cristallizzano nell'oggetto.

La Universal Audio produce strumentazioni per musicisti. Ma produce anche beni di consumo. Una parte della sua attività si dedica a replicare famosi apparecchi per l'elaborazione del suono in uso negli anni '60 e '70, ma realizza anche software che riproducono in forma digitale gli apparecchi analogici. Nei software, l'interfaccia grafica sullo schermo replica l'immagine dell'hardware, mentre gli algoritmi informano ogni parte dei circuiti fino al livello dei componenti. L'utente "manovra" con il mouse le immagini delle manopole per ruotarle. Anche se il software consente una maggiore flessibilità, la Universal Audio evita di introdurre nuove funzioni, a parte forse trasformare un'unità monofonica in un'unità stereofonica o aggiungere delle impostazioni predefinite. E se è indubbio che la Universal Audio è eccezionalmente attenta a restare fedele agli originali, è anche vero che non è la sola. Aziende come la Native Instruments di Berlino hanno cominciato la loro attività creando software che riproducono vecchi organi Hammond e sintetizzatori analogici. La Line6 di Los Angeles costruisce attrezzature per chitarristi basandosi su modelli classici di strumenti, amplificatori ed effetti, ma si concentra sul suono, più che sulle vecchie interfacce. Se acquistiamo un pacchetto completo per la registrazione, la composizione o l'esecuzione della musica, è probabile che il software contenga riproduzioni di vecchi strumenti e apparecchi per l'elaborazione dei suoni. Anche i suoi parametri operativi saranno scheumorfici, per cui l'utente si troverà di fronte a manopole e leve come se fosse seduto davanti a una gigantesca console di missaggio: un'immagine che offre un buon grado di leggibilità ma anche l'illusione del comando. Riflettendo sulle interfacce dei software, Wendy Chun afferma che sono "guidate dal sogno del controllo individuale: la manipolazione diretta dello schermo e quindi, per estensione, del sistema che lo schermo indicizza o rappresenta. ... Le interfacce ci offrono un rapporto immaginario con il nostro hardware."[3] Tra il click di un mouse e l'azione sullo schermo intervengono innumerevoli istruzioni digitali che vengono deliberatamente occultate all'utente finale, insieme al lavoro materiale impiegato nella costruzione del computer, del mouse e dello schermo, e agli elaborati parametri e protocolli che permettono ai componenti di interagire in modo coerente tra loro e con altri sistemi, pur restando invisibili al nostro immaginario utente finale. L'analisi di Chun riguarda espressamente le interfacce dei software, ma questa illusione del comando vale per *qualunque* tipo di interfaccia, qualunque tecnologia per la produzione dei suoni o della musica. Barry Blesser e Linda-Ruth Salter esplicitano questo nesso quando discutono dei riverberatori artificiali (strumenti che aggiungono un'eco di tipo ambientale) che Blesser elaborava negli anni '70: "quando un parametro spaziale è collegato a una manopola, un pulsante o un tasto [dalla prospettiva della persona che lo aziona], un riverberatore diventa di fatto indistinguibile da uno strumento musicale suonato in tempo reale da un musicista." Qui, i mezzi di comunicazione si sovrappongono agli strumenti, o meglio, la dimensione mediatica degli strumenti si rivela appieno.[4]

Siamo abituati a pensare al design di uno strumento come al design di un'interfaccia solo quando si tratta di uno strumento digitale, come per esempio un campionatore o un sintetizzatore. Ma in realtà tutti gli strumenti hanno delle interfacce. L'apparente immediatezza di uno strumento acustico nasconde altrettanto lavoro, altrettanta maestria e altrettanti processi di standardizzazione di quelli racchiusi in un'interfaccia informatica o in un elaboratore digitale del suono. Insieme a una serie di pedali, ogni tasto di un pianoforte aziona un martelletto, che articola e modula la dimensione percussiva dello stru mento. La pelle tesa sulla testa di una tabla si presenta come un recettore per il musicista, offrendo una varietà di toni a seconda del punto e del modo in cui le dita toccano la superficie. Le dita dovrebbero colpire una sola estremità dei tasti di uno mbira, non l'altra, proprio come i suonatori degli strumenti a corde occidentali sanno che, nella famiglia degli archi e delle chitarre, le corde devono essere suonate tra il ponticello e il capotasto (anche se molti hanno violato questa norma per ottenere effetti interessanti). Come nel caso del software, tutti questi meccanismi offuscano i processi decisionali e le fasi di lavorazione che implicano. L'accordatura del pianoforte è una delle professioni più antiche nella storia della musica ed è praticata tuttora. I compromessi del temperamento per costruire le scale di note incorporate nella tastiera di un pianoforte o di una chitarra derivano da un'incessante riflessione collettiva sulle discrepanze tra la fisica del suono e l'orecchio esperto dei musicisti e degli ascoltatori. Gli strumenti della famiglia degli archi e degli ottoni – o le percussioni nel "drumkit" – esistono in un mutuo rapporto di complementarietà. Persino strumenti come le chitarre e i bassi acustici sono dei sistemi. La catena delle cause e degli effetti fisici è più evidente all'osservatore inesperto, ma non è meno reale: se muoviamo il ponticello anche solo leggermente, è probabile che si debba sistemare anche il manico.[5] Le decisioni sottese a questi strumenti hanno ripercussioni reali sui musicisti, ripercussioni che vanno dalla finezza tecnica dell'esecuzione alle lesioni che possono patire se suonano troppo a lungo o in modo scorretto. Durante lo stesso viaggio nel nord della California parlo con Roger Linn; siamo nel suo soggiorno, seduti accanto a un pianoforte e non lontani da una chitarra elettrica collegata a un pedale e quindi a un computer. Linn si è distinto come inventore di strumenti elettronici. Dopo aver progettato una delle prime drum machine da campionamento – ovvero una batteria elettronica che usava i suoni registrati di altre batterie, invece di produrre suoni propri – ha sviluppato il concept delll'MPC (MIDI Producion Center) che, insieme al piatto, è diventato lo strumento più importante del rap e dell'hip hop. Linn ha faticato non poco per creare uno strumento digitale che fosse sufficientemente "espressivo". Per *espressivo* egli intende che abbia le qualità di uno strumento a corde, nel quale anche i gesti più lievi producono variazioni di altezza, timbro o volume del suono. Lanciando un'occhiata al pianoforte e alla chitarra, li critica per la concezione della loro interfaccia. Entrambi gli strumenti impongono al musicista di cambiare il gesto della mano ogni qualvolta vuole cambiare tasto o corda. In altre parole, non è possibile trasporre questa capacità. Dal punto di vista della moderna teoria dell'interfaccia, essi presentano inutili difficoltà per chi impara a utilizzarli e sono privi di ergonomia del suono, sostiene Linn, che pure ha appena mosso una critica agli strumenti digitali perché mancano di espressività, poiché non lasciano ai musicisti un controllo sufficiente a produrre sofisticate melodie, armonie e variazioni timbriche in tempo reale.[6] Dal punto di vista del musicista, la padronanza di uno strumento è la padronanza di un'interfaccia e del sistema a cui è collegata, sia che i trasferimenti di energia avvengano tra i componenti di metallo, legno e avorio, sia che si tratti di componenti elettronici e materiali plastici. Suonare uno strumento è una forma di conoscenza incarnata, una conoscenza che si acquisisce con ore e anni di pratica, una "seconda natura", come la definisce Pierre Bourdieu.[7] Quella conoscenza incarnata tempera la standardizzazione insita nell'interfaccia dello strumento. Quando imbraccio un basso elettrico, i movimenti delle mie mani sono completamente spontanei, nonostante abbiano richiesto grandi sforzi in passato. Dopo anni di pratica, il mio corpo sembra conformarsi allo strumento, anche se ho la sensazione che sia lo strumento a conformarsi al mio corpo e alla mia volontà. Dal punto di vista fenomenico, è un'azione non molto diversa dal digitare questa frase su una tastiera, un'altra operazione in cui il mio corpo si è conformato a uno standard; uno standard originariamente sviluppato per un sistema del tutto diverso da quello di un computer.[8]

Dall'esterno, gesti come questi possono apparire naturali a uno sguardo superficiale, perché il lavoro che implicano è oscurato dal design dello strumento e dalle capacità del musicista. Quando guardiamo suonare dei bravi musicisti (o persone abili in qualunque tipo di tecnica), possiamo dire che operano in una condizione di *strumentalità*. Ovvero, nel loro rapporto con lo strumento, l'intenzione e l'azione si conformano l'una all'altra, e certe azioni e conoscenze acquisite sono relegate in secondo piano, cosicché il musicista possa "sostenere una certa direzione", per citare le parole di Sara Ahmed.[9] Ma che cosa accade quando ci troviamo di fronte agli strumenti musicali in momenti diversi dell'esibizione? Possono essere esperiti come strumenti per l'apprendimento, o come oggetti parziali, ma anche come oggetti magici di per sé.

Nella sua classica analisi etnografica di uno studio di registrazione sudafricano, Louise Meintjes racconta la storia di una cantante, Joana, la quale ha scambiato un orologio MIDI, l'oggetto che sincronizza tutta la strumentazione dello studio, per un contenitore di suoni. "L'orologio MIDI non contiene davvero i suoni a cui Joana si riferisce. ... Per lei c'è un intero mondo sonico racchiuso in quella macchina lucente [...] È un mondo che Joana può sfiorare, ma al quale non può accedere. È invisibile, ma viene percepito come un mondo di enormi proporzioni".[10] Ancora una volta, abbiamo un caso specifico che illustra una condizione generale: quando l'osservatore non ha accesso al mondo interiore dello strumento – per conoscenze, esperienze, possibilità, abitudini – è portato ad attribuire a quell'oggetto un vasto e complesso mondo interiore.

In parte questo risultato è voluto e deriva da un aspetto specifico della progettazione, un aspetto molto antico. Il mio insegnante di registrazione, Mark Rubel, definisce questa dimensione degli strumenti "psicocosmetica". Le chiocciole elaborate degli archi; le palette delle chitarre su cui è impresso il marchio; la superficie levigata e lucente di un pianoforte; i led che occhieggiano in una drum machine: sono tutti elementi che alludono alla magia racchiusa nello strumento, alle azioni incorporate nella cosa, proprio come le punte delle dita di un essere umano. Raccontando dello studio di registrazione, Meintjes coglie l'aspetto futuristico che caratterizza molti strumenti elettronici dell'era contemporanea. Per lei, lo spazio dello studio è "costruito ed esperito come uno spazio magico, ed è percepito come un feticcio dai musicisti e dai tecnici che ci lavorano. Connotandolo come uno spazio magico, intendo dire che è lontano dalla sfera ordinaria e che, grazie all'arte dell'illusione e alla capacità d'immaginazione, sembra custodire una forza naturale ... che, invocata, produce un'arte ammaliante. Pensando allo studio come feticcio, lo reifico in un oggetto che può procurare, a coloro che vi accedono, i servizi di quella forza, o 'spirito', che racchiude al suo interno."[11]

Appesi alle pareti dei negozi di musica, appoggiati alle sedie sul palco, sistemati sugli scaffali o conservati nei musei in base a una classificazione organologica, gli strumenti possono evocare una sorta di spirito, quando sono lontani dal momento del loro utilizzo. Forse è più affascinante credere che la magia risieda negli strumenti, più che nel lavoro delle persone, per la mistificazione che investe la musica e la figura del musicista separandole dalla vita quotidiana, per l'apparente naturalezza che caratterizza le esibizioni di alto livello e per la distanza che si frappone tra l'osservatore e lo strumento nella maggior parte delle sue esperienze vissute.

Nelle vite fantastiche dei musicisti e degli artisti possiamo anche scorgere un legame tra la strumentalità e lo spirito o la magia. Una breve ricerca in google della frase "è quasi come uno strumento" genera una litania monocorde di rievocazioni nostalgiche legate alle attrezzature. Come nei parametri assegnati da Blesser a una manopola, la magia si manifesta nella macchina nel momento in cui la regolazione dei parametri e l'efficiente utilizzo da parte dell'utente generano ispirazione e offrono all'arte risorse nascoste. A esprimere questo genere di commenti sono comunità di musicisti che discutono circa la strumentazione, a volte come utenti, a volte come recensori. Il recensore di un microfono scrive: "È un piacere cantare in questo microfono, è come uno strumento che sta al cantante come la chitarra sta al chitarrista." L'utente di un riverberatore digitale ValhallaShimmer, scrive: "Vedo lo Shimmer come un effetto speciale; ha un tale carattere che è quasi come uno strumento." Un altro musicista, parlando di un sequenziatore (che controlla altri strumenti, ma di per sé non produce suoni), scrive: "È quasi come uno strumento, quando cominci a fare cose tipo destinare le manopole A + B al controllo quando succedono cose relative ad altre cose."[12]

Anche gli artisti usano questa terminologia. Commentando l'Aleph Null, un programma interattivo per l'arte visiva, Jim Andrews scrive: "Ci vuole pratica per tirarne fuori qualcosa di buono. In questo senso è come uno strumento". Notiamo che il qualcosa di buono viene "tirato fuori" dall'utente, ma risiede nella cosa in sé. In un'intervista rilasciata da Janet Cardiff e George Bures Miller ad ArtIT, un'installazione è descritta in modo analogo:

"ART iT: [commentando *Experiment in F# Minor* (2013)] Se non sai dove sono i sensori, viene naturale esplorare l'opera attraverso il movimento, per vedere se puoi controllare l'intensità muovendo le braccia in una direzione o nell'altra. L'opera ti spinge a un comportamento performativo. Anche quando cammini, agisce un meccanismo inconscio che induce a entrare in diverse zone di comportamento.

Janet Cardiff: Scopri quanto può essere illimitato. È come uno strumento.

George Bures Miller: È questo il nostro problema. Scopriamo sempre cose che possono essere illimitate.

Cardiff: Per *Pandemonium* (2005) abbiamo installato delle bacchette da percussione robotiche nelle celle dell'Eastern State Penitentiary Museum di Philadelphia, e abbiamo scoperto che anche quello era come uno strumento. Potevi suonarci qualunque tipo di musica. Potevi invitare un percussionista e dire, ecco, questo è uno strumento, che cosa puoi farci? Ci sono tutte queste propaggini che sarebbe fantastico sviluppare. Il nostro problema è che abbiamo troppe idee e poco tempo."[13]

"Scopri quanto può essere illimitato": questa frase segna il momento in cui la descrizione invoca il feticismo. L'illimitatezza – "Potevi suonarci qualunque tipo di musica" – è la fantasia di quella "forza naturale che produce buona arte" a cui accenna Meintjes.

Non è un caso che l'attribuzione di questa magia avvenga in un momento lontano da quello della strumentalità: nel momento dell'osservazione a distanza, in cui la conoscenza pratica non è pienamente disponibile, o nel momento della riflessione a distanza di tempo. In *Queer Phenomenology*, Sarah Ahmed scrive di ciò che si rivela quando "le tecnologie non sono più disponibili all'azione." L'esperienza dell'impossibilità di usare una cosa – o la distanza dal momento dell'uso – induce ad attribuire delle proprietà alla cosa in sé. Analizzando il fallimento, e ricorrendo al celebre esempio del martello di Heidegger, Ahmed scrive che il fallimento "potrebbe portare 'l'oggetto' a *vedersi attribuire* proprietà, qualità e valori. In altre parole, il fattore che entra in gioco nei momenti di fallimento non è l'accesso a certe proprietà ma l'attribuzione di certe proprietà, un processo che dipende da come *approcciamo* l'oggetto. [...] Il momento del 'non uso' è [...] il momento in cui gli oggetti possono essere giudicati poiché si rendono inadeguati a un compito, è il momento in cui 'rimproveriamo lo strumento'".[14]

Tuttavia può essere anche il momento in cui diamo credito allo strumento e ci riteniamo inadeguati a usarlo. La critica di Christopher Small alle culture musicali d'Occidente cade a proposito: l'idea che alcune persone posseggano innate doti musicali e altre no.[15] L'intima sensazione che sia lo strumento a infondere magia o potere al musicista, e non viceversa, è un residuo di questa visione generalizzata.

I circa cinque secoli di strumenti catalogati in "Art or Sound" esprimono entrambi gli aspetti di questo fenomeno. Quale miglior esempio del potere dell'oggetto di consumo di un'esposizione di strumenti, offerti all'osservazione e all'esperienza sensoriale, consacrati come opere d'arte in sé, separati dal contesto della pratica e dell'esecuzione musicale, e ricontestualizzati come oggetti di solenne contemplazione, come opere d'arte separate dalla vita? Allo stesso tempo, incorniciati come opere d'arte in un contesto museale, gli strumenti sembrano commentarsi da sé: dalla squisita maestria artigianale racchiusa in una viola o una chitarra [50, 52], del Seicento, alle esplicite riflessioni sulla natura della strumentazione moderna che esprimono i laptop acustici di Tore Honoré Bøe [436], o le decostruzioni delle moderne batterie proposte da Yoshihiko Satoh e Manuel Rocha Iturbide [418]. Quando chiunque può avvicinarsi e interagire con l'oggetto consacrato, l'idea che gli strumenti siano appannaggio di alcune persone e non di altre comincia a erodersi; l'invito a toccare, sentire, confrontarsi o interagire infrange la separazione tra esecutore e pubblico. La raccolta in sé offre un commento sugli strumenti intesi come oggetti che racchiudono una magia. Quando si esperisce nella sua totalità, la raccolta, con la sua ampiezza e vastità, testimonia il lavoro umano e la creatività sedimentati negli strumenti, il lavoro occulto che si annida dentro il feticcio.

Ringrazio Carrie Rentschler

1
Matt Ward, intervista rilasciata all'autore, 17 febbraio 2012.

2
Wendy Hui Kyong Chun, *Programmed Visions: Software and Memory*, The MIT Press, Cambridge (MA) 2011, p. 50.

3
Ibid., pp. 62, 66.

4
Barry Blesser, Linda-Ruth Salter, *Spaces Speak, Are You Listening?: Experiencing Aural Architecture*, The MIT Press, Cambridge (MA) 2007, p. 91; Paul Théberge, *Any Sound You Can Imagine: Making Music/Consuming Technology*, Wesleyan University Press, Hanover (NH) 1997.

5
Myles Jackson, *Harmonious Triads: Physicists, Musicians and Instrument-Makers in Nineteenth-Century Germany*, The MIT Press, Cambridge (MA) 2006.

6
Roger Linn, intervista rilasciata all'autore, 13 febbraio 2012.

7
Pierre Bourdieu, *The Logic of Practice*, trad. di Richard Nice, Stanford University Press, Stanford (CA) 1990.

8
Sarah S. Lochlann Jain, *Injury: the Politics of Product Design and Safety Law in the United States*, Princeton University Press, Princeton (NJ) 2006.

9
Sara Ahmed, *Queer Phenomenology: Orientations, Objects, Others*, Duke University Press, Durham (NC) 2006, p. 31.

10
Louise Meintjes, *Sound of Africa!: Making Music Zulu in a South African Studio*, Duke University Press, Durham (NC) 2003, p. 89.

11
Ibid., pp. 73-74.

12
Pete Pell, recensione del microfono a condensatore modello Sterling Audio ST66 Large Diaphragm Tube, sul sito commerciale *Musician's Friend*, musiciansfriend.com/tube-microphones/sterling-audio-st66-large-diaphragm-tube-condenser-microphone; Aniston, commento dell'utente nella discussione "Valhalla Shimmer o EOS?", nel forum del sito *KRVAudio* dedicato al software KVRAudio, kvraudio.com/forum/viewtopic.php?t=302426; Gosh, commento dell'utente nella discussione "Sequentix Cirklon", nel forum del sito *Muff Wiggler*, muffwiggler.com/forum/viewtopic.php?p=933082&sid=c75363c58507ef9674cb8b3bac00fc8c (tutti consultati il 2 marzo 2014).

13
Jim Andrews, profilo dell'artista su sul sito *Rhizome*, rhizome.org/profile/jimandrews/?page=2; intervista "Janet Cardiff and George Bures Miller: Pt II" sul sito *ART iT*, art-it.asia/u/admin_ed_itv_e/k15N0UpRnFSy6i3caPT9; (entrambi consultati il 2 marzo 2014).

14
Sara Ahmed, *op. cit.*, p. 49.

15
Christopher Small, *Music-Society-Education*, John Calder, London 1977.

PHOTO CREDITS

While all efforts have been made to contact the relevant copyright holders with respect to the materials included in this publication, we invite any person whose rights have not been cleared as yet to contact the Publisher for necessary arrangements.

© For their images: Terry Allen, Arman, Richard Artschwager, François Baschet, Giacomo Balla, Wladimir Baranoff Rossiné, Joseph Beuys, Céleste Boursier-Mougenot, Alexander Calder, Maurizio Cattelan, Salvador Dalí, Thomas Demand, Marcel Duchamp, Loris Gréaud, Maurice Henry, Rebecca Horn, Milan Knížák, Jannis Kounellis, Man Ray, Eliseo Mattiacci, Robert Morris, Bruce Nauman, Henrik Neugeboren, Carsten Nicolai, Roman Opalka, Michelangelo Pistoletto, Robert Rauschenberg, Takako Saito, Andreas Slominski, Alberto Tadiello, Takis, Jean Tinguely, Günther Uecker, Max Vandervorst, Wolf Vostell, Tom Wesselmann by SIAE 2014.

Courtesy Alias Editorial; photo Rodolfo Díaz Cervantes: p. 417.

Courtesy A.N. Scriabin Memorial Museum, Moscow: p. 137.

© Archivio Fotografico della Soprintendenza Speciale per il Patrimonio Storico, Artistico ed Etnoantropologico e per il Polo Museale della città di Roma; photo Maximiliano Massaroni, Massimo Taruffi e Gianfranco Zecca: pp. 55, 69, 119, 163.

Courtesy Archivio Maurice Henry, Milan; photo Paolo Vandrasch: p. 185.

Courtesy Arman Studio Archives, New York: pp. 219, 345.

Courtesy Athanasios Argianas and Max Wigram Gallery: p. 413.

© Atelier Leitner: p. 289.

Courtesy Tarek Atoui and Galerie Chantal Crousel; photo Alex Guirkinger: pp. 426–27.

Courtesy Audio Visual Arts: p. 273.

Courtesy Marco Bagnoli: p. 349.

Courtesy Bauhaus-Archiv Berlin; photo Hermann Kiessling, Berlin: p. 179.

Courtesy Riccardo Beretta; photo Vincenzo Caccia: p. 425.

Beyer Zurich Clock and Watch Museum, Zurich: pp. 49, 63, 117.

Courtesy René Block Archive, Berlin: p. 311.

Courtesy Tore Honoré Bøe and the Origami Republika Archive. Under Creative Common License: p. 437.

© Céleste Boursier-Mougenot. Courtesy Galerie Xippas; photo Frédéric Lanternier: p. 429.

Courtesy Ken Butler: pp. 299–301, 307.

Calder Foundation/Art Resource, New York © 2014; Calder Foundation, New York/SIAE, Rome: p. 187.

Courtesy Janet Cardiff & George Bures Miller and Luhring Augustine, New York: p. 431.

Courtesy Leo Castelli Gallery, New York: p. 241.

Courtesy Maurizio Cattelan Archive; photo Michael Bodycomb: p. 369.

Courtesy Cittadellarte – Fondazione Pistoletto, Biella; photo A. Lacirasella: pp. 250–51.

© Martin Creed. Courtesy the Artist and Hauser & Wirth: pp. 338–39.

Courtesy Di Rosa Collection, Napa: p. 276.

Collection Musée de la Musique, Paris; photo Albert Giordan: p. 47.

Collection Musée de la Musique, Paris: photo Jean-Marc Anglès: pp. 57, 191, 193; photo Claude Germaine: p. 101; photo Thierry Ollivier: p. 111.

Collection Nouveau Musée National de Monaco, Monaco; photo E. Mathon/ Miti: p. 121.

Collection Claes Oldenburg and Coosje van Bruggen, © 2005 Claes Oldenburg and Coosje van Bruggen: p. 379.

Collection of Dennis Oppenheim Estate, Courtesy Galleria Fumagalli, Milan: p. 283.

Collezione Peggy Guggenheim, Venezia (Fondazione Solomon R. Guggenheim, New York); photo David Heald: p. 167.

© Thomas Demand, VG Bild-Kunst, Bonn/SIAE, Rome: p. 407.

Courtesy Paul DeMarinis: p. 281.

Courtesy Brian Dewan and Pierogi Gallery: p. 375.

Courtesy Ruth Ewan; photo Stephan Baumann, bild_raum: p. 371.

© Estate of Nam June Paik; photo Moderner Kunst Stiftung Ludwig Wien, Vienna; former Hahn Collection, Cologne: p. 223.

Courtesy Neil Feather: p. 359.

Courtesy Friedrich Christian Flick Collection im Hamburger Bahnhof, Berlin; photo Stephan Altenburger, Zurich: p. 249.

Courtesy Fondazione Bonotto: pp. 287, 293.

Courtesy Fondazione Marconi, Milan: photo Filippo Armellin: p. 173; photo Gianni Ummarino: p. 177; photo X Ray Trust /Apice, Milan: p. 175.

Courtesy Fondazione Prada, Milan; photo Attilio Maranzano: p. 367.

Courtesy Galerie EIGEN + ART Leipzig/Berlin and The Pace Gallery; photo Uwe Walter: p. 351.

Photo Gallery Paule Anglim, San Francisco: p. 317.

Photo Isabelle Gardner: pp. 414–15.

Courtesy Loris Gréaud: pp. 372–73.

Courtesy Subodh Gupta: pp. 376–77.

Courtesy Hauser & Wirth; photo Alex Delfanne: p. 225.

Courtesy Alanna Heiss: p. 291.

Courtesy Bart Hopkin: p. 347, 353.

© Joe Jones; courtesy Moderner Kunst Stiftung Ludwig Wien, Vienna; former Hahn Collection, Cologne, p. 235.

© Martin Kersels; courtesy Mitchell-Innes & Nash, New York: p. 419.

Courtesy Doug Aitken, Galerie Eva Presenhuber, Zurich and 303 Gallery, New York; Photo Stefan Altenburger Photography, Zurich: p. 421.

Courtesy Walter Kitundu: p. 403.

Courtesy Kronos Quartet, San Francisco; photo Walter Kitundu: p. 405.

Courtesy Kunsthistorisches Museum, Vienna: p. 45.

Courtesy Kurimanzutto, Mexico City; photo © 2011 Anri Sala: p. 409.

Courtesy Los Angeles County Museum of Art, Los Angeles: p. 245.

Courtesy Macedonian Museum of Contemporary Art, Thessaloniki: p. 285.

Courtesy Walter Marchetti; photo Fabrizio Garghetti: pp. 314–15.

© Christian Marclay. Courtesy Paula Cooper Gallery, New York: p. 361.

Courtesy MART, Museo di arte moderna e contemporanea di Trento e Rovereto, Rovereto. © Archivio Fotografico MART: p. 161.

Courtesy Eliseo Mattiacci; photo Claudio Abate: p. 309.

© Philippe Migeat – Centre Pompidou, Musée national d'art moderne, Centre de création Industrielle (diffusion RMN), © Adagp, Paris: p. 171.

Courtesy Ministero dei Beni e delle Attività Culturali e del Turismo, Soprintendenza per i Beni Storici Artistici ed Etnoantropologici di Milano; Archivio fotografico: p. 41.

Courtesy Ministero dei Beni e delle Attività culturali e del Turismo; photo Archivio fotografico Soprintendenza per i Beni Storici Artistici ed Etnoantropologici di Modena e Reggio Emilia: pp. 51, 53.

Courtesy Haroon Mirza: p. 401.

Courtesy Haroon Mirza and Lisson Gallery, London: p. 435.

© Moderna Museet, Stockholm/© Succession Marcel Duchamp/SIAE: 2014 p. 165.

© Moderner Kunst Stiftung Ludwig Wien, Vienna; former Hahn Collection, Cologne: p. 229.

Musée d'art contemporain de Lyon, Lyon. © Collection du macLYON © Blaise Adilon: p. 297.

Musée des Instruments de Musique, 4e Dép. des Musées royaux d'Art et d'Histoire, Brussels; photo Anne Deknock: pp. 59, 71–73, 77–79, 103, 105, 107, 109, 113, 127, 129, 135, 181.

Musée Historique de Strasbourg; photo M. Bertola: p. 115.

© Museo della Città di Bologna: pp. 133, 169, 183.

Museo di Castelvecchio, Verona; photo Archivio fotografico, Ottica Nodari di Caliaro Nereo: p. 43.

© Museum für Musikautomaten, Seewen: pp. 61, 75.

Museum Tinguely, Basel; photo Christian Baur, © Siae 2014: p. 227.

Music Division, The New York Public Library for the Performing Arts, Astor, Lenox and Tilden Foundations: pp. 215, 217.

Novumichi Tosa Maywa Denki: p. 341.

Courtesy Oldenburg van Bruggen Studio, © 2005 Claes Oldenburg and Coosje van Bruggen; photo Kerry Ryan McFate, courtesy Pace Gallery, NY: p. 381.

Courtesy Marie-Madeleine G. Opalka and Galleria Michela Rizzo, Venice; photo Maurice Aeschlimann, Genève: p. 243.

Courtesy Ed Potokar; photo Amy Rosenfeld: p. 357.

Courtesy Private Collection: p. 303.

Courtesy Private Collection, Düsseldorf; photo Frank Oleski: p. 247.

Private Collection, Courtesy Het Noordbrabants Museum, 's-Hertogenbosch; photo Peter Cox: p. 189.

© Pedro Reyes; Courtesy Lisson Gallery and Alumnos 47; photo Ken Adlard: p. 433.

Courtesy Seattle Art Museum, Gift of Mr. and Mrs. Bagley Wright, 82.190, © Robert Morris, Photo Elizabeth Mann: p. 221.

© Service de la documentation photographique du Musée national d'art moderne – Centre Pompidou, MNAM-CCI (diffusion RMN) ; © Robert Rauschenberg/Adagp, Paris: pp. 230–31.

© Speelklok Museum, Utrecht: p. 123, 131; photo Albertine Dijkema: p. 125.

Courtesy Gian Enzo Sperone, Switzerland: pp. 362–63.

Courtesy Sarah Hoe Sterling: p. 275.

Courtesy T293, Napoli – Roma: p. 423.

Courtesy The Estate of Terry Adkins and Salon 94, New York: p. 365.

The Gilbert and Lila Silverman Fluxus Collection Gift, 2008. © 2014; Digital image, The Museum of Modern Art, New York/Scala, Firenze: p. 295.

© The State Hermitage Museum: photo Natalia Antonova, Inessa Regentova: pp. 66–67. photo Alexander Lavrentyev: p. 65.

Courtesy The Wolf Vostell Estate: p. 253.

Courtesy Max Vandervorst; photo Ecole de Photographie de la Ville de Bruxelles: p. 313.

Courtesy Vehbi Koç Foundation, Contemporary Art Collection, Istanbul: p. 271; photo Uwe Walter, Berlin: p. 305.

Courtesy Claire Wesselmann; photo Jeffrey Sturges: p. 239.

Courtesy Ethan Wiley and Ochi Gallery, Ketchum: p. 277.

Courtesy Würth Collection; photo Philipp Schönborn, Munich: p. 237.

© 2009 Guido van der Werve. Courtesy the artist and Luhring Augustine, New York: p. 411.

© 2014 Rebecca Horn, VG Bild Kunst; photo Attilio Maranzano: p. 355.

© 2014 Digital image, The Museum of Modern Art, New York/Scala, Firenze; photo Jonathan Muzikar: pp. 278–79.

© 2014 Digital image, The Museum of Modern Art, New York/Scala, Firenze: p. 233.

© 2014 photo Scala, Firenze/BPK, Bildagentur für Kunst, Kultur und Geschichte: p. 343.

ISBN 978-88-87029-56-7

9 788887 029567